HANDBOOK OF METAL LIGAND HEATS

HANDBOOK OF METAL LIGAND HEATS

and Related Thermodynamic Quantities

Second Edition, Revised and Expanded

JAMES J. CHRISTENSEN
DEPARTMENT OF CHEMICAL ENGINEERING SCIENCE
BRIGHAM YOUNG UNIVERSITY
PROVO, UTAH

DELBERT J. EATOUGH
CENTER FOR THERMOCHEMICAL STUDIES
BRIGHAM YOUNG UNIVERSITY
PROVO, UTAH

REED M. IZATT
DEPARTMENT OF CHEMISTRY
BRIGHAM YOUNG UNIVERSITY
PROVO, UTAH

MARCEL DEKKER, INC. New York

Contribution No. 66 from the Center for Thermochemical Studies,
Brigham Young University, Provo, Utah

MARCEL DEKKER, INC.
270 Madison Avenue, New York, New York 10016

LIBRARY OF CONGRESS CATALOG CARD NUMBER: 75-4480

ISBN: 0-8247-6317-3

Current printing (last digit):
10 9 8 7 6 5 4 3 2 1

PRINTED IN THE UNITED STATES OF AMERICA

To

V.B.C., J.P.E., and H.F.I.

PREFACE

We have undertaken the compilation of heats of metal*-ligand interactions in solution to aid those working in the fields of chemistry, physics, biology, medicine, engineering, etc., where these data are necessary and useful. We have found that an enormous amount of such data has been collected and in some cases partly tabulated, but that up to now there has been no one reasonably complete source of this information.

This book is a compilation of heats of interaction, ΔH, together with the related thermodynamic quantities, log K, ΔS, and ΔCp, where available. The book consists of a Table in which are summarized the published literature values through 1973 under the headings of the various ligands and four indexes. The first edition published in 1970 covered the literature up to mid-1969. This edition contains 1083 separate ligand entries, more than double the 514 entries in the first edition. The criterion for including a metal-ligand pair in the Table was that ΔH values must have been reported in the published literature. Values appearing in theses, technical reports, books, or other *nonrefereed* sources were not included. In addition to the ΔH, log K, ΔS, and ΔCp values, the following information is included in the Table: the appropriate reaction, temperature, method, and conditions of measurement of ΔH, original and additional literature references, and pertinent supplemental information.

Four indexes — Empirical Formula, Element, Synonym, and Reference — are included in the book. The Empirical Formula Index contains the empirical formula of each ligand together with the location of the ligand in the Table. The Element Index contains an alphabetical listing of all elements appearing in the Metal Ion column of the Table together with their valence states and locations in the Table. The Synonym Index is an alphabetical listing of the common synonyms of the ligands appearing in the Table with their locations in the Table. The Reference Index includes a year by year alphabetical listing of all references cited in the Table.

The compilers have read the original of each paper, if possible, and data from other sources are reported only if the original paper could not be obtained through any of the means available including interlibrary loans. All values are those reported by the original investigators except when only log K or ΔS was reported in addition to ΔH and in such cases the third quantity, either log K or ΔS, was calculated from the other two. Values calculated in this manner are indicated by being underlined. Where sufficient data for a given reaction were available a critical evaluation was made by the compilers and the value or values reported are those judged to be most reliable. The authors

*In the compilation the word *metal* includes many elements from the right side of the periodic table, i.e., I, Br, Se, etc.

do not claim to be infallible with respect to the selection of the most correct value in these cases and are solely responsible for any errors which may have been made in either compiling or critically evaluating the data. Comments concerning errors and omitted data would be most appreciated.

This book is intended as a companion to the following published Tables.

"Heats of Proton Ionization and Related Thermodynamic Quantities" by R. M. Izatt and J. J. Christensen in *Handbook of Biochemistry and Selected Data for Molecular Biology,* ed. H. Sober, The Chemical Rubber Publishing Company, Cleveland, Ohio, Second Edition, 1970, pp. J58-J173.

"Stability Constants of Metal-Ion Complexes," Special Publication No. 17, The Chemical Society, Burlington House, London, 1971, by L. G. Sillen and A. E. Martell.

"Stability Constants of Metal-Ion Complexes, Supplement No. 1," Special Publication No. 25, The Chemical Society, Burlington House, London, 1971, by L. G. Sillen and A. E. Martell.

"Instability Constants of Complex Compounds," Translated from Russian, Consultants Bureau, New York, 1960, by K. B. Yatsimirskii and V. P. Vasil'ev.

The authors express their gratitude to Robert Reeder for editorial assistance and to Susan Belnap for technical typing of the manuscript.

James J. Christensen
Delbert J. Eatough
Reed M. Izatt

Brigham Young University
Provo, Utah
January 1975

TABLE OF CONTENTS

HANDBOOK OF

METAL LIGAND HEATS

USE OF TABLE AND INDEXES

A. Ligands

1. The ligands are ordered alphabetically and each is assigned a letter-number combination, i.e., A1, A2, A3, etc. In general, the nomenclature follows that in the 53rd Edition of the Handbook of Chemistry and Physics, The Chemical Rubber Co., 1972-1973. One exception is the naming of the synthetic macrocyclic compounds containing O and S atoms for which the term "crown" is used to represent the central ring following the nomenclature devised by C. J. Pedersen, *J. Amer. Chem. Soc.*, *89*, 7017 (1967).

2. The charge on the ligand is indicated by a superscript on the symbol *L* which is located directly after the formula of the compound/ion. In those cases where the coordinating form of the ligand is variable, the form is noted in the Conditions column if it has been indicated by the author.

B. Formula, Synonyms

1. Empirical and structural formulas of each ligand are given. Empirical formulas are also given in the Empirical Formula Index where they are arranged with the elements in the order C, H, O, N, Cl, Br, I, F, S, P followed by other elements in the alphabetical order of their symbols according to Beilsteins system.

2. The neutral form of each ligand is given except in a few cases, i.e., Cl^-, Br^-, I^-, CrO_4^{2-}, etc.

3. Synonyms for each ligand are given alphabetically in the Synonym Index.

C. Metals

1. The metals under each ligand follow the order given in L. G. Sillen and A. E. Martell, "Stability Constants," Chem. Soc. Special Publ. No. 17, 1964, which is given below. Where an element has several valencies, a lower oxidation state precedes a higher one.

group IA:	H, Li, Na, K, Rb, Cs, Fr.
group IIA:	Be, Mg, Ca, Sr, Ba, Ra.
group IIIB + 4f:	Sc, Y, La, Ce, Pr, Nd, Pm, Sm, Eu, Gd, Tb, Dy, Ho, Er, Tm, Yb, Lu, Ac.
group IVB:	Ti, Zr, Hf, Th.

group VB:	V, Nb, Ta, Pa.
group VIB + 5f:	Cr, Mo, W, U, Np, Pu, Am, Cm, Bk, Cf, Es, Fm, Md, No, Lw.
group VIIB:	Mn, Tc, Re.
group VIII:	Fe, Co, Ni, Ru, Rh, Pd, Os, Ir, Pt.
group IB:	Cu, Ag, Au.
group IIB:	Zn, Cd, Hg.
group IIIA:	B, Al, Ga, In, Tl.
groups IVA:	C, Si, Ge, Sn, Pb.
group VA:	N, P, As, Sb, Bi.
group VIA:	O, S, Se, Te, Po.
group VIIA:	F, Cl, Br, I, At.

2. When the metal is part of a complex molecule or ion consisting of the metal and one or more ligands, the formula of the complex is given in the Metal Ion column. Ligands attached to the metal, except oxygen, are indicated by letters of the alphabet, e.g., A, B, etc., and each ligand is identified in the Conditions column, i.e., A = Cl, B = C_2H_4, etc.

3. Enclosure of the metal or metal ion in brackets, [], indicates that the metal is part of a complex molecule or ion which is named in the Conditions column.

4. All elements in the Metal Ion column are listed alphabetically in the Metal Index together with their oxidation states and locations in the Table.

D. <u>ΔH</u>

1. The first criterion for construction of the Table was to include all reliable ΔH data. When only one ΔH value was reported for a given reaction, this value is given in the Table irrespective of its apparent validity and can be identified by the single reference. Where sufficient data for a given reaction were available to make a critical evaluation the value or values judged by the authors to be most reliable are the ones reported. In those cases where values reported by different workers disagree widely with no apparent reason all values are given.

2. The order of the thermodynamic quantities is as follows. All values for the first consecutive reaction are presented before the second, etc. Values in water are presented first with those for other solvents following, the solvents being arranged alphabetically. For a given solvent, values are arranged in order of increasing ionic strength. For a given ionic strength, values are arranged in order of increasing temperature. This order is given on the following page in outline form.

a. Reaction

(1) Consecutive reactions beginning with first

(2) Overall reactions

b. Solvent

(1) Water

(2) Other solvents arranged alphabetically

c. Ionic Strength, μ

(1) $\mu = 0$

(2) Increasing μ values

d. Temperature - lowest temperature first

3. The ΔH values are valid at the temperature or over the temperature range listed in the T, °C column.

4. Uncertainties of the ΔH values are not reported when less than 10%; however, the number of significant figures given is that reported in the original article.

E. Log K

1. Each log K value is valid for the reaction designated by the italicized number(s) which is in parentheses immediately following the log K value. These numbers have the following meanings.

Consecutive reactions -- (*1*): ($M + L = ML$), (*2*): ($ML + L = ML_2$), etc.

Overall reactions -- (*1-2*): ($M + 2L = ML_2$), (*1-3*): ($M + 3L = ML_3$), (*2-3*): ($2M + 3L = M_2L_3$), etc.

If no number appears after the log K value see Conditions or Remarks column for the reaction.

2. Each log K value is valid at the temperature listed in the T, °C column, unless otherwise specified in parentheses immediately following the log K value.

3. Uncertainties of the log K values are not reported; however, the number of significant figures given is that reported in the original article if log K is reported directly. If ΔG or K is reported in the original article then log K is usually reported with one additional significant figure.

F. ΔS

1. The ΔS values are valid at the temperatures listed in the T, °C column if the ΔH and log K values are valid at the same temperature. If the ΔH and log K values are not valid at the same temperature, no attempt has been made to designate the temperature at which ΔS is valid.

2. Uncertainties of the ΔS values are not reported: however, the number of significant figures given is that reported in the original article.

G. Values Calculated by Compilers

1. In those cases where ΔH and only one of the other two thermodynamic quantities, log K or ΔS, are given, the compilers have taken the liberty to calculate the third quantity. This calculated value is underlined in each case.

H. T, °C

1. Where ΔH values were determined calorimetrically, they are valid at the stated temperatures.

2. Where ΔH values were determined by the temperature variation method, the temperature or temperature range is that given by the author(s).

I. Method

1. The method of ΔH determination is designated by C (calorimetric) or T (temperature variation).

J. Conditions

1. The experimental conditions under which ΔH was determined are given. Supporting electrolyte, when present, is identified by its composition and molarity (M). Composition of non-aqueous solvents, when present, is indicated by the symbol, S: followed by the composition of the solvent. In a solvent mixture if the second component is not stipulated it is understood to be water. The letters A, B, etc., indicate ligands attached to the metal as described under heading C.2.

2. If the reaction is not stipulated in parentheses immediately following the log K values (see heading E.1.), it is indicated either in the Conditions column by R: followed by the chemical reaction or in the Remarks column by an alphabetic letter.

3. Abbreviations used are defined under heading N.

4. The units of ΔCp are cal./°C-gram-mole.

K. References

1. The references given are those from which the data were taken.

2. All references in the Table are listed in the Reference Index according to the year of publication. Under each year the references are in order alphabetically according to the authors. In the Reference Index all journal titles are abbreviated in

accordance with Chemical Abstracts usage as given in ACCESS, 1969 edition.

L. Additional References

1. These references also contain ΔH values, but the values were not included in the Table since in the opinion of the compilers equally reliable, more reliable or more extensive data were available for a given set of conditions. The opinions of the compilers, however, are not considered by them to be infallible in every case. These references appear with the first data listings for each metal and no attempt was made to indicate the reactions for which they are valid.

M. Remarks

1. When the same comment was applicable to several data entries that comment was given an alphabetic letter. These letters have the following meanings.

   a. ΔH values are calculated from log K values determined at only two temperatures.

   b. ΔH values are average values over the indicated temperature range, resulting in a combination of thermodynamic data valid at different temperatures.

   c. No correction has been made for other species prior to calculation of ΔH. Species, if known, given in parentheses.

   d. Numbers in the Conditions column refer to the pH of the solution at the ionic strength indicated by the numbers following the letter, e.g., if all the pH measurements were made at $\mu = 0.05$ then d:0.05 appears in the Remarks column.

   e. Questionable whether ΔH value refers to overall or stepwise reaction.

   f. The ΔH value for the stepwise reaction was obtained by dividing the ΔH value for the overall reaction by the number of consecutive reactions, e.g., f:6 where 6 is the number of consecutive reactions.

   g. The salt of the ligand was used. This is indicated by placing the cation formula after the letter, e.g., if the sodium salt of the ligand was used then g:Na appears in the Remarks column.

   h. Indicates temperature range (°C) over which ΔH was assumed constant, e.g., if temperature range was 2-45°C then h:2-45° appears in the Remarks column.

   i. ΔH value was computed from earlier work, e.g., if earlier work has a reference number of 50Ha then i:50Ha appears in the Remarks column.

   j. Reaction not specified or some question as to what reaction applies to ΔH.

   k. ΔH value was calculated from K values where K was in mole fraction units.

l. K is in mole fraction units.

m. $K = \frac{[I_2L]}{[I_2]L}$ where $[I_2L]$ and $[I_2]$ are in concentration units and L is in mole fraction units.

n. ΔH value appears questionable.

o. The reaction took place in a two phase system, CCl_4 and H_2O. The reaction was $Ag^+(H_2O) + L(CCl_4) = complex^+(H_2O)$ where the phase is indicated in parentheses after each reactant and product.

p. Data are not internally consistent.

q. The author has placed a large uncertainty on ΔH, e.g., if the uncertainty was reported as ±3 kcal./mole then q:±3 appears in the Remarks column.

r. Temperature at which K is valid is not given or is uncertain.

s. Reaction consists of the complex changing from a tetrahedral to an octahedral configuration.

t. Reaction consists of the complex changing from a planar to a tetrahedral configuration.

u. Temperature at which ΔH is valid is not given. Most probable temperature as estimated by the compilers is given, e.g., if the most probable temperature is 25°C then u:25° appears in the Remarks column.

v. The author has placed a large uncertainty on Log K, e.g., if the uncertainty was reported as ±0.1 Log K unit then v:±0.1 appears in the Conditions column.

w. Indicates counter ion accompanying positively charged ligand, e.g., if the counter ion was I^- then w:I appears in the Remarks column.

x. The author has calculated the value of ΔH by combining his data with data in the literature. If other work has a reference number of 50Ha then x:50Ha appears in the Remarks column.

y. The sign on ΔH was changed to make value conform to other reported values.

z. Product of reaction is a solvent separated complex.

aa. Reaction consists of the complex changing from an octahedral to a tetrahedral configuration.

bb. Reaction consists of the complex changing from a <u>cis</u> to a <u>trans</u> configuration.

cc. ΔH value was calculated from K values where K was in pressure units (mm^{-1}).

N. Abbreviations

1. The following abbreviations were used in the Table primarily in the Conditions column.

L = ligand	$\bar{n}$ = average number of ligands bound per metal atom
S = solvent	C_M = concentration of metal in moles/liter
R = reaction	C_L = concentration of ligand in moles/liter
v = volume	μ = ionic strength
w = weight	D = dielectric constant
M = moles/liter	P = product of reaction
	Cp = heat capacity in cal./deg.-mole.

O. Heats of Proton Ionization and Related Thermodynamic Quantities

1. In many cases it is desirable to know related thermodynamic values for proton ionization. A table containing heat of proton ionization data and related thermodynamic quantities for many of the ligands contained in the present Table has been published by the compilers in the Handbook of Biochemistry and Selected Data for Molecular Biology, ed. H. Sober, The Chemical Rubber Co., Second Edition, 1970, pp. J58-J173.

Metal Ion	ΔH, kcal /mole	Log K	ΔS, cal /mole °K	T, °C	M	Conditions	Ref.	Remarks

A

A1 ACETIC ACID			$C_2H_4O_2$			CH_3CO_2H		L^-
Mg^{2+}	-1.52	1.24(25°,1)	0.6	25-35	T	μ=0	56Na	b
Ca^{2+}	0.91	1.24(25°,1)	8.7	25-35	T	μ=0	56Na	b
Sr^{2+}	0.74	1.16(25°,1)	7.8	25-35	T	μ=0	56Na	b
Ba^{2+}	-2.32	1.15(25°,1)	-2.5	25-35	T	μ=0	56Na	b
Y^{3+}	3.262	1.57(1)	18.1	25	C	μ=2.0($NaClO_4$)	64Gb	
	5.390	2.73(1-2)	30.6	25	C	μ=2.0($NaClO_4$)	64Gb	
	5.243	3.44(1-3)	33.3	25	C	μ=2.0($NaClO_4$)	64Gb	
La^{3+}	2.181	1.58(1)	14.6	25	C	μ=2.0($NaClO_4$)	64Gb	
	3.28	4.733(1)	32.7	25	C	0.5M$NaClO_4$. S: 95 v % CH_3OH	67Gc	
	2.60	3.571(2)	25.0	25	C	0.5M$NaClO_4$. S: 95 v % CH_3OH	67Gc	
	1.08	2.589(3)	15.5	25	C	0.5M$NaClO_4$. S: 95 v % CH_3OH	67Gc	
	0.41	0.85(4)	5.2	25	C	0.5M$NaClO_4$. S: 95 v % CH_3OH	67Gc	
	3.786	2.52(1-2)	24.2	25	C	μ=2.0($NaClO_4$)	64Gb	
	4.62	3.03(1-3)	29.4	25	C	μ=2.0($NaClO_4$)	64Gb	
Ce^{3+}	2.2	1.708(25°,1)	15.2	2.1-46.6	T	μ=2.0M(NaClO)	70Cg	b, q: ±0.2
	2.092	1.71(1)	14.8	25	C	μ=3.0($NaClO_4$)	64Gb	
	3.664	2.73(1-2)	24.8	25	C	μ=3.0($NaClO_4$)	64Gb	
	5.11	3.20(1-3)	31.8	25	C	μ=3.0($NaClO_4$)	64Gb	
Pr^{3+}	1.719	1.83(1)	14.2	25	C	μ=2.0($NaClO_4$)	64Gb	
	2.89	5.114(1)	33.1	25	C	0.5M$NaClO_4$. S: 95 v % CH_3OH	67Gc	
	2.64	3.821(2)	26.3	25	C	0.5M$NaClO_4$. S: 95 v % CH_3OH	67Gc	
	1.56	2.609(3)	17.2	25	C	0.5M$NaClO_4$. S: 95 v % CH_3OH	67Gc	
	0.66	0.92(4)	6.5	25	C	0.5M$NaClO_4$. S: 95 v % CH_3OH	67Gc	
	4.154	2.86(1-2)	27.0	25	C	μ=2.0($NaClO_4$)	64Gb	
	3.61	3.33(1-3)	27.3	25	C	μ=2.0($NaClO_4$)	64Gb	
Nd^{3+}	1.708	1.93(1)	14.6	25	C	μ=2.0($NaClO_4$)	64Gb	
	2.20	5.230(1)	31.3	25	C	0.5M$NaClO_4$. S: 95 v % CH_3OH	67Gc	
	2.21	3.963(2)	25.6	25	C	0.5M$NaClO_4$. S: 95 v % CH_3OH	67Gc	
	1.94	2.601(3)	18.4	25	C	0.5M$NaClO_4$. S: 95 v % CH_3OH	67Gc	
	1.71	1.30(4)	0.3	25	C	0.5M$NaClO_4$. S: 95 v % CH_3OH	67Gc	
	3.486	3.06(1-2)	25.7	25	C	μ=2.0($NaClO_4$)	64Gb	
	4.36	3.52(1-3)	30.7	25	C	μ=2.0($NaClO_4$)	64Gb	

Metal Ion	ΔH, kcal /mole	Log K	ΔS, cal /mole °K	T, °C	M	Conditions	Ref.	Remarks
Al, cont.								
Sm^{3+}	1.453	2.03(1)	14.2	25	C	μ=2.0($NaClO_4$)	64Gb	
	2.77	5.137(1)	32.0	25	C	0.5M$NaClO_4$. S: 95 v % CH_3OH	67Gc	
	1.54	4.059(2)	23.7	25	C	0.5M$NaClO_4$. S: 95 v % CH_3OH	67Gc	
	1.93	2.694(3)	18.8	25	C	0.5M$NaClO_4$. S: 95 v % CH_3OH	67Gc	
	-0.44	1.11(4)	3.6	25	C	0.5M$NaClO_4$. S: 95 v % CH_3OH	67Gc	
	2.876	3.29(1-2)	24.7	25	C	μ=2.0($NaClO_4$)	64Gb	
	3.773	3.90(1-3)	30.5	25	C	μ=2.0($NaClO_4$)	64Gb	
Eu^{3+}	1.4	1.903(25°,1)	13.4	0-55	T	μ=2.0M($NaClO_4$)	70Cq	b, q: ±0.1
Gd^{3+}	1.868	1.86(1)	14.8	25	C	μ=2.0($NaClO_4$)	64Gb	
	3.38	4.977(1)	34.1	25	C	0.5M$NaClO_4$. S: 95 v % CH_3OH	67Gc	
	1.71	4.000(2)	24.0	25	C	0.5M$NaClO_4$. S: 95 v % CH_3OH	67Gc	
	1.59	2.687(3)	17.7	25	C	0.5M$NaClO_4$. S: 95 v % CH_3OH	67Gc	
	-0.34	1.16(4)	4.1	25	C	0.5M$NaClO_4$. S: 95 v % CH_3OH	67Gc	
	3.245	3.16(1-2)	25.3	25	C	μ=2.0($NaClO_4$)	64Gb	
	3.543	3.76(1-3)	29.1	25	C	μ=2.0($NaClO_4$)	64Gb	
Dy^{3+}	2.925	1.71(1)	17.6	25	C	μ=2.0($NaClO_4$)	64Gb	
	4.09	5.026(1)	36.7	25	C	0.5M$NaClO_4$. S: 95 v % CH_3OH	67Gc	
	2.72	3.987(2)	27.4	25	C	0.5M$NaClO_4$. S: 95 v % CH_3OH	67Gc	
	1.00	2.777(3)	16.0	25	C	0.5M$NaClO_4$. S: 95 v % CH_3OH	67Gc	
	0.26	1.39(4)	7.3	25	C	0.5M$NaClO_4$. S: 95 v % CH_3OH	67Gc	
	4.439	3.03(1-2)	28.7	25	C	μ=2.0($NaClO_4$)	64Gb	
	4.309	3.84(1-3)	32.0	25	C	μ=2.0($NaClO_4$)	64Gb	
Ho^{3+}	3.167	1.67(1)	18.3	25	C	μ=3.0($NaClO_4$)	64Gb	
	5.009	2.92(1-2)	30.2	25	C	μ=3.0($NaClO_4$)	64Gb	
	4.533	3.80(1-3)	32.6	25	C	μ=3.0($NaClO_4$)	64Gb	
Er^{3+}	3.276	1.64(1)	18.5	25	C	μ=2.0($NaClO_4$)	64Gb	
	4.58	3.150(1)	38.9	25	C	0.5M$NaClO_4$. S: 95 v % CH_3OH	67Gc	
	2.92	3.965(2)	27.9	25	C	0.5M$NaClO_4$. S: 95 v % CH_3OH	67Gc	
	1.88	2.760(3)	19.0	25	C	0.5M$NaClO_4$. S: 95 v % CH_3OH	67Gc	
	-3.5	1.19(4)	-6.2	25	C	0.5M$NaClO_4$. S: 95 v % CH_3OH	67Gc	
	5.509	2.90(1-2)	31.8	25	C	μ=2.0($NaClO_4$)	64Gb	
	5.166	3.72(1-3)	34.3	25	C	μ=2.0($NaClO_4$)	64Gb	

Metal Ion	ΔH, kcal /mole	Log K	ΔS, cal /mole °K	T, °C	M	Conditions	Ref.	Remarks
Al, cont.								
Yb^{3+}	3.507	1.68(1)	19.5	25	C	μ=2.0($NaClO_4$)	64Gb	
	4.84	5.530(1)	41.5	25	C	0.5M$NaClO_4$. S: 95 v % CH_3OH	65Gc	
	3.50	4.472(2)	32.2	25	C	015M$NaClO_4$. S: 95 v % CH_3OH	67Gc	
	4.17	3.368(3)	29.4	25	C	0.5M$NaClO_4$. S: 95 v % CH_3OH	67Gc	
	-0.11	2.15(4)	9.4	25	C	0.5M$NaClO_4$. S: 95 v % CH_3OH	67Gc	
	6.065	2.91(1-2)	33.7	25	C	μ=2.0($NaClO_4$)	64Gb	
	6.58	3.63(1-3)	38.7	25	C	μ=2.0($NaClO_4$)	64Gb	
UO_2^{2+}	2.519	2.420(1)	19.5	25	C	μ=1.0($NaClO_4$)	71Ac	
	2.31	1.995(2)	16.9	25	C	μ=1.0($NaClO_4$)	71Ac	
	-0.96	1.983(3)	5.86	25	C	μ=1.0($NaClO_4$)	71Ac	
Am^{3+}	4.3	1.954(25°,1)	23.4	0-55	T	μ=2.0M($NaClO_4$)	70Cg	b, q: ±0.3
Cm^{3+}	4.3	2.033(25°,1)	23.8	0-55	T	μ=2.0M($NaClO_4$)	70Cg	b
Bk^{3+}	4.4	2.029(25°,1)	24.1	0.7-39.8	T	μ=2.0M($NaClO_4$)	70Cg	b
Cf^{3+}	3.8	2.124(25°,1)	22.4	0-55.2	T	μ=2.0M($NaClO_4$)	70Cg	b, q: ±0.3
Cu^{2+}	1.04	1.88(1)	12.1	25	C	3M$NaClO_4$	67Gb	e
	0.41	1.24(2)	7.1	25	C	3M$NaClO_4$	67Gb	e
	0.07	0.47(3)	2.4	25	C	3M$NaClO_4$	67Gb	e
CuL_2	-5	5.00(25°)	6	25&70	T	S: HL with 0.2MH_2O R: 2CuL_2= Cu_2L_4	71Gi	a,b, q: ±1
	-7	3.00(25°)	-11	25&70	T	S: E + OH with 0.1MH_2O R: 2CuL_2 = Cu_2L_4	71Gi	a,b, q: ±2
Zn^{2+}	2.04	0.946(1)	11.2	25	C	3M$NaClO_4$	67Gb	e
	3.22	0.396(2)	12.5	25	C	3M$NaClO_4$	67Gb	e
	1.15	0.506(3)	6.1	25	C	3M$NaClO_4$	67Gb	e
Cd^{2+}	1.69	1.27(1)	11.5	25	C	μ=0.25($NaClO_4$)	68Gd 67Gb	
	1.76	1.19(1)	11.3	25	C	μ=0.5($NaClO_4$)	68Gd	
	1.62	1.17(1)	10.8	25	C	μ=1($NaClO_4$)	68Gd	
	1.50	1.24(1)	10.7	25	C	μ=2($NaClO_4$)	68Gd	
	1.46	1.30(1)	10.9	25	C	μ=3($NaClO_4$)	68Gd	
	1.47	0.71(2)	8.2	25	C	μ=0.5($NaClO_4$)	68Gd	
	1.23	0.64(2)	7.1	25	C	μ=1($NaClO_4$)	68Gd	
	1.08	0.74(2)	7.0	25	C	μ=2($NaClO_4$)	68Gd	
	0.84	0.97(2)	7.3	25	C	μ=3($NaClO_4$)	68Gd	
	-0.60	0.22(3)	-1.0	25	C	μ=1($NaClO_4$)	68Gd	
	-0.07	0.15(3)	0.5	25	C	μ=2($NaClO_4$)	68Gd	
	0.41	0.14(3)	2.0	25	C	μ=3($NaClO_4$)	68Gd	
CdA	-18	3.63(25°)	-44	25-45	T	L=0.02M S: Ethanol	71Bh	b

Metal Ion	ΔH, kcal /mole	Log K	ΔS, cal /mole °K	T, °C	M	Conditions	Ref.	Remarks
Al, cont.								
CdA, cont.						A=Phaeophytin R: CdA + 2HL = $CdL_2 + H_2A$		
Pb^{2+}	-0.4	1.83(25°,1)	7.03	15-35	T	μ=0.75	67Hi	b
	-0.8	2.037(25°,1)	6.64	15-35	T	μ=1.0	67Hi	b
	-1.1	2.238(25°,1)	6.55	15-35	T	μ=2.0	67Hi	b
	-0.06	2.33(1)	10.5	25	C	$3M NaClO_4$	67Gb	e
	-1.0	2.740(25°,2)	9.18	15-35	T	μ=0.75	67Hi	b
	-0.8	2.914(25°,2)	10.6	15-35	T	μ=1.0	67Hi	b
	-0.7	3.299(25°,2)	12.7	15-35	T	μ=2.0	67Hi	b
	-0.09	1.27(2)	5.5	25	C	$3M NaClO_4$	67Gb	e
	-1.4	3.243(25°,3)	10.1	15-35	T	μ=0.75	67Hi	b
	-2.0	3.581(25°,3)	9.65	15-35	T	μ=1.0	67Hi	b
	-2.1	3.956(25°,3)	11.1	15-35	T	μ=2.0	67Hi	b
	-0.96	-0.015(3)	-3.3	25	C	$3M NaClO_4$	67Gb	e
A2 ACETIC ACID, amide			C_2H_5ON			CH_3CONH_2		L
I_2	-4.61	-0.215(22°,1)	-16.6	0&22	T	C_M=0.35. S: CH_2Cl_2	61Ta	a,b
A3 ACETIC ACID, amide, N-acetyl-N-phenyl-			$C_{10}H_{12}O_2N$			$(CH_3CO)_2NC_6H_5$		L^-
Be^{2+}	-1.7	8.69(25°,1)	35	10-40	T	S: 50% Dioxane	69Ha	b
	0	7.23(25°,2)	37	10-40	T	S: 50% Dioxane	69Ha	b
A4 ACETIC ACID, amide, N-benzoyl-N-(m-nitrophenyl)-			$C_{15}H_{11}O_4N_2$			$C_6H_5CON(C_6H_5NO_2)COCH_3$		HL
Fe^{3+}	-4.97	6.97(25°)	15.22	15-35	T	S: 60 w % Ethanol μ=0.1 pH=1.8-4.2 R: Fe^{3+} + 2HL = $FeL_2^+ + 2H^+$	72Ta	b,j
A5 ACETIC ACID, amide, N,N-diethyl			$C_6H_{13}ON$			$CH_3CON(C_2H_5)_2$		L
I_2	-5.4	1.27(25°,1)	-12.3	5-25	T	S: Heptane	71Te	b
A6 ACETIC ACID, amide, N,N-dimethyl			C_4H_9ON			$CH_3CON(CH_3)_2$		L
CuA_2	-8.0	3.217(1)	-12.8	25	C	S: CCl_4 A=Hexafluoroacetylacetone	70Pa	
	-6.6	(1)	--	25	C	S: O-Dichlorobenzene A=Hexafluoroacetylacetone	70Pa	
	-5.9	(1)	--	25	C	S: Methylene chloride A=Hexafluoroacetylacetonate	74Da	
	-5.2	0.964(2)	-13.0	25	C	S: CCl_4 A=Hexafluoroacetylacetone	70Pa	
CuA_2B	-2.1	--	--	25	C	S: CCl_4 A=Hexafluoroacetylacetone B=Ethyl acetate R: CuA_2B + L = CuA_2L + B	72Nd	

Metal Ion	ΔH, kcal /mole	Log K	ΔS, cal /mole °K	T, °C	M	Conditions	Ref.	Remarks
A6, cont.								
CuA_2B, cont.								
	-1.9	--	--	25	C	S: o-$Cl_2C_6H_4$ A=Hexafluoroacetylacetone B=Ethyl acetate R: CuA_2B + L = CuA_2L + B	72Nd	
SnA_3B	-7.9	0.59(1)	-24	26	C	S: CCl_4. A=CH_3 B=Cl	66Bg	
	-7.7	0.54(1)	-23	27	C	S: CCl_4. A=CH_3 B=Cl	65Bb	
	-9.4	0.193(1)	-30	26	C	S: N,N-Dimethyl-acetamide. A=CH_3 B=Cl	66Bh	
SbA_5	-27.80	(1)	--	25	C	C_M<.01. S: Ethylene chloride A=Cl	64Ob	
Br_2	-1.6	0.255(25°,1)	-4	5-40	T	S: CCl_4	62Db	b
I_2	-4.14	0.710(38°,1)	10	25.2-39.5	T	S: Benzene	66Dc 60Sf	b
	-4.0	0.839(25°,1)	9.4	29.4-46.7	T	S: CCl_4	61Db	b
	-5.3	1.18(25°,1)	-12.4	5-25	T	S: Heptane	71Te	b
IA	-9.5	3.042(25°,1)	-17	28-46	T	$C_M=10^{-2}$ $C_L=5x10^{-2}$. S: CCl_4 A=Cl	62Db	b
A7 ACETIC ACID, amide, N-phenyl-N-benzoyl		$C_{15}H_{13}O_2N$				$C_6H_5CON(C_6H_5)COCH_3$		L
Fe^{3+}	-2.48	8.67(25°,1-3)	31.34	20-35	T	S: 60% Ethanol pH=3.0	71Tc	b
A8 ACETIC ACID, amide, thiono		C_2H_5NS				CH_3CSNH_2		L
I_2	-8.2	4.398(25°,1)	-9.5	2.5 &17	T	$C_M=1x10^{-4}$ $C_L=10^{-4}-3x10^{-4}$ S: CH_2Cl_2	62La	a,b
A9 ACETIC ACID, amide, thiono N,N-dimethyl		C_4H_9NS				$CH_3CSN(CH_3)_2$		L
SnA_3B	-5.9	-0.322(1)	-21	26	C	S: CCl_4. A=CH_3 B=Cl	66Bg	
I_2	-9.5	3.076(25°,1)	-17.8	25-40	T	$C_M=10^{-3}$ $C_L=10^{-3}$. S: CCl_4	64Na	b
	-9.29	3.18(25°,1)	-16.6	--	T	S: CCl_4	70Gj	b,u: 25
A10 ACETIC ACID, amide, thiono, N-methyl-		C_3H_7NS				$CH_3C(S)NHCH_3$		
I_2	-8.4	2.732(25°,1)	-15.6	0-45	T	S: Methylenedichloride	66Bc	b, q: ±1

Metal Ion	ΔH, kcal /mole	Log K	ΔS, cal /mole °K	T, °C	M	Conditions	Ref.	Remarks
A11	ACETIC ACID, bromide		C_2H_3OBr			CH_3COBr		L
I_2	-0.9	-0.89(25°,1)	-7	5-25	T	S: Heptane	71Te	b
A12	ACETIC ACID, tert-butyl ester		$C_6H_{12}O_2$			$CH_3CO_2C(CH_3)_3$		L
SbA_5	-35.9	(1)	--	25	C	Excess $SbCl_5$ C_L low, variable, ∿0.02. S: 1,2-Dichloroethane. A=Cl	64Oa	
A13	ACETIC ACID, chloride		C_2H_3OCl			CH_3COCl		L
I_2	-1.1	-0.92(25°,1)	-7.8	5-25	T	S: Heptane	71Te	b
A14	ACETIC ACID, ethenyl ester		$C_4H_6O_2$			$CH_3CO_2CH:CH_2$		
NiAB	-13.7	2.49(27°)	-34.3	18-42	T	S: Tetrahydrofuran A=Dipyridyl B=Tetrahydrofuran R: NiAB + L = NiAL + B	71Ya	b
A15	ACETIC ACID, ethyl ester		$C_4H_8O_2$			$CH_3CO_2CH_2CH_3$		L
CuA_2	-5.9	2.290(1)	-9.3	25	C	S: CCl_4 A=Hexafluoroacetylacetone	70Pa	
	-4.7	1.545(1)	-8.7	25	C	S: o-Dichlorobenzene A=Hexafluoroacetylacetone	72Nd	q: ±0.3
SbA_5	-15.5	(1)	--	25	C	S: CCl_4 A=Cl	72Lb	
	-15.9	(1)	--	25	C	S: CCl_4 A=Cl	73Oa	
	-16.6	(1)	--	25	C	S: 1,2-Dichloroethane A=Cl	72Lb	
	-17.08	(1)	--	25	C	S: 1,2-Dichloroethane A=Cl	63Oa	
I_2	-2.05	(1)	--	--	C	S: Ethyl acetate	50Ha 09Ha	
	-3.2	-0.03(25°,1)	-10.5	5-25	T	S: Heptane	71Te	b
IA	-6.1	0.769(1)	-16.88	26	C	C_M=0.03. C_L= 0.1-0.9. S: CCl_4 A=Cl	68Bb	
A16	ACETIC ACID, isopropenyl ester		$C_5H_8O_2$			$CH_3CO_2C(CH_3):CH_2$		L
NiAB	-8.7	1.07(27°)	-24.1	17-42	T	S: Tetrahydrofuran A=Dipyridyl B=Tetrahydrofuran R: NiAB + L = NiAL + B	71Ya	b
A17	ACETIC ACID, methyl ester		$C_3H_6O_2$			$CH_3CO_2CH_3$		L

Metal Ion	ΔH, kcal /mole	Log K	ΔS, cal /mole °K	T, °C	M	Conditions	Ref.	Re-marks
A17, cont.								
SbA_5	-16.38	($\underline{1}$)	--	25	C	S: 1,2-Dichloro-ethane. A=Cl	63Oa	
I_2	-2.5	-0.284(25°,$\underline{1}$)	$\underline{-9.73}$	25-45	T	C_M=1.9x10^{-3}. C_L= 0.5. S: CCl_4	64Ma	b
	-3.1	-0.12(25°,$\underline{1}$)	-11	5-25	T	S: Heptane	71Te	b
	-2.4	($\underline{1}$)	--	--	C	S: Methyl acetate	50Ha	
A18 ACETIC ACID, nitrile			C_2H_3N			CH_3CN		L
CoA_2L_2	-14.6	-1.506	--	-38-29	T	S: Acetonitrile C_M=0.2-0.4M A=Cl R: CoL_2A_2 + 2L = CoL_4A_2	72Bg	b,l
	-18.2	-2.821	--	-34-49	T	S: Acetonitrile C_M=0.2-0.4M A=Br R: CoL_2A_2 + 2L = CoL_4A_2	72Bg	b,l
	-17.9	-2.966	--	-33-50	T	S: Acetonitrile C_M=0.2-0.4M A=I R: CoL_2A_2 + 2L = CoL_4A_2	72Bg	b,l
	14.6	-0.420	--	-38-29	T	S: Acetonitrile C_M=0.2-0.4M A=Cl R: $2CoL_2A_2$ + 2L = CoL_6^{2+} + CoA_4^{2-}	72Bg	b,l
	11.2	-0.468	--	-34-49	T	S: Acetonitrile C_M=0.2-0.4M A=Br R: $2CoL_2A_2$ + 2L = CoL_6^{2+} + CoA_4^{2-}	72Bg	b,l
	7.65	-0.409	--	-33-50	T	S: Acetonitrile C_M=0.2-0.4M A=I R: $2CoL_2A_2$ + 2L = CoL_6^{2+} + CoA_4^{2-}	72Bg	b
BA_3	-21	($\underline{1}$)	--	25	C	S: Acetonitrile A=I	73Bi	q: ±2.4
SuA_4	-6.0	2.20($\underline{1}$)	-10.1	25	C	S: Benzene A=Cl	70Gh 68Gf	
	-6.1	($\underline{1}$)	--	--	C	S: Benzene C_M=0.05-0.08M A=Cl	68Ge	
$SnAB_3$	-4.8	-0.260($\underline{1}$)	$\underline{-17.2}$	26	C	S: CCl_4. A=Cl, B=CH_3	66Bg	
SbA_5	-14.6	($\underline{1}$)	--	25	C	S: $C_2H_4Cl_2$ A=Cl	72Lb	

Metal Ion	ΔH, kcal /mole	Log K	ΔS, cal /mole °K	T, °C	M	Conditions	Ref.	Remarks
A18, cont.								
IA	-4.9	0.924(25°,1)	-12.3	0-40	T	S: CCl_4. A=Cl	63Pc	b
	-4.1	0.170(25°,1)	-12.8	0-40	T	S: CCl_4. A=Br	63Pc	b
I_2	-1.9	-0.38(25°,1)	-8.3	0-40	T	S: CCl_4	63Pc	b
	-2.92	-0.336(25°)	-11.31	--	T	S: CCl_4 C_L=0.2-1.0M C_M=2x10^{-3}	72Ma	b,j
A19	ACETIC ACID, 2-propyl ester		$C_5H_{10}O_2$			$CH_3CO_2CH_2CH_2CH_3$		L
SbA_5	-17.53	(1)	--	25	C	S: 1,2-Dichloroethane. A=Cl	63Oa	
A20	ACETIC ACID, 4-tolyl ester		$C_9H_{10}O_2$			$(CH_3)C_6H_4(CO_2CH_3)$		L
TiA_4	-8.6	2.602(25°,1)	-17	25&60	T	μ=0 A=Cl	66Gc	a,b,q: ±1.8
A21	ACETIC ACID, chloro-, amide, N,N- dimethyl-		C_4H_8ONCl			$ClCH_2CON(CH_3)_2$		L
I_2	-3.3	-0.114(25°,1)	-10.5	25-43	T	C_M=1.7x10^{-3} C_L=0.1-0.2 S: CCl_4	62Dc	b
A22	ACETIC ACID, chloro-, ethyl ester		$C_4H_7O_2Cl$			$ClCH_2CO_2CH_2CH_3$		L
SbA_5	-12.8	(1)	--	25	C	S: CCl_4 A=Cl	73Oa	
	-12.80	2.18(1)	-32.9	25	C	S: Ethylene chloride. A=Cl	67Od	
A23	ACETIC ACID, chloro-, nitrile		C_2H_2NCl			$ClCH_2CN$		L
SbA_5	-9.6	(1)	--	25	C	S: CCl_4 A=Cl	72Lb	
IA	-5.3	0.382(25°,1)	-16	0-40	T	S: CCl_4. A=Cl	63Pc	b
	-3.1	-0.260(25°,1)	-11.4	0-40	T	S: CCl_4. A=Br	63Pc	b
I_2	-1.5	-0.72(25°,1)	-8.4	0-40	T	S: CCl_4. A=Br	63Pc	b
A24	ACETIC ACID, dichloro-, ethyl ester		$C_4H_6O_2Cl_2$			$Cl_2CHCO_2CH_2CH_3$		L
SbA_5	-9.35	0.30(1)	-30.0	25	C	S: Ethylene chloride. A=Cl	67Od	
A25	ACETIC ACID, dichloro-, nitrile		C_2HNCl_2			Cl_2CHCN		L
IA	-3.7	-0.161(25°,1)	-13.3	0-40	T	S: CCl_4. A=Cl	63Pc	b
	-2.0	-0.638(25°,1)	-9.7	0-40	T	S: CCl_4. A=Br	63Pc	b
A26	ACETIC ACID, (ethylthio) hydroxide-		$C_4H_9O_3S$			$CH_3CH_2SCH_2(OH)CO_2H$		L^-
Tl^+	-1.6	1.708(30°,1)	2.4	20&30	T	μ=0.5M($NaClO_4$)	72Sd	a,b,j

Metal Ion	ΔH, kcal /mole	Log K	ΔS, cal /mole °K	T, °C	M	Conditions	Ref.	Remarks
A26, cont.								
Tl^{+}, cont.	-0.95	1.67(30°,1)	4.52	20&30	T	μ=2M($NaClO_4$)	74Ma	a,b
	-0.2	0.246(30°,2)	-0.1	20&30	T	μ=0.5M($NaClO_4$)	72Sd	a,b,j
	-0.35	0.72(30°,2)	2.17	20&30	T	μ=2M($NaClO_4$)	74Ma	a,b,i: 72Sd
	-0.2	1.350(30,3)	5.7	20&30	T	μ=0.5M($NaClO_4$)	74Ma	a,b,i: 72Sd
	-0.31	0.55(30°,3)	1.48	20&30	T	μ=2M($NaClO_4$)	74Ma	a,b,i: 72Sd
Pb^{2+}	5.3	2.00(30°,1)	27	20&30	T	μ=0.5M($NaClO_4$)	71Sa 72Sc	a,b,j
	3.2	4.477(30°,1-2)	31	20&30	T	μ=0.5M($NaClO_4$)	71Sa 72Sc	a,b,j
	2.8	6.832(30°,1-3)	41	20&30	T	μ=0.5M($NaClO_4$)	71Sa 72Sc	a,b,j
A27 ACETIC ACID, hydroxy-			$C_2H_4O_3$			$HOCH_2CO_2H$		L^-
Y^{3+}	-0.15	2.36(1)	10.3	25	C	μ=2.0($NaClO_4$)	66Cc	
	-0.073	2.47(1)	11.1	25	C	μ=2.0($NaClO_4$)	64Gb	
	-0.25	1.99(2)	8.3	25	C	μ=2.0($NaClO_4$)	66Cc	
	-0.100	1.93(2)	8.5	25	C	μ=2.0($NaClO_4$)	64Gb	
	-1.22	1.23(3)	1.5	25	C	μ=2.0($NaClO_4$)	66Cc	
	-0.722	1.29(3)	3.4	25	C	μ=2.0($NaClO_4$)	64Gb	
	-0.025	0.55(4)	2.5	25	C	μ=2.0($NaClO_4$)	64Gb	
La^{3+}	-0.85	2.14(1)	7	25	C	μ=2.0($NaClO_4$)	66Cc	
	-0.633	2.18(1)	7.9	25	C	μ=2.0($NaClO_4$)	64Gb	
	-0.23	1.80(2)	9	25	C	μ=2.0($NaClO_4$)	66Cc	
	-0.422	1.56(2)	5.7	25	C	μ=2.0($NaClO_4$)	64Gb	
	-0.625	1.05(3)	2.7	25	C	μ=2.0($NaClO_4$)	64Gb	
	-0.51	0.27(4)	-0.5	25	C	μ=2.0($NaClO_4$)	64Gb	
Ce^{3+}	-0.810	2.34(1)	8.0	25	C	μ=2.0($NaClO_4$)	64Gb 61Ca	
	-0.789	1.66(2)	4.9	25	C	μ=2.0($NaClO_4$)	64Gb	
	-0.669	1.12(3)	2.9	25	C	μ=2.0($NaClO_4$)	64Gb	
	-0.72	0.38(4)	-0.7	25	C	μ=2.0($NaClO_4$)	64Gb	
Nd^{3+}	-0.96	2.40(1)	7.8	25	C	μ=2.0($NaClO_4$)	66Cc	
	-1.193	2.49(1)	7.4	25	C	μ=2.0($NaClO_4$)	64Gb	
	-0.48	1.96(2)	7.4	25	C	μ=2.0($NaClO_4$)	66Cc	
	-0.995	1.83(2)	5.0	25	C	μ=2.0($NaClO_4$)	64Gb	
	-1.291	1.21(3)	1.2	25	C	μ=2.0($NaClO_4$)	64Gb	
	-0.52	0.47(4)	0.4	25	C	μ=2.0($NaClO_4$)	64Gb	
Sm^{3+}	-0.93	2.40(1)	7.9	25	C	μ=2.0($NaClO_4$)	66Cc	
	-1.039	2.54(1)	8.1	25	C	μ=2.0($NaClO_4$)	64Gb	
	-0.80	2.07(2)	6.8	25	C	μ=2.0($NaClO_4$)	66Cc	
	-0.365	1.96(2)	4.4	25	C	μ=2.0($NaClO_4$)	64Gb	
	-1.282	1.31(3)	1.7	25	C	μ=2.0($NaClO_4$)	64Gb	
	-1.25	0.57(4)	-1.6	25	C	μ=2.0($NaClO_4$)	64Gb	
Eu^{3+}	-0.84	2.38(1)	8.1	25	C	μ=2.0($NaClO_4$)	66Cc	
	-0.53	2.52(25°,1)	9.7	0.5-52	T	μ=2.0($NaClO_4$)	72Cd	b
	-0.90	2.08(2)	6.5	25	C	μ=2.0($NaClO_4$)	66Cc	

Metal Ion	ΔH, kcal /mole	Log K	ΔS, cal /mole °K	T, °C	M	Conditions	Ref.	Remarks
A27, cont.								
Gd^{3+}	-0.62	2.38(1)	8.8	25	C	μ=2.0($NaClO_4$)	66Cc	
	-0.614	2.47(1)	9.3	25	C	μ=2.0($NaClO_4$)	64Gb	
	-1.19	2.01(2)	5.2	25	C	μ=2.0($NaClO_4$)	66Cc	
	-1.120	1.94(2)	5.1	25	C	μ=2.0($NaClO_4$)	64Gb	
	-1.781	1.34(3)	0.1	25	C	μ=2.0($NaClO_4$)	64Gb	
	-0.75	0.61(4)	0.3	25	C	μ=2.0($NaClO_4$)	64Gb	
Tb^{3+}	-0.09	2.40(1)	10.7	25	C	μ=2.0($NaClO_4$)	66Cc	
	-0.83	2.05(2)	6.6	25	C	μ=2.0($NaClO_4$)	66Cc	
Dy^{3+}	-0.08	2.48(1)	11.1	25	C	μ=2.0($NaClO_4$)	66Cc	
	-0.164	2.53(1)	11.0	25	C	μ=2.0($NaClO_4$)	64Gb	
	-0.97	1.96(2)	5.8	25	C	μ=2.0($NaClO_4$)	66Cc	
	-0.444	1.94(2)	7.4	25	C	μ=2.0($NaClO_4$)	64Gb	
	-1.129	1.41(3)	2.7	25	C	μ=2.0($NaClO_4$)	64Gb	
	-0.84	0.57(4)	0.2	25	C	μ=2.0($NaClO_4$)	64Gb	
Ho^{3+}	-0.08	2.42(1)	10.8	25	C	μ=2.0($NaClO_4$)	66Cc	
	-0.40	2.12(2)	8.4	25	C	μ=2.0($NaClO_4$)	66Cc	
	-2.40	1.22(3)	-2.5	25	C	μ=2.0($NaClO_4$)	66Cc	
Er^{3+}	-0.09	2.64(1)	11.8	25	C	μ=2.0($NaClO_4$)	66Cc	
	-0.19	2.60(1)	11.3	25	C	μ=2.0($NaClO_4$)	64Gb	
	-0.39	1.88(2)	7.3	25	C	μ=2.0($NaClO_4$)	66Cc	
	-0.41	1.98(2)	7.6	25	C	μ=2.0($NaClO_4$)	64Gb	
	-0.68	1.42(3)	2.4	25	C	μ=2.0($NaClO_4$)	64Gb	
	-0.15	0.48(4)	3.9	25	C	μ=2.0($NaClO_4$)	64Gb	
Tm^{3+}	-0.29	2.62(1)	11.1	25	C	μ=2.0($NaClO_4$)	66Cc	
	-0.57	2.03(2)	7.4	25	C	μ=2.0($NaClO_4$)	66Cc	
	-0.84	1.46(3)	3.9	25	C	μ=2.0($NaClO_4$)	66Cc	
Yb^{3+}	-0.36	2.59(1)	10.6	25	C	μ=2.0($NaClO_4$)	66Cc	
	-0.288	2.71(1)	11.4	25	C	μ=2.0($NaClO_4$)	64Gb	
	-0.75	2.18(2)	7.4	25	C	μ=2.0($NaClO_4$)	66Cc	
	-0.480	2.10(2)	8.0	25	C	μ=2.0($NaClO_4$)	64Gb	
	-1.15	1.43(3)	2.7	25	C	μ=2.0($NaClO_4$)	66Cc	
	-0.89	1.49(3)	3.9	25	C	μ=2.0($NaClO_4$)	64Gb	
	1.07	0.47(4)	5.7	25	C	μ=2.0($NaClO_4$)	64Gb	
Lu^{3+}	-0.65	2.67(1)	10.0	25	C	μ=2.0($NaClO_4$)	66Cc	
	-0.59	2.09(2)	7.6	25	C	μ=2.0($NaClO_4$)	66Cc	
Am^{3+}	-1.33	2.57(25°,1)	7.3	0.5-52.6	T	μ=2.0($NaClO_4$)	72Cd	b
Cm^{3+}	-0.93	2.57(25°,1)	8.6	0.5-52	T	μ=2.0($NaClO_4$)	72Cd	b
Bk^{3+}	-1.21	2.65(25°,1)	8.1	0.5-52.8	T	μ=2.0($NaClO_4$)	72Cd	b
In^{3+}	3.1	3.00(35°,1)	24	25-45	T	μ=0.2M($NaClO_4$)	73Sc	b,q: ±0.5
	2.2	2.58(35°,2)	19	25-45	T	μ=0.2M($NaClO_4$)	73Sc	b,q: ±0.5

A28 ACETIC ACID, hydroxy (phenyl)- $C_8H_8O_3$ $C_6H_5CH(OH)CO_2H$ L^-

Metal Ion	ΔH, kcal /mole	Log K	ΔS, cal /mole °K	T, °C	M	Conditions	Ref.	Remarks
La^{3+}	-1.40	1.93(1)	4.1	25	C	μ=2.0M($NaClO_4$)	72Da	

Metal Ion	ΔH, kcal /mole	Log K	ΔS, cal /mole °K	T, °C	M	Conditions	Ref.	Remarks
A28, cont.								
Ce^{3+}	-1.26	2.17(1)	5.7	25	C	μ=2.0M($NaClO_4$)	72Da	
Pr^{3+}	-1.52	2.30(1)	5.5	25	C	μ=2.0M($NaClO_4$)	72Da	
Nd^{3+}	-1.43	2.43(1)	6.3	25	C	μ=2.0M($NaClO_4$)	72Da	
Sm^{3+}	-1.46	2.47(1)	6.4	25	C	μ=2.0M($NaClO_4$)	72Da	
Eu^{3+}	-1.39	2.25(1)	5.6	25	C	μ=2.0M($NaClO_4$)	72Da	
Gd^{3+}	-1.04	2.42(1)	7.6	25	C	μ=2.0M($NaClO_4$)	72Da	
Tb^{3+}	-0.93	2.52(1)	8.4	25	C	μ=2.0M($NaClO_4$)	72Da	
Dy^{3+}	-1.05	2.57(1)	8.3	25	C	μ=2.0M($NaClO_4$)	72Da	
Ho^{3+}	-1.02	2.54(1)	8.2	25	C	μ=2.0M($NaClO_4$)	72Da	
Er^{3+}	-1.25	2.68(1)	8.1	25	C	μ=2.0M($NaClO_4$)	72Da	
Tm^{3+}	-1.15	2.71(1)	8.6	25	C	μ=2.0M($NaClO_4$)	72Da	
Yb^{3+}	-1.03	2.72(1)	9.0	25	C	μ=2.0M($NaClO_4$)	72Da	
Lu^{3+}	-1.23	2.77(1)	8.6	25	C	μ=2.0M($NaClO_4$)	72Da	
A29 ACETIC ACID, iminodi-			$C_4H_7O_4N$			$HN(CH_2CO_2H)_2$		L^{2-}
Mg^{2+}	2.94	2.94(1)	23.5	20	C	0.1M(KNO_3)	64Ab	
Ca^{2+}	0.3	2.59(1)	12.7	20	C	0.1M(KNO_3)	64Ab	
Sr^{2+}	0.1	2.23(1)	10.5	20	C	0.1M(KNO_3)	64Ab	
Ba^{2+}	0.1	1.67(1)	8.0	20	C	0.1M(KNO_3)	64Ab	
Y^{3+}	-6.53	21.32	73.3	20	C	μ=0.1M(KNO_3) R: $Y^{3+} + EDTA^{4-} + L^{2-} = Y(EDTA)L^{3-}$	71Ga	x: 62Ma
YA^-	-6.53	3.24(1)	-7.4	20	C	μ=0.1M(KNO_3) A=EDTA	71Ga	
La^{3+}	0.17	5.70(1)	25.6	20	C	0.1M(KNO_3)	64Ab	
	-0.33	3.97(2)	18.0	20	C	0.1M(KNO_3)	64Ab	
	-4.49	18.31	68.5	20	C	μ=0.1M(KNO_3) R: $La^{3+} + EDTA^{4-} + L^{2-} = La(EDTA)L^{3-}$	71Ga	x: 62Ma
LaA^-	-1.56	2.80(1)	7.6	20	C	μ=0.1M(KNO_3) A=EDTA	71Ga	
Ce^{3+}	-4.54	19.08	71.8	20	C	μ=0.1M(KNO_3) R: $Ce^{3+} + EDTA^{4-} + L^{2-} = Ce(EDTA)L^{3-}$	71Ga	x: 62Ma
CeA^-	-1.60	3.11(1)	8.8	20	C	μ=0.1M(KNO_3) A=EDTA	71Ga	
Pr^{3+}	-0.867	6.074(1)	24.1	25	C	μ=1.0M($NaClO_4$)	72Ge	q: ±0.065
	-1.77	4.599(2)	15.0	25	C	μ=1.0M($NaClO_4$)	72Ge	
	-2.37	3.099(3)	6.4	25	C	μ=1.0M($NaClO_4$)	72Ge	
	-7.52	10.741	23.9	25	C	μ=1.0M($NaClO_4$) R: $M^{3+} + H^+ + L^{2-} = MHL^{2+}$	72Ge	

Metal Ion	ΔH, kcal /mole	Log K	ΔS, cal /mole °K	T, °C	M	Conditions	Ref.	Remarks
A29, cont.								
Pr^{3+}, cont.	-11.00	12.873	22.0	25	C	μ=1.0M($NaClO_4$) R: $M^{3+} + 2H^+ + L^{2-} = MH_2L^{3+}$	72Ge	
	-6.25	19.88	69.6	20	C	μ=0.1M(KNO_3) R: $Pr^{3+} + EDTA^{4-} + L^{2-} = Pr(EDTA)L^{3-}$	71Ga	x: 62Ma
PrA^-	-3.05	3.48(1)	5.5	20	C	μ=0.1M(KNO_3) A=EDTA	71Ga	
Nd^{3+}	-6.28	20.30	71.4	20	C	μ=0.1M(KNO_3) R: $Nd^{3+} + EDTA^{4-} + L^{2-} = Nd(EDTA)L^{3-}$	71Ga	x: 62Ma
NdA^-	-2.66	3.68(1)	7.8	20	C	μ=0.1M(KNO_3) A=EDTA	71Ga	
Sm^{3+}	-4.26	4.11(1)	4.3	20	C	μ=0.1M(KNO_3) A=EDTA	71Ga	
SmA^-	-7.61	21.26	71.3	20	C	μ=0.1M(KNO_3) R: $Sm^{3+} + EDTA^{4-} + L^{2-} = Sm(EDTA)L^{3-}$	71Ga	x: 62Ma
Eu^{3+}	-0.889	6.458(1)	26.5	25	C	μ=1.0M($NaClO_4$)	62Ge	q: ±0.094
	-3.057	5.200(2)	13.6	25	C	μ=1.0M($NaClO_4$)	72Ge	
	-3.977	4.129(3)	5.5	25	C	μ=1.0M($NaClO_4$)	72Ge	
	-7.366	10.814	24.8	25	C	μ=1.0M($NaClO_4$) R: $M^{2+} + H^+ + L^{2-} = MHL^{2+}$	72Ge	
	-10.53	12.908	23.7	25	C	μ=1.0M($NaClO_4$) R: $M^{2+} + 2H^+ + L^{2-} = MH_2L^{3+}$	72Ge	
	-8.13	21.60	71.0	20	C	μ=0.1M(KNO_3) R: $Eu^{3+} + EDTA^{4-} + L^{2-} = Eu(EDTA)L^{3-}$	71Ga	x: 62Ma
EuA^-	-5.57	4.23(1)	0.3	20	C	μ=0.1M(KNO_3) A=EDTA	71Ga	
Gd^{3+}	-8.34	21.69	70.7	20	C	μ=0.1M(KNO_3) R: $Gd^{3+} + EDTA^{4-} + L^{2-} = Gd(EDTA)L^{3-}$	71Ga	x: 62Ma
GdA^-	-6.61	4.30(1)	-2.9	20	C	μ=0.1M(KNO_3) A=EDTA	71Ga	
Tb^{3+}	-9.21	21.91	68.8	20	C	μ=0.1M(KNO_3) R: $Tb^{3+} + EDTA^{4-} + L^{2-} = Tb(EDTA)L^{3-}$	71Ga	
TbA^-	-8.10	3.98(1)	-9.4	20	C	μ=0.1M(KNO_3) A=EDTA	71Ga	
Dy^{3+}	-8.67	22.14	71.7	20	C	μ=0.1M(KNO_3) R: $Dy^{3+} + EDTA^{4-} + L^{2-} = Dy(EDTA)L^{3-}$	71Ga	x: 62Ma
DyA^-	-7.46	3.84(1)	-7.9	20	C	μ=0.1M(KNO_3) A=EDTA	71Ga	

Metal Ion	ΔH, kcal /mole	Log K	ΔS, cal /mole °K	T, °C	M	Conditions	Ref.	Re-marks
A29, cont.								
Ho^{3+}	0.35	6.642(1)	31.5	25	C	μ=1.0M($NaClO_4$)	72Ge	q: ±0.1
	-1.82	5.325(2)	18.2	25	C	μ=1.0M($NaClO_4$)	72Ge	
	-6.72	4.316(3)	-2.6	25	C	μ=1.0M($NaClO_4$)	72Ge	
	-6.83	10.677	25.8	25	C	μ=1.0M($NaClO_4$) R: $M^{3+} + H^+ + L^{2-} = MHL^{2+}$	72Ge	
	-11.08	12.829	21.5	25	C	μ=1.0M($NaClO_4$) R: $M^{3+} + 2H^+ + L^{2-} = MH_2L^{3+}$	72Ge	q: ±0.94
	-8.50	22.28	73.0	20	C	μ=0.1M(KNO_3) R: $Ho^{3+} + EDTA^{4-} + L^{2-} = Ho(EDTA)L^{3-}$	71Ga	x: 62Ma
HoA^-	-7.14	3.54(1)	-8.2	20	C	μ=0.1M(KNO_3) A=EDTA	71Ga	
Er^{3+}	-8.70	22.11	71.4	20	C	μ=0.1M(KNO_3) R: $Er^{3+} + EDTA^{4-} + L^{2-} = Er(EDTA)L^{3-}$	71Ga	x: 62Ma
ErA^-	-6.99	3.25(1)	-9.0	20	C	μ=0.1M(KNO_3) A=EDTA	71Ga	
Tm^{3+}	-8.37	22.17	72.9	20	C	μ=0.1M(KNO_3) R: $Tm^{3+} + EDTA^{4-} + L^{2-} = Tm(EDTA)L^{3-}$	71Ga	x: 62Ma
TmA^-	-6.50	2.86(1)	-9.1	20	C	μ=0.1M(KNO_3) A=EDTA	71Ga	
Yb^{3+}	-7.85	22.08	74.2	20	C	μ=0.1M(KNO_3) R: $Yb^{3+} + EDTA^{4-} + L^{2-} = Yb(EDTA)L^{3-}$	71Ga	x: 62Ma
YbA^-	-5.54	2.55(1)	-7.2	20	C	μ=0.1M(KNO_3) A=EDTA	71Ga	
Lu^{3+}	-0.13	7.064(1)	31.8	25	C	μ=1.0M($NaClO_4$)	72Ge	q: ±0.09
	-2.00	5.810(2)	20.1	25	C	μ=1.0M($NaClO_4$)	72Ge	
	-5.60	3.275(3)	-4.1	25	C	μ=1.0M($NaClO_4$)	72Ge	
	-6.69	10.601	26.0	25	C	μ=1.0M($NaClO_4$) R: $M^{3+} + H^+ + L^{2-} = MHL^{2+}$	72Ge	
	-11.82	12.707	18.4	25	C	μ=1.0M($NaClO_4$) R: $M^{3+} + 2H^+ + L^{2-} = MH_2L^{3+}$	72Ge	q: ±1.10
	-6.51	22.34	80.0	20	C	μ=0.1M(KNO_3) R: $Lu^{3+} + EDTA^{4-} + L^{2-} = Lu(EDTA)L^{3-}$	71Ga	x: 62Ma
LuA^-	-4.00	2.51(1)	-2.2	20	C	μ=0.1M(KNO_3) A=EDTA	71Ga	
ThA	0.84	6.11(1)	30.8	25	C	μ=0.10(KNO_3) A=1,2-Diaminocyclohexane- N,N,N',N'-tetraacetic acid	70K6	

Metal Ion	ΔH, kcal /mole	Log K	ΔS, cal /mole °K	T, °C	M	Conditions	Ref.	Re-marks
A29, cont.								
ThA, cont.	2.39	6.71(1)	38.7	25	C	μ=0.10(KNO_3) A=EDTA	70Kb	
	8.75	--	--	25	C	μ=0.10(KNO_3) A=1,2-Diaminocyclo-hexane- N,N,N',N'-tetraacetic acid R: ThA + H_2A = $ThAL^{2-}$ +$2H^+$	70Kb	
	10.3	--	--	25	C	μ=0.10(KNO_3) A=EDTA R: ThA + H_2A = $ThAL^{2-}$ + $2H^+$	70Kb	
Am^{3+}	6.9	(1)	13.1	15-90	T	C_M = $8.4x10^{-4}$M C_L = $5x10^{-3}$M	71Bg	b
Co^{2+}	-2.14	6.97(1)	24.6	20	C	0.1M(KNO_3)	64Ab	
	-3.86	5.34(2)	11.2	20	C	0.1M(KNO_3)	64Ab	
Ni^{2+}	-5.05	8.19(1)	20.0	20	C	0.1M(KNO_3)	64Ab	
	-5.4	4.11(1)	0.7	25	C	μ=0.1M(KNO_3)	71Ld	
	-4.45	6.11(2)	13.1	20	C	0.1M(KNO_3)	64Ab	
	-4.2	3.21(2)	0.6	25	C	μ=0.1M(KNO_3)	71Ld	
Cu^{2+}	-4.5	10.63(1)	33.3	20	C	0.1M(KNO_3)	64Ab	
	-6.4	6.05(2)	5.9	20	C	0.1M(KNO_3)	64Ab	
Zn^{2+}	-2.2	7.27(1)	25.7	20	C	0.1M(KNO_3)	64Ab	
	-3.7	5.33(2)	11.8	20	C	0.1M(KNO_3)	64Ab	
Cd^{2+}	-1.46	5.73(1)	21.3	20	C	0.1M(KNO_3)	64Ab	
	-4.02	4.46(2)	6.6	20	C	0.1M(KNO_3)	64Ab	
Pb^{2+}	-3.34	7.45(1)	22.7	20	C	0.1M(KNO_3)	64Ab	
A30	ACETIC ACID, iminodi, N-(2-carboxyethyl)-		$C_7H_{11}O_6N$			$CO_2HCH_2CH_2N(CH_2CO_2H)_2$		L^{3-}
Mg^{2+}	3.35	5.32(1)	35.6	25	C	μ=0.10(KNO_3)	69Va	
Ca^{2+}	-0.25	5.04(1)	22.2	25	C	μ=0.10(KNO_3)	69Va	
Sr^{2+}	-0.46	3.87(1)	16.2	25	C	μ=0.10(KNO_3)	69Va	
Ba^{2+}	-0.11	3.40(1)	15.2	25	C	μ=0.10(KNO_3)	69Va	
Mn^{2+}	1.1	7.33(1)	37.2	25	C	μ=0.1M(KNO_3)	70Mb	
Co^{2+}	-1.7	10.00(1)	40.1	25	C	μ=0.1M(KNO_3)	70Mb	
Ni^{2+}	-3.6	11.37(1)	39.9	25	C	μ=0.1M(KNO_3)	70Mb	
	-3.6	11.37(1)	39.9	25	C	μ=0.1M(KNO_3)	69Va	
Cu^{2+}	-4.1	13.22(1)	46.7	25	C	μ=0.1M(KNO_3)	70Mb	
Zn^{2+}	-1.1	9.98(1)	42.1	25	C	μ=0.1M(KNO_3)	70Mb	
A31	ACETIC ACID, iminodi, N-(o-carboxyphenyl)		$C_{11}H_{11}O_6N$			$(CO_2HC_6H_4)N(CH_2CO_2H)_2$		L^{3-}
Mg^{2+}	7.64	4.01(1)	44.0	25	C	μ=0.1(KNO_3)	69Va	

Metal Ion	ΔH, kcal /mole	Log K	ΔS, cal /mole °K	T, °C	M	Conditions	Ref.	Re-marks
A31, cont.								
Ca^{2+}	1.34	5.08(1)	27.7	25	C	μ=0.1(KNO_3)	69Va	
Sr^{2+}	1.36	3.93(1)	22.5	25	C	μ=0.1(KNO_3)	69Va	
Ba^{2+}	1.28	3.59(1)	20.7	25	C	μ=0.1(KNO_3)	69Va	
Mn^{2+}	4.7	3.85(1)	42.4	25	C	μ=0.1M(KNO_3)	70Mb	
Co^{2+}	3.3	8.42(1)	49.6	25	C	μ=0.1M(KNO_3)	70Mb	
Ni^{2+}	1.5	9.48(1)	48.3	25	C	μ=0.1M(KNO_3)	70Mb	
	1.46	9.48(1)	48.3	25	C	μ=0.1(KNO_3)	69Va	
Cu^{2+}	1.8	10.93(1)	56.1	25	C	μ=0.1M(KNO_3)	70Mb	
Zn^{2+}	3.1	8.42(1)	48.9	25	C	μ=0.1M(KNO_3)	70Mb	
A32 ACETIC ACID, iminodi, N-(2,5-dicarboxyphenyl)-			$C_{12}H_{11}O_8N$			$(CO_2HC_6H_3CO_2H)N(CH_2CO_2H)_2$		
Mg^{2+}	8.3	4.14(1)	46.8	25	C	μ=0.1M(KNO_3)	70Mc	
Ca^{2+}	1.6	5.05(1)	28.6	25	C	μ=0.1M(KNO_3)	70Mc	
Sr^{2+}	1.4	4.12(1)	23.7	25	C	μ=0.1M(KNO_3)	70Mc	
Co^{2+}	3.8	8.83(1)	53.3	25	C	μ=0.1M(KNO_3)	70Mc	
Ni^{2+}	1.8	9.95(1)	50.3	25	C	μ=0.1M(KNO_3)	70Mc	
Cu^{2+}	2.2	10.85(1)	56.9	25	C	μ=0.1M(KNO_3)	70Mc	
Zn^{2+}	3.4	8.86(1)	52.1	25	C	μ=0.1M(KNO_3)	70Mc	
A33 ACETIC ACID, iminodi, N-hydroxethyl-			$C_6H_{11}O_5N$			$HOCH_2CH_2N(CH_2COOH)_2$		L^{2-}
UO_2^{2+}	2.2	8.34(25°,1)	46	20-40	T	μ=0.1M(KNO_3)	69Sq	b
A34 ACETIC ACID, iminotri-			$C_6H_9O_6N$			$N(CH_2CO_2H)_3$		L^{3-}
Mg^{2+}	2.5	6.310(1)	38.0	0	T	μ=0	56Hb 56Ma	
	3.4	6.396(1)	41.2	10	T	μ=0	56Hb	
	4.3	6.500(1)	44.3	20	T	μ=0	56Hb	
	4.8	6.612(1)	46.0	30	T	μ=0	56Hb	
	4.44	5.41(1)	39.8	20	C	μ=0.1(KNO_3)	64Aa	
	4.070	(1)	--	20	C	μ=0.1	64Pc	
	1.97	5.58(1)	32	25	T	μ=0.10(KNO_3)	65Ma	
Ca^{2+}	-2.1	7.704(1)	27.5	0	T	μ=0	56Hb 56Ma	
	-1.7	7.652(1)	29.0	10	T	μ=0	56Hb	
	-1.0	7.608(1)	31.4	20	T	μ=0	56Hb	
	0	7.595(1)	34.6	30	T	μ=0	56Hb	
	-1.36	6.30(1)	24.7	20	C	μ=0.1	64Hc	
	-1.36	6.41(1)	24.2	20	C	μ=0.1(KNO_3)	64Aa	
	-0.81	6.57(1)	27.4	25	T	μ=0.1(KNO_3)	62Mf	
	-6.44	2.49(2)	-10.2	20	C	μ=0.1(KNO_3)	64Aa	
Sr^{2+}	-0.53	4.98(1)	17.5	20	C	μ=0.1(KNO_3)	64Aa	
	-0.54	6.68(1)	20.9	20	C	μ=0.1	64Hc	
	-2.66	5.16(1)	15	25	T	μ=0.10(KNO_3)	65Ma	

Metal Ion	ΔH, kcal /mole	Log K	ΔS, cal /mole °K	T, °C	M	Conditions	Ref.	Re-marks
A34, cont.								
Ba^{2+}	-2.25	5.968(1)	19.1	0	T	μ=0	56Hb	
	-1.70	5.914(1)	21.0	10	T	μ=0	56Hb	
	-1.20	5.875(1)	22.8	20	T	μ=0	56Hb	
	-1	5.86(1)	24	25	T	μ=0	56Ma	
	-0.50	5.587(1)	25.1	30	T	μ=0	56Hb	
	-1.44	4.82(1)	17.1	20	C	μ=0.1(KNO_3)	64Aa	
	-1.45	4.94(1)	18	25	T	μ=0.10(KNO_3)	65Ma	
Y^{3+}	1.027	11.458(1)	56	20	C	μ=0.1	64Pc	
	2.69	11.48(1)	61.6	25	T	μ=0.1(KNO_3)	62Mf	
	-4.19	9.02(2)	26.9	25	T	μ=0.1(KNO_3)	62Mf	
	-7.59	21.82	74.0	20	C	μ=0.1M(KNO_3) R: Y^{3+} + EDTA + L^{3-} = $Y(EDTA)L^{4-}$	71Ga	x: 62Ma
YA^-	-7.00	3.74(1)	-6.8	20	C	μ=0.1M(KNO_3) A=EDTA	71Ga	
La^{3+}	0.320	10.370(1)	48.8	20	C	μ=0.1	64Pc 61La	
	2.05	10.36(1)	54.3	25	T	μ=0.1(KNO_3)	62Mf	
	-2.43	7.24(2)	25.0	25	T	μ=0.1(KNO_3)	62Mf	
	-9.01	20.30	62.1	20	C	μ=0.1M(KNO_3) R: La^{3+} + $EDTA^{4-}$ + L^{3-} = $La(EDTA)L^{4-}$	71Ga	x: 62Ma
LaA^-	-6.08	4.79(1)	1.2	20	C	μ=0.1M(KNO_3) A=EDTA	71Ga	
Ce^{3+}	-0.215	10.824(1)	48.8	20	C	μ=0.1	64Pc 61La	
	1.25	10.83(1)	53.8	25	T	μ=0.1(KNO_3)	62Mf	
	0.40	10.26(1)	48.0	25	C	μ=0.2	69Mc	g:Na, q:±0.1
	-3.06	7.84(2)	25.6	25	T	μ=0.1(KNO_3)	62Mf	
	-3.00	7.99(2)	26.3	25	C	μ=0.2	69Mc	g:Na
	-8.66	20.70	65.1	20	C	μ=0.1M(KNO_3) R: Ce^{3+} + $EDTA^{4-}$ + L^{3-} = $Ce(EDTA)L^{4-}$	71Ga	x: 62Ma
CeA^-	-5.72	4.73(1)	2.1	20	C	μ=0.1M(KNO_3) A=EDTA	71Ga	
Pr^{3+}	-0.502	11.063(1)	48.9	20	C	μ=0.1	64Pc 61La	
	0.45	11.07(1)	52.2	25	T	μ=0.1(KNO_3)	62Mf	
	0.40	10.47(1)	49.0	25	C	μ=0.2	69Mc	g:Na, q:±0.1
	-3.72	8.18(2)	24.9	25	T	μ=0.1(KNO_3)	62Mf	
	-3.40	8.15(2)	27.1	25	C	μ=0.2	69Mc	g:Na
	-7.85	21.08	69.7	20	C	μ=0.1M(KNO_3) R: Pr^{3+} + $EDTA^{4-}$ + L^{3-} = $Pr(EDTA)L^{4-}$	71Ga	x: 62Ma
PrA^-	-4.65	4.67(1)	5.5	20	C	μ=0.1M(KNO_3) A=EDTA	71Ga	
Nd^{3+}	-0.803	11.249(1)	48.8	20	C	μ=0.1	64Pc 61La	

Metal Ion	ΔH, kcal /mole	Log K	ΔS, cal /mole °K	T, °C	M	Conditions	Ref.	Remarks
A34, cont.								
Nd^{3+}, cont.	0.68	11.26(1)	53.8	25	T	μ=0.1(KNO_3)	62Mf	
	0.20	10.67(1)	49.2	25	C	μ=0.2	69Mc	g:Na, q: ±0.05
	-3.78	8.47(2)	26.1	25	T	μ=0.1(KNO_3)	62Mf	
	-2.80	8.47(2)	29.2	25	C	μ=0.2	69Mc	g:Na
	-7.77	21.38	71.4	20	C	μ=0.1M(KNO_3) R: Nd^{3+} + $EDTA^{4-}$ + L^{3-} = Nd(EDTA)L^{4-}	71Ga	x: 62Ma
NdA^-	-4.15	4.77(1)	7.7	20	C	μ=0.1M(KNO_3) A=EDTA	71Ga	
Sm^{3+}	-1.047	11.50(1)	49.1	20	C	μ=0.1	64Pc 61La	
	0.40	11.53(1)	54.1	25	T	μ=0.1(KNO_3)	62Mf	
	-0.40	10.91(1)	48.0	25	C	μ=0.2	69Mc	g:Na, q: ±0.10
	-4.37	9.00(2)	26.5	25	T	μ=0.1(KNO_3)	62Mf	
	-3.90	9.18(2)	28.6	25	C	μ=0.2	69Mc	g:Na
	-9.58	22.15	68.6	20	C	μ=0.1M(KNO_3) R: Sm^{3+} + $EDTA^{4-}$ + L^{3-} = Sm(EDTA)L^{4-}	71Ga	x: 62Ma
SmA^-	-6.23	5.00(1)	1.6	20	C	μ=0.1M(KNO_3) A=EDTA	71Ga	
Eu^{3+}	-1.029	11.49(1)	49.1	20	C	μ=0.1	64Pc 61La	
	0.93	11.52(1)	55.8	25	T	μ=0.1(KNO_3)	62Mf	
	-5.08	9.18(2)	25.0	25	T	μ=0.1(KNO_3)	62Mf	
	-9.74	22.40	69.2	20	C	μ=0.1M(KNO_3) R: Eu^{3+} + $EDTA^{4-}$ + L^{3-} = Eu(EDTA)L^{4-}	71Ga	x: 62Ma
EuA^-	-7.18	5.03(1)	-1.5	20	C	μ=0.1M(KNO_3) A=EDTA	71Ga	
Gd^{3+}	-0.626	11.54(1)	50.7	20	C	μ=0.1	64Pc 61La	
	1.02	11.54(1)	56.2	25	T	μ=0.1(KNO_3)	62Mf	
	-0.40	11.00(1)	48.7	25	C	μ=0.2	69Mc	g:Na, q: ±0.10
	-5.80	9.27(2)	22.9	25	T	μ=0.1(KNO_3)	62Mf	
	-4.80	9.31(2)	26.4	25	C	μ=0.2	69Mc	g:Na
	-9.65	22.25	68.8	20	C	μ=0.1M(KNO_3) R: Gd^{3+} + $EDTA^{4+}$ + L^{3-} = Gd(EDTA)L^{4-}	71Ga	x: 62Ma
GdA^-	-7.92	4.86(1)	-4.7	20	C	μ=0.1M(KNO_3) A=EDTA	71Ga	
Tb^{3+}	-0.061	11.58(1)	52.8	20	C	μ=0.1	64Pc 61La	
	1.71	11.60(1)	58.8	25	T	μ=0.1(KNO_3)	62Mf	

Metal Ion	ΔH, kcal /mole	Log K	ΔS, cal /mole °K	T, °C	M	Conditions	Ref.	Remarks
A34, cont.								
Tb^{3+}, cont.	-5.21	9.38(2)	25.4	25	T	μ=0.1(KNO_3)	62Mf	
	-9.60	22.50	70.6	20	C	μ=0.1M(KNO_3) R: Tb^{3+} + $EDTA^{4-}$ + L^{3-} = $Tb(EDTA)L^{4-}$	71Ga	x: 62Ma
TbA^{-}	-8.49	4.65(1)	-7.7	20	C	μ=0.1M(KNO_3) A=EDTA	71Ga	
Dy^{3+}	0.350	11.70(1)	54.8	20	C	μ=0.1	64Pc 61La	
	2.25	11.74(1)	61.3	25	T	μ=0.1(KNO_3)	62Mf	
	-5.61	9.41(2)	24.2	25	T	μ=0.1(KNO_3)	62Mf	
	-9.66	22.59	70.4	20	C	μ=0.1M(KNO_3) R: Dy^{3+} + $EDTA^{4-}$ + L^{3-} = $Dy(EDTA)L^{4-}$	71Ga	x: 62Ma
DyA^{-}	-8.45	4.29(1)	-9.2	20	C	μ=0.1M(KNO_3) A=EDTA	71Ga	
Ho^{3+}	0.543	11.846(1)	56.1	20	C	μ=0.1	64Pc 61La	
	2.63	11.90(1)	63.3	25	T	μ=0.1(KNO_3)	62Mf	
	-5.15	9.35(2)	25.5	25	T	μ=0.1(KNO_3)	62Mf	
	-9.42	22.70	71.7	20	C	μ=0.1M(KNO_3) R: Ho^{3+} + $EDTA^{4-}$ + L^{3-} = $Ho(EDTA)L^{4-}$	71Ga	x: 62Ma
HoA^{-}	-8.06	3.95(1)	-9.4	20	C	μ=0.1M(KNO_3) A=EDTA	71Ga	
Er^{3+}	0.593	11.995(1)	56.9	20	C	μ=0.1	64Pc 61La	
	2.51	12.04(1)	63.5	25	T	μ=0.1(KNO_3)	62Mf	
	-3.87	9.26(2)	29.4	25	T	μ=0.1(KNO_3)	62Mf	
	-8.94	22.38	71.9	20	C	μ=0.1M(KNO_3) R: Er^{3+} + $EDTA^{4-}$ + L^{3-} = $Er(EDTA)L^{4-}$	71Ga	x: 62Ma
ErA^{-}	-7.32	3.52(1)	-8.5	20	C	μ=0.1M(KNO_3) A-EDTA	71Ga	
Tm^{3+}	0.585	12.196(1)	57.8	20	C	μ=0.1	64Pc 61La	
	1.90	12.23(1)	62.3	25	T	μ=0.1(KNO_3)	62Mf	
	-2.36	9.24(2)	34.3	25	T	μ=0.1(KNO_3)	62Mf	
	-8.62	22.44	73.3	20	C	μ=0.1M(KNO_3) R: Tm^{3+} + $EDTA^{4-}$ + L^{3-} = $Tm(EDTA)L^{4-}$	71Ga	x: 62Ma
TmA^{-}	-6.75	3.13(1)	-8.7	20	C	μ=0.1M(KNO_3) A=EDTA	71Ga	
Yb^{3+}	0.400	12.368(1)	58.0	20	C	μ=0.1	64Pc 61La	
	2.09	12.40(1)	63.7	25	T	μ=0.1(KNO_3)	62Mf	
	-1.86	9.29(2)	36.3	25	T	μ=0.1(KNO_3)	62Mf	
	-7.56	22.38	76.5	20	C	μ=0.1M(KNO_3) R: Yb^{3+} + $EDTA^{4-}$ + L^{3-} = $Yb(EDTA)L^{4-}$	71Ga	x: 62Ma

Metal Ion	ΔH, kcal /mole	Log K	ΔS, cal /mole °K	T, °C	M	Conditions	Ref.	Remarks
A34, cont.								
YbA^-	-5.25	2.85(1)	-4.9	20	C	μ=0.1M(KNO_3) A=EDTA	71Ga	
Lu^{3+}	0.180	12.465(1)	57.7	20	C	μ=0.1	64Pc 61La	
	1.98	12.49(1)	63.8	25	T	μ=0.1(KNO_3)	62Mf	
	-1.11	9.42(2)	39.4	25	T	μ=0.1(KNO_3)	62Mf	
	-6.11	22.67	82.8	20	C	μ=0.1M(KNO_3) R: Lu^{3+} + $EDTA^{4-}$ + L^{3-} = $Lu(EDTA)L^{4-}$	71Ga	x: 62Ma
LuA^-	-3.60	2.81(1)	0.6	20	C	μ=0.1M(KNO_3) A=EDTA	71Ga	
Mn^{2+}	∿0	8.527(1)	39.1	0	T	μ=0	56Hb	a
	0.8	8.534(1)	41.9	10	T	μ=0	56Hb	a
	2.0	8.573(1)	46.0	20	T	μ=0	56Hb	a
	3.5	8.644(1)	51.1	30	T	μ=0	56Hb	a
	1.14	7.44(1)	37.9	20	C	μ=0.1	64Hc	
	1.44	7.44(1)	38.9	20	C	μ=0.1(KNO_3)	64Aa	
	-5.58	3.55(2)	-1.7	20	C	μ=0.1(KNO_3)	64Aa	
Fe^{2+}	1.5	6.06	-23	25	T	μ=1.0($NaClO_4$) A=NH_3 R: Fe^{2+} + CoA_5L = CoA_5LFe^{2+}	74Ca	
Fe^{3+}	3.2	15.87(1)	83.6	20	C	μ=0.1(KNO_3)	64Aa	
Co^{2+}	-0.15	10.40(1)	47.1	20	C	μ=0.1	64Hc	
	-0.07	10.38(1)	47.2	20	C	μ=0.1(KNO_3)	64Aa	
	-4.76	4.01(2)	2.1	20	C	μ=0.1(KNO_3)	64Aa	
Ni^{2+}	-2.53	11.53(1)	44.1	20	C	μ=0.1	64Hc	
	-2.56	11.53(1)	43.9	20	C	μ=0.1(KNO_3)	64Aa	
	-5.57	4.88(2)	7.0	20	C	μ=0.1(KNO_3)	64Aa	
Cu^{2+}	-1.84	12.96(1)	53.0	20	C	μ=0.1	64Hc	
	-1.87	12.96(1)	52.8	20	C	μ=0.1(KNO_3)	64Aa	
	-1.1	13.10(1)	56.3	25	T	μ=0.1(KNO_3)	62Mf	
	-6.03	4.47(2)	-3.5	20	C	μ=0.1(KNO_3)	64Aa	
Zn^{2+}	-0.84	10.67(1)	46.0	20	C	μ=0.1	64Hc	
	-0.87	10.67(1)	45.5	20	C	μ=0.1(KNO_3)	64Aa	
	-2.81	3.62(2)	7.4	20	C	μ=0.1(KNO_3)	64Aa	
Cd^{2+}	3.97	9.82(1)	31.3	20	C	μ=0.1(KNO_3)	64Aa	
	-13.02	4.79(2)	4.7	20	C	μ=0.1(KNO_3)	64Aa	
Pb^{2+}	-3.81	11.39(1)	39.1	20	C	μ=0.1(KNO_3)	64Aa	
A35	ACETIC ACID, mercapto-		$C_2H_4O_2S$			$HSCH_2CO_2H$		L^-
La^{3+}	1.486	1.41(1)	11.5	25	C	μ=2.0($NaClO_4$)	64Gb	
	2.40	0.70(2)	11.2	25	C	μ=2.0($NaClO_4$)	64Gb	
Pr^{3+}	2.91	2.742(1)	22.3	25	C	μ=1.00M($NaClO_4$)	73Dd	
	1.80	1.643(2)	13.6	25	C	μ=1.00M($NaClO_4$)	73Dd	q: ±0.3

Metal Ion	ΔH, kcal /mole	Log K	ΔS, cal /mole °K	T, °C	M	Conditions	Ref.	Remarks
A35, cont.								
Pr^{3+}, cont.	1.53	5.447	30.1	25	C	μ=1.00M($NaClO_4$) R: M^{3+} + L^{2-} + H^+ = MHL^{2+}	73Dd	q: ±0.6
	2.29	2.190	17.7	25	C	μ=1.00M($NaClO_4$) R: MHL^{2+} + L^{2-} = MHL_2	73Dd	q: ±1.2
Sm^{3+}	2.72	2.903(1)	22.4	25	C	μ=1.00M($NaClO_4$)	73Dd	
	1.335	1.80(1)	12.7	25	C	μ=2.0($NaClO_4$)	64Gb	
	2.21	1.773(2)	15.6	25	C	μ=1.00M($NaClO_4$)	73Dd	
	1.97	0.91(2)	10.8	25	C	μ=2.0($NaClO_4$)	64Gb	
	1.29	5.592	29.9	25	C	μ=1.00M($NaClO_4$) R: M^{3+} + L^{2-} +H^+ = MHL^{2+}	73Dd	q: ±0.14
	2.17	2.225	17.4	25	C	μ=1.00M($NaClO_4$) R: MHL^{2+} + L^{2-} = MHL_2	73Dd	q: ±0.35
Tb^{3+}	4.11	2.521(1)	25.3	25	C	μ=1.00M($NaClO_4$)	73Dd	
	2.17	1.606(2)	14.6	25	C	μ=1.00M($NaClO_4$)	73Dd	
	3.08	5.205	34.2	25	C	μ=1.00M($NaClO_4$) R: M^{3+} + L^{2-} + H^+ = MHL^{2+}	73Dd	q: ±0.6
	1.70	2.085	15.3	25	C	μ=1.00M($NaClO_4$) R: MHL^{2+} + L^{2-} = MHL_2	73Dd	q: ±1.4
Er^{3+}	5.01	2.360(1)	27.6	25	C	μ=1.00M($NaClO_4$)	73Dd	
	2.380	1.28(1)	13.8	25	C	μ=2.0($NaClO_4$)	64Gb	
	2.90	1.489(2)	16.6	25	C	μ=1.00M($NaClO_4$)	73Dd	
	3.06	0.90(2)	14.4	25	C	μ=2.0($NaClO_4$)	64Gb	
	3.66	5.139	35.8	25	C	μ=1.00M($NaClO_4$) R: M^{3+} + L^{2-} + H^+ = MHL^{2+}	73Dd	q: ±0.55
	2.37	2.062	17.2	25	C	μ=1.00M($NaClO_4$) R: MHL^{2+} + L^{2-}	73Dd	q: ±0.91
Yb^{3+}	5.35	2.360(1)	28.7	25	C	μ=1.00M($NaClO_4$)	73Dd	
	3.28	1.400(2)	17.4	25	C	μ=1.00M($NaClO_4$)	73Dd	
	3.51	5.095	35.1	25	C	μ=1.00M($NaClO_4$) R: M^{3+} + L^{2-} + H^+ = MHL^{2+}	73Dd	q: ±0.26
	2.84	2.176	19.4	25	C	μ=1.00M($NaClO_4$) R: MHL^{2+} + L^{2-}	73Dd	q: ±0.38
Ni^{2+}	-3.5	13.041(1-2)	48	25	T	μ=0.11(KCl) L=L^{2-}	60Lb	
	-31	49.845(4-6)	124	25	T	μ=0.11(KCl) L=L^{2-}	60Lb	
In^{3+}	-6.1	11.98(35°,1)	35	25-45	T	μ=0.2M($NaClO_4$)	73Sc	b,q: ±0.5
	-6.3	10.19(35°,2)	26	25-45	T	μ=0.2M($NaClO_4$)	73Sc	b,q: ±0.5
	-7.0	6.16(35°,3)	5	25-45	T	μ=0.2M($NaClO_4$)	73Sc	b,q: ±0.5
A36 ACETIC ACID, methyliminodi-			$C_5H_9O_4N$			$CH_3N(CH_2CO_2H)_2$		L^{2-}
Mg^{2+}	2.85	3.48(1)	25.6	25	C	μ=0.1	68Na	
	-0.67	2.32(2)	8.4	25	C	μ=0.1	68Na	

Metal Ion	ΔH, kcal /mole	Log K	ΔS, cal /mole °K	T, °C	M	Conditions	Ref.	Remarks
A36, cont.								
Ca^{2+}	-1.19	3.85(1)	13.6	25	C	μ=0.1	68Na	
	-0.74	2.81(2)	10.0	25	C	μ=0.1	68Na	
Sr^{2+}	-0.83	2.95(1)	10.7	25	C	μ=0.1	68Na	
Ba^{2+}	-1.06	2.61(1)	8.4	25	C	μ=0.1	68Na	
$UO_2{}^{2+}$	1.0	9.71(25°,1)	48	20-40	T	μ=0.1M(KNO_3)	69Sg	b
A37 ACETIC ACID, oxydi-			$C_4H_6O_5$			$O(CH_2CO_2H)_2$		L^{2-}
Sc^{3+}	-0.47	8.28(1)	36.3	25	C	μ=1.0($NaClO_4$)	69Ge	
	-2.55	4.48(2)	12.0	25	C	μ=1.0($NaClO_4$)	69Ge	
Y^{3+}	1.732	5.26(1)	30.0	25	C	μ=1.0	63Gc	
	-1.243	4.51(2)	16.4	25	C	μ=1.0	63Gc	
	-4.198	3.23(3)	0.8	25	C	μ=1.0	63Gc	
La^{3+}	-0.070	4.93(1)	22.3	25	C	μ=1.0	63Gc	
	-0.758	3.47(2)	13.3	25	C	μ=1.0	63Gc	
	0.354	1.84(3)	9.6	25	C	μ=1.0	63Gc	
Ce^{3+}	-0.401	5.15(1)	22.2	25	C	μ=1.0	63Gc	
	-0.869	3.76(2)	14.3	25	C	μ=1.0	63Gc	
	-0.494	2.30(3)	8.9	25	C	μ=1.0	63Gc	
Pr^{3+}	-1.280	(1)	--	5	C	μ=1.0M($NaClO_4$)	72Gf	
	-0.757	(1)	--	20	C	μ=1.0M($NaClO_4$)	72Gf	
	-0.680	5.33(1)	22.1	25	C	μ=1.0	63Gc	
	-0.284	(1)	--	35	C	μ=1.0M($NaClO_4$)	72Gf	
	0.234	(1)	--	50	C	μ=1.0M($NaClO_4$)	72Gf	
		(1)				$\Delta H=A + BT + CT^2 + DT^3$		
						A=-137.928	72Gf	
						B=1.306856		
						C=-4.237067×10^{-3}		
						D=4.691358×10^{-6}		
						ΔCp(25°,1) = 31.4		
	-1.362	(2)	--	5	C	μ=1.0M($NaClO_4$)	72Gf	
	-1.039	(2)	--	20	C	μ=1.0M($NaClO_4$)	72Gf	
	-1.032	3.88(2)	14.3	25	C	μ=1.0	63Gc	
	-0.791	(2)	--	35	C	μ=1.0M($NaClO_4$)	72Gf	
	-0.494	(2)	--	50	C	μ=1.0M($NaClO_4$)	72Gf	
		(2)				$\Delta H=A + BT + CT^2 + DT^3$		
						A=-174.819	72Gf	
						B=1.694176		
						C=-5.5552124×10^{-3}		
						D=6.123457×10^{-6}		
						ΔCp(25°,2) = 16.4		
	-0.335	(3)	--	5	C	μ=1.0M($NaClO_4$)	72Gf	
	-0.331	(3)	--	20	C	μ=1.0M($NaClO_4$)	72Gf	
	-0.788	2.39(3)	8.3	25	C	μ=1.0	63Gc	
	-0.452	(3)	--	35	C	μ=1.0M($NaClO_4$)	72Gf	
	-0.545	(3)	--	50	C	μ=1.0M($NaClO_4$)	72Gf	
		(3)				$\Delta H=A + BT + CT^2 + DT^3$		
						A=-212.924	72Gf	
						B=2.105303		
						C=-6.922738×10^{-3}		
						D=7.555556×10^{-6}		

Metal Ion	ΔH, kcal /mole	Log K	ΔS, cal /mole °K	T, °C	M	Conditions	Ref.	Remarks
A37, cont.								
Pr^{3+}, cont.						ΔCp(25°,3) = -7.8		
Nd^{3+}	-0.848	3.44(1)	22.1	25	C	μ=1.0	63Gc	
	-1.255	4.04(2)	14.3	25	C	μ=1.0	63Gc	
	-0.897	2.65(3)	9.0	25	C	μ=1.0	63Gc	
Sm^{3+}	-1.673	(1)	--	5	C	μ=1.0M($NaClO_4$)	72Gf	
	-1.160	(1)	--	20	C	μ=1.0M($NaClO_4$)	72Gf	
	-1.048	5.53(1)	21.8	25	C	μ=1.0	63Gc	
	-0.658	(1)	--	35	C	μ=1.0M($NaClO_4$)	72Gf	
	-0.081	(1)	--	50	C	μ=1.0M($NaClO_4$)	72Gf	
		(1)				$\Delta H=A + BT + CT^2 + DT^3$		
						A=-119.900	72Gf	q:
						B=1.142185		±0.016
						$C=-3.759520x10^{-3}$		
						$D=4.246914x10^{-6}$		
						ΔCp(25°,1) = 33.0		
	-2.163	(2)	--	5	C	μ=1.0M($NaClO_4$)	72Gf	
	-1.835	(2)	--	20	C	μ=1.0M($NaClO_4$)	72Gf	
	-1.830	4.33(2)	13.7	25	C	μ=1.0	63Gc	
	-1.593	(2)	--	35	C	μ=1.0M($NaClO_4$)	72Gf	
	-1.361	(2)	--	50	C	μ=1.0M($NaClO_4$)	72Gf	
		(2)				$\Delta H=A + BT + CT^2 + DT^3$		
						A=-159.179	72Gf	
						B=1.505553		
						$C=-4.839253x10^{-3}$		
						$D=5.234568x10^{-6}$		
						ΔCp(25°,2) = 15.9		
	-1.033	(3)	--	5	C	μ=1.0M($NaClO_4$)	72Gf	
	-1.070	(3)	--	20	C	μ=1.0M($NaClO_4$)	72Gf	
	-1.413	2.89(3)	8.4	25	C	μ=1.0	63Gc	
	-1.225	(3)	--	35	C	μ=1.0M($NaClO_4$)	72Gf	
	-1.370	(3)	--	50	C	μ=1.0M($NaClO_4$)	72Cf	
		(3)				$\Delta H=A + BT + CT^2 + DT^3$		
						A=-180.570	72Gf	
						B=1.775654		
						$C=-5.821404x10^{-3}$		
						$D=6.320988x10^{-6}$		
						ΔCp(25°,3) =-10.0		
Eu^{3+}	-0.781	5.52(1)	22.6	25	C	μ=1.0	63Gc	
	2.162	4.48(2)	13.3	25	C	μ=1.0	63Gc	
	1.564	3.15(3)	9.1	25	C	μ=1.0	63Gc	
Gd^{3+}	-0.976	(1)	--	5	C	μ=1.0M($NaClO_4$)	72Gf	
	-0.476	5.490(1)	23.4	20	C	μ=1.0M($NaClO_4$)	72Gf	
	-0.360	5.40(1)	23.5	25	C	μ=1.0	63Gc	
	0.034	(1)	--	35	C	μ=1.0M($NaClO_4$)	72Gf	
	0.602	(1)	--	50	C	μ=1.0M($NaClO_4$)	72Gf	
		(1)				$\Delta H=A + BT + CT^2 + DT^3$		
						A=-68.001	72Gf	
						B=0.631253		
						$C=-2.062471x10^{-3}$		q:
						$D=2.370370x10^{6}$		±0.016
						ΔCp(25°,1) = 33.5		

Metal Ion	ΔH, kcal /mole	Log K	ΔS, cal /mole °K	T, °C	M	Conditions	Ref.	Re-marks
A37, cont.								
Gd^{3+}, cont.	-2.706	(2)	--	5	C	μ=1.0M($NaClO_4$)	72Gf	
	-2.361	4.543(2)	12.7	20	C	μ=1.0M($NaClO_4$)	72Gf	
	-2.312	4.50(2)	12.8	25	C	μ=1.0	63Gc	
	-2.127	(2)	--	35	C	μ=1.0M($NaClO_4$)	72Gf	
	-1.759	(2)	--	50	C	μ=1.0M($NaClO_4$)	72Gf	
		(2)				$\Delta H=A + BT + CT^2 + DT^3$		
						A=-333.248	72Gf	
						B=3.280608		
						$C=-10.887279\times10^{-3}$		
						$D=12.098765\times10^{-6}$		
						ΔCp(25°,2) = 15.0		
	-1.339	(3)	--	5	C	μ=1.0M($NaClO_4$)	72Gf	
	-1.482	3.187(3)	3.7	20	C	μ=1.0M($NaClO_4$)	72Gf	
	-1.918	3.08(3)	7.7	25	C	μ=1.0	63Gc	
	-1.714	(3)	--	35	C	μ=1.0M($NaClO_4$)	72Gf	
	-1.955	(3)	--	50	C	μ=1.0M($NaClO_4$)	72Gf	
		(3)				$\Delta H=A + BT + CT^2 + DT^3$		
						A=-114.090	72Gf	
						B=1.121153		
						$C=-3.672267\times10^{-3}$		
						$D=3.950617\times10^{-6}$		
						ΔCp(25°,3) = -15.1		
Tb^{3+}	0.094	(1)	--	5	C	μ=1.0M($NaClO_4$)	72Gf	
	0.544	5.422(1)	23.0	20	C	μ=1.0M($NaClO_4$)	72Gf	
	0.765	5.34(1)	27.0	25	C	μ=1.0	63Gc	
	0.984	(1)	--	35	C	μ=1.0M($NaClO_4$	72Gf	
	1.479	(1)	--	50	C	μ=1.0M($NaClO_4$)	72Gf	
		(1)				$\Delta H=A + BT + CT^2 + DT^3$		
						A=-90.724	72Gf	q:
						B=0.869571		±0.009
						$C=-2.845244\times10^{-3}$		
						$D=3.209877\times10^{-6}$		
						ΔCp(25°,1) = 29.0		
	-3.049	(2)	--	5	C	μ=1.0M($NaClO_4$)	72Gf	
	-2.625	4.657(2)	12.4	20	C	μ=1.0M($NaClO_4$)	72Gf	
	-2.652	4.61(2)	12.2	25	C	μ=1.0	63Gc	
	-2.224	(2)	--	35	C	μ=1.0M($NaClO_4$)	72Gf	
	-1.711	(2)	--	50	C	μ=1.0M($NaClO_4$)	72Gf	
		(2)				$\Delta H=A + BT + CT^2 + DT^3$		
						A=-182.606	72Gf	
						B=1.774823		
						$C=-5.914311\times10^{-3}$		
						$D=6.666667\times10^{-6}$		
						ΔCp(25°,2) = 26.0		
	-1.982	(3)	--	5	C	μ=1.0M($NaClO_4$)	72Gf	
	-2.251	3.329(3)	7.6	20	C	μ=1.0M($NaClO_4$)	72Gf	
	-2.577	3.25(3)	6.2	25	C	μ=1.0	63Gc	
	-2.651	(3)	--	35	C	μ=1.0M($NaClO_4$)	72Gf	
	-3.057	(3)	--	50	C	μ=1.0M($NaClO_4$)	72Gf	
		(3)				$\Delta H=A + BT + CT^2 + DT^3$		
						A=-175.850	72Gf	
						B=1.738528		
						$C=-5.720000\times10^{-3}$		

Metal Ion	ΔH, kcal /mole	Log K	ΔS, cal /mole °K	T, °C	M	Conditions	Ref.	Remarks
A37, cont.								
Tb^{3+},						$D=6.172840\times10^{-6}$		
cont.						$\Delta Cp(25°,3) = 26.1$		
Dy^{3+}	0.574	(1)	--	5	C	$\mu=1.0M(NaClO_4)$	72Gf	
	1.084	(1)	--	20	C	$\mu=1.0M(NaClO_4)$	72Gf	
	1.323	5.33(1)	28.8	25	C	$\mu=1.0$	63Gc	
	1.468	(1)	--	35	C	$\mu=1.0M(NaClO_4)$	72Gf	
	2.040	(1)	--	50	C	$\mu=1.0M(NaClO_4)$	72Gf	
		(1)				$\Delta H=A + BT + CT^2 + DT^3$		
						$A=-63.978$	72Gf	
						$B=0.604455$		
						$C=-1.956622\times10^{-3}$		
						$D=2.222222\times10^{-6}$		
						$\Delta Cp(25°,1) = 30.3$		
	-2.833	(2)	--	5	C	$\mu=1.0M(NaClO_4)$	72Gf	
	-2.397	(2)	--	20	C	$\mu=1.0M(NaClO_4)$	72Gf	
	-2.404	4.64(2)	13.2	25	C	$\mu=1.0$	63Gc	
	-1.830	(2)	--	35	C	$\mu=1.0M(NaClO_4)$	72Gf	
	-1.424	(2)	--	50	C	$\mu=1.0M(NaClO_4)$	72Gf	
		(2)				$\Delta H=A + BT + CT^2 + DT^3$		
						$A=-14.249$	72Gf	
						$B=0.052939$		
						$C=-0.056764\times10^{-3}$		
						$D=0.049383\times10^{-6}$		
						$\Delta Cp(25°,2) = 32.3$		
	-2.598	(3)	--	5	C	$\mu=1.0M(NaClO_4)$	72Gf	
	-2.976	(3)	--	20	C	$\mu=1.0M(NaClO_4)$	72Gf	
	-3.332	3.34(3)	4.1	25	C	$\mu=1.0$	63Gc	
	-3.450	(3)	--	35	C	$\mu=1.0M(NaClO_4)$	72Gf	
	-3.893	(3)	--	50	C	$\mu=1.0M(NaClO_4)$	72Gf	
		(3)				$\Delta H=A + BT + CT^2 + DT^3$		
						$A=-170.854$	72Gf	
						$B=1.712268$		
						$C=-5.729084\times10^{-3}$		
						$D=6.271605\times10^{-6}$		
						$\Delta Cp(25°,3) = -31.5$		
Ho^{3+}	0.953	(1)	--	5	C	$\mu=1.0M(NaClO_4)$	72Gf	
	1.382	5.396(1)	29.4	20	C	$\mu=1.0M(NaClO_4)$	72Gf	
	1.591	5.30(1)	29.6	25	C	$\mu=1.0$	63Gc	
	1.788	(1)	--	35	C	$\mu=1.0M(NaClO_4)$	72Gf	
	2.222	(1)	--	50	C	$\mu=1.0M(NaClO_4)$	72Gf	
		(1)				$\Delta H=A + BT + CT^2 + DT^3$		
						$A=-74.458$	72Gf	
						$B=0.706580$		
						$C=-2.266098\times10^{-3}$		
						$D=2.518519\times10^{-6}$		
						$\Delta Cp(25°,1) = 26.9$		
	-2.468	(2)	--	5	C	$\mu=1.0M(NaClO_4)$	72Gf	
	-1.866	4.676(2)	15.0	20	C	$\mu=1.0M(NaClO_4)$	72Gf	
	-1.852	4.64(2)	15.0	25	C	$\mu=1.0$	63Gc	
	-1.340	(2)	--	35	C	$\mu=1.0M(NaClO_4$	72Gf	
	-0.864	(2)	--	50	C	$\mu=1.0M(NaClO_4)$	72Gf	
		(2)				$\Delta H=A + BT + CT^2 + DT^3$		
						$A=-59.668$	72Gf	

Metal Ion	ΔH, kcal /mole	Log K	ΔS, cal /mole °K	T, °C	M	Conditions	Ref.	Re-marks
A37, cont.								
Ho^{3+}, cont.						B=0.467373		
						C=-1.298098x10^{-3}		
						D=1.283951x10^{-6}		
						ΔCp(25°,2) = 35.7		
	-3.350	(3)	--	5	C	μ=1.0M($NaClO_4$)	72Gf	
	-3.793	3.392(3)	2.6	20	C	μ=1.0M($NaClO_4$)	72Gf	
	-4.123	3.32(3)	1.4	25	C	μ=1.0	63Gc	
	-4.260	(3)	--	35	C	μ=1.0M($NaClO_4$)	72Gf	
	-4.670	(3)	--	50	C	μ=1.0M($NaClO_4$)	72Gf	
		(3)				ΔH=A + BT + CT^2 + DT^3		
						A=-100.000	72Gf	
						B=1.031351		
						C=-3.571253x10^{-3}		
						D=4.000000x10^{-6}		
						ΔCp(25°,3) = 31.5		
Er^{3+}	1.027	(1)	--	5	C	μ=1.0M($NaClO_4$)	72Gf	
	1.466	5.470(1)	30.0	20	C	μ=1.0M($NaClO_4$)	72Gf	
	1.660	5.36(1)	30.1	25	C	μ=1.0	63Gc	
	1.868	(1)	--	35	C	μ=1.0M($NaClO_4$)	72Gf	
	2.278	(1)	--	50	C	μ=1.0M($NaClO_4$)	72Gf	
		(1)				ΔH=A + BT + CT^2 + DT^3		
						A=-69.661	72Gf	
						B=0.648694		
						C=-2.036622x10^{-3}		
						D=2.222222x10^{-6}		
						ΔCp(25°,1) = 26.9		
	-1.658	(2)	--	5	C	μ=1.0M($NaClO_4$)	72Gf	
	-1.120	4.688(2)	17.6	20	C	μ=1.0M($NaClO_4$)	72Gf	
	-0.961	4.67(2)	18.1	25	C	μ=1.0	63Gc	
	-0.730	(2)	--	35	C	μ=1.0M($NaClO_4$)	72Gf	
	-0.374	(2)	--	50	C	μ=1.0M($NaClO_4$)	72Gf	
		(2)				ΔH=A + BT + CT^2 + DT^3		
						A=-179.921	72Gf	
						B=1.673979		
						C=-5.280036x10^{-3}		
						D=5.629630x10^{-6}		
						ΔCp(25°,2) = 26.8		
	-4.140	(3)	--	5	C	μ=1.0M($NaClO_4$)	72Gf	
	-4.449	3.248(3)	-0.31	20	C	μ=1.0M($NaClO_4$)	72Gf	
	-4.901	3.15(3)	-2.0	25	C	μ=1.0	63Gc	
	-4.815	(3)	--	35	C	μ=1.0M($NaClO_4$)	72Gf	
	-5.111	(3)	--	50	C	μ=1.0M($NaClO_4$)	72Gf	
		(3)				ΔH=A + BT + CT^2 + DT^3		
						A=-166.338	72Gf	
						B=1.667354		
						C=-5.642418x10^{-3}		
						D=6.271605x10^{-6}		
						ΔCp(25°,3) = -24.7		
Tm^{3+}	1.574	5.51(1)	30.5	25	C	μ=1.0	63Gc	
	-0.409	4.73(2)	20.3	25	C	μ=1.0	63Gc	
	-4.984	3.00(3)	-3.0	25	C	μ=1.0	63Gc	
Yb^{3+}	0.816	(1)	--	5	C	μ=1.0M($NaClO_4$)	72Gf	

Metal Ion	ΔH, kcal /mole	Log K	ΔS, cal /mole °K	T, °C	M	Conditions	Ref.	Remarks
A37, cont.								
Yb^{3+}, cont.	1.311	(1)	--	20	C	μ=1.0M(NaClO$_4$)	72Gf	
	1.423	5.57(1)	30.2	25	C	μ=1.0	63Gc	
	1.683	(1)	--	35	C	μ=1.0M(NaClO$_4$)	72Gf	
	2.158	(1)	--	50	C	μ=1.0M(NaClO$_4$)	72Gf	
		(1)				$\Delta H=A + BT + CT^2 + DT^3$		
						A=-101.812	72Gf	
						B=0.984906		
						C=-3.217138×10^{-3}		
						D=3.604938×10^{-6}		
						ΔCp(25°,1) = 27.9		
	-0.538	(2)	--	5	C	μ=1.0M(NaClO$_4$)	72Gf	
	-0.360	(2)	--	20	C	μ=1.0M(NaClO$_4$)	72Gf	
	-0.377	4.81(2)	20.8	25	C	μ=1.0	63Gc	
	-0.136	(2)	--	35	C	μ=1.0M(NaClO$_4$)	72Gf	
	0.040	(2)	--	50	C	μ=1.0M(NaClO$_4$)	72Gf	
		(2)				$\Delta H=A + BT + CT^2 + DT^3$		
						A=-88.148	72Gf	
						B=0.836912		
						C=-2.686880×10^{-3}		q:
						D=2.913580×10^{-6}		±0.015
						ΔCp(25°,2) = 11.7		q:
								±0.019
	-4.507	(3)	--	5	C	μ=1.0M(NaClO$_4$)	72Gf	
	-4.489	(3)	--	20	C	μ=1.0M(NaClO$_4$)	72Gf	
	-4.903	2.75(3)	-3.9	25	C	μ=1.0	63Gc	
	-4.792	(3)	--	35	C	μ=1.0M(NaClO$_4$)	72Gf	
	-5.048	(3)	--	50	C	μ=1.0M(NaClO$_4$)	72Gf	
		(3)				$\Delta H=A + BT + CT^2 + DT^2$		
						A=-519.676	72Gf	
						B=5.090126		
						C=-16.695982×10^{-3}		
						D=18.172840×10^{-6}		
						ΔCp(25°,3) = -19.4		
Lu^{3+}	0.753	(1)	--	5	C	μ=1.0M(NaClO$_4$)	72Gf	
	1.176	5.793(1)	30.5	20	C	μ=1.0M(NaClO$_4$)	72Gf	
	1.230	5.65(1)	30.0	25	C	μ=1.0	63Gc	
	1.599	(1)	--	35	C	μ=1.0M(NaClO$_4$)	72Gf	
	2.050	(1)	--	50	C	μ=1.0M(NaClO$_4$)	72Gf	
		(1)				$\Delta H=A + BT + CT^2 + DT^3$		
						A=-41.837	72Gf	
						B=0.384392		
						C=-1.216071×10^{-3}		
						D=1.382716×10^{-6}		
						ΔCp(25°,1) = 28.0		
	-0.734	(2)	--	5	C	μ=1.0M(NaClO$_4$)	72Gf	
	-0.430	4.960(2)	--	20	C	μ=1.0M(NaClO$_4$)	72Gf	
	-0.449	4.91(2)	20.9	25	C	μ=1.0	63Gc	
	-0.230	(2)	--	35	C	μ=1.0M(NaClO$_4$)	72Gf	
	-0.101	(2)	--	50	C	μ=1.0M(NaClO$_4$)	72Gf	
		(2)				$\Delta H=A + BT + CT^2 + DT^3$		
						A=-66.168	72Gf	
						B=0.572103		
						C=-1.664338×10^{-3}		

Metal Ion	ΔH, kcal /mole	Log K	ΔS, cal /mole °K	T, °C	M	Conditions	Ref.	Remarks
A37, cont.								
Lu^{3+}, cont.						$D=1.629630x10^{-6}$		q:
						ΔCp(25°,2) = 14.2		±0.016
	-4.091	(3)	--	5	C	μ=1.0M($NaClO_4$)	72Gf	
	-4.102	2.669(3)	--	20	C	μ=1.0M($NaClO_4$)	72Gf	
	-4.600	2.55(3)	-2.7	25	C	μ=1.0	63Gc	
	-4.210	(3)	--	35	C	μ=1.0M($NaClO_4$)	72Gf	
	-4.131	(3)	--	50	C	μ=1.0M($NaClO_4$)	72Gf	
		(3)				$ΔH=A + B + CT^2 + DT^3$ A=-373.891 B=3.735225 $C=-12.549991x10^{-3}$ $D=14.024691x10^{-6}$ ΔCp(25°,3) = -8.2	72Gf	
A38 ACETIC ACID, thio-, S-(3-butenyl)-			$C_6H_{10}O_2S$			$CH_2=CH(CH_2)_2SCH_2CO_2H$		L^-
Ag^+	-11.02	4.77(25°,1)	-15.15	0-40	T	μ=0.20M pH=5.8	71Bd 71Bc	b,q: ±1.4
	-19.1	2.26(25°,2)	-53.7	0-40	T	μ=0.20M pH=5.8	71Bd 71Bc	b,q: ±2.4
	1.9	2.07(25°)	16.0	0-40	T	μ=0.20M R: $AgL + Ag^+ = Ag_2L^+$ pH=5.8	71Bd	b,q: ±1.4
A39 ACETIC ACID, thio, S-(n-butyl)			$C_6H_{12}O_2S$			$CH_3(CH_2)_3SCH_2CO_2H$		L^-
Ag^+	-6.81	3.92(25°,1)	-4.89	0-40	T	μ=0.20M pH=5.8	71Bd 71Bc	b,q: ±1.2
	-7.17	3.56(25°,2)	-8.02	0-40	T	μ=0.20M pH=5.8	71Bd 71Bc	b,q: ±0.7
A40 ACETIC ACID, thio, S-(4-bentyl)-			$C_7H_{12}O_2S$			$CH_2:CH(CH_2)_3SCH_2CO_2H$		
Ag^+	-8.39	4.19(25°,1)	-8.98	0-40	T	μ=0.20M pH=5.8	71Bd 71Bc	b,q: ±0.7
	-4.78	2.82(25°,2)	-3.21	0-40	T	μ=0.20M pH=5.8	71Bd 71Bc	b,q: ±2.4
	-7.41	1.91(25°)	-16.03	0-40	T	μ=0.20M R: $AgL + Ag^+ = Ag_2L^+$ pH=5.8	71Bd	b,q: ±0.7
A41 ACETIC ACID, thio-, S-(allyl)-			$C_5H_8O_2S$			$CH_2:CHCH_2SCH_2CO_2H$		L^-
Ag^+	-6.91	3.75(25°,1)	-6.01	0-40	T	μ=0.20M pH=5.8	71Bd	b,q: ±1
	-4.78	2.96(25°,2)	-2.40	0-40	T	μ=0.20M pH=5.8	71Bd	b,q: ±0.7
A42 ACETIC ACID, thiolo- S-methyl ester			C_3H_6OS			CH_3COSCH_3		L
I_2	-3.15	-0.143(25°,1)	-11.2	24-41	T	$C_M=3.5x10^{-3}$ $C_L=0.33$. S: CCl_4	64Ma	b

Metal Ion	ΔH, kcal /mole	Log K	ΔS, cal /mole °K	T, °C	M	Conditions	Ref.	Remarks
A43	ACETIC ACID, 2,2,2-trichloro-, amide, N,N-dimethyl-		$C_4H_6ONCl_3$			$CCl_3CON(CH_3)_2$		L
I_2	-2.5	-0.523(25°,<u>1</u>)	-10.7	14-43	T	C_M=3x10^{-3}-7x10^{-3} C_L=4x10^{-2}-0.18 S: CCl_4	62Da	b
A44	ACETIC ACID, 2,2,2-trichloro-, ethyl ester		$C_4H_5O_2Cl_3$			$CCl_3CO_2CH_2CH_3$		L
SbA_5	-3.05	-1.17(<u>1</u>)	-15.6	25	C	S: Ethylene chloride. A=Cl	67Od	
A45	ACETIC ACID, trichloro-, nitrile		C_2NCl_3			Cl_3CCN		L
IA	-5.4	-0.70(25°,<u>1</u>)	-21.3	0.40	T	S: CCl_4. A=Cl	63Pc	b
A46	ACETIC ACID, 2,2,2-trifluoro-, amide, N,N-dimethyl-		$C_4H_6ONF_3$			$CF_3CON(CH_3)_2$		L
SbA_5	-16.58	(<u>1</u>)	--	25	C	S: Ethylene chloride. A=Cl	67Oa	
A47	ACETIC ACID, 2,2,2-trifluoro-, methyl ester		$C_3H_3O_2F_3$			$CF_3CO_2CH_3$		L
SbA_5	-2.7	-0.222(<u>1</u>)	-16.4	25	C	S: Ethylene chloride. A=Cl	67Oe	
A48	ACETOPHENONE		C_8H_8O			$CH_3COC_6H_5$		L
I_2	-3.4	0.060(25°,<u>1</u>)	-11	5-25	T	S: Heptane	71Te	b
A49	ACRIDINE		$C_{13}H_9N$			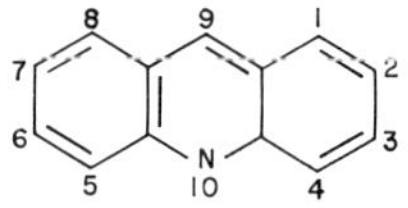		L
CoA_2	-1.0	2.27(25°,<u>1</u>)	7.0	25	C	S: Acetone A=Cl	73Lb	q: ±0.1
A50	ADENOSINE-5'-diphosphoric acid		$C_{10}H_{15}O_{10}N_5P_2$			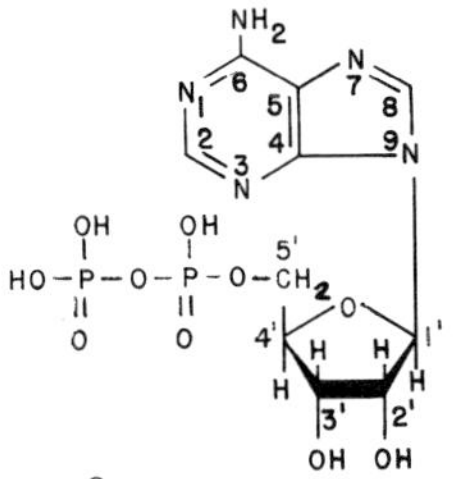		L^{3-}
Mg^{2+}	4.3	4.27(<u>1</u>)	33.9	25	T	μ=0	66Pe <u>67Ka</u>	
	5.8	4.10(<u>1</u>)	38	25	T	μ=0	63Ga	i: K. Burton <u>Biochem</u> <u>J.</u>,71, 388 (1959)

Metal Ion	ΔH, kcal /mole	Log K	ΔS, cal /mole °K	T, °C	M	Conditions	Ref.	Re-marks
A50, cont.								
Mg^{2+}, cont.	0.9	2.45(1)	14.3	25	T	μ=0. L=HL^{2-}	66Pe 63Ga	
	2.0	2.00(1)	15.9	25	T	μ=0.1. L=HL^{2-}	66Pe	
	3.6	3.17(25°,1)	26.6	5-35	T	μ=0.1M(KNO_3)	73Bb	b,g:Na
	3.9	1.64(25°,1)	20.6	5-35	T	μ=0.1M(KNO_3) L=HL^{3-}	73Bb	b,g:Na
	3.15	3.695(1)	27.3	30	C	μ=0.2 (tetramethyl-ammonium chloride) pH=8.5 Conc. triethanolamine buffer = 0.15M	69Bf	g:Na
	3.18	3.66(1)	27.3	30	C	μ=0.2 (tetramethyl-ammonium chloride) pH=8.50	73Sa	
Ca^{2+}	-1.2	2.86(25°,1)	9.1	0-40	T	μ=0.1(KNO_3)	67Ka	b,c g:Na
	-0.6	1.58(25°,1)	5.2	0-40	T	μ=0.1(KNO_3) L=HL^{2-}	67Ka	b,c g:Na
Sr^{2+}	-2.68	2.54(25°,1)	5.5	0-40	T	μ=0.1(KNO_3)	67Ka	b,c g:Na
	-1.2	1.52(25°,1)	3.0	0-40	T	μ=0.1(KNO_3) L=HL^{2-}	67Ka	b,c g:Na
Ba^{2+}	-2.9	2.36(25°,1)	1.1	0-40	T	μ=0.1(KNO_3)	67Ka	b,c g:Na
	-1.8	1.44(25°,1)	0.6	0-40	T	μ=0.1(KNO_3) L=HL^{2-}	67Ka	b,c g:Na
Mn^{2+}	2.6	4.982(24.8°,1)	31	1.5-47.2	T	C_L=2x10^{-5}M - 1.0x10^{-4}M C_M≈1x10^{-4}M 50mM N-Ethyl Morpholine buffer	70Ja	b
	-2.4	4.16(25°,1)	11.0	0-40	T	μ=0.1(KNO_3)	67Ka	b,c g:Na
	-1.9	1.88(25°,1)	2.2	0-40	T	μ=0.1(KNO_3) L=HL^{2-}	67Ka	b,c g:Na
Co^{2+}	-2.0	4.19(25°,1)	12.5	0-40	T	μ=0.1(KNO_3)	67Ka	b,c g:Na
	-1.9	2.01(25°,1)	2.8	0-40	T	μ=0.1(KNO_3) L=HL^{2-}	67Ka	b,c g:Na
Ni^{2+}	-1.9	4.49(25°,1)	14.1	0-40	T	μ=0.1(KNO_3)	67Ka	b,c g:Na
	-2.1	2.29(25°,1)	3.3	0-40	T	μ=0.1(KNO_3) L=HL^{2-}	67Ka	b,c g:Na
Cu^{2+}	-4.1	5.89(25°,1)	13.0	0-40	T	μ=0.1(KNO_3)	67Ka	b,c g:Na
	-2.7	2.62(25°,1)	3.0	0-40	T	μ=0.1(KNO_3) L=HL^{2-}	67Ka	b,c g:Na
Ag^{+}	-5.49	2.02(25°,1)	-9.1	10-40	T	μ=0	68Pb	b,q: ±0.5
	-3.66	1.84(25°,2)	-3.8	10-40	T	μ=0	68Pb	b,q: ±1.8

Metal Ion	ΔH, kcal /mole	Log K	ΔS, cal /mole °K	T, °C	M	Conditions	Ref.	Re-marks
A50, cont.								
Zn^{2+}	-2.0	4.27(25°,1)	12.5	0-40	T	μ=0.1(KNO_3)	67Ka	b,c g:Na
	-1.9	2.04(25°,1)	3.0	0-40	T	μ=0.1(KNO_3) L=HL^{2-}	67Ka	b,c g:Na

A51 ADENOSINE-2'-monophosphoric acid $C_{10}H_{14}O_7N_5P$ L^{2-}

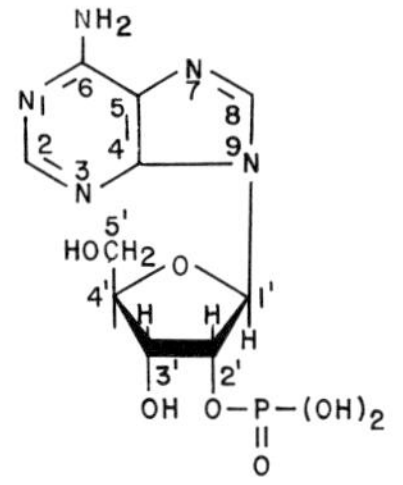

Metal Ion	ΔH, kcal /mole	Log K	ΔS, cal /mole °K	T, °C	M	Conditions	Ref.	Re-marks
Mg^{2+}	3.5	1.93(25°,1)	20.5	0-40	T	μ=0.1(KNO_3)	67Ka	b
Ca^{2+}	-0.6	1.82(25°,1)	6.5	0-40	T	μ=0.1(KNO_3)	67Ka	b
Sr^{2+}	-1.0	1.74(25°,1)	4.5	0-40	T	μ=0.1(KNO_3)	67Ka	b
Ba^{2+}	-2.0	1.71(25°,1)	1.2	0-40	T	μ=0.1(KNO_3)	67Ka	b
Mn^{2+}	-1.0	2.37(25°,1)	7.5	0-40	T	μ=0.1(KNO_3)	67Ka	b
Co^{2+}	-0.7	2.34(25°,1)	8.5	0-40	T	μ=0.1(KNO_3)	67Ka	b
Ni^{2+}	-1.0	2.81(25°,1)	9.5	0-40	T	μ=0.1(KNO_3)	67Ka	b
Cu^{2+}	-1.9	3.16(25°,1)	8.0	0-40	T	μ=0.1(KNO_3)	67Ka	b
Zn^{2+}	-1.2	2.64(25°,1)	8.0	0-40	T	μ=0.1(KNO_3)	67Ka	b
A52 ADENOSINE-3'-monophosphoric acid			$C_{10}H_{14}O_7N_5P$					L^{2-}
Mg^{2+}	3.5	1.89(25°,1)	20.5	0-40	T	μ=0.1(KNO_3)	67Ka	b
Ca^{2+}	-0.6	1.80(25°,1)	6.5	0-40	T	μ=0.1(KNO_3)	67Ka	b
Sr^{2+}	-0.9	1.71(25°,1)	4.5	0-40	T	μ=0.1(KNO_3)	67Ka	b
Ba^{2+}	-1.9	1.68(25°,1)	1.2	0-40	T	μ=0.1(KNO_3)	67Ka	b
Mn^{2+}	-0.9	2.28(25°,1)	7.6	0-40	T	μ=0.1(KNO_3)	67Ka	b
Co^{2+}	-0.6	2.24(25°,1)	8.5	0-40	T	μ=0.1(KNO_3)	67Ka	b
Ni^{2+}	-1.0	2.78(25°,1)	9.6	0-40	T	μ=0.1(KNO_3)	67Ka	b
Cu^{2+}	-1.7	2.95(25°,1)	8.0	0-40	T	μ=0.1(KNO_3)	67Ka	b
Zn^{2+}	-1.1	2.59(25°,1)	8.2	0-40	T	μ=0.1(KNO_3)	67Ka	b
A53 ADENOSINE-5'-monophosphoric acid			$C_{10}H_{14}O_7N_5P$					L^{2-}
Mg^{2+}	3.4	1.96(25°,1)	20.4	0-40	T	μ=0.1(KNO_3)	67Ka	b
	1.77	1.79(1)	14.0	30	C	μ=0.2 (Tetramethylammonium chloride) pH=8.50	73Sa	
	1.78	1.806(1)	14.1	30	C	μ=0.20	69Bf	

Metal Ion	ΔH, kcal /mole	Log K	ΔS, cal /mole °K	T, °C	M	Conditions	Ref.	Remarks
A53, cont.								
Mg^{2+}, cont.						(Tetramethylammonium chloride) pH=8.5 Conc. triethanolamine buffer=0.15M		
Ca^{2+}	-0.6	1.85(25°,1)	6.4	0-40	T	μ=0.1(KNO_3)	67Ka	b
Sr^{2+}	-1.4	1.79(25°,1)	4.4	0-40	T	μ=0.1(KNO_3)	67Ka	b
Ba^{2+}	-2.0	1.73(25°,1)	1.2	0-40	T	μ=0.1(KNO_3)	67Ka	b
Mn^{2+}	-1.0	2.40(25°,1)	7.6	0-40	T	μ=0.1(KNO_3)	67Ka	b
Co^{2+}	-1.1	2.40(25°,1)	8.4	0-40	T	μ=0.1(KNO_3)	67Ka	b
Ni^{2+}	-1.0	2.84(25°,1)	9.6	0-40	T	μ=0.1(KNO_3)	67Ka	b
Cu^{2+}	-2.0	3.17(25°,1)	8.0	0-40	T	μ=0.1(KNO_3)	67Ka	b
Zn^{2+}	-1.2	2.71(25°,1)	8.2	0-40	T	μ=0.1(KNO_3)	67Ka	b

A54 ADENOSINE-5'-triphosphoric acid $C_{10}H_{16}O_{13}N_5P_3$ L^{4-}

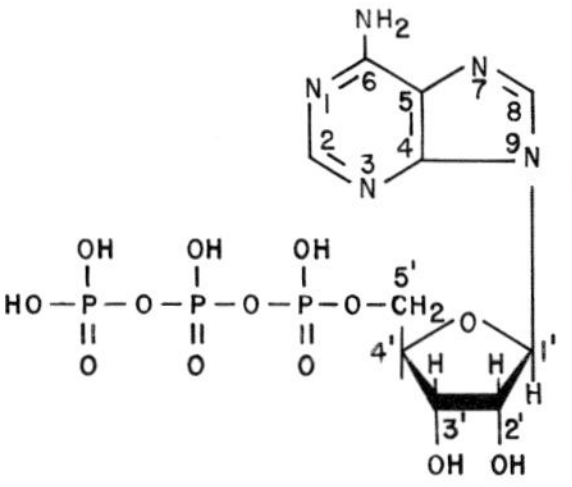

Metal Ion	ΔH, kcal /mole	Log K	ΔS, cal /mole °K	T, °C	M	Conditions	Ref.	Remarks
Mg^{2+}	5.1	5.83(1)	43.7	25	T	μ=0	66Pe 66Ka	
	4.8	5.72(1)	42	25	T	μ=0	63Ga	i: K. Burton, Biochem J. 71, 388 (1959)
	2.2	3.59(1)	23.8	25	T	μ=0 L=HL^{3-}	66Pe 63Ga 57Nc	
	1.9	2.86(1)	19.4	25	T	μ=0.1 L=HL^{3-}	66Pe	
	2.6	4.22(25°,1)	28.0	5-35	T	μ=0.1M(KNO_3)	73Bb	b,g:Na
	3.4	2.24(25°,1)	21.6	5-35	T	μ=0.1M(KNO_3)	73Bb	b,g:Na
	4.120	3.343(25°,1)	29.1	1-43	T	μ=0.15(NaCl) pH=8.8	57Nc	b
	4.47	4.693(1)	36.2	30	C	μ=0.2 (Tetramethylammonium chloride) pH=8.5 Conc. triethanolamine buffer=0.15M	69Bf	g:Na
	4.46	4.67(1)	36.1	30	C	μ=0.2 (Tetramethylammonium chloride) pH=8.50	73Sa	
Ca^{2+}	-0.9	3.97(25°,1)	12	0-40	T	0.1M(KNO_3)	66Ka 57Nc	b,c

A54, cont.

Metal Ion	ΔH, kcal /mole	Log K	ΔS, cal /mole °K	T, °C	M	Conditions	Ref.	Re-marks
Ca^{2+},	-0.3	2.17(25°,1)	9	0-40	T	0.1M(KNO_3) L=HL^{3-}	66Ka	b,c
cont.	4.560	2.932(25°,1)	28.7	1-43	T	μ=0.15(NaCl) pH=8.8	57Nc	b
Sr^{2+}	-3.0	3.54(25°,1)	6	0-40	T	0.1M(KNO_3)	66Ka	b,c
	-1.6	2.05(25°,1)	4	0-40	T	0.1M(KNO_3) L=HL^{3-}	66Ka	b,c
Ba^{2+}	-3.9	3.29(25°,1)	2	0-40	T	0.1M(KNO_3)	66Ka	b,c
	-2.1	1.85(25°,1)	2	0-40	T	0.1M(KNO_3) L=HL^{3-}	66Ka	b,c
YA^-	-4.5	5.3(25°,1)	9	2-45	T	μ=0.10M(KNO_3) A=$EDTA^{2-}$	73Kf	b,g:Na, q:±2.0
LaA^-	-3.4	4.5(25°,1)	9	2-45	T	μ=0.10M(KNO_3) A=$EDTA^{2-}$	73Kf	b,g:Na, q:±1.4
CeA^-	-2.4	4.3(25°,1)	11	2-45	T	μ=0.10M(KNO_3) A=$EDTA^{2-}$	73Kf	b,g:Na, q:±0.5
PrA^-	-2.4	4.4(25°,1)	12	2-45	T	μ=0.10M(KNO_3) A=$EDTA^{2-}$	73Kf	b,g:Na, q:±0.4
NdA^-	-2.4	4.4(25°,1)	13	2-45	T	μ=0.10M(KNO_3) A=$EDTA^{2-}$	73Kf	b,g:Na, q:±0.4
SmA^-	-3.5	4.9(25°,1)	11	2-45	T	μ=0.10M(KNO_3) A=$EDTA^{2-}$	73Kf	b,g:Na, q:±1.1
EuA^-	-2.4	4.7(25°,1)	13	2-45	T	μ=0.10M(KNO_3) A=$EDTA^{2-}$	73Kf	b,g:Na, q:±0.5
GdA^-	-2.7	4.3(25°,1)	11	2-45	T	μ=0.10M(KNO_3) A=$EDTA^{2-}$	73Kf	b,g:Na, q:±0.6
TbA^-	-3.3	5.4(25°,1)	14	2-45	T	μ=0.10M(KNO_3) A=$EDTA^{2-}$	73Kf	b,g:Na, q:±1.0
DyA^-	-4.7	5.7(25°,1)	10	2-45	T	μ=0.10M(KNO_3) A=$EDTA^{2-}$	73Kf	b,g:Na, q:±1.8
HoA^-	-4.6	5.6(25°,1)	10	2-45	T	μ=0.10M(KNO_3) A=$EDTA^{2-}$	73Kf	b,g:Na, q:±0.8
ErA^-	-3.6	6.1(25°,1)	16	2-45	T	μ=0.10M(KNO_3) A=$EDTA^{2-}$	73Kf	b,g:Na, q:±0.5
TmA^-	-4.5	6.5(25°,1)	14	2-45	T	μ=0.10M(KNO_3) A=$EDTA^{2-}$	73Kf	b,g:Na, q:±2.0
Mn^{2+}	-4.9	5.518(24.8°,1)	9	1.5- 47.2	T T	C_L=2x10^{-5}-1.0x10^{-4}M C_M≃1x10^{-4}M 50mM N-Ethyl Morpholine buffer	70Ja	b
	-3.0	4.78(25°,1)	12	0-40	T	0.1M(KNO_3)	66Ka	b,c
	-2.3	2.39(25°,1)	3	0-40	T	0.1M(KNO_3) L=HL^{3-}	66Ka	b,c
[Fe]	-10.4	3.778(1)	-20	5	C	C_{Cl^-}=0 pH(initial)=7.32 [Fe]=Horse Oxyhemoglobin	72He	q: ±1.4
	-5.2	4.799(1)	4.5	25	C	C_{Cl^-}=0 pH(initial)=6.93 [Fe]=Horse Oxyhemoglobin	72He	

Metal Ion	ΔH, kcal /mole	Log K	ΔS, cal /mole °K	T, °C	M	Conditions	Ref.	Re-marks
A54, cont.								
[Fe], cont.	-10.3	5.114(1)	-14	5	C	C_{Cl^-}=0.005 pH(initial)=6.50 [Fe]=Horse Oxyhemoglobin	72He	
	-4.5	2.903(1)	-1.5	25	C	C_{Cl^-}=0.10 pH(initial)=7.05 [Fe]=Horse Oxyhemoglobin	72He	q: ±0.5
	-6.75	2.167(14°)	-13.6	5-21	T	pH=6.2 [Fe]=Human Methaemoglobin R: [Fe] + 10L = $[Fe]L_{10}$	70Jb	b,j: Probably per mole L bound
	-6.85	2.201(17°)	-14.4	5&17	T	pH=7.2 [Fe]=Human Methaemoglobin R: [Fe] + 10L = $[Fe]L_{10}$	70Jb	b,j: Probably per mole L bound
Co^{2+}	-2.2	4.66(25°,1)	14	0-40	T	0.1M(KNO_3)	66Ka	b,c
	-2.1	2.32(25°,1)	4	0-40	T	0.1M(KNO_3) L=HL^{3-}	66Ka	b,c
Ni^{2+}	-2.5	5.02(25°,1)	15	0-40	T	0.1M(KNO_3)	66Ka	b,c
	-2.4	2.72(25°,1)	4	0-40	T	0.1M(KNO_3) L=HL^{3-}	66Ka	b,c
Cu^{2+}	-4.3	6.13(25°,1)	14	0-40	T	0.1M(KNO_3)	66Ka	b,c
	-3.0	3.12(25°,1)	4	0-40	T	0.1M(KNO_3) L=HL^{3-}	66Ka	b,c
Zn^{2+}	-2.7	4.85(25°,1)	13	0-40	T	0.1M(KNO_3)	66Ka	b,c
	-2.4	2.67(25°,1)	4	0-40	T	0.1M(KNO_3) L=HL^{3-}	66Ka	b,c
A55 ADRENALINE			$C_9H_{13}O_3N$					L
Be^{2+}	-25.0	9.65(25°,1)	-40	0-45	T	μ=0.12M(KCl) $C_M \simeq 2x10^{-3}M$ $C_L \simeq 8x10^{-3}M$	69Cb	b
	-23.8	6.31(25°,2)	-51	0-45	T	μ=0.12M(KCl) $C_M \simeq 2x10^{-3}M$ $C_L \simeq 8x10^{-3}M$	69Cb	b
A56 α-ALANINE			$C_3H_7O_2N$			$CH_3CH(NH_2)CO_2H$		L^-
Eu^{3+}	1.20	0.743	7.4	25	C	μ=2.0($NaClO_4$) R: $HL^{\pm}$ + Eu^{3+} = $EuHL^{3+}$	71Ai	
Mn^{2+}	-2.5	3.08(20°,1)	5	20-60	T	μ=0.1(KNO_3)	73Bf	b
	-1.2	3.00(20°,2)	9	20-60	T	μ=0.1(KNO_3)		b
Co^{2+}	-1.3	4.354(25°,1)	15	27	C	μ=0.05M(KCl)	71Gb	
	-2.0	4.354(25°,1)	13	20-35	T	μ=0.05(KCl)	71Gb 70Ga 70Gb	b,q: ±0.5
	-2.6	4.41(15°,1)	11.0	15-40	T	μ=0.2	65Sb 64Se	b
	-2.3	3.505(25°,2)	9	27	C	μ=0.05M(KCl)	71Gb	
	-2.3	3.505(25°,2)	8	20-35	T	μ=0.05(KCl)	71Gb 70Ga 70Gb	b,q: ±0.8
	-3.1	3.27(15°,2)	4.0	15-40	T	μ=0.2	65Sb	b
	-5.9	7.7(1-2)	20.0	21.8	C	0.1M(KNO_3)	67Sc	
	-5.7	7.68(15°,1-2)	15.0	15-40	T	μ=0.2	65Sb	b

Metal Ion	ΔH, kcal /mole	Log K	ΔS, cal /mole °K	T, °C	M	Conditions	Ref.	Remarks
A56, cont.								
Ni^{2+}	-3.4	5.93(1)	15.1	10	C	μ=0	67Aa	
	-3.3	5.81(1)	15.5	25	C	μ=0	67Aa	
	-3.6	5.69(1)	15.4	40	C	μ=0	67Aa	
	-3.6	5.463(25°,1)	13	27	C	μ=0.05M(KCl)	71Gb	
	-3.4	5.463(25°,1)	14	20-35	T	μ=0.05(KCl)	70Ga, 70Gb	b,q: ±0.2
	-3.35	3.09(1)	2.85	25	C	0.1M(KNO_3)	62Sc, 67Sc	
	-4.4	5.65(15°,1)	10.4	15-40	T	μ=0.2	65Sb, 64Se	b
	-4.6	4.87(2)	5.9	10	C	μ=0	67Aa	
	-3.9	4.73(2)	8.7	25	C	μ=0	67Aa	
	-2.6	4.50(2)	11.4	40	C	μ=0	67Aa	
	-3.8	4.467(25°,2)	8	27	C	μ=0.05M(KCl)	71Gb	
	-3.9	4.467(25°,2)	7	20-35	T	μ=0.05(KCl)	70Ga, 70Gb	b,q: ±0.3
	-3.96	2.47(2)	-1.98	25	C	0.1M(KNO_3)	62Sc	
	-4.8	4.57(15°,2)	4.3	15-40	T	μ=0.2	65Sb	b
	-2.60	1.44(3)	-2.15	25	C	0.1M(KNO_3)	62Sc	
	-9.2	10.22(15°,1-2)	14.7	15-40	T	μ=0.2	65Sb	b
Cu^{2+}	-5.38	8.70(1)	20.8	10	C	μ=0	66Ab, 65Ib	
	-4.51	8.54(1)	23.9	25	C	μ=0	66Ab, 64Se	
	-3.99	8.32(1)	25.3	40	C	μ=0	66Ab	
	-5.6	8.51(25°,1)	20.3	0-40	T	μ=0	61Ib	b
	-4.9	8.174(25°,1)	21	27	C	μ=0.05M(KCl)	71Gb	
	-4.7	8.174(25°,1)	22	20-35	T	μ=0.05(KCl)	70Ga, 70Gb	b,q: ±0.4
	-5.6	8.15(1)	18.5	25	C	μ=0.1M(KNO_3)	72Ia	
	-6.3	8.19(1)	16.3	25	C	μ=0.1M($NaClO_4$)	72Ia	
	-4.94	8.07(1)	20	25	C	μ=0.2M(KCl)	73Ga	
	-4.9	8.40(15°,1)	21.2	15-40	T	μ=0.2	65Sb	b
	-5.47	7.26(2)	12.9	10	C	μ=0	66Ab	
	-5.24	6.98(2)	9.0	25	C	μ=0	66Ab	
	-5.65	6.76(2)	10.7	40	C	μ-0	66Ab	
	-6.2	6.86(25°,2)	10.9	0-40	T	μ=0	61Ib	b
	-5.5	6.779(25°,2)	13	27	C	μ=0.05M(KCl)	71Gb	
	-6.0	6.779(25°,2)	10	20-35	T	μ=0.05(KCl)	70Ga, 70Gb	b,q: ±0.7
	-6.15	6.67(2)	9.9	25	C	μ=0.1M(KNO_3)	72Ia	
	-5.4	6.75(2)	12.8	25	C	μ=0.1M($NaClO_4$)	72Ia	
	-7.06	6.72(2)	7	25	C	μ=0.2M(KCl)	73Ga	
	-5.8	6.86(15°,2)	11.3	15-40	T	μ=0.2	65Sb	b
	-11.9	15.0(1-2)	28.4	22.3	C	0.1M(KNO_3)	67Sc	
	-12.4	14.9(1-2)	26.8	25	C	μ=0.1M	71Be	
	-12.3	14.9(1-2)	27.0	25	C	μ=0.1M	71Be	
	-10.7	15.26(15°,1-2)	32.5	15-40	T	μ=0.2	65Sb	
	-12.02	15.77	31.8	25	C	μ=0 A=α-Aminoisobutric acid R: $Cu^{2+} + L^- + A^- = CuLA$	72Ya	

Metal Ion	ΔH, kcal /mole	Log K	ΔS, cal /mole °K	T, °C	M	Conditions	Ref.	Re-marks
A56, cont.								
Cu^{2+}, cont.	-11.7	14.91	29.0	25	C	μ=0.1M(KNO_3) R: Cu + L + Serine = [CuL(Serine)]	72Ia	
	-11.78	15.20	30.0	25	C	μ=0.1M(KNO_3) R: Cu + L + Valine = [CuL(Valine)]	72Ia	
	0.92	1.77(25°)	11.2	-75&25	T	S: 50 v % Methanol R: Cu + CuL_2 = 2CuL	73Ya	a,b, ±1.27
Zn^{2+}	-1.5	4.604(25°,1)	17	27	C	μ=0.05M(KCl)	71Gb	
	-2.3	4.604(25°,1)	13	20-35	T	μ=0.05(KCl)	70Ga 70Gb	b,q: ±0.3
	-3.0	4.98(15°,1)	12.4	15-40	T	μ=0.2	65Sb 64Se	b
	-2.8	4.069(25°,2)	10	27	C	μ=0.05M(KCl)	71Gb	
	-1.7	4.069(25°,2)	13	20-35	T	μ=0.05(KCl)	70Ga 70Gb	b,q: ±0.2
	-4.3	4.12(15°,2)	4.0	15-40	T	μ=0.2	65Sb	b
	-7.3	9.10(15°,1-2)	16.4	15-40	T	μ=0.2	65Sb	b
In^{3+}	3.5	2.57(35°,1)	23	25-45	T	μ=0.2M($NaClO_4$)	73Sc	b,q: ±0.5
A57 β-ALANINE			$C_3H_7O_2N$			$H_2NCH_2CH_2CO_2H$		L^-
Mn^{2+}	-2.1	2.13(20°,1)	3	20-60	T	μ=0.1(KNO_3)	73Bf	b
Co^{2+}	-3.32	4.21(1)	8.1	25	C	μ=0	67Bc 64Se	
	-2.6	3.69(15°,1)	7.8	15-40	T	μ=0.2	65Sb	b
	-2.3	2.59(15°,2)	3.8	15-40	T	μ=0.2	65Sb	b
	-4.9	6.28(15°,1-2)	11.6	15-40	T	μ=0.2	65Sb	b
Ni^{2+}	-3.81	5.01(1)	10.2	25	C	μ=0	67Bc 64Se	
	-4.0	4.80(15°,1)	8.2	15-40	T	μ=0.2	65Sb	b
	-5.0	3.54(15°,2)	-1.0	15-40	T	μ=0.2	65Sb	b
	-6.1	8.0(1-2)	15.9	21.7	C	0.1M(KNO_3)	67Sc	
	-9.0	8.34(15°,1-2)	7.2	15-40	T	μ=0.2	65Sb	b
Cu^{2+}	-4.0	7.16(15°,1)	19.6	15-40	T	μ=0.2	67Bc 64Se	b
	-6.0	5.59(15°,2)	4.4	15-40	T	μ=0.2	65Sb	b
	-10.9	12.6(1-2)	20.3	22.0	C	0.1M(KNO_3)	67Sc	
	-10.0	12.75(15°,1-2)	24.0	15-40	T	μ=0.2	65Sb	b
	-0.11	1.36(25°)	5.9	-75&25	T	S: 50 v % Methanol R: Cu + CuL_2 = 2CuL	73Ya	a,b,q: ±0.46
Zn^{2+}	-1.17	4.274(1)	15.4	5	T	μ=0.1(KCl)	73Rb	q:±0.71
	-1.47	4.205(1)	14.3	25	T	μ=0.1(KCl)	73Rb	q:±0.71
	-1.82	4.130(1)	13.2	45	T	μ=0.1(KCl)	73Rb	q:±0.71
	-4.1	(1-2)	--	21.4	C	0.1M(KNO_3)	67Sc	
In^{3+}	8.0	5.51(35°,1-2)	51	25-45	T	μ=0	72Sa	b,q: ±1.5
	7.3	5.42(35°,1-2)	48	25-45	T	μ=0.01M($NaClO_4$)	72Sa	b,q: ±1.5
	3.0	2.77(35°,1)	22	25-45	T	μ=0.20M($NaClO_4$)	73Sc	b

Metal Ion	ΔH, kcal /mole	Log K	ΔS, cal /mole °K	T, °C	M	Conditions	Ref.	Remarks
A58	ALANINE, dehydro, N-acetyl- (Polymer)		$(C_5H_7O_3N)_n$			$[H_2C{:}C(CO_2H)NHCOCH_3]_n$		
Co^{2+}	-1.332	--	--	25	C	μ<0.1M R: $(C_5H_7O_3N)_n$ + $0.5nCoCl_2$ = $[C_5H_6O_3N(0.5Co)]_n$ + nHCl	67Ia	
A59	ALANINE, N-glycyl-		$C_5H_{10}O_3N_2$			$H_2NCH_2CONHCH(CH_3)CO_2H$		
Cu^{2+}	-6.6	5.80(1)	4.4	25	C	μ=0.1M(KNO_3)	72Bl	
A60	α-ALANINE, 3-phenyl-		$C_9H_{11}O_2N$			$C_6H_5CH_2CH(NH_2)CO_2H$		L-
Mn^{2+}	-1.1	2.39(20°,1)	7	20-60	T	μ=0.1(KNO_3)	73Bf	b
Co^{2+}	-1.5	4.02(25°,1)	14	27	C	μ=0.05M(KCl)	71Gb	
	-1.1	4.02(25°,1)	15	20-35	T	μ=0.05M(KCl)	71Gb	b
	-1.27	4.450(1)	16.1	25	C	μ=3.0($NaClO_4$)	72Wd	
	-0.3	3.43(25°,2)	15	27	C	μ=0.05M(KCl)	71Gb	
	-2.2	3.43(25°,2)	8	20-35	T	μ=0.05M(KCl)	71Gb	b
	-1.93	3.991(2)	11.8	25	C	μ=3.0($NaClO_4$)	72Wd	
Ni^{2+}	-3.4	5.61(1)	13.8	10	C	μ=0	67Aa	
	-3.2	5.56(1)	14.8	25	C	μ=0	67Aa	
	-2.6	5.52(1)	17.0	40	C	μ=0	67Aa	
	-2.7	5.13(25°,1)	15	27	C	μ=0.05M(KCl)	71Gb	b
	-2.2	5.13(25°,1)	16	20-35	T	μ=0.05M(KCl)	71Gb	
	-2.35	5.356(1)	16.6	25	C	μ=3.0($NaClO_4$)	72Wd	
	-3.6	4.95(2)	10.0	10	C	μ=0	67Aa	
	-3.3	4.66(2)	10.3	25	C	μ=0	67Aa	
	-2.9	4.39(2)	10.7	40	C	μ=0	67Aa	
	-1.9	4.36(25°,2)	14	27	C	μ=0.05M(KCl)	71Gb	
	-2.3	4.36(25°,2)	12	20-35	T	μ=0.05M(KCl)	71Gb	b
	-3.52	5.126(2)	11.7	25	C	μ=3.0($NaClO_4$)	72Wd	
Cu^{2+}	-6.0	8.48(1)	17.7	10	C	μ=0	66Aa	
	-5.3	8.25(1)	19.8	25	C	μ=0	66Aa	
	-5.1	8.30(25°,1)	20.7	0-40	T	μ=0	65Ib	b
	-5.1	7.70(30°,1)	20.7	0-40	T	μ=0	61Ib	b
	-4.85	8.13(1)	21.8	40	C	μ=0	66Aa	
	-4.7	7.90(25°,1)	20	27	C	μ=0.05M(KCl)	71Gb	
	-4.6	7.90(25°,1)	21	20-35	T	μ=0.05M(KCl)	71Gb	b
	-4.58	8.250(1)	22.4	25	C	μ=3.0($NaClO_4$)	72Wd	
	-6.4	7.43(2)	10.3	10	C	μ=0	66Aa	
	-6.4	7.13(2)	13.7	25	C	μ=0	66Aa	
	-6.4	7.01(25°,2)	10.7	0-40	T	μ=0	65Ib	b
	-6.4	6.94(30°,2)	10.7	0-40	T	μ=0	61Ib	b
	-5.5	6.94(3)	14.2	40	C	μ=0	66Aa	
	-6.9	6.91(25°,2)	9	27	C	μ=0.05M(KCl)	71Gb	
	-6.4	6.91(25°,2)	10	20-35	T	μ=0.05M(KCl)	71Gb	b
	-9.36	7.304(2)	2.0	25	C	μ=3.0($NaClO_4$)	72Wd	
Zn^{2+}	-0.9	4.31(25°,1)	17	27	C	μ=0.05M(KCl)	71Gb	
	-1.8	4.31(25°,1)	13	20-35	T	μ=0.05M(KCl)	71Gb	b
	-0.24	4.477(1)	19.6	5	T	μ=0.1(KCl)	73Rb	q: ±0.71
	-0.7	4.453(1)	18.0	25	T	μ=0.1(KCl)	73Rb	q: ±0.71

Metal Ion	ΔH, kcal /mole	Log K	ΔS, cal /mole °K	T, °C	M	Conditions	Ref.	Re-marks
A60, cont.								
Zn^{2+}, cont.	-1.3	4.408(1)	16.1	45	T	μ=0.1(KCl)	73Rb	q: ±0.71
	-1.7	4.09(25°,2)	13	27	C	μ=0.05M(KCl)	71Gb	
	-2.8	4.09(25°,2)	9	20-35	T	μ=0.05M(KCl)	71Gb	b
A61 β-ALANINE, 3-phenyl-			$C_9H_{11}O_2N$			$C_6H_5CH(NH_2)CH_2CO_2H$		L^-
Mn^{2+}	-1.7	2.06(20°,1)	4	20-60	T	μ=0.1(KNO_3)	73Sr	b
A62 AMINE, allyl-			C_3H_7N			CH_2:$CHCH_2NH_2$		L^+
PtA_4^{2-}	-7.1	3.540(25°)	-7.6	25-45	T	1.90M(NaCl) 0.1M(HCl) A=Cl R: PtA_4^{2-} + L^+ = PtA_3L + A^-	68Mc	
	-7.1	3.240(44°)	-7.6	30-60	T	μ=2M(NaCl + HCl) A=Cl R: PtA_4^{2-} + L^+ = PtA_3L + A^-	67Da	b
A63 AMINE, allyl diethyl-			$C_7H_{15}N$			CH_2:$CHCH_2NH(C_2H_5)_2$		L^+
PtA_4^{2-}	-5.6	2.740(45.3°)	-5.0	30-60	T	μ=2M(NaCl + HCl) A=Cl R: PtA_4^{2-} + L^+ = PtA_3L + A^-	67Da	b
A64 AMINE, allyl ethyl-			$C_5H_{12}N$			CH_2:$CHCH_2NHC_2H_5$		
PtA_4^{2-}	-4.8	2.382(25°)	-5.3	0-35	T	μ=2M(KBr + HBr) A=Br R: PtA_4^{2-} + L^+ = PtA_3L + A^-	67Dc	b
	-5.9	3.091(44°)	-4.4	30-60	T	μ=2M(NaCl + HCl) A=Cl R: RtA_4^{2-} + L^+ = PtA_3L + A^-	67Da	b
A65 AMINE, allyl methyl-			C_4H_9N			CH_2:$CHCH(CH_3)NH_2$		L^+
PtA_4^{2-}	-6.7	3.108(45.3°)	-6.9	30-60	T	μ=2M(NaCl + HCl) A=Cl R: PtA_4^{2-} + L^+ = PtA_3L + A^-	67Db	b
A66 AMINE, 2-butenyl-(trans)			C_4H_9N			CH_3CH:$CHCH_2NH_2$		L^+
PtA_4^{2-}	-5.1	2.483(44.5°)	-4.6	30-60	T	μ=2M(NaCl + HCl) A=Cl R: PtA_4^{2-} + L^+ = RtA_3L + A^-	67Db	b
A67 AMINE, 3-butenyl-			C_4H_9N			CH_2:$CHCH_2CH_2NH_2$		L^+
PtA_4^{2-}	-5.1	3.480(44.5°)	-0.2	30-60	T	μ=2M(NaCl + HCl) A=Cl R: PtA_4^{2-} + L^+ = PtA_3L + A^-	67Da	b

Metal Ion	ΔH, kcal /mole	Log K	ΔS, cal /mole °K	T, °C	M	Conditions	Ref.	Remarks
A68 AMINE, cyclohexyl dimethyl-			$C_8H_{17}N$			$C_6H_{13}N(CH_3)_2$		L
BA_3C	-17.0	--	--	25	C	S: Benzene A=F $C=(CH_3CH_2CH_2CH_2)_2O$ R: $BA_3C + L = BA_3L + C$	68Rd	
	-20.2	--	--	25	C	S: Benzene C_M=0.03-0.04M A=F C=Tributylphosphine R: $BA_3C + L = BA_3L + C$	73Rd	
	-19.5	--	--	25	C	S: Benzene C_M=0.03-0.04M A=F C=Trihexylphosphine R: $BA_3C + L = BA_3L + C$	73Rd	
	-7.7	--	--	25	C	S: Benzene C_M=0.03-0.04M A=F C=Triphenylphosphine R: $BA_3C + L = BA_3L + C$	73Rd	
Al_2A_6	-32.7	--	--	25	C	S: Benzene C_M=0.07M A=Br R: $0.5(Al_2A_6) + L = AlA_3L$	68Rc	
A69 AMINE, 2,2'-diacetic acid methyl-			C_5H_9N			$CH_3N(CH_2CO_2H)_2$		L^{2-}
Mg^{2+}	-3.12	3.44(1)	26.4	20	C	0.1M(KNO_3)	65Ab 56Ma	
Ca^{2+}	-1.64	3.75(1)	11.58	20	C	0.1M(KNO_3)	65Ab	
Sr^{2+}	-1.23	2.85(1)	8.85	20	C	0.1M(KNO_3)	65Ab 56Ma	
Ba^{2+}	-0.79	2.59(1)	9.17	20	C	0.1M(KNO_3)	65Ab	
Mn^{2+}	0.56	5.4(1)	26.6	20	C	0.1M(KNO_3)	65Ab 56Ma	
	0.23	9.56(1-2)	44.55	20	C	0.1M(KNO_3)	65Ab	
Co^{2+}	-1.85	7.62(1)	28.58	20	C	0.1M(KNO_3)	65Ab	
	-5.48	13.91(1-2)	45.0	20	C	0.1M(KNO_3)	65Ab	
Ni^{2+}	-4.7	8.73(1)	23.9	20	C	0.1M(KNO_3)	65Ab	
	-7.65	15.95(1-2)	46.9	20	C	0.1M(KNO_3)	65Ab	
Cu^{2+}	-3.84	11.09(1)	37.7	20	C	0.1M(KNO_3)	65Ab	
	-12.11	17.92(1-2)	40.75	20	C	0.1M(KNO_3)	65Ab	
Zn^{2+}	-2.17	7.66(1)	27.7	20	C	0.1M(KNO_3)	65Ab	
	-5.83	14.09(1-2)	44.40	20	C	0.1M(KNO_3)	65Ab	
Cd^{2+}	-1.89	6.77(1)	24.52	20	C	0.1M(KNO_3)	65Ab	
	-7.27	12.52(1-2)	32.5	20	C	0.1M(KNO_3)	65Ab	
Pb^{2+}	-3.56	8.02(1)	24.6	20	C	0.1M(KNO_3)	65Ab	

Metal Ion	ΔH, kcal /mole	Log K	ΔS, cal /mole °K	T, °C	M	Conditions	Ref.	Remarks
A70	AMINE, di(2-aminoethyl)-		$C_4H_{13}N$			$HN(CH_2CH_2NH_2)_2$		L
La^{3+}	-23.8	(1)	--	23	C	$\mu=0.024M[La(ClO_4)_3]$	71Fd	
	-19.8	(2)	--	23	C	$\mu=0.024M[La(ClO_4)_3]$	71Fd	
	-13.1	(3)	--	23	C	$\mu=0.024M[La(ClO_4)_3]$	71Fd	
Pr^{3+}	-24.6	(1)	--	23	C	$\mu=0.024M[La(ClO_4)_3]$	71Fd	
	-20.8	(2)	--	23	C	$\mu=0.024M[La(ClO_4)_3]$	71Fd	
	-12.6	(3)	--	23	C	$\mu=0.024M[La(ClO_4)_3]$	71Fd	
Nd^{3+}	-25.6	(1)	--	23	C	$\mu=0.024M[La(ClO_4)_3]$	71Fd	
	-20.9	(2)	--	23	C	$\mu=0.024M[La(ClO_4)_3]$	71Fd	
	-11.5	(3)	--	23	C	$\mu=0.024M[La(ClO_4)_3]$	71Fd	
Sm^{3+}	-25.7	(1)	--	23	C	$\mu=0.024M[La(ClO_4)_3]$	71Fd	
	-21.2	(2)	--	23	C	$\mu=0.024M[La(ClO_4)_3]$	71Fd	
	-10.7	(3)	--	23	C	$\mu=0.024M[La(ClO_4)_3]$	71Fd	
Gd^{3+}	-26.1	(1)	--	23	C	$\mu=0.024M[La(ClO_4)_3]$	71Fd	
	-21.5	(2)	--	23	C	$\mu=0.024M[La(ClO_4)_3]$	71Fd	
	-9.9	(3)	--	23	C	$\mu=0.024M[La(ClO_4)_3]$	71Fd	
Dy^{3+}	-27.2	(1)	--	23	C	$\mu=0.024M[La(ClO_4)_3]$	71Fd	
	-22.2	(2)	--	23	C	$\mu=0.024M[La(ClO_4)_3]$	71Fd	
	-6.9	(3)	--	23	C	$\mu=0.024M[La(ClO_4)_3]$	71Fd	
Er^{3+}	-28.1	(1)	--	23	C	$\mu=0.024M[La(ClO_4)_3]$	71Fd	
	-23.5	(2)	--	23	C	$\mu=0.024M[La(ClO_4)_3]$	71Fd	
	-6.2	(3)	--	23	C	$\mu=0.024M[La(ClO_4)_3]$	71Fd	
Yb^{3+}	27.4	(1)	--	23	C	$\mu=0.024M[La(ClO_4)_3]$	71Fd	
	22.7	(2)	--	23	C	$\mu=0.024M[La(ClO_4)_3]$	71Fd	
	-6.1	(3)	--	23	C	$\mu=0.024M[La(ClO_4)_3]$	71Fd	
Mn^{2+}	-4	3.99(30°,1)	5.1	30-40	T	$\mu\simeq2(KCl, KNO_3)$	52Ja	a,b
	-5	2.83(30°,2)	-3.5	30-40	T	$\mu\simeq2(KCl, KNO_3)$	52Ja	a,b
	-6.95	(1)	--	25	C	0.1KCl	61Cb	
Fe^{2+}	-9	6.23(30°,1)	-1.2	30-40	T	$\mu\simeq2(KCl, KNO_3)$	52Ja	a,b
	-8	4.13(30°,2)	-7.5	30-40	T	$\mu\simeq2(KCl, KNO_3)$	52Ja	a,b
	-12.95	(1-2)	--	25	C	0.1KCl	61Cb	
Co^{2+}	-8.15	7.99(1)	9.0	25	C	$\mu=0.1(KCl)$	61Cb	
	-9	8.47(30°,1)	9.1	30-40	T	$\mu\simeq2(KCl, KNO_3)$	52Ja	a,b
	-10.25	5.86(2)	-7.5	25	C	$\mu=0.1(KCl)$	61Cb	
	-10	6.07(30°,2)	-5.2	30-40	T	$\mu\simeq2(KCl, KNO_3)$	52Ja	a,b
	-18.40	13.85(1-2)	1.5	25	C	$\mu=0.1(KCl)$	61Cb	
Ni^{2+}	-11.85	10.59(1)	8.5	25	C	$\mu=0.1(KCl)$	61Cb 50Jb	
	-12	10.81(30°,1)	9.9	30-40	T	$\mu\simeq2(KCl, KNO_3)$	52Ja	a,b
	-13.45	7.99(2)	-8.5	25	C	$\mu=0.1(KCl)$	61Cb	
	-13	8.14(30°,2)	-5.6	30-40	T	$\mu\simeq2(KCl, KNO_3)$	52Ja	a,b
	-25.30	18.58(1-2)	0.0	25	C	$\mu=0.1(KCl)$	61Cb	
Cu^{2+}	-18.9	15.43(30°,1)	8	10-40	T	$\mu=0$	59Ma 50Jb	b
	-18.00	15.80(1)	12.0	25	C	$\mu=0.1(KCl)$	61Cb	
	-20	16.11(30°,1)	7.8	30-40	T	$\mu\simeq2(KCl, KNO_3)$	52Ja	a,b
	-6.3	4.68(30°,2)	1	10-40	T	$\mu=0$	59Ma	b
	-8.15	5.20(2)	-3.5	25	C	$\mu=0.1(KCl)$	61Cb	

Metal Ion	ΔH, kcal /mole	Log K	ΔS, cal /mole °K	T, °C	M	Conditions	Ref.	Remarks
A70, cont.								
Cu^{2+}, cont.	-26.15	21.00(1-2)	8.5	25	C	μ=0.1(KCl)	61Cb	
Zn^{2+}	-6.4	8.57(30°,1)	18	10-40	T	μ=0	59Ma	b
	-6.45	8.80(1)	18.5	25	C	μ=0.10(KCl)	61Cb	
	-8	9.14(30°,1)	15.4	30-40	T	μ≃2(KCl, KNO_3)	52Ja	a,b
	-10.15	5.50(2)	-9.0	25	C	μ=0.10(KCl)	61Cb	
	-16.60	14.29(1-2)	9.5	25	C	μ=0.10(KCl)	61Cb	
Hg^{2+}	-36	25.03(1-2)	-7	25	T	μ=0.1(KNO_3)	61Rc	
A71	AMINE, di(3-aminopropyl)-		$C_6H_{17}N$			$HN(CH_2CH_2CH_2NH_2)_2$		L
Co^{2+}	-7.77	6.92(1)	5.6	25	C	0.1M(KCl)	66Pa	
	-9	6.63(1)	1	30	T	1M(KNO_3)	56Ha	
Ni^{2+}	-10.56	9.18(1)	6.6	25	C	0.1M(KCl)	66Pa	
	-10	9.09(1)	9	30	T	1M(KNO_3)	56Ha	
	-7.08	3.55(2)	-7.5	25	C	0.1M(KCl)	66Pa	
Cu^{2+}	-16.09	14.20(1)	11.0	25	C	0.1M(KCl)	66Pa	
	-16	14.25(1)	12	30	T	1M(KNO_3)	56Ha	
Zn^{2+}	-5.44	7.92(1)	18	25	C	0.1M(KCl)	66Pa	
A72	AMINE, dibutyl-		$C_8H_{19}N$			$HN(CH_2CH_2CH_2CH_3)_2$		L
VOA_2	-9.75	-1.47(1)	-27.8	25	C	μ≃0. S: Nitrobenzene. A=Acetylacetonate ion	65Ca	
NiA	-13.1	3.03(1)	-30.0	25	T	$C_M=10^{-4}$ S: Benzene A=Diacetylbisbenzoyl hydrazine	60Sa	
ZnA_2	-8.9	2.463(25°,1)	-19	18.5-65	T	S: Toluene A=[$S_2CN(CH_3)_2$]	65Ch	b
A73	AMINE, di(2-butyl)-		$C_8H_{19}N$			$HN[CH(CH_3)CH_2CH_3]_2$		
NiA	-12.6	0.56(1-2)	-39.7	25	T	$C_M=10^{-4}$ S: Benzene A=Diacetylbisbenzoyl hydrazine	60Sa	
A74	AMINE, diethyl		$C_4H_{11}N$			$HN(CH_2CH_3)_2$		L
VOA_2	-9.50	-1.36(1)	-27.3	25	C	μ≃0. S: Nitrobenzene. A=Acetylacetonate ion	65Ca	
NiA	-13.5	3.72(1-2)	-28.4	25	T	$C_M=10^{-4}$. S: Benzene A=Diacetylbisbenzoyl hydrazine	60Sa	
NiA_2	-7.53	(1)	--	25	C	S: Benzene A=2,2,6,6-Tetramethyl-3,5-heptanedione	73Ce	

Metal Ion	ΔH, kcal /mole	Log K	ΔS, cal /mole °K	T, °C	M	Conditions	Ref.	Remarks
A74, cont.								
NiA_2, cont.	-11.66	1.569(2)	-32.2	25	C	S: Benzene A=2,2,6,6-Tetramethyl-3,5-heptanedione	73Ce	
[Ni]	-5.3	1.176(20°,1)	-13	-12-32	T	S: Toluene [Ni]=Bis(0,0'-diethyldithiophosphato)nickel(II)	71Cc	b,q: ±0.5
	-4.8	-0.377(20°,2)	-18	-12-32	T	S: Toluene [Ni]=Bis(0,0'-diethyldithiophosphato)nickel(II)	71Cc	b,q: ±0.3
Ag^+	-10.65	-8.70(1-2)	-6.5	25	C	--	55Fa	
AlA_3	-27.33	(1)	--	25-28	C	S: n-Hexane A=CH_3	68Hc	
I_2	-8.8	3.853(20°,1)	-1.2	22	C	low μ S: n-Heptane	64Sd 60Ya	
A75 AMINE, diisobutyl-			$C_8H_{19}N$			$HN[CH_2CH(CH_3)_2]_2$		L
NiA	-13.2	1.69(1-2)	-36.6	25	T	$C_M=10^{-4}$ S: Benzene A=Diacetylbisbenzoyl hydrazine	60Sa	
A76 AMINE, dimethyl-			C_2H_7N			$HN(CH_3)_2$		L
NiA	-14.4	6.46(1-2)	-19.1	25	T	$C_M=10^{-4}$ S: Benzene A=Diacetylbisbenzoyl hydrazine	60Sa	
Ag^+	-11.25	5.77(1-2)	-14.5	2	T	μ=0	65Le	h: 2-45°
	-11.25	5.22(1-2)	-13.9	25	T	μ=0	65Le	h: 2-45°
	-11.25	4.94(1-2)	-14.0	35	T	μ=0	65Le	h: 2-45°
	-11.25	4.68(1-2)	-14.0	45	T	μ=0	65Le	h: 2-45°
	-9.7	5.36(1-2)	-8.0	25	C	--	55Fa	
	-11.25	6.28(1-2)	-12.2	2	T	μ=0 S: 50 v % CH_3OH	65Le	h: 2-45°
	-11.25	5.54(1-2)	-12.4	25	T	μ=0 S: 50 v % CH_3OH	65Le	h: 2-45°
	-11.25	5.30(1-2)	-12.2	35	T	μ=0 S: 50 v % CH_3OH	65Le	h: 2-45°
	-11.25	4.97(1-2)	-12.7	45	T	μ=0 S: 50 v % CH_3OH	65Le	h: 2-45°
AlA_3	-30.84	(1)	--	25-28	C	S: n-Hexane A=CH_3	68Hc	
A77 AMINE, diphenyl methyl-			$C_{13}H_{13}N$			$(C_6H_5)_2NCH_3$		

Metal Ion	ΔH, kcal /mole	Log K	ΔS, cal /mole °K	T, °C	M	Conditions	Ref.	Re-marks
A77, cont.								
Al_2A_6	-17.4	--	--	25	C	S: Benzene C_M=0.07M A=Br R: 0.5(Al_2A_6) + L = AlA_3L	68Rc	
A78 AMINE, dipropyl-			$C_6H_{15}N$			$HN(CH_2CH_2CH_3)_2$		L
NiA	-13.3	3.36(1-2)	-29.3	25	T	$C_M=10^{-4}$ S: Benzene A=Diacetylbisbenzoyl hydrazine	60Sa	
A79 AMINE, di(2-propyl)-			$C_6H_{15}N$			$HN[CH(CH_3)_2]_2$		L
NiA	-12.6	1.80(1-2)	-34.5	25	T	$C_M=10^{-4}$ S: Benzene A=Diacetylbisbenzoyl hydrazine	60Sa	
A80 AMINE, methyl bis-(3-aminopropyl)-			$C_7H_{19}N_3$			$CH_3N(CH_2CH_2CH_2NH_2)_2$		L
Ni^{2+}	-7.85	7.05(30°,1)	6	10-40	T	μ=0	59Gc	b
Cu^{2+}	-14.1	12.28(30°,1)	10	10-40	T	μ=0	59Gc	b
Cd^{2+}	-4.65	5.87(30°,1)	12	10-40	T	μ=0	59Gc	b
A81 AMINE, tri(2-aminoethyl)-			$C_6H_{18}N_4$			$N(CH_2CH_2NH_2)_3$		L
Ca^{2+}	-5.54	8.43(25°,1)	19.9	15-40	T	μ=0.11(KNO_3)	61Mf	b
Pr^{3+}	-62.5	(1-2)	--	23	C	S: Acetonitrile	71Fc	
Gd^{3+}	-60.3	(1-2)	--	23	C	S: Acetonitrile	71Fc	
Mn^{2+}	-3.00	5.80(1)	16.5	25	C	μ=0.1(KCl)	63Pb	
Fe^{2+}	-6.30	8.6(1)	18.5	25	C	μ=0.1(KCl)	63Pb 60Pa	
Co^{2+}	-10.65	12.61(1)	22.0	25	C	μ=0.1(KCl)	63Pb 60Pa	
Ni^{2+}	-15.15	14.59(1)	16.0	25	C	μ=0.1(KCl)	63Pb 60Pa	
Cu^{2+}	-15.7	18.40(30°,1)	32	10-40	T	μ=0	58Ba 60Pa	b
	-20.40	18.84(1)	18.0	25	C	μ=0.1(KCl)	63Pb 61Mf	
Zn^{2+}	-13.85	14.48(1)	19.5	25	C	μ=0.1(KCl)	63Pb 60Pa	
A82 AMINE, tri(3-aminopropyl)-			$C_9H_{24}N_4$			$N(CH_2CH_2CH_2NH_2)_3$		L
Ni^{2+}	-9.19	8.702(1)	9.0	25	C	0.1M(KCl)	68Va	
Cu^{2+}	-14.67	13.113(1)	10.8	25	C	0.1M(KCl)	68Va	

Metal Ion	ΔH, kcal /mole	Log K	ΔS, cal /mole °K	T, °C	M	Conditions	Ref.	Re-marks
A82, cont.								
Cu^{2+}, cont.	-12.83	10.752(1)	6.1	25	C	0.1M(KCl) L=HL^+	68Va	
Zn^{2+}	-8.89	10.696(1)	19.1	25	C	0.1M(KCl)	68Va	
A83 AMINE, tribenzyl-			$C_{21}H_{21}N$			$(C_6H_5CH_2)_3N$		
I_2	-2.3	(1)	--	25	C	S: Benzene	66Ad	
A84 AMINE, tribenzyl-N-oxide			$C_{21}H_{21}NO$			$(C_6H_5CH_2)_3NO$		L
I_2	-10.5	3.540(20°,1)	-19.6	4-20	T	C_M=4.6x10^{-4} C_L=4x10^{-4}-1.4x10^{-3} S: CH_2Cl_2	65Kc	b
A85 AMINE tributyl-			$C_{12}H_{27}N$			$(CH_3CH_2CH_2CH_2)_3N$		L
NiA	-13.1	0.51(1-2)	-41.8	25	T	C_M=10^{-4} S: Benzene A=Diacetylbisbenzoyl hydrazine	60Sd	
HgA_2	-10.35	2.322(1)	-23.7	30	C	S: Benzene A=Cl C_M=1-5x10^{-3}M	73Fa	
	-10.47	2.539(1)	-22.9	30	C	S: Benzene A=Br C_M=1-5x10^{-3}M	73Fa	
	-10.68	2.348(1)	-24.6	30	C	S: Benzene A=I C_M=1-5x10^{-3}M	73Fa	
I_2	-11.6	3.20(20°,1)	-25	22	C	low μ S: n-Heptane	64Sd 60Ya	b
A86 AMINE, triethyl-			$C_6H_{15}N$			$(CH_3CH_2)_3N$		L
NiA	-13.1	1.36(1-2)	-38.2	25	T	C_M=10^{-4} S: Benzene A=Diacetylbisbenzoyl hydrazine	60Sd	
NiA_2	-7.49	1.918(1)	-16.3	25	C	S: Benzene A=2,2,6,6-Tetramethyl-3,5-heptanedione	73Ce	
	-2.47	(2)	--	25	C	S: Benzene A=2,2,6,6-Tetramethyl-3,5-heptanedione	73Ce	
AlA_3	-26.47	(1)	--	25-28	C	S: n-Hexane A=CH_3	68Hc	
Al_2A_6	-31.5	--	--	25	C	S: Benzene C_M=0.07M A=Br R: 0.5(Al_2A_6) + L = AlA_3L	68Rc	
CuA_2	-16.9	3.699(1)	-39.6	25	C	S: CCl_4 A=Hexa-fluoroacetylacetone	70Pa	
	-15.1	4.400(1)	-30.5	25	C	S: Cyclohexane A=Hexafluoroacetyl-acetone	72Nd	

Metal Ion	ΔH, kcal /mole	Log K	ΔS, cal /mole °K	T, °C	M	Conditions	Ref.	Remarks
A86, cont.								
CuA_2, cont.	-13.4	3.201(1)	-30.2	25	C	S: o-Dichlorobenzene A=Hexafluoroacetylacetone	72Nd	
CuA_2B	-6.8	--	--	25	C	S: $o-Cl_2C_6H_4$ A=Hexafluoroacetylacetone B=Dimethylacetate R: $CuA_2B + L = CuA_2L + B$	72Nd	
	-6.2	--	--	25	C	S: $o-Cl_2C_6H_4$ A=Hexafluoracetylacetone B=Dimethyl sulfoxide R: $CuA_2B + L = CuA_2L + B$	72Nd	
	-8.7	--	--	25	C	S: $o-Cl_2C_6H_4$ A=Hexafluoroacetylacetone B=Ethyl acetate R: $CuA_2B + L = CuA_2L + B$	72Nd	
	-7.1	--	--	25	C	S: Cyclohexane A=Hexafluoroacetylacetone B=Dimethylacetate R: $CuA_2B + L = CuA_2L + B$	72Nd	
	-6.6	--	--	25	C	S: Cyclohexane A=Hexafluoracetylacetone B=Dimethyl sulfoxide R: $CuA_2B + L = CuA_2L + B$	72Nd	
	-9.2	--	--	25	C	S: Cyclohexane A=Hexafluoroacetylacetone B=Ethyl acetate R: $CuA_2B + L = CuA_2L + B$	72Nd	
BA_3C	-14.6	(1)	--	25	C	S: Benzene A=F $C=(CH_3CH_2)_2O$	68Rd	
	-17.2	(1)	--	25	C	S: Benzene A=F $C=(CH_3CH_2CH_2CH_2)_2O$	68Rd	
GaA_3	-30.2	4.91(1)	-78.8	25	C	S: Benzene A=Cl	70Gh	
I_2	-11.8	3.671(25°,1)	-22.8	22	C	low μ S: n-Heptane	64Sd 58Na	b
A87 AMINE, trimethyl-			C_3H_9N			$(CH_3)_3N$		L
NiA	-13.9	3.39(1-2)	-31.0	25	T	$C_M=10^{-4}$ S: Benzene A=Diacetylbisbenzoyl hydrazine	60Sd	
AlA_3	-29.96	(1)	--	25-28	C	S: n-Hexane $A=CH_3$	68Hc	
I_2	-12.1	4.083(20°,1)	-22.6	20-40	T	$C_M=10^{-5}$ $C_L=10^{-4}-10^{-3}$ S: n-Heptane	60Ya	b

Metal Ion	ΔH, kcal /mole	Log K	ΔS, cal /mole °K	T, °C	M	Conditions	Ref.	Remarks
A88 AMINE, trimethyl, oxide			C_3H_9ON			$(CH_3)_3NO$		L
VAB_2	-37.6	(1)	--	25	C	S: CH_2Cl_2 A=O B=2,4-Pentanedione	69Pd	
CuA_2	-11.4	(1)	--	25	C	S: o-Dichloro-benzene A=Hexa-fluoroacetylacetone	70Pa	
	-11.4	(1)	--	25	C	S: o-Dichloro-benzene A=Hexa-fluoroacetylacetone	72Nd 72Nd	
I_2	-10.0	3.753(22°,1)	-16.9	6.5-22	T	C_M=5.2x10^{-4} C_L=1.8x10^{-4}-6.5x10^{-4} S: CH_2Cl_2	65Kc	b
A89 AMINE, trioctyl-			$C_{24}H_{51}N$			$[CH_2(CH_2)_6CH_3]_3N$		
I_2	-12.2	(1)	--	25	C	S: Benzene	66Ad	
A90 AMINE, tripentyl			$C_{15}H_{33}N$			$[CH_3(CH_2)_4]_3N$		L
BA_3C	-12.1	(1)	--	25	C	S: Benzene A=F C=$(CH_3CH_2)_2O$	68Rd	
Al_2A_6	-31.5	--	--	25	C	S: Benzene C_M=0.07M A=Br R: 0.5(Al_2A_6) + L = AlA_3L	68Rc	
I_2	-12.1	(1)	--	22	C	low μ S: n-Heptane	64Sd	
A91 AMINE, triphenyl-			$C_{18}H_{15}N$			$(C_6H_5)_3N$		L
I_2	-1.1	(1)	--	25	C	S: Benzene	66Ad	
A92 AMINE, tripropyl			$C_9H_{21}N$			$(CH_2CH_2CH_3)_3N$		L
NiA	-13.2	0.88(1-2)	-40.3	25	T	C_M=10^{-4}. S: Benzene A=Diacetylbisbenzoyl hydrazine	60Sd	
BA_3C	-14.4	(1)	--	25	C	S: Benzene A=F C=$(CH_3CH_2)_2O$	68Rd	
	-16.8	(1)	--	25	C	S: Benzene A=F C=$(CH_3CH_2CH_2CH_2)_2O$	68Rd	
Al_2A_6	-33.4	--	--	25	C	S: Benzene C_M=0.07M A=Br R: 0.5(Al_2A_6) + L = AlA_3L	68Rc	
I_2	-12.0	3.143(25°,1)	-25.8	22	C	low μ S: n-Heptane	64Sd 60Ya	
A93 AMINE, tri(2-hydroxyethyl)-			$C_6H_{15}O_3N$			$(CH_2CH_2OH)_3N$		L
Ni^{2+}	-3.86	2.27(1)	-2.5	25	C	μ=0.5	63Sa	
	-0.51	0.82(2)	2.1	25	C	μ=0.5	63Sa	
	-4.37	3.09(1-2)	-0.4	25	C	μ=0.5	63Sa	
A94 AMMONIA			NH_3			NH_3		L
Li^+	-0.5	2.47(15°)	-12.6	14-17	C	Very qualitative calorimetry. Not	52Fa	

Metal Ion	ΔH, kcal /mole	Log K	ΔS, cal /mole °K	T, °C	M	Conditions	Ref.	Remarks
A94, cont.								
Li^{+}, cont.						known how many NH_3 coordinate-thought to be 3		
Mg^{2+}	-0.1	3.40(15°)	-19.0	14-17	C	Very qualitative calorimetry. Not known how many NH_3 coordinate-thought to be 6	52Fa	
Co^{2+}	-5.26	(1-3)	--	--	C	--	36Ca	
Co^{3+}	-56.8	29.68(1-6)	-53.9	25	C	--	61Kb	
	-48.8	32.80	-13.7	25	C	R: $Co^{3+} + 5NH_3 + H_2O = [Co(NH_3)_5H_2O]^{3+}$	61Kb	
Ni^{2+}	-3.5	2.77(1)	0.9	25	C	1M(KNO_3)	55Pb 57Pa 46Ca 52Fa	f:6
	-4.0	2.80(1)	-0.5	26.8	C	--	57Yc	
	-3.5	2.22(2)	-1.6	25	C	1M(KNO_3)	55Pb	f:6
	-4.0	2.24(2)	-3.1	26.8	C	--	57Yc	
	-3.5	1.71(3)	-3.9	25	C	1M(KNO_3)	55Pb	f:6
	-4.0	1.73(3)	-5.4	26.8	C	--	57Yc	
	-3.5	1.17(4)	-6.4	25	C	1M(KNO_3)	55Pb	f:6
	-4.0	1.19(4)	-7.9	26.8	C	--	57Yc	
	-3.5	0.73(5)	-8.4	25	C	1M(KNO_3)	55Pb	f:6
	-4.3	0.75(5)	-10.9	26.8	C	--	57Yc	
	-3.5	0.01(6)	-11.7	25	C	1M(KNO_3)	55Pb	f:6
	-4.3	0.03(6)	-14.2	26.8	C	--	57Yc	
	-21.0	8.61(1-6)	-31.1	25	C	1M(KNO_3)	55Pb	f:6
NiA	-14.6	2.10(25°,1-2)	-39.2	10-45	T	$C_M=10^{-4}$ S: Benzene A=Diacetylbisbenzoyl hydrazine	60Sb	b
PtL_2A_2	-3.0	--	--	25	C	9.4% NH_3 solution A=I R: PtL_2A_2(cis) = PtL_2A_2(trans)	73Pa	
	-3.8	--	--	25	C	9.4% NH_3 solution A=I R: PtL_2A_2(cis) = PtL_2A_2(trans)	73Pa	
	-7.2	--	--	25	C	9.4% NH_3 solution A=I R: PtL_2A_2(cis) = PtL_2A_2(trans)	73Pa	
Cu^{2+}	-5.0	4.14(1)	2.1	25	C	1M(KNO_3) (Bouzat, Ann. Chim. Phys., 29, 312 (1903))	55Pb 52Fa	i:see cond. colm.
	-5.6	4.15(1)	0.3	26.8	C	--	57Yc	
	-5.0	3.49(2)	-0.8	25	C	1M(KNO_3 (Bouzat, Ann. Chim. Phys., 29, 312 (1903))	55Pb	i:see cond. colm.

Metal Ion	ΔH, kcal /mole	Log K	ΔS, cal /mole °K	T, °C	M	Conditions	Ref.	Remarks
A94, cont.								
Cu^{2+},	-5.5	3.50(2)	-2.3	26.8	C	--	57Yc	
cont.	-5.0	2.88(3)	-3.6	25	C	1M(KNO_3) (Bouzat, Ann. Chim. Phys., 29, 312 (1903))	55Pb	i:see cond. colm.
	-5.6	2.89(3)	-5.4	26.8	C	--	57Yc	
	-5.0	2.12(4)	-7.1	25	C	1M(KNO_3) (Bouzat, Ann. Chim. Phys., 29, 312 (1903))	55Pb	i:see cond. colm.
	-5.3	2.13(4)	-7.9	26.8	C	--	57Yc	
	-3.2	-0.55(5)	-13.3	25	C	1M(KNO_3) (Bouzat, Ann. Chim. Phys., 29, 312 (1903))	55Pb	i:see cond. colm.
	-5.1	-0.52(5)	-19.4	26.8	C	--	57Yc	
	-12.0	7.865(1-2)	-4.36	25	T	μ=2	53Sb	
	-20.0	12.63(1-4)	-9.4	25	C	1M(KNO_3) (Bouzat, Ann. Chim. Phys., 29, 312 (1903))	55Pb	i:see cond. colm.
	-23.6	13.05(1-4)	-19.8	25	T	μ=2	53Sb	
Ag^+	-13.40	7.225(1-2)	-11.9	25	C	μ=0	37Sa, 52Fa	
	-13	7.204(1-2)	-10.7	0-45	T	μ≈0.15	49Ja, 55Fa, 66Ma, 57Sb	b
Zn^{2+}	-2.6	2.35(1)	2.1	26.8	C	C_L=0.1-10	57Yb, 61Me, 52Fa	
	-3.1	2.45(2)	0.9	26.8	C	C_L=0.1-10	57Yb	
	-3.9	2.51(3)	-1.5	26.8	C	C_L=0.1-10	57Yb	
	-5.2	2.15(4)	-7.5	26.8	C	C_L=0.1-10	57Yb	
ZnA_4^{2-}	-21.0	(1-4)	--	--	C	S: 25 w % NH_4Cl A=Cl	57Ta	
Cd^{2+}	-3.5	2.66(1)	0.5	26.8	C	C_L=0.1-10	57Yb, 52Fa	f:6
	-3.5	2.09(2)	-2.1	26.8	C	C_L=0.1-10	57Yb	f:6
	-3.5	1.43(3)	-5.1	26.8	C	C_L=0.1-10	57Yb	f:6
	-3.5	0.93(4)	-7.4	26.8	C	C_L=0.1-10	57Yb	f:6
	-3.5	-0.29(5)	-13.0	26.8	C	C_L=0.1-10	57Yb	f:6
	-3.5	-2.42(6)	-19.4	26.8	C	C_L=0.1-10	57Yb	f:6
	-7.12	4.95(1-2)	-1.24	25	T	μ=2	53Sa	
	-12.7	7.44(1-4)	-34.08	25	T	μ=2	53Sa	
Hg^{2+}	-3.3	1.00(3)	-6.4	26.8	C	C_L=0.1-10	57Yb, 52Fa	
	-3.6	0.80(4)	-8.3	26.8	C	C_L=0.1-10	57Yb	
	-24.7	17.5(1-2)	-2.3	26.8	C	C_L=0.1-10	57Yb	
AlA_3	-27.55	(1)	--	25-28	C	S: n-Hexane A=CH_3	68Hc	
I_2	-4.8	1.83(20°,1)	-8.0	10&20	T	$C_M=10^{-5}$ $C_L=10^{-4}-10^{-3}$ S: n-Heptane	60Ya	a,b

Metal Ion	ΔH, kcal /mole	Log K	ΔS, cal /mole °K	T, °C	M	Conditions	Ref.	Re-marks
A95 AMMONIUM ION, allyl triethyl-			$C_9H_{20}N$			$CH_2:CHCH_2N(Et)_3$		L^+
PtA_4^{2-}	-4.9	2.185(45°)	-4.6	30-60	T	μ=2M(NaCl + HCl) A=Cl R: $PtA_4^{2-} + L^+ = PtA_3L + A^-$	67Da	b
A96 AMMONIUM ION, allyl trimethyl-			$C_6H_{14}N$			$CH_2:CHCH_2N(Me)_3$		L^+
PtA_4^{2-}	-5.3	2.238(44.5°)	-6.6	30-60	T	μ=2M(NaCl + HCl) A=Cl R: $PtA_4^{2-} + L^+ = PtA_3L + A^-$	67Da	b
A97 AMMONIUM ION, 3-butenyl triethyl-			$C_{10}H_{22}N$			$CH_2:CHCH_2CH_2N(Et)_3$		L^+
PtA_4^{2-}	-3.8	3.772(44.8°)	5.4	30-60	T	μ=2M(NaCl + HCl) A=Cl R: $PtA_4^{2-} + L^+ = PtA_3L + A^-$	67Da	b,q: ±0.4
A98 AMYLOSE			$(C_6H_{10}O_5)_n$			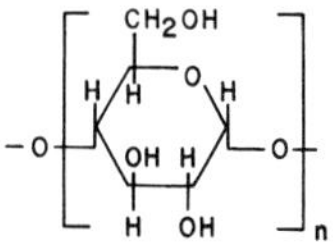		L
I_2	-15.5	9.34	-11.0	16	T	0.05M(KI)	60Ka	j
	-20.8	--	--	50	T	0.05M(KI)	60Ka	j
A99 ANILINE			C_6H_7N			$C_6H_5NH_2$		
CoA_2	-7.48	4.897(1-2)	-6.04	25	C	S: Acetone C_M=1x10^{-3}M C_L=0.2M A=Cl	71Zc	q: ±0.6
	-7.48	4.163(1-2)	-6.04	25	C	S: Acetone A=Cl	71Ze	q: ±0.6
	-13.70	3.577(1-2)	-29.60	25	C	S: Acetonitrile A=Cl	71Ze	q: ±1.5
	-9.93	3.188(1-2)	-18.70	25	C	S: n-Butanol A=Cl	71Ze	
	-15.2	1.832(1-2)	-42.6	25	C	S: t-Butyl alcohol C_M=1x10^{-3}M C_L=0.15M A=Cl	71Zd	
	-14.82	4.354(1-2)	-29.80	25	C	S: Cyclohexanone A=Cl	71Ze	
	-4.25	4.178(1-2)	4.76	25	C	S: Dimethyl-formamide A=Cl	71Ze	
	-12.98	2.030(1-2)	-34.27	25	C	S: Methanol A=Cl	71Ze	
A100 ANILINE, 4-bromo-N,N-dimethyl-N-oxide-			$C_8H_{10}ONBr$			$C_6H_4N(Br)(CH_3)_2(\rightarrow O)$		
VAB_2	-33.9	(1)	--	25	C	S: CH_2Cl_2 A=O B=2,4-Pentanedione	69Pd	

Metal Ion	ΔH, kcal /mole	Log K	ΔS, cal /mole °K	T, °C	M	Conditions	Ref.	Re-marks
A101	ANILINE, 2-bromo-N-methylazoxy		$C_7H_7ON_3Br$			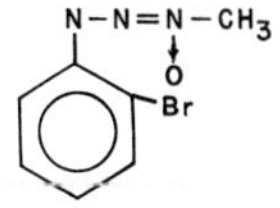		
NiL_2	1	--	0.4	21-50	T	S: Benzene R: NiL_2(planar, S=0) = NiL_2(pseudooctrahedral, S=1)	71Za	b,q: ±0.3
A102	ANILINE, 4-chloro-		C_6H_6NCl			C_6H_5NHCl		
CoA_2	-9.75	3.848(1-2)	-15.20	25	C	S: Acetone C_M=1x10^{-3}M C_L=0.2M A=Cl	71Zc	
A103	ANILINE, 2-chloro-N-methylazo, 3'-N-oxide		$C_7H_7ON_3Cl$			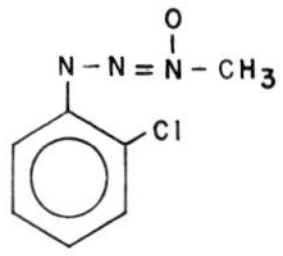		
NiL_2	2.0	--	3	20-60	T	S: Benzene R: NiL_2(planar, S=0) = NiL_2(pseudooctrahedral, S=1)	71Za	b,q: ±0.2
A104	ANILINE, N,N-diethyl-		$C_{10}H_{15}N$			$C_6H_5N(C_2H_5)_2$		L
CoA_2	-3.82	4.545(1-2)	7.95	25	C	S: Acetone C_M=1x10^{-3}M C_L=0.2M A=Cl	71Zc	q: ±0.9
A105	ANILINE, N,N-dimethyl-		$C_8H_{11}N$			$C_6H_5N(CH_3)_2$		L
CoA_2	-6.42	4.347(1-2)	-1.64	25	C	S: Acetone C_M=1x10^{-3}M C_L=0.2M A=Cl	71Zc	q: ±0.9
BA_3C	-7.3	(1)	--	25	C	S: Benzene A=F C=$(CH_3CH_2CH_2CH_3)_2O$	68Rd	
AlA_6	-24.2	--	--	25	C	S: Benzene C_M=0.07M A=Br R: 0.5(Al_2A_6) + L = AlA_3L	68Rc	
I_2	-8.2	1.24(27°,1)	-21.6	11-40	T	C_M=2.9x10^{-5} C_L=0.01-0.04 S: n-Heptane	60Tc	b
A106	ANILINE, 2,3-dimethyl-N-(2'-hydroxybenzylidine)-		$C_{15}H_{15}ON$					

Metal Ion	ΔH, kcal /mole	Log K	ΔS, cal /mole °K	T, °C	M	Conditions	Ref.	Remarks
A106, cont.								
Co^{2+}	-11.58	11.77(25°,1-2)	14.62	--	T	S: 80 v % Methanol	71De	b,j, u:25
Cu^{2+}	-13.0	15.91(25°,1-2)	29	15-45	T	S: 80 v % Methanol μ=0.1 $C_M=4x10^{-4}M$	71Kb	b
	-13.57	15.86(25°,1-2)	26.28	--	T	S: 80 v % Methanol	71De	b,j, u:25
A107	ANILINE, 1,2-dimethyl-N-(2'-hydroxy-3'-methoxy-benzylidene)	$C_{16}H_{17}O_2N$				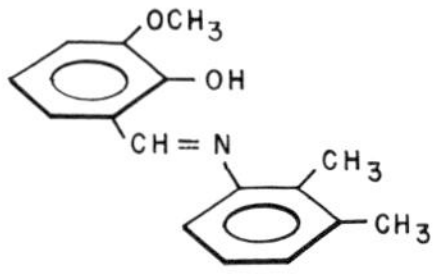		
Co^{2+}	-9.96	11.16(25°,1-2)	21.25	--	T	S: 80 v % Methanol	71De	b,j, u:25
Cu^{2+}	-11.2	16.86(25°,1-2)	38	15-45	T	S: 80 v % Methanol μ=0.1 $C_M=4x10^{-4}M$	71Kb	b
	-10.49	16.52(25°,1-2)	38.59	--	T	S: 80 v % Methanol	71De	b,j, u:25
A108	ANILINE, 2-fluoro-N-methylazo, 3'-N-oxide	$C_7H_7ON_3F$				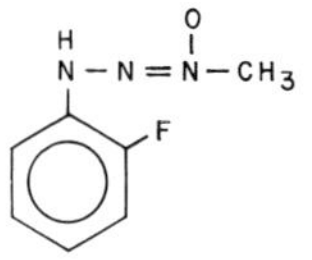		
NiL_2	2.4	--	1.5	30-70	T	S: Benzene R: NiL_2(planar, S=0) = NiL_2(pseudooctrahedral, S=1)	71Za	b,q: ±0.2
A109	ANILINE, N-(2'-hydroxy-benzylidene)-	$C_{13}H_{11}ON$				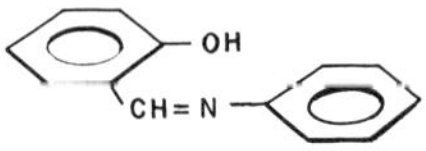		L
Co^{2+}	-12.34	10.75(25°,1-2)	7.42	--	T	S: 80 v % Methanol	71De	b,j, u:25
Cu^{2+}	-15.8	14.07(25°,1-2)	11	15-45	T	S: 80 v % Methanol μ=0.1 $C_M=4x10^{-4}M$	71Kb	b
	-15.83	14.10(25°,1-2)	11.46	--	T	S: 80 v % Methanol	71De	b,j, u:25
A110	ANILINE, N-(2'-hydroxy-3'methoxybenzylidene)	$C_{14}H_{13}O_2N$				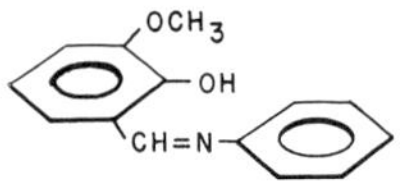		

Metal Ion	ΔH, kcal /mole	Log K	ΔS, cal /mole °K	T, °C	M	Conditions	Ref.	Remarks
A110, cont.								
Co^{2+}	-11.16	10.94(25°,1-2)	12.38	--	T	S: 80 v % Methanol	71De	b,j, u:25
Cu^{2+}	-14.3	15.32(25°,1-2)	21	15-45	T	S: 80 v % Methanol μ=0.1 C_M=4x10^{-4}M	71Kb	b
	-14.33	15.36(25°,1-2)	21.38	--	T	S: 80 v % Methanol	71De	b,j, u:25
A111	ANILINE, 2-methoxy-		C_7H_9ON			$C_6H_4NH_2(OCH_3)$		
CoA_2	-3.67	3.878(1-2)	5.50	25	C	S: Acetone C_M=1x10^{-3}M C_L=0.2M A=Cl	71Zc	q: ±0.5
A112	ANILINE, 4-methoxy-		C_7H_9ON			$C_6H_4NH_2(OCH_3)$		
CoA_2	-15.55	4.083(1-2)	-34.16	25	C	S: Acetone C_M=1x10^{-3}M C_L=0.2M A=Cl	71Zc	q: ±1.5
A113	ANILINE, 4-methyl-N-(2'-hydroxybenzylidene)-		$C_{14}H_{13}ON$			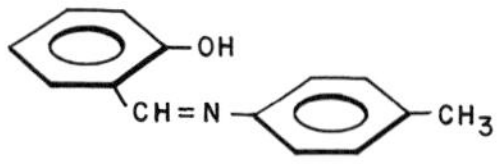		
Co^{2+}	-11.93	11.14(25°,1-2)	14.91	--	T	S: 80 v % Methanol	71De	b,j, u:25
Cu^{2+}	-15.3	15.98(25°,1-2)	20	15-45	T	S: 80 v % Methanol μ=0.1 C_M=4x10^{-4}M	71Kb	b
	-15.32	16.00(25°,1-2)	20.74	--	T	S: 80 v % Methanol	71De	b,j, u:25
A114	ANILINE, 4-methyl-N-(2'-hydroxy-3'-methoxy-benzylidene)-		$C_{15}H_{15}O_2N$			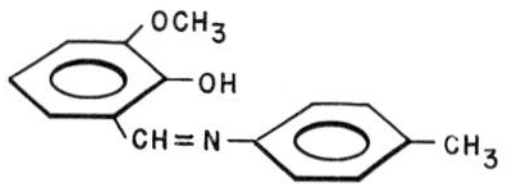		
Co^{2+}	-12.48	11.96(25°,1-2)	21.61	--	T	S: 80 v % Methanol	71De	b,j, u:25
Cu^{2+}	-12.4	16.20(25°,1-2)	33	15-45	T	S: 80 v % Methanol μ=0.1 C_M=4x10^{-4}M	71Kb	b
	-12.45	16.19(25°,1-2)	33.51	--	T	S: 80 v % Methanol	71De	b,j, u:25
A115	ANILINE, 3-nitro-		$C_6H_6O_2N_2$			$C_6H_4NH_2(NO_2)$		
CoA_2	-13.87	2.566(1-2)	-34.80	25	C	S: Acetone C_M=5x10^{-4} C_L=7.5x10^{-4} A=Cl	71Zc	q:±1

Metal Ion	ΔH, kcal /mole	Log K	ΔS, cal /mole °K	T, °C	M	Conditions	Ref.	Remarks
A116	ANILINE, 4-nitro-		$C_6H_6O_2N_2$			$C_6H_4NH_2(NO_2)$		
CoA_2	-14.72	2.155(1-2)	-39.53	25	C	S: Acetone C_M=5x10^{-4} C_L=7.5x10^{-4} A=Cl	71Zc	q:±1.1
A117	ANILINE, 2,4,6-trimethyl-N-(2'-hydroxybenzylidene)-		$C_{16}H_{17}ON$			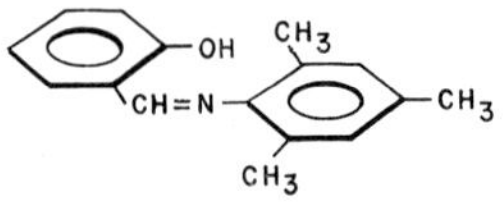		
Cu^{2+}	-18.5	15.83(25°,1-2)	26	15-45	T	S: 80 v % CH_3OH μ=0.1 C_M=4x10^{-4}M	71Kb	b
A118	ANILINE, 2,4,6-trimethyl-N-(2'-hydroxy-3'-methoxy-benzilidene)-		$C_{17}H_{19}O_2N$			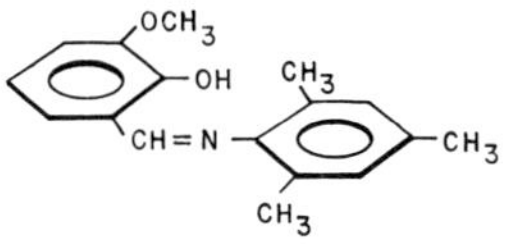		
Cu^{2+}	-10.5	16.49(25°,1-2)	38	15-45	T	S: 80 v % CH_3OH μ=0.1 C_M=4x10^{-4}M	71Kb	b
A119	ANTAZOLINE		$C_{17}H_{19}N_3$			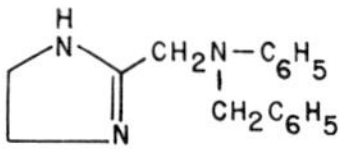		L
Be^{2+}	-5.7	7.44(25°,1)	15	0-45	T	μ=0.12M(KCl) C_M≈2x10^{-3}M C_L≈8x10^{-3}M	69Cb	b
	-3.3	6.20(25°,2)	17	0-45	T	μ=0.12M(KCl) C_M≈2x10^{-3}M C_L≈8x10^{-3}M	69Cb	b
Co^{2+}	11.0	7.54(25°,1-2)	72	0&25	T	0.06M(KCl)	62Aa	a,b
Ni^{2+}	9.1	7.70(25°,1-2)	66	0&25	T	0.06M(KCl)	62Aa	a,b
Cu^{2+}	-12.5	12.30(25°,1-2)	14.4	0&25	T	0.06M(KCl)	62Aa	a,b
Cd^{2+}	-17.7	8.73(25°,1-2)	-19	0-45	T	μ=0.1(KNO_3)	64Ad	b
A120	ANTHRACENE		$C_{14}H_{10}$			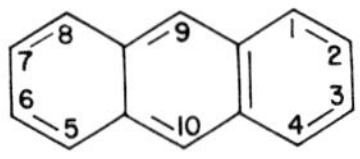		L^-
Li^+	0.0 1.9	5.149(1) 5.305(1)	24 31	-65 15	T T	S: Tetrahydrofuran S: Tetrahydrofuran	68Nb 68Nb	q:±0.4 q:±0.4
Na^+	-4.6	--	-23	-92-12	T	S: 2-Methyltetra-hydrofuran C_L=10^{-5}-10^{-4}M R: Tight ion pair = Loose ion pair	67He	b,q: ±0.3
	-6.2	--	-25	-40-30	T	S: 50% Tetrahydro-furan 50% 2-Methyl-	67He	b,q: ±0.5

Metal Ion	ΔH, kcal /mole	Log K	ΔS, cal /mole °K	T, °C	M	Conditions	Ref.	Re-marks
A120, cont.								
Na^+, cont.						tetrahydrofuran $C_L=10^{-5}-10^{-4}M$ R: Tight ion pair = Loose ion pair		
	0	4.54(1)	21	-65	T	$C_M=C_L=10^{-6}-10^{-3}$ S: Tetrahydrofuran	66Cb	q:±0.4
	6.1	5.19(1)	45	15	T	$C_M=C_L=10^{-6}-10^{-3}$ S: Tetrahydrofuran	66Cb	q:±0.4
A121 APOCARBONIC ANHYDRASE B, bovine								
Zn^{2+}	3.86	10.48	60.9	25	C	pH=5.5	69Hc	j
	9.84	12.02	88.0	25	C	pH=7.0	69Hc	j
A122 L-(+)-ARABINOSE			$C_5H_{10}O_5$					L
$H_5TeO_6^-$	-1.3	1.67	3	25	T	μ=0.1 R: $H_5TeO_6^-$ + L = $HTeO_4L^-$ + $2H_2O$	62Ea	
A123 ARGININE			$C_6H_{14}O_2N_4$			$H_2NC(:NH)NH(CH_2)_3CH(NH_2)CO_2H$		L^-
Ni^{2+}	-3.75	4.92(25°,1)	9.9	17-40	T	--	60Pe	b
	-4.75	4.20(25°,2)	3.3	17-40	T	--	60Pe	b
	-6.50	3.08(25°,3)	-7.7	17-40	T	--	60Pe	b
Cu^{2+}	-9.25	7.34(25°,1)	2.5	17-40	T	--	60Pe	b
	-7.85	6.42(25°,2)	3.0	17-40	T	--	60Pe	b
Ag^+	-5.76	3.18(1)	-4.78	25	T	μ=0.024	59Da	
	-7.24	3.65(2)	-7.64	25	T	μ=0.024	59Da	
Zn^{2+}	-1.40	4.19(25°,1)	14.5	17-40	T	--	60Pe	b
	-3.35	3.93(25°,2)	6.7	17-40	T	--	60Pe	b
Cd^{2+}	-2.10	3.27(25°,1)	7.9	17-40	T	--	60Pe	b
	-6.35	3.18(25°,2)	-6.7	17-40	T	--	60Pe	b
A124 ARSINE, triphenyl-			$C_{18}H_{15}As$			$(C_6H_5)_3As$		L
HgA_2	-12.9	2.583(1)	-30.8	30	C	S: Benzene A=Cl $C_M=1-5x10^{-3}M$	73Fa	
	-16.0	2.120(1)	-43.2	30	C	S: Benzene A=Br $C_M=1-5x10^{-3}M$	73Fa	
A125 ARSONIUM ION, allyl triethyl-			$C_9H_{20}As$			$CH_2:CHCH_2As(C_2H_5)_3$		L^+
PtA_4^{2-}	-5.9	3.121(45°)	-4.3	30-60	T	μ=2M(NaCl + HCl) A=Cl R: PtA_4^{2-} + L^+ = PtA_3L + A^-	67Da	b,q: ±0.6
A126 ARSONIUM ION, 3-butenyl triethyl			$C_{10}H_{22}As$			$CH_2:CHCH_2CH_2As(C_2H_5)_3$		L^+
PtA_4^{2-}	-3.3	3.846(44.8°)	7.2	30-60	T	μ=2M(NaCl + HCl) A=Cl R: PtA_4^{2-} + L^+ = PtA_3L + A^-	67Da	b,q: ±0.35

Metal Ion	ΔH, kcal /mole	Log K	ΔS, cal /mole °K	T, °C	M	Conditions	Ref.	Remarks
A127	ASPARAGINE		$C_4H_8O_3N_2$			$H_2NCOCH_2CH(NH_2)CO_2H$		L^-
Cu^{2+}	-11.35	14.4(2)	28.0	25	C	μ=0.1M	70Bd 71Be	y:70Bd
	-11.27	14.4(2)	28.2	25	C	μ=0.1M	70Bd 71Be	y:70Bd
Zn^{2+}	-0.53	6.000(1)	25.5	5	T	μ=0.1(KCl)	73Rb	q:±0.71
	-1.83	5.940(1)	21.0	25	T	μ=0.1(KCl)	73Rb	q:±0.71
	-3.46	5.820(1)	15.8	45	T	μ=0.1(KCl)	73Rb	q:±0.71
A128	ASPARTIC ACID		$C_4H_7O_4N$			$HO_2CCH_2CH(NH_2)CO_2H$		L^{2-}
Nd^{3+}	-91.4	5.66(30°)	-275.8	30-50	T	μ=0.1M($NaClO_4$) R: $Nd^{3+} + H_2L = NdHL^{2+} + H^+$	71Te	b
Sm^{3+}	-57.12	7.83(30°)	-152.83	30-50	T	μ=0.1M($NaClO_4$) R: $Sm^{3+} + H_2L = SmHL^{2+} + H^+$	71Te	b
$UO_2{}^{2+}$	-25.966	8.34(30°)	-47.67	30-50	T	μ=0.1M($NaClO_4$) R: $UO_2{}^{2+} + HL^- = UO_2L + H^+$	71Td	b
Cu^{2+}	-6.1	8.80(1)	20	25	C	μ=0.2(KCl)	74Na	
	-6.1	8.80(25°,1)	20	20-35	T	μ=0.2(KCl)	74Na	b
	-6.3	6.96(2)	11	25	C	μ=0.2(KCl)	74Na	
	-6.0	6.96(25°,2)	12	20-35	T	μ=0.2(KCl)	74Na	b
In^{3+}	4.9	3.28(35°,1)	31	25-45	T	μ=0.20M($NaClO_4$)	73Sc	b
	9.99	6.40(35°,1-2)	61.7	25-45	T	μ=0	72Sb	b
A129	1,2-AZABOROLIDINE, 2-butyl-		$C_7H_{16}NB$			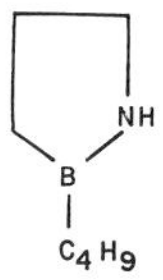		
[B]	15.2	0.535(100°)	43.2	30-110	T	[B]=L R: $2L=L_2$	71Te	b
A130	AZIDE ION		N_3			$N_3{}^-$		L^-
MnA	1.28	1.473(1)	11.2	25	T	μ=0.25 A=EDTA	67Sd	
	0.7	1.466(1)	9.2	25	T	μ=0.25 ΔH calculated from rate data. A=EDTA	67Sd	
Fe^{3+}	-1.1	4.84(25°,1)	18.3	15-35	T	μ=0.25	61Ba 52Ua	b
[Fe^{3+}]	-15.2	4.398	-31	25	C	Tris-cacodylate buffer (≈0.1M Tris) [Fe]=G. gouldii Hemerythrin=$HrFe_2$ R: $HrFe_2(H_2O)_m + L = HrFe_2L + mH_2O$	61Lb	g:Na q:±1.5
[Fe^{3+}]	-12.9	5.5(1)	-18.4	25	-	Hemoglobin	60Gb	
[Fe^{3+}]	-10.2	4.8(1)	-12.2	25	-	Chironomus Hemoglobin	60Gb	

Metal Ion	ΔH, kcal /mole	Log K	ΔS, cal /mole °K	T, °C	M	Conditions	Ref.	Re-marks
A130, cont.								
$[Fe^{3+}]$, cont.						$[Fe^{3+}]$=Ferrihemoglobin μ=0.05(NaCl)		d:0.05
	-14.1	(1)	--	25.9	C	5.65	73Ag	
	-14.1	(1)	--	25.9	C	5.99	73Ag	
	-14.1	5.290(1)	-23.19	25.9	C	6.00	73Ag	
	-14.63	5.209(1)	-25.20	25.9	C	6.40	73Ag	
	-14.8	(1)	--	25.9	C	6.41	73Ag	
	-17.0	(1)	--	25.9	C	6.82	73Ag	
	-17.73	5.055(1)	-36.27	25.9	C	7.00	73Ag	
	-16.8	(1)	--	25.9	C	7.24	73Ag	
	-16.8	(1)	--	25.9	C	7.60	73Ag	
	-16.3	(1)	--	25.9	C	7.82	73Ag	
	-16.0	(1)	--	25.9	C	8.01	73Ag	
	-15.41	5.231(1)	-27.71	25.9	C	8.47	73Ag	
	-14.5	(1)	--	25.9	C	8.83	73Ag	
	-12.6	(1)	--	25.9	C	9.06	73Ag	
	-11.5	(1)	--	25.9	C	9.21	73Ag	
						μ=0.05(NaCl) S: 5% Tert-butyl alcohol		d:0.05
	-13.8	(1)	--	25.9	C	5.8	73Ag	
	-14.8	(1)	--	25.9	C	6.0	73Ag	
	-14.7	(1)	--	25.9	C	6.41	73Ag	
	-15.07	4.791(1)	-28.58	25.9	C	6.50	73Ag	
	-15.2	(1)	--	25.9	C	6.64	73Ag	
	-15.1	(1)	--	25.9	C	6.80	73Ag	
	-15.54	4.696(1)	-30.58	25.9	C	7.0	73Ag	
	-16.6	(1)	--	25.9	C	7.25	73Ag	
	-16.60	4.549(1)	-34.80	25.9	C	7.52	73Ag	
	-16.6	(1)	--	25.9	C	7.55	73Ag	
	-14.70	4.630(1)	-28.07	25.9	C	7.94	73Ag	
	-15.2	(1)	--	25.9	C	8.02	73Ag	
	-14.2	(1)	--	25.9	C	8.40	73Ag	
	-14.30	4.813(1)	-25.90	25.9	C	8.52	73Ag	
	-14.3	(1)	--	25.9	C	8.56	73Ag	
	-13.7	(1)	--	25.9	C	8.80	73Ag	
	-12.41	4.769(1)	-19.78	25.9	C	9.00	73Ag	
	-12.4	(1)	--	25.9	C	9.01	73Ag	
$[Fe^{3+}]$						$[Fe^{3+}]$=Sperm Whale Melmyoglobin μ=0.05		b, d:0.05
	-12.3	4.79(20°,1)	-20.0	8.7-27.2	T	6.0	69Ba	
	-12.1	4.73(20°,1)	-20.0	8.7-27.2	T	6.4	69Ba	b
	-11.2	4.66(20°,1)	-16.8	8.7-27.2	T	6.8	69Ba	b
	-11.2	4.60(20°,1)	-17.2	8.7-27.2	T	7.2	69Ba	b
	-11.9	4.56(20°,1)	-19.7	8.7-27.2	T	7.6	69Ba	b
	-12.8	4.52(20°,1)	-23.0	8.7-27.2	T	8.0	69Ba	b
	-13.1	4.50(20°,1)	-24.1	8.7-27.2	T	8.4	69Ba	b

Metal Ion	ΔH, kcal /mole	Log K	ΔS, cal /mole °K	T, °C	M	Conditions	Ref.	Remarks
A130, cont.								
$[Fe^{3+}]$, cont.	-13.1	4.50(20°,1)	-24.1	8.7-27.2	T	8.8	69Ba	b
	-12.8	4.51(20°,1)	-23.0	8.7-27.2	T	9.2	69Ba	b
	-12.4	4.52(20°,1)	-21.6	8.7-27.2	T	9.6	69Ba	b
	-10.9	4.57(20°,1)	-16.3	8.7-27.2	T	10.0	69Ba	b
$[Fe^{3+}]$	Reaction: Pigeon methaemoglobin + L. ΔH values reported as a function of pH in graphical form.						68Ab	
$[Fe^{3+}]$	Reaction: Dog methaemoglobin + L. ΔH values reported as a function of pH in graphical form.						68Ab	
$[Fe^{3+}]$	Reaction: Guinea pig methaemoglobin + L. ΔH values reported as a function of pH in graphical form.						68Ab	
$[Fe^{3+}$	Reaction: Human methaemoglobin + L. ΔH values reported as a function of pH in graphical form.						66Ac	
Co^{2+}	1.37	1.279(25°,1)	18.18	20.5-36.5	T	$C_L=C_M=1.8x10^{-3}M$	73Aa	b
	4.2	3.743(27°,1-4)	31.5	23-33	T	S: Methanol	73Aa	b
$CoA_6{}^{3+}$	-3.01	2.01(1)	-1.0	25	T	μ=0 A=NH_3	55Na 53Ea	
$CoA_3{}^{3+}$	-5.2	1.06(1)	-13	25	T	μ=0.054 A=Ethylenediamine	53Ea	
$PtA_2{}^{2+}$	≈0	1.59(23°,1)	7.3	23-48	T	μ=0 S: CH_3OH A=o-Phenylenebis-dimethyl arsine	66Db	b
CuA^{2+}	3.32	2.48(1)	22.4	25	C	S: Methanol A=N,N'-bis(2-Aminoethyl) propane-1,3-diamine	72Bd	
	1.59	3.07(1)	19.9	25	C	S: Methanol A=N,N' bis(3-Aminopropyl) propane-1,3-diamine	72Bd	
	3.09	2.79(1)	23.1	25	C	S: Methanol A=Triethylene-tetramine	72Bd	
$CuA_2{}^{2+}$	3.48	2.48(1)	23.0	25	C	S: Methanol A=Ethylenediamine	72Bd	
	2.41	2.57(1)	19.8	25	C	S: Methanol A=Trimethylene-diamine	72Bd	
Cd^{2+}	-1.15	1.62(1)	3.5	25	C	$3.0M(NaClO_4)$	66Gb	
	-1.41	1.13(2)	0.2	25	C	$3.0M(NaClO_4)$	66Gb	
	-1.66	0.53(3)	-3.1	25	C	$3.0M(NaClO_4)$	66Gb	
	-1.27	0.56(4)	-1.7	25	C	$3.0M(NaClO_4)$	66Gb	
Tl^+	-1.33	0.39(1)	-2.7	25	T	μ=0	57Na	

Metal Ion	ΔH, kcal /mole	Log K	ΔS, cal /mole °K	T, °C	M	Conditions	Ref.	Re-marks

B

B1	BARBITURIC ACID, 5-amino-N,N-diacetic acid		$C_8H_9O_7N_3$					L^{3-}
Li^+	-7.0	4.70(27°,1)	-1	20-39	T	μ=0.1M[$(CH_3)_4NNO_3$]	63Ia	b
Na^+	-8.7	2.54(27°,1)	-18	20-34	T	μ=0.1M[$(CH_3)_4NNO_3$]	63Ia	b
K^+	-11.8	1.00(27°,1)	-35	20-39	T	μ=0.1M[$(CH_3)_4NNO_3$]	63Ia	b
Th^+	-15.4	5.76(27°,1)	-25	20-39	T	μ=0.1M[$(CH_3)_4NNO_3$]	63Ia	b
B2	BENZALDEHYDE		C_7H_6O			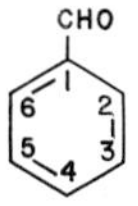		L
FeA_3	-12.8	1.48(1)	-36	25	C	S: Chlorobenzene A=Cl	49Db	
AlA_3	-24.7	1.78(1)	-75	25	C	S: Chlorobenzene A=Cl	49Db	
SnA_4	-28.0	5.56(1-2)	-68	25	C	S: Chlorobenzene A=Cl	49Db	
SbA_3	-3.9	0.95(1)	-9	25	C	S: Chlorobenzene A=Cl	49Db	
I_2	-3.0	-0.28(25°,1)	-11.3	5-25	T	S: Heptane	71Te	
B3	BENZALDEHYDE, oxime, N-methyl-		C_8H_9ON			$C_6H_5CH{:}N(CH_3)OH$		L
I_2	-5.25	1.554(15°,1)	-11.2	5-28	T	C_M=7.5x10^{-5} C_L=3.8x10^{-3}-1.5x10^{-2} S: CCl_4	65Kc	b
B4	BENZALDEHYDE, 5-bromo-2-hydroxy-		$C_7H_5O_2Br$			$C_6H_4O(Br)(OH)$		
Mg^{2+}	-2.8	3.18(25°,1)	5.6	15-50	T	S: 50% Dioxane μ=0.3M($NaClO_4$)	73Ca	b
	-3.3	2.68(25°,2)	1.3	15-50	T	S: 50% Dioxane μ=0.3M($NaClO_4$)	73Ca	b
Sr^{2+}	-1.3	3.09(25°,1)	9.9	15-50	T	S: 50% Dioxane μ=0.3M($NaClO_4$)	73Ca	b
	-2.3	2.64(25°,2)	4.1	15-50	T	S: 50% Dioxane μ=0.3M($NaClO_4$)	73Ca	b
Ni^{2+}	-4.9	4.28(25°,1)	3.3	15-50	T	S: 50% Dioxane μ=0.3M($NaClO_4$)	73Ca	b
	-3.5	3.36(25°,2)	3.5	15-50	T	S: 50% Dioxane μ=0.3M($NaClO_4$)	73Ca	b
Cu^{2+}	-6.3	6.37(25°,1)	7.8	15-50	T	S: 50% Dioxane μ=0.3M($NaClO_4$)	73Ca	b
	-6.5	4.75(25°,2)	0.0	15-50	T	S: 50% Dioxane μ=0.3M($NaClO_4$)	73Ca	b

Metal Ion	ΔH, kcal /mole	Log K	ΔS, cal /mole °K	T, °C	M	Conditions	Ref.	Remarks
B4, cont.								
Zn^{2+}	-2.5	3.81(25°,1)	9.1	15-50	T	S: 50% Dioxane μ=0.3M($NaClO_4$)	73Ca	b
	-2.5	3.13(25°,2)	5.7	15-50	T	S: 50% Dioxane μ=0.3M($NaClO_4$)	73Ca	b
B5 BENZALDEHYDE, 5-chloro-2-hydroxy-			$C_7H_5O_2Cl$			$C_6H_4O(Cl)(OH)$		
Mg^{2+}	-4.0	3.30(25°,1)	1.6	15-50	T	S: 50% Dioxane μ=0.3M($NaClO_4$)	73Ca	b
	-4.2	2.82(25°,2)	-1.1	15-50	T	S: 50% Dioxane μ=0.3M($NaClO_4$)	73Ca	b
Sr^{2+}	-1.3	3.06(25°,1)	9.7	15-50	T	S: 50% Dioxane μ=0.3M($NaClO_4$)	73Ca	b
	-0.7	2.59(25°,2)	-9.5	15-50	T	S: 50% Dioxane μ=0.3M($NaClO_4$)	73Ca	b
Ni^{2+}	-5.8	4.56(25°,1)	1.4	15-50	T	S: 50% Dioxane μ=0.3M($NaClO_4$)	73Ca	b
	-5.4	3.52(25°,2)	-2.0	15-50	T	S: 50% Dioxane μ=0.3M($NaClO_4$)	73Ca	b
Cu^{2+}	-5.7	6.21(25,1)	9.5	15-50	T	S: 50%Dioxane μ=0.3M($NaClO_4$)	73Ca	b
	-4.9	4.66(25°,2)	4.8	15-50	T	S: 50% Dioxane μ=0.3M($NaClO_4$)	73Ca	b
Zn^{2+}	-3.3	3.65(25°,1)	5.8	15-50	T	S: 50% Dioxane μ=0.3M($NaClO_4$)	73Ca	b
	-0.6	3.03(25°,2)	11.9	15-50	T	S: 50% Dioxane μ=0.3M($NaClO_4$)	73Ca	b
B6 BENZALDEHYDE, 2-hydroxy-			$C_7H_6O_2$			$C_6H_5O(OH)$		L
Fe^{3+}	2.0	8.8(25°,1)	47	15-35	T	μ=3.0	56Aa	b
B7 BENZALDEHYDE, 4-hydroxy			$C_7H_6O_2$			$C_6H_5O(OH)$		
Fe^{3+}	0.5	5.35(1)	26	25	T	μ=0.1	69Da	
B8 BENZALDEHYDE, 2-hydroxy-5-nitro-			$C_7H_5O_4N$			$C_6H_3O(OH)(NO_2)$		
Mg^{2+}	-7.2	3.18(25°,1)	-11.9	15-50	T	S: 50% Dioxane μ=0.3M($NaClO_4$)	73Ca	b
	-4.3	2.63(25°,2)	-3.0	15-50	T	S: 50% Dioxane μ=0.3M($NaClO_4$)	73Ca	b
Sr^{2+}	-10.3	3.16(25°,1)	-21.0	15-50	T	S: 50% Dioxane μ=0.3M($NaClO_4$)	73Ca	b
	-6.2	2.65(25°,2)	-9.3	15-50	T	S: 50% Dioxane μ=0.3M($NaClO_4$)	73Ca	b
Ni^{2+}	-10.2	3.80(25°,1)	-17.1	15-50	T	S: 50% Dioxane μ=0.3M($NaClO_4$)	73Ca	b
	-5.5	3.13(25°,2)	-4.3	15-50	T	S: 50% Dioxane μ=0.3M($NaClO_4$)	73Ca	b

Metal Ion	ΔH, kcal /mole	Log K	ΔS, cal /mole °K	T, °C	M	Conditions	Ref.	Remarks
B8, cont.								
Cu^{2+}	-7.9	4.39(25°,1)	-6.5	15-50	T	S: 50% Dioxane μ=0.3M($NaClO_4$)	73Ca	b
	-4.0	3.34(25°,2)	2.3	15-50	T	S: 50% Dioxane μ=0.3M($NaClO_4$)	73Ca	b
Zn^{2+}	-3.6	3.27(25°,1)	2.8	15-50	T	S: 50% Dioxane μ=0.3M($NaClO_4$)	73Ca	b
	-2.9	2.69(25°,2)	2.6	15-50	T	S: 50% Dioxane μ=0.3M($NaClO_4$)	73Ca	b

B9 BENZENE C_6H_6

Metal Ion	ΔH, kcal /mole	Log K	ΔS, cal /mole °K	T, °C	M	Conditions	Ref.	Remarks
VAB_3	-0.140	(1)	--	5-50	T	S: CCl_4 A=O B=Cl	71Ha	b
	-0.300	(1)	--	5-50	T	S: Cyclohexane A=O B=Cl	71Ha	b
NbA_5	-2.100	(1)	--	-30-0	T	S: CCl_4 A=Cl	71Ha	b,q: ±0.40
	-2.400	(1)	--	-30-0	T	S: Cyclohexane A=Cl	71Ha	b,q: ±0.40
TaA_5	-1.100	(1)	--	-30-0	T	S: CCl_4 A=Cl	71Ha	b
	-2.500	(1)	--	-30-0	T	S: Cyclohexane A=Cl	71Ha	b,q: ±0.40
WA_6	-0.150	(1)	--	-24-26	T	S: CCl_4 A=Cl	71Hd	b
	-0.050	(1)	--	-24-26	T	S: Perfluoromethyl-cyclohexane A=Cl	71Hd	b
WAB_4	0.130	(1)	--	-14-26	T	S: CCl_4 A=O B=Cl	71Hd	b
	0.060	(1)	--	-14-26	T	S: Cyclohexane A=O B=Cl	71Hd	b
OsA_4	-0.525	(1)	--	0-31.5	T	S: Cyclohexane A=O	61Hc	b,q: ±0.175
Ag^+	-2.86	0.041(1)	-9.40	25	T	μ=0.5 S: Equimolar H_2O - CH_3OH soln.	56Oa	
AgA	4.0	0.220(20°,1)	14.6	1.4&20	T	S: CH_3OH A=ClO_4^-	64Tb	a,b
GaA_3	-11.3	(1)	--	25	C	S: Benzene A=Cl	72Rd	
SO_2	-7.8	-0.252(20°,1)	-27.68	0-30	T	0.4M(NaOH) 0.2M(HCl) S: CCl_4	57Da	b
	-7.8	-0.252(20°,1)	-27.68	11-25	T	low μ. S: Ethanol	57Db	b
I_2	-1.4	(1)	--	--	C	S: Benzene	61Aa 55Kb 61Ja	i: 50Ha
	-0.85	(1)	--	--	C	S: Benzene	50Ha	
	-1.452	0.336(25°,1)	-3.3	0-44.5	T	Mole fraction of Benzene in CCl_4:1.00	50Ca	b
	-1.416	0.282(25°,1)	-3.5	0-44.5	T	Mole fraction of Benzene in CCl_4:0.620	50Ca	b
	-1.349	0.265(25°,1)	-3.3	0-44.5	T	Mole fraction of Benzene in CCl_4: 0.0434	50Ca	b

Metal Ion	ΔH, kcal /mole	Log K	ΔS, cal /mole °K	T, °C	M	Conditions	Ref.	Remarks
B9, cont.								
I_2, cont.	-1.317	0.281(25°,1)	-3.1	0-44.5	T	Mole fraction of Benzene in CCl_4: 0.0217	50Ca	b
	-1.3	(1)	--	--	-	S: Cyclohexane	60Bc	
	-1.300	0.083(1)	-4.0	25	T	S: n-Hexane	54Ka	1
	-1.14	0.649(25°)	-0.85	-20-40	T	S: Hexane A=Hexane R: $I_2A + L = I_2L + A$	73Si	b,q: ±0.15
IA	-2.54	0.74(25°,1)	-5.14	1.6&25	T	C_L=0.1-1 C=10^{-4}-10^{-3} S: CCl_4 A=Cl	550a	a,b
	-3.14	0.74(25°,1)	-7.14	25& 45.8	T	C_L=0.1-1 C=10^{-4}-10^{-3} S: CCl_4 A=Cl	550a	a,b
B10	BENZENE, 1-tert-butyl-4-nitro-		$C_{10}H_{13}O_2N$			$C_6H_4(NO_2)[C(CH_3)_3]$		L^-
Li^+	13.3	--	--	25	T	S: Hexamethyl-phosphoramide R: $Li^+ + L^- = LiL$	73Sq	q:0.8
B11	BENZENE, 1-carbamyl-4-nitro-		$C_7H_6O_3N_2$			$C_6H_4(NO_3)(CONH_2)$		
Li^+	7.0	--	--	25	T	S: Hexamethyl-phosphoramide R: $Li^+ + L^- = LiL$	73Sq	q: ±0.6
B12	BENZENE, chloro-		C_6H_5Cl			$C_6H_5(Cl)$		
I_2	-1.10	-0.174(1)	-4.4	25	T	S: n-Hexane	54Ka	1
B13	BENZENE, 1-chloro-2-cyclo-propyl-		C_9H_9Cl			$C_6H_4(Cl)(CHCH_2CH_2)$		L
HgA_2	9.77	-0.777(27°)	29.0	35-20	T	Large excess of L S: CH_3OH R: $HgA_2 + L = HgAl + A$. A=$OCOCH_3$	64Pb	b
B14	BENZENE, 4-chloro-1,2-dihydroxy-		$C_6H_5O_2Cl$			$C_6H_3(Cl)(OH)_2$		L^{2-}
Cu^{2+}	-4.0	0.92	-9.1	25	C	μ=0.100M(KNO_3) R: $Cu^{2+} + HL^- = CuL + H^+$	72Jb	
	2.85	-7.602	-25.1	25	C	μ=0.100M(KNO_3) R: $Cu^{2+} + H_2L = CuL + 2H^+$	72Jb	
B15	BENZENE, 1-chloro-4-nitro-		$C_6H_4O_2NCl$			$C_6H_4(NO_2)(Cl)$		L^-
Li^+	10.8	(1)	--	25	T	S: Hexamethyl-phosphoramide	73Sq	

Metal Ion	ΔH, kcal /mole	Log K	ΔS, cal /mole °K	T, °C	M	Conditions	Ref.	Remarks
B16	BENZENE, cyclopropyl		C_9H_{10}			$C_6H_5(CHCH_2CH_2)$		L
HgA_2	5.70	-0.356(11°)	18.4	-14-20	T	Large excess of L S: CH_3OH R: HgA_2 + L = HgAL + A. A=$OCOCH_3$	64Pb	b
B17	BENZENE, 1-cyclopropyl-2-nitro-		$C_9H_9O_2N$			$C_6H_4(CHCH_2CH_2)(NO_2)$		L
HgA_2	10.61	-0.914(27°)	31.2	15-35	T	Large excess of L S: CH_3OH R: HgA_2 + L = HgAL + A A=$OCOCH_3$	64Pb	b
B18	BENZENE, 1,2-dichloro-		$C_6H_4Cl_2$			$C_6H_4(Cl)_2$		L
I_2	-1.00	-0.180(1)	-4.3	25	T	S: n-Hexane	54Ka	1
B19	BENZENE, 1,2-dihydroxy-		$C_6H_6O_2$			$C_6H_4(OH)_2$		L^{2-}
ThA	12.8	--	--	25	C	μ=0.10(KNO_3) A=1,2-Diaminocyclohexane-N,N',N'-tetraacetic acid R: ThA + H_2L = $ThAL^{2-}$ + $2H^+$	70Kb	
	12.2	--	--	25	C	μ=0.10(KNO_3) A=EDTA R: ThA + H_2L = $ThAL^{2-}$ + $2H^+$	70Kb	
Cu^{2+}	-10.0	13.827(1)	-31.0	25	C	μ=0.100M(KNO_3)	72Ja	
	-5.0	0.85	-12.9	25	C	μ=0.100M(KNO_3) R: Cu^{2+} + HL^- = CuL + H^+	72Ja	
	-3.20	-8.345	-27.5	25	C	μ=0.100M(KNO_3) R: Cu^{2+} + H_2L = CuL + $2H^+$	72Ja	
B20	BENZENE, 1,4-dihydroxy		$C_6H_6O_2$			$C_6H_4(OH)_2$		L
SO_2	-14.0	0.121(14.5°,1)	-42.45	4.5-20	T	Low μ. S: C_2H_5OH	57Db	b
B21	BENZENE, (1,2-dihydroxyethyl)		$C_8H_{10}O_2$			$C_6H_5[CH(OH)(CH_2OH)]$		L
$H_5TeO_6^-$	-3.6	1.63	-5	25	T	μ=0.1 R: $H_5TeO_6^-$ + L = $HTeO_4L^-$ + $2H_2O$	62Ea	
B22	BENZENE, 1,3-dimethyl-		C_8H_{10}			$C_6H_4(CH_3)_2$		L
AgA	4.5	0.783(20°,1)	18.9	1.4&20	T	S: CH_3OH A=ClO_4^-	64Tb	a,b
B23	BENZENE, 1,4-dimethyl-		C_8H_{10}			$C_6H_4(CH_3)_2$		L
I_2	-2.00	0.471(1)	-4.9	25	T	S: n-Hexane Orientation (o, m, or p) not specified	54Ka	1

Metal Ion	ΔH, kcal /mole	Log K	ΔS, cal /mole °K	T, °C	M	Conditions	Ref.	Remarks
B23, cont.								
IA	-3.60	1.29(25°,1)	-6.18	1.6&25	T	$C_M=10^{-4}-10^{-3}$ A=Cl $C_L=0.1-1$ S. CCl_4	550a 60Dc 55Kb	a,b,k
	-3.82	1.29(25°,1)	-6.91	25& 45.8	T	$C_M=10^{-4}-10^{-3}$ A=Cl $C_L=0.1-1$ S: CCl_4	550a	a,b,k
B24	BENZENE, 1,3-dimethyl-2-(N,N-dimethylamino)-		$C_{10}H_{15}N$			$C_6H_3(CH_3)_2[N(CH_3)_2]$		L
I_2	-1.7	0.42(33°,1)	-3.6	16&33	T	$C_M \approx 10^{-5}$. $C_L \approx 0.05-0.13$. S: n-Heptane	60Tc	a,b
B25	BENZENE, ethoxy-		$C_8H_{10}O$			$C_6H_5OCH_2CH_3$		L
I_2	-0.5	(1)	--	25	C	S: Octane	66Ad	
B26	BENZENE, ethynyl-		C_8H_6			$C_6H_5C{:}CH$		L
AlA_3	0.74	-1.19(27°,1)	--	-5-27	T	S: Cyclohexane R: $1/2Al_2A_6 + L = AlA_3L$ A=Et	73Ae	b,q: ±.2
B27	BENZENE, hexaethyl-		$C_{18}H_{30}$			$C_6(CH_2CH_3)_6$		L
I_2	-1.79	0.58(25°,1)	-3.4	1.6-45.8	T	$C_L=0.01-0.15$mole fraction. S: CCl_4	55Kb	b,m
IA	-3.59	1.25(25°,1)	-6.30	1.6&25	T	$C_M=10^{-4}-10^{-3}$ $C_L=0.1-1$ S: CCl_4 A=Cl	550a	a,b,k
	-3.35	1.25(25°,1)	-5.50	25& 45.8	T	$C_M=10^{-4}-10^{-3}$ $C_L=0.1-1$ S: CCl_4 A=Cl	550a	a,b,k
B28	BENZENE, hexamethyl-		$C_{12}H_{18}$			$C_6(CH_3)_6$		L
Ag^+	-3.45	-0.201(1)	-12.5	25	T	μ=0.5 S: Equimolar H_2O - CH_3OH soln.	560a	
PA_5	0.590	(1)	--	-30-31	T	S: Cyclohexane A=Cl	71Hb	b
I_2	-3.73	1.20(25°,1)	-7.1	1.6-45.8	T	$C_L=0.01-0.15$ mole fraction S: CCl_4	55Kb	b,m
IA	-5.25	2.13(25°,1)	-7.86	1.6&25	T	$C_M=10^{-4}-10^{-3}$ $C_L=0.1-1$ S: CCl_4 A=Cl	550a	a,b,k
	-5.39	2.13(25°,1)	-8.32	25& 45.8	T	$C_M=10^{-4}-10^{-3}$ $C_L=0.1-1$ S: CCl_4 A=Cl	550a	a,b,k
B29	BENZENE, 1-isopropyl-4-nitro-		$C_9H_{11}O_2N$			$(CH_3)_2CHC_6H_5NO_2$		L^-
Li^+	12.5	(1)	--	25	T	S: Hexamethyl-phosphoramide	73Sq	q: ±1.2

Metal Ion	ΔH, kcal /mole	Log K	ΔS, cal /mole °K	T, °C	M	Conditions	Ref.	Re-marks
B30	BENZENE, α-ketoethyl-2-hydroxy-		$C_8H_8O_2$			$C_6H_4(COCH_3)(OH)$		L
Fe^{3+}	0.8	10.6(25°,1)	51	15-35	T	μ=3.0	56Aa	b
B31	BENZENE, methoxy-		C_7H_8O			$C_6H_5OCH_3$		L
Al_2A_6	-15.5	--	--	25	C	S: Benzene C_M=0.07M A=Br R: $0.5(Al_2A_6) + L = AlA_3L$	68Rc	
	-14.6	--	--	25	C	S: Cyclohexane C_M=0.07M A=Br R: $0.5(Al_2A_6) + L = AlA_3L$	68Rc	
GaA_3	-21.1	(1)	--	25	C	S: Benzene A=Cl	72Rd	
	-19.5	(1)	--	25	C	S: Cyclohexane A=Cl	72Rd	
	-21.9	(1)	--	25	C	S: Mesitylene A=Cl	72Rd	
B32	BENZENE, 1-methoxy-4-(S-methylmercapto)		$C_8H_{10}OS$			$C_6H_4(OCH_3)(SCH_3)$		L
I_2	-6.2	2.339(25°,1)	-7.8	25-55	T	C_L<0.02 mole fraction. S: CCl_4	65Lb	b,1
B33	BENZENE, methylthio-		C_7H_8S			$C_6H_5(SCH_3)$		L
PtA_2L_2	3.6	0.398(28°)	14	30-60	T	S: Chloroform A=Cl R: PtL_2A_2(cis) = PtL_2A_2(trans)	73Re	b
Al_2A_6	-13.1	--	--	25	C	S: Benzene C_M=0.07M A=Br R: $0.5(Al_2A_6) + L = AlA_3L$	68Rc	
I_2	-6.1	0.964(25°,1)	-16.1	25-40	T	$C_M=10^{-3}$. C_L=0.16 S: CCl_4	64Na	b
	-5.5	1.083(25°,1)	-13	15-35	T	S: Cyclohexane	63Va	b
B34	BENZENE, methylthio-4-bromo-		C_7H_7BrS			$C_6H_4(SCH_3)(Br)$		L
I_2	-5.8	0.771(25°,1)	-16	15-35	T	S: Cyclohexane	63Va	b
B35	BENZENE, methylthio-3-chloro-		C_7H_7ClS			$C_6H_5(SCH_3)(Cl)$		L
I_2	-5.8	0.663(25°,1)	-16	15-35	T	S: Cyclohexane	63Va	b
B36	BENZENE, methylthio-4-chloro-		C_7H_7ClS			$C_6H_4(SCH_3)(Cl)$		L
I_2	-5.8	0.748(25°,1)	-16	15-35	T	S: Cyclohexane	63Va	b
B37	BENZENE, methylthio-3-fluoro-		C_7H_7FS			$C_6H_4(SCH_3)(F)$		L
I_2	-5.3	0.633(25°,1)	-15	15-35	T	S: Cyclohexane	63Va	b
B38	BENZENE, methylthio-4-fluoro-		C_7H_7FS			$C_6H_4(SCH_3)(F)$		L

Metal Ion	ΔH, kcal /mole	Log K	ΔS, cal /mole °K	T, °C	M	Conditions	Ref.	Remarks
B38, cont.								
I_2	-6.5	0.949(25°,1)	-17	15-35	T	S: Cyclohexane	73Va	b
B39	BENZENE, methylthio-4-methyl-		$C_8H_{10}S$			$C_6H_4(SCH_3)(CH_3)$		L
I_2	-6.3	1.292(25°,1)	-15	15-35	T	S: Cyclohexane	63Va	b
B40	BENZENE, nitro-		$C_6H_5O_2N$			$C_6H_5(NO_2)$		L
Li^+	9.0	(1)	--	25	T	S: Hexamethyl-phosphoramide	73Sq	q: ±0.8
SbA_5	-8.6	(1)	--	25	C	S: CCl_4 A=Cl	73Oa	q:±2
I_2	-2.75	(1)	--	--	C	S: Nitrobenzene	50Ha	
B41	BENZENE, 4-nitro-1,2-dihydroxy-		$C_6H_5O_5N$			$C_6H_3(NO_3)(OH)_2$		L^{2-}
Cu^{2+}	-4.0	0.81	-10.0	25	C	μ=0.100M(KNO_3) R: $Cu^{2+} + HL^- = CuL + H^+$	72Jb	
	1.71	-5.888	-21.2	25	C	μ=0.100M(KNO_3) R: $Cu^{2+} + H_2L = CuL + 2H^+$	72Jb	
B42	BENZENE, pentamethyl-		$C_{11}H_{16}$			$C_6H(CH_3)_5$		
IA	-5.20	1.82(25°,1)	-9.10	1.6&25	T	$C_M=10^{-4}-10^{-3}$ C_L=0.1-1.0 S: CCl_4 A=Cl	55Oa	a,b,k
	-4.97	1.82(25°,1)	-8.32	25& 45.8	T	$C_M=10^{-4}-10^{-3}$ C_L=0.1-1 S: CCl_4 A=Cl	55Oa	a,b,k
B43	BENZENE, 1,2,4,5-tetramethyl-		$C_{10}H_{14}$			$C_6H_2(CH_3)_4$		L
I_2	-2.78	0.814(25°,1)	-5.6	1.6-45.8	T	C_L=0.01-0.15 mole fraction S: CCl_4	55Kb	b,m
B44	BENZENE, 1,2,4-triethyl-		$C_{12}H_{18}$			$C_6H_3(CH_2CH_3)_4$		L
IA	-4.22	1.40(25°,1)	-7.76	1.6&25	T	$C_M=10^{-4}-10^{-3}$ C_L=0.1-1. S: CCl_4 A=Cl	55Oa	a,b,k
	-4.46	1.40(25°,1)	-8.55	25& 45.8	T	$C_M=10^{-4}-10^{-3}$ C_L=0.1-1. S: CCl_4 A=Cl	55Oa	a,b,k
B45	BENZENE, 1,3,5-triethyl		$C_{12}H_{18}$			$C_6H_2(CH_2CH_3)_3$		L
I_2	-2.64	0.72(25°,1)	-5.6	1.6-45.8	T	C_L=0.01-0.15 mole fraction S: CCl_4	55Kb	b,m
B46	BENZENE, 1,3,5-trimethyl-		C_9H_{12}			$C_6H_3(CH_3)_3$		L

Metal Ion	ΔH, kcal /mole	Log K	ΔS, cal /mole °K	T, °C	M	Conditions	Ref.	Remarks
B46, cont.								
NbA_5	-1.200	(1)	--	-30-0	T	S: CCl_4 A=Cl	71Ha	b
OsA_4	-0.770	(1)	--	0-31.5	T	S: CCl_4 A=O	71Hc	b,q: ±0.11
Ag^+	-2.93	-0.097(1)	-10.3	25	T	μ=0.5 S: Equimolar H_2O - CH_3OH soln.	56Oa	
	0.70	-0.292(20°,1)	--	1.4&20	T	S: CH_3OH. A=ClO_4	64Tb	a,b
AlA_3	-15.4	(1)	--	25	C	S: Mesitylene A=Cl	72Rd	
GaA_3	-15.2	(1)	--	25	C	S: Mesitylene A=Cl	72Rd	
I_2	-2.86	0.780(25°,1)	-6.0	1.6-45.8	T	C_L=0.01-0.15 mole fraction S: CCl_4	55Kb	b,m
IA	-4.45	0.936(1)	-10.6	25	C	S: CCl_4 A=Cl	73Sf	
	-4.75	1.58(25°,1)	-8.69	1.6& 45.8	T	$C_M=10^{-4}-10^{-3}$ C_L=0.1-1. S: CCl_4 A=Cl	55Oa	a,b,k
	-4.68	1.58(25°,1)	-8.45	25& 45.8	T	$C_M=10^{-4}-10^{-3}$ C_L=0.1-1. S: CCl_4 A=Cl	55Oa	a,b,k
B47 BENZENE, 1,3,5-tris-(1,1-dimethylethyl)-		$C_{18}H_{30}$				$C_6H_3[C(CH_3)_3]_3$		L
Ag^+	-3.29	-0.495(1)	-13.3	25	T	μ=0.5 S: Equimolar H_2O - CH_3OH soln.	56Oa	a
I_2	-2.18	0.46(25°,1)	-5.2	1.6-45.8	T	C_L=0.01-0.15 mole fraction	55Kb	b,m
B48 1,3-BENZENEDISULFONIC ACID, 4,5-dihydroxy-, disodium salt		$C_6H_4O_8S_2Na_2$				$C_6H_2(SO_3Na)_2(OH)_2$		L^{2-}
ThA	3.78	--	--	25	C	μ=0.10(KNO_3) A=EDTA R: ThA + H_2L = $ThAL^{2-}$ + $2H^+$	70Kb	g:Na
	6.79	--	--	25	C	μ=0.10(KNO_3) A=1,2-Diaminocyclohexane-N,N,N',N'-tetraacetic acid R: ThA + H_2L = $ThAL^{2-}$ + $2H^+$	70Kb	g:Na
VO^{2+}	-2.5	16.74(25°,1)	67	25-35	T	μ=0.10(KNO_3)	66Mg	b,q: ±0.5
Ga^{3+}	0.392	4.62(1)	21.3	25	T	μ=0.1	65Da	g:Na
In^{3+}	1.43	3.77(29°,1)	22.2	20-45	T	μ=0.1	65Nb	b,g:Na
B49 BENZIMIDAZOLE, 2-acetic acid		$C_9H_8O_2N_2$				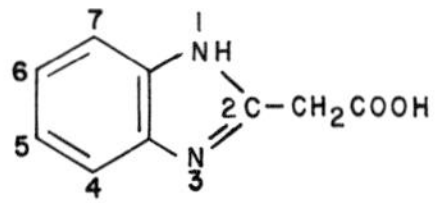		L^-

Metal Ion	ΔH, kcal /mole	Log K	ΔS, cal /mole °K	T, °C	M	Conditions	Ref.	Re-marks
B49, cont.								
Mn^{2+}	-4.5	5.40(25°,1)	9.6	3-40	T	S: 50 v % dioxane	59La	b
	-8.4	3.95(25°,2)	10.1	3-40	T	S: 50 v % dioxane	59La	b
Co^{2+}	-7.3	5.09(25°,1)	-1.2	3-40	T	S: 50 v % dioxane	59La	b
	-6.4	4.37(25°,1)	-1.5	3-40	T	S: 50 v % dioxane	59La	b
Ni^{2+}	-9.7	6.98(25°,1)	-0.60	25&40	T	S: 50 v % dioxane	59La	b
	-8.3	6.50(25°,2)	1.91	3-40	T	S: 50 v % dioxane	59La	b
Cu^{2+}	-11.7	7.85(25°,1)	-3.4	25&40	T	S: 50 v % dioxane	59La	b
	-8.9	6.72(25°,2)	0.91	3-40	T	S: 50 v % dioxane	59La	b
Zn^{2+}	-9.8	7.59(25°,1)	1.88	3-40	T	S: 50 v % dioxane	59La	b
	-9.3	6.95(25°,2)	0.60	3-40	T	S: 50 v % dioxane	59La	b
Pb^{2+}	-3.0	2.76(25°,1)	2.6	3&25	T	S: 50 v % dioxane	59La	b
	-3.2	2.45(25°,2)	0.47	3&25	T	S: 50 v % dioxane	59La	b
B50	BENZIMIDAZOLE, 2-aminomethyl-		$C_8H_9N_3$			$C_7H_5N_2(CH_2NH_2)$		L
Cu^{2+}	-19.2	14.23(25°,1-2)	0.67	15-45	T	S: 50 v % dioxane	63La	b
B51	BENZIMIDAZOLE, 2-(butylaminomethyl)-		$C_{12}H_{17}N_3$			$C_7H_5N_2[CH_2NH(CH_2)_3CH_3]$		L
Cu^{2+}	-11.6	11.45(25°,1-2)	13.5	15-45	T	S: 50 v % dioxane	63La	b
B52	BENZIMIDAZOLE, 2-(ethylaminomethyl)-		$C_{10}H_{13}N_3$			$C_7H_5N_2(CH_2NHCH_2CH_2)_3$		L
Cu^{2+}	-12.8	11.56(25°,1-2)	10.0	15-45	T	S: 50 v % dioxane	63La	b
B53	BENZIMIDAZOLE, 2-(1-hydroxy-butyl)-		$C_{11}H_{14}ON_2$			$C_7H_5N_2[CHOH(CH_2)_2CH_3]$		L
Cu^{2+}	-11.1	17.54(25°,1-2)	43.0	15-45	T	S: 50 v % dioxane	63La	b
B54	BENZIMIDAZOLE, 2-(1-hydroxy-ethyl)-		$C_9H_{10}ON_2$			$C_7H_5N_2(CHOHCH_3)$		L
Cu^{2+}	-13.5	17.34(25°,1-2)	34.1	15-45	T	S: 50 v % dioxane	63La	b
B55	BENZIMIDAZOLE, 2-hydroxy-methyl-		$C_8H_8ON_2$			$C_7H_5N_2(CH_2OH)$		L
Cu^{2+}	-16.7	17.68(25°,1-2)	24.9	15-45	T	S: 50 v % dioxane	63La	b
B56	BENZIMIDAZOLE, 2-(1-hydroxy-pentyl)-		$C_{12}H_{16}ON_2$			$C_7H_5N_2[CHOH(CH_2)_3CH_3]$		L
Cu^{2+}	-10.1	17.86(1-2)	47.8	15-45	T	S: 50 v % dioxane	63La	b
B57	BENZIMIDAZOLE, 2-(1-hydroxy-propyl)-		$C_{10}H_{12}ON_2$			$C_7H_5N_2(CHOHCH_2CH_3)$		L
Cu^{2+}	-11.3	17.60(25°,1-2)	42.6	15-45	T	S: 50 v % dioxane	63La	b
B58	BENZIMIDAZOLE, 2-(N-methyl-aminoethyl)		$C_{10}H_{13}N_3$			$C_7H_5N_2(CH_2CH_2NHCH_3)$		L

Metal Ion	ΔH, kcal /mole	Log K	ΔS, cal /mole °K	T, °C	M	Conditions	Ref.	Remarks
B58, cont.								
Cu^{2+}	-13.7	11.68(25°,1-2)	7.5	15-45	T	S: 50 v % dioxane	63La	b
B59	BENZIMIDAZOLE, 2-propanoic acid-		$C_{10}H_{10}O_2N_2$			$C_7H_5N_2(CH_2CH_2CO_2H)$		L^-
Mn^{2+}	-4.5	3.61(25°,1)	1.4	3-40	T	S: 50 v % dioxane	59La	b
	-4.0	3.21(25°,2)	1.3	3-40	T	S: 50 v % dioxane	59La	b
Co^{2+}	-7.3	3.79(25°,1)	-7.1	3-40	T	S: 50 v % dioxane	59La	b
	-4.2	3.37(25°,2)	1.3	3-40	T	S: 50 v % dioxane	59La	b
Ni^{2+}	-6.5	4.11(25°,1)	-3.0	3-40	T	S: 50 v % dioxane	59La	b
	-4.2	3.27(25°,2)	0.87	3-40	T	S: 50 v % dioxane	59La	b
Cu^{2+}	-8.1	5.25(25°,1)	-3.2	3-40	T	S: 50 v % dioxane	59La	b
	-2.1	4.15(25°,2)	11.9	3-40	T	S: 50 v % dioxane	59La	b
Zn^{2+}	-18.0	4.65(25°,1)	-39.1	3-40	T	S: 50 v % dioxane	59La	b
	-2.5	3.95(25°,2)	9.7	3-40	T	S: 50 v % dioxane	59La	b
B60	BENZIMIDAZOLE, 2-(N-propyl-aminomethyl)-		$C_{11}H_{15}N_3$			$C_7H_5N_2(CH_2NHCH_2CH_2CH_3)$		L
Cu^{2+}	-11.4	12.15(25°,1-2)	17.4	15-45	T	S: 50 v % dioxane	63La	b
B61	BENZOIC ACID		$C_7H_6O_2$			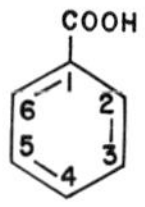		L^-
$[Fe^{3+}]$	-2.1	2.78(1)	5.7	25	T	Myoglobin	60Gb	
Cu^{2+}	0.0	4.04(30°,1)	18.2	25-35	T	μ≈0.03. S: 50 v % dioxane	62Mb	b,j
Pb^{2+}	7.12	3.299(30°,1-2)	38.6	30&40	T	=1.0($NaClO_4$)		a,b
B62	BENZOIC ACID, amide, N,N-dimethyl-		C_9H_4ON			$C_6H_5[CON(CH_3)_2]$		L
I_2	-4.0	0.58(25°,1)	-10.7	25-45	T	$C_M=1.6x10^{-3}$ $C_L=0.038-0.18$ S: CCl_4	62Ce	b
	-5.0	0.91(25°,1)	-11.9	5-25	T	S: Heptane	71Te	b
B63	BENZOIC ACID, bromide		C_7H_5OBr			C_6H_5COBr		L
I_2	-0.9	-1.0(25°,1)	-7.3	5-25	T	S: Heptane	71Te	b
B64	BENZOIC ACID, chloride		C_7H_5OCl			C_6H_5COCl		L
I_2	-1.3	-1.0(25°,1)	-9.1	5-25	T	S: Heptane	71Te	b
B65	BENZOIC ACID, ethyl ester		$C_9H_{10}O_2$			$C_6H_5COOC_2H_5$		L
I_2	-2.6	-0.15(25°,1)	-9.2	5-25	T	S: Heptane	71Te	b

Metal Ion	ΔH, kcal /mole	Log K	ΔS, cal /mole °K	T, °C	M	Conditions	Ref.	Re- marks
B66	BENZOIC ACID. methyl ester		$C_8H_8O_2$			$C_6H_5COOCH_3$		L
I_2	-2.5	-0.21(25°,1)	-9.6	5-25	T	S: Heptane	71Te	b
B67	BENZOIC ACID, nitrile		C_7H_5N			C_6H_5CN		L
TiA_4	-11.3	(1)	--	--	C	S: Benzene C_M=0.05-0.08M A=Cl	68Ge	μ:25
SnA_4	-5.1	(1)	--	--	C	S: Benzene C_M=0.05-0.08M A=Cl	68Ge	μ:25
SbA_5	-12.7	(1)	--	25	C	S: CCl_4 A=Cl	72Lb	
	-13.0	(1)	--	25	C	S: $C_2H_4Cl_2$ A=Cl	72Lb	
B68	BENZOIC ACID, 2-amino-		$C_7H_7O_2N$			$C_6H_4(NH_2)(CO_2H)$		L^-
In^{3+}	-1.9	5.24(30°,1)	17	20-40	T	S: 75 v % Ethanol μ=0.2($NaClO_4$)	73Sb	b,j,q: ±0.5
	-3.2	4.27(30°,2)	9	20-40	T	S: 75 v % Ethanol μ=0.2($NaClO_4$)	73Sb	b,j,q: ±1.0
B69	BENZOIC ACID, 4-amino-2-hydroxy-, amide		$C_7H_8O_2N_2$			$C_6H_3(CONH_2)(OH)(NH_2)$		L
Fe^{3+}	0.4	10.8(25°,1)	51	15-35	T	μ=3.0	56Aa	b
B70	BENZOIC ACID, 2-amino-5-sulpho-		$C_7H_7O_5NS$			$C_6H_3(CO_2H)(NH_2)(HSO_3)$		
Fe^{3+}	2.29	3.647(22°,1)	24.4	22-41	T	μ=0.1M(KCl) pH<2	69Bi	b
B71	BENZOIC ACID, 3-bromo-		$C_7H_5O_2Br$			$C_6H_4(CO_2H)(Br)$		L^-
Cu^{2+}	21.4	3.72(30°,1)	87.4	25-35	T	μ≃0.03. S: 50 v % dioxane	62Mb	b,j
B72	BENZOIC ACID, 2-chloro-, amide, N,N-dimethyl-		$C_9H_{10}ONCl$			$C_6H_4[CON(CH_3)_2](Cl)$		L
I_2	-4.1	0.481(25°,1)	-11.0	25-45	T	C_M=1.5x10^{-3} C_L=0.08-0.13 S: CCl_4	63Ca	b
B73	BENZOIC ACID, 3-chloro-		$C_7H_5O_2Cl$			$C_6H_4(CO_2H)(Cl)$		L^-
Cu^{2+}	10.9	3.88(30°,1)	53.4	25-35	T	μ≃0.03. S: 50 v % dioxane	62Mb	b,j
B74	BENZOIC ACID, 4-chloro-, amide, N,N-dimethyl-		$C_9H_{10}ONCl$			$C_6H_4[CON(CH_3)_2](Cl)$		L
I_2	-3.8	0.447(25°,1)	-10.7	25-45	T	C_M=1.4x10^{-3} C_L=0.08-0.15 S: CCl_4	63Ca	b

Metal Ion	ΔH, kcal /mole	Log K	ΔS, cal /mole °K	T, °C	M	Conditions	Ref.	Re-marks
B75 BENZOIC ACID, 4-chloro-, nitrile			C_7H_4NCl			$C_6H_4(CN)(Cl)$		L
SnA_4	-7.1	0.556(25°,1)	-21.3	20-35	T	S: Benzene A=Cl	64La	b
B76 BENZOIC ACID, 3,4-dichloro-, amide, N,N-dimethyl-			$C_9H_9ONCl_2$			$C_6H_3[CON(CH_3)_2](Cl)_2$		L
I_2	-3.5	0.230(25°,1)	-10.9	25-45	T	$C_M=1.2x10^{-3}$ $C_L=0.09-0.43$ S: CCl_4	63Ca	b
B77 BENZOIC ACID, 2,3-dihydroxy-			$C_7H_6O_4$			$C_6H_3(CO_2H)(OH)_2$		L^-
UO_2^{2+}	-10.84	5.76(30,1)	-9.41	20-50	T	μ=0.1 pH-5.5	73Pc	b
B78 BENZOIC ACID, 2,4-dihydroxy-			$C_7H_6O_4$			$C_6H_3(CO_2H)(OH)_2$		L^-
Be^{3+}	1.49	4.330(30°,1)	24.7	10-60	T	μ=0.01 pH=5.0	69Gf	b
Cu^{2+}	1.5	2.770(30°,1)	17.7	10-60	T	μ=0.01 pH=5.5	69Gh	b
B79 BENZOIC ACID, 2,5-dihydroxy-			$C_7H_6O_4$			$C_6H_3(CO_2H)(OH)_2$		L^-
UO_2^{2+}	-2.18	3.20(30°,1)	4.84	20-50	T	μ=0.1 pH=4.7	73Pc	b
B80 BENZOIC ACID, 2,6-dihydroxy-			$C_7H_6O_4$			$C_6H_3(CO_2H)(OH)_2$		L^-
UO_2^{2+}	-2.49	3.02(30°,1)	5.60	20-50	T	μ=0.1 pH=4.9	73Pc	b
B81 BENZOIC ACID, 2-hydroxy-			$C_7H_6O_3$			$C_6H_4(CO_2H)(OH)$		L^-
Be^{2+}	1.67	3.47(25)	21.5	20-45	T	μ=0.2 pH=4.0 R: Be + HL = BeL + H	64Da	b,q: ±0.15
ThA	-4.64	10.7(1)	33.5	25	C	$μ=0.10(KNO_3)$ A=EDTA	70Kb	
	8.60	--	--	25	C	$μ=0.10(KNO_3)$ A=1,2-Diaminocyclo-hexane-N,N',N'-tetraacetic acid R: ThA + H_2L = $ThAL^{2-}$ + $2H^+$	70Kb	
	8.56	--	--	25	C	$μ=0.10(KNO_3)$ A=EDTA R: ThA + H_2L = $ThAL^{2-}$ + $2H^+$	70Kb	
VO^{2+}	-2.1	13.38(25°,1)	54	25-35	T	$μ=0.10(KNO_3)$	66Mg	b,q: ±0.5
Fe^{3+}	-7.950	17.80(1)	53.4	10	T	μ=0	73Vb	
	-6.760	17.57(1)	57.7	25	T	μ=0	73Vb	Cp=85
	-5.850	17.46(1)	60.9	35	T	μ=0	73Vb	
	-4.570	17.18(1)	64.5	50	T	μ=0	73Vb	

Metal Ion	ΔH, kcal /mole	Log K	ΔS, cal /mole °K	T, °C	M	Conditions	Ref.	Remarks
B81, cont.								
Fe^{3+}, cont.	-9.050	14.44(1)	34.1	10	T	μ=1.0	73Vb	
	0.050	14.13(1)	37.6	25	T	μ=1.0	73Vb	Cp=69
	-7.360	13.97(1)	40.0	35	T	μ=1.0	73Vb	
	-6.300	13.70(1)	43.2	50	T	μ=1.0	73Vb	
	-9.390	14.32(1)	32.3	10	T	μ=1.5	73Vb	
	-8.420	13.98(1)	35.7	25	T	μ=1.5	73Vb	Cp=66
	-7.760	13.81(1)	38.0	35	T	μ=1.5	73Vb	
	-6.750	13.55(1)	41.1	50	T	μ=1.5	73Vb	
	-9.680	14.17(1)	30.6	10	T	μ=2.0	73Vb	
	-8.740	13.83(1)	33.9	25	T	μ=2.0	73Vb	Cp=64
	-8.100	13.64(1)	36.1	35	T	μ=2.0	73Vb	
	-7.130	13.37(1)	39.1	50	T	μ=2.0	73Vb	
	-5.3	15.9(25°,1)	55	15-35	T	μ=3.0	56Aa	b
	-11.17	5.73(25°)	-11.2	15-35	T	S: 70 w % Dioxane R: $Fe^{3+} + HL^- = FeL^+ + H^+$	69Bj	b
	-13.92	7.29(25°)	-13.3	15-35	T	S: 82 w % Dioxane R: $Fe^{3+} + HL^- = FeL^+ + H^+$	69Bj	b
In^{3+}	2.1	2.59(30°,1)	19	20-40	T	S: 75 v % Ethanol μ=0.2($NaClO_4$)	73Sb	b,j,q: ±0.5
B82 BENZOIC ACID, 2-hydroxy-, amide		$C_7H_7O_2H$				$C_6H_4(CONH_2)(OH)$		L
Fe^{3+}	0.2	10.1(25°,1)	47	15-35	T	μ=3.0	56Aa	b
B83 BENZOIC ACID, 2-hydroxy-, methyl ester		$C_8H_8O_3$				$C_6H_4(CO_2CH_3)(OH)$		L
Fe^{3+}	1.6	9.82(25°,1)	51	15-35	T	μ=3.0	56Aa	b
B84 BENZOIC ACID, 2-hydroxy-5-sulfo-		$C_7H_6O_6S$				$C_6H_3(CO_2H)(OH)(SO_3H)$		L^{3-}
Be^{2+}	1.40	3.56(25)	21.0	20-45	T	μ=0.2 pH=4.0 R: Be + HL = BeL + H	64Da	b,q: ±0.15
ThA	7.58	--	--	25	C	μ=0.10(KNO_3) A=EDTA R: $ThA + H_3L = ThAL^{3-} + 3H^+$	70Kb	
	8.36	--	--	25	C	μ=0.10(KNO_3) A=1,2-Diaminocyclo-hexane-N,N',N'-tetraacetic acid R: $ThA + H_3L = ThAL^{3-} + 3H^+$	70Kb	
	-1.7	11.71(25°,1)	45	25-35	T	μ=0.10(KNO_3)	66Mg	b,q: ±0.5
Fe^{3+}	-2.7	14.5(25°,1)	58	15-35	T	μ=3.0	56Aa	b
	-7.17	5.85(25°)	2.7	15-35	T	S: 70 w % Dioxane R: $Fe^{3+} + HL^- = FeL^+ + H^+$	69Bj	b

Metal Ion	ΔH, kcal /mole	Log K	ΔS, cal /mole °K	T, °C	M	Conditions	Ref.	Remarks
B84, cont.								
Fe^{3+}, cont.	-12.30	6.55(25°)	-11.3	15-35	T	S: 82 w % Dioxane R: $Fe^{3+} + HL^- = FeL^+ + H^+$	69Bj	b
B85	BENZOIC ACID, 3-iodo-		$C_7H_5O_2I$			$C_6H_4(CO_2H)(I)$		L^-
Cu^{2+}	24.4	3.88(30°,1)	98.0	25-35	T	μ=0.03 S: 50 v % Dioxane	62Mb	b,j
B86	BENZOIC ACID, 2-mercapto-		$C_7H_6O_2S$			$C_6H_4(CO_2H)(SH)$		L^{2-}
In^{3+}	-2.3	11.10(30°,1)	43	20-40	T	S: 75 v % Ethanol μ=0.2($NaClO_4$)	73Sb	b,j,q: ±0.5
	-4.6	8.90(30°,2)	25	20-40	T	S: 75 v % Ethanol μ=0.2($NaClO_4$)	73Sb	b,j,q: ±1.0
	-6.8	5.96(30°,3)	5	20-40	T	S: 75 v % Ethanol μ=0.2($NaClO_4$)	73Sb	b,j,q: ±1.0
B87	BENZOIC ACID, 4-methoxy-, amide, N,N-dimethyl-		$C_{10}H_{13}O_2N$			$C_6H_4[CON(CH_3)_2](OCH_3)$		L
I_2	-4.6	0.724(25°,1)	-12.0	25-45	T	$C_M=1.3x10^{-3}$ C_L=0.05-0.25 S: CCl_4	63Ca	b
B88	BENZOIC ACID, 4-methoxy-, nitrile		C_8H_7ON			$C_6H_4(CN)(OCH_3)$		L
SnA_4	-12.5	1.19(30°,1)	-35.8	30-40	T	S: Benzene A=Cl	64La	b
B89	BENZOIC ACID, 4-methyl-, amide, N,N-dimethyl-		$C_{10}H_{13}ON$			$C_6H_4[CON(CH_3)_2](CH_3)$		
I_2	-4.5	0.681(25°,1)	-12.0	25-45	T	$C_M=1.5x10^{-3}$ C_L=0.069 S: CCl_4	63Ca	b
B90	BENZOIC ACID, 4-nitro, methyl ester		$C_8H_7O_4N$			$O_2NC_6H_4CO_2CH_3$		L^-
Li^+	1.1	(1)	--	25	T	S: Hexamethyl-phosphoramide	73Sq	q:±0.2
B91	BENZOIC ACID, 4-nitro-, nitrile		$C_7H_4O_2N_2$			$C_6H_4(NO_2)(CN)$		L^-
Li^+	0	(1)	--	25	T	S: Hexamethyl-phosphoramide	73Sq	
B92	BENZOIC ACID, 2-phosphonooxy-		$C_7H_7O_6P$			$C_6H_4(PO_4H_2)(CO_2H)$		L^{3-}
VO^{2+}	-0.5	5.81(25°,1)	25	25-35	T	μ=0.10(KNO_3)	66Mg	b,q: ±0.5

Metal Ion	ΔH, kcal /mole	Log K	ΔS, cal /mole °K	T, °C	M	Conditions	Ref.	Re-marks
B93	9,10-BENZOPHENANTHRENE		$C_{18}H_{12}$			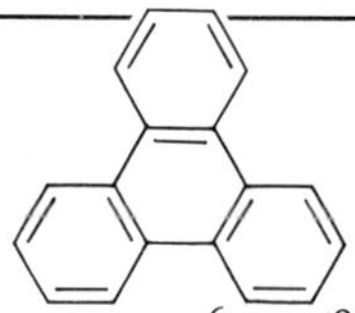		L
Na^+	0	4.830(1)	22	-55	T	$C_M=C_L=10^{-6}-10^{-3}$ S: Dimethoxyethane	66Cb	
	2.4	5.201(15°,1)	32.5	20	T	$C_M=C_L=10^{-6}-10^{-3}$ S: Dimethoxyethane	66Cb	
	0.4	4.534(1)	23	-65	T	$C_M=C_L=10^{-6}-10^{-3}$ S: Tetrahydrofuran	66Cb	
	5.2	5.130(1)	42	15	T	$C_M=C_L=10^{-6}-10^{-3}$ S: Tetrahydrofuran	66Cb	
B94	BENZOPHENONE		$C_{13}H_{10}O$			$C_6H_5COC_6H_5$		L
I_2	-2.8	-0.06(25°,1)	-10	5-25	T	S: Heptane	71Te	b
B95	1,4-BENZOQUINONE		$C_6H_4O_2$					L^-
Na^+	1.05	0.824	7.3	0-35	T	S: Hexamethyl-phosphoramide	73Sp	b,q: ±0.08, z
K^+	3.68	1.444	19	-10-30	T	S: Hexamethyl-phosphoramide	73Sp	b,z
B96	1,4-BENZOQUINONE, 2,5-dichloro-3,6-dihydroxy-		$C_6H_2O_4Cl_2$			$C_6H_4O_2(OH)_2(Cl)_2$		L^{2-}
Fe^{3+}	-3.36	5.813(25°,1)	-15.4	15-35	T	μ=0.15 pH=4.77 Complex in configuration 1	67Ca	b
	-14.7	5.513(25°,1)	24.4	15-35	T	μ=0.15 pH=4.77 Complex in configuration 2	67Ca	b,p
	0.7	4.030(25°,2)	-20.8	15-35	T	μ=0.15 pH=4.77 Complex in configuration 1	67Ca	b,p
	4.35	3.954(25°,2)	-31.8	15-35	T	μ=0.15 pH=4.77 Complex in configuration 2	67Ca	b,p
Ni^{2+}	-1.66	4.041(25,1)	-12.9	15-35	T	μ=0.15 pH=7.56	67Ca	b
B97	BICYCLO [2,2,1]-hepta-2,5-diene		C_7H_8			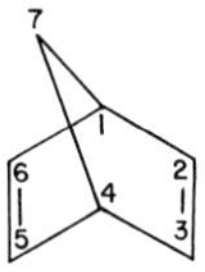		L

Metal Ion	ΔH, kcal /mole	Log K	ΔS, cal /mole °K	T, °C	M	Conditions	Ref.	Remarks
B97, cont.								
PdA_2B_2	-13.3	--	--	25	C	S: CH_2Cl A=Cl $B=C_6H_5CN$ R: $PdA_2B_2 + L = PdA_2L + 2B$ $C_L=3.5x10^{-3}M$ $C_M=1x10^{-4}M$	72Pc	
Cu^+	-15	4.11(1)	-30.7	30	T	S: 1M($LiClO_4$) in 2-propanol	69Hb	
B98 BICYCLO [2,2,1]-heptadiene			C_7H_8					L
Ag^+	-11.18	-0.243	-38.7	25	T	1M(KNO_3) Location of double bonds not specified. Two phase system: CCl_4 & H_2O. R: $Ag^+(H_2O) + L(CCl_4) = Complex^+(H_2O)$	59Ta	
B99 BICYCLO [2,2,1]-2-heptene			C_7H_{10}			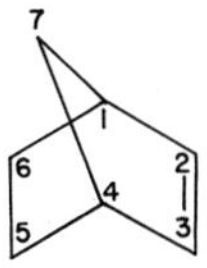		L
Cu^+	-10.8	4.26(1)	-16.1	30	T	S: 1M($LiClO_4$) in 2-propanol	69Hb	
Ag^+	-6.68	-0.572	-25.0	25	T	1M(KNO_3). Two phase system: CCl_4 & H_2O. R: $Ag^+(H_2O) + L(CCl_4) = Complex^+(H_2O)$		
B100 BICYCLO [2,2,2]-2-octene			C_8H_{12}			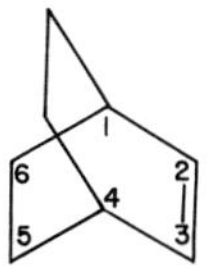		L
Ag^+	-4.94	-1.02(1)	-21.2	25	T	1M(KNO_3). Two phase system: CCl_4& H_2O. R: $Ag^+(H_2O) + L(CCl_4) = Complex^+(H_2O)$	59Ta	
B101 BIPHENYL			$C_{12}H_{10}$			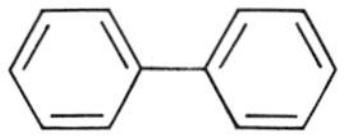		L
Li^+	0.4	5.119(1)	26	-65	T	S: Tetrahydrofuran	68Nb	q:±0.4
	2.5	5.328(1)	33	15	T	S: Tetrahydrofuran	68Nb	q:±0.4
Na^+	0	4.769(1)	22.0	-55	T	$C_M=C_L=10^{-6}-10^{-3}$ S: Dimethoxyethane	66Cb	

Metal Ion	ΔH, kcal /mole	Log K	ΔS, cal /mole °K	T, °C	M	Conditions	Ref.	Remarks
B101, cont.								
Na^+, cont.	2.1	5.207(15°,1)	31.5	20	T	$C_M=C_L=10^{-6}-10^{-3}$ S: Dimethoxyethane	66Cb	
	3.4	9.75(25°,1)	56	-70-25	T	low μ ∿10^{-4} S: 2-Methyltetrahydrofuran	67Sb	b
	1.6	4.530(1)	28	-65	T	$C_M=C_L=10^{-6}-10^{-3}$ S: Tetrahydrofuran	66Cb 68Nb	
	7.3	5.824(1)	52	15	T	$C_M=C_L=10^{-6}-10^{-3}$ S: Tetrahydrofuran	66Cb 68Nb	
B102 2,2'-BIPYRIDYL			$C_{10}H_8N_2$			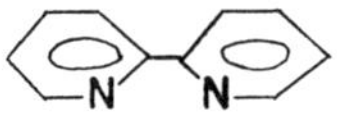		L
LaA_3	-6.62	3.25(1)	-7.46	25	C	S: Chloroform A=Thenoyltrifluoroacetone	72Kb	
	-5.85	2.25(2)	-9.19	25	C	S: Chloroform A=Thenoyltrifluoroacetone	72Kb	
NdA_3	-7.90	3.61(1)	-9.97	25	C	S: Chloroform A=Thenoyltrifluoroacetone	72Kb	
	-10.54	3.26(2)	-20.45	25	C	S: Chloroform A=Thenoyltrifluoroacetone	72Kb	
GdA_3	-9.78	3.71(1)	-15.63	25	C	S: Chloroform A=Thenoyltrifluoroacetone	72Kb	
	-13.40	3.79(2)	-27.79	25	C	S: Chloroform A=Thenoyltrifluoroacetone	72Kb	
LuA_3	-14.32	4.75(1)	-26.40	25	C	S: Chloroform A=Thenoyltrifluoroacetone	72Kb	
	-14.82	3.00(2)	-35.88	25	C	S: Chloroform A=Thenoyltrifluoroacetone	72Kb	
Mn^{2+}	-3.5	2.6(1)	0	20	C	μ=0.1($NaNO_3$)	63Ac 62Ab	
	-2	2.7(25°,1)	6	-46-0	T	μ=0.2($NaClO_4$) S: Methanol	73Be	b,q
	-5.73	2.54(1)	-7.3	30.3	C	1M(KNO_3)	65Db	
	-6.11	4.40(1-2)	0	30.3	C	1M(KNO_3)	65Db	
	-6.23	5.90(1-3)	6.5	30.3	C	1M(KNO_3)	65Db	
Fe^{2+}	-31.35	17.45(1-3)	-27.0	20	C	μ=0.1($NaNO_3$)	63Ac 50Bb 50Ba 64Lc 49Kc	
	-28.0	17.16(1-3)	-13.8	30.3	C	μ=1.0(KNO_3)	65Db	
Fe^{3+}	-1.2	11.3(1-3)	47.5	25	T	μ=0	59Ga	

Metal Ion	ΔH, kcal /mole	Log K	ΔS, cal /mole °K	T, °C	M	Conditions	Ref.	Remarks
B102, cont.								
Co^{2+}	-8.2	6.06(1)	-0.35	20	C	μ=0.1($NaNO_3$)	63Ac	
	-7.20	-7.95(1)	2.5	30.3	C	1M(KNO_3)	65Db	
	-15.2	11.42(1-2)	0.35	20	C	μ=0.1($NaNO_3$)	63Ac	
	-14.40	-15.46(1-2)	3.5	30.3	C	1M(KNO_3)	65Db	
	-21.3	16.02(1-3)	1.4	20	C	μ=0.1($NaNO_3$)	63Ac	
	-19.66	-22.13(1-3)	8.1	30.3	C	1M(KNO_3)	65Db	
Ni^{2+}	-9.6	7.13(1)	0	20	C	μ=0.1($NaNO_3$)	63Ac, 62Ab	
	-8.90	9.65(1)	2.5	30.3	C	1M(KNO_3)	65Db	
	-19.0	14.01(1-2)	-0.7	20	C	μ=0.1($NaNO_3$)	63Ac	
	-17.80	19.14(1-2)	4.4	30.3	C	1M(KNO_3)	65Db	
	-28.2	20.54(1-3)	-2.1	20	C	μ=0.1($NaNO_3$)	63Ac	
	-26.70	27.95(1-3)	4.1	30.3	C	1M(KNO_3)	65Db	
NiA_2	-18.2	(1)	--	30	C	S: Benzene A=Diethyldithiophospato	71Dc	q:±2
Cu^{2+}	-11.9	8.0(1)	-4.1	20	C	μ=0.1($NaNO_3$)	63Ac, 62Ab	
	-10.16	8.39(1)	4.9	30.3	C	1M(KNO_3)	65Db, 62Cb	
	-17.3	13.6(1-2)	3.1	20	C	μ=0.1($NaNO_3$)	63Ac	
	-19.02	14.03(1-2)	1.5	30.3	C	1M(KNO_3)	65Db	
	-23.8	17.08(1-3)	-3.1	20	C	μ=0.1($NaNO_3$)	63Ac	
	-21.62	17.66(1-3)	9.5	30.3	C	1M(KNO_3)	65Db	
CuA_2	-19.1	(1)	--	30	C	S: Benzene A=1,1,1,5,5,5-Hexafluoropentane-2,4-dione	74Gb	
	-11.95	3.004(1)	-25.8	30	C	S: Benzene A=1,1,1-Trifluoropentane-2,4-dione	74Gb	
	-13.6	3.164(1)	--	30	C	S: Benzene A=1,1,1-Trifluoro-4-phenylbutane-2,4-dione	74Gb	
	-13.0	(1)	--	30	C	S: Benzene A=1,1,1-Trifluoro-4-(2-thienyl) butane-2,4-dione	74Gb	
	-13.09	6.07(1)	-16.14	25	C	S: Chloroform A=1,1,1,5,5,5-Hexafluoro-2,4-pentanedione	72Ka	
	-0.35	1.02(1)	3.52	25	C	S: Chloroform A=2,4-Pentanedione	72Ka	
	-9.65	2.65(1)	-20.43	25	C	S: Chloroform A=1,1,1-Trifluoro-2,4-pentanedione	72Ka	
	-10.32	3.67(1)	-17.78	25	C	S: Chloroform A=4,4,4-Trifluoro-1-(2-thienyl)-1,3-butanedione	72Ka	

Metal Ion	ΔH, kcal /mole	Log K	ΔS, cal /mole °K	T, °C	M	Conditions	Ref.	Remarks
B102, cont.								
Ag^+	-7.654	3.70(25°,1)	-8.74	20-40	T	0.1M(KNO_3) S: 50 v % 95% C_2H_5OH	58Cc 58Cb	b
	-4.086	3.52(25°,2)	2.40	20-40	T	0.1M(KNO_3) S: 50 v % 95% C_2H_5OH	58Cc	b
	-11.740	7.22(25°,1-2)	-6.34	20-40	T	0.1M(KNO_3) S: 50 v % 95% C_2H_5OH	58Cc	b
Zn^{2+}	-7.1	5.3(1)	0	20	C	μ=0.1($NaNO_3$)	63Ac 62Ab	
	-7.0	5.16(25°,1)	0.1	10-40	T	μ=0.1M(KNO_3)	62Ca	b
	-6.25	5.26(1)	3.5	30.3	C	1M(KNO_3)	65Db	
	-7.6	4.88(25°,1)	-3.1	10-40	T	μ=0.1M(KNO_3) S: 41.5 w % EtOH	62Ca	b
	-6.7	4.46(25°,2)	-2.1	10-40	T	μ=0.1M(KNO_3)	62Ca	b
	-7.4	4.08(25°,2)	-6.1	10-40	T	μ=0.1M(KNO_3) S: 41.5 w % EtOH	62Ca	b
	-5.8	3.74(25°,3)	-2.3	10-40	T	μ=0.1M(KNO_3)	62Ca	b,q: ±0.6
	-5.2	3.05(25°,3)	-3.5	10-40	T	μ=0.1M(KNO_3) S: 41.5 w % EtOH	62Ca	b
	-12.5	9.83(1-2)	2.4	20	C	μ=0.1($NaNO_3$)	63Ac	
	-11.75	9.83(1-2)	6.2	30.3	C	1M(KNO_3)	65Db	
	-17.5	13.63(1-3)	2.7	20	C	μ=0.1($NaNO_3$)	63Ac	
	-15.92	13.78(1-3)	10.6	30.3	C	1M(KNO_3)	65Db	
ZnA_2	-14.9	-2.572(1)	-37.2	30	C	S: Benzene A=Diisobutyldithio-phosphate	71Db	q:±1.4
	-5.69	1.95(1)	-10.17	25	C	S: Chloroform A=4,4,4-Trifluoro-1-(2-thienyl)-1,3-butanedione	72Ka	
	-17.6	2.866(1)	-46	25	T	S: Toluene A=CH_3	67Ra	
	-35.1	3.651(1)	-103	25	T	S: Toluene A=C_2H_5	67Ra	
	-15.7	3.269(1)	-59	25	T	S: Toluene A=C_3H_7	67Ra	
	-10.8	2.925(1)	-23	25	T	S: Toluene A=C_4H_9	67Ra	
	-13.0	3.028(1)	-30	25	T	S: Toluene A=C_5H_{11}	67Ra	
	-12.6	3.079(1)	-28	25	T	S: Toluene A=C_6H_{13}	67Ra	
	-15.6	3.137(1)	-38	25	T	S: Toluene A=C_7H_{15}	67Ra	
	-15.0	3.064(1)	-36	25	T	S: Toluene A=C_8H_{17}	67Ra	
	-16.0	3.079(1)	-39	25	T	S: Toluene A=C_9H_{19}	67Ra	
Cd^{2+}	-5.08	4.28(25°,1)	2.54	11-40	T	0.04M(KNO_3)	58Ca 58Cb	b
	-5.1	4.25(1)	2.1	20	C	μ=0.1($NaNO_3$)	63Ac	
	-5.86	3.99(25°,1)	-1.4	20-40	T	0.05M(KNO_3) S: 50 v % 95% C_2H_5OH	58Ca	b
	-3.79	3.51(25°,2)	3.34	11-40	T	0.04M(KNO_3)	58Ca	b
	-4.20	3.05(25°,2)	-0.1	20-40	T	0.05M(KNO_3) S: 50 v % 95% C_2H_5OH	58Ca	b
	-3.07	2.69(25°,3)	2.01	11-40	T	0.04M(KNO_3)	58Ca	b
	-2.72	2.10(25°,3)	0.5	20-40	T	0.05M(KNO_3) S: 50 v % 95% C_2H_5OH	58Ca	b
	-9.4	7.85(1-2)	3.7	20	C	μ=0.1($NaNO_3$)	63Ac	
	-11.94	10.48(25°,1-3)	7.89	11-40	T	0.04M(KNO_3)	58Ca	b

Metal Ion	ΔH, kcal /mole	Log K	ΔS, cal /mole °K	T, °C	M	Conditions	Ref.	Remarks
B102, cont.								
Cd^{2+},	-14.0	10.55(<u>1-3</u>)	0.3	20	C	μ=0.1($NaNO_3$)	63Ac	
cont.	-12.78	9.14(<u>25°</u>,<u>1-3</u>)	-1.4	20-40	T	0.05M(KNO_3) S: 50 v % 95% C_2H_5OH	58Ca	b
CdA_2	-14.6	(<u>1</u>)	--	30	C	S: Benzene A=Diisobutyldithiophosphate	71Db	q:±1.4
Cd_2A_4	-10.8	-2.308(<u>1</u>)	<u>-25.1</u>	30	C	S: Benzene A=Diisobutyldithiophosphato R: $1/2Cd_2A_4$ + L = CdA_2L	71Db	q:±1
HgA_2	-4.19	1.283(<u>1</u>)	-8.2	25	C	S: C_6H_6 A=C_6H_5	73Pl	
	-7.28	1.744(<u>1</u>)	-15.4	25	C	S: CCl_4 A=C_6F_5	73Pl	
Tl^{3+}	-10.1	9.38(<u>1</u>)	-9	25	C	μ=1.0	62Kc	
	-20.1	16.1(<u>1-2</u>)	-6	25	C	μ=1.0	62Kc	
	-6.1	4.69	-1	25	C	μ=1.0 R: Tl^{3+} + HL^+ = TlL^{3+} + H^+	62Kc	
	-12.1	6.74	10	25	C	μ=1.0 R: Tl^{3+} + $2HL^+$ = TlL_2^{3+} + $2H^+$	62Kc	
SnA_4	-6.18	>6.96(<u>1</u>)	>11.3	25	C	μ≈0. S: Acetonitrile. A=Cl	66Mb	
$SnAB_3$	-20.43	>6.96(<u>1</u>)	>-36.6	25	C	μ≈0. S: Acetonitrile. A=n-C_4H_9 B=Cl	66Mb	
	-26.85	(<u>1</u>)	--	25	C	μ≈0. S: Acetonitrile. A=n-C_4H_9 B=Cl. $SnAB_3$ was initially dissolved in CCl_4	66Mb	
SnA_2B_2	-11.88	3.30(<u>1</u>)	-24.8	25	C	μ≈0. S: Acetonitrile. A=CH_3 B=Cl	66Mb	
	-17.25	(<u>1</u>)	--	25	C	μ≈0. S: Acetonitrile. A=CH_3 B=Cl. SnA_2B_2 was initially dissolved in CCl_4	66Mb	
	-12.43	3.47(<u>1</u>)	-25.8	25	C	μ≈0. S: Acetonitrile. A=C_2H_5 B=Cl	66Mb	
	-17.01	(<u>1</u>)	--	25	C	μ≈0. S: Acetonitrile. A=C_2H_5 B=Cl. SnA_2B_2 was initially dissolved in CCl_4	66Mb	
	-13.12	3.07(<u>1</u>)	-30.0	25	C	μ≈0. S: Acetonitrile. A=n-C_3H_7 B=Cl	66Mb	
	-13.04	3.03(<u>1</u>)	-29.9	25	C	μ≈0. S: Acetonitrile. A=n-C_4H_9 B=Cl	66Mb	

Metal Ion	ΔH, kcal /mole	Log K	ΔS, cal /mole °K	T, °C	M	Conditions	Ref.	Remarks
B102, cont.								
SnA_2B_2 cont.	-15.78	(1)	--	25	C	$\mu \approx 0$. S: Acetonitrile. A=n-C_4H_9 B=Cl. SnA_2B_2 was initially dissolved in CCl_4	66Mb	
	-16.8	3.96(35°,1)	-38.2	30-40	T	S: Benzene A=C_4H_9 B=SCN	71Wb	b
	-12.7	1.32(35°,1)	-34.8	30-40	T	S: Benzene A=C_8H_{17} B=Cl	71Wb	b
	-17.2	4.01(35°,1)	-38.4	30-40	T	S: Benzene A=C_8H_{17} B=SCN	71Wb	b
	-12.6	1.24(35°,1)	-35.1	30-40	T	S: Toluene A=C_8H_{17} B=Cl	71Wb	b
B103 4,4'-BIPYRIDYL			$C_{10}H_8N_2$			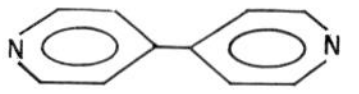		L
CrA_3^{2+}	12	-2.149(25°)	33	13.5-31.5	T	μ=0.1(NaCl) A=α,α'-Bipyridy R: CrA_3^{2+} + HL^+ = CrA_2HL^{3+} + A	71Ua	b
CuA_2	-6.71	2.51(1)	-11.03	25	C	S: Chloroform A=4,4,4-Trifluoro-1-(2-thienyl)-1,3-butanedione	72Ka	
ZnA_2	-4.89	1.71(1)	-8.57	25	C	S: Chloroform A=4,4,4-Trifluoro-1-(2-thienyl)-1,3-butanedione	72Ka	
B104 BISMUTHATE(III) ION, hexachloro			Cl_6Bi			$BiCl_6^{3-}$		L^{3-}
Na^+	≃-0.9	2.64(25°)	-5.0	25-65	T	3M(Li^+, H^+, Na^+, Cl^-). R: xNa + L = Na_xL	63Mg	b
K^+	-1.2	0.19(25°)	-3.2	25-65	T	3M(Li^+, H^+, K^+, Cl^-). R: xK + L = K_xL	63Mg	b
Rb^+	-1.3	0.30(25°)	-3.0	25-65	T	3M(Li^+, H^+, Rb^+, Cl^-). R: xRb + L = Rb_xL	63Mg	b
Cs^+	-1.4	0.40(25°)	-2.9	25-65	T	3M(Li^+,H^+,Cs^+, Cl^-). R: xCs + L = Cs_xL	63Mg	b
NH_4^+	5.3	0.26(25°,1)	19	15-65	T	μ=1	63Md	b
B105 BORIC ACID, tris (pyrazoly)			$Fe[HB(pz)_3]_2$					L^-
FeL_2	3.85	-0.328(25°)	11.4	--	T	S: Methanol-dichloromethane R: Transition from high to low spin isomer. $^1A \rightleftharpoons {}^5A$	73Bd	u:25

Metal Ion	ΔH, kcal /mole	Log K	ΔS, cal /mole °K	T, °C	M	Conditions	Ref.	Remarks
B106 BOVINE SERIUM ALBUMIN								
Cu^{2+}	4.0	3.739(1)	31	20	C	μ=0.1(NaAc) pH=4.80 Note: For the reactions with Cu^{2+} one or more hydrogen ions may have been released during binding of the Cu^{2+}.	73Rc	
	2.780	4.331(25°,1)	29.2	0&25	T	Buffer: 0.036M(HAc) 0.064M(NaAc) C_L=2% protein	48Ka	a,b
	3.7	3.313	28	20	C	μ=0.1(NaAc) pH=4.80 R: CuL + Cu = Cu_2L	73Rc	
	2.840	3.958(25°)	27.6	0&25	T	Buffer: 0.036M(HAC) 0.064M(NaAc) C_L=2% protein R: CuL + Cu = Cu_2L	48Ka	a,b
	3.4	2.961	25	20	C	μ=0.1(NaAc) pH=4.80 R: Cu_2L + Cu = Cu_3L	73Rc	
	2.900	3.708(25°)	26.7	0&25	T	Buffer: 0.036M(HAc) 0.064M(NaAc) C_L=2% protein R: Cu_2L + Cu = Cu_3L	48Ka	a,b
	3.2	2.940	24	20	C	μ=0.1(NaAc) pH=4.80 R: Cu_3L + Cu = Cu_4L	73Rc	
	2.960	3.507(25°)	26.0	0&25	T	Buffer: 0.036M(HAc) 0.064M(NaAc) C_L=2% protein R: Cu_3L + Cu = CuL_4	48Ka	a,b
	3.0	2.616	22	20	C	μ=0.1(NaAc) pH=4.80 R: Cu_4L + Cu = Cu_5L	73Rc	
	3.020	3.331(25°)	25.4	0&25	T	Buffer: 0.036M(HAc) 0.064M(NaAc) C_L=2% protein R: Cu_4L + Cu = Cu_5L	48Ka	a,b
	2.8	2.535	21	20	C	μ=0.1(NaAc) pH=4.80 R: Cu_5L + Cu = Cu_6L	73Rc	
	3.080	3.170(25°)	24.8	0&25	T	Buffer: 0.036M(HAc) 0.064M(NaAc) C_L=2% protein R: Cu_5L + Cu = Cu_6L	48Ka	a,b
	2.6	2.417	20	20	C	μ=0.1(NaAc) pH=4.80 R: Cu_6L + Cu = Cu_7L	73Rc	
	3.140	3.018(25°)	24.3	0&25	T	Buffer: 0.036M(HAc) 0.064M(NaAc) C_L=2% protein R: Cu_6L + Cu = Cu_7L	48Ka	a,b

Metal Ion	ΔH, kcal /mole	Log K	ΔS, cal /mole °K	T, °C	M	Conditions	Ref.	Remarks
B106, cont.								
Cu^{2+}, cont.	2.4	2.267	19	20	C	μ=0.1(NaAc) pH=4.80 R: $Cu_7L + Cu = Cu_8L$	73Rc	
	3.200	2.870(25°)	23.8	0&25	T	Buffer: 0.036M(HAc) 0.064M(NaAc) C_L=2% protein R: $Cu_7L + Cu = Cu_8L$	48Ka	a,b
	2.3	2.144	18	20	C	μ=0.1(NaAc) pH=4.80 R: $Cu_8L + Cu = Cu_9L$	73Rc	
	3.260	2.723(25°)	23.4	0&25	T	Buffer: 0.036M(HAc) 0.064M(NaAc) C_L=2% protein R: $Cu_8L + Cu = Cu_9L$	48Ka	a,b
	2.2	2.037	17	20	C	μ=0.1(NaAc) pH=4.80 R: $Cu_9L + Cu = Cu_{10}L$	73Rc	
	3.320	2.575(25°)	22.9	0&25	T	Buffer: 0.036M(HAc) 0.064M(NaAc C_L=2% protein R: $Cu_9L + Cu = Cu_{10}L$	48Ka	a,b
	3.380	2.423(25°)	22.4	0&25	T	Buffer: 0.036M(HAc) 0.064M(NaAc C_L=2% protein R: $Cu_{10}L + Cu = Cu_{11}L$	48Ka	a,b
	3.440	2.262(25°)	21.8	0&25	T	Buffer: 0.036M(HAc) 0.064M(NaAc) C_L=2% protein R: $Cu_{11}L + Cu = Cu_{12}L$	48Ka	a,b
	3.500	2.086(25°)	21.2	0&25	T	Buffer: 0.036M(HAc) 0.064M(NaAc) C_L=2% protein R: $Cu_{12}L + Cu = Cu_{13}L$	48Ka	a,b
	3.560	1.885(25°)	20.5	0&25	T	Buffer: 0.036M(HAc) 0.064M(NaAc) C_L=2% protein R: $Cu_{13}L + Cu = Cu_{14}L$	48Ka	a,b
	3.620	1.635(25°)	19.6	0&25	T	Buffer: 0.036M(HAc) 0.064M(NaAc) C_L=2% protein R: $Cu_{14}L + Cu = Cu_{15}L$	48Ka	a,b
	3.680	1.262(25°)	18.0	0&25	T	Buffer: 0.036M(HAc) 0.064M(NaAc) C_L=2% protein R: $Cu_{15}L + Cu = Cu_{16}L$	48Ka	a,b
B107	BROMATE ION		O_3Br			BrO_3^-		L^-
K^+	-9.540	1.24(<u>1</u>)	-26.3	25	T	μ=0	34La	
B108	BROMIDE ION		Br			Br^-		L^-
Li^+	4.70	14.69(<u>1</u>)	90	25	T	$C_M \approx 10^{-4}$ S: 12.65 w % Butanol 87.35 w % Hexane D=2.16	61Rd	

Metal Ion	ΔH, kcal /mole	Log K	ΔS, cal /mole °K	T, °C	M	Conditions	Ref.	Remarks
B108, cont.								
Li^+, cont.	15.30	11.65(1)	105	25	T	$C_M \simeq 10^{-4}$ S: 25 w % Butanol 75 w % Hexane D=2.85	61Rd	
	9.20	13.44(1)	92	25	T	$C_M \simeq 10^{-4}$ S: 80 w % Butanol 20 w % Hexane D=2.51	61Rd	
Na^+	0.044	-3.101(1)	-14.0	25	T	μ=0	36La 34La	
K^+	-4.910	-1.133(1)	-21.7	25	T	μ=0	34La	
	-7.4	-0.824	-29	25&35	T	μ=4M(Li,K,ClO_4,Br) R: $K^+ + CdL_4^{2-} = K^+CdL_4^{2-}$	64Mc	a,b
Rb^+	-7.0	-0.658	-27	25&35	T	μ=4M(Li,Rb,ClO_4,Br) R: $Rb^+ + CdL_4^{2-} = Rb^+CdL_4^{2-}$	64Mc	a,b
Cs^+	-6.2	-0.456	-23	25&35	T	μ=4M(Li,Rb,ClO_4,Br) R: $Cs^+ + CdL_4^{2-} = Cs^+CdL_4^{2-}$	64Mc	a,b
Sc^{3+}	-0.5	1.17(1)	4	25	T	μ=0.5($NaClO_4$)	62Pa	
	-0.6	0.5(2)	0	25	T	μ=0.5($NaClO_4$)	62Pa	
Y^{3+}	-0.9	0.45(1)	-1	25	T	μ=0.5($NaClO_4$)	62Pa	
Cr^{3+}	7.5	--	--	45-75	T	μ=0.50-2.0M (Hydrobromic acid solution) A=H_2O R: $CrA_6^{3+} + L^- = CrA_5L^{2+} + A$	71Zf	b,q: ±0.5
	5.1	-2.648(25°,1)	4.9	0-45	T	μ=2.00($NaClO_4$)	60Ea	b
	6.5	--	--	45-75	T	μ=4.0M (Hydrobromic acid solution) A=H_2O R: $CrA_6^{3+} + L^- = CrA_5L^{2+} + A$	71Zf	b,q: ±0.5
	5.9	--	--	45-75	T	μ=8.0M (Hydrobromic acid solution) A=H_2O R: $CrA_6^{3+} + L^- = CrA_5L^{2+} + A$	71Zf	b,q: ±0.5
	3.0	--	--	45-75	T	μ=4.0M (Hydrobromic acid solution) A=H_2O R: $CrA_5L^{2+} + L = CrA_4L_2^{+1} + A$	71Zf	b,q: ±0.5
	2.7	--	--	45-75	T	μ=8.0M (Hydrobromic acid solution) A=H_2O R: $CrA_5L^{2+} + L = CrA_4L_2^{+1} + A$	71Zf	b,q: ±0.5
$HCrO_4^-$	-2.25	1.139(26°)	-2.3	26&38	T	μ=2.0M($NaClO_4$) R: $HCrO_4^- + 2H^+ + Br^- = HCrO_3Br + H_2O$	71Yc	a,b
Fe^{3+}	6.1	0.60(1)	23	25	T	μ=0	42Ra 52Ua	

Metal Ion	ΔH, kcal /mole	Log K	ΔS, cal /mole °K	T, °C	M	Conditions	Ref.	Re-marks
B108, cont.								
Co^{2+}	0.14	-0.114(1)	-0.07	25	T	μ=2.0	66Kb	
	0.15	-0.108(1)	0.0	40	C	μ=2.0	66Kb	
	11.22	2.39(20°,1)	14.36	20-40	T	$C_M=5x10^{-4}-5x10^{-3}$ $C_L=1x10^{-3}-3.5x10^{-1}$ S: C_2H_5OH	61Sd	b
	11.70	1.49(20°,2)	13.77	20-40	T	$C_M=5x10^{-4}-5x10^{-3}$ $C_L=1x10^{-3}-3.5x10^{-1}$ S: C_2H_5OH	61Sd	b
	14.58	1.04(20°,3)	15.82	20-40	T	$C_M=5x10^{-4}-5x10^{-3}$ $C_L=1x10^{-3}-3.5x10^{-1}$ S: C_2H_5OH	61Sd	b
	7.64	0.865(20°,4)	9.67	20-40	T	$C_M=5x10^{-4}-5x10^{-3}$ $C_L=1x10^{-3}-3.5x10^{-1}$ S: C_2H_5OH	61Sd	b
	13.78	-0.34(20°,5)	13.33	20-40	T	$C_M=5x10^{-4}-5x10^{-3}$ $C_L=1x10^{-3}-3.5x10^{-1}$ S: C_2H_5OH	61Sd	b
CoA_6^{3+}	2.83	2.38(1)	20	25	T	μ=0. A=NH_3	55Na 53Ea	a
CoA_2	0.6	(1)	--	--	C	A=1,2-Bis(Diphenyl-phosphino) Ethane	72Mi	q:±0.2 u:25
CoA_3^{3+}	1.96	1.33(1)	13	25	T	μ=0.054 A=Ethylenediamine	53Ea	a
cis-$CoABC_2$	-1.27	2.839(1)	8.8	30	T	μ=0. A=Br. B=Cl. C=Ethylenediamine	66La	
	7.2	3.964(1)	42	30	T	μ=0. A=Dimethyl-formamide. B=Cl C=Ethylenediamine	66La	
Ni^{2+}	0.08	-0.119(1)	-0.27	25	C	μ=2.0	66Kb	
	0.07	-0.114(2)	-0.29	40	C	μ=2.0	66Kb	
NiA_2	0.0	(1)	--	--	C	A=1,2-Bis(Diphenyl-phosphino) Ethane	72Mi	q:±0.2 u:25
	0.638	0.155(25°)	2.85	10-59	T	S: Chloroform A=[(p-C_6H_5)(p-C_6H_5)MeP] R: NiA_2L_2(planar) = NiA_2L_2(tetrahedral)	70Pc	b
	1.150	-0.119(25°)	3.31	4-60	T	S: Chloroform A=[(p-C_6H_5)(p-ClC_6H_4)MeP] R: NiA_2L_2(planar) = NiA_2L_2(tetrahedral)	70Pc	b
	0.497	0.310(25°)	3.09	9-59	T	S: Chloroform A=[(p-C_6H_5)(p-MeC_6H_4)MeP] R: NiA_2L_2(planar) = NiA_2L_2(tetrahedral)	70Pc	b
	0.383	0.375(25°)	3.00	13-70	T	S: Chloroform A=[(p-C_6H_5)(p-$MeOC_6H_4$)MeP] R: NiA_2L_2(planar) = NiA_2L_2(tetrahedral)	70Pc	b
	0.136	0.675(25°)	3.55	-5-59	T	S: Chloroform	70Pc	b

Metal Ion	ΔH, kcal /mole	Log K	ΔS, cal /mole $^\circ$K	T, $^\circ$C	M	Conditions	Ref.	Re- marks
B108, cont.								
NiA$_2$, cont.						A=[(p-C$_6$H$_5$)(p-Me$_2$NC$_6$H$_4$) MeP] R: NiA$_2$L$_2$(planar) = NiA$_2$L$_2$(tetrahedral)		
	1.860	-0.548(25°)	3.73	-8-56	T	S: Chloroform A=[(p-ClC$_6$H$_4$)(p-ClC$_6$H$_4$) MeP] R: NiA$_2$L$_2$(planar) = NiA$_2$L$_2$(tetrahedral)	70Pc	b
	0.492	0.336(25°)	3.19	10-59	T	S: Chloroform A=[(p-MeC$_6$H$_4$)(p-MeC$_6$H$_4$) MeP] R: NiA$_2$L$_2$(planar) = NiA$_2$L$_2$(tetrahedral)	70Pc	b
	0.0063	0.516(25°)	2.38	-8-59	T	S: Chloroform A=[(p-MeOC$_6$H$_4$) (p-MeOC$_6$H$_4$)MeP] R: NiA$_2$L$_2$(planar) = NiA$_2$L$_2$(tetrahedral)	70Pc	b
NiAL$_2$	3.42	0.00(27°)	11.5	17-157	T	S: Nitrobenzene A=(C$_6$H$_5$)$_2$P(CH$_2$)$_3$P(C$_6$H$_5$)$_2$ R: NiAL$_2$(planar) = NiAL$_2$(tetrahedral)	66Va	b
NiA$_4$L$_2$	12.13	--	--	25-69	T	μ=4.30M(LiBr) S: CH$_3$OH A=CH$_3$OH R: NiL$_2$A$_4$ + L$^-$ = NiL$_3$A$^-$ + 3A (octahedral→tetrahedral)	67Sa	b
	17.11	--	--	25-69	T	μ=2.3M(LiBr) S: C$_2$H$_5$OH A=C$_2$H$_5$OH R: NiL$_2$A$_4$ + L$^-$ = NiL$_3$A$^-$ + 3A (octahedral→tetrahedral)	67Sa	b
	17.03	--	--	25-69	T	μ=1.152M(LiBr) S: n-C$_4$H$_9$OH A=n-C$_4$H$_9$OH R: NiL2A$_4$ = NiL$_2$A$_2$ + 2A (octahedral→tetrahedral)	67Sa	b
	14.8	--	--	54-111	T	μ=1.152M(LiBr) S: n-C$_4$H$_9$OH A=n-C$_4$H$_9$OH R: NiL2A$_4$ + L$^-$ = NiL$_3$A$^-$ + 3A (octahedral→tetrahedral)	67Sa	b
RhA$_2$BC	-3.2	3.222	5	50	T	μ=0.2M A=Ethylenediamine B=Cl C=H$_2$O R: <u>trans</u>-RhA$_2$BC^{2+} + L$^-$ = <u>trans</u>-RhA$_2$BL$^+$ + C	67Bb	q: ±0.4
	-2.9	3.000	5	50	T	μ=0.2M A=Ethylenediamine B=Br C=H$_2$O	67Bb	q: ±0.4

Metal Ion	ΔH, kcal /mole	Log K	ΔS, cal /mole °K	T, °C	M	Conditions	Ref.	Remarks
B108, cont.								
RhA_2BC, cont.	-4.5	3.222	2	50	T	R: trans-RhA_2BC^{2+} + L^- = trans-RhA_2BL^+ + C μ=0.2M A=Ethylenediamine B=I C=H_2O R: trans-RhA_2BC^{2+} + L^- = trans-RhA_2BL^+ + C	67Bb	
RhA_5B	0.7	--	--	--	-	A=NH_3 B=H_2O R: RhA_5B^{3+} + L^- = RhA_5L^{2+} + B	67Pa	q:±0.4 u:25
Pd^{2+}	-5.10	5.17(1)	6.5	25	C	μ=1M($HClO_4$)	72Re	
	-4.3	2.3(25°,4)	-3.5	10-45	T	μ=1.0	66Sb	b
	-13.11	(1-4)	--	25	C	μ=0.10	67Ic	
PdA_2	-2.2	2.283(1)	3.3	25	T	μ=0 A=o-Phenylene-bisdimethylarsine Thermodynamic values are average values as reported in Ref. 67Ed	67Ec 67Ed	
	-2.5	2.057(35°,1)	1.3	25-55	T	μ=0.01($NaNO_3$) A=o-Phenylenebisdi-methylarsine	67Ed	b
	-2.2	1.832(35°,1)	1.3	25-55	T	μ=0.06($NaNO_3$) A=o-Phenylenebisdi-methylarsine	67Ed	b
$PdAB^+$	-1.2	0.471(27°)	-1.8	20-63	T	μ=0.5 A=Diethylenetriamine B=Cl R: $PdAB^+$ + L^- = $PdAL^+$ + B^-	67Hd	b,q: ±0.1
	0.6	0.114(27°)	3.0	20-63	T	μ=0.5 A=1,1,7,7-Tetra-ethyldiethylenetriamine B=Cl R: $PdAB^+$ + L^- = $PdAL^+$ + B^-	67Hd	b,q: ±0.1
PtA_2^{2+}	-3.1	1.732(1)	-2.6	25	T	μ=0 A=o-Phenylene-bisdimethylarsine	67Ec 67Ed	
	-1.02	3.81(23°,1)	14	23-56	T	μ=0. S: CH_3OH A=o-Phenylene-bisdimethylarsine	66Db	b
	-3.4	1.152(49°,1)	-5.3	25-55	T	μ=0.06($NaNO_3$) A=o-Phenylene-bisdimethylarsine	67Ed	b
	-0.52	0.663(1)	1.2	25	C	μ=1M(NaF) C_M=0.040-0.050M A=Ethylenediamine	73Ma	q: ±0.05
PtA_3^{4+}	1.3	1.18(1)	9.7	25	T	μ→0. A=Ethylene-diamine	63Gb	
	3.1	-0.95(1)	6	25	C	3.0M($NaClO_4$ + NaCl) A=Ethylenediamine	72Mf	q: ±0.02

Metal Ion	ΔH, kcal /mole	Log K	ΔS, cal /mole °K	T, °C	M	Conditions	Ref.	Remarks
B108, cont.								
PtA_4^{2+}	-0.84	0.505(1)	-2.9	25	C	μ=1M(NaF) C_M=0.040-0.050M A=NH_3	73Ma	q: ±0.15
PtA_6^{2-}	4	1.90(1-6)	22	25	T	μ=0.5. A=Cl	60Pf	
Cu^{2+}	0.90	-0.071(1)	2.75	25	C	μ=2.0	66Kb 60Ld	
	1.00	-0.036(1)	3.03	40	C	μ=2.0	66Kb	
CuA_2^{2+}	0.83	3.04(1)	16.7	25	C	S: Methanol A=N,N-Dimethyl-ethylenediamine	72Bf	
	1.38	3.13(1)	18.9	25	C	S: Methanol A=N,N'-Dimethyl-ethylenediamine	72Bf	
	2.86	1.91(1)	18.3	25	C	S: Methanol A=Ethylenediamine	72Bf	
	1.61	2.68(1)	17.6	25	C	S: Methanol A=N-Methylethylene-diamine	72Bf	
AuA_4^-	-11.0	--	--	25	C	μ=2.4M R: AuA_4^- + $4L^-$ = $4A^-$ + AuL_4^-	69Pb	
Ag^+	-5.94	1.97(376°,1)	-0.14	376& 415	T	μ=0. $NaNO_3$-KNO_3 Eutectic mixture	61Dc 60Ja 50Fa	a,b
	-13.9	1.52(376°,2)	-14.4	376& 415	T	μ=0. $NaNO_3$-KNO_3 Eutectic mixture	61Dc 38Oa	a,b
	-12.6	9.021(20°)	-1.7	20-70	T	R: AgBr + $2Br^-$ = $AgBr_3^{2-}$	54Pc	b
	-20.4	9.365(20°)	-26.7	20-70	T	R: AgBr + $4Br^-$ = $AgBr_5^{4-}$	54Pc	b
Zn^{2+}	0.35	-0.57(1)	-1.4	25	C	μ=3M($NaClO_4$)	69Gb	q: ±0.03
	5.1	(1)	8.7	20-95	T	C_M + C_L=4M	70Ma	b
	0	0.601(1)	-2.75	25	C	μ=4.5	56Sa	
	10	-0.83(2)	30	25	C	μ=3M($NaClO_4$)	69Gb	q:±2
	-3.1	(2)	-3.8	20-95	T	C_M + C_L = 4M	70Ma	b
	-2	0.51(3)	-4	25	C	μ=3M($NaClO_4$)	69Gb	q:±2
	-0.8	(3)	-3.5	20-95	T	C_M + C_L = 4M	70Ma	b
	-0.3	(4)	-2.5	20-95	T	C_M + C_L = 4M	70Ma	b
	0	1.026(1-2)	-4.70	25	C	μ=4.5	56Sa	
	26.400	1.680(1-3)	80.8	25	C	μ=4.5	56Sa	n
	-2.000	1.320(1-4)	-12.7	25	C	μ=4.5	56Sa	
Cd^{2+}	-0.320	(1)	--	25	T	μ=0	59Aa	a
	-0.800	2.054(25°,1)	6.7	5-35	T	μ=0	73Sl	b,q: ±0.3
	-0.98	1.76(1)	4.7	25	C	μ=3.0($NaClO_4$)	66Ga	
	2.7	(1)	3.7	20-95	T	C_M + C_L = 4M	70Ma	b
	2.3	1.68(1)	15.4	25	T	μ=4.5	57Sc	
	-4.0	2.75(40°,1)	-0.2	40-70	T	S: Melt of NH_4NO_3	72Nb	b
	22.00	-0.110(25°,2)	73	5-35	T	μ=0	73Sl	b,q: ±2

Metal Ion	ΔH, kcal /mole	Log K	ΔS, cal /mole °K	T, °C	M	Conditions	Ref.	Remarks
B108, cont.								
Cd^{2+}, cont.	-0.57	0.58(2)	0.8	25	C	μ=3.0($NaClO_4$)	66Ga	
	0.1	(2)	6.0	20-95	T	$C_M + C_L$ = 4M	70Ma	b
	-4.3	0.733(2)	-11.1	25	T	μ=4.5	57Sc	
	1.72	0.98(3)	10.2	25	C	μ=3.0($NaClO_4$)	66Ga	
	3.8	(3)	8.4	20-95	T	$C_M + C_L$ = 4M	70Ma	b
	2.0	0.806(3)	10.4	25	T	μ=4.5	57Sc	
	0.30	0.35(4)	2.7	25	C	μ=3.0($NaClO_4$)	66Ga	
	-0.4	(4)	0.7	20-95	T	$C_M + C_L$ = 4M	70Ma	b
	2.6	0.513(4)	11.1	25	T	μ=4.5	57Sc	
Hg^{2+}	-11.1	9.48(1)	4.0	8	C	μ=0	64Cc, 58Sc	
	-10.2	9.01(1)	7.1	25	C	μ=0	64Cc, 61Ma, 63Bb	
	-10.0	8.75(1)	7.9	40	C	μ=0	64Cc	
	-10.1	(1)	--	25	C	0.5M($NaClO_4$)	65Ae	
	-9.57	9.40(1)	10.9	25	C	3M($NaClO_4$)	65Ae	
	-11.5	8.78(2)	-0.6	8	C	μ=0	64Cc	
	-11.0	8.28(2)	1.1	25	C	μ=0	64Cc	
	-11.4	7.77(2)	-0.9	40	C	μ=0	64Cc	
	-10.7	(2)	--	25	C	0.5M($NaClO_4$)	65Ae	
	-9.61	8.58(2)	7.1	25	C	3M($NaClO_4$)	65Ae	
	-2.08	2.26(3)	3.4	25	C	dil soln.	63Ba	
	-3.0	(3)	--	25	C	0.5M($NaClO_4$)	65Ae	
	-2.58	2.76(3)	4.0	25	C	3M($NaClO_4$)	65Ae	
	-3.46	1.38(4)	-5.3	25	C	dil soln.	63Ba	
	-4.1	(4)	--	25	C	0.5M($NaClO_4$)	65Ae	
	-4.44	1.49(4)	-8.1	25	C	3M($NaClO_4$)	65Ae	
	-22.60	18.26(1-2)	3.4	8	C	μ=0	64Cc	
	-21.23	17.29(1-2)	8.2	25	C	μ=0	64Cc	
	-21.40	16.52(1-2)	7.0	40	C	μ=0	64Cc	
	-20.8	(1-2)	--	25	C	0.5M($NaClO_4$)	65Ae	
	-19.18	(1-2)	--	25	C	3M($NaClO_4$)	65Ae	
	-27.9	(1-4)	--	25	C	0.5M($NaClO_4$)	65Ae	
	-27.7	20.95(1-4)	3.0	25	C	μ=0.5	60Ga	
	-26.0	(1-4)	--	25	C	3M($NaClO_4$)	65Ae	
	-5.54	3.64	-1.9	25	C	dil soln. R: $HgBr_2 + 2Br^- = HgBr_4$	63Ba	
Hg(II)	-0.5	0.60(25°)	1.1	5-60	T	$\mu \simeq 10^{-3}$. R: 1/2 $HgBr_2$ + 1/2 $HgCl_2$ = HgBrCl	64Ea	b,r
	-0.4	0.55(25°)	1.2	5-60	T	$\mu \simeq 10^{-3}$. R: 1/2 $HgBr_2$ + 1/2 HgI_2 = HgBrI	64Ea	b,r
HgA^+	-9.9	6.635(1)	-3.6	20	C	μ=0.1(KNO_3) A=CH_3	65Sa	
Ga^{3+}	13.1	-2.01(1-4)	27.7	80	T	μ=6.0	62Nb	
	11.4	-3.00(1-4)	18.4	80	T	μ=7.5	62Nb	
	10.4	-3.08(1-4)	15.3	80	T	μ=9.0	62Nb	
	9.6	-3.11(1-4)	12.6	80	T	μ=9.9	62Nb	
	9.3	-3.10(1-4)	12.2	80	T	μ=10.4	62Nb	

Metal Ion	ΔH, kcal /mole	Log K	ΔS, cal /mole °K	T, °C	M	Conditions	Ref.	Remarks
B108, cont.								
Ga^{3+}, cont.	8.8	-3.36(1-4)	9.7	80	T	μ=11.0	62Nb	
	9.2	-3.40(1-4)	10.5	80	T	μ=11.5	62Nb	
	7.9	-3.30(1-4)	7.1	80	T	μ=13.6	62Nb	
	8.4	-3.51(1-4)	7.7	80	T	μ=14.9	62Nb	
In^{3+}	0.47	1.98(1)	10.6	25	C	μ=2.00M($NaClO_4$)	69Rd	
	1.35	0.60(2)	7.3	25	C	μ=2.00M($NaClO_4$)	69Rd	
Tl^{+}	-2.45	0.88(1)	-4.2	25	T	μ=0	57Na	
Tl^{3+}	-8.96	9.50(1)	13.4	25	C	3M($HClO_4$) + 1M($NaClO_4$)	64Lf	
	-6.09	7.37(2)	13.3	25	C	3M($HClO_4$) + 1M($NaClO_4$)	64Lf	
	-4.58	5.44(3)	9.7	25	C	3M($HClO_4$) + 1M($NaClO_4$)	64Lf	
	-2.14	4.12(4)	11.7	25	C	3M($HClO_4$) + 1M($NaClO_4$)	64Lf	
Sn^{2+}	1.380	0.729(1)	8.0	25	T	3M($NaClO_4$)	52Va	
	2.730	1.140(1-2)	14.4	25	T	3M($NaClO_4$)	52Va	
	2.365	1.338(1-3)	14.1	25	T	3M($NaClO_4$)	52Va	
Pb^{2+}	0.300	(1)	--	25	C	μ=0	59Aa 55Na	
	0.048	(1)	--	25	C	μ=0.15	59Aa	
	-1.35	1.342(1)	1.3	5	T	μ=3M($LiClO_4$)	68Fb	
	-0.88	1.310(1)	3.0	15	T	μ=3M($LiClO_4$)	68Fb	
	-0.52	1.292(1)	4.1	25	T	μ=3M($LiClO_4$)	68Fb	
	-0.20	1.286(1)	5.2	35	T	μ=3M($LiClO_4$)	68Fb	q: ±0.05
	0.08	1.283(1)	6.1	45	T	μ=3M($LiClO_4$)	68Fa	q: ±0.05
	0.54	1.290(1)	7.6	55	T	μ=3M($LiClO_4$)	68Fb	q: ±0.10
	0.75	1.305(1)	8.2	65	T	μ=3M($LiClO_4$)	68Fb	q: ±0.20
	-0.52	1.283(25°,1)	4	18-65	T	μ=3M($LiClO_4$)	68Fa	b
	0.800	(1)	--	25	C	μ=0. S: 20.86 mole % CH_3OH	59Aa	
	0.638	(1)	--	25	C	μ=0.015. S: 20.86 mole % CH_3OH	59Aa	
	1.450	(1)	--	25	C	μ=0. S: 40.16 mole % CH_3OH	59Aa	
	1.178	(1)	--	25	C	μ=0.015. S: 40.16 mole % CH_3OH	59Aa	
	2.070	(1)	--	25	C	μ=0. S: 61.01 mole % CH_3OH	59Aa	
	1.778	(1)	--	25	C	μ=0.009. S: 61.01 mole % CH_3OH	59Aa	
	-2.6	2.09(40°,1)	1.2	40-70	T	S: Melt of NH_4NO_3	72Nb	b
	-1.62	0.984(2)	-1.3	5	T	μ=3M($LiClO_4$)	68Fb	
	-1.39	0.940(2)	-0.6	15	T	μ=3M($LiClO_4$)	68Fb	
	-0.86	0.918(2)	1.4	25	T	μ=3M($LiClO_4$)	68Fb	q: ±0.12
	-0.27	0.896(2)	3.3	35	T	μ=3M($LiClO_4$)	68Fb	q: ±0.15

Metal Ion	ΔH, kcal /mole	Log K	ΔS, cal /mole °K	T, °C	M	Conditions	Ref.	Remarks
B108, cont.								
Pb^{2+}, cont.	0.69	0.905(2)	6.1	45	T	μ=3M($LiClO_4$)	68Fb	q: ±0.20
	1.42	0.925(2)	8.6	55	T	μ=3M($LiClO_4$)	68Fb	q: ±0.12
	2.54	0.962(2)	11.8	65	T	μ=3M($LiClO_4$)	68Fb	q: ±0.40
	-0.86	0.931(25°,2)	1.5	18-65	T	μ=3M($LiClO_4$)	68Fa	b
	-0.83	0.593(3)	-0.3	5	T	μ=3M($LiClO_4$)	68Fb	q:±0.5
	-0.40	0.582(3)	1.3	15	T	μ=3M($LiClO_4$)	68Fb	q:±0.4
	-0.09	0.568(3)	2.3	25	T	μ=3M($LiClO_4$)	68Fb	q: ±0.17
	0.20	0.581(3)	3.5	35	T	μ=3M($LiClO_4$)	68Fb	q: ±0.20
	0.43	0.583(3)	4.2	45	T	μ=3M($LiClO_4$)	68Fb	q: ±0.13
	0.64	0.598(3)	4.6	55	T	μ=3M($LiClO_4$)	68Fb	q: ±0.18
	0.98	0.614(3)	6.0	65	T	μ=3M($LiClO_4$)	68Fb	q: ±0.40
	-0.36	0.579(25°,3)	2.5	18-65	T	μ=3M($LiClO_4$)	68Fa	b
	-2.2	0.271(4)	-6.7	5	T	μ=3M($LiClO_4$)	68Fb	q:±0.6
	-2.13	0.209(4)	-6.7	15	T	μ=3M($LiClO_4$)	68Fb	q:±0.4
	-2.23	0.151(4)	-6.8	25	T	μ=3M($LiClO_4$)	68Fb	q: ±0.23
	-2.63	0.088(4)	-8.0	35	T	μ=3M($LiClO_4$)	68Fb	q: ±0.30
	-3.3	0.007(4)	-10.4	45	T	μ=3M($LiClO_4$)	68Fb	q: ±0.20
	-3.8	-0.050(4)	-11.8	55	T	μ=3M($LiClO_4$)	68Fb	q: ±0.40
	-5.07	-0.149(4)	-16	65	T	μ=3M($LiClO_4$)	68Fb	
	-1.96	0.132(25°,4)	-7	18-65	T	μ=3M($LiClO_4$)	68Fa	b
Bi^{3+}	3.300	3.28(1)	26.1	25	C	μ=0	67Va	
	-0.557	2.64(1)	10.0	10	C	μ=4.0($HClO_4$) ΔCp=34.4	67Va	
	-0.355	2.64(1)	11.0	18	C	μ=4.0($HClO_4$) ΔCp=33.9	67Va	
	-0.019	2.64(1)	11.8	25	C	μ=4.0($HClO_4$) ΔCp=33.4	67Va	
	0.097	2.64(1)	12.4	30	C	μ=4.0($HClO_4$) ΔCp=33.1	67Va	
	0.250	2.64(1)	12.9	35	C	μ=4.0($HClO_4$) ΔCp=32.8	67Va	
	0.751	2.64(1)	14.4	50	C	μ=4.0($HClO_4$) ΔCp=31.8 ΔCp=52.44-0.0638T	67Va	
	-1.520	2.80(1)	10.8(1)	25	C	μ=5.0($HClO_4$)	67Va	
TeA_4	-39.9	--	--	25	C	S: Benzene A=Cl R: TeA_4 + AlL_3 = $TeA_4 \cdot AlL_3$	73Gb	
	-26.3	--	--	25	C	S: Benzene A=Cl R: TeA_4 + GaL_3 = $TeA_4 \cdot GaL_3$	73Gb	

Metal Ion	ΔH, kcal /mole	Log K	ΔS, cal /mole °K	T, °C	M	Conditions	Ref.	Re-marks
B108, cont.								
Br_2	∿0	1.22(1)	∿0	25	T	μ=0.5. 63Db reports ΔH=3.06 as the value given in 58Sb	58Sb	
	-3.75	1.373(25°,1)	6.3	18-39	T	--	60Db	b
	-3.4	2.00(2°,1)	3.3	2&25	T	μ=0.1. S: 75% Acetic acid	60Db	a,b,n
	-5.79	2.31(18°,1)	9.4	-15-25	T	S: CH_3OH	63Db	b
B109 BROMINE			Br_2			Br_2		L
KA	-7.110	1.435(1)	15.4	50.00	T	μ=0.2(Lithium ethane sulfate) S: Glacial acetic acid. A=Br	42Na	a
PbA_4	-86.55	--	--	--	C	S: $CHCl_3$ A=Phenyl R: $PbA_4 + 2L_2 = PbL_2A_2 + 2LA$	70Bg	u
PbA_3L	-40.86	--	--	--	C	S: $CHCl_3$ A=Phenyl R: $PbA_3L + L_2 = PbA_2L_2 + LA$	70Bg	u
SeA	-13.1	3.04(1)	-30	25	T	Dilute solution C_M→0. S: CCl_4 A=4-chlorodiphenyl	60Ma 49Ma	n
	-9.7	3.44(1)	-17	25	T	Dilute solution C_M→0. S: CCl_4 A=Diphenyl	60Ma	n
	-11.3	3.65(1)	-21	25	T	Dilute solution C_M→0. S: CCl_4 A=3-Methoxydiphenyl	60Ma	n
	-14.4	4.21(1)	-29	25	T	Dilute solution C_M→0. S: CCl_4 A=4-Methoxydiphenyl	60Ma	n
SeA_2	-7.8	2.68(1)	-14	25	T	S: CCl_4 A=p-Chlorophenyl	45Ma	
	-9.7	2.55(1)	-18	25	T	S: CCl4 A=p-Bromophenyl	45Ma	
	-8.4	4.44(1)	-8	25	T	S: CCl_4 A=p-Methylphenyl	45Ma	
B110 1,3-BUTADIENE			C_4H_6			CH_2=CHCH=CH_2		L
Ag^+	3.2	0.653(25°,1)	-7.7	0-40	T	S: Ethyleneglycol	65Ci	b
B111 1,3-BUTADIENE, 2,3-dimethyl-			C_6H_{10}			CH_2:CC(CH_3)C(CH_3):CH_2		
NiA_2	-17.1	4.041(25°,1)	-38	--	T	S: Hydrocarbons, CCl_4. $C_L=C_M=2\times10^{-4}$M A=cis(1,2-perfluoromethyl-ethene-1,2-dithiol)	71Bb	u:25
B112 BUTANE, 1-amino-			$C_4H_{11}N$			$CH_3(CH_2)_3NH_2$		L
VOA_2	-10.42	(1)	--	25	C	μ→0. S: Nitro-benzene A=Acetylacetone	65Ca	

Metal Ion	ΔH, kcal /mole	Log K	ΔS, cal /mole °K	T, °C	M	Conditions	Ref.	Remarks
B112, cont.								
NiA	-16.9	6.01(1-2)	-29.1	25	T	$C_M=10^{-4}$ S: Benzene A=Diacetyl-bisbenzoyl hydrazine	60Sb	
NiA_2	-6	2.700(35°,1)	-7	25-45	T	S: Toluene A=0,0'-diethyldithiophosphate	71Cc	b,q: ±1
	-5	2.725(35°,2)	-4	25-45	T	S: Toluene A=0,0'-diethyldithiophosphate	71Cc	b,q: ±1
Ag^+	-4	3.43(1)	3.4	25	C	$\mu=0$	71He	q:±2
	-9	4.05(2)	-10.0	25	C	$\mu=0$	71He	q:±2
	-12.57	7.48(1-2)	-8.0	25	C	$\mu=0$	71He	
ZnA_2	-7.5	3.159(25°,1)	-11	18.5-65	T	S: Toluene A=$[S_2CN(CH_3)_2]$	65Ch	b
[B]	-11.9	-0.6284(100°,1)	-35.6	40-120	T	[B]=2-Hexyl-7-methyl-1,2-oxaborepane 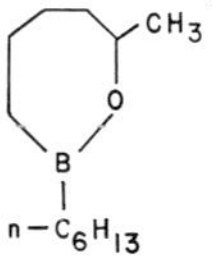	72Vc	b
I_2	-1.48	1.173(25°,1)	0.40	15-40	T	$\mu<0.02M$	60Da	b
	-1.37	1.021(1)	0	20	T	--	64Eb	
	-8.4	3.090(20°,1)	-14.8	20-40	T	$C_M=10^{-5}$ $C_L=10^{-4}-10^{-3}$ S: n-Heptane	60Ya	b
B113	BUTANE, 2-amino-		$C_4H_{11}N$			$CH_3CH_2CH(NH_2)CH_3$		L
NiA	-16.5	4.08(1-2)	-36.5	25	T	$C_M=10^{-4}$ S: Benzene A=Diacetyl-bisbenzoyl hydrazine	60Sb	
B114	BUTANE, 1-amino-3-methyl-		$C_5H_{13}N$			$(CH_3)_2CHCH_2CH_2NH_2$		L
NiA	-17.2	5.96(1-2)	-30.3	25	T	$C_M=10^{-4}$ S: Benzene A=Diacetyl-bisbenzoyl hydrazine	60Sb	
B115	BUTANE, 1-amino-3-thia-		C_3H_9NS			$CH_3SCH_2CH_2NH_2$		L
Ni^{2+}	-5	3.23(30°,1)	-3	0-50	T	$\mu=1.0(KCl$ or $KNO_3)$	54Gb	b
	-6	2.79(30°,2)	-8	0-50	T	$\mu=1.0(KCl$ or $KNO_3)$	54Gb	b
	-4	1.73(30°,3)	-7	0-50	T	$\mu=1.0(KCl$ or $KNO_3)$	54Gb	b
	-15	7.75(30°,1-3)	-17	0-50	T	$\mu=1.0(KCl$ or $KNO_3)$	54Gb	b
Cu^{2+}	-7.2	5.41(30°,1)	1	10-40	T	$\mu=0$	59Ma	b
B116	BUTANE, 1,2-diamino-		$C_4H_{12}N_2$			$NH_2CH_2CH(NH_2)CH_2CH_3$		L
Cu^{2+}	-11.76	10.047(1)	6.5	25	C	$\mu=0$	67Pc	

Metal Ion	ΔH, kcal /mole	Log K	ΔS, cal /mole °K	T, °C	M	Conditions	Ref.	Re-marks
B116, cont.								
Cu^{2+}, cont.	-12.04	9.069(2)	1.1	25	C	μ=0	67Pc	
	-23.8	19.12(1-2)	7.6	25	C	μ=0	67Pc	
B117	BUTANE, 2,3-diamino-2,3-dimethyl-		$C_6H_{16}N_2$			$CH_3C(CH_3)(NH_2)C(CH_3)(NH_2)CH_3$		L
Ni^{2+}	-14.1	14.7(1-2)	20	25	C	μ=0.5(KNO_3)	54Ba	
Cu^{2+}	-24.4	21.9(1-2)	19	25	C	μ=0.5(KNO_3)	54Ba	
Zn^{2+}	1.449	-1.083(20°,1-2)	0	0-40	T	μ=0	68Pc	b
Cd^{2+}	1.264	-1.039(20°,1-2)	0.4	0-40	T	μ=0	68Pc	b
B118	BUTANE, 1,4-diamino-N,N,N',N'-tetraacetic acid		$C_{12}H_{20}O_8N_2$			$(CH_2CO_2H)_2N(CH_2)_4N(CH_2CO_2H)_2$		L^{4-}
Mg^{2+}	8.5	6.22(1)	54.0	20	C	μ=0.1(KNO_3)	64Ab	
Ca^{2+}	-0.9	5.66(1)	29.7	20	C	μ=0.1(KNO_3)	64Ab	
La^{3+}	1.88	9.13(1)	48.2	20	C	μ=0.1(KNO_3)	64Ab	
Mn^{2+}	3.41	9.53(1)	55.2	20	C	μ=0.1(KNO_3)	64Ab	
Co^{2+}	-1.6	15.66(1)	66.3	20	C	μ=0.1(KNO_3)	64Ab	
Ni^{2+}	-6.95	17.36(1)	55.5	20	C	μ=0.1(KNO_3)	64Ab	
Cu^{2+}	-6.52	17.33(1)	57.0	20	C	μ=0.1(KNO_3)	64Ab	
Zn^{2+}	-3.48	15.01(1)	56.8	20	C	μ=0.1(KNO_3)	64Ab	
Cd^{2+}	-2.88	12.02(1)	45.2	20	C	μ=0.1(KNO_3)	64Ab	
Hg^{2+}	-19.1	20.99(1)	30.8	20	C	μ=0.1(KNO_3)	64Ab	
Pb^{2+}	-4.85	10.53(1)	31.6	20	C	μ=0.1(KNO_3)	64Ab	
B119	BUTANE, 1,2-diethoxy-		$C_8H_{18}O_2$			$CH_3CH_2CH(OC_2H_5)CH_2OC_2H_5$		L
SnA_4	-12.3	2.71(1)	-28.9	25	C	S: Benzene A=Cl	70Gh 68Gf	
B120	BUTANE, 2,3-dihydroxy-(ℓ)		$C_4H_{10}O_2$			$CH_3CH(OH)CH(OH)CH_3$		L
BA_4^-	-4.30	1.568(25°,1)	-7.2	0-35	T	C_L=0-0.9M A=OH^-	67Cc	b,q: ±0.4
	-5.48	2.215(25°,1-2)	-8.2	0-35	T	C_L=0-0.9M A=OH^-	67Cc	b,q: ±0.6
B121	BUTANE, 2,3-dihydroxy-,(meso)		$C_4H_{10}O_2$			$CH_3CH(OH)CH(OH)CH_3$		L
BA_4^-	-3.43	0.431(25°,1)	-9.5	0-35	T	C_L=0-0.9M A=OH^-	67Cc	b,q: ±0.4
	-7.19	0.663(25°,1-2)	-21	0-35	T	C_L=0-0.9M A=OH^-	67Cc	b,q: ±0.7
$H_5TeO_6^-$	-3.2	1.22	-5	25	T	μ=0.1. R: $H_5TeO_6^-$ + L = $HTeO_4L$ + $2H_2O$	62Ea	
B122	BUTANE, 2,3-dihydroxy-2,3-dimethyl-		$C_6H_{14}O_2$			$CH_3C(OH)(CH_3)C(OH)(CH_3)CH_3$		L

Metal Ion	ΔH, kcal /mole	Log K	ΔS, cal /mole °K	T, °C	M	Conditions	Ref.	Remarks
B122, cont.								
BA_4^-	-7.09	1.041(25°,1)	-19	0-35	T	C_L=0-0.9M A=OH^-	67Cc	b,q: ±0.7
	-9.00	2.334(25°,1-2)	-20	0-35	T	C_L=0-0.9M A=OH^-	67Cc	b,q: ±0.9
$H_5TeO_6^-$	-2.9	0.59	-7	25	T	μ=0.1R: $H_5TeO_6^-$ + L = $HTeO_4L^-$ + $2H_2O$	62Ea	
B123	1,4-BUTANEDIOL, diethyl ether	$C_8H_{18}O_2$				$C_2H_5O(CH_2)_4OC_2H_5$		
SnA_4	-12.3	(1)	--	--	C	S: Benzene C_M=0.05-0.08M A=Cl	68Ge	u
B124	2,3-BUTANEDIOL, 2-methyl-	$C_5H_{12}O_2$				$CH_3CH(OH)C(OH)(CH_3)_2$		L
BA_4^-	-4.90	1.255(25°,1)	-11	0-35	T	C_L=0-0.9M A=OH^-	67Cc	b,q: ±0.5
	-8.00	2.322(25°,1-2)	-16	0-35	T	C_L=0-0.9M A=OH^-	67Cc	b,q: ±0.8
B125	1,3-BUTANEDIONE, 1-phenyl-	$C_{10}H_{10}O_2$				$C_6H_5C(O)CH_2C(O)CH_3$		
Co^{3+}	-0.17	-0.421(56°)	-2.5	56-104	T	S: Chlorobenzene R: CoL_3(trans) = CoL_3(cis)	70Gg	b,q: ±0.33
Ni^{2+}	-2.6	9.13(25°,1)	33	15-40	T	S: 75 v % Dioxane μ=0	71Rb	b,q: ±0.5
	-2.5	7.76(25°,2)	27	15-40	T	S: 75 v % Dioxane μ=0	71Rb	b,q: ±0.5
NiL_2	53.78	6.0	204	30	C	S: Benzene C_M=0.006-0.062M R: $3NiL_2$ = Ni_3L_6	73Af	q
Cu^{2+}	-6.8	12.42(25°,1)	34	15-40	T	S: 75 v % Dioxane μ=0	71Rb	b
	-6.6	10.81(25°,2)	27	15-40	T	S: 75 v % Dioxane μ=0	71Rb	b
B126	1-BUTANESULPHONIC ACID, 4-amino-	$C_4H_{11}O_3NS$				$NH_2(CH_2)_4SO_3H$		L^-
Ag^+	-5.51	3.379(1)	-3.0	25	C	μ=0.5M(KNO_3)	72Va	
	-7.16	3.702(2)	-7.1	25	C	μ=0.5M(KNO_3)	72Va	
B127	BUTANOIC ACID, amide, 3-oxo-N-(o-chlorophenyl)-	$C_{10}H_{10}O_2NCl$				$C_6H_4(Cl)[NHC(O)CH_2C(O)CH_3]$		L
Fe^{3+}	-2.86	5.91(25°)	17.47	15-35	T	S: 50 w % Ethanol μ=0.1 pH=1.5-3.8 R: Fe^{3+} + 2HL = FeL_2^+ + $2H^+$	72Ta	b,j
B128	BUTANOIC ACID, amide, 3-oxo-4-(2,4-dimethylphenyl)-	$C_{12}H_{15}O_2N$				$C_6H_3(CH_3)_2[NHC(O)CH_2C(O)CH_3]$		L

Metal Ion	ΔH, kcal /mole	Log K	ΔS, cal /mole °K	T, °C	M	Conditions	Ref.	Remarks
B128, cont.								
Fe^{3+}	-3.85	5.86(25°)	13.89	15-35	T	S: 50 w % Ethanol μ=0.1 pH=1.8-3.8 R: Fe^{3+} + 2HL = FeL_2^+ + $2H^+$	72Ta	b,j
B129	BUTANOIC ACID, amide, 3-oxo-N-(o-methoxyphenyl)-		$C_{11}H_{13}O_3N$			$C_6H_4(OCH_3)[NHC(O)CH_2C(O)CH_3]$		L
Fe^{3+}	-4.43	5.81(25°)	11.71	15-35	T	S: 50 w % Ethanol μ=0.1 pH=1.5-3.8 R: Fe^{3+} + 2HL = FeL_2^+ + $2H^+$	72Ta	b,j
B130	BUTANOIC ACID, amide 3-oxo-N-(o-tolyl)-		$C_{11}H_{13}O_2N$			$C_6H_4(CH_3)[NHC(O)CH_2C(O)CH_3]$		L
Fe^{3+}	-1.64	2.31(25°)	5.07	15-35	T	S: 50 w % Ethanol μ=0.1 pH=1.5-3.8 R: Fe^{3+} + HL = FeL^{2+} + H^+	72Ta	b,j
B131	BUTANOIC ACID, ethyl ester		$C_6H_{12}O_2$			$CH_3CH_2CH_2CO_2CH_2CH_3$		L
SbA_5	-16.76	(1)	--	25	C	S: 1,2-Dichloro-ethane. A=Cl	630a	
B132	BUTANOIC ACID, nitrile		C_4H_7N			$CH_3CH_2CH_2CN$		L
I_2	-1.72	-0.284(25°,1)	-7.1	--	T	S: Carbon disulfide	72Ma	b,j, u:25
	-2.83	-0.294(25°,1)	-10.9	--	T	S: CCl_4 C_L=0.2-1.0M C_M=2x10^{-3} Data analyzed by method 1.	72Ma	j,u: 25
	-2.97	-0.294(25°,1)	-11.31	--	T	S: CCl_4 C_L=0.2-1.0M C_M=2x10^{-3} Data analyzed by method 2.	72Ma	j,u: 25
	-2.89	-0.032(25°,1)	-9.8	--	T	S: Cyclohexane	72Ma	b,j, u:25
	-3.20	0.005(25°,1)	-10.7	--	T	S: n-Heptane C_L=0.2-1.0M C_M=2x10^{-3} Data analyzed by method 1.	72Ma	j,u: 25
	-3.91	0.005(25°,1)	-13.07	--	T	S: n-Heptane C_L=0.2-1.0M C_M=2x10^{-3} Data analyzed by method 2.	72Ma	j,u: 25
	-4.24	-0.032(25°,1)	-14.4	--	T	S: n-Hexane	72Ma	b,j, u:25

Metal Ion	ΔH, kcal /mole	Log K	ΔS, cal /mole °K	T, °C	M	Conditions	Ref.	Remarks
B132, cont.								
I_2, cont.	-3.69	-0.387(25°,1)	-14.1	--	T	S: Tetrachloro-ethylene	72Ma	b,j, μ:25
B133	BUTANOIC ACID, 2-amino-		$C_4H_9O_2N$			$CH_3CH_2CHNH_2CO_2H$		L^-
Co^{2+}	-1.2	4.21(1)	15	25	C	μ=0.05M(KCl)	72Gb, 72Ga	
	-2.5	4.31(15°,1)	11.1	15&40	T	μ=0.2	65Sb, 64Se	a,b
	-2.2	3.50(2)	7	25	C	μ=0.05M(KCl)	72Gb, 72Ga	
	-3.0	3.19(15°,2)	4.3	15&40	T	μ=0.2	65Sb	a,b
	-5.5	7.50(15°,1-2)	15.4	15&40	T	μ=0.2	65Sb	a,b
Ni^{2+}	-2.7	5.35(1)	15	25	C	μ=0.05M(KCl)	72Gb, 72Ga	
	-4.8	5.46(25°,1)	8.9	15-40	T	μ=0.2	65Sb, 64Se	b
	-2.8	4.41(2)	11	25	C	μ=0.05M(KCl)	72Gb, 72Ga	
	-4.8	4.36(25°,2)	3.9	15-40	T	μ=0.2	65Sb	b
	-9.6	9.82(25°,1-2)	12.7	15-40	T	μ=0.2	65Sb	b
Cu^{2+}	-4.4	8.01(1)	22	25	C	μ=0.05M(KCl)	72Gb, 72Ga	
	-5.40	8.02(1)	19	25	C	μ=0.2M(KCl)	73Ga	
	-5.4	8.21(25°,1)	19.2	15-40	T	μ=0.2	65Sb, 64Se	b
	-5.4	6.75(1)	13	25	C	μ=0.05M(KCl)	72Gb, 72Ga	
	-6.6	6.70(2)	9	25	C	μ=0.2M(KCl)	73Ga	
	-5.3	6.72(25°,2)	13.0	15-40	T	μ=0.2	65Sb	b
	-10.7	14.93(25°,1-2)	32.1	15-40	T	μ=0.2	65Sb	b
Zn^{2+}	-2.1(1)	4.50	14	25	C	μ=0.05M(KCl)	72Gb, 72Ga	
	-3.1	4.78(25°,1)	11.4	15-40	T	μ=0.2	65Sb, 65Se	b
	-1.6(2)	4.15	13	25	C	μ=0.05M(KCl)	72Gb, 72Ga	
	-3.3	3.90(25°,2)	6.7	15-40	T	μ=0.2	65Sb	b
	-6.4	8.68(25°,1-2)	18.0	15-40	T	μ=0.2	65Sb	b
B134	BUTANOIC ACID, 3-amino-		$C_4H_9O_2N$			$CH_3CH(NH_2)CH_2CO_2H$		L^-
Co^{2+}	-1.7	3.52(25°,1)	10.5	15-40	T	μ=0.2	65Sb, 64Se	b
	-3.3	2.36(25°,1)	-0.2	15-40	T	μ=0.2	65Sb	b
	-5.0	5.88(25°,1-2)	10.2	15-40	T	μ=0.2	65Sb	b
Ni^{2+}	-2.70	4.56(1)	11.8	25	C	μ=0.16	70Lb	
	-3.8	4.60(25°,1)	8.4	15-40	T	μ=0.2	65Sb, 64Se	b
	-2.89	3.30(2)	5.4	25	C	μ=0.16	70Lb	
	-4.8	3.32(25°,2)	1.0	15-40	T	μ=0.2	65Sb	b
	-8.6	7.92(25°,1-2)	7.3	15-40	T	μ=0.2	65Sb	b

Metal Ion	ΔH, kcal /mole	Log K	ΔS, cal /mole °K	T, °C	M	Conditions	Ref.	Re-marks
B134, cont.								
Cu^{2+}	-4.94	7.12(1)	16.0	25	C	μ=0.16	70Lb	
	-5.0	7.18(25°,1)	16.2	15-40	T	μ=0.2	65Sb 64Se	b
	-5.58	4.73(2)	7.5	25	C	μ=0.16	70Lb	
	-4.6	5.59(25°,1)	9.8	15-40	T	μ=0.2	65Sb	b
	-9.6	12.77(25°,1-2)	25.9	15-40	T	μ=0.2	65Sb	b
B135	BUTANOIC ACID, 4-amino-N-glycyl-		$C_6H_{12}O_3N_2$			$NH_2CH_2C(O)NHCH_2CH_2CH_2CO_2H$		
Ni^{2+}	-4.6	4.11(1)	3.4	25	C	μ=0.1M(KNO_3)	71Ld	
	-4.2	3.38(2)	1.7	25	C	μ=0.1M(KNO_3)	71Ld	
Cu^{2+}	-6.1	5.66(1)	5.4	25	C	μ=0.1M(KNO_3)	72Bl	
B136	BUTANOIC ACID, 2-amino-4-hydroxy-		$C_4H_9O_3N$			$HOCH_2CH_2CH(NH_2)CO_2H$		L^-
Ni^{2+}	-3.92	5.51(1)	12.1	25	C	μ=0.16	70Lb	
	-4.48	4.60(2)	6.0	25	C	μ=0.16	70Lb	
	-5.49	3.26(3)	-3.5	25	C	μ=0.16	70Lb	
Cu^{2+}	-5.45	8.00(1)	18.3	25	C	μ=0.16	70Lb	
	-5.94	6.69(2)	10.7	25	C	μ=0.16	70Lb	
B137	BUTANOIC ACID, 4-amino-3-hydroxy-		$C_4H_9O_3N$			$NH_2CH_2CH(OH)CH_2CO_2H$		L^-
Ni^{2+}	-2.60	3.99(1)	9.4	25	C	μ=0.16	70Lb	
	-2.71	3.18(2)	5.5	25	C	μ=0.16	70Lb	
Cu^{2+}	-4.87	6.48(1)	13.3	25	C	μ=0.16	70Lb	
	-5.55	6.06(2)	9.1	25	C	μ=0.16	70Lb	
B138	BUTANOIC ACID, 2-(dihydroxyamino)-		$C_4H_9O_4N$			$CH_3CH_2CH[N(OH)_2]CO_2H$		L^-
Ce^{3+}	-20.5	12.26(30,1-3)	-10	30-50	T	μ=0.1M($NaClO_4$)	73Kb	b
Ce^{4+}	-24.6	14.35(30,1-3)	-15	30-50	T	μ=0.1M($NaClO_4$)	73Kb	b
B139	2-BUTANONE		C_4H_8O			$CH_3COCH_2CH_3$		L
SbA_5	-17.43	(1)	--	25	C	Excess $SbCl_5$. C_L low, variable, ∿0.02. S: 1,2-Dichloroethane A=Cl	64Oa	
TeA_4	-5.4	(1)	--	25	C	S: Benzene A=Cl	72Pe	
I_2	-3.5	0.013(25°,1)	-11.4	5-25	T	S: Heptane	71Te	b
B140	3-BUTANONE, 2-amino-2-methyl-, oxime		$C_5H_{12}ON_2$			$CH_3C(CH_3)(NH_2)C(NOH)CH_3$		
Ni^{2+}	-8.5	6.08	-1	25	C	μ<0.06. R: Ni^{2+} + 2L = $[NiL_2\text{-}H]^+$ + H^+	64Wb	
	-16.9	--	--	25	C	μ<0.06. R: $[NiL_2\text{-}H]^+$ + $3H^+$ = Ni^{2+} + $2LH^+$	64Wb	

Metal Ion	ΔH, kcal /mole	Log K	ΔS, cal /mole °K	T, °C	M	Conditions	Ref.	Remarks
B140, cont.								
Cu^{2+}	-11.3	7.84	-2	25	C	μ<0.06. R: Cu^{2+} + 2L = $[CuL_2-H^+]$ + H^+	64Wb	
	-14.1	--	--	25	C	μ<0.06. R: $[CuL_2-H]^+$ + $3H^+$ = Cu^{2+} + $2LH^+$.	64Wb	
B141	2-BUTANONE, 3,3-dimethyl-		$C_6H_{12}O$			$CH_3COC(CH_3)_3$		L
SbA_5	-16.95	(<u>1</u>)	--	25	C	Excess $SbCl_5$ C_L low, variable, ∿0.02. S: 1,2-Dichloroethane. A=Cl	64Oa	
I_2	-3.6	0.057(25°,<u>1</u>)	-11.8	5-25	T	S: Heptane	71Te	b
B142	3-BUTANONE, 2-ethylamino-2-methyl-, oxime		$C_7H_{16}ON_2$			$CH_3C(CH_3)(NHCH_2CH_3)C(:NOH)CH_3$		L
Ni^{2+}	2.6	--	--	25	C	μ<0.06. R: Ni^{2+} + 2L = $[NiL_2-H]^+$ + H^+	64Wb	
	-23.8	--	--	25	C	μ<0.06. R: $[NiL_2-H]^+$ + $3H^+$ = Ni^{2+} + $2HL^+$	64Wb	
B143	2-BUTANONE, 3-methyl-		$C_5H_{10}O$			$CH_3COCH(CH_3)_2$		L
SbA_5	-17.07	(<u>1</u>)	--	25	C	Excess $SbCl_5$ C_L low, variable, ∿0.02. S: 1,2-Dichloroethane. A=Cl	64Oa	
I_2	-3.6	0.033(25°,<u>1</u>)	-11.9	5-25	T	S: Heptane	71Te	b
B144	3-BUTANONE, 2-methylamino-2-methyl-, oxime		$C_6H_{14}ON_2$			$CH_3C(CH_3)(NHCH_3)C(:NOH)CH_3$		L
Ni^{2+}	-0.7	3.59	14	25	C	μ<0.06. R: Ni^{2+} + 2L = $[NiL_2-H]^+$ + H^+	64Wb	
	-19.9	--	--	25	C	μ<0.06. R: $[NiL_2-H]^+$ + $3H^+$ = Ni^{2+} + $2HL^+$	64Wb	
Cu^{2+}	-8.7	6.96	3	25	C	μ<0.06. R: Cu^{2+} + 2L = $[CuL_2-H]^+$ + H^+	64Wb	
	-11.9	--	--	25	C	μ<0.06. R: $[CuL_2-H]^+$ + $3H^+$ = Cu^{2+} + $2HL^+$	64Wb	
B145	3-BUTANONE, 2-methyl-2-(n-propylamino)-, oxime		$C_8H_{18}ON_2$			$CH_3C(CH_3)(NHCH_2CH_2CH_3)-C(:NOH)CH_3$		L
Ni^{2+}	2.0	1.1	12	25	C	μ<0.06. R: Ni^{2+} + 2L = $[NiL_2-H]^+$ + H^+	64Wb	
	-23.6	--	--	25	C	μ<0.06. R: $[NiL_2-H]^+$ + $3H^+$ = Ni^{2+} + $2HL^+$	64Wb	

Metal Ion	ΔH, kcal /mole	Log K	ΔS, cal /mole °K	T, °C	M	Conditions	Ref.	Remarks
B146 1-BUTENE			C_4H_8			$CH_3CH_2CH{:}CH_2$		L
$RhAB_2$	1.0	-1.036(25°)	-1.7	0&25	T	S: Toluene A=2,4-Pentanedionato B=Ethylene R: L + $RhAB_2$ = RhABL + B	67Cd	a,b,q: ±1.4
PdA_4^{2-}	0	1.093(20°)	5	5-25	T	μ=4.0. R: PdA_4^{2-} + L=PbA_3L^- + A^- A=Cl	65Pb	b
	0	0.53(20°)	2.5	5-25	T	μ=4.0. R: PdA_4 + L + H_2O = $PdA_2L(OH)_2$ + 2A. A=Cl	65Pb	b
Ag^+	-5.95	2.08(1)	-10.36	25	T	μ=0.1M(KNO_3)	72Pf	
	3.7	0.944(25°,1)	-8.0	0-40	T	S: Ethyleneglycol	65Ci	b
AgAB	-0.3	-0.301(25°)	-2.0	0&25	T	S: Toluene A=NO_3. B=Ethylene R: L + AgAB = AbAL + B	67Cd	a,b,i: 65Ci
B147 2-BUTENE(cis)			C_4H_8			$CH_3CH{:}CHCH_3$		L
$RhAB_2$	1.8	-2.383(25°)	-4.9	0&25	T	S: Toluene A=2,4-Pentanedionato B=Ethylene R: L + $RhAB_2$ = RhABL + B	67Cd	a,b,q: ±0.8
Ag^+	3.4	0.690(25°,1)	-8.2	0-40	T	S: Ethyleneglycol	65Ci	b
	-0.1	-0.553(25°)	-2.2	0-25	T	S: Toluene A=NO_3. B=Ethylene R: L + AgAB = AgAL + B	67Cd	a,b,i: 65Ci
B148 2-BUTENE(trans)			C_4H_8			$CH_3CH{:}CHCH_3$		
$RhAB_2$	1.9	-2.699(25°)	-6.1	0&25	T	S: Toluene A=2,4-Pentanedionato B=Ethylene R: L + $RhAB_2$ = RhABL + B	67Cd	a,b,q: ±0.7
Ag^+	2.6	0.204(25°,1)	-7.7	0-40	T	S: Ethyleneglycol	65Ci	b
B149 2-BUTENE, 2,3-dimethyl-			C_6H_{12}			$(CH_3)_2C{:}C(CH_3)_2$		L
Ag^+	1.9	-0.468(25°,1)	-8.5	0-40	T	S: Ethyleneglycol	65Ci	b
B150 1-BUTENE, 2-methyl-			C_5H_{10}			$CH_3CH_2C(CH_3){:}CH_2$		L
Ag^+	3.5	0.644(25°,1)	-8.6	0-40	T	S: Ethyleneglycol	65Ci	b
B151 1-BUTENE, 3-methyl-			C_5H_{10}			$CH_3CH(CH_3)CH{:}CH_2$		
Ag^+	3.9	0.903(25°,1)	-8.8	0-40	T	S: Ethyleneglycol	65Ci	b
B152 2-BUTEN-1-ol(cis)			C_4H_8O			$CH_3CH{:}CHCH_2OH$		L
CuA_2B	-3.13	0.992(25°)	-6.0	0-30	T	S: 2-Butenyl alcohol. A=trans-2-Butenyl alcohol. B=H_2O R: CuA_2B^+ + 2L = CuL_2A^+ + 2A	72Ib	b

Metal Ion	ΔH, kcal /mole	Log K	ΔS, cal /mole °K	T, °C	M	Conditions	Ref.	Remarks
B153	3-BUTEN-2-ONE		C_4H_6O			CH_2:CHC(O)CH_3		L
NiAB	-13.7	2.58(27°)	-33.9	24-41	T	S: Tetrahydrofuran A=Dipyridyl B=Tetrahydrofuran R: NiAB + L = NiAL + B	71Ya	b
B154	3-BUTEN-2-ONE, 4-amino-3-methyl-		C_5H_9ON			$CH_3COC(CH_3)$:$CHNH_2$		L^-
Co^{2+}	1.91	-0.493(25°)	4.15	-33-37	T	S: 15 v % Tetramethylsilane in $CHCl_3$	68Ea	b,t
B155	2-BUTEN-1-ONE, 3-amino-1-phenyl-		$C_{10}H_{11}ON$			C_6H_5COCH:$C(NH_2)CH_3$		L^-
Co^{2+}	1.64	0.100(25°)	5.96	-38-57	T	S: 15% v% Tetramethylsilane in $CHCl_3$	68Ea	b,t
B156	2-BUTEN-1-ONE, 3-methylamino-1-(β-naphthyl)-		$C_{15}H_{15}ON$			β-$C_{10}H_7COCH$:$C(NHCH_3)CH_3$		L^-
Ni^{2+}	2.300	1.35(25°)	1.56	57-102	T	S: $CDCl_3$	68Ea	b,t
B157	2-BUTEN-1-ONE, 3-methylamino-1-phenyl		$C_{11}H_{13}ON$			C_6H_5COCH:$C(NHCH_3)CH_3$		L^-
Ni^{2+}	4.81	-1.730(25°)	8.22	-33-87	T	S: $CDCl_3$	68Ea	b,t
B158	3-BUTEN-2-ONE, 3-methyl-4-methylamino-		$C_6H_{11}O$			$CH_3COC(CH_3)$:$CHNHCH_3$		L^-
Ni^{2+}	5.62	-3.26(25°)	3.92	67-127	T	S: $CDCl_3$	68Ea	b,t
B159	BUTYROHYDROXAMIC ACID, N-phenyl		$C_5H_{11}O_3N$			$CH_3CH_2CH_2CH_2COONOH$		
Zn^{2+}	2.95	7.85(25°,<u>1</u>)	-26.1	25-35	T	μ=0 S: 50 v % Dioxane	71Aa	b
	4.21	6.40(25°,<u>2</u>)	-14.7	25-35	T	μ=0 S: 50 v % Dioxane	71Aa	b
			C					
C1	CADMIUMATE(II) ION, tetrachloro-		Cl_4Cd			$CdCl_4^{2-}$		L^{2-}
Na^+	-7.3	0.30(<u>1</u>)	26	25	T	4M(Li,Na,ClO_4,Cl)	63Mf	
K^+	-6.9	0.15(<u>1</u>)	23	25	T	4M(Li,Na,ClO_4,Cl)	63Mf	
Rb^+	-6.8	0.26(<u>1</u>)	22	25	T	4M(Li,Na,ClO_4,Cl)	63Mf	
Cs^+	-6.3	0.34(<u>1</u>)	20	25	T	4M(Li,Na,ClO_4,Cl)	63Mf	
NH_4^+	-5.5	0.04(25°,<u>1</u>)	-18	15-65	T	μ=1	63Md	b
C2	CAMPHOR(<u>d</u>)		$C_{10}H_{16}O$			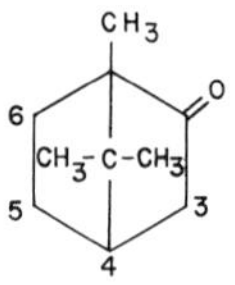		L

Metal Ion	ΔH, kcal /mole	Log K	ΔS, cal /mole °K	T, °C	M	Conditions	Ref.	Remarks
C2, cont.								
[Fe]	8.7	5.47(4.6°,1)	56.4	0-13	T	μ=0.1M(KCl) pH=7.4 [Fe]=Ferric Cytochrome P-450	72Gg	b
	0	5.72(21.0°,1)	26.2	13-40	T	μ=0.1M(KCl) pH=7.4 [Fe]=Ferric Cytochrome P-450	72Gg	b
C3 CAMPHOR, thiono-			$C_{10}H_{16}S$			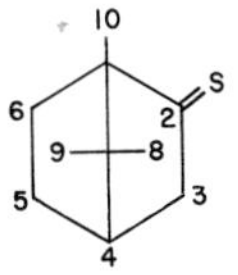		
I_2	-11.0	1.978(25°,1)	-27.8	0-45	T	S: Cyclohexane	66Bc	b,q:±1
C4 CARBAMIC ACID, N,N-dimethyl-, chloride			C_3H_6ONCl			$(CH_3)_2NCOCl$		L
SbA_2	-17.22	(1)	--	25	C	S: Ethylene chloride. A=Cl	67Oa	
C5 CARBAMIC ACID, N,N-dimethyl-, ethyl ester			$C_5H_{11}O_2N$			$(CH_3)_2NC(O)OC_2H_5$		L
SbA_2	-22.37	(1)	--	25	C	S: Ethylene chloride. A=Cl	67Oa	
SbA_5	-20.0	(1)	--	25	C	S: CCl_4. A=Cl	73Oa	
C6 CARBAMIC ACID, N-isopropyl-, thiono-, methyl ester			$C_5H_{11}ONS$			$(CH_3)_2CHNHC(S)OCH_3$		L
I_2	-7.8	2.204(25°,1)	-16.1	0-45	T	S: Cyclohexane	66Bc	b,q:±1
C7 CARBAMIC ACID, N-methyl-N-benzyl-dithio-			$C_9H_{11}NS_2$			$C_6H_5CH_2N(CH_3)C(S)SH$		
FeL_3A	-0.15	0.477(25°)	1.7	-54-36	T	S: CH_2Cl_2 A=BF_4 R: Cis→Trans	73Ea	b
C8 CARBAMIC ACID, N-methyl-N-isopropyl-dithio-			$C_5H_{11}NS_2$			$(CH_3)_2CHN(CH_3)C(O)SH$		
FeL_3	0.72	0.613(25°)	5.2	-54-34	T	S: CH_2Cl_2 R: Cis→Trans	73Ea	b
FeL_3A	0.17	0.462(25°)	2.7	-56-43	T	S: CH_2Cl_2 A=BF_4 R: Cis→Trans	73Ea	b
C9 CARBAMIC ACID, N-methyl, thiono-, methyl ester			C_3H_7ONS			$CH_3NHC(S)OCH_3$		L
I_2	-7.3	3.910(25°,1)	-6.7	0-45	T	S: Cyclohexane	66Bc	b,q:±1
C10 CARBONATE ION			CO_3			CO_3^{2-}		L^{2-}

Metal Ion	ΔH, kcal /mole	Log K	ΔS, cal /mole °K	T, °C	M	Conditions	Ref.	Remarks
C10, cont.								
CrA_3	4.59	-0.046(1)	15.2	25	C	μ=3.0($NaClO_4$) A=Ethylenediamine	73Ua	
	3.30	-0.070(2)	10.8	25	C	μ=3.0($NaClO_4$) A=Ethylenediamine	73Ua	
CoA_3	-1.15	0.46(1)	-6.0	25	C	μ=3.0M($NaClO_4$ + Na_2L) A=Ethylenediamine	73Md	q:±0.1
	-0.30	0.146(2)	-2	25	C	μ=3.0M($NaClO_4$ + Na_2L) A=Ethylenediamine	73Md	q:±0.1
	1.45	-0.40(3)	7	25	C	μ=3.0M($NaClO_4$ + Na_2L) A=Ethylenediamine	73Md	q:±0.3
C11 CARBONIC ACID, ethyl ester			$C_3H_6O_3$			$HCO_3CH_2CH_3$		L
SbA_5	-15.98	(1)	--	25	C	S: 1,2-Dichloroethane. A=Cl	63Oa	
C12 CARBONIC ACID, methyl ester			$C_2H_4O_3$			HCO_3CH_3		L
SbA_5	-15.17	(1)	--	25	C	S: 1,2-Dichloroethane. A=Cl	63Oa	
C13 CARBONIC ACID, chloro-, ethyl ester			$C_3H_5O_2Cl$			$ClCO_2CH_2CH_3$		L
SbA_5	-21.4	(1)	--	25	C	Excess $SbCl_5$ C_L low, variable, ∿0.02. S: Ethylene chloride. A=Cl	67Oa	
C14 CARBON MONOXIDE			CO			CO		L
$[MoAL_3]_2$	0.44	0.418(25°)	3.38	-15-62	T	S: Acetone A=(h^5-C_5H_5) R: Gauche⇌Trans Transformation	73Ab	b,q: ±0.15
[Fe]	-21.2	(1)	-38.5	0-30	T	0.001M THAM buffer pH=8.5 [Fe]=Sperm whale myoglobin. Standard state for L = unit mole fraction in organic solvent.	71Ka	b
$RhABC_2$	-2	1.477(20°,1)	-0.1	10-60	T	S: Chlorobenzene A=I. B=CO. C=Ph_3P	71Vb	b
$IABC_2$	-1	3.653(20°,1)	13.3	10-60	T	S: Chlorobenzene A=I. B=CO. C=Ph_3P	71Vb	b
C15 CHLORATE ION			ClO_3			ClO_3^-		L^-
K^+	-10.120	-1.277(1)	-29.7	25	T	μ=0	34La	

Metal Ion	ΔH, kcal /mole	Log K	ΔS, cal /mole °K	T, °C	M	Conditions	Ref.	Remarks
C16 CHLORIDE ION			Cl			Cl^-		L^-
Li^+	14.7	12.06(1)	105	25	T	$C_M \simeq 10^{-4}$. S: 25 w% Butanol-75 w% Hexane	61Rd	
Na^+	-1.019	-1.850(1)	-11.9	25	T	$\mu=0$	34La	
	16.9	12.70(100°)	102.9	25-300	T	S: Dioxane-Water mixtures R: $Na^+ + L^- = NaL + 7.8H_2O$	72La	b
	7.0	17.31(500°)	88.3	300-800	T	R: $Na^+ + L^- = NaL + 10.2H_2O$		
K^+	-4.119	-0.885(1)	-17.9	25	T	$\mu=0$	34La	
Sc^{3+}	-1.5	1.08(1)	0	25	T	$\mu=0.5(NaClO_4)$	62Pa	
	-1.7	1.03(2)	-1	25	T	$\mu=0.5(NaClO_4)$	62Pa	
Y^{3+}	-0.3	0.36(1)	-1	25	T	$\mu=0.5(NaClO_4)$	62Pa	
Ce^{3+}	5.4	0.0(1)	18	25	C	--	67Ab 62Mi	
Eu^{3+}	-0.05	-0.05(1)	-0.4	25	T	$\mu=1.0$. Data in 65Ce recalculated from 63Cb which contains an error in the unitary entropy value	65Ce 63Cb	
VAB_4^{2+}	7.5	-1.92(27°)	16.3	20-130	T	$\mu=0$. A=0. $B=H_2O$ R: $VAB_4^{2+} + L^- = VAB_3L^+ + B$	71Zb	b
VAB_3L^+	2.6	-3.00(27°)	-5.0	--	T	$\mu=0$. A=0. $B=H_2O$ R: $VAB_3L^+ + L^- = VAB_2L_2 + B$	71Zb	b
UO_2^{2+}	3.8	-0.06(1)	12	25	T	$\mu=2.0(NaClO_4)$	54Db	
NpO_2^{2+}	-8.7	-0.72(0°,1)	-32.2	0-9.84	T	$\mu=3.0$. Difference in ΔH due to mechanism assumed to obtain rate constant	55Cb	b
	-6.9	0.51(0°,1)	-21	0-9.84	T	$\mu=3.0$. Difference in ΔH due to mechanism assumed to obtain rate constant	55Cb	b
	3.5	1.02(0°,2)	16	0-9.84	T	$\mu=3.0$	55Cb	b
Pu^{3+}	4.6	0.57(1)	18	25	C	$\mu=0.1$	58Ma	
Pu^{4+}	1.9	-0.25(1)	5	25	T	1M(HCl)	55Ra	
PuO_2^{2+}	3.3	-0.25(20.2°,1)	9.9	2.4-29.6	T	$\mu=2.0$	61Ra	b
Cr^{3+}	-0.43	-0.056(25°,1)	-1.7	10-50	T	$\mu=1.0(HClO_4)$	68Md	b,q: ±0.23
	6.1	-0.710(1)	17.2	25	T	$\mu \simeq 4.4$. $[Cl^-]=0$	58Ga	
	4.5	-0.982(1)	10.6	25	T	$\mu \simeq 4.4$. $[Cl^-]=4.4$	58Ga	
	6.6	-0.66(1)	19	25	C	$\mu=5.1(NaClO_4)$	67Ab 58Sd	

Metal Ion	ΔH, kcal /mole	Log K	ΔS, cal /mole °K	T, °C	M	Conditions	Ref.	Remarks
C16, cont.								
Cr^{3+}, cont.	4.3	-1.54(2)	7.5	25	T	μ≃4.4	58Ga	
	5.0	(2)	--	25	C	μ=5.1($NaClO_4$)	58Gd	
$CrA_6{}^{3+}$	7.4	-0.745(43.6°,1)	20.0	30-60	T	0 M % CH_3OH in water at μ = 0.418. A=H_2O	64Bb	b
	7.4	-0.623(43.6°,1)	20.5	30-60	T	0.048M % CH_3OH in water at μ = 0.418. A=H_2O	64Bb	b
	7.4	-0.426(43.6°,1)	21.5	30-60	T	0.128M % CH_3OH in water at μ = 0.418. A=H_2O	64Bb	b
	7.4	-0.267(43.6°,1)	22.1	30-60	T	0.190M % CH_3OH in water at μ = 0.418. A=H_2O	64Bb	b
	7.4	-0.011(43.6°,1)	23.1	30-60	T	0.306M % CH_3OH in water at μ = 0.418. A=H_2O	64Bb	b
	7.4	0.193(43.6°,1)	24.2	30-60	T	0.391M % CH_3OH in water at μ = 0.418. A=H_2O	64Bb	b
	7.4	0.415(43.6°,1)	25.3	30-60	T	0.505M % CH_3OH in water at μ = 0.418. A=H_2O	64Bb	b
	7.4	0.785(43.6°,1)	27.0	30-60	T	0.625M % CH_3OH in water at μ = 0.418. A=H_2O	64Bb	b
	7.4	1.04(43.6°,1)	28.1	30-60	T	0.706M % CH_3OH in water at μ = 0.418. A=H_2O	64Bb	b
	6.0	-0.27(60°)	15.7	40-80.5	T	0.98M($HClO_4$). R: $CrA_6{}^{3+} + L^- = CrA_5L^{2+} + A$. A=$H_2O$	67Ha	b
	5.6	-0.054(60°)	15.7	40-80.5	T	4.00M($HClO_4$). R: $CrA_6{}^{3+} + L^- = CrA_5L^{2+} + A$. A=$H_2O$		
	3.6	0.26(60°)	13.0	40-80.5	T	6.70M($HClO_4$). R: $CrA_6{}^{3+} + L^- = CrA_5L^{2+} + A$. A=$H_2O$	67Ha	b
	3.1	0.54(60°)	13.8	40-80.5	T	10.0M($HClO_4$). R: $CrA_6{}^{3+} + L^- = CrA_5L^{2+} + A$. A=$H_2O$	67Ha	b
$CrAL_2$	-8.20	--	--	--	C	0.1M($CaCl_2$). R: $CrAL_2 + HL = CrL_3 + H_2O$. A=OH	10Ba	
	-9.60	--	--	--	T	0.1M($CaCl_2$). R: $CrAL_2 + HL = CrL_3 + H_2O$. A=OH	10Ba	
CrA_2L	-8.060	--	--	--	T	0.1M($CaCl_2$). R:	10Ba	

Metal Ion	ΔH, kcal /mole	Log K	ΔS, cal /mole °K	T, °C	M	Conditions	Ref.	Re-marks
C16, cont.								
CrA_2L, cont.						$CrA_2L + HL = CrAL_2 + H_2O$. A=OH		
$HCrO_4^-$	1.13	1.05	8.6	15-35	T	μ=1.0. R: $H^+ + L^- + HCrO_4^- = CrO_3L^- + H_2O$	66Ta	b
	-5.20	1.515(25°)	-11.62	26&38	T	μ=2.0M($NaClO_4$) R: $HCrO_4^- + 2H^+ + L = HCrO_3L + H_2O$	71Yc	b
CrA_5B	6	0.030	19	45	T	μ=0.106M. A=NH_3. B=H_2O R: $CrA_5B + L^- = CrA_5L^{2+} + B$	67De	
Fe^{3+}	4.640	(1)	22.1	10	C	μ=0	67Vb	ΔCp= 61
	5.140	(1)	23.6	18	C	μ=0	67Vb	ΔCp= 58
	5.600	1.38(1)	25.1	25	C	μ=0	67Vb	ΔCp= 58
	8.5	1.48(1)	35	25	T	μ=0	42Ra 52Ua	
	6.160	(1)	26.9	35	C	μ=0	67Vb	ΔCp= 56
	6.680	(1)	28.6	45	C	μ=0	67Vb	ΔCp= 54
	7.200	(1)	30.2	55	C	μ=0	67Vb	ΔCp= 53
	5.0	0.64(1)	24	25	T	μ=0.5	61Se	
	4.2	0.46(1)	16.6	25	C	μ=1.0($HClO_4$)	62Wa	
	6.0	0.83(25°,1)	23.9	22-45	-	μ=1.382	59Cb	b
	3.300	(1)	13.5	10	C	μ=2.0M($HClO_4$)	67Vb	ΔCp= 43
	3.670	(1)	14.7	18	C	μ=2.0M($HClO_4$)	67Vb	ΔCp= 41
	3.980	0.53(1)	15.7	25	C	μ=2.0M($HClO_4$)	67Vb	ΔCp= 41
	4.260	(1)	17.0	35	C	μ=2.0M($HClO_4$)	67Vb	ΔCp= 39
	4.690	(1)	18.2	45	C	μ=2.0M($HClO_4$)	67Vb	ΔCp= 38
	5.050	(1)	19.6	55	C	μ=2.0M($HClO_4$)	67Vb	ΔCp= 37
	2.900	(1)	13.1	10	C	μ=3.0M($HClO_4$)	67Vb	ΔCp= 38
	3.190	(1)	14.1	18	C	μ=3.0M($HClO_4$)	67Vb	ΔCp= 37
	3.330	0.76(1)	15.0	25	C	μ=3.0M($HClO_4$)	67Vb	ΔCp= 36
	3.760	(1)	16.1	35	C	μ=3.0M($HClO_4$)	67Vb	ΔCp= 34
	3.8	0.73(25°,1)	16	25-45	T	μ=3.0($HClO_4$)	59Ca	b
	4.240	(1)	17.2	45	C	μ=3.0M($HClO_4$)	67Vb	ΔCp= 33

Metal Ion	ΔH, kcal /mole	Log K	ΔS, cal /mole °K	T, °C	M	Conditions	Ref.	Re-marks
C16, cont.								
Fe^{3+}, cont.	4.480	(1)	18.2	55	C	μ=3.0M($HClO_4$)	67Vb	ΔCp= 32
	2.440	(1)	13.0	10	C	μ=4.0M($HClO_4$)	67Vb	ΔCp= 32
	2.640	(1)	13.9	18	C	μ=4.0M($HClO_4$)	67Vb	ΔCp= 31
	2.890	1.10(1)	14.6	25	C	μ=4.0M($HClO_4$)	67Vb	ΔCp= 30
	3.200	(1)	15.6	35	C	μ=4.0M($HClO_4$)	67Vb	ΔCp= 29
	3.490	(1)	16.6	45	C	μ=4.0M($HClO_4$)	67Nb	ΔCp= 28
	3.810	(1)	17.5	55	C	μ=4.0M($HClO_4$)	67Vb	ΔCp= 27
	1.840	(1)	13.0	10	C	μ=5.0M($HClO_4$)	67Vb	ΔCp= 25
	2.050	(1)	13.7	18	C	μ=5.0M($HClO_4$)	67Vb	ΔCp= 24
	2.140	1.50(1)	14.2	25	C	μ=5.0M($HClO_4$)	67Vb	ΔCp= 24
	2.330	(1)	15.0	35	C	μ=5.0M($HClO_4$)	67Vb	ΔCp= 23
	2.640	(1)	15.8	45	C	μ=5.0M($HClO_4$)	67Vb	ΔCp= 23
	2.870	(1)	16.5	55	C	μ=5.0M($HClO_4$)	67Vb	ΔCp= 22
	-0.220	(1)	--	0	C	μ=7.0M($HClO_4$)	67Vb	q: ±0.03
	-0.002	(1)	11.2	10	C	μ=7.0M($HClO_4$)	67Vb	q: ±0.025 ΔCp= 23
	0.227	(1)	12.0	18	C	μ=7.0M($HClO_4$)	67Vb	q: ±0.03 ΔCp= 22
	0.398	2.46(1)	12.6	25	C	μ=7.0M($HClO_4$)	67Vb	ΔCp= 22
	0.553	(1)	13.1	35	C	μ=7.0M($HClO_4$)	67Vb	ΔCp= 21
	0.788	(1)	13.9	45	C	μ=7.0M($HClO_4$)	67Vb	ΔCp= 20
	0.986	(1)	14.5	55	C	μ=7.0M($HClO_4$)	67Vb	ΔCp= 20
	-0.769	3.13(1)	11.7	25	C	μ=8.0M($HClO_4$)	67Vb	q: ±0.08
	-2.3	3.44(25°,1)	8.2	25-45	T	μ=8.5($HClO_4$)	59Ca	b
	4.2	0.94(1)	18.4	25	T	μ=0.5. S: D_2O	61Se	
Co^{2+}	2.9	-0.770(1)	6.1	27	T	μ=0.5-16M(HCl) R: $Co(H_2O)_6^{2+} + L^- = CoL(H_2O)_5^+ + H_2O$	68Za	q: ±1.0
	0.52	0.137(1)	1.11	25	C	μ=2.0	66Kb	
	1.0	-0.176(25°,1)	2.5	12-40	T	μ=2.0	60Lb	b

Metal Ion	ΔH, kcal /mole	Log K	ΔS, cal /mole °K	T, °C	M	Conditions	Ref.	Remarks
C16, cont.								
Co^{2+}, cont.	0.50	0.119(1)	1.05	40	C	μ=2.0	66Kb	
	2.1	-2.770(2)	-5.6	27	T	μ=0.5-16M(HCl) R: $CoL(H_2O)_5^+ + L^- = CoL_2(H_2O)_2 + 3H_2O$	68Za	q: ±1.0
	11.3	-2.507(3)	26.2	27	T	μ=0.5-16M(HCl) R: $CoL_2(H_2O)_2 + L^- = CoL_3(H_2O)^- + H_2O$	68Za	q: ±1.3
	0.80	-2.056(4)	-6.7	27	T	μ=0.5-16M(HCl) R: $CoL_3(H_2O)^- + L^- = CoL_4^{2-} + H_2O$	68Za	q: ±0.4
CoA_6^{3+}	4.32	2.59(1)	26	25	T	μ=0. A=NH_3	55Na 53Ea	
	0.6	(1)	--	25	C	0-0.4M(NaCl). A=NH_3	62Fa 53Ya	
	5.8	-1.185(1)	11	25	T	0-0.9M(NaCl). A=NH_3	62Fa	
	5.6	-0.29(1)	18	25	T	1M($NaClO_4$ & NaCl) A=NH_3	62Me	
cis-$CoAB_2S$	0.93	4.179	22.1	30	T	μ=0. S: N,N-Dimethylformamide R: cis-$CoAB_2S$ + L = $CoALB_2^+$ + S. A=Cl. B=Ethylenediamine. S=Solvent	66La	
	-3.8	6.083	15.4	30	T	μ=0. S: N,N-Dimethylformamide R: cis-$CoAB_2S$ + 2L = $CoAL_2B_2$ + S. A=Cl B=Ethylenediamine S=Solvent	66La	
cis-$CoA_2B_2^+$	-2.3	3.712(25°,1)	9.4	20-30	T	μ=0. S: N,N-Dimethylformamide A=Ethylenediamine B=Cl	66Md	b
CoA_4L_2	13.57	--	--	25-63	T	μ=0.354M(LiCl) S: CH_3OH. A=CH_3OH R: $CoL_2A_4 + L^- = CoL_3A^- + 3A$	67Sa	b,aa
	11.70	--	--	25-69	T	μ=8.47M(LiCl) A=H_2O R: $CoL_2A_4 + L^- = CoL_3A^- + 3A$	67Sa	b,aa
CoA_5B	0.4	0.176(25°)	1.6	25-86	T	$C_M \approx 2x10^{-3}M$ A=NH_3. B=H_2O R: CoA_5B^{3+}---L^- = CoA_5L^{2+}---H_2O (ligand interchange between inner and outer coordination spheres)	67Lb	b

Metal Ion	ΔH, kcal /mole	Log K	ΔS, cal /mole °K	T, °C	M	Conditions	Ref.	Re-marks
C16, cont.								
Ni^{2+}	0.50	-0.174(1)	0.87	25	C	μ=2.0	66Kb	
	1.25	0.569(25°,1)	6.8	12-40	T	μ=2.0	60Ld	b
	0.44	-0.155(1)	0.61	40	C	μ=2.0	66Kb	
	2.5-2.6	≈-0.48(4)	--	19-70	T	3.6M(Ni^{2+})	56Ka	b,r
NiA_2L_2	1.640	-0.394(25°)	3.70	-18-52	T	S: Chloroform A=[(p-C_6H_5)(p-C_6H_5)MeP]	70Pc	b,t
	2.140	-0.626(25°)	4.31	-18-52	T	S: Chloroform A=[p-C_6H_5)(p-ClC_6H_4)MeP]	70Pc	b,t
	1.830	-0.356(25°)	4.49	-13-54	T	S: Chloroform A=[(p-C_6H_5)(p-MeC_6H_4)MeP]	70Pc	b,t
	1.730	-0.260(25°)	4.63	0-70	T	S: Chloroform A=[(p-C_6H_5)(p-$MeOC_6H_4$)MeP]	70Pc	b,t
	1.410	-0.0088(25°)	4.68	-5-59	T	S: Chloroform A=[(p-C_6H_5)(p-$Me_2NC_6H_4$)MeP]	70Pc	b,t
	1.820	-0.274(25°)	4.86	-7-50	T	S: Chloroform A=[(p-MeC_6H_4)(p-MeC_6H_4)MeP]	70Pc	b,t
	1.380	-0.194(25°)	3.76	-8-59	T	S: Chloroform A=[(p-$MeOC_6H_4$)(p-$MeOC_6H_4$)MeP]	70Pc	b,t
NiA_4L_2	14.64	--	--	40-85	T	μ=2.4M(LiBr) S: C_2H_5OH A=C_2H_5OH R: $NiL_2A_4 + L^- = NiL_3A^- + 3A$	67Sa	b,aa
	13.3	1.176(100°)	41	63-160	T	S: Dimethyl Sulphoxide A=Dimethyl Sulphoxide R: $NiA_4L_2 + L^- = NiAL_3^- + 3A$	70Gr	b,aa
RuA_2L^+	-2.2	4.137(25°,1)	11.7	25-49	T	S: N,N-Dimethyl-acetamide. μ=0 A=Triphenylphosphine	74Ja	b
RhA_2BC	0.6	2.699	14	50	T	μ=0.2M A=Ethylenediamine B=Cl . C=H_2O R: trans-RhA_2BC^{2+} + L^- = trans-RhA_2BL^+ + C	67Bb	q: ±0.4
	-1.2	3.222	11	50	T	μ=0.2M A=Ethylenediamine B=Br . C=H_2O R: trans-RhA_2BC^{2+} + L^- = trans-RhA_2BL^+ + C	67Bb	q: ±0.7
	-1.6	3.000	8	50	T	μ=0.2M A=Ethylenediamine B=I . C=H_2O R: trans-RhA_2BC^{2+} + L^- = trans-RhA_2BL^+ + C	67Bb	q: ±0.5

Metal Ion	ΔH, kcal /mole	Log K	ΔS, cal /mole °K	T, °C	M	Conditions	Ref.	Remarks
Cl6, cont.								
RuA_5B^{3+}	1.7	2.272(64°)	15	35-90	T	μ=0.1 + 0.1M p-toluenesulfonic acid. K values good to ± 30%. R: RuA_5B + L = RuA_5L + B. $A=NH_3$. $B=H_2O$	64Be	b
	3.5	--	27	77-97.5	T	μ=0.01-0.07M $A=NH_3$. $B=H_2O$ R: RhA_5B^{3+} + L^- = RhA_5L^{2+} + B	68Lb	b
	1.3	--	--	--	-	$A=NH_3$. $B=H_2O$ R: RhA_5B^{3+} + L^- = RhA_5L^{2+} + B	67Pa	u,q: ±0.6
Pd^{2+}	-3.03	4.47(1)	10.3	25	C	μ=1M($HClO_4$)	72Re	
	-2.59	3.29(2)	6.4	25	C	μ=1M($HClO_4$)	72Re	
	-2.56	2.41(3)	2.4	25	C	μ=1M($HClO_4$)	72Re	
	-3.41	1.37(4)	-5.2	25	C	μ=1M($HClO_4$)	72Re	
	-2.8	1.42(25°,4)	-2.9	10-45	T	μ=1.0	66Sb 57De	b
	-2.8	1.59(25°,4)	-2.1	15-40	T	μ=2.0($LiClO_4$) $C_M=1x10^{-4}M$	68Le	b
	-5.5	(1-4)	--	25	C	μ=0.10	67Ic	
PdA_2B_2	2.8	-0.28(32°)	7.9	--	T	S: Acetone $A=PhP(CH_3)_2$ B=Cl	73Ra	q:±0.5 u:32, bb
	3.1	-0.26(32°)	9.0	--	T	S: Deuterio-chloroform $A=PhP(CH_3)_2$ B=Cl	73Ra	q:±0.5 u:32, bb
	4.6	1.12(32°)	20.2	--	T	S: Deuterio-chloroform $A=Ph_2PCH_3$ B=Cl	73Ra	q:±0.5 u:32, bb
	4.4	0.36(32°)	15.1	--	T	S: m-Dichloro-benzene $A=PhP(CH_3)_2$ B=Cl	73Ra	q:±0.5 u:32, bb
	4.6	-0.24(32°)	14.0	--	T	S: o-Dichloro-benzene $A=PhP(CH_3)_2$ B=Cl	73Ra	q:±0.5 u:32, bb
	7.7	-1.12(32°)	20.2	--	T	S: Nitrobenzene $A=PhP(CH_3)_2$ B=Cl	73Ra	q:±0.5 u:32, bb
	8.1	-0.19(32°)	25.7	--	T	S: Nitrobenzene $A=Ph_2PCH_3$ B=Cl	73Ra	q:±0.5 u:32, bb
	4.4	-0.99(32°)	16.2	--	T	S: Nitromethane $A=PhP(CH_3)_2$ B=Cl	73Ra	q:±0.5 u:32, bb
	5.7	-0.83(32°)	14.9	--	T	S: Nitromethane $A=Ph_2PCH_3$ B=Cl	73Ra	q:±0.5 u:32, bb
	5.8	-0.81(32°)	15.3	--	T	S: sym-Tetrachloro-ethane $A=PhP(CH_3)_2$ B=Cl	73Ra	q:±0.5 u:32 bb

Metal Ion	ΔH, kcal /mole	Log K	ΔS, cal /mole °K	T, °C	M	Conditions	Ref.	Remarks
C16, cont.								
PdA_2B_2, cont.	6.4	-0.46(32°)	18.9	--	T	S: Tetrachloroethane A=Ph_2PCH_3 B=Cl	73Ra	q:±0.5 u:32 bb
PtA_2^{2+}	≈0	2.52(23°,1)	11.5	23-54	T	μ=0. S: CH_3OH. A=o-Phenylenebisdimethylarsine	66Db	b
PtA_3^{4+}	3.9	-0.21(1)	12	25	C	3.0M($NaClO_4$ + NaCl) A=Ethylenediamine	72Mf	q:±0.5
	1.6	1.24(25°,1)	11.0	10-40	T	μ→0. A=Ethylenediamine	63Gb	b
PtA_2L_2	1.975	1.470(25°)	13.3	20-35	T	S: Benzene A=PPr_3	56Cb	b,bb
	1.180	2.243(25°)	14.2	15-70	T	S: Benzene A=$AsEt_3$	52Cb	b,bb
	2.410	0.281(25°)	9.4	15-35	T	S: Benzene A=$SbEt_3$	52Cb	b,bb
	2.200	0.600(25°)	10.1	20-35	T	S: Benzene A=$SbPr_3$	56Cb	b,bb
Cu^+	-4.1	5.845(50°,1-2)	14.1	25-80	T	μ=10M(NH_4NO_3)	69Si	b,q: ±0.5
	-8.7	5.447(50°,1-3)	-1.9	25-80	T	μ=10M(NH_4NO_3)	69Si	b,q: ±0.5
	-11.6	4.863(50°,1-3)	-13.2	25-80	T	μ=10M(NH_4NO_3)	69Si	b
	-13.3	12.079(50°,2-3)	13.5	25-80	T	μ=10M(NH_4NO_3)	69Si	b,q: ±1.0
	-16.0	11.845(50°,2-4)	3.9	25-80	T	μ=10M(NH_4NO_3)	69Si	b
	-17.6	11.602(50°,2-5)	-1.8	25-80	T	μ=10M(NH_4NO_3)	69Si	b
	-16.9	11.301(50°,2-6)	-1.5	25-80	T	μ=10M(NH_4NO_3)	69Si	b
	-21.4	18.176(50°,3-4)	15.9	25-80	T	μ=10M(NH_4NO_3)	69Si	b,q: ±1.5
	-22.4	18.041(50°,3-5)	12.6	25-80	T	μ=10M(NH_4NO_3)	69Si	b,q: ±1.5
	-26.7	17.699(50°,3-6)	-2.7	25-80	T	μ=10M(NH_4NO_3)	69Si	b
	-33.4	24.477(50°,4-5)	7	25-80	T	μ=10M(NH_4NO_3)	69Si	b,q: ±2.5
	-35.2	24.255(50°,4-6)	0.6	25-80	T	μ=10M(NH_4NO_3)	69Si	b,q: ±2.5
Cu^{2+}	0.60	0.114(25°,1)	2.5	25.2& 46.9	T	μ=1.0($NaClO_4$)	50Ma 60Ld 52Ua	a,b
	1.58	0.086(1)	5.67	25	C	μ=2.0	66Kb	
	1.98	0.149(1)	6.99	40	C	μ=2.0	66Kb	
	3.1	-0.346(1)	8	25	C	3.0N($LiClO_4$ + LiCl) Values are for formation of inner sphere complexes	70Bf	
	2.4	-0.346(1)	6.5	25	C	Values are for formation of outer sphere complexes	70Bf	
	3.68	0.512(3)	14.7	25	T	C_L→0	65Ta	
CuA_2^{2+}	2.88	2.13(1)	19.4	25	C	S: Methanol A=Ethylenediamine	72Bf	
	2.19	2.65(1)	19.5	25	C	S: Methanol A=N-methylethylenediamine	72Bf	
	2.08	3.01(1)	20.7	25	C	S: Methanol	72Bf	

Metal Ion	ΔH, kcal /mole	Log K	ΔS, cal /mole °K	T, °C	M	Conditions	Ref.	Re- marks
Cl6, cont.								
CuA_2^{2+}, cont.						A=N,N'-dimethyl-ethylenediamine		
	1.63	3.04(1)	19.4	25	C	S: Methanol	72Bf	
						A=N,N-dimethyl-ethylenediamine		
Ag^+	-2.7	3.3(25°,1)	6	15-35	T	μ=0	67Ab	b
	-7.96	1.70(370°,1)	-4.6	370& 436	T	μ=0. S: KNO_3	61Dc	a,b
	-4.40	1.548(333°,1)	-0.18	333& 374	T	μ=0. S: $NaNO_3$-KNO_3 Eutectic	61Dc	a,b
	-4.92	1.23(370°,2)	-2.0	370& 436	T	μ=0. S: KNO_3	61Dc	a,b
	-7.73	1.127(333°,2)	-7.6	333& 374	T	μ=0. S: $NaNO_3$-KNO_3 Eutectic	61Dc	a,b
	-3.9	5.3(25°,1-2)	11	15-35	T	μ=0	52Jc	b
	-14	5.90(25°,1-4)	-19.9	0&25	T	Sat. Sol. AgCl	41Ea	a,b
Zn^{2+}	3.1	-0.139(25°,1)	9.8	15-65	T	μ=1M($LiClO_4$)	70Fb	b
	1.55	-0.176(25°,1)	4.3	15-65	T	μ=2M($LiClO_4$)	70Fb	b
	1.32	-0.19(1)	3.6	25	C	μ=3M($NaClO_4$)	69Gb	
	3.57	0.00(25°,1)	11.9	15-65	T	μ=3M($LiClO_4$)	70Fb	b
	4.09	0.147(25,1)	14.2	15-65	T	μ=4M($LiClO_4$)	70Fb	b,q: ±0.48
	0	-0.322(1)	-1.5	25	C	μ=4.5	56Sa	n
	4.30	1.66(70°,1)	20.1	55-85	T	S: $NH_4NO_3\cdot 2H_2O$ melt	74Nb	b
	2.5	-0.132(25°,2)	7.7	15-65	T	μ=1M($LiClO_4$)	70Fb	b
	1.77	0.081(25°,2)	6.0	15-65	T	μ=2M($LiClO_4$)	70Fb	b,q: ±0.21
	9	-0.42(2)	27	25	C	μ=3M($NaClO_4$)	69Gb	q:±2
	2.41	-0.117(25°,2)	7.9	15-65	T	μ=3M($LiClO_4$)	70Fb	b,q: ±0.88
	2.83	0.418(25°,2)	12.4	15-65	T	μ=4M($LiClO_4$)	70Fb	b,q: ±0.34
	12.58	-0.601(25°,3)	40.7	15-65	T	μ=2M($LiClO_4$)	70Fb	b,q: ±1.9
	0	0.75(3)	2	25	C	μ=3M($NaClO_4$)	69Gb	q:±2
	4.12	0.476(25°,3)	15.2	15-65	T	μ=3M($LiClO_4$)	70Fb	b,q: ±1.0
	-0.82	0.212(25°,3)	-2.6	15-65	T	μ=4M($LiClO_4$)	70Fb	b,q: ±1.7
	5.100	-0.044(1-2)	16.8	25	C	μ=4.5	56Sa	n
	19.200	-0.300(1-3)	63	25	C	μ=4.5	56Sa	n
	0.120	-0.154(1-4)	-0.35	25	C	μ=4.5	56Sa	n
	7.0	(1-4)	--	--	C	S: 25 w % NH_4Cl	57Ta	
Cd^{2+}	0.160	1.92(1)	9.4	25	C	μ=0	72Ff	q: ±0.03
	0.8	(1)	--	25	C	μ=0	70Pd	q:±0.5
	0.60	1.98(1)	11.2	25	T	μ=0.0	53Va 67Ab	
	0.800	1.971(25°,1)	11.7	5-35	T	μ=0	69Sa	b
	0.21	1.44(1)	7.3	25	C	μ=0.25($NaClO_4$)	68Gc 49Ka	
	0.275	1.33(1)	7.0	25	C	μ=0.5M($LiClO_4$ & LiCl)	72Ff	q: ±0.02

Metal Ion	ΔH, kcal /mole	Log K	ΔS, cal /mole °K	T, °C	M	Conditions	Ref.	Remarks
Cl6, cont.								
Cd^{2+}, cont.	0.25	1.37(1)	7.1	25	C	μ=0.50($NaClO_4$)	68Gc	
	0.29	1.37(1)	7.2	25	T	μ=0.5($NaClO_4$)	52Ua 53Va	
	0.137	1.37(1)	6.7	25	C	μ=1M($LiClO_4$ & LiCl)	62Bc 72Ff	q: ±0.015
	0.1	(1)	--	25	C	μ=1M	70Pd	q:±0.5
	0.13	1.35(1)	6.6	25	C	μ=1.0($NaClO_4$)	68Gc	
	0.18	1.37(1)	6.9	25	T	μ=1.0($NaClO_4$)	59Aa 53Va	
	-0.109	1.46(1)	6.3	25	C	μ=2M($LiClO_4$ & LiCl)	72Ff	q: ±0.015
	0.00	1.42(1)	6.5	25	C	μ=2.0($NaClO_4$)	68Gc	
	0	1.43(1)	6.5	25	T	μ=2.0($NaClO_4$)	53Va	
	-0.340	1.51(1)	5.8	25	C	μ=3M($LiClO_4$ & LiCl)	72Ff	q: ±0.025
	-0.10	1.58(1)	6.9	25	C	μ=3.0($NaClO_4$)	68Gc	
	0.10	1.586(1)	6.9	25	C	μ=3.0($NaClO_4$)	66Ga	
	-0.045	1.54(1)	6.9	25	T	μ=3.0($NaClO_4$)	53Va	
	-0.680	1.77(1)	5.8	25	C	μ=4M($LiClO_4$ & LiCl)	72Ff	
	0	1.32(1)	6.1	25	T	μ=4.5	57Sc	
	-3.7	2.53(40°,1)	-0.2	40-70	T	S: Melt of NH_4NO_3	72Nb	b
	0.940	0.61(2)	5.9	25	C	μ=0	72Ff	q: ±0.09
	-0.6	(2)	--	25	C	μ=0	70Pd	q:±1.0
	1.450	0.538(25°,2)	7.3	5-35	T	μ=0	69Sa	b
	0.67	0.42(2)	4.2	25	C	μ=0.5($NaClO_4$)	68Gc	
	0.755	0.44(2)	4.5	25	C	μ=0.5M($LiClO_4$ & LiCl)	72Ff	q: ±0.08
	0.49	0.43(2)	3.6	25	C	μ=1($NaClO_4$)	68Gc	
	0.707	0.23(2)	3.4	25	C	μ=1M($LiClO_4$ & LiCl)	72Ff	
	0.6	(2)	--	25	C	μ=1M	70Pd	q:±0.5
	0.34	0.53(2)	3.6	25	C	μ=2($NaClO_4$)	68Gc	
	0.323	0.49(2)	3.3	25	C	μ=2M($LiClO_4$ & LiCl)	72Ff	q: ±0.03
	0.02	0.64(2)	3.0	25	C	μ=3($NaClO_4$)	68Gc	
	-0.02	1.644(2)	3.0	25	C	μ=3.0($NaClO_4$)	66Ga	
	-0.340	0.82(2)	2.6	25	C	μ=3M($LiClO_4$ & LiCl)	72Ff	q: ±0.05
	-0.640	0.70(2)	1.1	25	C	μ=4M($LiClO_4$ & LiCl)	72Ff	
	-0.300	0.902(2)	3.1	25	T	μ=4.5	57Sc	
	4.3	(3)	--	25	C	μ=0	70Pd	q:±1.0
	-0.920	-0.20(3)	-4.3	25	C	μ=0($LiClO_4$ & LiCl)	72Ff	q:±0.1
	1.8	-0.37(3)	4.3	25	C	μ=1($NaClO_4$)	68Gc	
	0.156	0.10(3)	1.0	25	C	μ=1($LiClO_4$ & LiCl)	72Ff	
	2.9	(3)	--	25	C	μ=1M	70Pd	q:±1.0
	1.93	-0.16(3)	5.7	25	C	μ=2($NaClO_4$)	68Gc	
	1.046	0.22(3)	4.5	25	C	μ=2($LiClO_4$ & LiCl)	72Ff	q: ±0.15
	1.85	0.18(3)	7.0	25	C	μ=3($NaClO_4$)	68Gc	
	-1.85	0.185(3)	7.0	25	C	μ=3.0($NaClO_4$)	66Ga	
	2.440	0.30(3)	9.5	25	C	μ=3($LiClO_4$ + LiCl)	72Ff	
	1.500	0.64(3)	7.9	25	C	μ=4($LiClO_4$ & LiCl)	72Ff	

Metal Ion	ΔH, kcal /mole	Log K	ΔS, cal /mole °K	T, °C	M	Conditions	Ref.	Remarks
Cl6, cont.								
Cd^{2+}, cont.	2.60	0.088(3)	9.2	25	T	μ=4.5	57Sc	
	5.920	-0.55(4)	17.3	25	C	μ=4.0($LiClO_4$ & LiCl)	72Ff	
	2.90	0.454(4)	7.7	25	T	μ=4.5	56Sc	
	1.20	2.70(1-2)	16.4	25	T	μ=0.0	53Va	
	0.78	1.82(1-2)	10.9	25	T	μ=0.5($NaClO_4$)	53Va	
	0.75	1.79(1-2)	10.7	25	T	μ=1.0($NaClO_4$)	53Va	
	0.70	1.96(1-2)	11.3	25	T	μ=2.0($NaClO_4$)	53Va	
	0.65	2.20(1-2)	12.2	25	T	μ=3.0($NaClO_4$)	53Va	
	3.85	2.11(1-3)	22.3	25	T	μ=0.0	53Va	
	3.30	1.49(1-3)	17.9	25	T	μ=1.0($NaClO_4$)	53Va	
	3.02	1.82(1-3)	18.4	25	T	μ=2.0($NaClO_4$)	53Va	
	2.84	2.30(1-3)	20.1	25	T	μ=3.0($NaClO_4$)	53Va	
Hg^{2+}	-4.8	7.32(1)	17	25	C	$\mu \to 0$	61Ma 67Ab	
	-6.75	7.2(1)	9.0	8	C	μ=0.50	64Cc	
	-5.5	6.7(1)	12.4	25	C	μ=0.50	64Cc	
	-5.9	6.74(1)	11.1	25	C	μ=0.50	60Ga	
	-5.6	6.6(1)	12.2	40	C	μ=0.50	64Cc	
	-5.79	7.07(1)	12.9	25	C	3M($NaClO_4$)	65Ae	
	-7.19	5.16(70°,1)	2.5	55-85	T	S: $NH_4NO_3 \cdot 2H_2O$ melt	74Nb	b
	-7.3	6.7(2)	4.7	8	C	μ=0.50	64Cc	
	-7.25	6.5(2)	5.3	25	C	μ=0.50	64Cc	
	-6.9	6.48(2)	6.3	25	C	μ=0.50	60Ga	
	-6.9	6.2(2)	6.2	40	C	μ=0.50	64Cc	
	-6.49	6.91(2)	9.9	25	C	3M($NaClO_4$)	65Ae	
	-2.2	0.95(3)	-3.0	25	C	μ=0.5	60Ga	
	-1.03	0.75(1.08)(3)	0(1.5)	25	C	3M($NaClO_4$). Log K and ΔS values in parentheses determined calorimetrically	65Ae	
	0.1	1.049(4)	5.1	25	C	μ=0.5	60Ga	
	-1.48	1.38(1.09)(4)	1.3(0)	25	C	3M($NaClO_4$). Log K and ΔS values in parentheses determined calorimetrically	65Ae	
	-14.05	13.9(1-2)	13.7	8	C	μ=0.50	64Cc	
	-12.75	13.2(1-2)	17.7	25	C	μ=0.50	64Cc	
	-12.50	12.7(1-2)	18.4	40	C	μ=0.50	64Cc	
	-12.28	(1-2)	--	25	C	3M($NaClO_4$)	65Ae	
	-14.9	15.22(1-4)	19.5	25	C	μ=0.50	60Ga	
	-14.79	(1-4)	--	25	C	3M($NaClO_4$)	65Ae	
HgA	-6.0	5.219(1)	3.6	20	C	μ=0.1(KNO_3) A=CH_3	65Sa	
	-0.5	0.60(25°)	1.1	5-60	T	$\mu \approx 10^{-3}$. A=Br R: $1/2HgL_2 + 1/2HgA_2$ = HgLA	64Ea	b,r
	-0.7	0.88(25°)	1.7	5-60	T	$\mu \approx 10^{-3}$. A=I R: $1/2HgL_2 + 1/2HgA_2$ = HgLA	64Ea	b,r
AlL_4^-	4.0	-4.0(250°)	-7.0	175-275	T	S: $AlCl_3$-LiCl melt R: $2AlL_4^- = L^- + Al_2L_7^-$	72Tc	b

Metal Ion	ΔH, kcal /mole	Log K	ΔS, cal /mole °K	T, °C	M	Conditions	Ref.	Remarks
Cl6, cont.								
AlL_4^-, cont.	13.6	-5.7(300°)	-2.4	175-450	T	S: $AlCl_3$-NaCl melt R: $2AlL_4^- = L^- + Al_2L_7^-$	72Tc	b
	22.9	-7.1(300°)	7.5	275-450	T	S: $AlCl_3$-KCl melt R: $2AlL_4^- = L^- + Al_2L_7^-$	72Tc	b
In^{3+}	1.23	2.10(1)	13.7	25	C	μ=2.00M($NaClO_4$)	69Rd	
	0.78	1.50(2)	9.5	25	C	μ=2.00M($NaClO_4$)	69Rd	
	8	-0.24(3)	26	25	C	μ=2.00M($NaClO_4$)	69Rd	
Tl^+	-1.4	0.68(1)	-1.6	25	C	μ=0	69Bk 34La	
	-1.43	0.68(1)	-1.7	25	T	μ=0	53Ba 63Kb	
	~0	0.67(25°,1)	3	0-45	T	μ=0	67Ab	
	-1.52	0.380(25°,1)	-1.0	10-60	T	μ=0	72Fd	b,q: ±0.10
	-0.15	0.107(25°,1)	-0.44	10-60	T	μ=0.5($LiClO_4$)	72Fd	b,q: ±0.01
	0.02	-0.013(25°,1)	4.0	10-60	T	μ=1.0($LiClO_4$)	72Fd	b,q: ±0.11
	0.10	-0.076(25°,1)	-0.5	10-60	T	μ=2.0($LiClO_4$)	72Fd	b,q: ±0.04
	0.27	-0.201(25°,1)	5.9	10-60	T	μ=3.0($LiClO_4$)	72Fd	b,q: ±0.03
	0.12	-0.092(25°,1)	4.6	10-60	T	μ=4.0($LiClO_4$)	72Fd	b,q: ±0.06
	-1.1	0.00(25°,1)	-4	15-80	T	4M(LiCl + $LiClO_4$)	67Ab	
	2.3	-0.81(25°,1-2)	4	15-80	T	4M(LiCl + $LiClO_4$)	63Kb	
Tl^{3+}	-6.04	7.49(1)	13.9	25	C	3M($HClO_4$)+1M($NaClO_4$)	64Lf	
	-5.45	7.724(1)	14.5	25	C	μ=3.0	64Wa	
	-7.8	7.1(1)	7	25	C	3M(HCl,$HClO_4$, LiCl,$LiClO_4$)	65Kd	
	-4.05	5.78(2)	12.9	25	C	3M($HClO_4$)+1M($NaClO_4$)	64Lf	
	-4.40	5.763(2)	10.1	25	C	μ=3.0	64Wa	
	-3.7	3.4(2)	12	25	C	3M(HCl,$HClO_4$, LiCl,$LiClO_4$)	65Kd	
	-1.08	3.39(3)	11.9	25	C	3M($HClO_4$)+1M($NaClO_4$)	64Lf	
	-1.1	3.000(3)	12.6	25	C	μ=3.0	64Wa	
	0	3.4(3)	16	25	C	3M(HCl,$HClO_4$, LiCl,$LiClO_4$)	65Kd	
	-5.1	1.470(25°,4)	-10	0&25	T	μ=0	64Nb	a,b
	-0.17	2.78(4)	12.2	25	C	3M($HClO_4$+1M($NaClO_4$)	64Lf	
	-0.3	1.815(4)	8.9	25	C	μ=3.0	64Wa	
	0	3.4(4)	6	25	C	3M(HCl,$HClO_4$, LiCl,$LiClO_4$)	65Kd	
	-11.5	12.5(1-2)	19	25	C	3M(HCl,$HClO_4$, LiCl,$LiClO_4$)	65Kd	
	-11.5	16.0(1-3)	35	25	C	3M(HCl,$HClO_4$, LiCl,$LiClO_4$)	65Kd	
	-11.5	18.5(1-4)	41	25	C	3M(HCl,$HClO_4$, LiCl,$LiClO_4$)	65Kd	
Sn^{2+}	2.600	1.15(1)	14.0	25	T	3.0M($NaClO_4$)	52Vb 67Ab	a
	3.200	1.171(1-2)	18.5	25	T	3.0M($NaClO_4$)	52Vb	a

Metal Ion	ΔH, kcal /mole	Log K	ΔS, cal /mole °K	T, °C	M	Conditions	Ref.	Remarks
C16, cont.								
Sn^{2+}, cont.	5.600	1.69(1-3)	26.5	25	T	3.0M($NaClO_4$)	52Vb	a
Pb^{2+}	1.250	(1)	--	25	C	μ=0	59Aa	
	4.38	1.60(1)	22.0	25	T	μ=0	55Na	a
	0.983	(1)	--	25	C	μ=0.15	59Aa	
	0.86	1.16(1)	8.2	25	T	μ=3.0($LiClO_4$,LiCl)	64Mb	
	-0.86	1.158(25°,1)	8	18-65	T	μ=3M($LiClO_4$)	68Fa	b
	-2.5	(1)	--	40-70	T	S: Melt of NH_4NO_3	72Nb	b
	1.9	1.81(2)	15	25	T	μ=3.0($LiClO_4$,LiCl)	64Mb	
	-1.04	0.652(25°,2)	7	18-65	T	μ=3M($LiClO_4$)	68Fa	b
	2.6	1.91(3)	18	25	T	μ=3.0($LiClO_4$,LiCl)	64Mb	
	-0.7	0.103(25°,3)	3	18-65	T	μ=3M($LiClO_4$)	68Fa	b
	0.6	1.20(4)	7.5	25	T	μ=3.0($LiClO_4$,LiCl)	64Mb	
	2.0	-0.711(25°,4)	-10	18-65	T	μ=3M($LiClO_4$)	68Fa	b
Bi^{3+}	4.000	3.42(1)	32.4	25	C	μ=0	67Va	
	0	2.2(25°,1)	10	25-65	T	μ=3(Li,H,ClO_4,Cl)	63Me	b
	-0.633	2.77(1)	10.3	-7.4	C	μ=4.0 ΔCp=35.6	67Va	
	-2.268	2.75(1)	11.6	0	C	μ=4.0 ΔCp=35.1	67Va	
	-0.140	2.82(1)	12.3	5	C	μ=4.0 ΔCp=34.8	67Va	
	0.115	2.71(1)	12.9	10	C	μ=4.0 ΔCp=34.5	67Va	
	0.342	2.78(1)	13.8	18	C	μ=4.0 ΔCp=34.0	67Va	
	0.521	2.78(1)	14.6	25	C	μ=4.0 ΔCp=33.5	67Va	
	1.032	2.79(1)	16.0	40	C	μ=4.0 ΔCp=32.6 ΔCp=52.61-0.0640T	67Va	
	-0.738	2.94(1)	12.6	25	C	μ=5.0	67Va	
	-0.261	3.12(1)	9.7	25	C	μ=6.0	67Va	
	4.5	3.5(25°,2)	31	25-65	T	μ=3(Li,H,ClO_4,Cl)	63Me	b
	5.7	5.8(25°,3)	46	25-65	T	μ=3(Li,H,ClO_4,Cl)	63Me	b
	4.5	6.7(25°,4)	46	25-65	T	μ=3(Li,H,ClO_4,Cl)	63Me	b
	4.0	7.29(25°,5)	47	25-65	T	μ=3(Li,H,ClO_4,Cl)	63Me	b
	0	7.33(25°,6)	34	25-65	T	μ=3(Li,H,ClO_4,Cl)	63Me	b
	-7.880	8.29(1-5)	11.0	15	C	μ=4.0($HClO_4$)	72Vb	Cp= 84
	-7.150	8.19(1-5)	13.5	25	C	μ=4.0($HClO_4$)	72Vb	ΔCp= 81
	-6.320	8.03(1-5)	16.2	35	C	μ=4.0($HClO_4$)	72Vb	ΔCp= 79
	-5.500	7.91(1-5)	18.9	45	C	μ=4.0($HClO_4$)	72Vb	ΔCp= 76
TeA_4	-21.3	--	--	25	C	S: Benzene A=Cl R: TeA_4 + GaL_3 = $TeA_4 \cdot GaL_3$	73Gb	
C17 CHROMATE ION			O_4Cr			$CrO_4^{=}$		L^{2-}
Th^{4+}	3.67	0.67(25°)	15.4	10-25	T	μ=0.20M($HClO_4$) R: Th^{4+} + HL^- = ThL^{2+} + H^+	72Bn	b
	3.91	0.28(25°)	14.4	10-25	T	μ=1.00M($HClO_4$) R: Th^{4+} + HL^- = ThL^{2+} + H^+	72Bn	b,q: ±0.4
Np^{4+}	1.17	1.81(25°)	12.2	10-25	T	μ=0.20M($HClO_4$)	72Bn	b

Metal Ion	ΔH, kcal /mole	Log K	ΔS, cal /mole °K	T, °C	M	Conditions	Ref.	Remarks
C17, cont.								
Np^{4+}, cont.						R: $Np^{4+} + HL^- = NpL^{2+} + H^+$		
Fe^{3+}	2.88	0.45(25°)	11.7	1-25	T	μ=0.084($HClO_4$) R: $Fe^{3+} + HL^- = FeL^+ + H^+$	72Bn	b
	3.09	0.29	11.7	1-25	T	μ=0.20M($HClO_4$) R: $Fe^{3+} + HL^- = FeL^+ + H^+$	72Bn	b
	2.09	0.11(25°)	7.5	1-25	T	μ=1.00M($HClO_4$) R: $Fe^{3+} + HL^- = FeL^+ + H^+$	72Bn	b
C18 CHROMIUM(III) ION, hexaaquo-			$O_6H_{12}Cr$			$Cr(H_2O)_6{}^{3+}$		L^{3+}
Np^{5+}	-3.3	0.418(25°,<u>1</u>)	-9.0	25-50	T	μ=8.0	64Sf	b
C19 CITRIC ACID			$C_6H_8O_7$			$HO_2CCH_2C(OH)(CO_2H)CH_2CO_2H$		L^{3-}
Ca^{2+}	-2.480	3.10(28.2°,<u>1</u>)	5.9	19.6-65.0	T	μ=0.2M(NH_4Cl)	71Rd	b
Sr^{2+}	-5.850	2.72(20°,<u>1</u>)	-7.6	10-25	T	μ=0.2M(NH_4Cl)	71Rd	b
	0.934	2.59(40°,<u>1</u>)	14.8	30-55	T	μ=0.2M(NH_4Cl)	71Rd	b
Ba^{2+}	-5.060	2.30(20°,<u>1</u>)	-6.9	10-25	T	μ=0.2M(NH_4Cl)	71Rd	b
	4.280	2.27(40°,<u>1</u>)	24.0	30-55	T	μ=0.2M(NH_4Cl)	71Rd	b
Ra^{2+}	0	2.04(37°,<u>1</u>)	9	7-37	T	μ=0.15(NaCl)	50Sa	b
C20 COBALTATE(II) ION, hexacyano-			C_6H_6Co			$Co(CN)_6{}^{4-}$		L^{4-}
Al^{3+}	4.30	-1.301	8.4	10-40	T	μ=0 R: $Al(OH_2)L = AlL + H_2O$ (Outer Sphere) = (Inner Sphere)	71Ke	b,q: ±1
C21 COBALTATE(III) ION, hexacyano			C_6N_6Co			$Co(CN)_6{}^{3-}$		L^{3-}
La^{3+}	1.33	3.762(25°,<u>1</u>)	21.7	25	T	μ=0	50Ja	a
C22 COPPER(II) CHLORIDE			Cl_2Cu			$CuCl_2$		L
LiA	1.6	(<u>2</u>-<u>1</u>)	--	55	T	S: Acetic acid sol. A=Cl	67Eb	
	3.0	--	--	55	T	S: Acetic acid sol. R: $CuCl_2 + 2LiA = CuA_2 + 2LiCl$ A=Acetate	67Eb	
C23 15-CROWN-5, dibenzo-			$C_{18}H_{20}O_5$			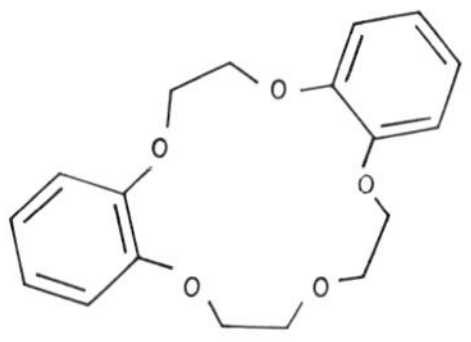		L

Metal Ion	ΔH, kcal /mole	Log K	ΔS, cal /mole $^\circ$K	T, $^\circ$C	M	Conditions	Ref.	Re-marks
C23, cont.								
Na$^+$	−1.77	0.72(1)	−2.6	25	C	μ=0 S: 20 w % CH$_3$OH	74Cd	
	−2.63	1.17(1)	−3.5	25	C	μ=0 S: 40 w % CH$_3$OH	74Cd	
	−3.78	1.64(1)	−5.2	25	C	μ=0 S: 60 w % CH$_3$OH	74Cd	
	−3.82	1.99(1)	−3.7	25	C	μ=0 S: 70 w % CH$_3$OH	74Cd	
	−8.32	2.26(1)	−17.6	25	C	μ=0 S: 80 w % CH$_3$OH	74Cd	
K$^+$	−1.8	1.22(1)	−0.5	25	C	μ=0 S: 20 w % CH$_3$OH	74Cd	
	−2.51	1.92(1)	0.4	25	C	μ=0 S: 40 w % CH$_3$OH	74Cd	
	−3.52	2.54(1)	−0.2	25	C	μ=0 S: 60 w % CH$_3$OH	74Cd	
	−3.67	2.93(1)	1.1	25	C	μ=0 S: 70 w % CH$_3$OH	74Cd	
	−10.20	2.82(1)	−21.3	25	C	μ=0 S: 80 w % CH$_3$OH	74Cd	

C24 18-CROWN-6, di(4-aminobenzo-) (cis) $C_{20}H_{26}O_6N_2$

								L
Na$^+$	−6	−−	−7	10-40	T	S: Dimethoxy-fluorenyl Salt: NaSCN	73Sk	b

C25 18-CROWN-6, dibenzo- $C_{20}H_{24}O_6$

								L
Na$^+$	−3.9	3.728(20°,1)	3.9	10-40	T	S: Dimethoxyethane	73Sk	b
	−6	2.778(25°,1)	−7	0-40	T	S: N,N-Dimethyl-formamide C_M=10^{-3}M	71Sc	b
NaA	−3.4	3.684(20°,1)	5.4	10-40	T	S: Dimethoxyethane A=Tetraphenylboron B(Ph)$_4^-$	73Sk	b
K$^+$	−5.5	2.5(1)	−6.9	25	C	μ=0 S: Dimethyl Sulfoxide	74Cd	

C26 18-CROWN-6, dicyclohexyl-(Isomer A) $C_{20}H_{36}O_6$

L

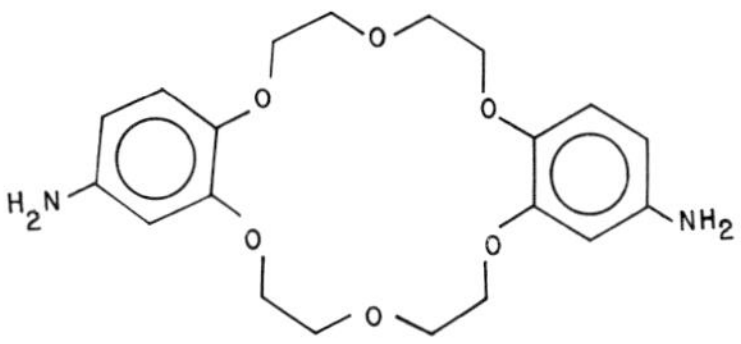

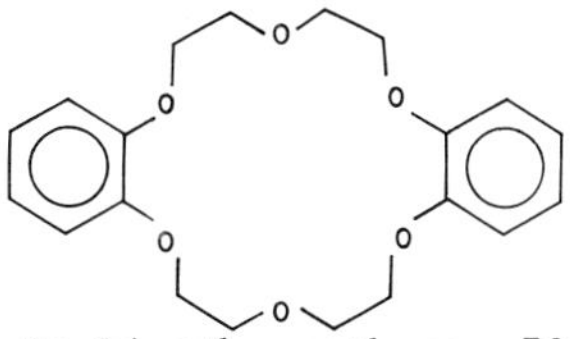

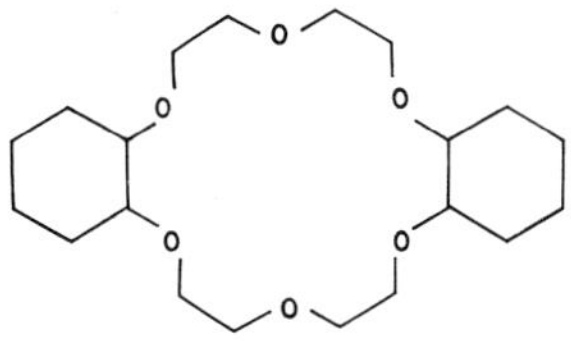

Metal Ion	ΔH, kcal /mole	Log K	ΔS, cal /mole °K	T, °C	M	Conditions	Ref.	Remarks
C26, cont.								
K^+	-4.14	2.15(1)	-4.8	10	C	μ=0	71Ih	
	-3.88	2.02(1)	-3.8	25	C	μ=0	71Ih	
	-3.89	2.01(1)	-3.8	25	C	μ=0	69Ic	
	-3.58	1.91(1)	-2.7	40	C	μ=0	71Ih	
	-10.12	(1)	--	25	C	S: Acetone. Salt:Kl	71Ag	
KAL	-3.9	0.255(25°)	-11.7	-17-25	T	S: Tetrahydrofuran A=Fluorenyl R: $A^-K^+L = A^-LK^+$	71Zd	b,q: ±0.3
Rb^+	-3.43	1.61(1)	-4.8	10	C	μ=0	71Ih	
	-3.33	1.52(1)	-4.2	25	C	μ=0	71Ih	
	-3.48	1.47(1)	-5.0	25	C	μ=0	69Ic	
	-3.29	1.40(1)	-4.1	40	C	μ=0	71Ih	
Cs^+	-2.40	1.00(1)	-3.9	10	C	μ=0	71Ih	
	-2.41	0.96(1)	-3.7	25	C	μ=0	71Ih	
	-2.00	1.07(1)	-1.8	25	C	μ=0	69Ic	
	-2.38	0.96(1)	-3.2	40	C	μ=0	71Ih	
Sr^{2+}	-3.68	3.43(1)	2.7	10	C	μ=0	71Ih	
	-3.68	3.24(1)	2.5	25	C	μ=0	71Ih	
	-3.70	3.16(1)	2.6	40	C	μ=0	71Ih	
Ba^{2+}	-4.97	3.84(1)	0.0	10	C	μ=0	71Ih	
	-4.92	3.57(1)	-0.2	25	C	μ=0	71Ih	
	-4.85	3.47(1)	0.4	40	C	μ=0	71Ih	
NH_4^+	-2.16	1.33(1)	-1.2	25	C	μ=0	71Ih	
C27	18-CROWN-6, dicyclohexyl- (Isomer B)		$C_{20}H_{36}O_6$					L
Na^+	-2.5	1.7(1)	-0.6	25	C	μ=0 S: Dimethyl Sulfoxide	74Cd 71Ih	
	∿-8.6	∿4.5(1)	--	25	C	μ=0 S: Ethanol	74Cd 71Ih	
	-5.6	3.68(1)	-1.9	25	C	μ=0 S: Methanol	74Cd 71Ih	
K^+	-5.78	1.79(1)	-12.2	10	C	μ=0	71Ih	
	-5.07	1.63(1)	-9.6	25	C	μ=0	71Ih	
	-5.18	1.60(1)	-10.1	25	C	μ=0	69Ic	
	-4.19	1.50(1)	-6.5	40	C	μ=0	71Ih	
	-11.06	(1)	--	25	C	S: Acetone Salt:KI	71Ag	
	-7.7	2.7(1)	-13.5	25	C	μ=0 S: Dimethyl Sulfoxide	74Cd 71Ih	
	∿-12.6	>6.0(1)	--	25	C	μ=0 S: Ethanol	74Cd 71Ih	
	-10.5	5.38(1)	-10.5	25	C	μ=0 S: Methanol	74Cd 71Ih	
KAL	-3.1	-0.097(25°)	-10.7	-17-25	T	S: Tetrahydrofuran A=Fluorenyl R: $A^+K^+L = A^-LK^+$	71Zd	b,q: ±0.3

Metal Ion	ΔH, kcal /mole	Log K	ΔS, cal /mole °K	T, °C	M	Conditions	Ref.	Re-marks
C27, cont.								
Rb^+	-4.6	0.95(1)	-11.9	10	C	μ=0	71Ih	
	-3.97	0.87(1)	-9.3	25	C	μ=0	71Ih	
	-3.30	0.86(1)	-6.6	40	C	μ=0	71Ih	
Sr^{2+}	-3.45	2.80(1)	0.6	10	C	μ=0	71Ih	
	-3.16	2.64(1)	1.5	25	C	μ=0	71Ih	
	-2.91	2.56(1)	2.4	40	C	μ=0	71Ih	
Ba^{2+}	-6.82	3.44(1)	-8.3	10	C	μ=0	71Ih	
	-6.20	3.27(1)	-5.8	25	C	μ=0	71Ih	
	-5.78	3.12(1)	-4.2	40	C	μ=0	71Ih	
Ag^+	-2.09	1.59(1)	0.3	25	C	μ=0	71Ih	
NH_4^+	-3.41	0.80(1)	-7.8	25	C	μ=0	71Ih	
C28	18-CROWN-6, dicyclohexyl- (Isomer A+B)		$C_{20}H_{36}O_6$					L
Li^+	-3.87	(1)	--	25	C	S: Acetone Salt:LiBr	71Ag	
	∿0	(1)	--	25	C	S: Dimethyl Sulfoxide	71Ag	
	∿0	(1)	--	25	C	S: Tetrahydrofuran	71Ag	
Na^+	-6.43	(1)	--	25	C	S: Acetone Salt:NaI	71Ag	
	-5.99	(1)	--	25	C	S: Acetone Salt:$NaB(C_6H_5)_4$	71Ag	
	-0.45	(1)	--	25	C	S: Dimethyl Sulfoxide Salt:NaI	71Ag	
	-4.78	(1)	--	25	C	S: Tetrahydrofuran Salt:$NaB(C_6H_5)_4$ Salt added to ligand	71Ag	
	-5.03	(1)	--	25	C	S: Tetrahydrofuran Salt:$NaB(C_6H_5)_4$ Ligand added to salt	71Ag	
K^+	-9.71	(1)	--	25	C	S: Acetone Salt:KI	71Ag	
	-9.29	(1)	--	25	C	S: Acetone Salt:$KB(C_6H_5)_4$	71Ag	
	-7.39	(1)	--	25	C	S: Dimethyl Sulfoxide Salt:KI Metal added to ligand	71Ag	
	-6.52	(1)	--	25	C	S: Dimethyl Sulfoxide Salt:KI Ligand added to metal	71Ag	
	-7.53	(1)	--	25	C	S: Dimethyl Sulfoxide Salt:KNO_3	71Ag	
	-7.45	(1)	--	25	C	S: Dimethyl Sulfoxide Salt:$KB(C_6H_5)_4$	71Ag	

Metal Ion	ΔH, kcal /mole	Log K	ΔS, cal /mole °K	T, °C	M	Conditions	Ref.	Remarks
C28, cont.								
Cs^+	-8.40	(1)	--	25	C	S: Acetone Salt: $CsB(C_6H_5)_4$	71Ag	
	-6.94	(1)	--	25	C	S: Dimethyl Sulfoxide Salt: CsI	71Ag	
NH_4^+	-9.98	(1)	--	25	C	S: Acetone Salt: NH_4I	71Ag	
	-4.15	(1)	--	25	C	S: Dimethyl Sulfoxide Salt: NH_4I	71Ag	
C29	18-CROWN-6, (4'-methylbenzo)-		$C_{17}H_{26}O_6$			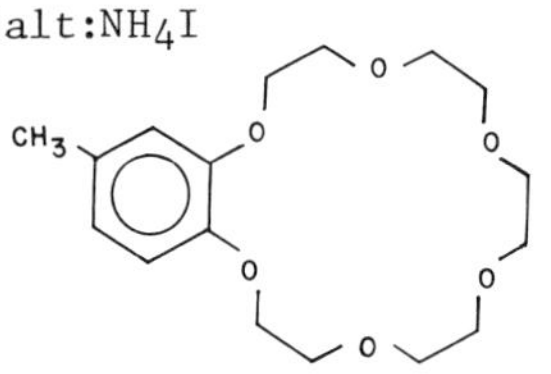		L
KAL	-3.0	-0.032(25°)	-10.4	-17-25	T	S: Tetrahydrofuran A=Fluorenyl R: $A^-K^+L = A^-LK^+$	71Ta	b,q: ±0.3
C30	30-CROWN-10, dibenzo-		$C_{28}H_{40}O_{10}$					L
Na^+	-4	2.114(25°,1)	-3.8	5-35	T	μ=0.15M (Tetra-butylammonium perchlorate) S: Methanol	72Cc	b,q: ±1
K^+	-11.5	4.568(25°,1)	-17.7	5-35	T	μ=0.15M(Tetra-butylammonium perchlorate) S: Methanol	72Cc	b,q: ±2
Rb^+	-12.7	4.643(25°,1)	-21.4	5-35	T	μ=0.15M(Tetra-butylammonium perchlorate) S: Methanol	72Cc	b,q: ±2
Cs^+	-11.2	4.230(25°,1)	-18.2	5-35	T	μ-0.15M(Tetra-butylammonium perchlorate) S: Methanol	72Cc	b,q: ±2
Tl^+	-11	4.505(25°,1)	-16.3	5-35	T	μ=0.15M(Tetra-butylammonium perchlorate) S: Methanol	72Cc	b,q: ±2
NH_4^+	-5.5	2.431(25°,1)	-7.3	5-35	T	μ=0.15M(Tetra-butylammonium perchlorate) S: Methanol	72Cc	b,q: ±1
C31	CRYPTATE, Diaza-Heptaoxa		$C_{20}H_{40}O_7N_2$			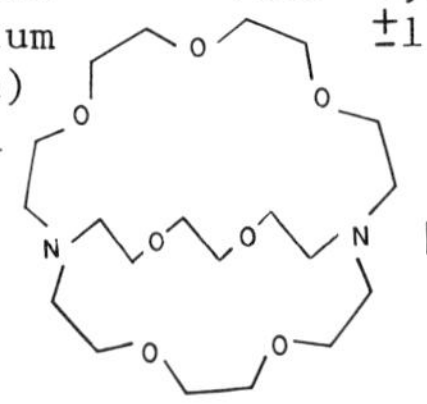		L

Metal Ion	ΔH, kcal /mole	Log K	ΔS, cal /mole °K	T, °C	M	Conditions	Ref.	Remarks
C31, cont.								
K^+	-6.2	2.1(1)	-11	25	C	4-8x10^{-2}M(LiCl)	74Cd	
Rb^+	-5.4	2.05(1)	-9	25	C	4-8x10^{-2}M(LiCl)	74Cd	
Cs^+	-6.5	1.8(1)	-14	25	C	4-8x10^{-2}M(LiCl)	74Cd	
C32 CRYPTATE, Diaza-Hexaoxa			$C_{18}H_{36}O_6N_2$			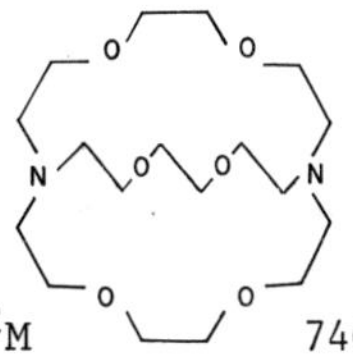		L
Na^+	-5.8	3.90(1)	-2	25	C	4-8x10^{-2}M [N$(CH_3)_4$Br]	74Cd 71Lc	
K^+	-11.1	5.30(1)	-13	25	C	4-8x10^{-2}M [N$(CH_3)_4$Br]	74Cd	
Rb^+	-10.5	4.35(1)	-15	25	C	4-8x10^{-2}M [N$(CH_3)_4$Br]	74Cd 71Lc	
Ca^{2+}	-0.15	4.40(1)	20	25	C	4-8x10^{-2}M [N$(CH_3)_4$Br]	74Cd 71Lc	
Ba^{2+}	-12.9	9.50(1)	0	25	C	4-8x10^{-2}M [N$(CH_3)_4$Br]	74Cd 71Lc	
Sr^{2+}	-8.1	8.00(1)	9	25	C	4-8x10^{-2}M [N$(CH_3)_4$Br]	74Cd 71Lc	
C33 CRYPTATE, Diaza-Pentaoxa			$C_{16}H_{32}O_5N_2$					L
Li^+	0.4	2.50(1)	13	25	C	4-8x10^{-2}M [N$(CH_3)_4$Br]	71Lc 74Cd	
Na^+	-3.2	5.40(1)	14	25	C	4-8x10^{-2}M [N$(CH_3)_4$Br]	71Lc	
	-4.7	5.30(1)	8	25	C	4-8x10^{-2}M [N$(CH_3)_4$Br]	74Cd	
K^+	-5.9	3.95(1)	-2	25	C	4-8x10^{-2}M [N$(CH_3)_4$Br]	71Lc 74Cd	
Rb^+	-3.2	2.55(1)	1	25	C	4-8x10^{-2}M [N$(CH_3)_4$Br]	71Lc 74Cd	
Ca^+	-1.7	6.95(1)	26	25	C	4-8x10^{-2}M [N$(CH_3)_4$Br]	71Lc	
	-2.7	6.95(1)	23	25	C	4-8x10^{-2}M [N$(CH_3)_4$Br]	74Cd	
Ba^{2+}	-5.1	6.30(1)	12	25	C	4-8x10^{-2}M [N$(CH_3)_4$Br]	71Lc 74Cd	
Sr^{2+}	-5.0	7.35(1)	17	25	C	4-8x10^{-2}M [N$(CH_3)_4$Br]	71Lc 74Cd	
C34 CRYPTATE, Diaza-Tetraoxa			$C_{14}H_{28}O_4N_2$			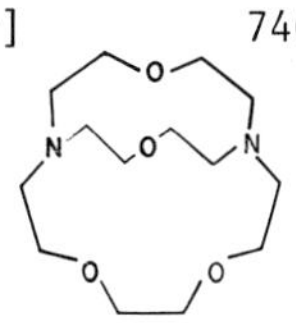		L

Metal Ion	ΔH, kcal /mole	Log K	ΔS, cal /mole °K	T, °C	M	Conditions	Ref.	Re-marks
C34, cont.								
Li^+	-4.6	4.30(1)	4	25	C	4-8x10^{-2}M [$N(CH_3)_4Br$]	71Lc 74Cd	
Na^+	-4.7	2.70(1)	-4	25	C	4-8x10^{-2}M [$N(CH_3)_4Br$]	74Cd	
Ca^{2+}	0	2.80(1)	13	25	C	4-8x10^{-2}M [$N(CH_3)_4Br$]	71Lc 74Cd	
C35 CYANAMIDE, dimethyl-			$C_3H_6N_2$			$(CH_3)_2NCN$		L
[Co^{3+}]	-12.7	--	--	24	C	S: Dichloromethane [Co]=Methylcobaloxime R: $[Co]_2$ + 2L = 2[Co]L	73Cf	
I_2	-2.8	0.249(25°,1)	-8.2	13-43	T	S: CCl_4	65Af	b
IA	-5.6	1.202(25°,1)	-13.3	9-40	T	S: CCl_4 A=Br	65Af	b
	-7.3	2.13(25°,1)	-15.1	13-44	T	S: CCl_4 A=Cl	65Af	b
C36 CYANIC ACID			CHON			HOCN		L^-
[Fe^{3+}]	-10.6	2.86(1)	-22.6	25	-	Chironomus hemo-globin	60Gb	
C37 CYANIDE ION			CN			CN^-		L^-
Ca^{2+}	1.4	2.631(25°)	17	15-35	T	μ=0 R: Ca^{2+} + $Fe(CN)_6{}^{3-}$ = $CaFe(CN)_6{}^-$	74Ha	b,q: ±0.3
	2.0	3.634(25°)	24	15-35	T	μ=0 R: Ca^{2+} + $Fe(CN)_6{}^{4-}$ = $CaFe(CN)_6{}^{2-}$	74Ha	b,q: ±0.3
	1.6	2.631	17	25	C	μ=0.07M R: Ca^{2+} + $Fe(CN)_6{}^{3-}$ = $CaFe(CN)_6{}^-$	74Ha	q: ±0.2
	2.1	3.634	24	25	C	μ=0.07M R: Ca^{2+} + $Fe(CN)_6{}^{4-}$ = $CaFe(CN)_6{}^{2-}$	74Ha	q: ±0.2
V^{2+}	-47.0	(1-6)	--	25	C	1M(KCN)	64Gc	
Cr^{2+}	-63.2	(1-6)	--	25	C	1M(KCN)	64Gc	
Mn^{2+}	-34.5	(1-6)	--	25	C	1M(KCN)	64Gc	
Fe^{2+}	-85.77	35.4(1-6)	-125.9	25	C	μ=0	65Wa 64Gc	
	-15.15	--	--	25	C	C_{Fe_T}=0 R: Fe^{2+} + $FeL_6{}^{3-}$ = Turnbull's blue	70Ia	
Fe^{3+}	-4.8	7.04(1)	16	25	-	--	60Gb 64Gc	
	-70.14	43.6(1-6)	-35.7	25	C	μ=0	65Wa	
	0.72	--	--	25	C	C_{Fe_T}=0 R: Fe^{3+} + $FeL_6{}^{4-}$ = Prussian blue	70Ia	q: ±0.18

Metal Ion	ΔH, kcal /mole	Log K	ΔS, cal /mole °K	T, °C	M	Conditions	Ref.	Remarks
C37, cont.								
[Fe^{3+}]	-18.6	8.36(1)	-24	25	-	Ferimyoglobin	60Gb 55Ga	
[Fe^{3+}]	1.1	6.10(24.6°,1)	31.3	11.7-30.4	T	Ferricytochrome. 0.075M borate buffer	55Ga 52Ga	b
[Fe^{3+}]	-16.50	9.102(1)	-14.6	20	T	Human Methaemoglobin A. 6.0	68Ac	d: 0.05
	-18.20	9.028(1)	-20.8	20	T	6.4	68Ac	
	-19.50	8.968(1)	-25.5	20	T	6.8	68Ac	
	-19.90	8.946(1)	-25.2	20	T	7.0	68Ac	
	-18.39	8.819(1)	-22.1	20	T	7.2	68Ac	
	-14.40	8.610(1)	-9.72	20	T	7.6	68Ac	
	-11.40	8.558(1)	0.27	20	T	7.8	68Ac	
[Fe^{3+}]	-14.50	9.073(1)	-7.95	20	T	Human Methaemoglobin C. 6.0	68Ac	d: 0.05
	-16.30	9.020(1)	-14.3	20	T	6.4	68Ac	
	-17.20	8.924(1)	-17.8	20	T	6.8	68Ac	
	-18.10	8.871(1)	-21.1	20	T	7.0	68Ac	
	-19.00	8.834(1)	-24.4	20	T	7.2	68Ac	
	-21.10	8.633(1)	-32.5	20	T	7.5	68Ac	
	-20.60	8.416(1)	-31.8	20	T	7.8	68Ac	
FeL_6^{3-}	-15	3.176	-35.8	12-25	T	μ=0.6M(HAC + Ac^-) pH=5.0 A=EDTA R: $Fe(III)L_6^{3-}$ + $Co(II)A^{2-}$ = $[ACo(III)\text{-}L\text{-}Fe(II)L_5]^{5-}$	67Hh	b,r
Co^{2+}	-61.5	(1-5)	--	25	C	μ=0	68Ic	
	-74.4	(1-6)	--	25	C	1M(KCN). Product may be $Co(CN)_5H_2O^{3-}$ or a dimer rather than $Co(CN)_6^{4-}$	64Gc	
CoA^{2-}	-8.6	1.580(1)	-21.7	25	T	μ=0.10($NaClO_4$) A=1,2-diaminocyclohexanetetraacetate	69Jb	
Ni^{2+}	-3.0	-0.721(25°,5)	-13.3	15-35	T	μ=1.34	60Mb	b
	-45.2	32.2(1-4)	-7.8	10	C	μ=0	71Ig	
	-43.2	30.1(1-4)	-7	25	C	μ=0	63Cc	
				25	C	μ=0	71Ig	ΔCp= 36
	-43.9	27.43(1-4)	-14.7	40	C	μ=0	71Ig	
Pd^{2+}	-0.2	2.8(5)	8	25	C	μ=0	67Ic	
	-92.3	42.2(1-4)	-116	25	C	μ=0	67Ic	
Cu^+	-11.1	5.30(3)	-13.4	25	C	μ=0. Log K value from ref. 50Va corrected using HCN, pK=9.21	67Ib	
	-11.2	1.5(4)	-31	25	C	μ=0. Log K value from ref. 50Va	67Ib	

Metal Ion	ΔH, kcal /mole	Log K	ΔS, cal /mole °K	T, °C	M	Conditions	Ref.	Remarks
C37, cont.								
Cu^{+}, cont.						corrected using HCN, pK=9.21		
	-29.1	23.94(1-2)	12	25	C	μ=0. Log K value from ref. 50Va corrected using HCN, pK=9.21	67Ib	
Ag^{+}	-0.6	1.55(3)	5	25	C	μ=0	67Ib	
	-32.9	20.44(1-2)	-16	25	C	μ=0	67Ib	
	-35.72	10.87(1-2)	-18.5	250	T	S: Equimolar molten salt of KNO_3 & $NaNO_3$. K in units of moles^2/ 1000gm^2	56Fa	
Zn^{2+}	-9.5	5.17(3)	-9.9	10	C	μ=0	71Ig	
	-9.2	4.98(3)	-8.0	25	C	μ=0	71Ig	ΔCp=5
	-8.4	4.98(3)	-5.4	25	C	μ=0	65Ia 67Ab	
	-9.3	4.50(3)	-9.1	40	C	μ=0	71Ig	
	-7.4	3.79(4)	-8.8	10	C	μ=0	71Ig	
	-7.7	3.57(4)	-9.5	25	C	μ=0	71Ig	ΔCp= -17
	-8.6	3.57(4)	-12.4	25	C	μ=0	65Ia	
	-7.9	3.10(4)	-14.2	40	C	μ=0	71Ig	
	-11.6	11.47(1-2)	11.5	10	C	μ=0	71Ig	
	-11.0	11.07(1-2)	13.7	25	C	μ=0	71Ig	ΔCp= 36
	-10.8	11.067(1-2)	14	25	C	μ=0	65Ia	
	-10.5	10.70(1-2)	15.4	40	C	μ=0	71Ig	
	-19.2	16.050(1-3)	9.0	25	C	μ=0	65Ia	
	-28.5	20.43(1-4)	-7.2	10	C	μ=0	71Ig	
	-27.9	19.62(1-4)	-3.8	25	C	μ=0	71Ig	ΔCp= 24
	-27.8	19.62(1-4)	-3.4	25	C	μ=0	65Ia	
	-27.7	18.30(1-4)	-4.7	40	C	μ=0	71Ig	
	-27.7	(1-4)	--	25	C	1M(KCN)	64Gc	
Cd^{2+}	-7.9	6.22(1)	0.6	10	C	μ=0	71Ig	
	-7.3	6.01(1)	3.0	25	C	μ=0	71Ig	ΔCp= 42
	-6.65	5.73(1)	5.0	40	C	μ=0	71Ig	
	-7.39	5.48(1)	0.3	25	C	3.0M($NaClO_4$)	66Gb	
	-6.89	5.38(2)	-0.3	10	C	μ=0	71Ig	
	-5.7	5.11(2)	4.3	25	C	μ=0	71Ig	ΔCp= 56
	-5.15	4.90(2)	6.0	40	C	μ=0	71Ig	
	-7.73	5.12(2)	-2.5	25	C	3.0M($NaClO_4$)	66Gb	
	-8.3	4.77(3)	-7.6	10	C	μ=0	71Ig	
	-8.56	4.53(3)	-8.0	25	C	μ=0	71Ig	ΔCp= -7
	-8.6	4.12(3)	-8.5	40	C	μ=0	71Ig	
	-7.12	4.54(3)	-3.1	25	C	3.0M($NaClO_4$)	66Gb	
	-5.2	2.52(4)	-7.0	10	C	μ=0	71Ig	
	-5.1	2.27(4)	-6.9	25	C	μ=0	71Ig	ΔCp= -7

Metal Ion	ΔH, kcal /mole	Log K	ΔS, cal /mole °K	T, °C	M	Conditions	Ref.	Remarks
C37, cont.								
Cd^{2+}, cont.	-5.42	2.12(4)	-7.6	40	C	μ=0	71Ig	
	-7.03	3.16(4)	-9.1	25	C	3.0M($NaClO_4$)	66Gb	
	-28.37	18.89(1-4)	-13.8	10	C	μ=0	71Ig	
	-26.76	17.92(1-4)	-8.0	25	C	μ=0	71Ig	ΔCp= 84
	-25.77	16.87(1-4)	-5.1	40	C	μ=0	71Ig	
	-29.3	18.79(1-4)	-12.3	25	C	μ=3	67Ab	
Hg^{2+}	-23.92	17.97(1)	-2.3	10	C	μ=0	71Ig	
	-23.2	17.00(1)	0.0	25	C	μ=0	71Ig	ΔCp= 43
	-23.0	17.00(1)	0.7	25	C	μ=0	65Cg	
	-22.6	16.26(1)	2.2	40	C	μ=0	71Ig	
	-24.04	16.74(2)	-8.3	10	C	μ=0	71Ig	
	-23.4	15.75(2)	-6.4	25	C	μ=0	71Ig	ΔCp= 37
	-25.5	15.75(2)	-13.4	25	C	μ=0	65Cg	
	-22.9	15.02(2)	-4.4	40	C	μ=0	71Ig	
	-7.36	3.81(3)	-8.5	10	C	μ=0	71Ig	
	-6.84	3.56(3)	-6.6	25	C	μ=0	71Ig	ΔCp= 21
	-7.6	3.56(3)	-9.0	25	C	μ=0	65Cg	
	-6.71	3.37(3)	-6.0	40	C	μ=0	71Ig	
	-5.26	2.81(4)	-5.7	10	C	μ=0	71Ig	
	-6.3	2.66(4)	-9.0	25	C	μ=0	71Ig	ΔCp= -61
	-7.2	2.66(4)	-12.1	25	C	μ=0	65Cg	
	-7.10	2.46(4)	-11.4	40	C	μ=0	71Ig	
	-60.59	41.33(1-4)	-24.9	10	C	μ=0	71Ig	
	-59.8	38.97(1-4)	-22.0	25	C	μ=0	71Ig	ΔCp= 41
	-59.3	37.11(1-4)	-19.6	40	C	μ=0	71Ig	
HgA^+	-22.1	14.016(1)	-11.4	20	C	μ=0.1(KNO_3) A=CH_3	65Sa	
C38 CYCLOBUTANE, methylene-			C_5H_8			$CH_2CH_2CH_2CH=CH_2$		L
Ag^+	-4.04	-1.046(1)	-18.4	25	T	1M(KNO_3)	59Ta	o
I_2	-0.84	0.444(1)	0.8	25	T	S: 2,2,4-Trimethylpentane	59Ta	
C39 1,5,9-CYCLODODECATRIENE			$C_{12}H_{18}$			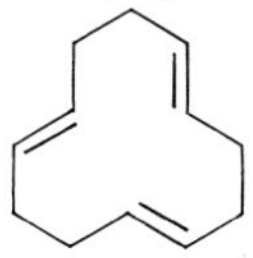		L
PdA_2B_2	-5.9	--	--	25	C	S: CH_2Cl_2 C_L=4.6x10^{-3}M C_M=2.6x10^{-4}M A=Cl B=C_6H_5CN R: PdA_2B_2 + L = PdA_2L + 2B	72Pc	
C40 CYCLOHEPTANE, 1-aza-			$C_6H_{13}N$					

Metal Ion	ΔH, kcal /mole	Log K	ΔS, cal /mole °K	T, °C	M	Conditions	Ref.	Re-marks
C40, cont.								
NiA_2	-7.8	2.037(24.6°,1)	-16.8	8-39.5	T	S: Benzene A=0,0' Diethyldithio phosphate	72Mc	b
	-7.9	2.561(24.6°,1)	-15	8-39.5	T	S: Benzene A=0,0'-Dimethyldithio-phosphate	72Mc	b
C41 CYCLOHEPTANE, 1,4-diaza			$C_5H_{12}N_2$			$CH_2NHCH_2CH_2NHCH_2CH_2$		L
Cu^{2+}	-5.6	7.72(30°,1)	17	10-40	T	$\mu \to 0$	61Pa	b
	-5.8	5.93(30°,2)	8	10-40	T	$\mu \to 0$	61Pa	b
C42 CYCLOHEPTANE, methylene-			C_8H_{14}			$CH_2CH_2CH_2CH_2CH_2CH_2C{=}CH_2$		L
Ag^+	-3.21	-1.140(1)	-16.1	25	T	1M(KNO_3)	59Ta	o
I_2	-2.88	0.4265(1)	7.7	25	T	S: 2,2,4-Trimethylpentane	59Ta	
C43 2,4,6-CYCLOHEPTATRIEN-1-one, 2-hydroxy-			$C_7H_6O_2$			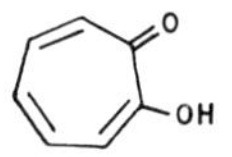		
Y^{3+}	-2.56	7.17(1)	24.2	25	C	μ=0.1M(KNO_3)	70Ca	
La^{3+}	-2.48	6.19(1)	20	25	C	μ=0.1M(KNO_3)	70Ca	
Ce^{3+}	-2.70	6.55(1)	21.0	25	C	μ=0.1M(KNO_3)	70Ca	
Pr^{3+}	-2.82	6.60(1)	20.7	25	C	μ=0.1M(KNO_3)	70Ca	
Nd^{3+}	-2.85	6.77(1)	21.8	25	C	μ=0.1M(KNO_3)	70Ca	
Sm^{3+}	-2.81	6.90(1)	22.2	25	C	μ=0.1M(KNO_3)	70Ca	
Eu^{3+}	-2.76	7.10(1)	23.2	25	C	μ=0.1M(KNO_3)	70Ca	
Gd^{3+}	-2.72	7.03(1)	23.1	25	C	μ=0.1M(KNO_3)	70Ca	
Tb^{3+}	-2.74	7.14(1)	23.5	25	C	μ=0.1M(KNO_3)	70Ca	
Dy^{3+}	-2.80	7.23(1)	23.7	25	C	μ=0.1M(KNO_3)	70Ca	
Ho^{3+}	-2.76	7.40(1)	24.6	25	C	μ=0.1M(KNO_3)	70Ca	
Er^{3+}	-2.79	7.54(1)	25.2	25	C	μ=0.1M(KNO_3)	70Ca	
Tm^{3+}	-2.94	7.59(1)	24.9	25	C	μ=0.1M(KNO_3)	70Ca	
Yb^{3+}	-3.12	7.84(1)	25.5	25	C	μ=0.1M(KNO_3)	70Ca	
Lu^{3+}	-2.90	7.69(1)	25.5	25	C	μ=0.1M(KNO_3)	70Ca	
C44 2,4,6-CYCLOHEPTATRIEN-1-one, 2-hydroxy-7-isopropenyl-			$C_{10}H_{10}O_2$			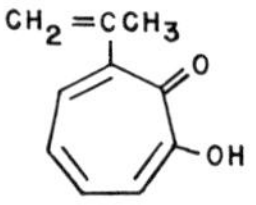		
CoL_3	-0.8	0.518(25°)	-0.3	-56-20	T	S: $CDCl_3$	72Ea	b,q: ±0.08 bb
AlL_3	-1.7	0.623(25°)	-2.7	-35-30	T	S: $C_2H_2Cl_4$	72Ea	b,q: ±0.2 bb

Metal Ion	ΔH, kcal /mole	Log K	ΔS, cal /mole °K	T, °C	M	Conditions	Ref.	Remarks
C44, cont.								
AlL_3, cont.	-0.20	0.602(25°)	2.2	-63-25	T	S: CH_2Cl_2	72Ea	b,q: ±0.02 bb
C45	2,4,6-CYCLOHEPTATRIEN-1-ONE, 2-hydroxy-		$C_{10}H_{11}O_2$			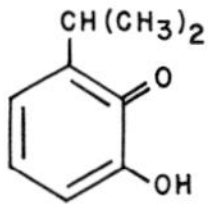		
CoL_3	-0.94	0.623(25°)	-0.2	-55- -20	T	S: $CHCl_3$	72Ea	b,q: ±0.40 bb
AlL_3	0.0	0.505(25°)	2.9	-55- -20	T	S: $C_2H_2Cl_4$	72Ea	b,q: ±0.5 bb
C46	CYCLOHEPTENE		C_7H_{12}			$CH_2CH_2CH_2CH_2CH_2CH{=}CH$ (ring)		L
Ag^+	-6.49	-1.674(1)	-29.4	25	T	1M(KNO_3)	59Ta	o
	-6.61	-1.66(1)	-29.8	25	T	μ=1.0($AgNO_3$)	59Ta	o
I_2	-1.95	0.484(1)	4.3	25	T	S: 2,2,4-Trimethylpentane	59Ta	
C47	1,4-CYCLOHEXADIENE		C_6H_8			$CH{=}CHCH{=}CHCH_2CH_2$ (ring)		L
Cu^+	-7.0	2.53(1)	-11.5	30	T	S: 1M($LiClO_4$) in 2-propanol	69Hb	
C48	CYCLOHEXANE, 1,2-diamino- (cis)		$C_6H_{14}N_2$			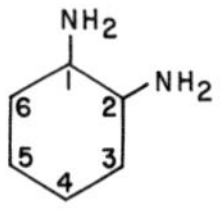		L
Ni^{2+}	-7.7	7.12(30°,1)	7	10-40	T	μ=0	58Ba	b
	-6.3	5.80(30°,2)	6	10-40	T	μ=0	58Ba	b
Cu^{2+}	-11.6	10.72(30°,1)	10	10-40	T	μ=0	58Ba	b
	-12.0	9.40(30°,2)	3	10-40	T	μ=0	58Ba	b
Zn^{2+}	-5.2	5.74(30°,1)	9	10-40	T	μ=0	58Ba	b
	-4.2	5.38(30°,2)	10	10-40	T	μ=0	58Ba	b
Cd^{2+}	-5.7	5.65(30°,1)	7	10-40	T	μ=0	58Ba	b
	-5.6	4.64(30°,2)	2	10-40	T	μ=0	58Ba	b
C49	CYCLOHEXANE, 1,2-diamino- (trans)		$C_6H_{14}N_2$			$C_6H_{10}(NH_2)_2$		L
Ni^{2+}	-8.5	7.82(30°,1)	8	10-40	T	μ=0	58Ba	b
	-7.9	6.71(30°,2)	5	10-40	T	μ=0	58Ba	b
	-9.1	4.67(30°,3)	9	10-40	T	μ=0	58Ba	b
Cu^{2+}	-13.6	10.96(30°,1)	5	10-40	T	μ=0	58Ba	b
	-12.3	9.54(30°,2)	3	10-40	T	μ=0	58Ba	b
Zn^{2+}	-5.0	6.14(30°,1)	12	10-40	T	μ=0	58Ba	b
	-5.2	5.31(30°,2)	7	10-40	T	μ=0	58Ba	b
Cd^{2+}	-5.6	5.74(30°,1)	8	10-40	T	μ=0	58Ba	b

Metal Ion	ΔH, kcal /mole	Log K	ΔS, cal /mole °K	T, °C	M	Conditions	Ref.	Re-marks
C49, cont.								
Cd^{2+}, cont.	-6.6	4.74(30°,2)	0	10-40	T	μ=0	58Ba	b
C50 CYCLOHEXANE, 1,2-diamino-N,N,N',N'-tetraacetic acid (trans)			$C_{14}H_{22}O_8N_2$			$C_6H_{10}[N(CH_2CO_2H]_2$		L^{4-}
Mg^{2+}	3.80	10.97(1)	63.1	20	C	0.1M(KNO_3)	63Ab	
	1.6	10.3(1)	52	25	C	μ=0.1(KNO_3)	65Wd	
	4.59	10.78(1)	64.7	25	C	μ=0.20(KNO_3)	69Ca	
	1.89	10.408(1)	54.1	20	C	μ=1M(KCl) C_L=0.01	69Sh	g:k
Ca^{2+}	-3.70	13.15(1)	47.5	20	C	0.1M(KNO_3)	63Ab	
	-6.2	12.2(1)	35	25	C	μ=0.1(KNO_3)	65Wd	
	-3.62	12.92(1)	47.0	25	C	μ=0.20(KNO_3)	69Ca	
	-5.55	12.712(1)	39.2	20	C	μ=1M(KCl) C_L=0.01	69Sh	g:k
Sr^{2+}	-0.74	10.5(1)	45.7	20	C	0.1M(KNO_3)	63Ab	
	-3.6	9.97(1)	34	25	C	μ=0.1(KNO_3)	65Wd	
	-0.73	10.36(1)	44.9	25	C	μ=0.20(KNO_3)	69Ca	
	-0.56	10.520(1)	46.2	20	C	μ=1M(KCl) C_L=0.01	69Sh	g:k
Ba^{2+}	0.33	8.64(1)	40.9	20	C	0.1M(KNO_3)	63Ab	
	-2.2	7.84(1)	29	25	C	μ=0.1(KNO_3)	65Wd	
	0.28	8.49(1)	39.6	25	C	μ=0.20(KNO_3)	69Ca	
	-0.91	8.112(1)	34.0	20	C	μ=1M(KCl) C_L=0.01	69Sh	g:k
Y^{3+}	4.23	19.41(1)	103.0	25	T	0.1M(KNO_3)	62Mg	
La^{3+}	1.60	16.91(1)	82.8	20	C	0.1M(KNO_3)	63Ab	
	3.55	16.35(1)	86.7	25	T	0.1M(KNO_3)	62Mg	
Pr^{3+}	5.00	17.23(1)	95.6	25	T	0.1M(KNO_3)	62Mg	
Nd^{3+}	5.00	17.69(1)	97.6	25	T	0.1M(KNO_3)	62Mg	
Sm^{3+}	5.00	18.63(1)	102	25	T	0.1M(KNO_3)	62Mg	
Eu^{3+}	5.54	18.77(1)	104.5	25	T	0.1M(KNO_3)	62Mg	
	-0.6	17.60(50°,1)	78.5	30-80	T	μ=1.0(KCl) C_M=0.1 Complex with Structure 1 as determined by absorption spectra	71Kc	b
	5.32	16.97(50°,1)	94.0	30-80	T	μ=1.0(KCl) C_M=0.1 Complex with Structure 2 as determined by absorption spectra	71Kc	b
Gd^{3+}	5.75	18.80(1)	105.3	25	T	0.1M(KNO_3)	62Mg	
Tb^{3+}	5.00	19.30(1)	105.1	25	T	0.1M(KNO_3)	62Mg	
Dy^{3+}	3.09	19.69(1)	100.5	25	T	0.1M(KNO_3)	62Mg	
Ho^{3+}	1.18	19.89(1)	95.0	25	T	0.1M(KNO_3)	62Mg	

Metal Ion	ΔH, kcal /mole	Log K	ΔS, cal /mole °K	T, °C	M	Conditions	Ref.	Remarks
C50, cont.								
Er^{3+}	0.12	20.20(1)	92.8	25	T	0.1M(KNO_3)	62Mg	
Tm^{3+}	-1.59	20.46(1)	88.3	25	T	0.1M(KNO_3)	62Mg	
Yb^{3+}	-4.49	20.80(1)	80.2	25	T	0.1M(KNO_3)	62Mg	
Lu^{3+}	-4.92	20.91(1)	79.2	25	T	0.1M(KNO_3)	62Mg	
Zr^{4+}	-4.21	20.68(1)	80.51	25	T	μ=0.1(KNO_3)	65Hd	
Th^{4+}	-0.53	23.78(1)	107.08	25	T	μ=0.1(KNO_3)	65Hd	
Mn^{2+}	-4.14	17.43(1)	65.60	20	C	0.1M(KNO_3)	63Ab	
	-7.1	16.6(1)	52	25	C	μ=0.1(KNO_3)	65Wd	
Fe^{2+}	-6.6	18.2(1)	61	25	C	μ=0.1(KNO_3)	65Wd	
	-0.9	9.322(28.1°)	39	18.2-40	T	μ=0.20 R: $Fe^{2+} + HL^{3-} = FeHL^-$	72Bk	b,q: ±0.3
	-1.1	5.301(28.1°)	20	18.2-40	T	μ=0.20 R: $Fe^{2+} + H_2L^{2-} = FeH_2L$	72Bk	b,q: ±0.9
Fe^{3+}	-14.2	29.146(28.1°,1)	86	18.2-40	T	μ=0.20	72Bk	b
Co^{2+}	-2.80	19.57(1)	80.0	20	C	0.1M(KNO_3)	63Ab	
	-5.4	18.8(1)	68	25	C	μ=0.1(KNO_3)	65Wd	
	-17.5	1.639(25°)	-51.2	17-30	T	μ=0.660(NaAc + HAc) pH=5.0 R: $CoL^{2-} + Fe(CN)_6{}^{3-} = [LCoCNFeCN_5{}^{5-}]$	73Hf	b
Ni^{2+}	-5.37	(1)	--	20	C	0.1M(KNO_3)	63Ab	
	-7.5	19.4(1)	63	25	C	μ=0.1(KNO_3)	65Wd	
Cu^{2+}	-6.07	21.95(1)	79.7	20	C	0.1M(KNO_3)	63Ab	
	-8.8	21.1(1)	67	25	C	μ=0.1(KNO_3)	65Wd	
	-7.18	20.94(25°,1)	71.7	10-40	T	μ=0.1(KNO_3)	67Hg	b
Zn^{2+}	-1.94	19.32(1)	81.8	20	C	0.1M(KNO_3)	63Ab	
	-7.7	18.5(1)	59	25	C	μ=0.1(KNO_3)	65Wd	
	-3.58	18.53(25°,1)	72.8	10-40	T	μ=0.1(KNO_3)	67Hg	b
	-1.94	18.99(1)	80.4	25	C	μ=0.20(KNO_3)	69Ca	
Cd^{2+}	-7.40	19.88(1)	65.7	20	C	0.1M(KNO_3)	63Ab	
	-11.2	19.1(1)	50	25	C	μ=0.1(KNO_3)	65Wd	
	-6.63	18.50(25°,1)	62.4	10-40	T	μ=0.1(KNO_3)	67Hg	b
	-6.95	19.54(1)	66.1	25	C	μ=0.20(KNO_3)	69Ca	
Hg^{2+}	-16.60	24.95(1)	59.0	20	C	0.1M(KNO_3)	63Ab 62Mg	
	-18.9	24.3(1)	48	25	C	μ=0.1(KNO_3)	65Wd 65Hd 66Mf	
	-16.05	24.53(1)	58.4	25	C	μ=0.20(KNO_3)	69Ca	
Al^{3+}	10.7	18.73(1)	121.7	25	T	μ=0.1(KNO_3)	66Mf	
Pb^{2+}	-11.36	20.33(1)	54.2	20	C	0.1M(KNO_3)	63Ab	
	-12.4	19.4(1)	47	25	C	μ=0.1(KNO_3)	65Wd	

Metal Ion	ΔH, kcal /mole	Log K	ΔS, cal /mole °K	T, °C	M	Conditions	Ref.	Remarks
C51 CYCLOHEXANE, 1,2-dihydroxy-(trans)			$C_6H_{12}O_2$			$C_6H_{10}(OH)_2$		L
$H_5TeO_6^-$	-3.2	1.16	-5	25	T	μ=0.1. R: $H_5TeO_6^-$ + L = $HTeO_4^-$ + $2H_2O$	62Ea	
C52 CYCLOHEXANE, 1,4-epoxy-			$C_6H_{10}O$					L
I_2	-4.74	0.517(25°,1)	-13.5	6.7-40	T	S: n-Heptane ΔH Determined from visible spectral data	64Ta	b
	-4.94	0.591(25°,1)	-13.8	6.7-40	T	S: n-Heptane ΔH Determined from ultraviolet spectral data	64Ta	b
C53 CYCLOHEXANE, methylene			C_7H_{12}			$C_6H_{11}(CH_2)$		L
Ag^+	-3.11	-1.223(1)	-16.0	25	T	1M(KNO_3)	59Ta	o
I_2	-1.78	0.562(1)	3.4	25	T	S: 2,2,4-Trimethylpentane	59Ta	
C54 1,2-CYCLOHEXANEDIOL(cis)			$C_6H_{12}O_2$			$C_6H_{10}(OH)_2$		L
BA_4^-	-4.00	0.000(25°,1)	-10	0-35	T	C_L=0-0.9M A=OH^-	67Cc	b,q: ±0.1
	-10.00	-0.155(25°,1-2)	-40	0-35	T	C_L=0-0.9M A=OH^-		b,q: ±2.5
C55 CYCLOHEXANONE			$C_6H_{10}O$			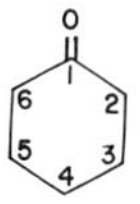		L
TiA_4	-13.72	(1)	--	25	C	S: CH_2Cl_2 A=Cl	72Eb	
	-13.60	(2)	--	25	C	S: CH_2Cl_2 A=Cl	72Eb	
BF_3	-31.3	(1)	--	25	C	S: CH_2Cl_2	71Pa	
$SiAB_3$	-3.65	1.45(30°,1)	-5.41	20-60	T	S: CCl_4 $X_L=5x10^{-2}$ $X_M=5x10^{-4}$ A=OH B=C_2H_5	69Lb	b
	-3.27	1.46(30°,1)	-4.13	10-45	T	S: CCl_4 $X_L=5x10^{-2}$ $X_M=5x10^{-4}$ A=OH B=CH_3	69Lb	b
	-3.94	1.74(30°,1)	-5.02	0-50	T	S: CCl_4 $X_L=5x10^{-2}$ $X_M=4x10^{-4}$ A=OH B=Phenyl	69Lb	b
I_2	-6.1	1.34(20°,1)	-15	-15-35	T	S: Freon	65Wc	b,k
I_2L	-20	3.95(20°)	-50	-15-35	T	S: Freon R: $2I_2L$ = $(I_2L)_2$	65Wc	b,k

Metal Ion	ΔH, kcal /mole	Log K	ΔS, cal /mole °K	T, °C	M	Conditions	Ref.	Remarks
C56	CYCLOHEXANONE, 2,6-dimethyl-		$C_8H_{14}O$			$C_6H_{10}O(CH_3)_2$		L
TiA_4	-9.76	(1)	--	25	C	S: CH_2Cl_2 A=Cl	72Eb	
	-14.4	(2)	--	25	C	S: CH_2Cl_2 A=Cl	72Eb	
BF_3	-23.5	(1)	--	25	C	S: CH_2Cl_2	71Pa	
C57	CYCLOHEXANONE, 2-methyl-		$C_7H_{12}O$			$C_6H_{10}O(CH_3)$		L
TiA_4	-10.73	(1)	--	25	C	S: CH_2Cl_2 A=Cl	72Eb	
	-9.36	(2)	--	25	C	S: CH_2Cl_2 A=Cl	72Eb	
BF_3	-25.0	(1)	--	25	C	S: CH_2Cl_2	71Pa	
C58	CYCLOHEXANONE, 3-methyl-		$C_7H_{12}O$			$C_6H_{10}O(CH_3)$		L
TiA_4	-11.44	(1)	--	25	C	S: CH_2Cl_2 A=Cl	72Eb	
	-9.16	(2)	--	25	C	S: CH_2Cl_2 A=Cl	72Eb	
BF_3	-25.5	(1)	--	25	C	S: CH_2Cl_2	71Pa	
C59	CYCLOHEXANONE, 4-methyl-		$C_7H_{12}O$			$C_6H_{10}O(CH_3)$		L
TiA_4	-12.24	(1)	--	25	C	S: CH_2Cl_2 A=Cl	72Eb	
	-8.92	(2)	--	25	C	S: CH_2Cl_2 A=Cl	72Eb	
BF_3	-27.0	(1)	--	25	C	S: CH_2Cl_2	71Pa	
C60	CYCLOHEXENE(cis)		C_6H_{10}			$CH_2CH_2CH_2CH_2CH{=}CH$ (ring)		L
Cu^+	-8.3	2.09(1)	-17.8	30	T	S: 1M($LiClO_4$) in 2-propanol	69Hb	
Ag^+	-5.74	-1.735(1)	-27.2	25	T	1M(KNO_3)	59Ta	o
	-5.58	-1.73(1)	-26.6	25	T	1.0M($AgNO_3$)	56Tc	o
	-5.80	-1.735(25°,1)	-27.4	0-25	T	μ=1.0($AgNO_3$,KNO_3)	38Wa	b,o
I_2	-2.18	0.518(1)	4.9	25	T	S: 2,2,4-Trimethylpentane	59Ta	
C61	2-CYCLOHEXEN-1-ONE		C_6H_8O			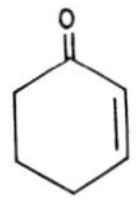		L
I_2	-5.4	1.49(20°,1)	-11	-15-35	T	S: Freon	65Wc	b,k
C62	CYCLOLEUCINE		$C_6H_{13}O_2N$			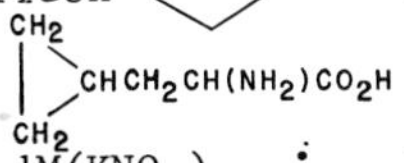		
CoA	-5.1	3.60(25°,1)	6.9	15-55	T	μ=0.1M(KNO_3) A=Methyliminodiacetate	72Ic	b
NiA	-4.7	4.50(25°,1)	4.8	15-55	T	μ=0.1M(KNO_3) A=Methyliminodiacetate	72Ic	b,q: ±0.8
CuA	-7.6	6.11(25°,1)	2.6	15-55	T	μ=0.1M(KNO_3) A=Methyliminodiacetate	72Ic	b,q: ±1.6
CdA	-4.8	3.41(25°,1)	1.2	15-55	T	μ=0.1M(KNO_3) A=Methyliminodiacetate	72Ic	b,q: ±0.5
C63	1,5-CYCLOOCTADIENE(cis,cis)		C_8H_{12}			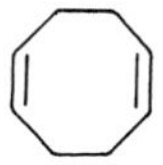		L

Metal Ion	ΔH, kcal /mole	Log K	ΔS, cal /mole °K	T, °C	M	Conditions	Ref.	Remarks
C63, cont.								
PdA_2B_2	-13.0	--	--	25	C	S: CH_2Cl_2 C_L=5x10^{-3}M C_M=1.3x10^{-4}M A=Cl B=C_6H_5CN R: PdA_2B_2 + L = PdA_2L + 2B	72Pc 73Pe	
C64 1,3,5,7-CYCLOOCTATETRAENE			C_8H_8			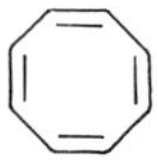		L
PdA_2B_2	-5.6	--	--	25	C	S: CH_2Cl_2 C_L=5x10^{-3}M C_M=2x10^{-4}M A=Cl B=C_6H_5CN R: PdA_2B_2 + L = PdA_2L + 2B	72Pc	
C65 CYCLOOCTENE(cis)			C_8H_{14}			$CH_2(CH_2)_5CH=CH$		L
Cu^+	-15.3	3.46(1)	-34.6	30	T	S: 1M($LiClO_4$) in 2-propanol	69Hb	
Ag^+	-4.11	-2.298(1)	-24.8	25	T	1M(KNO_3)	59Ta	o
I_2	-0.48	0.037(1)	1.4	25	T	S: 2,2,4-Trimethylpentane	59Ta	
C66 CYCLOPENTANE, 1,2-diamino-N,N,N',N'-tetraacetic acid (trans)			$C_{13}H_{20}O_8N_2$			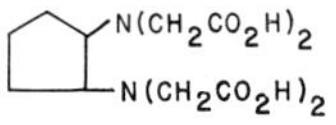		
Mg^{2+}	4.340	(1)	--	20	C	μ=0.1M($NaNO_3$)	71Ie	
	4.37	9.349(1)	57.7	20	C	μ=1M(HCl) C_L=0.01	69Sh	g:k
Ca^{2+}	-4.87	11.370(1)	35.4	20	C	μ=1M(HCl) C_L=0.01	69Sh	g:k
Sr^{2+}	-0.32	9.528(1)	42.5	20	C	μ=1M(HCl) C_L=0.01	69Sh	g:k
Ba^{2+}	-2.63	8.678(1)	30.7	20	C	μ=1M(HCl) C_L=0.01	69Sh	g:k
La^{3+}	-1.684	17.01(1)	72.1	20	C	μ=0.1M($NaNO_3$)	71Ie	
	-6.014	--	--	20	C	μ=0.1M($NaNO_3$) R: La^{3+} + MgL^{2-} = LaL^- + Mg^{2+}	71Ie	
Ce^{3+}	-1.633	17.28(1)	73.5	20	C	μ=0.1M($NaNO_3$)	71Ie	
	-5.963	--	--	20	C	μ=0.1M($NaNO_3$) R: Ce^{3+} + MgL^{2-} = CeL^- + Mg^{2+}	71Ie	
Pr^{3+}	-2.477	17.47(1)	71.5	20	C	μ=0.1M($NaNO_3$)	71Ie	
	-6.807	--	--	20	C	μ=0.1M($NaNO_3$) R: Pr^{3+} + MgL^{2-} = PrL^- + Mg^{2+}	71Ie	
Nd^{3+}	-2.722	17.72(1)	71.8	20	C	μ=0.1M($NaNO_3$)	71Ie	
	-7.052	--	--	20	C	μ=0.1M($NaNO_3$) R: Nd^{3+} + MgL^{2-} = NdL^- + Mg^{2+}	71Ie	

Metal Ion	ΔH, kcal /mole	Log K	ΔS, cal /mole °K	T, °C	M	Conditions	Ref.	Remarks
C66, cont.								
Sm^{3+}	-2.670	18.11(1)	73.8	20	C	μ=0.1M($NaNO_3$)	71Ie	
	-7.000	--	--	20	C	μ=0.1M($NaNO_3$) R: Sm^{3+} + MgL^{2-} = SmL^- + Mg^{2+}	71Ie	
Eu^{3+}	-1.855	18.21(1)	77.0	20	C	μ=0.1M($NaNO_3$)	71Ie	
	-6.185	--	--	20	C	μ=0.1M($NaNO_3$) R: Eu^{3+} + MgL^{2-} = EuL^- + Mg^{2+}	71Ie	
Gd^{3+}	-1.235	18.24(1)	79.3	20	C	μ=0.1M($NaNO_3$)	71Ie	
	-5.565	--	--	20	C	μ=0.1M($NaNO_3$) R: Gd^{3+} + MgL^{2-} = GdL^- + Mg^{2+}	71Ie	
Tb^{3+}	-0.348	18.64(1)	84.1	20	C	μ=0.1M($NaNO_3$)	71Ie	q: ±0.02
	-4.678	--	--	20	C	μ=0.1M($NaNO_3$) R: Tb^{3+} + MgL^{2-} = TbL^- + Mg^{2+}	71Ie	
Dy^{3+}	0.949	18.94(1)	89.9	20	C	μ=0.1M($NaNO_3$)	71Ie	
	-3.381	--	--	20	C	μ=0.1M($NaNO_3$) R: Dy^{3+} + MgL^{2-} = DyL^- + Mg^{2+}	71Ie	
Ho^{3+}	1.587	19.24(1)	93.5	20	C	μ=0.1M($NaNO_3$)	71Ie	
	-2.743	--	--	20	C	μ=0.1M($NaNO_3$) R: Ho^{3+} + MgL^{2-} = HoL^- + Mg^{2+}	71Ie	
Er^{3+}	1.203	19.49(1)	93.3	20	C	μ=0.1M($NaNO_3$)	71Ie	
	-3.127	--	--	20	C	μ=0.1M($NaNO_3$) R: Er^{3+} + MgL^{2-} = ErL^- + Mg^{2+}	71Ie	
Tm^{3+}	-0.263	19.71(1)	89.3	20	C	μ=0.1M($NaNO_3$)	71Ie	q: ±0.02
	-4.593	--	--	20	C	μ=0.1M($NaNO_3$) R: Tm^{3+} + MgL^{2-} = TmL^- + Mg^{2+}	71Ie	
Yb^{3+}	-1.861	19.95(1)	84.9	20	C	μ=0.1M($NaNO_3$)	71Ie	
	-6.191	--	--	20	C	μ=0.1M($NaNO_3$) R: Yb^{3+} + MgL^{2-} = YbL^- + Mg^{2+}	71Ie	
Lu^{3+}	0.214	20.20(1)	93.2	20	C	μ=0.1M($NaNO_3$)	71Ie	
	-4.016	--	--	20	C	μ=0.1M($NaNO_3$) R: Lu^{3+} + MgL^{2-} = LuL^- + Mg^{2+}	71Ie	

C67 CYCLOPENTANE, 1,3-diaza-1,3-dimethyl-2-oxo-2-thia- $C_4H_{12}ON_2S$

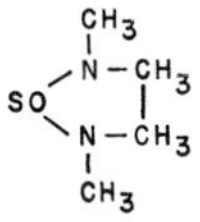

Metal Ion	ΔH, kcal /mole	Log K	ΔS, cal /mole °K	T, °C	M	Conditions	Ref.	Remarks
C67, cont.								
AlA_3	-24.32	--	--	28	C	S: Hexane A=Cl R: $1/2[AlA_3]_2 + L = AlLA_3$	68Hd	
C68	CYCLOPENTANE, methylene-		C_6H_{10}			$CH_2CH_2CH_2CH_2C{=}CH_2$ (ring)		L
Ag^+	-3.31	-1.174(<u>1</u>)	-16.5	25	T	1M(KNO_3)	59Ta	o
I_2	-0.96	0.415(<u>1</u>)	1.3	25	T	S: 2,2,4-Tri-methylpentane	59Ta	
C69	1,2-CYCLOPENTANEDIOL(<u>cis</u>)		$C_5H_{10}O_2$			$C_5H_8(OH)_2$		L
BA_4^-	-3.53	1.415(25°,<u>1</u>)	-5.4	0-35	T	C_L=0-0.9M A=OH^-	67Cc	b,q: ±0.4
	-6.06	2.152(25°,<u>1</u>-<u>2</u>)	-11	0-35	T	C_L=0-0.9M A=OH^-		b,q: ±0.6
C70	CYCLOPENTANONE		C_5H_8O			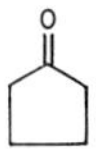		L
TiA_4	-13.58	(<u>1</u>)	--	25	C	S: CH_2Cl_2 A=Cl	72Eb	
	-13.50	(<u>2</u>)	--	25	C	S: CH_2Cl_2 A=Cl	72Eb	
BF_3	-28.5	(<u>1</u>)	--	25	C	S: CH_2Cl_2	71Pa	
I_2	-2.5	(<u>1</u>)	--	0-45	T	S: Cyclohexane $C_L \simeq$ 1M	66Bc	b,q: ±1
	-4.7	1.34(20°,<u>1</u>)	-10	-15-35	T	S: Freon	65Wc	b,k
I_2L	-16	3.88(20°)	-37	-15-35	T	S: Freon R: $2I_2L{=}(I_2L)_2$	65Wc	b,k
C71	CYCLOPENTANONE, 2,5-dimethyl-		$C_7H_{12}O$			$C_5H_8O(CH_3)_2$		L
TiA_4	-10.34	(<u>1</u>)	--	25	C	S: CH_2Cl_2 A=Cl	72Eb	
	-4.98	(<u>2</u>)	--	25	C	S: CH_2Cl_2 A=Cl	72Eb	
C72	CYCLOPENTANONE, 2-methyl-		$C_6H_{10}O$			$C_5H_8O(CH_3)$		L
TiA_4	-12.93	(<u>1</u>)	--	25	C	S: CH_2Cl_2 A=Cl	72Eb	
	-10.20	(<u>2</u>)	--	25	C	S: CH_2Cl_2 A=Cl	72Eb	
BF_3	-28.2	(<u>1</u>)	--	25	C	S: CH_2Cl_2	71Pa	
C73	CYCLOPENTANONE, 2,2,5,5-tetramethyl-		$C_9H_{16}O$			$C_5H_8O(CH_3)_4$		L
TiA_4	-9.35	(<u>1</u>)	--	25	C	S: CH_2Cl_2 A=Cl	72Eb	
	-4.82	(<u>2</u>)	--	25	C	S: CH_2Cl_2 A=Cl	72Eb	
BF_3	-22.2	(<u>1</u>)	--	25	C	S: CH_2Cl_2	71Pa	
C74	CYCLOPENTANONE, 2,2,4-trimethyl-		$C_8H_{14}O$			$C_5H_8O(CH_3)_3$		L
BF_3	-26.3	(<u>1</u>)	--	25	C	S: CH_2Cl_2	71Pa	

Metal Ion	ΔH, kcal /mole	Log K	ΔS, cal /mole °K	T, °C	M	Conditions	Ref.	Remarks
C75	CYCLOPENTANONE, 2,4,4-trimethyl-		$C_8H_{14}O$			$C_5H_8O(CH_3)_3$		L
TiA_4	-11.95	(1)	--	25	C	S: CH_2Cl_2 A=Cl	72Eb	
	-10.31	(2)	--	25	C	S: CH_2Cl_2 A=Cl	72Eb	
BF_3	-27.0	(1)	--	25	C	S: CH_2Cl_2	71Pa	
C76	CYCLOPENTENE		C_5H_8			$CH_2CH_2CH_2CH=CH$ (ring)		L
Cu^+	-12.7	2.86(1)	-28.7	30	T	S: 1M($LiClO_4$) in 2-Propanol	69Hb	
Ag^+	4.3	1.009(25°,1)	-9.9	0-40	T	S: Ethyleneglycol	65Ci	b
	-6.63	-0.925(1)	-26.5	25	T	1M(KNO_3)	59Ta	o
	-7.03	-0.945(1)	-27.9	25	T	1.0M($AgNO_3$)	56Tc	o
I_2	-0.43	0.447(1)	-0.6	25	T	S: 2,2,4-Trimethylpentane	59Ta	
C77	CYSTEINE		$C_3H_7O_2NS$			$HSCH_2CH(NH_2)CO_2H$		L^{2-}
Ni^{2+}	-13.7	19.77(25°,1-2)	45	15-40	T	μ=0.2M(KNO_3)	73Sn	b
C78	CYTIDINE-5'-diphosphoric acid		$C_{14}H_{24}O_{15}N_3P_2$					L^{4-}
Mg^{2+}	3.6	3.17(25°,1)	26.6	5-35	T	μ=0.1M(KNO_3)	73Bb	b,g:Na
	3.9	1.64(25°)	20.6	5-35	T	μ=0.1M(KNO_3) R: Mg^{2+} + HL^{3-} = $MgHL^-$	73Bb	b,g:Na
C79	CYTOCHROME C							
	-12.8	--	--	25	C	0.05M Buffer pH=6.00 R: Fe(II)L + Fe(III) = Fe(III)L + Fe(II)	69Wa	
	-12.6	--	--	25	C	0.05M Buffer pH=6.50 R: Fe(II)L + Fe(III) = Fe(III)L + Fe(II)	69Wa	
	-12.6	--	--	25	C	0.05M Buffer pH=7.00 R: Fe(II)L + Fe(III) = Fe(III)L + Fe(II)	69Wa	
	-11.6	--	--	25	C	0.05M Buffer pH=7.50 R: Fe(II)L + Fe(III) = Fe(III)L + Fe(II)	69Wa	
	-10.8	--	--	25	C	0.05M Buffer pH=8.00 R: Fe(II)L + Fe(III) = Fe(III)L + Fe(II)	69Wa	
	-8.92	--	--	25	C	0.05M Buffer pH=8.45 R: Fe(II)L + Fe(III) = Fe(III)L + Fe(II)	69Wa	
	-7.80	--	--	25	C	0.05M Buffer pH=8.90 R: Fe(II)L + Fe(III) = Fe(III)L + Fe(II)	69Wa	

Metal Ion	ΔH, kcal /mole	Log K	ΔS, cal /mole °K	T, °C	M	Conditions	Ref.	Remarks
C79, cont.								
	-8.87	--	--	25	C	0.05M Buffer pH=9.00 R: Fe(II)L + Fe(III) = Fe(III)L + Fe(II)	69Wa	
	-1.88	--	--	25	C	0.05M Buffer pH=9.75 R: Fe(II)L + Fe(III) = Fe(III)L + Fe(II)	69Wa	
	0.70	--	--	25	C	0.05M Buffer pH=10.00 R: Fe(II)L + Fe(III) = Fe(III)L + Fe(II)		
			D					
D1 DECANE, 1-amino-			$C_{10}H_{23}N$			$CH_3(CH_2)_9NH_2$		L
VOA_2	-10.59	(1)	--	25	C	μ→0. S: Nitrobenzene. A=Acetylacetonate	65Ca	
NiA	-17.2	6.13(1-2)	-29.6	25	T	$C_M=10^{-4}$. S: Benzene A=Diacetyl-bisbenzoyl hydrazine	60Sb	
D2 DECANE, 4,7-diaza-1,10-diamino-			$C_8H_{22}N_4$			$NH_2(CH_2)_3NH(CH_2)_2NH(CH_2)_3NH_2$		L
Ni^{2+}	-19.19	14.69(1)	2.9	25	C	μ=0.5M(KNO_3)	73Bc	
	-12.0	9.75	4.4	25	C	μ=0.5M(KNO_3) R: $Ni^{2+} + HL^+ = NiHL^{3+}$	73Bc	
Cu^{2+}	-24.8	21.69(1)	15.8	25	C	μ=0.10M	73Ha	
	-25.89	21.83(1)	13.1	25	C	μ=0.5M(KNO_3)	73Bc	
	-16.8	14.69	10.9	25	C	μ=0.10M R: $Cu^{2+} + HL^+ = CuHL^{3+}$	73Ha	
	-18.1	14.73	6.7	25	C	μ=0.5M(KNO_3) R: $Cu^{2+} + HL^+ = CuHL^{3+}$	73Bc	
Zn^{2+}	-10.62	11.25(1)	15.9	25	C	μ=0.5M(KNO_3)	73Bc	
	-7.7	7.18	7.0	25	C	μ=0.5M(KNO_3) R: $Zn^{2+} + HL^+ = ZnHL^{3+}$	73Bc	
D3 DEOXYRIBONUCLEIC ACID								
Fe^{3+}	≃0	5.114(25°,1)	23	25-45	T	$C_M<5\times10^{-4}$. pH=3.1	65Kb	b
D4 DICARBONYL, pentahaptocyclopentadienyl iron (dimer)								
Fe	0.90	--	3.5	-90- -70	T	S: 3:1 v/v $CS_2-C_6H_5CD_3$ mixture R: Complex Structure(I)→ complex structure(III)	72Bn	b
D5 α-DICYCLOPENTADIENE			$C_{10}H_{12}$			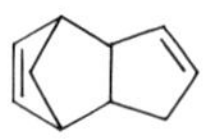		L

Metal Ion	ΔH, kcal /mole	Log K	ΔS, cal /mole °K	T, °C	M	Conditions	Ref.	Remarks
D5, cont.								
PdA_2B_2	-4.9	--	--	25	C	S: CH_2Cl_2 C_L=5x10^{-3}M C_M=1.9x10^{-4}M A=Cl B=C_6H_5CN R: PdA_2B_2 + L = PdA_2L + 2B	72Pc	
D6 2,12-DIMETHYL-3,7,11,17-tetraazabicyclo [11.3.1] heptadeca-1(17),2,11,13,15-pentaene		$C_{15}H_{22}N_3$				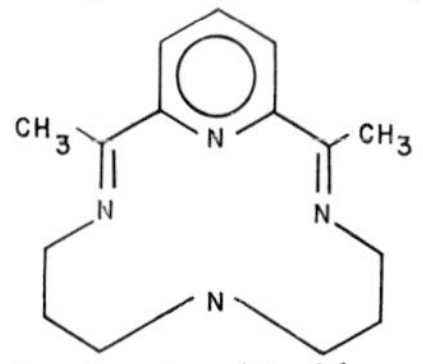		
$NiLA_2$	-4.28	--	-11.0	-43-84	T	S: Acetonitrile A=PF_6 R: $NiLA_2$(diamagnetic) = $NiLA_2$(paramagnetic)	71Re	b
	-6.97	--	-18.3	-3-84	T	A=BF_4 R: $NiLA_2$(diamagnetic) = $NiLA_2$(paramagnetic)	71Re	b
	-3.50	--	-6.7	-43-127	T	S: N,N-Dimethyl formamide A=PF_6 R: $NiLA_2$(diamagnetic) = $NiLA_2$(paramagnetic)	71Re	b
	-4.60	--	-9.2	21-127	T	S: Dimethyl sulfoxide A=PF_6 R: $NiLA_2$(diamagnetic) = $NiLA_2$(paramagnetic)	71Re	b
	-4.26	--	-11.5	-65-60	T	S: Methanol A=BF_4 R: $NiLA_2$(diamagnetic) = $NiLA_2$(paramagnetic)	71Re	b
D7 DINACTIN								
Na^+	-6.60	2.9(1)	-8.8	25	C	S: Methanol	73Za	
D8 1,4-DIOXANE			$C_4H_8O_2$			$OCH_2CH_2OCH_2CH_2$		L
CuA_2	-1.55	-0.155(1)	5.7	30	C	S: Benzene A=N-nitroso-N-Phenylhydroxylamine	71Gh	q: ±0.36
Al_2A_6	-20.9	--	--	25	C	S: Benzene C_M=0.07M A=Br R: 0.5(Al_2A_6) + L = AlA_3L	68Rc	
	-37.0	--	--	25	C	S: Benzene C_M=0.07M A=Br R: 0.5(Al_2A_6) + 2L = AlA_3L_2	68Rc	
I_2	-3.3	0.959(25°,1)	-6.7	--	T	S: CCl_4	57Dd 52Ka	1
	-2.70	0.29(25°,1)	-7.7	17-50	T	C_L=0.015 in mole fraction units S: CCl_4	51Ka	b,1

Metal Ion	ΔH, kcal /mole	Log K	ΔS, cal /mole °K	T, °C	M	Conditions	Ref.	Re-marks
D8, cont.								
I_2, cont.	-2.3	(1)	--	17-50	T	C_L=0-0.15 in mole fraction units S: CCl_4	51Ka	b,l
	-3.50	0.969(25°,1)	-7.3	25	T	S: n-Hexane	57Dd 52Ka 54Ka	b,l
	-4.3	1.367(25°,1)	-8.2	--	T	S: c-Hexane	57Dd 52Ka	l
	-3.6	(1)	--	--	C	S: Dioxane. Recalculation of data in 50 Ha ref.	61Aa	
	-3.25	(1)	--	--	C	S: Dioxane	50Ha	
D9 DISULFIDE, dibutyl-			$C_8H_{18}S_2$			$CH_3(CH_2)_3SS(CH_2)_3CH_3$		
I_2	-4.3	0.544(1)	-12.1	25	C	S: Ethylene chloride	66Nb	
D10 DISULFIDE, di-tert-butyl-			$C_8H_{18}S_2$			$(CH_3)_3CSSC(CH_3)_3$		L
I_2	-6.31	0.773(1)	-17.7	25	C	S: CCl_4 C_M=0.05M	72Na	
	-5.7	0.832(1)	-15.4	25	C	S: Ethylene chloride	66Nb	
D11 DISULFIDE, diethyl-			$C_4H_{10}S_2$			$C_2H_5SSC_2H_5$		L
I_2	-4.47	0.569(1)	-12.4	25	C	S: CCl_4 C_M=0.05M	72Na	
	-5.75	1.559(30°,1)	-11.8	20-40	T	S: CCl_4	61Ga	b
	-4.62	0.750(20°,1)	-12.3	0-30	T	C_M=10^{-3}. C_L=0.07-0.3. S: Dichloromethane	61Ta	b
	-4.3	0.491(1)	-12.1	25	C	S: Ethylene chloride	66Nb	
	-7.13	1.415(30°,1)	-17.0	20-40	T	S: n-Heptane	61Ga	b
D12 DISULFIDE, bis(diethylthiocarbamyl)-			$C_{10}H_{20}N_2S_4$			$(C_2H_5)_2NC(S)SSC(S)N(C_2H_5)_2$		L
I_2	-6.38	2.243(25°,1)	-11.4	10-30	T	S: CCl_4	69Gd	b,q: ±0.80
D13 DISULFIDE, diheptyl-			$C_{14}H_{30}S_2$			$C_7H_{15}SSC_7H_5$		L
I_2	-8.1	2.854(1)	-14.0	25	C	S: Octane	66Ad	
D14 DISULFIDE, diisopropyl-			$C_6H_{14}S_2$			$(CH_3)_2CHSSCH(CH_3)_2$		L
I_2	-4.73	0.639(1)	-13.0	25	C	S: CCl_4 C_M=0.05M	72Na	
	-4.4	0.544(1)	-12.4	25	C	S: Ethylene chloride	66Nb	
D15 DISULFIDE, dimethyl-			$C_2H_6S_2$			CH_3SSCH_3		L
I_2	-5.07	0.423(1)	-15.1	25	C	S: CCl_4 C_M=0.05M	72Na	

Metal Ion	ΔH, kcal /mole	Log K	ΔS, cal /mole °K	T, °C	M	Conditions	Ref.	Remarks
D15, cont.								
I_2, cont.	-4.4	0.380(1)	-13.1	25	C	S: Ethylene chloride	66Nb	
D16 DISULFIDE, bis(dimethylarsine)-			$C_4H_{12}S_2As_2$			$(CH_3)_2AsSSAs(CH_3)_2$		L
As	6.77	0.72(89°)	15.4	29-120	T	S: $C_2H_2Cl_4$ C_M=0.05-0.10M (Rearrangement reaction) R: $(CH_3)_2As(S)SAs(CH_3)_2$= $(CH_3)_2AsSSAs(CH_3)_2$	71Zf	
D17 DISULFIDE, bis(dimethylthiocarbamyl)-			$C_6H_{12}N_2S_4$			$(CH_3)_2NC(S)SSC(S)N(CH_3)_2$		L
I_2	-7.18	2.182(20°,1)	-14.4	10-30	T	S: CCl_4	69Gd	b,q: ±0.62
D18 DISULFIDE, dioctyl-			$C_{16}H_{34}S_2$			$H_{17}C_8SSC_8H_{17}$		L
I_2	-7.4	3.000(1)	-11.0	25	C	S: Octane	66Ad	
D19 DISULFIDE, dipropyl-			$C_6H_{14}S_2$			$CH_3CH_2CH_2SSCH_2CH_2CH_3$		L
I_2	-4.5	0.447(1)	-13.1	25	C	S: Ethylene chloride	66Nb	q: ±0.3
D20 DISULFIDE, 1-methyl-2-tert-butyl-			$C_5H_{12}S_2$			$CH_3SSC(CH_3)_3$		L
I_2	-5.00	0.559(1)	-14.2	25	C	S: CCl_4 C_M=0.05M	72Na	
D21 1,2-DITHIANE			$C_4H_6S_2$			CH_2:CHSSCH:CH_2		L
I_2	-5.74	0.997(1)	-14.7	25	C	S: CCl_4 C_M=0.037M	72Na	
D22 1,4-DITHIANE			$C_4H_8S_2$					L
I_2	-4.65	1.064(1)	-10.9	25	C	S: Ethylene chloride	66Nb	
D23 DODECANE, 5,8-diaza-, 2,11-			$C_{14}H_{34}N_4$			$H_2NCH(CH_3)CH_2C(CH_3)_2NH(CH_2)_2NHC(CH_3)_2CH_2CH(CH_3)NH_2$		L
Cu^{2+}	-24.86	22.41(1)	19.17	25	C	μ=0.10M(NaCl)	73Hc	
D24 DODECANE, 5,8-diaza-, 2,11-dionedioxime-,4,4,9,9-tetramethyl-			$C_{14}H_{30}O_2N_4$			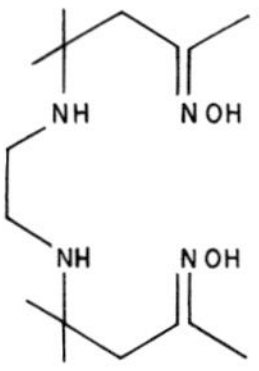		L^{2-}

Metal Ion	ΔH, kcal /mole	Log K	ΔS, cal /mole °K	T, °C	M	Conditions	Ref.	Remarks
D24, cont.								
	-13.10	13.24	16.7	25	C	μ=0.10M R: $Cu^{2+} + H_2L = CuH_2L^{2+}$	74Hb	
	5.66	-3.23	4.30	25	C	μ=0.10M R: $CuH_2L^{2+} = Cu(HL)^+ + H^+$	74Hb	
D25	DODECANE, 5,8-dithio-		$C_{10}H_{22}S_2$			$CH_3(CH_2)_3S(CH_2)_2S(CH_2)_3CH_3$		L
TiA_4	-25.2	3.66(1)	-67.8	25	C	S: Benzene A=Cl	68Ge 70Gh 68Gf	
SnA_4	-20.1	(1)	--	--	C	S: Benzene C_M=0.05-0.08M A=Br	68Ge	u
	-27.8	4.25(1)	-73.8	25	C	S: Benzene A=Cl	65Gb 68Ge 70Gh 68Gf	
TeA_4	-12.8	(1)	--	25	C	S: Benzene A=Cl	73Ph	
D26	DODECANE, 2,5,8,11-tetraoxa-		$C_8H_{18}O_4$			$CH_3OCH_2CH_2OCH_2CH_2OCH_2CH_2OCH_3$		L
NaA	-3.6	1.83(25°)	-3.6	-40-25	T	$C_L<4x10^{-2}$. S: 2-Methyl tetrahydrofuran. R: NaA + L = LNaA A=Biphenyl	67Sb	b
	-5.4	1.83(25°)	-9.7	-40-25	T	$C_L<4x10^{-2}$. S: 2-Methyl tetrahydrofuran. R: NaA + L = NaLA A=Biphenyl	67Sb	b
	-5.4	0.954(25°)	-14.5	0-50	T	S: Tetrahydrofuran A=Fluorenyl ion R: NaA + L = NaLA	70Ce	b
	-4.6	1.54(25°)	-8.4	-40-25	T	$C_L<4x10^{-2}$. S: Tetrahydropyran R: NaA + L = LNaA A=Biphenyl	67Sb	b
	-7.0	1.39(25°)	-17	-40-25	T	$C_L<4x10^{-2}$. S: Tetrahydropyran R: NaA + L = NaLA A=Biphenyl	67Sb	b
D27	DUROSEMIQUINONE ION							
Na^+	2.7	9.54(-75°,1)	38	--	T	S: 1,2-Dimethoxyethane	72Ac	b,u: (-78°--60°)
K^+	3.5	12.64(-75°,1)	50	--	T	S: 1,2-Dimethoxyethane	72Ac	b,u: (-78°--60°)

Metal Ion	ΔH, kcal /mole	Log K	ΔS, cal /mole °K	T, °C	M	Conditions	Ref.	Remarks
D27, cont.								
Rb^+	5.5	9.76(-75°,1)	48	--	T	S: 1,2-Dimethoxy-ethane	72Ac	b,u: (-78°--60°)
Cs^+	7.5	9.87(-75°,1)	55	--	T	S: 1,2-Dimethoxy-ethane	72Ac	b,u: (-78°--60°)

E

Metal Ion	ΔH, kcal /mole	Log K	ΔS, cal /mole °K	T, °C	M	Conditions	Ref.	Remarks
E1 EICOSANE, 5,16-dithio-			$C_{18}H_{38}S_2$			$CH_3(CH_2)_3S(CH_2)_{10}S(CH_2)_3CH_3$		L
SnA_4	-24.0	(1)	--	--	C	S: Benzene C_M=0.05-0.08M. A=Cl	65Gb	u
TeA_4	-12.6	3.14(1)	-27.9	25	C	S: Benzene A=Cl	73Ph	
	-7.2	(2)	--	25	C	S: Benzene A=Cl	73Ph	
E2 EPHEDRINE			$C_{10}H_{15}ON$			$C_6H_5CH(OH)CH(CH_3)NHCH_3$		L
Be^{2+}	-12.6	6.57(25°,1)	-13	0-45	T	μ=0.12M(KCl) $C_M \simeq 2 \times 10^{-3}M$ $C_L \simeq 8 \times 10^{-3}M$	69Cb	b
	-13.5	5.47(25°,2)	-21	0-45	T	μ=0.12M(KCl) $C_M \simeq 2 \times 10^{-3}M$ $C_L \simeq 8 \times 10^{-3}M$	69Cb	b
E3 ERIOCHROME BLACK T			$C_{20}H_{12}O_7N_3SNa$			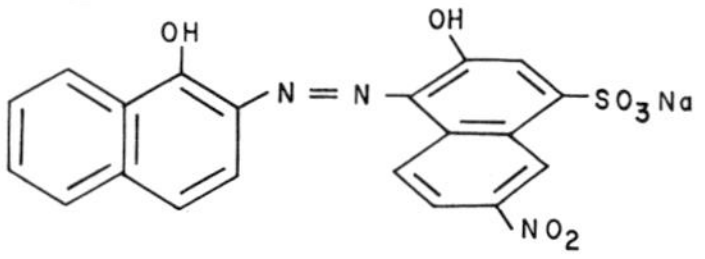		L^{3-}
Mg^{2+}	9.9	3.954	51	10-60	T	μ=0.010 R: $Mg^{2+} + HL^{2-} = MgL^- + H^+$	71Kd	b,r
	10.0	3.786	51	10-60	T	μ=0.029 R: $Mg^{2+} + HL^{2-} = MgL^- + H^+$	71Kd	b,r
	9.7	3.636	49	10-60	T	μ=0.060 R: $Mg^{2+} + HL^{2-} = MgL^- + H^+$	71Kd	b,r
	9.9	3.509	49	10-60	T	μ=0.100 R: $Mg^{2+} + HL^{2-} = MgL^- + H^+$	71Kd	b,r
E4 ETHANE, 1-amino-			C_2H_7N			$NH_2CH_2CH_3$		L
NiA	-16.5	4.65(1-2)	-33.3	25	T	$C_M=10^{-4}$ S: Benzene A=Diacetyl-bisbenzoyl hydrazine	60Sb	
Ag^+	-5	3.46(1)	-3.4	25	C	μ=0	71He	q:±2
	-7	3.90(2)	-6.7	25	C	μ=0	71He	q:±2
	-12.51	7.36(1-2)	-8.3	25	C	μ=0	71He	
	-13.0	7.32(1-2)	-10.1	25	C	--	55Fa	

Metal Ion	ΔH, kcal /mole	Log K	ΔS, cal /mole °K	T, °C	M	Conditions	Ref.	Remarks
E4, cont.								
I_2	-7.5	2.857(20°,1)	-12.5	22	C	low μ. S: n-Heptane	64Sd 60Ya	b
	-9.8	3.833(20°,1)	-15.9	20-40	T	$C_M=10^{-5}$ $C_L=10^{-4}-10^{-3}$ S: n-Heptane	60Ya	b
E5 ETHANE,1-amino-2-diphenyl-methoxy-N,N-dimethyl-			$C_{17}H_{21}ON$			$(C_6H_5)_2CHOCH_2CH_2N(CH_3)_2$		L^-
Be^{2+}	-16.3	6.30(25°,1)	-26	0-45	T	μ=0.12M(KCl) $C_M \approx 2x10^{-3}M$ $C_L \approx 8x10^{-3}M$	69Cb	b
	-2.5	5.04(25°,2)	14	0-45	T	μ=0.12M(KCl) $C_M \approx 2x10^{-3}M$ $C_L \approx 8x10^{-3}M$	69Cb	b
Co^{2+}	-1.8	6.54(25°,1-2)	24.0	0&25	T	0.06M(KCl)	62Aa	a,b
Ni^{2+}	-6.3	6.52(25°,1-2)	8.8	0&25	T	0.06M(KCl)	62Aa	a,b
Cu^{2+}	-7.8	10.50(25°,1-2)	22.0	0&25	T	0.06M(KCl)	62Aa	a,b
Cd^{2+}	-8.6	7.28(1)	4	25	T	μ=0.1(KNO_3)	64Ad	
E6 ETHANE, 1-amino-2-methoxy-			C_3H_9ON			$NHCH_2CH_2OCH_3$		L
Ag^+	-5.5	3.18(30°,1)	-3	25	T	μ=0	59Lb	b
	-5.3	3.37(30°,2)	-2	25	T	μ=0	59Lb	b
E7 ETHANE, 1,2-bis(methylthio)-			$C_4H_{10}S_2$			$CH_3SCH_2CH_2SCH_3$		L
CuA_2	-14.85	(1)	--	--	C	A=Cl	25Ma	
CuA	-5.88	(1)	--	--	C	A=I	25Ma	
ZnA_2	-15.35	(1)	--	--	C	A=Br	25Ma	
	-13.21	(1)	--	--	C	A=Cl	25Ma	
	-14.23	(1)	--	--	C	A=I	25Ma	
CdA_2	-11.27	(1)	--	--	C	A=Br	25Ma	
	-12.16	(1)	--	--	C	A=Cl	25Ma	
	-9.96	(1)	--	--	C	A=I	25Ma	
HgA_2	-11.33	(1)	--	--	C	A=Br	25Ma	
	-13.23	(1)	--	--	C	A=Cl	25Ma	
	-4.83	(1)	--	--	C	A=I	25Ma	
SnA_4	-19.40	(1)	--	--	C	A=Br	25Ma	
	-28.06	(1)	--	--	C	A=Cl	25Ma	
	-16.58	(1)	--	--	C	A=I	25Ma	
BiA_3	-12.0	(2-3)	--	--	C	A=I	25Ma	
E8 ETHANE, cyano-			C_3H_5N			CH_3CH_2CN		L
TiA_4	-11.5	3.59(1)	-22.1	25	C	S: Benzene A=Cl	70Gh	
	-11.9	3.96(1)	-21.8	25	C	S: Benzene A=Cl	70Gh 68Gf	
SnA_4	-5.1	(1)	--	25	C	S: Benzene A=Cl	70Gh 68Gf	

Metal Ion	ΔH, kcal /mole	Log K	ΔS, cal /mole °K	T, °C	M	Conditions	Ref.	Remarks
E9 ETHANE, 1,2-diamino-			$C_2H_8N_2$			$H_2NCH_2CH_2NH_2$		L
Y^{3+}	-20.0	(1)	--	23	C	S: Acetonitrile 0.026M $Y(ClO_4)_3$	69Fc	c
	-18.7	(2)	--	23	C	S: Acetonitrile 0.026M $Y(ClO_4)_3$	69Fc	c
	-12.9	(3)	--	23	C	S: Acetonitrile 0.026M $Y(ClO_4)_3$	69Fc	c
	-10.5	(4)	--	23	C	S: Acetonitrile 0.026M $Y(ClO_4)_3$	69Fc	c
La^{3+}	-17.3	9.5(1)	-15.1	23	C	S: Acetonitrile 0.026M $La(ClO_4)_3$	69Fc	c
	-13.9	(1)	--	23	C	S: Acetonitrile 0.026M $La(NO_3)_3$	69Fc	c
	-15.5	7.5(2)	-18.2	23	C	S: Acetonitrile 0.026M $La(ClO_4)_3$	69Fc	c
	-12.6	(2)	--	23	C	S: Acetonitrile 0.026M $La(NO_3)_3$	69Fc	c
	-13.8	6.2(3)	-18.4	23	C	S: Acetonitrile 0.026M $La(ClO_4)_3$	69Fc	c
	-11.0	3.3(4)	-22.1	23	C	S: Acetonitrile 0.026M $La(ClO_4)_3$	69Fc	c
Pr^{3+}	-18.7	(1)	--	23	C	S: Acetonitrile 0.026M $Pr(ClO_4)_3$	69Fc	c
	-14.3	(1)	--	23	C	S: Acetonitrile 0.026M $Pr(NO_3)_3$	69Fc	c
	-16.8	(2)	--	23	C	S: Acetonitrile 0.026M $Pr(ClO_4)_3$	69Fc	c
	-12.8	(2)	--	23	C	S: Acetonitrile 0.026M $Pr(NO_3)_3$	69Fc	c
	-13.6	(3)	--	23	C	S: Acetonitrile 0.026M $Pr(ClO_4)_3$	69Fc	c
	-10.8	(4)	--	23	C	S: Acetonitrile 0.026M $Pr(ClO_4)_3$	69Fc	c
Nd^{3+}	-18.8	(1)	--	23	C	S: Acetonitrile 0.026M $Nd(ClO_4)_3$	69Fc	c
	-14.7	(1)	--	23	C	S: Acetonitrile 0.026M $Nd(NO_3)_3$	69Fc	c
	-16.9	(2)	--	23	C	S: Acetonitrile 0.026M $Nd(ClO_4)_3$	69Fc	c
	-13.1	(2)	--	23	C	S: Acetonitrile 0.026M $Nd(NO_3)_3$	69Fc	c
	-13.8	(3)	--	23	C	S: Acetonitrile 0.026M $Nd(ClO_4)_3$	69Fc	c
	-10.9	(4)	--	23	C	S: Acetonitrile 0.026M $Nd(ClO_4)_3$	69Fc	c
Sm^{3+}	-19.3	(1)	--	23	C	S: Acetonitrile 0.026M $Sm(ClO_4)_3$	69Fc	c
	-18.0	(2)	--	23	C	S: Acetonitrile 0.026M $Sm(ClO_4)_3$	69Fc	c
	-13.5	(3)	--	23	C	S: Acetonitrile 0.026M $Sm(ClO_4)_3$	69Fc	c
	-9.9	(4)	--	23	C	S: Acetonitrile 0.026M $Sm(ClO_4)_3$	69Fc	c

Metal Ion	ΔH, kcal /mole	Log K	ΔS, cal /mole °K	T, °C	M	Conditions	Ref.	Re-marks
E9, cont.								
Eu^{3+},	-19.8	(1)	--	23	C	S: Acetonitrile 0.026M $Eu(ClO_4)_3$	69Fc	c
	-18.3	(2)	--	23	C	S: Acetonitrile 0.026M $Eu(ClO_4)_3$	69Fc	c
	-13.9	(3)	--	23	C	S: Acetonitrile 0.026M $Eu(ClO_4)_3$	69Fc	c
	-9.7	(4)	--	23	C	S: Acetonitrile 0.026M $Eu(ClO_4)_3$	69Fc	c
Gd^{3+}	-19.5	(1)	--	23	C	S: Acetonitrile 0.026M $Gd(ClO_4)_3$	69Fc	c
	-18.0	(2)	--	23	C	S: Acetonitrile 0.026M $Gd(ClO_4)_3$	69Fc	c
	-13.9	(3)	--	23	C	S: Acetonitrile 0.026M $Gd(ClO_4)_3$	69Fc	c
	-9.5	(4)	--	23	C	S: Acetonitrile 0.026M $Gd(ClO_4)_3$	69Fc	c
Tb^{3+}	-19.9	10.4(1)	-19.8	23	C	S: Acetonitrile 0.026M $Tb(ClO_4)_3$	69Fc	c
	-18.6	8.4(2)	-24.5	23	C	S: Acetonitrile 0.026M $Tb(ClO_4)_3$	69Fc	c
	-13.1	6.2(3)	-16.0	23	C	S: Acetonitrile 0.026M $Tb(ClO_4)_3$	69Fc	c
	-9.0	3.2(4)	-15.8	23	C	S: Acetonitrile 0.026M $Tb(ClO_4)_3$	69Fc	c
Dy^{3+}	-19.9	(1)	--	23	C	S: Acetonitrile 0.026M $Dy(ClO_4)_3$	69Fc	c
	-18.4	(2)	--	23	C	S: Acetonitrile 0.026M $Dy(ClO_4)_3$	69Fc	c
	-12.6	(3)	--	23	C	S: Acetonitrile 0.026M $Dy(ClO_4)_3$	69Fc	c
	-9.2	(4)	--	23	C	S: Acetonitrile 0.026M $Dy(ClO_4)_3$	69Fc	c
Ho^{3+}	-19.9	(1)	--	23	C	S: Acetonitrile 0.026M $Ho(ClO_4)_3$	69Fc	c
	-18.2	(2)	--	23	C	S: Acetonitrile 0.026M $Ho(ClO_4)_3$	69Fc	c
	-12.7	(3)	--	23	C	S: Acetonitrile 0.026M $Ho(ClO_4)_3$	69Fc	c
	-10.0	(4)	--	23	C	S: Acetonitrile 0.026M $Ho(ClO_4)_3$	69Fc	c
Er^{3+}	-20.1	(1)	--	23	C	S: Acetonitrile 0.026M $Er(ClO_4)_3$	69Fc	c
	-18.7	(2)	--	23	C	S: Acetonitrile 0.026M $Er(ClO_4)_3$	69Fc	c
	-13.1	(3)	--	23	C	S: Acetonitrile 0.026M $Er(ClO_4)_3$	69Fc	c
	-11.5	(4)	--	23	C	S: Acetonitrile 0.026M $Er(ClO_4)_3$	69Fc	c
Yb^{3+}	-20.1	11.5(1)	-15.4	23	C	S: Acetonitrile 0.026M $Yb(ClO_4)_3$	69Fc	c

Metal Ion	ΔH, kcal /mole	Log K	ΔS, cal /mole °K	T, °C	M	Conditions	Ref.	Remarks
E9, cont.								
Yb^{3+}, cont.	-18.8	9.3(2)	-21.1	23	C	S: Acetonitrile 0.026M $Yb(ClO_4)_3$	69Fc	c
	-14.4	6.5(3)	-19.0	23	C	S: Acetonitrile 0.026M $Yb(ClO_4)_3$	69Fc	c
	-12.8	3.8(4)	-25.9	23	C	S: Acetonitrile 0.026M $Yb(ClO_4)_3$	69Fc	c
Lu^{3+}	-20.0	(1)	--	23	C	S: Acetonitrile 0.026M $Lu(ClO_4)_3$	69Fc	c
	-18.6	(2)	--	23	C	S: Acetonitrile 0.026M $Lu(ClO_4)_3$	69Fc	c
	-14.3	(3)	--	23	C	S: Acetonitrile 0.026M $Lu(ClO_4)_3$	69Fc	c
	-12.8	(4)	--	23	C	S: Acetonitrile 0.026M $Lu(ClO_4)_3$	69Fc	c
Mn^{2+}	-2.80	2.75(1)	3.2	25	C	1M(KCl)	60Cb	
	-6.00	4.87(1-2)	2.2	25	C	1M(KCl)	60Ca/60Cb	
	-11.05	5.79(1-3)	-10.5	25	C	1M(KCl)	60Cb	
Fe^{2+}	-5.05	4.32(1)	2.9	25	C	1M(KCl)	60Cb	
	-10.40	7.66(1-2)	0.2	25	C	1M(KCl)	60Ca/60Cb	
	-15.85	9.71(1-3)	-8.7	25	C	1M(KCl)	60Cb	
Co^{2+}	-6.90	5.94(1)	4.0	25	C	1M(KCl)	60Cb	
	-13.95	10.66(1-2)	2.0	25	C	1M(KCl)	60Ca/60Cb	
	-22.15	13.67(1-3)	11.8	25	C	1M(KCl)	54Wa/60Cb	
CoA	-8.1	4.36(1)	-7.2	25	C	μ=0.10M(KNO_3) A=Ethylenediamine-diacetic acid	70Da	
Ni^{2+}	-9.5	7.27(30°,1)	2	10-40	T	μ=0	58Ba	b
	-15.00	7.32(1)	-17	25	C	0.1M(KCl)	60Cb/54Da	
	-9.25	7.49(1)	3.3	25	C	μ=0.30(ClO_4^-)	60Ca 67Hf	
	-9.01	7.51(1)	4.1	25	C	1M(KNO_3)	52Ba/55Pb	
	-9.05	7.51(1)	4	25	C	1M(KNO_3)	54Ba/60Ca	
	-9.07	(1)	--	25	C	μ=1M(KNO_3)	54Wa/68Pa	
	-7.5	6.11(30°,2)	3	10-40	T	μ=0	58Ba	b
	-9.19	6.44(2)	-1.3	25	C	μ=0.30(ClO_4^-)	55Cc/67Hf	
	-9.18	6.35(2)	-1.8	25	C	1M(KNO_3)	46Ca/55Pb	
	-9.24	(2)	--	25	C	μ=1M(KNO_3)	68Pa	
	-8.7	4.20(30°,3)	-10	10-40	T	μ=0	58Ba	b
	-9.46	4.11(3)	-12.9	25	C	μ=0. 0(ClO_4^-)	67Hf	
	-9.71	4.42(3)	-12.4	25	C	1M(KNO_3)	55Pb	

Metal Ion	ΔH, kcal /mole	Log K	ΔS, cal /mole °K	T, °C	M	Conditions	Ref.	Re-marks
E9, cont.								
Ni^{2+}, cont.	-9.66	(3)	--	25	C	μ=1M(KNO_3)	68Pa	
	-17.25	13.38(1-2)	3.4	25	C	0.1M(KCl)	54Da	
	-16.3	14.5(1-2)	7	0	C	0.50M(KNO_3)	54Bb	
	-17.90	13.80(1-2)	3.5	25	C	1M(KNO3)	60Ca	
	-28.01	17.48(1-3)	-14.0	25	C	0.1M(KCl)	54Da	
	-24.9	20.1(1-3)	1	0	C	0.50M(KNO_3)	54Bb	
	-27.90	18.28(1-3)	-10.1	25	C	1M(KNO_3)	55Pb	
	-27.40	18.25(1-3)	-8.5	25	C	1M(KNO_3)	60Ca	
NiA	-9.86	6.31(1)	-4.2	25	C	μ=0.10M(KNO_3) A=Ethylenediamine-diacetic acid	70Da	
	-9.46	5.51(1)	-6.5	25	C	μ=0.10M(KNO_3) A=Ethylenediamine-dipropionic acid	70Da	
NiL_3	5.8	-1.69(25°)	11.8	0.2-75.0		S: DMSO R: δ→λ Ring Inversion	73Cg	b
	7.6	-1.61(25°)	18.2	20.1-103.5		S: D_2O R: δ→λ Ring Inversion	73Cg	b
Cu^{2+}	-12.6	10.48(1)	5.7	25	C	μ→0	67Pc 52Ba	
	-12.8	10.36(30°,1)	5	10-40	T	μ=0	58Ba 54Ba	b
	-11.9	10.67(25°,1)	9.0	0-50	T	μ=0.15	55Cc 54Wa	b
	-12.45	10.44(1)	6.0	25	C	μ=0.30(ClO_4^-)	67Hf 48Ba	
	-12.1	10.57(1)	6.5	15-40	T	μ=0.3M($NaClO_4$)	72Md	b
	-12.56	10.582(1)	6.3	25	C	0.5M(KNO_3)	72Be	
	-13.0	10.72(1)	5.4	25	C	1M(KNO_3)	55Pb	
	-14.6	11.023(1)	1.34	25	T	μ=2	53Sb	
	-12.6	9.07(2)	-0.7	25	C	μ→0	67Pc	
	-12.3	8.93(30°,2)	0	10-40	T	μ=0	59Ma	b
	-11.3	9.10(25°,2)	3.7	0-50	T	μ=0.15	55Cc	b
	-12.28	8.98(2)	-0.1	25	C	μ=0.30(ClO_4^-)	67Hf	
	-12.5	9.19(2)	0.2	15-40	T	μ=0.3M($NaClO_4$)	72Md	b
	-12.64	9.151(2)	-0.5	25	C	0.5M(KNO_3)	72Be	
	-12.4	9.31(2)	0.9	25	C	1M(KNO_3)	55Pb	
	-25.2	19.56(1-2)	5.0	25	C	μ→0	67Pc	
	-25.16	20.03(1-2)	7.1	25	C	0.1M(KCl)	54Da	
	-24.6	21.3(1-2)	7	0	C	0.5M(KNO_3)	54Bb	
	-25.4	20.03(1-2)	6.3	25	C	1M(KNO_3)	55Pb	
	-28.4	20.61(1-2)	-0.7	25	T	μ=2	53Sb	
	-0.51	1.42(25°)	4.8	-75&25	T	S: 50 v % Methanol R: Cu + CuL_2 = 2CuL	73Ya	a,b,q: ±0.24
CuA	-13.3	6.66(1)	-14.1	25	C	μ=0.1M(KNO_3) A=Ethylenediamine-diacetic acid	70Da	
CuL_2	-0.86	1.44(20°)	3.65	20&50	T	S: 50 v % Methanol A=L-Alanine R: CuL_2 + CuA_2 = 2CuAL	71Yb 71Yc	a,b,q: ±0.6

Metal Ion	ΔH, kcal /mole	Log K	ΔS, cal /mole °K	T, °C	M	Conditions	Ref.	Remarks
E9, cont.								
CuL_2, cont.	0.01	0.78(20°)	3.62	-100-50	T	S: 50 v % Methanol A=β-Alanine R: $CuL_2 + CuA_2 = 2CuAL$	71Yb 71Yc	b,q: ±0.2
	1.02	0.14(20°)	4.12	-100-50	T	S: 50 v % Methanol A=N,N-Dimethyl-glycine R: $CuL_2 + CuA_2 = 2CuAL$	71Yb 71Yc	q: ±0.3
Zn^{2+}	-5.0	5.55(30°,1)	9	10-40	T	μ=0	58Ba 60Cb	b
	-9.8	5.50(1)	-8	25	C	0.1M(KCl)	54Da 59Ma	
	-6.65	5.79(1)	4.2	25	C	1M(KCl)	60Ca 54Wa	
	-5.2	5.88(1)	9.6	27	C	1M(KNO_3)	61Me	
	-6.6	6.148(1)	6.0	25	T	μ=2	53Sb	
	-5.2	4.89(30°,2)	5	10-40	T	μ=0	58Ba	b
	-5.2	5.12(2)	6.1	27	C	1M(KNO_3)	61Me	
	-1.8	3.22(30°,3)	9	10-40	T	μ=0	58Ba	b
	-5.2	1.85(3)	-8.8	27	C	1M(KNO_3)	61Me	
	-11.45	10.05(1-2)	7.6	25	C	0.1M(KCl)	54Da	
	-13.75	10.55(1-2)	2.2	25	C	1M(KCl)	60Cb	
	-12.5	11.49(1-2)	10.7	25	T	μ=2	53Sb	
	-18.46	11.40(1-3)	-9.8	25	C	0.1M(KCl)	54Da	
	-20.70	12.31(1-3)	-13.1	25	C	1M(KCl)	60Cb	
ZnA	-7.1	4.44(1)	-3.5	25	C	μ=0.10(KNO_3) A=Ethylenediamine-diacetic acid	70Da	
	-6.6	4.03(1)	-3.7	25	C	μ=0.10M(KNO_3) A=Ethylenediamine-dipropionic acid	70Da	
Cd^{2+}	-6.2	5.34(30°,1)	4	10-40	T	μ=0	58Ba	b
	-1.20	5.2(1)	20	25	C	0.1M(KCl)	54Da	
	-6.2	5.6(1)	5	27	C	1M(KNO_3)	62Md	
	-7.03	5.836(1)	3.1	25	T	μ=2	53Sa	
	-7.5	4.38(30°,2)	-5	10-40	T	μ=0	58Ba	b
	-13.33	9.4(1-2)	-1.7	25	C	0.1M(KCl)	54Da	
	-12.4	10.2(1-2)	5.2	27	C	1M(KNO_3)	62Md	
	-13.5	10.62(1-2)	3.3	25	T	μ=2	53Sa	
	-19.70	11.1(1-3)	-15.3	25	C	0.1M(KCl)	54Da	
	-18.6	12.2(1-3)	-6.0	27	C	1M(KNO_3)	62Md	
CdA	-7.1	4.33(1)	-4.0	25	C	μ=0.10M(KNO_3). A= Ethylenediamine-diacetic acid	70Da	
HgA_2	-32.9	23.10(1-2)	-5	25	T	0.1M(KNO_3)	61Rc	
	-8.66	5.54	-3.7	25	C	μ=0. R: $HgA_2 + L = HgAL^+ + A^-$. A=Cl	66Pd	
	-9.03	4.19	-11.1	25	C	μ=0. R: $HgAL^+ + L = HgL_2^{2+} + A^-$. A=Cl		
E10 ETHANE, 1,2-diamino-N-n-butyl- $C_6H_{16}N_2$						$CH_3CH_2CH_2CH_2NHCH_2CH_2NH_2$		L
Ni^{2+}	-7.8	7.25(1)	4.8	0	T	μ=0.5(KNO_3)	52Ba	
	-6.1	5.97(2)	5.1	0	T	μ=0.5(KNO_3)	52Ba	
	-8.8	2.79(3)	-19.4	0	T	μ=0.5(KNO_3)	52Ba	

Metal Ion	ΔH, kcal /mole	Log K	ΔS, cal /mole °K	T, °C	M	Conditions	Ref.	Re-marks
E10, cont.								
Cu^{2+}	-8.0	10.47(1)	19	0	T	μ=0.5(KNO_3)	52Ba	
	-8.2	8.82(2)	12	0	T	μ=0.5(KNO_3)	52Ba	
E11 ETHANE, 1,2-diamino-N,N'-diacetic acid			$C_6H_{12}O_4N_2$			$HO_2CCH_2NHCH_2CH_2NHCH_2CO_2H$		L^{2-}
La^{3+}	-1.12	6.955(1)	28.0	25	C	μ=1.00($NaClO_4$)	74Gc	c
	-2.04	4.598(2)	14.3	25	C	μ=1.00($NaClO_4$)	74Gc	
Ce^{3+}	-1.72	(1)	28.0	25	C	μ=1.00($NaClO_4$)	74Gc	
	-1.37	(2)	17.6	25	C	μ=1.00($NaClO_4$)	74Gc	
Pr^{3+}	-2.09	(1)	28.4	25	C	μ=1.00($NaClO_4$)	74Gc	
	-1.71	(2)	18.0	25	C	μ=1.00($NaClO_4$)	74Gc	
Nd^{3+}	-2.51	7.976(1)	28.0	25	C	μ=1.00($NaClO_4$)	74Gc	
	-2.08	5.613(2)	18.8	25	C	μ=1.00($NaClO_4$)	74Gc	
Sm^{3+}	-2.92	8.259(1)	28.0	25	C	μ=1.00($NaClO_4$)	74Gc	
	-3.10	6.180(2)	17.9	25	C	μ=1.00($NaClO_4$)	74Gc	
Gd^{3+}	-2.12	8.245(1)	30.4	25	C	μ=1.00($NaClO_4$)	74Gc	
	-3.84	6.173(2)	16.0	25	C	μ=1.00($NaClO_4$)	74Gc	
Tb^{3+}	-1.14	8.263(1)	33.9	25	C	μ=1.00($NaClO_4$)	74Gc	
	-4.15	6.786(2)	17.2	25	C	μ=1.00($NaClO_4$)	74Gc	
Dy^{3+}	-0.85	(1)	35.4	25	C	μ=1.00($NaClO_4$)	74Gc	
	-3.89	(2)	19.3	25	C	μ=1.00($NaClO_4$)	74Gc	
Ho^{3+}	-0.48	(1)	36.6	25	C	μ=1.00($NaClO_4$)	74Gc	
	-2.86	(2)	25.5	25	C	μ=1.00($NaClO_4$)	74Gc	
Er^{3+}	-0.33	8.508(1)	37.8	25	C	μ=1.00($NaClO_4$)	74Gc	
	-3.11	7.430(2)	23.6	25	C	μ=1.00($NaClO_4$)	74Gc	
Tm^{3+}	-0.39	(1)	38.2	25	C	μ=1.00($NaClO_4$)	74Gc	
	-3.36	(2)	24.9	25	C	μ=1.00($NaClO_4$)	74Gc	
Yb^{3+}	-0.77	8.827(1)	37.8	25	C	μ=1.00($NaClO_4$)	74Gc	
	-3.81	8.110(2)	24.3	25	C	μ=1.00($NaClO_4$)	74Gc	
UO_2^{2+}	1.5	11.41(25°,1)	57	20-40	T	μ=0.1M(KNO_3)	69Sg	b
Mn^{2+}	-0.85	7.05(1)	29.4	25	C	μ=0.10M(KNO_3)	70Da	
Co^{2+}	-5.76	11.25(1)	32.2	25	C	μ=0.10M(KNO_3)	70Da	
Ni^{2+}	-10.0	13.65(1)	28.9	25	C	μ=0.10M(KNO_3)	70Da	
Zn^{2+}	-6.1	11.22(1)	31.0	25	C	μ=0.10M(KNO_3)	70Da	
Cd^{2+}	-5.38	9.40(1)	25.0	25	C	μ=0.10M(KNO_3)	70Da	
E12 ETHANE, 1,2-diamino-N,N'-di-(2-amino ethyl)-			$C_6H_{18}N_4$			$H_2N(CH_2)_2NH(CH_2)_2NH(CH_2)_2NH_2$		L
Mn^{2+}	-2.30	4.91(1)	15	25	C	0.1M(KCl)	60Sc 61Sa	b
	-4	5.43(30°,1)	11.3	30-40	T	μ≈2(KCl + KNO_3)	52Ja	b
Fe^{2+}	-6.05	7.69(1)	15	25	C	0.1M(KCl)	60Sc 61Sa	

Metal Ion	ΔH, kcal /mole	Log K	ΔS, cal /mole °K	T, °C	M	Conditions	Ref.	Remarks
E12, cont.								
Fe^{2+}, cont.	-9	8.31(30°,1)	7.8	30-40	T	μ≃2(KCl + KNO_3)	52Ja	b
Co^{2+}	-10.65	10.92(1)	14.5	25	C	0.1M(KCl)	60Sc 61Sa	
	-9	11.21(30°,1)	21.6	30-40	T	μ≃2(KCl + KNO_3)	52Ja	b
Ni^{2+}	-14.00	13.78(1)	16	25	C	0.1M(KCl)	60Sc 61Sa	
	-13.9	14.1(1)	18	25	C	μ=0.1(KNO_3)	65Wd	
	-13.0	14.3(30°,1)	22.4	30&40	T	1M(KNO_3)	51Jc	a,b
	-13	14.34(30°,1)	22.8	30&40	T	μ=2(KCl + KNO_3)	52Ja	a,b
	-21.7	20.1(30°,1-2)	20.4	30&40	T	1M(KNO_3)	51Jc	a,b
	-9	5.63(30°,2-3)	-4.0	30&40	T	μ=2(KCl + KNO_3)	52Ja	a,b
	-8.7	5.6(30°,3-2)	-3.0	30&40	T	1M(KNO_3)	51Jc	a,b
Cu^{2+}	-21.55	20.08(1)	19.5	25	C	0.1M(KCl)	60Sc 61Sa	
	-21.4	20.2(1)	21	25	C	μ=0.1(KNO_3)	65Wd	
	-22	20.62(30°,1)	21.8	30&40	T	μ=2(KCl + KNO_3)	52Ja	a,b
	-21.7	20.6(30°,1-2)	22.8	30&40	T	μ=1(KNO_3)	51Jc	a,b
Zn^{2+}	-8.90	12.02(1)	25.0	25	C	0.1M(KCl)	60Sc 61Sa	
	-8.3	11.9(1)	27	25	C	μ=0.1(KNO_3)	65Wd	
	-4	11.94(30°,1)	41	30&40	T	μ=2(KCl + KNO_3)	52Ja	a,b
Cd^{2+}	-9.2	10.8(1)	19	25	C	μ=0.1(KNO_3)	65Wd	
	-4	10.92(30°,1)	36.8	30&40	T	μ=2(KCl + KNO_3)	52Ja	a,b
E13	ETHANE, 1,2-diamino-N,N'-diethyl-		$C_6H_{16}N_2$			$CH_3CH_2NHCH_2CH_2NHCH_2CH_3$		L
Ni^{2+}	-7.8	12.24(1-2)	27	0	C	0.5M(KNO_3)	54Bb	
Cu^{2+}	-7.5	8.01(1)	11.4	15-40	T	μ=0.3M($NaClO_4$)	72Md	b
	-6.4	5.70(2)	4.6	15-40	T	μ=0.3M($NaClO_4$)	72Md	b
	-17.5	18.6(1-2)	21	0	C	0.5M(KNO_3)	54Bb	
	-2.35	1.99(25°)	1.2	-75-25	T	S: 50 v % Methanol R: Cu + CuL_2 = 2CuL	72Hb	a,b,q: ±0.6
E14	ETHANE, 1,2-diamino-N,N-dimethyl-		$C_4H_{12}N_2$			$NH_2CH_2CH_2N(CH_3)_2$		L
Cu^{2+}	-9.82	9.280(1)	9.5	25	C	0.5M(KNO_3)	72Be	
	-9.48	7.028(2)	0.4	25	C	0.5M(KNO_3)	72Be	
E15	ETHANE, 1,2-diamino-N,N'-dimethyl-		$C_4H_{12}N_2$			$CH_3NHCH_2CH_2NHCH_3$		
Cu^{2+}	-11.10	10.06(1)	9.0	25	C	0.5M(KNO_3)	72Be	
	-9.67	7.151(2)	0.3	25	C	0.5M(KNO_3)	72Be	
	-1.68	2.10(25°)	4.0	-75-25	T	S: 50 v % Methanol R: Cu + CuL_2 = 2CuL	73Ya	a,b,q: ±0.43
E16	ETHANE, 1,2-diamino-N,N'-dimethyl-N,N'-diacetic acid-		$C_8H_{16}N_2$			$CH_3N(CH_2CO_2H)CH_2CH_2N$-$(CH_2CO_2H)CH_3$		L^{2-}

Metal Ion	ΔH, kcal /mole	Log K	ΔS, cal /mole °K	T, °C	M	Conditions	Ref.	Remarks
E16, cont.								
Mg^{2+}	1	5.7(1-2)	31	25	T	μ=0	56Ma	
E17	ETHANE, 1,2-diamino-1,2-dimethyl-N,N,N',N'-tetraacetic acid		$C_{12}H_{20}O_8N_2$			$(HO_2CH_2C)_2NCH(CH_3)CH(CH_3)$-$N(CH_2CO_2H)_2$		L^{4-}
Mg^{2+}	3.80	8.462(1)	51.7	20	C	μ=1M(HCl) C_L=0.01	69Sh	g:k
Cu^{2+}	-3.54	9.200(1)	30.0	20	C	μ=1M(HCl) C_L=0.01	69Sh	g:k
Sr^{2+}	0.37	7.210(1)	31.7	20	C	μ=1M(HCl) C_L=0.01	69Sh	g:k
Ba^{2+}	-1.83	6.032(1)	21.4	20	C	μ=1M(HCl) C_L=0.01	69Sh	g:k
E18	ETHANE, 1,2-diamino-N-ethyl-		$C_4H_{12}N_2$			$CH_3CH_2NHCH_2CH_2NH_2$		L
Ni^{2+}	-6.1	7.19(1)	10.6	0	T	μ=0.5(KNO_3)	52Ba	
	-7.2	5.78(2)	0.4	0	T	μ=0.5(KNO_3)	52Ba	
	-7.6	2.51(3)	-16.5	0	T	μ=0.5(KNO_3)	52Ba	
Cu^{2+}	-5.4	10.55(1)	29	0	T	μ=0.5(KNO_3)	52Ba	
	-6.4	8.81(2)	17	0	T	μ=0.5(KNO_3)	52Ba	
E19	ETHANE, 1,2-diamino-N-(2-hydroxyethyl)-N,N',N'-triacetic acid-		$C_{10}H_{18}O_7N_2$			$(CH_2CO_2H)_2NCH_2CH_2N(CH_2CO_2H)$-$CH_2CH_2OH$		L^{3-}
Mg^{2+}	3.4	6.96(1)	43	25	C	μ=0.1	65Wd	
	-3.80	6.92(1)	19	25	T	μ=0.10(KNO_3)	65Ma	
Ca^{2+}	-6.5	8.06(1)	15	25	C	μ=0.1(KNO_3)	61Mf	
	-5.4	8.44(1)	20	25	T	0.1M(KNO_3)	65Wd	
	-5.44	8.12(1)	19	25	T	μ=0.10(KNO_3)	65Ma	
Sr^{2+}	-5.2	6.82(1)	14	25	C	μ=0.1(KNO_3)	65Wd	
	-5.57	6.62(1)	11	25	T	μ=0.10(KNO_3)	65Ma	
Ba^{2+}	-5.4	6.16(1)	10	25	C	μ=0.1(KNO_3)	65Wd	
	-6.44	5.89(1)	5	25	T	μ=0.10(KNO_3)	65Ma	
Y^{3+}	-2.43	14.49(1)	58.2	25	C	μ=0.1(KCl)	68Fe	
	-0.29	14.6(1)	66.1	25	T	0.1M(KNO_3)	61Mf	
La^{3+}	-4.17	13.22(1)	46.5	25	C	μ=0.1(KCl)	68Fe	
	-2.20	13.47(1)	54.2	25	T	0.1M(KNO_3)	61Mf	
Ce^{3+}	-5.21	14.08(1)	47.0	25	C	μ=0.1(KCl)	68Fe	
	-3.06	14.12(1)	54.4	25	T	0.1M(KNO_3)	61Mf	
Pr^{3+}	-5.62	14.39(1)	45.5	25	C	μ=0.1(KCl)	68Fe	
	-4.45	14.62(1)	52.0	25	T	0.1M(KNO_3)	61Mf	
Nd^{3+}	-6.16	14.71(1)	46.7	25	C	μ=0.1(KCl)	68Fe	
	-4.25	14.87(1)	53.8	25	T	0.1M(KNO_3)	61Mf	
Sm^{3+}	-7.03	15.15(1)	45.7	25	C	μ=0.1(KCl)	68Fe	
	-4.65	15.30(1)	54.4	25	T	0.1M(KNO_3)	61Mf	
Eu^{3+}	-7.04	15.21(1)	46.0	25	C	μ=0.1(KCl)	68Fe	
	-4.81	15.36(1)	54.1	25	T	0.1M(KNO_3)	61Mf	
Gd^{3+}	-6.76	15.10(1)	46.4	25	C	μ=0.1(KCl)	68Fe	

Metal Ion	ΔH, kcal /mole	Log K	ΔS, cal /mole °K	T, °C	M	Conditions	Ref.	Re-marks
E19, cont.								
Gd^{3+}, cont.	-4.66	15.23(1)	54.1	25	T	0.1M(KNO_3)	61Mf	
Tb^{3+}	-5.92	15.10(1)	49.2	25	C	μ=0.1(KCl)	68Fe	
	-3.39	15.33(1)	58.8	25	T	0.1M(KNO_3)	61Mf	
Dy^{3+}	-4.81	15.08(1)	52.9	25	C	μ=0.1(KCl)	68Fe	
	-2.12	15.31(1)	62.8	25	T	0.1M(KNO_3)	61Mf	
Ho^{3+}	-3.42	15.06(1)	57.5	25	C	μ=0.1(KCl)	68Fe	
	-1.14	15.33(1)	66.3	25	T	0.1M(KNO_3)	61Mf	
Er^{3+}	-2.40	15.17(1)	61.4	25	C	μ=0.1(KCl)	68Fe	
	-0.32	15.44(1)	69.4	25	T	0.1M(KNO_3)	61Mf	
Tm^{3+}	-1.81	15.38(1)	64.3	25	C	μ=0.1(KCl)	68Fe	
	0.92	15.60(1)	74.5	25	T	0.1M(KNO_3)	61Mf	
Yb^{3+}	-2.11	15.64(1)	64.5	25	C	μ=0.1(KCl)	68Fe	
	0.36	15.89(1)	74.0	25	T	0.1M(KNO_3)	61Mf	
Lu^{3+}	-2.20	15.79(1)	64.9	25	C	μ=0.1(KCl)	68Fe	
	0.22	15.89(1)	73.4	25	T	0.1M(KNO_3)	61Mf	
Mn^{2+}	-5.2	10.8(1)	32	25	C	μ=0.1(KNO_3)	65Wd	
Fe^{2+}	-6.0	11.6(1)	33	25	C	μ=0.1(KNO_3)	65Wd	
Co^{2+}	-6.5	14.4(1)	44	25	C	μ=0.1(KNO_3)	65Wd	
Ni^{2+}	-10.3	17.1(1)	45	25	C	μ=0.1(KNO_3)	65Wd	
Cu^{2+}	-9.4	17.4(1)	48	25	C	μ=0.1(KNO_3)	65Wd	
	-6.8	17.41(1)	57	25	T	0.1M(KNO_3)	61Mf	
Zn^{2+}	-8.4	14.4(1)	38	25	C	μ=0.1(KNO_3)	65Wd	
Cd^{2+}	-10.3	13.0(1)	25	25	C	μ=0.1(KNO_3)	65Wd	
Hg^{2+}	-20.0	20.0(1)	25	25	C	μ=0.1(KNO_3)	65Wd	
	-7.1	19.38(1)	64.8	25	T	μ=0.1(KNO_3)	66Mf	
Al^{3+}	9.4	12.54(1)	89.1	25	T	μ=0.1(KNO_3)	66Mf	
Pb^{2+}	-12.6	15.5(1)	29	25	C	μ=0.1(KNO_3)	65Wd	
E20 ETHANE, 1,2-diamino-N-methyl-			$C_3H_{10}N_2$			$CH_3NHCH_2CH_2NH_2$		L
Ni^{2+}	-9.3	6.97(30°,1)	1	10-40	T	μ=0	59Ma	b
	-8.8	7.95(1)	4.0	0	T	μ=0.5(KNO_3)	52Ba	
	-7.8	5.40(30°,2)	-1	10-40	T	μ=0	59Ma	b
	-6.1	6.15(2)	5.9	0	T	μ=0.5(KNO_3)	52Ba	
	-4.1	2.68(30°,3)	-1	10-40	T	μ=0	59Ma	b
	-6.0	2.41(3)	-11.0	0	T	μ=0.5(KNO_3)	52Ba	
	-17.0	-17.2(1-2)	1	0	C	0.5KNO_3	54Bb	
Cu^{2+}	-12.2	10.06(30°,1)	6	10-40	T	μ=0	59Ma	b
							52Ba	
	-11.53	10.417(1)	9.0	25	C	0.5M(KNO_3)	72Be	
	-12.5	8.38(30°,2)	-3	10-40	T	μ=0	59Ma	b
	-12.2	8.703(2)	-1.0	25	C	0.5M(KNO_3)	72Be	
	-23.0	20.2(1-2)	8	0	C	0.5KNO_3)	54Bb	

Metal Ion	ΔH, kcal /mole	Log K	ΔS, cal /mole °K	T, °C	M	Conditions	Ref.	Re-marks
E20, cont.								
Zn^{2+}	-5.1	5.29(30°,1)	1	10-40	T	μ=0	59Ma	b
	-3.4	4.23(30°,2)	1	10-40	T	μ=0	59Ma	b
E21	ETHANE, 1,2-diamino-N-(1-methylethyl)-		$C_5H_{14}N_2$			$H_2NCH_2CH_2NHCH(CH_3)_2$		L
Ni^{2+}	-6.7	5.62(1)	1.1	0	T	μ=0.5(KNO_3)	52Ba	
	-5.5	3.84(2)	-2.6	0	T	μ=0.5(KNO_3)	52Ba	
Cu^{2+}	-5.8	9.46(1)	22	0	T	μ=0.5(KNO_3)	52Ba	
	-8.2	8.00(2)	7	0	T	μ=0.5(KNO_3)	52Ba	
E22	ETHANE, 1,2-diamino-1-phenyl-		$C_8H_{12}N_2$			$C_6H_5CH(NH_2)CH_2NH_2$		
Cu^{2+}	-13.1	10.25(1)	2.9	15-40	T	μ=0.3M($NaClO_4$)	72Md	b
	-13.5	9.01(2)	-4.0	15-40	T	μ=0.3M($NaClO_4$)	72Md	b
E23	ETHANE, 1,2-diamino-N-n-propyl-		$C_5H_{14}N_2$			$CH_3CH_2CH_2NHCH_2CH_2NH_2$		L
Ni^{2+}	-7.5	7.10(1)	5.1	0	T	μ=0.5(KNO_3)	52Ba	
	-9.9	5.82(2)	-9.5	0	T	μ=0.5(KNO_3)	52Ba	
	-3.9	2.26(3)	-4.0	0	T	μ=0.5(KNO_3)	52Ba	
Cu^{2+}	-7.6	10.49(1)	20	0	T	μ=0.5(KNO_3)	52Ba	
	-8.0	8.70(2)	11	0	T	μ=0.5(KNO_3)	52Ba	
E24	ETHANE, 1,2-diamino-N,N,N',N'-tetraacetic acid		$C_{10}H_{16}O_8N_2$			$(CH_2CO_2H)_2NCH_2CH_2N(CH_2CO_2H)_2$		L^{4-}
Li^+	0.1	2.8(1)	13	25	C	μ≦0.1	54Cb	
Na^+	-1.4	1.7(1)	3	25	C	μ≦0.1	54Cb	
Mg^{2+}	5.5	(1)	--	25	C	$C_M=10^{-2}$	57Jc, 54Ca	
	3.1	9.09(1)	52	25	C	μ<0.1	54Cb, 56Ma	
	3.14	8.69(1)	50.5	20	C	μ=0.17	56Ca, 64Ab	
	3.49	8.69(1)	51.0	20	C	0.1M(KNO_3)	63Ab	
Ca^{2+}	-3	11.0(1)	42	25	T	μ=0	56Ma, 54Ca	
	-5.7	(1)	--	25	C	$C_M=10^{-2}$	57Jc, 64Ab	
	-5.74	11.18(1)	31.9	25	C	μ=0.060	59Yb	
	-5.69	11.06(1)	31.5	25	C	μ=0.075	59Yb	
	-5.8	10.97(1)	31	25	C	μ≦0.1	54Cb	
	-6.55	10.7(1)	26.6	20	C	0.1M(KNO_3)	63Ab	
	-5.22	10.58(1)	30.9	25	C	μ=0.15	59Yb	
	-6.45	10.69(1)	26.9	20	C	μ=0.17(KNO_3)	56Ca	
	-5.10	10.45(1)	30.7	25	C	μ=0.22	59Yb	
	-4.72	10.13(1)	30.5	25	C	μ=0.54	59Yb	
	-4.65	10.15(1)	30.8	25	C	μ=0.75	59Yb	
	-4.82	10.16(1)	30.3	25	C	μ=0.80	59Yb	
	-4.73	10.17(1)	30.7	25	C	μ=0.86	58Ya	
	-4.97	10.29(1)	30.4	25	C	μ=1.08	59Yb	

Metal Ion	ΔH, kcal /mole	Log K	ΔS, cal /mole °K	T, °C	M	Conditions	Ref.	Remarks
E24, cont.								
Ca^{2+}, cont.	-5.27	10.61(1)	30.9	25	C	μ=1.46	59Yb	
	-5.52	10.73(1)	30.6	25	C	μ=1.58	59Yb	
Sr^{2+}	-4	8.72(1)	26	25	T	μ=0	54Ca 56Ma	
	-4.08	8.63(1)	25.6	20	C	0.1M(KNO_3)	63Ab	
	-4.2	8.72(1)	26	25	C	μ≦0.1	54Cb	
	-4.11	8.62(1)	25.4	20	C	μ=0.17	56Ca	
Ba^{2+}	-4	7.70(1)	22	25	T	μ=0	54Ca 56Ma	
	-5.1	7.72(1)	18	25	C	μ≦0.1	54Cb	
	-4.93	7.76(1)	18.7	20	C	0.1M(KNO_3)	63Ab	
	-5.3	7.84(1)	18	25	C	μ=0.1(KNO_3)	65Wd	
	-4.83	7.75(1)	19.0	20	C	μ=0.17	56Ca	
Y^{3+}	0.32	18.085(1)	83.8	20	C	μ=0.1	58Se	
	-0.588	17.372(1)	77.5	25	C	μ=0.1(KNO_3)	62Ma	
	-1.6	17.11(1)	73	25	C	μ=0.44	65Ka	
La^{3+}	-2.80	15.50(1)	61.4	20	C	0.1M(KNO_3)	63Ab 64Ab	
	-1.15	15.88(1)	68.8	25	C	μ=0.1	65Fb 59Ba	
	-2.926	15.19(1)	59.7	25	C	μ=0.1(KNO_3)	62Ma 62Ya	
	-4.1	-14.51(1)	53	25	C	μ=0.44	65Ka	
	-5.285	--	--	10	C	μ=1.0M($NaClO_4$) R: La^{3+} + MgL^{2-} = LaL^- + Mg^{2+}	73Oc	
	-5.141	--	--	15	C	μ=1.0M($NaClO_4$) R: La^{3+} + MgL^{2-} = LaL^- + Mg^{2+}	73Oc	
	-5.020	--	--	20	C	μ=1.0M($NaClO_4$) R: La^{3+} + MgL^{2-} = LaL^- + Mg^{2+}	73Oc	
	-4.918	--	--	25	C	μ=1.0M($NaClO_4$) R: La^{3+} + MgL^{2-} = LaL^- + Mg^{2+}. ΔCp=20.9	73Oc	
	-4.806	--	--	30	C	μ=1.0M($NaClO_4$) R: La^{3+} + MgL^{2-} = LaL^- + Mg^{2+}	73Oc	
	-4.688	--	--	35	C	μ=1.0M($NaClO_4$) R: La^{3+} + MgL^{2-} = LaL^- + Mg^{2+}	73Oc	
	-4.538	--	--	40	C	μ=1.0M($NaClO_4$) R: La^{3+} + MgL^{2-} = LaL^- + Mg^{2+}	73Oc	
Ce^{3+}	-2.43	15.98(1)	64.8	20	C	μ=0.1	58Se 59Ba	
	-2.938	15.44(1)	60.8	25	C	μ=0.1(KNO_3)	62Ma	
	-4.8	15.00(1)	53	25	C	μ=0.44	65Ka	
Pr^{3+}	-3.198	15.75(1)	61.4	25	C	μ=0.1(KNO_3)	62Ma 59Ba	

Metal Ion	ΔH, kcal /mole	Log K	ΔS, cal /mole °K	T, °C	M	Conditions	Ref.	Re-marks
E24, cont.								
Pr^{3+}, cont.	-4.7	15.42(1)	55	25	C	μ=0.44	65Ka	
	-5.455	--	--	10	C	μ=1.0M($NaClO_4$) R: Pr^{3+} + MgL^{2-} = PrL^- + Mg^{2+}	73Oc	
	-5.231	--	--	20	C	μ=1.0M($NaClO_4$) R: Pr^{3+} + MgL^{2-} = PrL^- + Mg^{2+}	73Oc	
	-5.145	--	--	25	C	μ=1.0M($NaClO_4$) R: Pr^{3+} MgL^{2-} = PrL^- + Mg^{2+}. ΔCp=16.5	73Oc	
	-5.062	--	--	30	C	μ=1.0M($NaClO_4$) R: Pr^{3+} + MgL^{2-} = PrL^- + Mg^{2+}	73Oc	
	-4.864	--	--	40	C	μ=1.0M($NaClO_4$) R: Pr^{3+} + MgL^{2-} = PrL^- + Mg^{2+}	73Oc	
Nd^{3+}	-2.98	16.60(1)	65.8	20	C	μ=0.1	58Se 59Ba	
	-3.623	16.04(1)	61.3	25	C	μ=0.1(KNO_3)	62Ma 63Pd	
	-4.8	16.16(1)	55	25	C	μ=0.44	65Ka	
	-5.829	--	--	10	C	μ=1.0M($NaClO_4$) R: Nd^{3+} + MgL^{2-} = NdL^- + Mg^{2+}	73Oc	
	-5.570	--	--	20	C	μ=1.0M($NaClO_4$) R: Nd^{3+} + MgL^{2-} = NdL^- + Mg^{2+}	73Oc	
	-5.478	--	--	25	C	μ=1.0M($NaClO_4$) R: Nd^{3+} + MgL^{2-} = NdL^- + Mg^{2+}. ΔCp=17.7	73Oc	
	-5.392	--	--	30	C	μ=1.0M($NaClO_4$) R: Nd^{3+} + MgL^{2-} = NdL^- + Mg^{2+}	73Oc	
	-5.180	--	--	40	C	μ=1.0M($NaClO_4$) R: Nd^{3+} + MgL^{2-} = NdL^- + Mg^{2+}	73Oc	
Sm^{3+}	-3.349	16.52(1)	64.4	25	C	μ=0.1(KNO_3)	62Ma 59Ba	
	-4.6	16.16(1)	59	25	C	μ=0.44	65Ka 63Pd	
	-6.055	--	--	10	C	μ=1.0M($NaClO_4$) R: Sm^{3+} + MgL^{2-} = SmL^- + Mg^{2+}	73Oc	
	-5.723	--	--	20	C	μ=1.0M($NaClO_4$) R: Sm^{3+} + MgL^{2-} = SmL^- + Mg^{2+}	73Oc	
	-5.576	--	--	25	C	μ=1.0M($NaClO_4$) R: Sm^{3+} + MgL^{2-} = SmL^- + Mg^{2+}. ΔCp=28.9	73Oc	
	-5.429	--	--	30	C	μ=1.0M($NaClO_4$) R: Sm^{3+} + MgL^{2-} = SmL^- + Mg^{2+}	73Oc	

Metal Ion	ΔH, kcal /mole	Log K	ΔS, cal /mole °K	T, °C	M	Conditions	Ref.	Re-marks
E24, cont.								
Sm^{3+}, cont.	-5.104	--	--	40	C	μ=1.0M($NaClO_4$) R: Sm^{3+} + MgL^{2-} = SmL^- + Mg^{2+}	730c	
Eu^{3+}	-2.558	16.65(1)	67.6	25	C	μ=0.1(KNO_3)	62Ma 59Ba	
	-3.5	16.37(1)	63	25	C	μ=0.44	65Ka	
	-5.0	15.27(50°,1)	54.3	40-70	T	μ=1.0(KCl) C_M=0.1 Complex having Structure 1 as determined by absorption spectra	71Kc	b
	0	15.21(50°,1)	69.9	40-70	T	μ=1.0(KCl) C_M=0.1 Complex having Structure 2 as determined by absorption spectra	71Kc	b
	5.53	7.29(50°)	50.4	40-70	T	μ=1.0(KCl) C_M=0.1 R: Eu^{3+} +HL^{3-} = EuHL	71Kc	b
	-5.561	--	--	10	C	μ=1.0M($NaClO_4$) R: Eu^{3+} + MgL^{2-} = EuL^- + Mg^{2+}	730c	
	-5.387	--	--	15	C	μ=1.0M($NaClO_4$) R: Eu^{3+} + MgL^{2-} = EuL^- + Mg^{2+}	730c	
	-5.230	--	--	20	C	μ=1.0M($NaClO_4$) R: Eu^{3+} + MgL^{2-} = EuL^- + Mg^{2+}	730c	
	-5.077	--	--	25	C	μ=1.0M($NaClO_4$) R: Eu^{3+} + MgL^{2-} = EuL^- + Mg^{2+}	730c	Cp= 29.4
	-4.936	--	--	30	C	μ=1.0M($NaClO_4$) R: Eu^{3+} + MgL^{2-} EuL^- + Mg^{2+}	730c	
	-4.783	--	--	35	C	μ=1.0M($NaClO_4$) R: Eu^{3+} + MgL^{2-} = EuL^- + Mg^{2+}	730c	
	-4.629	--	--	40	C	μ=1.0M($NaClO_4$) R: Eu^{3+} + Mg^{2-} = EuL^- + Mg^{2+}	730c	
Gd^{3+}	-1.11	17.36(1)	75.7	20	C	μ=0.1	58Se 59Ba	
	-1.730	16.81(1)	71.2	25	C	μ=0.1(KNO_3)	62Ma 63Pd	
	-2.5	16.39(1)	66	25	C	μ=0.44	65Ka	
	-4.857	--	--	10	C	μ=1.0M($NaClO_4$) R: Gd^{3+} + MgL^{2-} = GdL^- + Mg^{2+}	730c	
	-4.573	--	--	20	C	μ=1.0M($NaClO_4$) R: Gd^{3+} + Mg^{2-} = GdL^- + Mg^{2+}	730c	

Metal Ion	ΔH, kcal /mole	Log K	ΔS, cal /mole °K	T, °C	M	Conditions	Ref.	Remarks
E24, cont.								
Gd^{3+}, cont.	-4.458	--	--	25	C	μ=1.0M($NaClO_4$) R: Gd^{3+} + Mg^{2+} = GdL^- + Mg^{2+}	730c	Cp= 22.4
	-4.345	--	--	30	C	μ=1.0M($NaClO_4$) R: Gd^{3+} + Mg^{2+} = GdL^- + Mg^{2+}	730c	
	-4.079	--	--	40	C	μ=1.0M($NaClO_4$) R: Gd^{3+} + Mg^{2+} = GdL^- + Mg^{2+}	730c	
Tb^{3+}	-1.114	17.31(1)	75.5	25	C	μ=0.1(KNO_3)	62Ma 59Ba	
	-3.981	--	--	10	C	μ=1.0M($NaClO_4$) R: Tb^{3+} + MgL^{2-} = TbL^- + Mg^{2+}	730c	
	-3.941	--	--	15	C	μ=1.0M($NaClO_4$) R: Tb^{3+} + MgL^{2-} = TbL^- + Mg^{2+}	730c	
	-3.916	--	--	20	C	μ=1.0M($NaClO_4$) R: Tb^{3+} + MgL^{2-} = TbL^- + Mg^{2+}	730c	
	-3.890	--	--	25	C	μ=1.0M($NaClO_4$) R: Tb^{3+} + MgL^- = TbL^- + Mg^{2+}	730c	Cp= 5.1
	-3.866	--	--	30	C	μ=1.0M($NaClO_4$) R: Tb^{3+} + MgL^- = TbL^- + Mg^{2+}	730c	
	-3.808	--	--	35	C	μ=1.0M($NaClO_4$) R: Tb^{3+} + MgL^- = TbL^- + Mg^{2+}	730c	
	-3.740	--	--	40	C	μ=1.0M($NaClO_4$) R: Tb^{3+} + MgL^- = TbL^- + Mg^{2+}	730c	
Dy^{3+}	-1.211	17.77(1)	77.3	25	C	μ=0.1(KNO_3)	62Ma 59Ba	
	-3.0	17.32(1)	69	25	C	μ=0.44	65Ka	
Ho^{3+}	-1.356	18.038(1)	78.0	25	C	μ=0.1(KNO_3)	62Ma 59Ba	
	-3.1	17.76(1)	71	25	C	μ=0.44	65Ka	
	-4.023	--	--	10	C	μ=1.0M($NaClO_4$) R: Ho^{3+} + MgL^{2-} = HoL^- + Mg^{2+}	730c	
	-4.046	--	--	20	C	μ=1.0M($NaClO_4$) R: Ho^{3+} + MgL^{2-} = HoL^- + Mg^{2+}	730c	
	-4.066	--	--	25	C	μ=1.0M($NaClO_4$) R: Ho^{3+} + MgL^{2-} = HoL^- + Mg^{2+}	730c	Cp= -3.4
	-4.075	--	--	30	C	μ=1.0M($NaClO_4$) R: Ho^{3+} + MgL^{2-} = HoL^- + Mg^{2+}	730c	
	-4.015	--	--	40	C	μ=1.0M($NaClO_4$) R: Ho^{3+} + MgL^{2-} = HoL^- + Mg^{2+}	730c	

Metal Ion	ΔH, kcal /mole	Log K	ΔS, cal /mole °K	T, °C	M	Conditions	Ref.	Remarks
E24, cont.								
Er^{3+}	-1.708	18.368(1)	78.3	25	C	μ=0.1(KNO_3)	62Ma 59Ba	
	-3.0	17.87(1)	72	25	C	μ=0.44	65Ka	
Tm^{3+}	-1.870	18.647(1)	79.1	25	C	μ=0.1(KNO_3)	62Ma 59Ba	
Yb^{3+}	-2.310	18.992(1)	79.2	25	C	μ=0.1(KNO_3)	62Ma 59Ba	
	-2.7	18.53(1)	76	25	C	μ=0.44	65Ka	
	-4.774	--	--	10	C	μ=1.0M($NaClO_4$) R: Yb^{3+} + MgL^{2-} = YbL^- + Mg^{2+}	73Oc	
	-4.781	--	--	20	C	μ=1.0M($NaClO_4$) R: Yb^{3+} + MgL^{2-} = YbL^- + Mg^{2+}	73Oc	
	-4.806	--	--	25	C	μ=1.0M($NaClO_4$) R: Yb^{3+} + MgL^{2-} = YbL^- + Mg^{2+}	73Oc	Cp= -5.3
	-4.829	--	--	30	C	μ=1.0M($NaClO_4$) R: Yb^{3+} + MgL^{2-} = YbL^- + Mg^{2+}	73Oc	
	-4.813	--	--	40	C	μ=1.0M($NaClO_4$) R: Yb^{3+} + MgL^{2-} = YbL^- + Mg^{2+}	73Oc	
Lu^{3+}	-2.512	19.138(1)	79.1	25	C	μ=0.1(KNO_3)	62Ma 59Ba	
	-3.2	18.84(1)	75	25	C	μ=0.44	65Ka	
UO_2^{2+}	3.2	7.40(25°)	45	20-40	T	μ=0.1M(KNO_3) R: UO_2^{2+} + HL = UO_2HL	69Sg	b
Pu^{3+}	-4.23	18.07(1)	68.6	25	C	μ=0.1	65Fb	
Am^{3+}	-4.67	18.16(1)	67.5	25	C	μ=0.1	65Fb	
Mn^{2+}	-4.56	13.80(1)	47.6	20	C	0.1M(KNO_3)	63Ab 64Ab	
	-5.2	12.6(1)	41	25	C	μ≤0.1	54Cb	
Fe^{2+}	-4.0	14.1(1)	51	25	C	μ=0.1(KNO_3)	65Wd	
	-0.41	6.903(20°)	30	10-40	T	μ=0.20 R: Fe^{2+} + HL^{3-} = $FeHL^-$	72Bk	b,q: ±0.06
	-0.80	3.322(20°)	12	10-40	T	μ=0.20 R: Fe^{2+} + H_2L^{2-} = FeH_2L	72Bk	b,q: ±0.4
Fe^{3+}	-12.2	25.447(20°,1)	75	10-40	T	μ=0.20	72Bk	b
Co^{2+}	-4.20	16.31(1)	60.3	20	C	0.1M(KNO_3)	63Ab 64Ab	
	-4.2	(1)	--	25	C	--	57Jc	
	-4.1	15.7(1)	58	25	C	μ≤0.1	54Cb	
	-16.7	2.920(25°)	-42.6	15-30	T	μ=0.660(NaAc+HAc) pH=5.0 R: CoL^{2-} + $Fe(CN)_6^{3-}$ = $[LCoCNFeCN_5^{5-}]$	73Hf	b

Metal Ion	ΔH, kcal /mole	Log K	ΔS, cal /mole °K	T, °C	M	Conditions	Ref.	Remarks
E24, cont.								
Ni^{2+}	-7.4	(1)	--	25	C	$C_M=10^{-2}$	57Jc 64Ab	
	-7.55	18.62(1)	59.4	20	C	0.1M(KNO_3)	63Ab	
	-7.6	17.6(1)	55	25	C	μ≦0.1	54Cb	
	-8.5	18.7(1)	57	25	C	μ=0.1(KNO_3)	65Wd	
	-8.35	18.61(1)	56.7	20	C	μ=0.17(KNO_3)	56Ca	
	-6.88	18.11(1)	59.8	25	C	μ=0.22	59Yb	
	-6.80	17.79(1)	58.6	25	C	μ=0.59	59Yb	
	-6.69	17.78(1)	58.9	25	C	μ=0.75	59Yb	
	-7.09	18.25(1)	59.7	25	C	μ=1.45	59Yb	
	-7.38	18.28(1)	58.9	25	C	μ=1.50	59Yb	
Cu^{2+}	-8.2	(1)	--	25	C	$C_M=10^{-2}$	57Jc 64Ab	
	-8.15	18.80(1)	58.2	20	C	0.1M(KNO_3)	63Ab	
	-8.2	17.88(1)	55	25	C	μ≦0.1	54Cb	
	-8.8	18.62(1)	56	25	C	0.1M(KNO_3)	65Wd	
	-8.67	18.79(1)	56.4	20	C	μ=0.17(KNO_3)	56Ca	
Zn^{2+}	-4.6	(1)	--	25	C	$C_M=10^{-2}$	57Jc 64Ab	
	-4.85	16.50(1)	59.0	20	C	0.1M(KNO_3)	63Ab	
	-4.5	15.3(1)	55	25	C	μ≦0.1	54Cb	
	-5.6	16.3(1)	57	25	C	μ=0.1(KNO_3)	65Wd	
	-5.61	16.49(1)	56.3	20	C	μ=0.17(KNO_3)	56Ca	
Cd^{2+}	-9.2	(1)	--	25	C	$C_M=10^{-2}$	57Jc 64Ab	
	-9.05	16.46(1)	44.4	20	C	0.1M(KNO_3)	63Ab	
	-9.1	15.0(1)	38	25	C	μ≦0.1	54Cb	
	-10.1	16.34(1)	42	25	C	0.1M(KNO_3)	65Wd	
	-10.08	16.45(1)	40.9	20	C	μ=0.17(KNO_3)	56Ca	
Hg^{2+}	-18.90	21.80(1)	35.5	20	C	0.1M(KNO_3)	63Ab 64Ab	
	-19.2	22.1(1)	37	25	C	μ=0.1(KNO_3)	65Wd 66Mf	
Al^{3+}	2.58	16.10(1)	116.6	20	C	μ=0.1	58Se	
	..1	16.78(1)	117.4	25	T	μ=0.1(KNO_3)	66Mf	
	..2.1	15.61(1)	112.0	25	C	μ=0.22	58Ya	
In^{3+}	-7.23	24.94(1)	89.5	20	C	μ=0.1	58Se	
Pb^{2+}	-12.8	(1)	--	25	C	$C_M=10^{-2}$	57Jc 54Cb	
	-13.20	18.04(1)	37.5	20	C	0.1M(KNO_3)	63Ab 64Ab	
	-13.1	17.3(1)	35	25	C	μ=0.1(KNO_3)	65Wd	
	-14.08	18.02(1)	34.5	20	C	μ=0.17(KNO_3)	56Ca	
E25	ETHANE, 1,2-diamino-N,N,N',N' tetra(2-aminoethyl)-	$C_{10}H_{28}N_2$				$(NH_2CH_2CH_2)_2NCH_2CH_2N(CH_2CH_2NH_2)_2$		L
Mn^{2+}	-8.85	9.30(1)	12.5	25	C	μ=0.1(KCl)	64Sc 60Cc	

Metal Ion	ΔH, kcal /mole	Log K	ΔS, cal /mole °K	T, °C	M	Conditions	Ref.	Remarks
E25, cont.								
Fe^{2+}	-9.65	11.05(1)	18.5	25	C	μ=0.1(KCl)	64Sc 60Cc	
Co^{2+}	-14.75	11.55(1)	3.4	25	C	μ=0.1(KCl)	64Cc 60Cc	
	-14.00	12.40	10.0	25	C	μ=0.1(KCl) R: $Co^{2+} + HL^{+} = CoHL^{3+}$	64Sc	
Ni^{2+}	-19.65	19.05(1)	21.5	25	C	μ=0.1(KCl)	64Sc 60Cc	
	-18.35	15.65	10.0	25	C	μ=0.1(KCl) R: $Ni^{2+} + HL^{+} = NiHL^{3+}$	64Sc	
Cu^{2+}	-24.50	22.15(1)	19.0	25	C	μ=0.1(KCl)	64Sc 60Cc	
	-24.80	20.15	9.0	25	C	μ=0.1(KCl) R: $Cu^{2+} + HL^{+} = CuHL^{3+}$	64Sc	
Zn^{2+}	-14.50	16.05(1)	25.0	25	C	μ=0.1(KCl)	64Sc 60Cc	
	-14.65	14.00	15.0	25	C	μ=0.1(KCl) R: $Zn^{2+} + HL^{+} = ZnHL^{3+}$	64Sc	
E26	ETHANE, 1,2-diamino-N,N,N',N'-tetrakis (2-hydroxyethyl)-	$C_{10}H_{24}O_4N_2$				$(HOCH_2CH_2)_2NCH_2CH_2N(CH_2CH_2OH)_2$		
Ni^{2+}	-4.5	6.40(25°,1)	14	10-55	T	μ=0.5($NaClO_4$)	69Pa	b
Cu^{2+}	-6.5	8.60(25°,1)	18	10-55	T	μ=0.5($NaClO_4$)	69Pa	b
Zn^{2+}	-2.5	4.78(25°,1)	14	10-55	T	μ=0.5($NaClO_4$)	69Pa	b
E27	ETHANE, 1,2-diamino-N,N,N',N'-tetramethyl-	$C_6H_{16}N_2$				$(CH_3)_2NCH_2CH_2(CH_3)_2$		L
Cu^{2+}	-6.15	7.376(1)	13.1	25	C	0.5M(KNO_3)	72Ad	
	3.3	-0.66	8.0	25	C	0.5M(KNO_3) R: $Cu^{2+} + L + H_2O = CuOHL^{+} + H^{+}$	72Ad	
	15.00	-10.91	0.4	25	C	0.5M(KNO_3) R: $Cu^{2+} + L + 2H_2O = Cu(OH)_2L + 2H^{+}$	72Ad	
	4.70	2.59	27.6	25	C	0.5M(KNO_3) R: $2Cu^{2+} + 2L + 2H_2O = [Cu_2(OH)_2L_2]^{2+} + 2H^{+}$	72Ad	
	17.7	-8.15	22.1	25	C	0.5M(KNO_3) R: $3Cu^{2+} + 2L + 4H_2O = [Cu_3(OH)_4L_2]^{2+} + 4H^{+}$	72Ad	
E29	ETHANE, 1,2-bis(dimethyl-amino)-	$C_6H_{16}N_2$				$(H_3C)_2NCH_2CH_2N(CH_3)_2$		L
PdA_2B_2	-27.3	--	--	25	C	S: CH_2Cl_2 $C_L=4x10^{-4}M$	72Pc	

Metal Ion	ΔH, kcal /mole	Log K	ΔS, cal /mole °K	T, °C	M	Conditions	Ref.	Remarks
E29, cont.								
PdA_2B_2, cont.						$C_M=2x10^{-5}M$ A=Cl B=C_6H_5CN R: PdA_2B_2 + L = PdA_2L + 2B		
E30 ETHANE, 1,2-bis(diphenyl-arsino)-			$C_{26}H_{24}As_2$			$(C_6H_5)_2AsCH_2CH_2As(C_6H_5)_2$		L
PdA_2B_2	-35.2	--	--	25	C	S: CH_2Cl_2 $C_L=1x10^{-4}M$ $C_M=2x10^{-4}M$ A=Cl B=C_6H_5CN R: PdA_2B_2 + L = PdA_2L + 2B	72Pc	
E31 ETHANE, 1,2-bis(diphenyl-phosphino)-			$C_{26}H_{24}P_2$			$(C_6H_5)_2PCH_2CH_2P(C_6H_5)_2$		L
PdA_2B_2	-50.5	--	--	25	C	S: CH_2Cl_2 $C_L=6x10^{-5}M$ $C_M=3x10^{-4}M$ A=Cl B=C_6H_5CN R: PdA_2B_2 + L = PdA_2L + 2B	72Pc	
E32 ETHANE, 1,1-diphosphonic acid			$C_2H_8O_6P_2$			$CH_3CH(PO_3H_2)_2$		L^{4-}
Li^+	-0.36	3.12(1)	13.1	25	C	0.5M$(CH_3)_4NCl$	68Ca	
Na^+	1.7	1.51(1)	12.6	25	C	0.5M$(CH_3)_4NCl$	68Ca	
K^+	0.8	1.20(1)	8.1	25	C	0.5M$(CH_3)_4NCl$	68Ca	
Ca^{2+}	2.2	5.22(1)	31	25	C	0.5M$(CH_3)_4NCl$	68Ca	
E33 ETHANE, 1-hydroxy, 1,1-diphosphonic acid			$C_2H_8O_7P_2$			$CH_3C(OH)(PO_3H_2)_2$		L^{4-}
Li^+	-0.57	3.35(1)	13.4	25	C	0.5M$(CH_3)_4NCl$	68Ca	
Na^+	0.4	2.07(1)	10.9	25	C	0.5M$(CH_3)_4NCl$	68Ca	
K^+	0.4	1.79(1)	9.4	25	C	0.5M$(CH_3)_4NCl$	68Ca	
Ca^{2+}	-0.52	5.75(1)	24.6	25	C	0.5M$(CH_3)_4NCl$	68Ca	
E34 ETHANE, 1,2-bis(3-mercapto-propionoxy)-			$C_8H_{14}O_4S_2$			$HSCH_2CH_2C(O)OCH_2CH_2OC(O)CH_2CH_2SH$		
Tl^+	-1.6	0.740(40°,1)	-1.8	30&40	T	μ=0.5M$(NaClO_4)$	72Se	a,b,j
	-4.3	0.903(40°,1-2)	-9.7	30&40	T	μ=0.5M$(NaClO_4)$	72Se	a,b,j
	-5.2	2.845(40°,1-3)	-3.5	30&40	T	μ=0.5M$(NaClO_4)$	72Se	a,b,j
E35 ETHANE, 1,1,2,2-tetramercapto-S,S',S",S"'-tetraacetic acid			$C_{10}H_{14}O_8S_4$			$(HO_2CCH_2S)_2CHCH(SCH_2CO_2H)_2$		L^{4-}
Cu^{2+}	1.68	5.00(1)	28.5	25	C	μ=0.10(KNO_3)	70Cd	
	2.49	7.33(2-1)	41.9	25	C	μ=0.10(KNO_3)	70Cd	

Metal Ion	ΔH, kcal /mole	Log K	ΔS, cal /mole °K	T, °C	M	Conditions	Ref.	Re-marks
E36	1,2-ETHANEDIOL		$C_2H_6O_2$			$HOCH_2CH_2OH$		L
BA_4^-	-2.70	0.332(25°,1)	-7.5	0-35	T	C_L=0-0.9M A=OH^-	67Cc	b,q: ±0.3
	-1.97	0.061(25°,1-2)	-6.3	0-35	T	C_L=0-0.9M A=OH^-		b,q: ±0.2
$H_5TeO_6^-$	-4.6	1.17	-10	25	T	μ=0.1 R: H_5TeO_6 + L = $HTeO_4L^-$ + $2H_2O$	62Ea	
E37	1,2-ETHANEDIOL, di(2'-amino-ethyl)ether		$C_6H_{16}O_2N_2$			$H_2N(CH_2CH_2O)_2CH_2CH_2NH_2$		L
Cu^{2+}	-7.9	7.82(30°,1)	10	10-40	T	μ=0	59Lb	b
Ag^+	-13.2	7.71(30°,1)	-8	10-40	T	μ=0	59Lb	b
E38	1,2-ETHANEDIOL, dibutyl ether		$C_{10}H_{22}O_2$			$CH_3(CH_2)_3OCH_2CH_2O(CH_2)_3CH_3$		
TiA_4	-25.7	(1)	--	25	C	S: Benzene C_M=0.05-0.08M A=Cl	68Ge 70Gh	
SnA_4	-21.9	4.03(1)	-55.0	25	C	S: Benzene C_M=0.05-0.08M A=Cl	68Ge 70Gh 68Gf	
E39	1,2-ETHANEDIOL, diethyl ether		$C_6H_{14}O_2$			$C_2H_5OCH_2CH_2OC_2H_5$		L
TiA_4	-21.2	(1)	--	25	C	S: Benzene C_M=0.05-0.08M A=Br	68Ge 68Gf 70Gh	
	-23.8	3.96(1)	-61.7	25	C	S: Benzene C_M=0.05-0.08M A=Cl	68Ge 70Gh 68Gf	
SnA_4	-22.8	(1)	--	25	C	S: Benzene C_M=0.05-0.08M A=Cl	68Ge 70Gh 68Gf	
E40	1-ETHANESULPHONIC ACID, 2-amino-		$C_2H_7O_3NS$			$NH_2(CH_2)_2SO_3H$		L^-
Ag^+	-4.50	2.969(1)	-1.5	25	C	μ=0.5M(KNO_3)	72Va	
	-7.050	3.181(2)	-9.1	25	C	μ=0.5M(KNO_3)	72Va	
E41	ETHANETHIOL, 2-(diisopropyl-amino)-		$C_8H_{19}NS$			$(C_3H_7)_2NC_2H_4SH$		L^-
UO_2^{2+}	16.46	15.75(30°)	128	30-40	T	μ=0.1M(KNO_3) R: UO_2^{2+} + 2HL + $2OH^-$ = UO_2L_2 + $2H_2O$	73Pb	a,b
E42	ETHANETHIOL, 2-(dimethyl-amino)-		$C_4H_{11}NS$			$(CH_3)_2NCH_2CH_2SH$		L^-
UO_2^{2+}	15.39	15.20(25°,1-2)	122.17	20&35	T	μ=0.25M(KNO_3)	73Me	a,b
Cu^{2+}	-4.11	19.20(25°,1-2)	73.88	20&35	T	μ=0.25M(KNO_3)	73Me	a,b
Zn^{2+}	10.39	13.44(25°,1-2)	97.04	20&35	T	μ=0.25M(KNO_3)	73Me	a,b

Metal Ion	ΔH, kcal /mole	Log K	ΔS, cal /mole °K	T, °C	M	Conditions	Ref.	Re-marks
E42, cont.								
Cd^{2+}	-26.36	15.98(25°,1-2)	-16.75	20&35	T	μ=0.25M(KNO_3)	73Me	a,b
Hg^{2+}	-5.22	20.28(25°,1-2)	75.13	20&35	T	μ=0.25M(KNO_3)	73Me	a,b
Pb^{2+}	-1.37	14.48(25°,1-2)	61.90	20&35	T	μ=0.25M(KNO_3)	73Me	a,b
E43 ETHANOL			C_2H_6O			CH_3CH_2OH		L
SO_2	-3.64	0.387(20°,1)	-10.80	0-30	T	0.4F(NaOH) 0.2F(HCl) S: CCl_4	57Da	b
I_2	-3.5	0.88(23°,1)	-8.1	23-45	T	Dilute solution	56Ab	b
	-2.10	0.61(25°,1)	-4.09	7-25	T	C_L=0.06-0.96 mole fraction S: CCl_4	57Dc	b
	-2.6	(1)	--	--	C	S: C_2H_5OH Recalculation of data in Ref 50Ha	61Aa	
	-2.1	(1)	--	--	C	S: C_2H_5OH	50Ha	
E44 ETHANOL, 2-amino-			C_2H_7ON			$NH_2CH_2CH_2OH$		L
Ag^+	-5.5	3.07(30°,1)	-4	10-40	T	μ=0	59Lb 67Ce	b
	-6.4	3.57(30°,2)	-5	10-40	T	μ=0	59Lb	b
Hg^{2+}	-10.2	8.56(1)	5.0	25	C	μ=0	70Ea	
	-8.3	8.77(2)	12.3	25	C	μ=0	70Ea	
E45 ETHANOL, 1,1-dimethyl-			$C_4H_{10}O$			$(CH_3)_3COH$		L
I_2	-3.4	1.05(25°,1)	-6.7	1.6-45.8	T	C_L=0.01-0.15 mole fraction S: CCl_4	55Kb	b,m
E46 ETHANOL, 2-mercapto-			C_2H_6OS			$HOCH_2CH_2SH$		L^-
HgA^+	-19.8	16.104(1)	6.2	20	C	μ=0.1(KNO_3) A=CH_3	65Sa	
E47 ETHENE			C_2H_4			CH_2:CH_2		L
$PdA_4{}^{2-}$	-1.5	1.182(20°)	0	5-25	T	μ=4.0. R: $PdA_4{}^{2-}$ + L = PdA_3L^- + A^- A=Cl	65Pb	b
	-11.5	0.634(20°)	-36	5-25	T	μ=4.0. R: $PdA_4{}^{2-}$ + L + H_2O = $PdA_2L(OH)_2$ + $2A^-$. A=Cl	65Pb	b
$IrABC_2$	-12	0.431(20°,1)	-39.0	10-60	T	S: Chlorobenzene A=I B=CO C=Ph_3P	71Vb	b
Ag^+	-5.50	1.929(25°,1)	-9.08	0-25	T	μ=0	70Wa	b
	-7.37	1.973(25°,1)	-15.7	17.5-32.5	T	1M Sodium trifluoroacetate	59Bb	b
Ag^+	3.5	1.243(25°,1)	-6.0	0-40	T	S: Ethyleneglycol	65Ci	b

Metal Ion	ΔH, kcal /mole	Log K	ΔS, cal /mole °K	T, °C	M	Conditions	Ref.	Remarks
E48	ETHENE, n-butoxy-		$C_6H_{12}O$			CH_2:$CHOC_4H_9$		L
Ag^+	-3.7	0.890(20°,1)	-8.8	10-30	T	S: Ethyleneglycol	68Fd	b,q: ±0.2
E49	ETHENE, chloro-		C_2H_3Cl			CH_2:CHCl		L
$RhAB_2$	0.8	-0.770	-0.8	0&25	T	S: Toluene A=2,4-Pentanedionato B=Ethylene R: L + $RhAB_2$ = RhABL + B	67Cd	a,b,q: ±0.8
E50	ETHENE, cyano-		C_3H_3N			CH_2:CHCN		L
TiA_4	-11.2	3.22(1)	-22.8	25	C	S: Benzene A=Cl	70Gh	
E51	ETHENE, 1,1-dimethyl-2-ethoxy-		$C_6H_{12}O$			$(CH_3)_2C$:$CHOC_2H_5$		L
Ag^+	-3.5	-0.633(20°,1)	-15	10-30	T	S: Ethyleneglycol	68Fd	b,q: ±0.2
E52	ETHENE, 1,2-diphenyl-(cis)		$C_{14}H_{12}$			C_6H_5CH:CHC_6H_5		L
I_2	-2.40	0.161(1)	-7.32	25	T	$C_M=10^{-4}$ $C_L=0.1$ S: CCl_4	59Ya	
	-7.54	0.707(1)	-22.1	25	T	$C_M=10^{-4}$ $C_L=0.1$ S: n-Hexane	59Ya	
E53	ETHENE, 1,2-diphenyl-(trans)		$C_{14}H_{12}$			C_6H_5CH:CHC_6H_5		L
I_2	-2.54	0.295(1)	-7.18	25	T	$C_M=10^{-4}$ $C_L=0.1$ S: n-Hexane	59Ya	
E54	ETHENE, ethoxy-		C_4H_8O			CH_2:$CHOC_2H_5$		L
Ag^+	-4.4	0.827(20°,1)	-11.0	10-30	T	S: Ethyleneglycol	68Fd	b,q: ±0.2
E55	ETHENE, 1-ethoxy-2-ethyl-(cis)		$C_6H_{12}O$			C_2H_5CH:$CHOC_2H_5$		L
Ag^+	-4.5	0.627(20°,1)	-12.6	10-30	T	S: Ethyleneglycol	68Fd	b,q: ±0.2
E56	ETHENE, 1-ethoxy-2-ethyl-(trans)		$C_6H_{12}O$			C_2H_5CH:$CHOC_2H_5$		L
Ag^+	-3.3	-0.137(20°,1)	-12.0	10-30	T	S: Ethyleneglycol	68Fd	b,q: ±0.2
E57	ETHENE, 1-ethoxy-2-isopropyl-(cis)		$C_7H_{14}O$			C_3H_7CH:$CHOC_2H_5$		L
Ag^+	-4.7	0.649(20°,1)	-12.9	10-30	T	S: Ethyleneglycol	68Fd	b,q: ±0.2
E58	ETHENE, 1-ethoxy-2-isopropyl-(trans)		$C_7H_{14}O$			C_3H_7CH:$CHOC_2H_5$		L
Ag^+	-3.5	-0.229(20°,1)	-12.9	10-30	T	S: Ethyleneglycol	68Fd	b,q: ±0.2

Metal Ion	ΔH, kcal /mole	Log K	ΔS, cal /mole °K	T, °C	M	Conditions	Ref.	Re- marks
E59	ETHENE, 1-ethoxy-2-methyl- (cis)		$C_5H_{10}O$			$CH_3CH:CHOC_2H_5$		L
Ag^+	-3.9	0.496(20°,1)	-11.0	10-30	T	S: Ethyleneglycol	68Fd	b,q: ±0.2
E60	ETHENE, 1-ethoxy-2-methyl- (trans)		$C_5H_{10}O$			$CH_3CH:CHOC_2H_5$		L
Ag^+	-3.2	-0.268(20°,1)	-12.1	10-30	T	S: Ethyleneglycol	68Fd	b,q: ±0.2
E61	ETHENE, fluoro-		C_2H_3F			$CH_2:CHF$		L
$RhAB_2$	-1.6	-0.495(25°)	-7.5	0&25	T	S: Toluene A=2,4-Pentanedionate B=Ethylene R: L + $RhAB_2$ = RhABL + B	67Cd	a,b,q: ±1.1
E62	ETHENE, isobutoxy-		$C_6H_{12}O$			$CH_2:CHOC_4H_9$		L
Ag^+	-4.0	0.834(20°,1)	-9.8	10-30	T	S: Ethyleneglycol	68Fd	b,q: ±0.2
E63	ETHENE, 1-isobutoxy-2-methyl- (cis)		$C_7H_{14}O$			$CH_3CH:CHOC_4H_9$		L
Ag^+	-3.7	0.456(20°,1)	-10.4	10-30	T	S: Ethyleneglycol	68Fd	b,q: ±0.2
E64	ETHENE, 1-isobutoxy-2-methyl- (trans)		$C_7H_{17}O$			$CH_3CH:CHOC_4H_9$		L
Ag^+	-2.8	-0.222(20°,1)	-10.6	10-30	T	S: Ethyleneglycol	68Fd	b,q: ±0.2
E65	ETHENE, 1-methoxy-1-methyl-		C_4H_8O			$CH_2:C(CH_3)OCH_3$		L
Ag^+	-4.7	0.866(20°,1)	-12	10-30	T	S: Ethyleneglycol	68Fd	b,q: ±0.2
E66	ETHENE, tetracyano-		C_6N_4			$(CN)_2C:C(CN)_2$		L
Na^+	-7.1	--	--	-95- -50	T	S: 2-Methyl-tetrahydrofuran ΔH Determined by ESR R: 2ML = M_2L_2	72Ie	b
	-7.6	4.699(-104°)	-11.8	-119- -88	T	S: 2-Methyl-tetrahydrofuran ΔH Determined by ESR R: 2ML = M_2L_2	72Ie	b
	-5.7	4.556(-104°)	-6.5	-129- -98	T	S: 5.4% Tetrahydro-furan R: 2ML = M_2L_2	72Ie	b
	-3.4	4.176(-104°)	-0.5	-126- -104	T	S: 9.6% Tetrahydro-furan R: 2ML = M_2L_2	72Ie	b

Metal Ion	ΔH, kcal /mole	Log K	ΔS, cal /mole °K	T, °C	M	Conditions	Ref.	Remarks
E66, cont.								
Na^+, cont.	-1.8	4.000(104°)	3.7	-126- -104	T	S: 14.5% Tetrahydrofuran R: 2ML = M_2L_2	72Ie	b
K^+	-4.8	--	--	-115- -85	T	S: 2-Methyltetrahydrofuran R: 2ML = M_2L_2	72Ie	b
Rb^+	-4.3	--	--	-105- -65	T	S: 2-Methyltetrahydrofuran R: 2ML = M_2L_2	72Ie	b
Cs^+	-4.0	--	--	-115- -85	T	S: 2-Methyltetrahydrofuran R: 2ML = M_2L_2	72Ie	b
E67	ETHENE, tetrakis(4-methoxyphenyl)		$C_{30}H_{28}O_4$			$[C_6H_4(OCH_3)]_2C:[C_6H_4(OCH_3)]_2$		L
Br_2	-9.56	3.94(1)	-14	26	T	$C_M \approx 10^{-3}$ S: Ethylene chloride	58Bc	
E68	ETHENE, tetraphenyl-		$C_{26}H_{20}$			$(C_6H_5)_2C:C(C_6H_5)_2$		L
LiL	15.8	2.613(30°)	64	0-50	T	S: Tetrahydrofuran R: 2LiL = Na_2L + L	63Ea	b
NaL	11.0	3.447(30°)	52	0-50	T	S: Tetrahydrofuran R: 2NaL = Na_2L + L	63Ea	b
	-13	0.143(0°)	-45	-37-20	T	S: Tetrahydrofuran R: NaL + L^- = NaL^- + L	65Ra	b,g:Na, q:±2
Na_2L	-19	-1.591(0°)	-75	-37-20	T	S: Tetrahydrofuran R: Na_2L + L = 2NaL	65Ra	b,g:Na, q:±2
E69	ETHENE, 1,1,2-trimethyl-		C_5H_{10}			$CH_3CH:C(CH_3)_2$		L
Ag^+	-6.01	-2.212(25°)	30.2	0-25	T	μ=1.0($AgNO_3$, KNO_3)	38Wa	b,o
	2.4	0.0043(25°,1)	-7.8	0-40	T	S: Ethyleneglycol	65Ci	b
E70	ETHER, allyl ethyl-		$C_5H_{10}O$			$C_2H_5OCH_2CH:CH_2$		L
	-12.0	(1)	--	25	C	S: Benzene C_M=0.05-0.08M A=Cl	68Ge	
E71	ETHER, dibutyl-		$C_8H_{18}O$			$CH_3(CH_2)_3O(CH_2)_3CH_3$		L
TiA_4	-6.0	--	--	25	C	S: Benzene C_M=0.05-0.08M A=Cl	70Gh 68Ge	u
Al_2A_6	-23.2	--	--	25	C	S: Benzene C_M=0.07M A=Br R: 0.5(Al_2A_6) + L = AlA_3L	68Rc	
	-21.5	--	--	25	C	S: Cyclohexane C_M=0.07M A=Br	68Rc	

Metal Ion	ΔH, kcal /mole	Log K	ΔS, cal /mole °K	T, °C	M	Conditions	Ref.	Remarks
E71, cont.								
Al_2A_6, cont						R: $0.5(Al_2A_6) + L = AlA_3L$		
GaA_3	-16.0	3.66(1)	-37	25	C	S: Benzene A=Cl	70Gh	
TeA_4	-1.5	(1)	--	25	C	S: Benzene A=Cl	72Pe	
E72 ETHER, diethyl-			$C_4H_{10}O$			$CH_3CH_2OCH_2CH_3$		L
AlA_3	-20.21	--	--	25-28	C	S: n-Hexane $A=CH_3$	68Hc	
Al_2A_6	-23.2	--	--	25	C	S: Benzene C_M=0.07M A=Br R: $0.5(Al_2A_6) + L = AlA_3L$	68Rc	
TeA_4	-4.2	(1)	--	25	C	S: Benzene A=Cl	72Pe	
I_2	-4.2	0.82(25°,1)	-10.3	4-25	T	S: n-Heptane	60Ta 60Bc	b
	-4.30	0.96(25°,1)	-9.86	4-25	T	S: CCl_4	60Ta 60Bc	b
	-4.06	(1)	--	2-25	T	C_L=0.00-0.16 S: CCl_4	61Da	
	-4.30	0.96(25°,1)	-9.86	7-25	T	C_L in mole fraction = 0.06-0.95 S: CCl_4	57Dc	b,1
	-0.94	(1)	--	2-25	T	C_L=0.00-0.16 S: $CHCl_3$	61Da	
	-3.62	(1)	--	10-25	T	C_L=0.00-0.16 S: Cyclohexane	61Da	
	-4.9	(1)	--	--	C	S: $(C_2H_5)_2O$	61Aa	i: 50Ha
	-4.4	(1)	--	--	C	S: $(C_2H_5)_2O$	50Ha	
	-2.70	(1)	--	2-25	T	C_L=0.00-0.16 S: n-Heptane	61Da	
	-3.76	(1)	--	2-25	T	C_L=0.00-0.16 S: n-Hexane	61Da	
E73 ETHER, diethyl, 2,2'-diamino-			$C_4H_{12}O$			$O(CH_2CH_2NH_2)_2$		L
Ni^{2+}	-7.1	5.54(30°,1)	2	10-40	T	μ=0	59Lb	b
	-7.5	3.19(30°,2)	-10	10-40	T	μ=0	59Lb	b
Cu^{2+}	-11.0	8.58(30°,1)	3	10-40	T	μ=0	59Lb	b
	-3.7	4.35(30°,2)	8	10-40	T	μ=0	59Lb	b
Ag^+	-14.5	5.31(30°,1)	-24	10-40	T	μ=0	59Lb	b
E74 ETHER, diethyl, 2,2'-diamino-N,N,N',N'-tetraacetic acid			$C_{12}H_{20}O_9N_2$			$(CH_2CO_2H)_2N(CH_2)_2O(CH_2)_2-N(CH_2CO_2H)_2$		L^{4-}
Mg^{2+}	3.51	8.32(1)	50.04	20	C	μ=0.1(KNO_3)	64Ab	
Ca^{2+}	-6.85	10.0(1)	22.4	20	C	μ=0.1(KNO_3)	64Ab	
La^{3+}	-3.35	16.6(1)	64.5	20	C	μ=0.1(KNO_3)	64Ab	
Mn^{2+}	-5.9	13.76(1)	45.6	20	C	μ=0.1(KNO_3)	64Ab	

Metal Ion	ΔH, kcal /mole	Log K	ΔS, cal /mole °K	T, °C	M	Conditions	Ref.	Remarks
E74, cont.								
Co^{2+}	-6.35	15.27(1)	48.2	20	C	μ=0.1(KNO_3)	64Ab	
Ni^{2+}	-4.74	15.07(1)	52.8	20	C	μ=0.1(KNO_3)	64Ab	
Cu^{2+}	-9.82	18.1(1)	49.0	20	C	μ=0.1(KNO_3)	64Ab	
Zn^{2+}	-5.99	15.3(1)	49.6	20	C	μ=0.1(KNO_3)	64Ab	
Cd^{2+}	-9.42	16.2(1)	42.0	20	C	μ=0.1(KNO_3)	64Ab	
Hg^{2+}	-20.5	23.9(1)	35.7	20	C	μ=0.1(KNO_3)	64Ab	
Pb^{2+}	-13.15	15.03(1)	23.9	20	C	μ=0.1(KNO_3)	64Ab	
E75	ETHER, diethyl, 2,2'-dimethoxy-		$C_6H_{14}O_3$			$CH_3O(CH_2)_2O(CH_2)_2OCH_3$		L
NaA	-4.5	(1)	-8.6	-40-25	T	$C_L<4x10^{-2}$. S: Tetrahydropyran R: NaA + L = LNaA A=Biphenyl	67Sb	r,b
	0	(1)	--	-40-25	T	$C_L<4x10^{-2}$ S: Tetrahydropyran R: NaA + L = NaLA A=Biphenyl	67Sb	
	-2.8	0.146(25°)	-9	0-50	T	S: Tetrahydrofuran A=Fluorenyl ion R: NaA + L = NaLA	70Ce	b
SnA_4	-20.3	(1)	--	--	C	S: Benzene C_M=0.05-0.08M A=Cl	68Ge	u
E76	ETHER, diisopropyl-		$C_6H_{14}O$			$(CH_3)_2CHOCH(CH_3)_2$		L
I_2	-6.11	0.647(25°,1)	--	--	T	S: n-Hexane	57Dd 52Ka	1
E77	ETHER, dimethyl-		C_2H_6O			CH_3OCH_3		L
AlA_3	-20.29	--	--	25-28	C	S: n-Hexane A=CH_3	68Hc	
E78	ETHER, dioctyl-		$C_{16}H_{34}O$			$C_8H_{17}OC_8H_{17}$		L
TiA_4	-6.3	(1)	--	25	C	S: Benzene C_M=0.05-0.08M A=Cl	68Ge 70Gh	
GaA_3	-17.4	3.81(1)	-40.9	25	C	S: Benzene A=Cl	70Gh	
E79	ETHER, diphenyl-		$C_{12}H_{10}O$			$C_6H_5OC_6H_5$		L
AlA_3	-3.90	(1)	-10.2	29-110	T	S: L R: $1/2Al_2A_6$ + L = AlA_3L A=Et	72Ab	b,r
Al_2A_6	-12.0	--	--	25	C	S: Benzene C_M=0.07M A=Br R: $0.5(Al_2A_6)$ + L = AlA_3L	68Rc	
E80	ETHER, diphenyl, 4,4'-dibromo-		$C_{12}H_8OBr_2$			$C_6H_4(Br)OC_6H_4(Br)$		L

Metal Ion	ΔH, kcal /mole	Log K	ΔS, cal /mole °K	T, °C	M	Conditions	Ref.	Re- marks

E80, cont.

Al_2A_6	-9.5	--	--	25	C	S: Benzene C_M=0.07M A=Br R: $0.5(Al_2A_6)$ + L = AlA_3L	68Rc	

E81 ETHER, diphenyl, 4,4'- dimethyl- $C_{14}H_{14}O$ $C_6H_4(CH_3)OC_6H_4(CH_3)$ L

Al_2A_6	-13.0	--	--	25	C	S: Benzene C_M=0.07M A=Br R: $0.5(Al_2A_6)$ + L = AlA_3L	68Rc	

E82 ETHER, methyl propyl- $C_4H_{10}O$ $CH_3OCH_2CH_2CH_3$ L

SbA_5	-18.09	(1)	--	25	C	S: Ethylene chloride. A=Cl	68Oa	

E83 ETHENE, tetrachloro- C_2Cl_4 $Cl_2C{:}CCl_2$ L

I_2A	-0.17	0.364(25°)	1.09	-20-40	T	S: Hexane A=Hexane R: I_2A + L = I_2L + A	73Si	b,q: ±0.03

E84 ETHYLENETRITHIOCARBONATE $C_3H_4S_3$

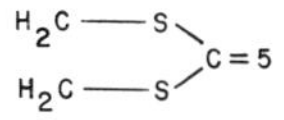

$$\begin{array}{c} H_2C \!-\! S \\   \diagdown \\   C=S \\ \diagup \\ H_2C \!-\! S \end{array}$$ L

I_2	-8.5	1.653(25°,1)	-21.0	0-45	T	S: Cyclohexane	66Bc	b,q:±1

E85 ETHYNE C_2H_2 CH:CH L

$IrABC_2$	-9.3	0.079(20°,1)	-31.4	10-60	T	S: Chlorobenzene A=I B=CO C=Ph_3P	71Vb	b
Ag^+	-13.20	1.60(1)	-36.90	25	T	P: $AgC_2H_2^+$ C_2H_2(g) is passed into soln. containing Ag_2SO_4 + H_2SO_4 ΔH_T=ΔH_{298} + 30.0 (T-298)	62Tc	
	6.86	1.38(1)	29.4	25	T	P: $AgC{\equiv}CH^+$ C_2H_2(g) is passed into soln. containing Ag_2SO_4 + H_2SO_4 ΔH_T=ΔH_{298} - 6.0 (T-298)	62Tc	

F

F1 FERRATE(II) ION, hexacyano C_6N_6Fe $Fe(CN)_6^{4-}$ L^{4-}

K^+	1.0	2.35(1)	13	25	C	μ=0	65Ea	
La^{3+}	0.85	3.702(1)	19.8	25	C	μ=0	73So	
Pr^{3+}	0.87	3.636(1)	19.5	25	C	μ=0	73So	

Metal Ion	ΔH, kcal /mole	Log K	ΔS, cal /mole °K	T, °C	M	Conditions	Ref.	Remarks
F1, cont.								
Nd^{3+}	0.80	3.768(1)	19.9	25	C	μ=0	73So	
Sm^{3+}	0.91	3.724(1)	20.1	25	C	μ=0	73So	
Eu^{3+}	0.98	3.651(1)	20.0	25	C	μ=0	73So	
Gd^{3+}	1.04	3.592(1)	19.9	25	C	μ=0	73So	
Tb^{3+}	0.94	3.760(1)	20.4	25	C	μ=0	73So	
Dy^{3+}	1.02	3.680(1)	20.3	25	C	μ=0	73So	
Ho^{3+}	1.06	3.658(1)	20.3	25	C	μ=0	73So	
Er^{3+}	1.04	3.680(1)	20.3	25	C	μ=0	73So	
Tm^{3+}	1.05	3.673(1)	20.3	25	C	μ=0	73So	
Yb^{3+}	1.04	3.658(1)	20.2	25	C	μ=0	73So	
Lu^{3+}	1.01	3.695(1)	20.3	25	C	μ=0	73So	
Tl^{+}	1.1	3.0(1)	17.4	25	T	μ=0	58Pa	
	-1.78	0.82(1)	-2.2	25	C	3M($LiClO_4$)	67Mc	
F2 FERRATE(III) ION, hexacyano			C_6N_6Fe			$Fe(CN)_6^{3-}$		L^{3-}
Na^{+}	-4.0	-0.77(1)	-17	25	C	μ=3	66Me	
K^{+}	0.5	1.46(1)	8	25	C	μ=0	67Ea	
	-5.4	-0.42(1)	-20	25	C	μ=3	66Me	
Mg^{2+}	-3.4	-1.03(1)	-16	25	C	μ=3	66Me	
Ba^{2+}	-3.7	-0.60(1)	-15	25	C	μ=3	66Me	
La^{3+}	2.02	3.73(25°,1)	23.9	20-35	T	μ=0	52Jb 48Da	b
Tl^{+}	0.82	3.22(1)	17.4	25	T	μ=0	53Ba	
F3 FLUORENE			$C_{13}H_{10}$			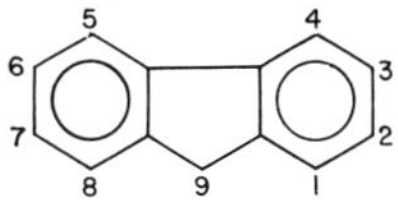		L^{-}
LiL	3.0	--	16	--	T	S: 3,4-Dihydropyran R: Solvent separated ion pair→contact ion pair $Li^{+}\|\|L^{-}$ = $Li^{+}L^{-}$ + Solvent	72Sh	u
	5.6	--	15	-40-80	T	S: 1,2-Dimethoxy-ethane. R: Solvent separated ion pair→contact ion pair $Li^{+}\|\|Li^{-}$ = $Li^{+}L^{-}$ + Solvent	72Gi	b
	2.9	--	16	-55-30	T	S: 2,5-Dimethoxy-tetrahydrofuran R: Solvent separated ion pair→contact ion pair $Li^{+}\|\|L^{-}$ = $Li^{+}L^{-}$ + Solvent	72Sh	

Metal Ion	ΔH, kcal /mole	Log K	ΔS, cal /mole °K	T, °C	M	Conditions	Ref.	Re-marks
F3, cont.								
LiL, cont.	2.0	--	14	-55-30	T	S: 2,5-Dimethyl-tetrahydrofuran R: Solvent separated ion pair→contact ion pair $Li^+||L^- = Li^+L^-$ + Solvent	72Sh	
	3.6	--	14	-55-30	T	S: Dioxane-tetra-hydrofuran(50%) R: Solvent separated ion pair→contact ion pair $Li^+||L^- = Li^+L^-$ + Solvent	72Sh	
	3.5	--	1.7	-55-30	T	S: Dioxolane R: Solvent separated ion pair→contact ion pair $Li^+||L^- = Li^+L^-$ + Solvent	72Sh	
	7.5	--	27	-55-30	T	S: 2-Methyltetra-hydrofuran R: Solvent separated ion pair→contact ion pair $Li^+||L^- = Li^+L^-$ + Solvent	72Sh	
	4.3	--	16	-55-30	T	S: Oxepane R: Solvent separated ion pair→contact ion pair $Li^+||L^- = Li^+L^-$ + Solvent	72Sh	
	7.5	--	22	-55-30	T	S: Tetrahydrofuran R: Solvent separated ion pair→contact ion pair $Li^+||L^- = Li^+L^-$ + Solvent	72Sh	
	6.6	--	28	-55-30	T	S: Tetrahydropyran R: Solvent separated ion pair→contact ion pair $Li^+||L^- = Li^+L^-$ + Solvent	72Sh	
LiA	-0.290	0.613(20°)	1.8	-10-30	T	S: Cyclohexylamine A=2,3-Benzfluorene R: CuA + L = CuL + A	72Si	b,q: ±0.06
NaL	5.4	--	17	-40-80	T	S: 1,2-Dimethoxy-ethane R: Solvent separated ion pair→contact ion pair $Na^+||L^- = Na^+L^-$ + Solvent	72Gi	b
	6.7	--	27	-40-80	T	S: Tetrahydrofuran R: Solvent separated ion pair→contact ion pair $Na^+||L^- = Na^+L^-$ + Solvent	72Gi	b

Metal Ion	ΔH, kcal /mole	Log K	ΔS, cal /mole °K	T, °C	M	Conditions	Ref.	Remarks
F3, cont.								
NaL, cont.	7.6	1.194(24.2°)	30.9	-63.0-24.2	T	S: Tetrahydrofuran R: Solvent separated ion pair→contact ion pair $Na^+\|\|L^- = Na^+L^-$ + Solvent	66Hc	b
KL	4.1	--	16	-40-80	T	S: 1,2-Dimethoxyethane R: Solvent separated ion pair→contact ion pair $K^+\|\|L^- = K^+L^-$ + Solvent	72Gi	b
Sr^{2+}	6.2	(1)	27	-80-20	T	S: $C_2H_5NH_2$	74Pc	b,j
F4 FLUORENE, 9-tert-butyl-			$C_{17}H_{18}$			$C_{13}H_9[C(CH_3)_3]$		L^-
LiA	1.140	-1.678(20°)	-3.8	-10-30	T	S: Cyclohexylamine A=2,3-Benzfluorene R: LiA + L = LiL + A	72Si	b
CsA	1.320	-1.086(20°)	0.6	-10-30	T	S: Cyclohexylamine A=2,3-Benzfluorene R: CsA + L = CsL + A	72Si	b
F5 FLUORENE, 2-fluoro-			$C_{13}H_9F$			$C_{13}H_9(F)$		L^-
Li^+	2.9	5.208(15°,1)	33	25	T	S: 1,2-Dimethoxyethane	69Ea	
	3.5	5.523(1)	37	25	T	S: Tetrahydrofuran	69Ea	
Na^+	3.9	5.260(1)	37	25	T	S: 1,2-Dimethoxyethane	69Ea	
	8.3	6.174(1)	56	25	T	S: Tetrahydrofuran	69Ea	
K^+	5.9	5.538(20°,1)	45	25	T	S: 1,2-Dimethoxyethane	69Ea	
	4.7	6.796(1)	47	25	T	S: Tetrahydrofuran	69Ea	
Cs^+	4.6	6.602(15°,1)	45	25	T	S: 1,2-Dimethoxyethane	69Ea	
	2.9	7.854(1)	46	25	T	S: Tetrahydrofuran	69Ea	
F6 FLUORENE, 9-ethyl-			$C_{15}H_{14}$			$C_{13}H_9(C_2H_5)$		L
LiA	-0.770	0.196(20°)	-1.8	-10-30	T	S: Cyclohexylamine A=2,3-Benzfluorene R: LiA + L = LiL + A	72Si	b
CsA	-0.650	0.580(20°)	0.4	-10-30	T	S: Cyclohexylamine A=2,3-Benzfluorene R: CsA + L = CsL + A	72Si	b,q: ±0.05
F7 FLUORENE, 9-isopropyl-			$C_{16}H_{16}$			$C_{13}H_9[CH(CH_3)_2]$		L^-
LiA	0.180	-0.585(20°)	-2.1	-10-30	T	S: Cyclohexylamine A=2,3-Benzfluorene R: LiA + L = LiL + A	72Si	b,q: ±0.04
CsA	0.400	-0.086(20°)	1.0	-10-30	T	S: Cyclohexylamine	72Si	b,q:

Metal Ion	ΔH, kcal /mole	Log K	ΔS, cal /mole °K	T, °C	M	Conditions	Ref.	Remarks
F7, cont.								
CsA, cont.						A=2,3-Benzfluorene R: CsA + L = CsL + A		±0.04
F8 FLUORENE, 9-methyl-			$C_{14}H_{12}$			$C_{13}H_9(CH_3)$		L^-
LiA	-0.780	0.573(20°)	0.0	-10-30	T	S: Cyclohexylamine A=2,3-Benzfluorene R: LiA + L = LiL + A	72Si	b
CsA	-0.740	0.724(20°)	0.7	-10-30	T	S: Cyclohexylamine A=2,3-Benzfluorene R: CsA + L = CsL + A	72Si	b
F9 FLUORENE, 9-(1-methylpentyl)			$C_{19}H_{22}$			$C_{13}H_9[CH(CH_3)C_4H_9]$		L^-
LiL	8.2	--	33	-55-30	T	S: 3,4-Dihydropyran R: Solvent separated ion pair→contact ion pair $Li^+\|\|L^- = Li^+L^-$ + Solvent	72Sh	
	10.0	--	50	-55-30	T	S: 2,5-Dimethyl-tetrahydrofuran R: Solvent separated ion pair→contact ion pair $Li^+\|\|L^- = Li^+L^-$ + Solvent	72Sh	
	11.0	--	36	-55-30	T	S: Dioxane-tetra-hydrofuran(28%) R: Solvent separated ion pair→contact ion pair $Li^+\|\|L^- = Li^+L^-$ + Solvent	72Sh	
	9.8	--	32	-55-30	T	S: 2-Methyltetra-hydrofuran R: Solvent separated ion pair→contact ion pair $Li^+\|\|L^- = Li^+L^-$ + Solvent	72Sh	
F10 9-FLUORENONE			$C_{13}H_8O$					L
SnA_4	-7.6	6.6(30°,$\underline{1}$)	-15.1	30-40	T	S: Benzene ΔS calculated using ΔG in atm^{-1}. A=Cl	64La	b
F11 FLUORIDE ION			F			F^-		L^-
Be^{2+}	-0.2	5.99($\underline{1}$)	27	25	T	μ=0	67Ab	
	-0.4	5.04($\underline{1}$)	22	25	T	μ=0.5	67Ab	
	-0.35	4.900($\underline{1}$)	21.2	25	T	1.0M(NaCl)	69Mb	
	-1.16	3.762($\underline{2}$)	13.3	25	T	1.0M(NaCl)	69Mb	
	-0.29	2.788($\underline{3}$)	11.8	25	T	1.0M(NaCl)	69Mb	
	-0.46	1.426($\underline{4}$)	5.0	25	T	1.0M(NaCl)	69Mb	

Metal Ion	ΔH, kcal /mole	Log K	ΔS, cal /mole °K	T, °C	M	Conditions	Ref.	Re-marks
F11, cont.								
Mg^{2+}	4	1.301(25°,1)	19	15-25	T	μ=0.50($NaClO_4$)	54Ce 68Tb	b
	3.2	1.32(1)	16.8	25	C	μ=1.0($NaClO_4$)	68Tb	
Ca^{2+}	3.5	0.63(1)	15.0	25	C	μ=1.0($NaClO_4$)	68Tb	
	3.8	0.63(1)	15.8	25	T	μ=1.0($NaClO_4$)	68Tb	
Sr^{2+}	4	0.15(1)	14.0	25	C	μ=1.0($NaClO_4$)	68Tb	
	4	0.15(1)	14.0	25	T	μ=1.0($NaClO_4$)	68Tb	
Ba^{2+}	0	-0.22(1)	-1.0	25	T	μ=1.0($NaClO_4$)	68Tb	
Sc^{3+}	0.6	6.16(25°,1)	30	15-35	T	μ=0.5	67Ab	b
	-2.3	3.28(25°)	7	15-35	T	μ=0.5. R: Sc^{3+} + HF = ScF^{2+} + H^+	59Kb	b
	-4.1	2.37(25°)	-3	15-35	T	μ=0.5. R: ScF^{2+} + HF = $ScF_2{}^+$ + H^+	59Kb	b
	-3.4	1.16(25°)	-6	15-35	T	μ=0.5. R: $ScF_2{}^+$ + HF = ScF_3 + H^+	59Kb	b
Y^{3+}	2.3	4.79(1)	30	25	T	μ=0	67Ab 61Pb	
	2.2	3.91(1)	25	25	T	μ=0.5	67Ab	
	8.32	3.599(1)	44.5	25	C	μ=1M($NaClO_4$)	67Wa	
La^{3+}	4.00	2.668(1)	25.6	25	C	μ=1M($NaClO_4$)	67Wa	
Ce^{3+}	4.82	2.815(1)	29.0	25	C	μ=1M($NaClO_4$)	67Wa	
Pr^{3+}	5.74	3.013(1)	33.1	25	C	μ=1M($NaClO_4$)	67Wa	
Nd^{3+}	6.83	3.086(1)	36.9	25	C	μ=1M($NaClO_4$)	67Wa	
Sm^{3+}	9.39	3.116(1)	45.8	25	C	μ=1M($NaClO_4$)	67Wa	
Eu^{3+}	9.22	3.189(1)	45.4	25	C	μ=1M($NaClO_4$)	67Wa	
Gd^{3+}	8.90	3.313(1)	45.1	25	C	μ=1M($NaClO_4$)	67Wa	
Tb^{3+}	7.51	3.423(1)	40.8	25	C	μ=1M($NaClO_4$)	67Wa	
Dy^{3+}	7.03	3.460(1)	39.4	25	C	μ=1M($NaClO_4$)	67Wa	
Ho^{3+}	7.26	3.519(1)	40.4	25	C	μ=1M($NaClO_4$)	67Wa	
Er^{3+}	7.43	3.541(1)	41.2	25	C	μ=1M($NaClO_4$)	67Wa	
Tm^{3+}	8.66	3.563(1)	45.4	25	C	μ=1M($NaClO_4$)	67Wa	
Yb^{3+}	9.56	3.585(1)	48.5	25	C	μ=1M($NaClO_4$)	67Wa	
Lu^{3+}	9.53	3.614(1)	48.5	25	C	μ=1M($NaClO_4$)	67Wa	
Th^{4+}	-1.2	8.44(25°,1)	39	5-40	T	μ=0	70Be	b
	-0.8	6.62(25°,2)	31	5-40	T	μ=0	70Be	b
	-0.8	4.75(25°,3)	19	5-40	T	μ=0	70Be	b
	-0.9	3.36(25°,4)	12	5&25	T	μ=0	70Be	a,b
VO^{2+}	1.88	3.371(1)	21.7	25	C	μ=1.0($NaClO_4$)	71Ab	
	1.55	2.369(2)	16.0	25	C	μ=1.0($NaClO_4$)	71Ab	
	1.39	1.554(3)	11.8	25	C	μ=1.0($NaClO_4$)	71Ab	
	1.46	0.785(4)	8.46	25	C	μ=1.0($NaClO_4$)	71Ab	
$CrA_6{}^{3+}$	1.3	1.346(86°)	6.2	77.2-94.7	T T	μ=1. R: $CrA_6{}^{3+}$ + HF = CrA_5F^{2+} + H^+ + A A=H_2O	65Se	b

Metal Ion	ΔH, kcal /mole	Log K	ΔS, cal /mole °K	T, °C	M	Conditions	Ref.	Re-marks
F11, cont.								
UO_2^{2+}	0.406	4.535(1)	22.1	25	C	μ=1.0($NaClO_4$)	71Ab	
	≃-2	4.4(25°,1)	≃13	10-40	T	μ=2. Data recalculated from those given in 54Db	67Ab 54Db	b
	0.096	3.442(2)	16.1	25	C	μ=1.0($NaClO_4$)	71Ab	
	0.060	2.437(3)	11.4	25	C	μ=1.0($NaClO_4$)	71Ab	
	-0.492	1.477(4)	5.11	25	C	μ=1.0($NaClO_4$)	71Ab	
Fe^{3+}	3.4	6.03(25°,1)	39	15-35	T	μ=0	67Ab	b
	2.35	5.17(1)	31.5	25	C	μ=0.5	67Ab	
	3.3	5.210(25°,1)	34.9	0-25	T	μ=0.5($NaClO_4$)	53Ha	b
	0.8	3.949(25°,2)	20.8	0-25	T	μ=0.5($NaClO_4$)	53Ha	b
	-0.65	2.26	8.2	25	T	μ=0.5. R: Fe^{3+} + HF = FeF^{2+} + H^+	56Cd	
	-1.2	1.01	0.6	25	T	μ=0.5. R: FeF^{2+} + HF = FeF_2^+ + H^+	56Cd	
[Fe^{3+}]	-1.5	1.46(1)	1.8	25	-	Ferrimyglobin	55Ga 60Gb	
[Fe^{3+}]	-2.5	1.76(1)	-0.6	25	-	Ferrihaemoglobin	55Ga 60Gb	
[Fe^{3+}]	ΔH, obtained by temperature variation of K, is reported in graphical form as a function of pH in the reactions: Hb(H_2O) + L = HbL + H_2O					Dog Methaemoglobin	69Be	
[Fe^{3+}]	ΔH, obtained by temperature variation of K, is reported in graphical form as a function of pH in the reactions: Hb(H_2O) + L = HbL + H_2O					Guinea Pig Methaemoglobin	69Be	
[Fe^{3+}]	-1.38	2.01(1)	4.5	20	T	Human Methaemoglobin A. 6.0	68Ac 69Be	d: 0.05
	-2.18	1.94(1)	1.5	20	T	6.5	68Ac	
	-2.89	1.86(1)	-1.3	20	T	7.0	68Ac	
	-2.38	1.77(1)	0	20	T	7.5	68Ac	
	-1.94	1.69(1)	1.1	20	T	8.0	68Ac	
	-1.38	1.68(1)	3.0	20	T	8.5	68Ac	
	-1.24	1.62(1)	3.2	20	T	9.0	68Ac	
[Fe^{3+}]	-1.26	2.13(1)	5.46	20	T	Human Methaemoglobin C. 6.0	68Ac 69Be	d: 0.05
	-2.03	2.06(1)	2.5	20	T	6.5	68Ac	
	-2.67	1.99(1)	0	20	T	7.0	68Ac	
	-3.21	1.89(1)	-2.3	20	T	7.5	68Ac	
	-2.67	1.78(1)	-1.1	20	T	8.0	68Ac	
	-1.72	1.67(1)	1.8	20	T	8.5	68Ac	
	-1.00	1.60(1)	3.79	20	T	9.0	68Ac	
[Fe^{3+}]	ΔH, obtained by temperature variation of K, is reported in graphical form as a function of pH in the reactions: Hb(H_2O) + L = HbL + H_2O					Pigeon Methaemoglobin	69Be	
[Fe^{3+}]	ΔH, obtained by temperature variation of K, is reported in graphical form					Sperm Whale Metmyoglobin	69Be	

Metal Ion	ΔH, kcal /mole	Log K	ΔS, cal /mole °K	T, °C	M	Conditions	Ref.	Re-marks
F11, cont.								
$[Fe^{3+}]$, cont.	as a function of pH in the reactions: $Hb(H_2O) + L = HbL + H_2O$							
$[Fe^{3+}]$	-0.7	1.90(1)	6.4	25	-	Chironomus Hemoglobin	60Gb	
Cu^{2+}	0.9	0.70(1)	6	25	T	μ=0.5	58Cd	
Ag^{+}	-2.8	-0.167(1)	-10	25	T	μ=0.50	61Cc 61Az	q: ±1.0
Zn^{2+}	3.80	1.16(1)	18.0	25	C	μ=0	71Ce	
	1.5	0.73(1)	8	25	T	μ=0.5	67Ab	
	1.99	0.748(1)	10.1	25	C	μ=1M($NaClO_4$)	69Gb	
	1.83	0.845(1)	10.0	25	C	μ=3M($NaClO_4$)	69Gb	
Cd^{2+}	1.02	0.568(1)	6.0	25	C	μ=3($NaClO_4$)	67Ab 66Ga	
Hg^{2+}	≃0.9	1.02(1)	≃8	25	T	μ=0.5	67Ab	
Al^{3+}	1.06	6.61(1)	33.8	25	C	μ=0.07	59Ka	
	1.150	6.14(1)	32	25	C	μ=0.06-0.2	53La	
	0.74	6.114(1)	30.5	25	C	μ=1.0M($NaClO_4$)	71Wa	q: ±0.08
	0.92	5.36(2)	27.6	25	C	μ=0.07	59Ka	
	0.780	5.02(2)	26	25	C	μ=0.06-0.2	53La	
	0.18	4.06(3)	19.2	25	C	μ=0.07	59Ka	
	0.190	3.85(3)	18	25	C	μ=0.06-0.2	53La	
	0.04	2.68(4)	12.4	25	C	μ=0.07	59Ka	
	0.280	2.74(4)	3	25	C	μ=0.06-0.2	53La	
	-0.36	1.33(5)	4.9	25	C	μ=0.07	59Ka	
	-0.750	1.63(5)	5	25	C	μ=0.06-0.2	53La	
	-1.550	0.469(6)	-3	25	C	μ=0.06-0.2	53La	
	0.100	19.842(1-6)	91	25	C	μ=0.06-0.2	53La	
Ga^{3+}	2.51	4.362(1)	28.4	25	C	μ=1.0M($NaClO_4$)	71Wa	
In^{3+}	3.5	4.62(1)	33	25	T	μ=0	67Ab	
	2.4	3.74(1)	25	25	T	μ=0.5	67Ab	
	2.20	3.69(1)	24.2	25	C	μ=1.00M($NaClO_4$)	69Rd	
	2.98	3.729(1)	27.2	25	C	μ=1.0M($NaClO_4$)	71Wa	q: ±0.15
	1.83	2.83(2)	19.1	25	C	μ=1.00M($NaClO_4$)	69Rd	
	3.3	2.11(3)	20.7	25	C	μ=1.00M($NaClO_4$)	69Rd	
	-0.508	0.839	2.1	25	T	μ=0.5. R: $In^{3+} + HF = InF^{2+} + H^+$	54Ha	
	1.0	-0.30	2.0	25	T	μ=0.5. R: $InF^{2+} + HF = InF_2^{+} + H^+$	54Ha	
Sn^{2+}	10.35	8.415(25°,1-3)	73.2	25-60	T	μ=0.85	68Ha	b
F12 FORMIC ACID			CH_2O_2			HCO_2H		L^-
Mg^{2+}	-1.77	1.42(25°,1)	0.6	25-35	T	μ=0	56Na	b
Ca^{2+}	0.98	1.42(25°,1)	10	25-35	T	μ=0	56Na	b
Sr^{2+}	0.59	1.38(25°,1)	8.3	25-35	T	μ=0	56Na	b
Ba^{2+}	-1.89	1.37(25°,1)	-0.3	25-35	T	μ=0	56Na	b

Metal Ion	ΔH, kcal /mole	Log K	ΔS, cal /mole °K	T, °C	M	Conditions	Ref.	Remarks
F12, cont.								
Cr^{2+}	3.16	1.928(25°,1)	19	25&35	T	μ=1.0($HClO_4$)	73Ta	a,b,y: ΔS
	2.72	2.000(35°,1)	18	35&50	T	μ=1.0($HClO_4$)	73Ta	a,b,y: ΔS
	1.32	0.674(25°,2)	8	25&35	T	μ=1.0($HClO_4$)	73Ta	a,b,y: ΔS
	4.70	0.709(35°,2)	18	35&50	T	μ=1.0($HClO_4$)	73Ta	a,b,y: ΔS
	6.42	1.283(25°,3)	27	25&35	T	μ=1.0($HClO_4$)	73Ta	a,b,y: ΔS
	-6.42	1.433(35°,3)	-14	35&50	T	μ=1.0($HClO_4$)	73Ta	a,b,y: ΔS
Mn^{3+}	11.88	-0.046(49.8°)	36.8	40.8 59.6	T	μ=4.0M R: Mn^{3+} + HL = MnL^{2+} + H^+	72Wa	b
[Fe^{3+}]	-6.3	1.46(1)	-14.6	25	-	Hemoglobin	60Gb	
[Fe^{3+}]	-6.30	1.51(1)	-14.6	20	T	Methemoglobin	55Sa	
F13 FORMIC ACID, amide, N,N-dimethyl-			C_3H_7ON			$HCON(CH_3)_2$		L
[Co]	-7.9	1.63(0°,1)	-21	-23-23	T	S: Toluene [Co]=Cobalt(II) protoporphyrin IX dimethyl ester	73Sr	b,q: ±0.6
I_2	-3.7	0.462(25°,1)	-10.4	8.6-28.2	T	C_M=5.5x10^{-3} C_L=4.9x10^{-3} S: CCl_4	62Dc	b
	-4.03	0.158(10°,1)	-13.5	0-20	T	C_M=5x10^{-4} C_L=0.2-0.7 S: CH_2Cl_2	61Ta	b
F14 FRUCTOSE			$C_6H_{12}O_6$			$CH_2(CHOH)_3COHCH_2OH$ (O ring bridge)		L
$H_5TcO_6^-$	-2.7	2.46	?	25	T	μ=0.1. R: $H_5TeO_6^-$ + L = $HTeO_4L^-$ + $2H_2O$	62Ea	
F15 FUMARIC ACID			$C_4H_4O_4$			$HO_2CCH:CHCO_2H$		
La^{3+}	2.69	2.74(1)	21.6	25	C	μ=0.10M($NaClO_4$)	73Cc	
Ce^{3+}	3.16	2.80(1)	23.4	25	C	μ=0.10M($NaClO_4$)	73Cc	
Pr^{3+}	3.24	2.84(1)	23.8	25	C	μ=0.10M($NaClO_4$)	73Cc	
Nd^{3+}	3.67	2.74(1)	24.8	25	C	μ=0.10M($NaClO_4$)	73Cc	
Sm^{3+}	3.46	2.83(1)	24.5	25	C	μ=0.10M($NaClO_4$)	73Cc	
Eu^{3+}	3.44	2.86(1)	24.6	25	C	μ=0.10M($NaClO_4$)	73Cc	
Gd^{3+}	3.84	2.88(1)	25.9	25	C	μ=0.10M($NaClO_4$)	73Cc	
Tb^{3+}	3.64	2.77(1)	24.9	25	C	μ=0.10M($NaClO_4$)	73Cc	
Dy^{3+}	3.81	2.80(1)	25.5	25	C	μ=0.10M($NaClO_4$)	73Cc	

Metal Ion	ΔH, kcal /mole	Log K	ΔS, cal /mole °K	T, °C	M	Conditions	Ref.	Remarks
F15, cont.								
Ho^{3+}	3.61	2.80(1)	24.9	25	C	μ=0.10M($NaClO_4$)	73Cc	
Er^{3+}	3.88	2.80(1)	25.8	25	C	μ=0.10M($NaClO_4$)	73Cc	
Tm^{3+}	3.48	2.81(1)	24.5	25	C	μ=0.10M($NaClO_4$)	73Cc	
Yb^{3+}	3.80	2.80(1)	25.5	25	C	μ=0.10M($NaClO_4$)	73Cc	
Lu^{3+}	3.81	2.81(1)	25.6	25	C	μ=0.10M($NaClO_4$)	73Cc	
F16	FURAN, 2,5-dimethyl-tetrahydro-		$C_6H_{12}O$			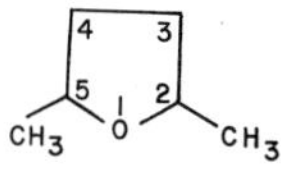		
AlA_3	-22.95	(1)	--	25-28	C	S: n-Hexane A=CH_3	68Hc	
F17	FURAN, 2-methyl-tetrahydro-		$C_5H_{10}O$			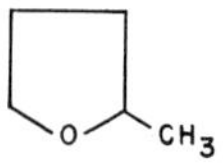		L
AlA_3	-22.94	(1)	--	25-28	C	S: n-Hexane A=CH_3	68Hc	
I_2	-6.2	1.35(25°,1)	-14.6	1-25	T	S: n-Heptane	60Bc 60Ta	b
F18	FURAN, tetrahydro-		C_4H_8O			$CH_2CH_2CH_2CH_2O$ (ring)		L
$[Co^{3+}]$	-15.8	--	--	24	C	S: Dichloromethane [Co]=Methylcobal-oxime R: $[Co]_2$ + 2L = 2[Co]L	73Cf	
HgA_2	-3.61	0.78(1)	-8.36	30	C	S: Benzene A=Cl C_M=1-5x10^{-3}M	73Fa	q: ±0.6
	-2.99	0.70(1)	-6.69	30	C	S: Benzene A=Br C_M=1-5x10^{-3}M	73Fa	q: ±0.6
AlA_3	-22.90	(1)	--	25-28	C	S: n-Hexane A=CH_3	68Hc	
SbA_5	-19.3	(1)	--	25	C	S: CCl_4 A=Cl	73Oa	
I_2	-5.3	1.25(25°,1)	-11.6	6-25	T	S: n-Heptane	60Bc 60Ta	b
F19	FURFURAL, oxime(syn)		$C_5H_5O_2N$			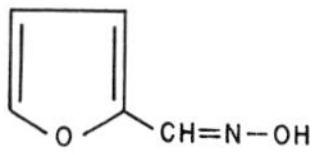		L
Cr^{3+}	-5.35	10.64(25°,1)	30.74	15-35	T	μ=0 S: 70 v % dioxane	63Aa	b
	-4.07	10.04(25°,2)	32.28	15-35	T	μ=0 S: 70 v % dioxane	63Aa	b
	-5.81	9.42(25°,3)	23.62	15-35	T	μ=0 S: 70 v % dioxane	63Aa	b
Fe^{3+}	-22.88	12.64(25°,1)	-18.93	15-35	T	μ=0 S: 70 v % dioxane	63Aa	b

Metal Ion	ΔH, kcal /mole	Log K	ΔS, cal /mole °K	T, °C	M	Conditions	Ref.	Remarks
F19, cont.								
Fe^{3+}, cont.	-51.70	10.80(25°,2)	-124	15-35	T	μ=0 S: 70 v % dioxane	63Aa	b
	-114.30	8.52(25°,3)	-344	15-35	T	μ=0 S: 70 v % dioxane	63Aa	b
Ni^{2+}	-20.36	7.60(25°,1)	-33.6	15-35	T	μ=0 S: 70 v % dioxane	63Aa	b
	-10.98	7.52(25°,2)	-2.42	15-35	T	μ=0 S: 70 v % dioxane	63Aa	b
Cu^{2+}	-9.86	10.28(25°,1)	13.96	15-35	T	μ=0 S: 70 v % dioxane	63Aa	b
	-13.98	9.88(25°,2)	-1.71	15-35	T	μ=0 S: 70 v % dioxane	63Aa	b
			G					
G1 GALACTOSE			$C_6H_{12}O_6$			$CH_2OHCH(CHOH)_3CHOH$ (ring bridged by O)		L
$H_5TeO_6^-$	-1.4	1.66(1)	3	25	T	μ=0.1 R: $H_5TeO_6^-$ + L = $HTeO_4L^-$ + $2H_2O$	62Ea	
G2 GLUCOSE			$C_6H_{12}O_6$			$CH_2OHCH(CHOH)_3CHOH$ (ring bridged by O)		L
BA_4^-	-3.52	2.130(25°,1)	-2.0	0-35	T	C_L=0-0.9M A=OH^-	67Cc	b,q: ±0.4
	-0.15	2.940(25°,1-2)	13	0-35	T	C_L=0-0.9M A=OH^-	67Cc	b,q: ±0.02
$H_5TeO_6^-$	-0.7	1.11(1)	3	25	T	μ=0.1 R: $H_5TeO_6^-$ + L = $HTeO_4L^-$ + $2H_2O$	62Ea	
G3 GLUCOSE, 2-amino-(D,β)			$C_6H_{13}O_5N$					
Co^{2+}	-3.0	4.35(1)	9.7	25	C	μ=0.05(KCl)	72Gd 72Gc	
	-7.0	8.04(1-2)	13	25	C	μ=0.05M(KCl)	72Gd 72Gc	
Ni^{2+}	-4.2	5.39(1)	10.7	25	C	μ=0.05M(KCl)	72Gd 72Gc	
	-8.6	9.85(1-2)	16	25	C	μ=0.05M(KCl)	72Gd 72Gc	
Cu^{2+}	-6.6	8.07(1)	9.7	25	C	μ=0.05M(KCl)	72Gd 72Gc	
	-12.8	14.76(1-2)	26	25	C	μ=0.05M(KCl)	72Gd 72Gc	
Zn^{2+}	-1.8	4.63(1)	15.0	25	C	μ=0.05M(KCl)	72Gd 72Gc	
	-5.4	8.64(1-2)	22	25	C	μ=0.05M(KCl)	72Gc	

Metal Ion	ΔH, kcal /mole	Log K	ΔS, cal /mole °K	T, °C	M	Conditions	Ref.	Re-marks
G4 GLUCOSE, 1-phosphate			$C_6H_{13}O_9P$			$CH_2OHCH(CHOH)_3CHOPO_3H_2$ (ring through O)		L
Mg^{2+}	2.913	2.479(1)	21.1	25	T	μ=0 ΔCp=59	54Cd	
	3.640	2.678(1)	23.5	37	T	μ=0 ΔCp=61.5	54Cd	
Ca^{2+}	2.403	2.495(1)	19.5	25	T	μ=0 ΔCp=63	56Cc	
	3.170	2.573(1)	22.0	37	T	μ=0 ΔCp=65	56Cc	
G5 GLUTAMIC ACID (D)			$C_5H_9O_4N$			$HO_2CCH_2CH_2CH(NH_2)CO_2H$		
Ni^{2+}	-7.39	9.81(2)	20.3	25	C	μ=0.1M	70Bd 71Be	y:ΔH (70Bd)
Cu^{2+}	-11.39	14.1(2)	26.5	25	C	μ=0.1M	70Bd 71Be	y:ΔH (70Bd)
G6 GLUTAMIC ACID (L)			$C_5H_9O_4N$			$HO_2CCH_2CH_2CH(NH_2)CO_2H$		L^-
Ni^{2+}	-7.36	9.81(2)	20.1	25	C	μ=0.1M	70Bd 71Be	y:ΔH (70Bd)
Cu^{2+}	-11.31	14.1(2)	26.8	25	C	μ=0.1M	70Bd 71Be	y:ΔH (70Bd)
G7 GLUTAMIC ACID (DL)			$C_5H_9O_4N$			$HO_2CCH_2CH_2CH(NH_2)CO_2H$		L^{2-}
Ni^{2+}	-7.34	9.71(1-2)	19.8	25	C	μ=0.1M	70Bd 71Be	y:ΔH (70Bd)
Cu^{2+}	-5.0	8.27(1)	21	25	C	μ=0.2(KCl)	74Na	
	-6.7	6.47(2)	7	25	C	μ=0.2(KCl)	74Na	
	-11.44	14.1(1-2)	26.3	25	C	μ=0.1M	70Bd 71Be	y:ΔH (70Bd)
G8 GLUTAMINE			$C_5H_{10}O_3N_2$			$HO_2CCH(NH_2)CH_2CH_2CONH_2$		
Zn^{2+}	-1.64	5.199(1)	17.9	5	T	μ=0.1(KCl)	73Rb	q: ±0.71
	-1.60	5.113(1)	18.0	25	T	μ=0.1(KCl)	73Rb	q: ±0.71
	-1.48	5.042(1)	18.4	45	T	μ=0.1(KCl)	73Rb	q: ±0.71
G9 GLUTAMINE SYNTHETASE FROM ESCHERICHIA COLI								
Mn^{2+}	-6.3	--	--	37	C	Hepes (N-2-hydroxy-elhylpiperazine-N'-2-ethanesulfonic acid) buffer pH=7.14	72Hf	
	-19.7	--	--	37	C	Tris-chloride buffer pH=7.19 R: $Mn^{2+} + H_2L = MnL + 2H^+$	72Hf	
G10 GLYCEROL			$C_3H_8O_3$			$HOCH_2CH(OH)CH_2OH$		L
BA_4^-	-2.85	1.146(25°,1)	-4.3	0-35	T	C_L=0-0.9M A=OH^-	67Cc	b,q: ±0.3
	-3.88	1.763(25°,1-2)	-4.9	0-35	T	C_L=0-0.9M A=OH^-	67Cc	b,q: ±0.4

Metal Ion	ΔH, kcal /mole	Log K	ΔS, cal /mole °K	T, °C	M	Conditions	Ref.	Remarks
G10, cont.								
$H_5TeO_6^-$	-3.8	1.90	-4	25	T	μ=0.1 R: $H_5TeO_6^-$ + L = $HTeO_4L^-$ + $2H_2O$	62Ea	
G11 GLYCEROL, 2-phosphoric acid			$C_3H_9O_6P$			$HOCH_2CH(OPO_3H)CH_2OH$		
Mg^{2+}	3.437	2.486(1)	22.9	25	T	μ=0 ΔCp=56	54Cd	
	4.110	2.700(1)	25.1	37	T	μ=0 ΔCp=58	54Cd	
G12 GLYCINE			$C_2H_5O_2N$			$H_2NCH_2CO_2H$		L^-
Ce^{3+}	3.3	0.531(25°,1)	14	0-55	T	μ=2.0M($NaClO_4$) pH=3.64	68Ta	b,q: ±0.4
Pm^{3+}	3.5	0.672(25°,1)	15	0-55	T	μ=2.0M($NaClO_4$) pH=3.64	68Ta	b,q: ±0.4
Eu^{3+}	2.3	0.699(25°,1)	11	0-55	T	μ=2.0M($NaClO_4$) pH=3.64	68Ta	b,q: ±0.4
Am^{3+}	2.9	0.690(25°,1)	13	0-55	T	μ=2.0M($NaClO_4$) pH=3.64	68Ta	b,q: ±0.4
Cm^{3+}	3.3	0.806(25°,1)	15	0-55	T	μ=2.0M($NaClO_4$) pH=3.64	68Ta	b,q: ±0.4
Mn^{2+}	-0.4	3.23(1)	13	10	C	μ=0	72Ig	
	-0.3	3.21(1)	14	25	C	μ=0	72Ig	q:±0.1
	-0.29	3.17(1)	13.5	25	T	μ=0 ΔCp=16	64Bd	
	-0.04	3.15(1)	14	40	C	μ=0	72Ig	q: ±0.13
				10-40	C	μ=0 ΔCp=12	72Ig	
	-5	2.96(1)	-3	30	T	μ=0.09(KCl)	57Mb	
	-1.4	2.60(25°,1)	7.4	10&25	T	μ=0.65(KCl)	64Lg	a,b
	-1.9	1.98(25°,2)	2.3	10&25	T	μ=0.65(KCl)	64Lg	a,b
	-4.0	5.36(25°)	11.1	10&25	T	μ=0.65(KCl) R: Mn^{2+} + L^- + $(Pyruvate)^-$ = MnL (Pyruvate)	64Lg	a,b
	-11.8	9.79(25°)	5.4	10&25	T	μ=0.65(KCl) R: Mn^{2+} + $2L^-$ + $2(Pyruvate)^-$ = $MnL_2(Pyruvate)_2^{2-}$	64Lg	a,b
Fe^{2+}	-3.7	4.36(1)	6.8	10	C	μ=0	72Ig	
	-3.64	4.31(1)	7.5	25	C	μ=0	72Ig	
	-3.55	4.28(1)	8.2	40	C	μ=0	72Ig	
				10-40	C	μ=0 ΔCp=7	72Ig	
Co^{2+}	-3.00	5.16(1)	13	10	C	μ=0	72Ig	
	-2.48	4.61(1)	12.8	25	C	μ=0	67Bc 66Na 64Bd 64Se	
	-2.86	5.07(1)	13.6	25	C	μ=0	72Ig	
	-2.15	4.98(1)	15.9	40	C	μ=0	72Ig	
				10-40	C	μ=0 ΔCp=28	72Ig	
	-3.0	4.66(1)	11	25	C	μ=0.05M(KCl)	70Gc 71Gc	

G12, cont.

Metal Ion	ΔH, kcal /mole	Log K	ΔS, cal /mole °K	T, °C	M	Conditions	Ref.	Re-marks
Co^{2+}, cont.	-2.0	4.76(15°,1)	14.9	15&40	T	μ=0.2	65Sb	a,b
	-4.00	4.07(2)	4.5	10	C	μ=0	72Ig	
	-3.54	4.02(2)	6.5	25	C	μ=0	72Ig	
	-2.55	3.97(2)	9.6	25	C	μ=0	67Bc	
	-3.41	3.91(2)	7.0	40	C	μ=0	72Ig	
				10-40	C	μ=0 ΔCp=20		
	-4.4	3.98(2)	3	25	C	μ=0.05M(KCl)	70Gc	
							71Gc	
	-3.6	3.56(15°,2)	3.7	15&40	T	μ=0.2	65Sb	a,b
	-3.63	2.67(3)	-0.6	10	C	μ=0	72Ig	
	-3.41	2.54(3)	0.2	25	C	μ=0	72Ig	
	-3.01	2.45(3)	1.6	40	C	μ=0	72Ig	
				10-40	C	μ=0 ΔCp=21		
	-6.6	8.4(1-2)	16.1	20	C	0.1M(KNO_3)	67Sc	
	-5.6	8.32(15°,1-2)	18.6	15&40	T	μ=0.2	65Sb	a,b
CoA	-5.0	4.03(25°,1)	1.8	15-70	T	μ=0.1M(KNO_3)	72Id	b
						A=Methyliminodiacetate		
Ni^{2+}	-5.2	6.36(1)	10.7	10	C	μ=0	67Aa	
	-4.63	6.28(1)	12.4	10	C	μ=0	72Ig	
	-4.38	6.13(1)	13.4	25	C	μ=0	72Ig	
	-4.9	6.18(1)	11.9	25	C	μ=0	67Aa	
	-4.14	6.18(1)	14.4	25	C	μ=0	67Bc	
	-4.09	6.18(1)	14.5	25	T	μ=0 ΔCp=18	64Bd	
	-4.3	6.09(1)	14.2	40	C	μ=0	67Aa	
	-3.77	6.00(1)	15.4	40	C	μ=0	72Ig	
				10-40	C	μ=0 ΔCp=28	72Ig	
	-3.6	5.77(1)	14	25	C	μ=0.05M(KCl)	70Gc	
							71Gc	
	-5.0	5.79(1)	9.6	25	C	μ=0.1M(KNO_3)	71Ld	
	-4.3	6.04(15°,1)	12.7	15-40	T	μ=0.2	65Sb	b
	-1.9	5.66(25°,1)	19.8	10&25	T	μ=0.65(KCl)	64Lg	
	-5.30	5.14(2)	4.8	10	C	μ=0	72Ig	
	-5.8	5.29(2)	3.7	10	C	μ=0	67Aa	
	-4.97	4.92(2)	5.8	25	C	μ=0	72Ig	
	-4.7	5.07(2)	7.6	25	C	μ=0	67Aa	
	-4.69	4.95(2)	6.9	25	T	μ=0 ΔCp=28	64Bd	
	-4.7	4.92(2)	7.4	40	C	μ=0	67Aa	
	-4.60	4.76(2)	7.1	40	C	μ=0	72Ig	
				10-40	C	μ=0 ΔCp=23	72Ig	
	-4.8	4.88(2)	6	25	C	μ=0.05M(KCl)	70Gc	
							71Gc	
	-4.4	4.78(2)	7.0	25	C	μ=0.1M(KNO_3)	71Ld	
	-4.4	4.98(15°,2)	6.0	15-40	T	μ=0.2	65Sb	b
	-5.6	4.85(25°,2)	3.0	10&25	T	μ=0.65(KCl)	64Lg	a,b
	-5.64	3.51(3)	-3.9	10	C	μ=0	72Ig	
	-5.55	3.18(3)	-4.1	25	C	μ=0	72Ig	
	-5.50	3.00(3)	-3.8	40	C	μ=0	72Ig	
				10-40	C	μ=0 ΔCp=5	72Ig	
	-8.7	(1-2)	--	24.6	C	μ=0.01M($HClO_4$)	72Sg	
	-8.8	10.5(1-2)	18.0	20.0	C	0.1M(KNO_3)	67Sc	
	-8.7	11.02(15°,1-2)	18.7	15-40	T	μ=0.2	65Sb	b
	-14.0	13.22(1-2)	14.0	25	T	μ=0.004. S: 45% Dioxane	56Lb	

Metal Ion	ΔH, kcal /mole	Log K	ΔS, cal /mole °K	T, °C	M	Conditions	Ref.	Remarks
G12, cont.								
Ni^{2+}, cont.	-2.6	8.09(25°)	28.5	10&25	T	μ=0.65(KCl) R: Ni^{2+} + L^- + $(Pyruvate)^-$ = NiL (Pyruvate)	64Lg	a,b
	-11.9	15.29(25°)	30.2	10&25	T	μ=0.65(KCl) R: Ni^{2+} + $2L^-$ + $2(Pyruvate)^-$ = NiL_2 $(Pyruvate)_2{}^{2-}$	64Lg	a,b
NiA	-2.6	4.91(25°,1)	5.4	15-70	T	μ=0.1M(KNO_3) A=Methyliminodiacetate	72Id	b,q: ±0.3
	-6.8	4.34(25°,1)	-2.8	15-70	T	μ=0.1M(KNO_3) A=Triethylenetetramine	72Id	b
Cu^{2+}	-6.23	8.85(1)	16.9	10	C	μ=0	72Ig	
	-7.28	8.85(1)	14.8	10	C	μ=0	66Aa 56Lb	
	-5.82	8.57(1)	19.7	25	C	μ=0	72Ig	
	-6.22	8.58(1)	18.4	25	C	μ=0	66Aa	
	-6.0	8.59(20°,1)	19	25	C	μ=0	64Ia	
	-6.76	8.58(1)	16.6	25	C	μ=0	67Bc	
	-5.75	8.42(1)	20.2	40	C	μ=0	66Aa	
	-5.47	8.33(1)	20.6	40	C	μ=0	72Ig	
				10-40	C	μ=0 ΔCp=25		
	-6.8	8.18(1)	14	25	C	μ=0.05M(KCl)	70Gc 71Gc	
	-7.0	8.00(1)	13	30	T	μ=0.09(KCl)	57Mb	
	-6.0	9.56(1)	17.5	25	C	μ=0.1M($NaClO_4$)	72Ia	
	-5.87	8.07(1)	17	25	C	μ=0.2M(KCl)	73Ga	
	-4.8	8.54(15°,1)	22.4	15-40	T	μ=0.2	65Sb	b
	-7.20	7.52(2)	9.0	10	C	μ=0	72Ig	
	-6.92	7.36(2)	9.2	10	C	μ=0	66Aa	
	-6.96	7.09(2)	9.1	25	C	μ=0	66Aa	
	-6.4	7.24(20°,2)	11	25	C	μ=0	64Ia	
	-6.89	6.94(2)	8.7	25	C	μ=0	67Bc	
	-6.93	7.26(2)	10.0	25	C	μ=0	72Ig	
	-6.59	7.00(2)	11.0	40	C	μ=0	72Ig	
				10-40	C	μ=0 ΔCp=20		
	-7.33	6.85(2)	7.9	40	C	μ=0	66Aa	
	-6.8	7.00(15°,2)	11.4	15-40	T	μ=0.2	65Sb	b
	-6.2	6.87(2)	11	25	C	μ=0.05M(KCl)	70Gc 71Gc	
	-6.6	6.80(2)	8.9	25	C	μ=0.1M($NaClO_4$)	72Ia	
	-7.63	6.77(2)	6	25	C	μ=0.2M(KCl)	73Ga	
	-13.6	(1-2)	--	24.1	C	μ=0.01M($HClO_4$)	72Sg	
	-12.8	11.2(1-2)	25.2	20.5	C	0.1M(KNO_3)	67Sc	
	-11.6	15.54(15°,1-2)	33.8	15-40	T	μ=0.2	65Sb	b
	-12.35	15.81	30.9	25	C	μ=0 A=Alanine R: Cu^{2+} + L^- + A^- = CuLA	72Ya	
	-11.75	15.05	29.4	25	C	μ=0.1M(KNO_3) A=α-Alanine R: Cu^{2+} + L^- + A^- = CuLA	72Ia	
	-11.49	15.89	34.2	25	C	μ=0 A=α-Aminoiso-butric acid R: Cu^{2+} + L^- + A^- = CuLA	72Ya	

Metal Ion	ΔH, kcal /mole	Log K	ΔS, cal /mole °K	T, °C	M	Conditions	Ref.	Remarks
G12, cont.								
Cu^{2+}, cont.	-13.3	15.78	28	25	C	μ=0.2(KCl) A=Aspartic acid R: Cu^{2+} + A^- + L^- CuAL	74Na	
	-10.91	15.59	34.8	25	C	μ=0 A=Sarcosine R: Cu^{2+} + L^- + A^- = CuLA	72Ya	
	-11.7	14.66	27.8	25	C	μ=0.1M(KNO_3) A=Serine R: Cu^{2+} + L^- + A^- = CuLA	72Ia	
	-11.84	15.06	29.2	25	C	μ=0.1M(KNO_3) A=Valine R: Cu^{2+} + L^- + A^- = CuLA	72Ia	
CuA	-8.1	5.92(25°,1)	0.3	15-70	T	μ=0.1M(KNO_3) A=Methyliminodiacetate	72Id	b
Ag^+	1.65	3.24(1)	21.7	5	T	μ=0.01 ΔCp=-241	59Da	
	-0.81	3.36(1)	13.1	15	T	μ=0.01 ΔCp=-250	59Da	
	-3.34	3.43(1)	4.5	25	T	μ=0.01 ΔCp=-258	59Da	
	-5.97	3.43(1)	-4.3	35	T	μ=0.01 ΔCp=-267	59Da	
	-8.67	3.36(1)	-12.9	45	T	μ=0.01 ΔCp=-276	59Da	
	-11.5	3.24(1)	-21.5	55	T	μ=0.01 ΔCp=-284	59Da	
	-8.05	3.43(2)	-11.2	25	T	μ=0.01	59Da	
Zn^{2+}	-3.14	5.50(1)	14.2	10	C	μ=0	72Ig	
	-2.76	5.38(1)	15.4	25	C	μ=0	72Ig	
	-3.39	5.50(1)	13.8	25	C	μ=0	67Bc	
	-2.22	5.29(1)	17.1	40	C	μ=0	72Ig	
				10-40	C	μ=0 ΔCp=31		
	-2.0	5.06(1)	16	25	C	μ=0.05M(KCl)	70Gc 71Gc	
	-1.87	5.156(1)	16.8	5	T	μ=0.1(KCl)	73Rb	q: ±0.71
	-2.08	5.052(1)	16.1	25	T	μ=0.1(KCl)	73Rb	q: ±0.71
	-2.29	4.951(1)	15.4	45	T	μ=0.1(KCl)	73Rb	q: ±0.71
	-3.3	5.27(15°,1)	12.7	15-40	T	μ=0.2	65Sb	b
	-2.1	4.88(25°,1)	15.4	10&25	T	μ=0.65(KCl)	64Lg	a,b
	-3.73	4.57(2)	7.7	10	C	μ=0	72Ig	
	-3.22	4.43(2)	9.5	25	C	μ=0	72Ig	
	-2.90	4.29(2)	10.4	40	C	μ=0	72Ig	
				10-40	C	μ=0 ΔCp=28	72Ig	
	-3.2	4.38(2)	9	25	C	μ=0.05M(KCl)	70Gc 71Gc	
	-4.0	4.31(15°,2)	6.0	15-40	T	μ=0.2	65Sb	b
	-3.8	4.13(25°,2)	6.0	10&25	T	μ=0.65(KCl)	64Lg	a,b
	-3.53	2.63(3)	-0.4	10	C	μ=0	72Ig	
	-3.56	2.52(3)	-0.4	25	C	μ=0	72Ig	
	-3.64	2.40(3)	-0.6	40	C	μ=0	72Ig	
				10-40	C	μ=0 ΔCp=-4	72Ig	
	-6.3	9.5(1-2)	22.1	20	C	0.1M(KNO_3)	67Sc	
	-7.3	9.58(15°,1-2)	18.7	15-40	T	μ=0.2	65Sb	b

Metal Ion	ΔH, kcal /mole	Log K	ΔS, cal /mole °K	T, °C	M	Conditions	Ref.	Remarks
G12, cont.								
Zn^{2+}, cont.	-1.9	7.53(25°)	28.2	10&25	T	μ=0.65(KCl) A=Pyruvate R: $Zn^{2+} + L^- + A^- = ZnLA$	64Lg	a,b
	-7.5	14.25(25°)	40.3	10&25	T	μ=0.65(KCl) A=Pyruvate R: $Zn^{2+} + 2L^- + 2A^- = ZnL_2A_2^{2-}$	64Lg	a,b
ZnA	-4.6	4.41(25°,1)	6.5	15-70	T	μ=0.1M(KNO_3) A=Methyliminodiacetate	72Id	b
Cd^{2+}	-2.26	4.73(1)	13.7	10	C	μ=0	72Ig	
	-2.12	4.69(1)	14.4	25	C	μ=0	72Ig	
	-1.95	4.60(1)	14.8	40	C	μ=0	72Ig	q: ±0.10
				10-40	C	μ=0 ΔCp=10		
	-3.74	3.76(2)	4.0	10	C	μ=0	72Ig	
	-3.24	3.71(2)	6.1	25	C	μ=0	72Ig	
	-2.95	3.60(2)	7.0	40	C	μ=0	72Ig	
				10-40	C	μ=0 ΔCp=26	72Ig	
	-2.84	2.53(3)	1.5	10	C	μ=0	72Ig	
	-3.21	2.28(3)	-0.3	25	C	μ=0	72Ig	
	-3.39	2.00(3)	-1.7	40	C	μ=0	72Ig	
				10-40	C	μ=0 ΔCp=-18	72Ig	
CdA	-4.4	3.82(25°,1)	2.8	15-70	T	μ=0.1M(KNO_3) A=Methyliminodiacetate	72Id	b
HgA_2	-2.94	2.61	2.1	25	C	μ=0. R: $HgAL + L^- = HgL_2 + A^-$. A=Cl	66Pd	
	-6.10	3.42	-4.8	25	C	μ=0. R: $HgA_2 + L^- = HgAL + A^-$. A=Cl	66Pd	
In^{3+}	3.0	2.46(35°,1)	21	25-45	T	μ=0.2M($NaClO_4$)	73Sc	b,q: ±0.5
G13 GLYCINE, 2-alanyl-			$C_5H_{10}O_3N_2$			$CH_3CHNH_2COCH(NH_2)CO_2H$		
Cu^{2+}	-6.7	5.34(1)	1.9	25	C	μ=0.1M(KNO_3)	72Bl	
G14 GLYCINE, amide-			$C_2H_6ON_2$			$H_2NCH_2C(O)NH_2$		
	-5.5	5.40(1)	6.3	25	C	μ=0.1M(KNO_3)	72Bl	
	-6.4	5.43(1)	3.4	25	C	μ=0.10($NaClO_4$)	72Tb	
	-5.7	4.22(2)	0.2	25	C	μ=0.1M(KNO_3)	72Bl	
	-6.1	4.10(2)	-1.7	25	C	μ=0.10($NaClO_4$)	72Tb	
G15 GLYCINE, N,N-dimethyl-			$C_4H_9O_2N$			$(CH_3)_2NCH_2CO_2H$		L^-
Mg^{2+}	0	1.68(1)	16	25	T	μ=0	56Ma	
CoA	-4.4	3.30(25°,1)	0.3	15-70	T	μ=0.1M(KNO_3) A=Nitrilotriacetate	71Id	b
NiA	-4.0	4.42(25°,1)	6.5	15-70	T	μ=0.1M(KNO_3) A=Nitrilotriacetate	71Id	b
CuA	-10.8	5.34(25°,1)	-11.6	15-70	T	μ=0.1M(KNO_3) A=Nitrilotriacetate	71Id	b

Metal Ion	ΔH, kcal /mole	Log K	ΔS, cal /mole °K	T, °C	M	Conditions	Ref.	Remarks
G15, cont.								
ZnA	-4.5	3.28(25°,1)	-0.3	15-70	T	μ=0.1M(KNO_3) A=Nitrilotriacetate	71Id	b
CdA	-2.7	2.70(25°,1)	3.2	15-70	T	μ=0.1M(KNO_3) A=Nitrilotriacetate	71Id	b
G16 GLYCINE, N-glycyl-			$C_4H_8O_3N_2$			$NH_2CH_2CONHCH_2CO_2H$		L^-
Mn^{2+}	12.5	3.44(1)	6	30	T	μ=0.09(KCl)	57Mb	p
Ni^{2+}	-4.9	4.09(1)	2.3	25	C	μ=0.10($NaClO_4$)	72Tb	
	-3.9	3.10(2)	1.1	25	C	μ=0.10($NaClO_4$)	72Tb	
	-5.5	2.11(3)	-8.8	25	C	μ=0.10($NaClO_4$)	72Tb	
	-9.08	(1-2)	--	24	C	μ=0.01M($HClO_4$)	72Sg	
Cu^{2+}	-6.1	5.55(1)	4.97	25	C	μ=0.10($NaClO_4$)	68Bc	
	-6.1	(1)	--	--	-	μ=0.01M($HClO_4$)	72Sg	r,u
	7.9	-7.6	-6	0	T	μ=0.09(KCl) R: Cu^{2+} + HL = CuL + $2H^+$	57Mb	
	-9.67	--	--	--	-	μ=0.01M($HClO_4$) R: Cu^{2+} + $2L^-$ = CuA + HL A=L^{2-}	72Sg	r,u
CuL	-16.9	--	--	--	-	μ=0.01M($HClO_4$) R: Cu^{2+} + A^{2-} + L^- = $CuAL^-$ A=L^{2-}	72Sg	r,u
CuA	-6.7	2.99(1)	-8.8	25	C	μ=0.10($NaClO_4$) A=Glyclyglycine	73Pe	
Zn^{2+}	-3.1	3.44(1)	5.3	25	C	μ=0.1M(KNO_3)	72Bl	
	-4.9	2.85(2)	-3.4	25	C	μ=0.1M(KNO_3)	72Bl	
Cd^{2+}	-2.5	3.08(25°,1)	5.7	10-40	T	μ=0.06	66Vb 65Va	b
	-2.8	2.85(1)	3.7	25	C	μ=0.1M(KNO_3)	72Bl	
	-2.2	2.566(25°,2)	4.3	10-40	T	μ=0.06M	65Va	b
	-4.3	2.48(2)	-3.0	25	C	μ=0.1M(KNO_3)	72Bl	
	-4.7	5.64(25°,1-2)	10	10-40	T	μ=0.06	66Vc	b
G17 GLYCINE, N(N-glycylglycyl-glycine)			$C_8H_{14}O_5N_4$			$NH_2CH_2CONHCH_2CONHCH_2CONH$-$CH_2CO_2H$		L^-
Cu^{2+}	-6.0	5.13(1)	3.3	25	C	μ=0.1($NaClO_4$)	69Na	
G18 BLYOXAL, dioxime, dimethyl-			$C_4H_8O_2N_2$			HON:C(CH_3)C(CH_3):NOH)		L^-
Ni^{2+}	2.1	21.58(25°,1-2)	106.5	25&40	T	S: 50 v % Dioxane	54Cc	a,b
Cu^{2+}	-13.9	23.30(25°,1-2)	60.5	25&40	T	S: 50 v % Dioxane	54Cc	a,b
G19 GUANOSINE-5'-diphosphoric acid			$C_{10}H_{15}O_{10}N_5P_2$					L^{3-}
Mg^{2+}	3.41	3.40(1)	26.8	30	C	μ=0.2[$(CH_3)_4NCl$] pH=8.50	73Sa	
G20 GUANOSINE-5'-phosphoric acid			$C_{10}H_{14}O_8N_5P$					
Mg^{2+}	1.72	1.75(1)	13.7	30	C	μ=0.2[$(CH_3)_4NCl$] pH=8.50	73Sa	

Metal Ion	ΔH, kcal /mole	Log K	ΔS, cal /mole °K	T, °C	M	Conditions	Ref.	Remarks
G21 GUANOSINE-5'-triphosphoric acid			$C_{10}H_{16}O_{14}N_5P_3$					
Mg^{2+}	0.90	4.98(25°,1)	26	25-45	T	μ=0.10M(KNO_3)	73Ke	b,g:Na
	4.29	3.91(1)	32.1	30	C	μ=0.2[$(CH_3)_4NCl$] pH=8.50	73Sa	
Ca^{2+}	-1.6	4.92(25°,1)	17	25-45	T	μ=0.10M(KNO_3)	73Ke	b,g:Na q:±0.1
Mn^{2+}	-2.1	5.18(25°,1)	16	25-45	T	μ=0.10M(KNO_3)	73Ke	b,g:Na
Co^{2+}	-1.6	5.57(25°,1)	20	25-45	T	μ=0.10M(KNO_3)	73Ke	b,g:Na q:±0.1
Ni^{2+}	-1.9	5.78(25°,1)	20	25-45	T	μ=0.10M(KNO_3)	73Ke	b,g:Na q:±0.1
Cu^{2+}	-4.5	6.55(25°,1)	15	25-45	T	μ=0.10M(KNO_3)	73Ke	b,g:Na
Zn^{2+}	1.8	5.72(25°,1)	20	25-45	T	μ=0.10M(KNO_3)	73Ke	b,g:Na q:±0.1

H

Metal Ion	ΔH, kcal /mole	Log K	ΔS, cal /mole °K	T, °C	M	Conditions	Ref.	Remarks
H0 HENDECANE, 5,7-thio-			$C_9H_{20}S_2$			$CH_3(CH_2)_3SCH_2S(CH_2)_3CH_2$		
SnA_4	-14.5	(1)	--	25	C	S: Benzene C_M=0.05-0.08M A=Cl	68Ge 65Gb	
H1 HENDECANE, 3,6,9-triaza-1,11-diamino-			$C_8H_{23}N_5$			$NH_2(CH_2CH_2NH)_3CH_2CH_2NH_2$		L
Mn^{2+}	-3.70	6.56(1)	17.5	25	C	μ=0.1(KCl)	64Pa	
	-5.16	7.62(25°,1)	17.5	26.5	C	--	58Jb	b
Fe^{2+}	-8.70	9.86(1)	16.0	25	C	μ=0.1(KCl)	64Pa	
	-9.36	11.40(25°,1)	20.7	26.5	C	--	58Jb	b
Co^{2+}	-13.85	13.30(1)	14.5	25	C	μ=0.1(KCl)	64Pa	
Ni^{2+}	-18.4	17.4(1)	18.1	25	C	μ=0.1(KNO_3)	65Wd	
	-18.90	17.44(1)	16.5	25	C	μ=0.1(KCl)	64Pa	
	-11.35	(1)	--	25	C	μ=0.5 Anion: Cl^-	57Jb	
	-10.43	17.51(1)	-45.1	25	C	μ=0.5 Anion: ClO_4^-	57Jb	
Cu^{2+}	-24.95	22.80(1)	20.5	25	C	μ=0.1(KCl)	64Pa 57Ja	
	-24	23.0(1)	25	25	C	μ=0.1(KNO_3)	65Wd	
Zn^{2+}	-13.85	15.10(1)	22.5	25	C	μ=0.1(KCl)	64Pa	
	-14.0	15.5(1)	24	25	C	μ=0.1(KNO_3)	64Pa	
Cd^{2+}	-12.8	14.1(1)	22	25	C	μ=0.1(KNO_3)	65Wd	
H2 2-HENDECANONE			$C_{11}H_{22}O$			$CH_3(CH_2)_8COCH_3$		
$SiAB_3$	-2.98	1.43(30°,1)	-3.3	10-45	T	S: CCl_4 $X_L=5x10^{-2}$ $X_M=5x10^{-4}$ A=OH B=CH_3	69Lb	b
	-3.43	1.33(30°,1)	-5.2	20-60	T	S: CCl_4 $X_L=5x10^{-2}$	69Lb	b

Metal Ion	ΔH, kcal /mole	Log K	ΔS, cal /mole °K	T, °C	M	Conditions	Ref.	Re-marks
H2, cont.								
$SiAB_3$, cont.						$X_M=5x10^{-4}$ A=OH $B=C_2H_5$		
	-3.72	1.51(30°,1)	-5.35	0-50	T	S: CCl_4 $X_L=5x10^{-2}$ $X_M=4x10^{-4}$ A=OH B=Phenyl	69Lb	b
H3 HEPTANE, 1-amino-			$C_7H_{17}N$			$CH_3(CH_2)_6NH_2$		L
NiA	-16.9	6.12(1-2)	-28.8	25	T	$C_M=10^{-4}M$ S: Benzene A=Diacetylbisbenzoyl hydrazine	60Sb	
H4 HEPTANE, 4-oxa-			$C_6H_{14}O$			$(CH_3CH_2CH_2)_2O$		L
SbA_5	-16.4	(1)	--	25	C	S: CCl_4 A=Cl	730a	
	-17.84	(1)	--	25	C	S: Ethylene Chloride. A=Cl	680a	
H5 HEPTANEDIOIC ACID, 4-thio-			$C_6H_{10}O_4S$			$HO_2CCH_2CH_2SCH_2CH_2CO_2H$		L^{2-}
Ca^{2+}	-2.03	2.99(30°,1-3)	7.00	30&40	T	μ=1.2M(KNO_3)	72Ra	a,b
	-1.92	2.94(40°,1-3)	7.35	40&50	T	μ=1.2M(KNO_3)	72Ra	a,b
Pb^{2+}	0.89	3.27(40°,1-3)	12.1	30-50	T	μ=1.2M(KNO_3)	72Rb	b
Cd^{2+}	-2.71	1.63(40°,1)	-1.20	30-50	T	μ=1.2(KNO_3) P.C. Rowat & C.M. Gupta, J. Inorg. Nucl. Chem., 34, 951 (1973)	74Ma	b,i
	-0.12	0.69(40°,2)	2.80	30-50	T	μ=1.2(KNO_3) P.C. Rowat & C.M. Gupta, J. Inorg. Nucl. Chem., 34, 951 (1973)	74Ma	b,i
	-0.34	0.51(40°,3)	1.55	30-50	T	μ=1.2(KNO_3) P.C. Rowat & C.M. Gupta, J. Inorg. Nucl. Chem., 34, 951 (1973)	74Ma	b,i
H6 3,5-HEPTANEDIONE			$C_7H_{12}O_2$			$CH_3CH_2COCH_2COCH_2CH_3$		
Co^{2+}	-1.00	7.32(25°,1)	30.35	5-45	T	S: 50 v % Dioxane	73Aj	b
	-1.60	5.84(25°,2)	21.51	5-45	T	S: 50 v % Dioxane	73Aj	b
H7 3,5-HEPTANEDIONE, 2,6-dimethyl-			$C_9H_{16}O_2$			$CH_3CH(CH_3)COCH_2COCH(CH_3)CH_3$		
NiL_2	-5.02	3.301	-1.45	30	C	S: Benzene R: $3NiL_2 = Ni_3L_6$	74Ga	
H8 6-HEPTANOIC ACID, 3-seleno-			$C_6H_{12}O_2Se$			$CH_3(CH_2)_3SeCH_2CO_2H$		L^-
Ag^+	-7.79	4.57(25°,1)	-5.21	0-40	T	μ=0.20M pH=5.8	71Bd 71Bc	b,q: ±1.9
	-6.45	3.45(25°,2)	-5.60	0-40	T	μ=0.20M pH=5.8	71Bd 71Bc	b,q: ±0.5

Metal Ion	ΔH, kcal /mole	Log K	ΔS, cal /mole °K	T, °C	M	Conditions	Ref.	Remarks
H9 4-HEPTANONE, 2,6-dimethyl-			$C_9H_{18}O$			$(CH_3)_2CHCH_2COCH_2CH(CH_3)_2$		L
SnA_4	-13.6	1.529(30°,1)	-31.6	30-40	T	S: Benzene ΔS calculated using ΔG in units of atm^{-1} at 30° A=Cl	64La	b
H10 6-HEPTENOIC ACID, 3-seleno-			$C_6H_{10}O_2Se$			$CH_2{:}CH(CH_2)_2SeCH_2CO_2H$		L^-
Ag^+	-13.0	5.17(25°,1)	-20.0	0-40	T	μ=0.20M pH=5.8	71Bd 71Bc	b,q: ±1.9
	-24.1	2.78(25°,2)	-68.1	0-40	T	μ=0.20M pH=5.8	71Bd 71Bc	b,q: ±4.3
	-0.72	1.96(25°)	6.41	0-40	T	μ=0.20M R: $AgL + Ag^+ = Ag_2L^+$ pH=5.8	71Bd	b,q: ±1.9
H11 1-HEPTYNE			C_7H_{12}			$CH_3(CH_2)_4C{\vdots}CH$		L
Ag^+	-5.9	1.049(25°,1)	-15.3	10-40	T	μ=1M($NaNO_3$) S: 52 v % Ethanol	73St	b,q: ±0.8
	-0.9	1.025(25°,1)	-4.7	10-40	T	μ=1M($NaNO_3$) S: 75 v % Ethanol	73St	b,q: ±0.2
H12 HEXADECANE, 5,12-dithio-			$C_{14}H_{30}S_2$			$CH_3(CH_2)_3S(CH_2)_6S(CH_2)_3CH_3$		L
SnA_4	-27.4	(1)	--	--	C	S: Benzene C_M=0.05-0.08M A=Cl	65Gb	u
TeA_4	-12.3	3.00(1)	-27.5	25	C	S: Benzene A=Cl	73Ph	
	-7.3	(2)	--	25	C	S: Benzene A=Cl	73Ph	
H13 1,5-HEXADIENE			C_6H_{10}			$CH_2{:}CHCH_2CH_2CH{:}CH_2$		L
PdA_2B_2	-8.0	--	--	25	C	S: CH_2Cl_2 C_L=5x10^{-3}M C_M=1.7x10^{-4}M A=Cl B=C_6H_5CN R: PdA_2B_2 + L = PdA_2L + 2B	72Pc	
H14 HEXAETHYLENE GLYCOL, dimethyl ether			$C_{14}H_{30}O_7$			$CH_3OCH_2(CH_2OCH_2)_5CH_2OCH_3$		L
NaA	-9.0	2.903(25°)	-17	0-50	T	S: Tetrahydrofuran A=Fluorenyl ion R: NaA + L = NaLA	70Ce	b
H15 HEXANE, 1-amino-			$C_6H_{15}N$			$CH_3(CH_2)_5NH_2$		L
NiA	-16.8	6.09(1-2)	-28.6	25	T	C_M=10^{-4}M S: Benzene A=Diacetylbisbenzoyl hydrazine	60Sb	
CuA_2	-7.3	0.322(25°,1)	-23	10-50	T	S: Methylcyclohexane-n-Hexylamine A= di-n-Butyldithiocarbamate	71Cf	b,q: ±1.4
Ag^+	-6	3.54(1)	-3.4	25	C	μ=0	71He	q:±2

Metal Ion	ΔH, kcal /mole	Log K	ΔS, cal /mole °K	T, °C	M	Conditions	Ref.	Re-marks
H15, cont.								
Ag^+,	-7	4.02(2)	-3.4	25	C	μ=0	71He	q:±2
cont.	-12.68	7.56(1-2)	-8.0	25	C	μ=0	71He	
H16	HEXANE, 1,6-diamino-N,N,N',N'-tetraacetic acid	$C_{14}H_{24}O_8N_2$				$(CH_2CO_2H)_2N(CH_2)_6N(CH_2CO_2H)_2$		L^{4-}
Ca^{2+}	-1.3	3.69(1)	12.5	25	C	μ=0.50	58Ya	
	-1.4	3.69(1)	12.1	25	C	μ=0.90	58Ya	
Mn^{2+}	0.87	4.8(1)	44.2	20	C	μ=0.1(KNO_3)	64Ab	
Co^{2+}	-4.56	13.05(1)	44.1	20	C	μ=0.1(KNO_3)	64Ab	
Ni^{2+}	-8.5	13.82(1)	34.2	20	C	μ=0.1(KNO_3)	64Ab	
Zn^{2+}	-4.0	12.68(1)	44.4	20	C	μ=0.1(KNO_3)	64Ab	
Cd^{2+}	-4.26	11.9(1)	39.9	20	C	μ=0.1(KNO_3)	64Ab	
Hg^{2+}	-20.97	21.58(1)	27.3	20	C	μ=0.1(KNO_3)	64Ab	
Pb^{2+}	-7.53	--	--	20	C	μ=0.1(KNO_3) R: Pb^{2+} + L + H_2O = HLPbOH	64Ab	
H17	2,4-HEXANEDIONE, 5-methyl-1-phenyl-	$C_{13}H_{16}O_2$				$C_6H_5CH_2C(O)CH_2C(O)CH(CH_3)CH_3$		
Gd^{3+}	0.38	0.398(25°)	3.1	-20-50	T	S: Chlorobenzene R: AlL_3(Cis) = AlL_3(Trans)	71Hj	b,q: ±0.50
Al^{3+}	0.15	0.398(25°)	2.3	-20-80	T	S: Chlorobenzene R: AlL_3(Cis) = AlL_3(Trans)	71Hj	b,q: ±0.50
H18	HEXANOIC ACID, 2-amino-	$C_6H_{13}O_2N$				$CH_3(CH_2)_3CH(NH_2)CO_2H$		
Co^{2+}	-5.9	7.89(1-2)	16	25	C	μ=0.05M(KCl)	72Gd 72Gc	
Ni^{2+}	-7.5	9.74(1-2)	19	25	C	μ=0.05M(KCl)	72Gd 72Gc	
H19	3-HEXANONE, 2,2,5,5-tetramethyl-	$C_{10}H_{20}O$				$(CH_3)_3CC(O)CH_2C(CH_3)_3$		L
SbA_5	-11.78	2.423(1)	-34.8	25	C	S: Ethylene chloride. A=Cl	67Of	
H20	2-HEXEN-4-ONE, 2-amino-5,5-dimethyl-	$C_8H_{15}ON$				$C(CH_3)_3C(O)CH:C(NH_2)CH_3$		L
Co^{2+}	2.80	-0.385(25°)	7.63	-38-52	T	S: 15 v % Tetramethylsilane in $CHCl_3$	68Ea	b,t
H21	2-HEXEN-4-ONE, 5,5-dimethyl-2-methylamino-	$C_9H_{17}ON$				$(CH_3)_3CC(O)CH:C(NHCH_3)CH_3$		L
Ni^{2+}	5.08	-2.02(25°)	7.80	17-107	T	S: $CDCl_3$	68Ea	b,t

Metal Ion	ΔH, kcal /mole	Log K	ΔS, cal /mole °K	T, °C	M	Conditions	Ref.	Remarks
H22	3-HEXYNE		C_6H_{10}			$CH_3CH_2C{:}CCH_2CH_3$		L
Ag^+	-4.5	-0.883(25°)	11.1	25-35	T	μ=1.0 C_L-7x10^{-3} R: $Ag^+ + L_{(liquid)} \rightarrow AgL^+$	56Hb	b
H23	3-HEXYNE, 2,2,5,5-tetra-methyl-		$C_{10}H_{18}$			$CH_3C(CH_3)_2C{:}CC(CH_3)_2CH_3$		L
Ag^+	-5.0	-2.719(25°)	4.3	25-35	T	μ=1.0 C_L=1.5x10^{-4} R: $Ag^+ + L_{(liquid)} = AgL^+$	57Hb	b
H24	HISTAMINE		$C_5H_9N_3$			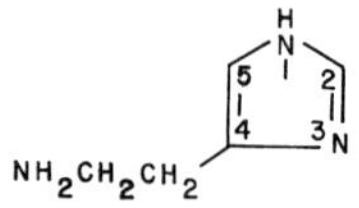		
Be^{2+}	-28.4	7.12(25°,1)	-62	0-45	T	μ=0.12M(KCl) C_M≃2x10^{-3}M C_L≃8x10^{-3}M	69Cb 70Cf	b
	-10.9	5.35(25°,2)	-12	0-45	T	μ=0.12M(KCl) C_M≃2x10^{-3}M C_L≃8x10^{-3}M	69Cb 70Cf	b
Mn^{2+}	-0.5	2.98(15°,1)	11.9	15&40	T	μ=0.2M(KNO_3)	71Ra	a,b
	-0.5	(2)	--	15&40	T	μ=0.2M(KNO_3)	71Ra	a,b
Co^{2+}	-6.9	5.08(30°,1)	-1	10-40	T	μ→0	61Nb	b
	-4.4	5.16(25°,1)	8.9	15-40	T	μ=0.2M(KNO_3)	71Ra	b
	-5	5.34(30°,1)	8	30-50	T	1M(KNO_3)	56Ha	b
	-4.8	3.76(30°,2)	1	10-40	T	μ→0	61Nb	b
	-4.64	3.64(25°,2)	1.2	15-40	T	μ=0.2M(KNO_3)	71Ra	b
	-10	3.75(30°,2)	-16	30-50	T	1M(KNO_3)	56Ha	b
	-7	1.88(30°,3)	-15	30-50	T	1M(KNO_3)	56Ha	b
Ni^{2+}	-7.1	6.69(1)	7	25	T	μ=0	59Sa	
	-10.9	6.77(30°,1)	-5	10-40	T	μ→0	61Nb	b
	-7.7	6.45(1)	5	37	T	μ=0	59Sa	
	-8.4	6.22(1)	3	50	T	μ=0	59Sa	
	-4	6.88(25°,1)	10	0&25	T	μ=0.135	55Ma	a,b, f:3
	-6.3	6.70(25°,1)	9.7	15-40	T	μ=0.2M(KNO_3)	71Ra	b
	-8	6.87(30°,1)	5	30-50	T	μ=0.2M(KNO_3)	71Ra	b
	-8.2	5.01(30°,2)	-4	10-40	T	μ→0	61Nb	b
	-4	5.03(25°,2)	10	0&25	T	μ=0.135	55Ma	a,b, f:3
	-6.6	5.03(25°,2)	0.9	15-40	T	μ=0.2M(KNO_3)	71Ra	b
	-4	4.96(30°,2)	10	30-50	T	1M(KNO_3)	56Ha	b
	-2.3	3.25(30°,3)	7	10-40	T	μ→0	61Nb	b
	-4	3.09(25°,3)	10	0&25	T	μ=0.135	55Ma	a,b, f:3
	-6	3.08(30°,3)	-6	30-50	T	1M(KNO_3)	56Ha	b
Cu^{2+}	-10.3	9.47(1)	9	25	T	μ=0	60Se	

Metal Ion	ΔH, kcal /mole	Log K	ΔS, cal /mole °K	T, °C	M	Conditions	Ref.	Re-marks
H24, cont.								
Cu^{2+}, cont.	-15.5	9.50(30°,1)	-1	10-40	T	μ→0	61Nb	b
	-11.2	9.20(1)	6	37	T	μ=0	60Se	
	-12.1	8.80(1)	3	50	T	μ=0	60Se	
	-7	9.55(25°,1)	12	0&25	T	μ=0.135	55Ma	a,b
	-13.4	9.60(1)	-1	25	C	μ=0.2M(KCl)	73Ga	
	-10.97	9.53(25°,1)	6.80	15-40	T	μ=0.2M(KNO_3)	71Ra	b
	-10.3	9.47(1)	8.7	25	T	--	56Sa	
	-11.2	9.20(1)	6.1	37	T	--	56Sa	
	-12.1	8.80(1)	2.9	50	T	--	56Sa	
	-11.6	6.45(30°,2)	-9	10-40	T	μ→0	61Nb	b
	-7	6.48(25°,2)	12	0&25	T	μ=0.135	55Ma	a,b
	-10.0	6.49(2)	-3	25	C	μ=0.2M(KCl)	73Ga	
	-10.20	6.21(25°,2)	-5.80	15-40	T	μ=0.2M(KNO_3)	71Ra	b
Zn^{2+}	-4.0	5.15(25°,1)	10.2	15-40	T	μ=0.2M(KNO_3)	71Ra	b
	-5.5	5.23(25°,1)	5	25	T	--	62Kb	
	-3.7	4.84(25°,2)	9.7	15-40	T	μ=0.2M(KNO_3)	71Ra	b
Cd^{2+}	-5.4	4.80(1)	4	25	T	μ=0	59Sa	
	-7.4	4.8(30°,1)	-3	10-40	T	μ→0	61Nb	b
	-5.9	4.62(1)	2	37	T	μ=0	59Sa	
	-6.4	4.44(1)	1	50	T	μ=0	59Sa	
	-11.8	8.57(1)	1	25	T	μ=0.1(KNO_3)	64Ad	
	-4.4	3.39(30°,2)	1	10-40	T	μ→0	61Nb	b

H25 HISTIDINE $C_6H_9O_2N_3$ L^-

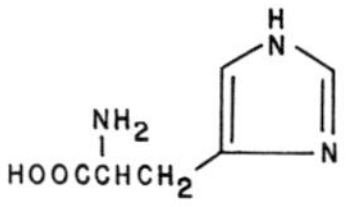

Metal Ion	ΔH, kcal /mole	Log K	ΔS, cal /mole °K	T, °C	M	Conditions	Ref.	Re-marks
Be^{2+}	-32.4	5.52(35°,1)	-79	25-40	T	μ=0.12(KCl) C_M≃2x10^{-3}M C_L≃9x10^{-3}M	70Cf	b
	-8.5	4.50(35°,2)	-7	25-40	T	μ=0.12(KCl) C_M≃2x10^{-3}M C_L≃9x10^{-3}M	70Cf	b
La^{3+}	-2.75	4.10(1)	9.49	25	C	μ=3.00M($NaClO_4$)	70Jc	
	-2.18	3.40(1)	8.5	37	C	μ=3.00M($NaClO_4$)	71Ja	
	7.4	1.30(2)	30.9	25	C	μ=3.00M($NaClO_4$)	70Jc	
	7.76	3.45(2)	40.8	37	C	μ=3.00M($NaClO_4$)	71Ja	
	-2.77	11.75	44.4	25	C	μ=3.00M($NaClO_4$) R: La^{3+} + HL = $LaHL^{3+}$	70Jc	
	-1.96	11.07	44.3	37	C	μ=3.00M($NaClO_4$) R: La^{3+} + HL = $LaHL^{3+}$	71Ja	q: ±0.21
Pr^{3+}	-2.46	4.36(1)	11.7	25	C	μ=3.00M($NaClO_4$)	70Jc	
	-2.04	3.69(1)	10.3	37	C	μ=3.00M($NaClO_4$)	71Ja	
	7.5	1.84(2)	33.5	25	C	μ=3.00M($NaClO_4$)	70Jc	
	8.11	4.09(2)	44.9	37	C	μ=3.00M($NaClO_4$)	71Ja	
	-2.94	11.77	44.0	25	C	μ=3.00M($NaClO_4$) R: Pr^{3+} + HL = $PrHL^{3+}$	70Jc	
	-2.11	11.04	43.7	37	C	μ=3.00M($NaClO_4$) R: Pr^{3+} + HL = $PrHL^{3+}$	71Ja	q: ±0.23
Nd^{3+}	-1.31	3.95(1)	13.8	37	C	μ=3.00M($NaClO_4$)	71Ja	q: ±0.15

Metal Ion	ΔH, kcal /mole	Log K	ΔS, cal /mole °K	T, °C	M	Conditions	Ref.	Re-marks
H25, cont.								
Nd^{3+}, cont.	-1.87	4.40(1)	13.9	25	C	μ=3.00M($NaClO_4$)	70Jc	
	0.77	2.19(2)	32.7	25	C	μ=3.00M($NaClO_4$)	70Jc	
	7.18	4.17(2)	42.3	37	C	μ=3.00M($NaClO_4$)	71Ja	
	-2.38	11.77	45.6	25	C	μ=3.00M($NaClO_4$) R: Nd^{3+} + HL = $NdHL^{3+}$	70Jc	q: ±0.22
	-1.52	11.20	46.3	37	C	μ=3.00M($NaClO_4$) R: Nd^{3+} + HL = $NdHL^{3+}$	71Ja	q: ±0.25
Sm^{3+}	-0.55	4.46(1)	18.5	25	C	μ=3.00M($NaClO_4$)	70Jc	q: ±0.12
	-0.06	4.37(1)	19.8	37	C	μ=3.00M($NaClO_4$)	71Ja	q: ±0.13
	4.16	4.25(2)	33.6	25	C	μ=3.00M($NaClO_4$)	70Jc	
	4.87	4.41(2)	35.9	37	C	μ=3.00M($NaClO_4$)	71Ja	
	-2.51	11.78	45.4	25	C	μ=3.00M($NaClO_4$) R: Sm^{3+} + HL = $SmHL^{3+}$	70Jc	
	-1.20	11.18	47.3	37	C	μ=3.00M($NaClO_4$) R: Sm^{3+} + HL = $SmHL^{3+}$	71Ja	q: ±0.18
Gd^{3+}	1.70	4.39(1)	25.8	25	C	μ=3.00M($NaClO_4$)	70Jc	
	2.16	4.94(1)	29.6	37	C	μ=3.00M($NaClO_4$)	71Ja	
	0.36	4.22(2)	20.6	25	C	μ=3.00M($NaClO_4$)	70Jc	q: ±0.19
	0.92	4.22(2)	22.2	37	C	μ=3.00M($NaClO_4$)	71Ja	
	-0.99	11.47	49.2	25	C	μ=3.00M($NaClO_4$) R: Gd^{3+} + HL = $GdHL^{3+}$	70Jc	q: ±0.17
	-0.18	11.30	51.1	37	C	μ=3.00M($NaClO_4$) R: Gd^{3+} + HL = $GdHL^{3+}$	71Ja	q: ±0.20
Dy^{3+}	2.20	4.40(1)	27.5	25	C	μ=3.00M($NaClO_4$)	70Jc	
	2.70	5.09(1)	31.8	37	C	μ=3.00M($NaClO_4$)	71Ja	
	-0.14	4.74(2)	21.3	25	C	μ=3.00M($NaClO_4$)	70Jc	q: ±0.24
	0.28	4.76(2)	22.7	37	C	μ=3.00M($NaClO_4$)	71Ja	q: ±0.26
	0.98	11.16	54.5	25	C	μ=3.00M($NaClO_4$) R: Dy^{3+} + HL = $DyHL^{3+}$	70Jc	q: ±0.19
	1.32	11.56	57.1	37	C	μ=3.00M($NaClO_4$) R: Dy^{3+} + HL = $DyHL^{3+}$	71Ja	q: ±0.21
Er^{3+}	1.69	4.49(1)	26.3	25	C	μ=3.00M($NaClO_4$)	70Jc	q: ±0.17
	2.03	4.99(1)	29.4	37	C	μ=3.00M($NaClO_4$)	71Ja	q: ±0.19
	5.09	4.50(2)	22.2	25	C	μ=3.00M($NaClO_4$)	70Jc	q: ±0.19
	1.07	4.92(2)	25.9	37	C	μ=3.00M($NaClO_4$)	71Ja	
	0.37	11.18	52.3	25	C	μ=3.00M($NaClO_4$) R: Er^{3+} + HL = $ErHL^{3+}$	70Jc	q: ±0.14
	1.19	11.40	56.0	37	C	μ=3.00M($NaClO_4$) R: Er^{3+} + HL = $ErHL^{3+}$	71Ja	q: ±0.19
Yb^{3+}	1.58	4.23(1)	24.6	25	C	μ=3.00M($NaClO_4$)	70Jc	q: ±0.17
	1.81	4.76(1)	27.6	37	C	μ=3.00M($NaClO_4$)	71Ja	q: ±0.20

Metal Ion	ΔH, kcal /mole	Log K	ΔS, cal /mole °K	T, °C	M	Conditions	Ref.	Remarks
H25, cont.								
Yb^{3+}, cont.	-0.36	5.60(2)	24.4	25	C	μ=3.00M($NaClO_4$)	70Jc	q: ±0.17
	0.35	5.55(2)	26.5	37	C	μ=3.00M($NaClO_4$)	71Ja	q: ±0.19
	0.29	11.40	53.3	25	C	μ=3.00M($NaClO_4$) R: Yb^{3+} + HL = $YbHL^{3+}$	70Jc	q: ±0.14
	1.13	11.60	56.7	37	C	μ=3.00M($NaClO_4$) R: Yb^{3+} + HL = $YbHL^{3+}$	71Ja	q: ±0.17
Mn^{2+}	-2.71	3.915(1)	8.8	25	C	μ=3.0($NaClO_4$)	70Wb	
	-2.52	2.691(2)	3.9	25	C	μ=3.0($NaClO_4$)	70Wb	
	-1.2	5.71(40°,1-2)	22.4	15&40	T	μ=0.2(KNO_3)	69Ra	a,b
Fe^{2+}	-4.37	5.883(1)	12.3	25	C	μ=3.0($NaClO_4$)	70Wb	
	-4.73	4.543(2)	4.9	25	C	μ=3.0($NaClO_4$)	70Wb	
	-4.0	8.50(40°,1-2)	26.3	15&40	T	μ=0.2(KNO_3)	69Ra	a,b
Co^{2+}	-8.0	7.10(15°,1)	4.86	0-40	T	μ=0.25(KCl)	65Ad	b
	-5.63	7.442(1)	15.2	25	C	μ=3.0($NaClO_4$)	70Wb	
	-12.6	5.62(15°,2)	-18.0	0-40	T	μ=0.25(KCl)	65Ad	b
	-6.62	6.036(2)	5.4	25	C	μ=3.0($NaClO_4$)	70Wb	
	-11.59	(1-2)	--	25	C	μ=0.02M	72Ph	
	-9.6	12.84(25°,1-2)	26.6	15-45	T	μ=0.2(KNO_3)	69Ra	b
Ni^{2+}	-11.4	8.75(15°,1)	0.7	0-40	T	μ=0.25(KCl)	65Ad	b
	-8.63	9.199(1)	13.1	25	C	μ=3.0($NaClO_4$)	70Wb	
	-13.7	7.08(15°,2)	-15.3	0-40	T	μ=0.25(KCl)	65Ad	b
	-10.26	7.454(2)	-0.3	25	C	μ=3.0($NaClO_4$)	70Wb	
	-16.6	15.5(1-2)	4.9	22	C	0.1M(KNO_3)	67Sc	
	-16.47	15.5(1-2)	15.5	25	C	μ=0.1M Histidine(-)	70Bd 71Be	y:ΔH (70Bd)
	-16.52	15.5(1-2)	15.5	25	C	μ=0.1M Histidine(+)(rac)	70Bd	y:ΔH (70Bd)
	-16.95	15.7(1-2)	14.8	25	C	μ=0.1M Histidine(+)(rac)	70Bd	y:ΔH (70Bd)
	-13.6	16.05(25°,1-2)	28.0	15-45	T	μ=0.2(KNO_3)	69Ra	b
Cu^{2+}	-10.49	10.086(1)	11.0	25	C	μ=3.0($NaClO_4$)	72Wc	
	-9.57	8.940(2)	8.82	25	C	μ=3.0($NaClO_4$)	72Wc	
	-21.3	18.8(1-2)	13.9	21.9	C	0.1M(KNO_3)	67Sc	
	-19.99	(1-2)	--	25	C	μ=0.1M Histidine(-)	70Bd	y:ΔH (70Bd)
	-19.74	(1-2)	--	25	C	μ=0.1M Histidine(+)(rac)	70Bd	y:ΔH (70Bd)
	-19.99	(1-2)	--	25	C	μ=0.1M Histidine(+)(rac)	70Bd	y:ΔH (70Bd)
	-22.1	19.40(25°,1-2)	14.5	15-40	T	μ=0.2(KNO_3)	69Ra	b
	-27.4	11.2(1-2)	-43.8	0-40	T	μ=0.25(KCl)	65Ad	b
	-16.01	15.615	17.7	25	C	μ=3.0($NaClO_4$) R: Cu^{2+} + L^- + H^+ = $CuLH^{2+}$	72Wc	
	-25.72	25.884	32.2	25	C	μ=3.0($NaClO_4$) R: Cu^{2+} + $2L^-$ + H^+ = CuL_2H^+	72Wc	
	-31.00	30.750	36.7	25	C	μ=3.0($NaClO_4$) R: Cu^{2+} + $2L^-$ + $2H^+$ = $CuL_2H_2{}^{2+}$	72Wc	

Metal Ion	ΔH, kcal /mole	Log K	ΔS, cal /mole °K	T, °C	M	Conditions	Ref.	Remarks
H25, cont.								
Cu^{2+}, cont	-6.36	5.97	6.0	25	C	μ=3.0($NaClO_4$) R: Cu^{2+} + HL = $CuHL^{2+}$	72Wc	
	-11.69	11.5	13.3	25	C	μ=3.0($NaClO_4$) R: Cu^{2+} + 2HL = $CuH_2L_2{}^{2+}$	72Wc	
CuL	-5.52	5.53	6.8	25	C	μ=3.0($NaClO_4$)	72Wc	
	-5.57	6.16	9.5	25	C	μ=3.0($NaClO_4$) R: CuL^+ + HL = $CuHL_2{}^+$	72Wc	
CuL_2	-5.66	6.86	12.4	25	C	μ=3.0($NaClO_4$) R: CuL_2 + H^+ = $CuHL_2{}^+$	72Wc	
CuAL	-9.70	10.28(1)	14.4	25	C	μ=3.0($NaClO_4$) A=H	72Wc	
	-5.33	5.50	7.3	25	C	μ=3.0($NaClO_4$) A=H R: $CuAL^{2+}$ + AL = $CuA_2L_2{}^{2+}$	72Wc	
$CuAL_2$	-5.28	5.58	4.5	25	C	μ=3.0($NaClO_4$) A=H R: $CuAL_2{}^+$ + A = $CuA_2L_2{}^{2+}$	72Wc	
Zn^{2+}	-6.9	6.78(15°,1)	6.9	0-40	T	μ=0.25(KCl)	65Ad	b
	-5.52	7.068(1)	13.8	25	C	μ=3.0($NaClO_4$)	70Wb	
	-6.23	5.673(2)	5.1	25	C	μ=3.0($NaClO_4$)	70Wb	
	-11.7	11.8(1-2)	14.6	21.3	C	0.1M(KNO_3)	67Sc	
	-11.40	12.05(1-2)	17.0	25	C	μ=0.1M Histidine(-)	70Bd 71Be	y:ΔH (70Bd)
	-11.44	12.05(1-2)	16.7	25	C	μ=0.1M Histidine(+)(rac)	70Bd 71Be	y:ΔH (70Bd)
	-11.76	12.16(1-2)	16.2	25	C	μ=0.1M Histidine(+)(rac)	70Bd 71Be	y:ΔH (70Bd)
	-8.3	12.44(25°,1-2)	29.2	15-40	T	μ=0.2(KNO_3)	69Ra	b
	-11.4	5.55(15°,1-2)	-14.2	0-40	T	μ=0.25(KCl)	65Ad	b
Cd^{2+}	-14.2	10.20(1)	0	25	T	μ=0.1(KNO_3)	64Ad	
H26 HISTIDINE, histidyl-		$C_{12}H_{16}O_4N_6$				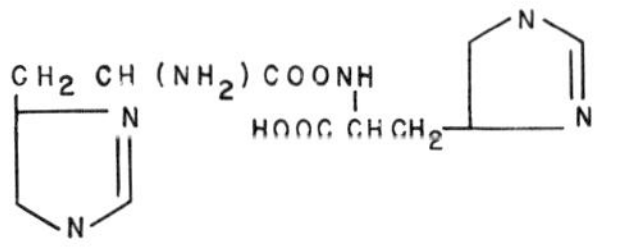		
Be^{2+}	-50.3	4.14(35°,1)	-143	25-45	T	μ=0.12(KCl) C_M≃2x10^{-3}M C_L≃9x10^{-3}M	70Cf	b
	-20.3	3.45(35°,2)	-50	25-45	T	μ=0.12(KCl) C_M≃2x10^{-3}M C_L≃9x10^{-3}M	70Cf	b
H27 HISTIDINE, methyl ester		$C_7H_{11}O_2N_3$				$C_3H_3N_2[CH_2CH(NH_2)COOCH_3]$		L^-
Be^{2+}	-14.1	4.50(35°,1)	-25	25-45	T	μ=0.12(KCl) C_M≃2x10^{-3}M C_L≃9x10^{-3}M	70Cf	b
	-10.1	3.22(35°,2)	-18	25-45	T	μ=0.12(KCl) C_M≃2x10^{-3}M C_L≃9x10^{-3}M	70Cf	b

Metal Ion	ΔH, kcal /mole	Log K	ΔS, cal /mole °K	T, °C	M	Conditions	Ref.	Re-marks
H27, cont.								
Co^{2+}	-16.0	5.00(15°,1)	-32.7	0-40	T	μ=0.25(KCl)	65Ad	b
	-16.0	3.57(15°,2)	-39.2	0-40	T	μ=0.25(KCl)	65Ad	b
Ni^{2+}	-16.0	6.65(15°,1)	-25.0	0-40	T	μ=0.25(KCl)	65Ad	b
	-16.0	5.14(15°,2)	-32.0	0-40	T	μ=0.25(KCl)	65Ad	b
	-17.2	2.76(15°,3)	-47.2	0-40	T	μ=0.25(KCl)	65Ad	b
Cu^{2+}	-12.3	8.52(25°,1)	-2.3	25-50	T	μ=0.1M	71Hf	b,q: ±0.7
	-18.3	9.12(15°,1)	-21.9	0-40	T	μ=0.25(KCl)	65Ad	b
	-9.0	5.98(25°,2)	-3.0	25-50	T	μ=0.1M	71Hf	b
	-16.0	6.54(15°,2)	-25.7	0-40	T	μ=0.25(KCl)	65Ad	b
Zn^{2+}	-9.2	4.82(15°,1)	-9.71	0-40	T	μ=0.25(KCl)	65Ad	b
	-9.2	3.93(15°,2)	-13.9	0-40	T	μ=0.25(KCl)	65Ad	b
Cd^{2+}	-11.2	7.42(1)	-3	25	T	μ=0.1(KNO_3)	64Ab	
H28	HYDRAZINE, 1,1,2-trimethyl-silyl-		$C_9H_{28}N_2S_3$			$[(CH_3)_3S]_2NNH[S(CH_3)_3]$		
Li^+	4.4	1.155(-50°)	25	-80- -20	T	S: Diethylether R: 2LiL = $(LiL)_2$ (monomer ⇄ dimer)	72Wb	b
H29	HYDROAZOIC ACID		HN_3			HN_3		L
Fe^{3+}	2.01	0.233	7.8	25	T	μ=0. R: Fe^{3+} + L = FeN_3^{2+} + H^+	61Wa	
	2.03	--	--	25	T	1M(HNO_3). R: Fe^{3+} + L = FeN_3^{2+} + H^+	61Wa	
	1.89	--	--	25	T	1M(HNO_3), 4M($NaNO_3$) R: Fe^{3+} + L = FeN_3^{2+} + H^+	61Wa	
	2.10	--	--	25	T	2.5M(HNO_3), 2.5M ($NaNO_3$). R: Fe^{3+} + L = FeN_3^{2+} + H^+	61Wa	
	2.42	--	--	25	T	5M(HNO_3). R: Fe^{3+} + L = FeN_3^{2+} + H^+	61Wa	
H30	HYDROBROMIC ACID		HBr			HBr		L
I_2	-1.602	1.079(25°,1)	5.1	15-45	T	0-3M(HBr)	36Lb	b
H31	HYDROGEN CHLORIDE		HCl			HCl		L
I_2	1.037	0.093(25°,1)	3.90	25-45	T	μ=0	37La	b
H32	HYDROGEN PEROXIDE		O_2H_2			H_2O_2		L^-
TiO^{2+}	-10.59	11.96(1)	17.98	15	C	μ=0.5N($HClO_4$)	70Va	
	-10.56	11.65(1)	17.91	25	C	μ=0.5N($HClO_4$)	70Va	
	-10.56	11.36(1)	17.69	35	C	μ=0.5N($HClO_4$)	70Va	
	-10.56	10.98(1)	17.07	45	C	μ=0.5N($HClO_4$)	70Va	
	-10.67	12.01(1)	17.98	15	C	μ=1N($HClO_4$)	70Va	
	-10.51	11.52(1)	17.44	25	C	μ=1N($HClO_4$)	70Va	
	-10.60	11.36(1)	17.60	35	C	μ=1N($HClO_4$)	70Va	
	-10.62	11.02(1)	17.03	45	C	μ=1N($HClO_4$)	70Va	

Metal Ion	ΔH, kcal /mole	Log K	ΔS, cal /mole °K	T, °C	M	Conditions	Ref.	Remarks
H32, cont.								
TiO^{2+}, cont.	-10.47	11.37(1)	16.78	25	C	μ=2N($HClO_4$)	70Va	
	-10.66	11.76(1)	16.80	15	C	μ=3N($HClO_4$)	70Va	
	-10.45	11.18(1)	16.10	25	C	μ=3N($HClO_4$)	70Va	
	-10.66	11.21(1)	16.68	35	C	μ=3N($HClO_4$)	70Va	
	-10.72	10.9(1)	16.20	45	C	μ=3N($HClO_4$)	70Va	
	-10.60	11.27(1)	16.00	25	C	μ=4N($HClO_4$)	70Va	
	-10.52	11.32(1)	15.27	15	C	μ=5N($HClO_4$)	70Va	
	-10.52	11.03	15.20	25	C	μ=5N($HClO_4$)	70Va	
	-10.55	10.77(1)	15.06	35	C	μ=5N($HClO_4$)	70Va	
	-10.66	10.65(1)	15.22	45	C	μ=5N($HClO_4$)	70Va	
Nb^{5+}	-7.0	2.73(25°)	-11	15-45	T	S: 63.51 m % H_2SO_4 R: Either $NbO_2^+ + L = NbO_2L^+$ or $NbOSO_4^+ + L = NbOSO_4L^+$	71Va	b,j,q: ±1.5
	-7.8	3.18(25°)	-11	15-45	T	S: 78.42 m % H_2SO_4 R: Either $NbO_2^+ + L = NbO_2L^+$ or $NbOSO_4^+ + L = NbOSO_4L^+$	71Va	b,j,q: ±1.0
	-8.3	3.56(25°)	-11	15-45	T	S: 88.71 m % H_2SO_4 R: Either $NbO_2^+ + L = NbO_2L^+$ or $NbOSO_4^+ + L = NbOSO_4L^+$	71Va	b,j,q: ±1.5
	-6.0	3.67(25°)	-3	15-45	T	S: 94.85 m % H_2SO_4 R: Either $NbO_2^+ + L = NbO_2L^+$ or $NbOSO_4^+ + L = NbOSO_4L^+$	71Va	b,j,q: ±0.9
TaABn	-4.18	2.40(25°,1)	-1	15-55	T	S: 63.5 w % H_2SO_4 A=O B=SO_4	73Vc	b,q: ±1.0
	-4.39	2.80(1)	-1.9	25	C	S: 78.4 w % H_2SO_4 A=O B=SO_4	73Vc	
	-4.50	2.73(1)	-2.1	35	C	S: 78.4 w % H_2SO_4 A=O B=SO_4	73Vc	
	-4.32	2.60(1)	-1.7	45	C	S: 78.4 w % H_2SO_4 A=O B=SO_4	73Vc	
	-4.35	2.80(25°,1)	-1.8	15-55	T	S: 78.4 w % H_2SO_4 A=O B=SO_4	73Vc	b,q: ±1.5
	-3.84	3.05(25°,1)	1	15-55	T	S: 88.71 w % H_2SO_4 A=O B=SO_4	73Vc	b,q: ±1.0
	-5.3	3.45(25°,1)	-2	15-55	T	S: 94.85 w % H_2SO_4 A=O B=SO_4	73Vc	b,q: ±1.0
Fe^{3+}	1.8	9.32(1)	49	20	T	μ=0	52Ua 49Ea	
Co^{3+}	-5	13.95(12.5°,1)	46	0-12.5	T	≃1M($HClO_4$)	57Bb	b
H33 HYDROGEN SULFIDE ION			HS			HS^-		L^-
[Fe^{3+}]	ΔH, obtained by temperature variation of K, is reported in graphical form as a function of pH for the reaction: $Hb(H_2O) + L^- = HbLH + H_2O$					Cow Methaemoglobin	69Be	
[Fe^{3+}]	ΔH, obtained by temperature variation					Dog Methaemoglobin	69Be	

Metal Ion	ΔH, kcal /mole	Log K	ΔS, cal /mole °K	T, °C	M	Conditions	Ref.	Re- marks
H33, cont.								
[Fe^{3+}], cont.	of K, is reported in graphical form as a function of pH in the reaction: $Hb(H_2O) + L^- = HbLH + H_2O$							
[Fe^{3+}]	ΔH, obtained by temperature variation of K, is reported in graphical form as a function of pH in the reaction: $Hb(H_2O) + L^- = HbLH + H_2O$					Guinea Pig Methaemoglobin	69Be	
[Fe^{3+}]	ΔH, obtained by temperature variation of K, is reported in graphical form as a function of pH in the reaction: $Hb(H_2O) + L^- = HbLH + H_2O$					Human Methaemo-globin	69Be	
[Fe^{3+}]	ΔH, obtained by temperature variation of K, is reported in graphical form as a function of pH in the reaction: $Hb(H_2O) + L^- = HbLH + H_2O$					Mouse Methaemo-globin	69Be	
[Fe^{3+}]	ΔH, obtained by temperature variation of K, is reported in graphical form as a function of pH in the reaction: $Hb(H_2O) + L^- = HbLH + H_2O$					Pigeon Methaemo-globin	69Be	
[Fe^{3+}]	ΔH, obtained by temperature variation of K, is reported in graphical form as a reaction of pH in the reaction: $Hb(H_2O) + L^- = HbLH + H_2O$					Sperm Whale Metmyoglobin	69Be	
H34 HYDROXIDE ION			OH			OH^-		L^-
Ca^{2+}	2.0	1.222(1)	12.3	25	C	$\mu=0$	65Hb	
	1.19	1.40(1)	10.4	25	T	$\mu=0$	53Ba	
Sr^{2+}	1.15	0.815(1)	7.6	25	T	$\mu=0$	54Ga	
Ba^{2+}	1.75	0.640(1)	8.8	25	T	$\mu=0$	54Ga	
V^{3+}	-3.7	12.6(1)	45.5	25	-	--	52Ua	
Fe^{2+}	-55.75	-6.63(25°,1)	-156	12-30	T	$\mu=1.0(NaClO_4)$	60Bb	b,n
Fe^{3+}	1.2	11.70(1)	50	25	T	$\mu=0$	60Gb 42Ra 52Ua	
A_5FeB	-16.2	6.18	-26.1	25	T	$\mu=1.0(KCl)$ R: $A_5FeB^{2-} + 2L^-$ $A_5FeB^{4-} + H_2O$ A=CN B=NO	66Se	
[Fe^{3+}]	-7.7	5.06(1)	-2.6	25	-	Ferrimyoglobin	55Ga 60Gb	
[Fe^{3+}]	-9.5	5.2(1)	-7.9	25	-	Ferrihaemoglobin	55Ga 60Gb	
[Fe^{3+}]	-9.6	5.86(1)	-5.0	25	-	Chironomus Hemo-globin	60Gb	
Co^{3+}	-3.5	12.16(1)	44	25	T	$\mu=1.0$	56Sd	

Metal Ion	ΔH, kcal /mole	Log K	ΔS, cal /mole °K	T, °C	M	Conditions	Ref.	Remarks
H34, cont.								
CoA_5B	-6.1	--	--	25	C	μ=0.1F(NaOH) A=NH_3 B=Cl R: CoA_5B^{2+} + L^- = CoA_5L^{2+} + B^-	69Hd 71Hh	
	-5.7	--	--	25	C	μ=0.1F(NaOH) A=NH_3 B=Br R: CoA_5B^{2+} + L^- = CoA_5L^{2+} + B^-	69Hd 71Hh	
	-3.65	--	--	25	C	μ=0.1F(NaOH) A=NH_3 B=I R: CoA_5B^{2+} + L^- = CoA_5L^{2+} + B^-	69Hd 71Hh	
	-3.85	--	--	25	C	μ=0.1F(NaOH) A=NH_3 B=NO_3 R: CoA_5B^{2+} + L^- = CoA_5L^{2+} + B^-	69Hd 71Hh	
RuA_6	5.1	<u>3.36</u>	32.5	25	T	A=NH_3 R: $RuA_6{}^{3+}$ + L^- = $RuA_5NH_2{}^{2+}$ + H_2O	71Wc	q: ±0.3
Cu^{2+}	-10.3	17.8(<u>2</u>-<u>2</u>)	47	25	C	3M($NaClO_4$)	68Ae	
	-15.44	16.45(25°)	23	5-35	T	μ=0.5($LiClO_4$) S: 50% C_2H_5OH R: Cu^{2+} + 2-carboxyphenyl-azo-β-naphthol + L = complex	58Ja	b
	-22.12	19.05(25°)	13	5-35	T	μ=0.5($LiClO_4$) S: 50% C_2H_5OH R: Cu^{2+} + L + 2-azo-β-napthylamine = complex	58Ja	b
	-37.37	22.34(25°)	-23	5-35	T	μ=0.5($LiClO_4$) S: 50% C_2H_5OH R: Cu^{2+} + L + o,o dihydroxyazobenzene = complex	58Ja	b
CuA^{2+}	-3.83	3.988(<u>1</u>)	5.4	25	C	0.1M(KCl) A=Tri(3-aminopropyl) amine	68Va	
	-2.28	4.14(<u>1</u>)	11.3	25	C	0.1M(KCl) A=3,3'-Diaminodipropyl amine	66Pa	
	-2.7	4.5(<u>1</u>)	11.5	25	C	μ=0.1M(KCl) A=Diethylenetriamine	68Pa	
	-3.7	4.65(<u>1</u>)	8.5	25	C	μ=0.1M(KCl) A=Triethylenetetraamine	63Pa	
	0	3.2(<u>1</u>)	14.5	25	C	μ=0.1M(KCl) A=Tetraethylenepenta-amine	63Pa	
	-5.9	8.789	20	25	C	0.5M(KNO_3) A=1,2-Diamino-N,N-dimethylethane	72Be	

Metal Ion	ΔH, kcal /mole	Log K	ΔS, cal /mole °K	T, °C	M	Conditions	Ref.	Re-marks
H34, cont.								
CuA^{2+}, cont.						R: $CuA^{2+} + 2L^- = Cu(L)_2A$		
	-7.4	8.306	13	25	C	0.5M(KNO_3) A=1,2-Diamino-N,N'-dimethylethane R: $CuA^{2+} + 2L^- = Cu(L)_2A$	72Be	
	-5.6	9.119	23.0	25	C	0.5M(KNO_3) A=1,2-Diamino-N,N,N',N'-tetramethylethane R: $CuA^{2+} + 2L^- = Cu(L)_2A$	72Ad 72Be	
	-11.2	15.416	33	25	C	0.5M(KNO_3) A=1,2-Diamino-N,N-dimethylethane R: $2CuA^{2+} + 2L^- = [Cu_2(L)_2A_2]^{2+}$	72Be	
	-11.2	15.020	31	25	C	0.5M(KNO_3) A=1,2-Diamino-N,N'-dimethylethane R: $2CuA^{2+} + 2L^- = [Cu_2(L)_2A_2]^{2+}$	72Be	
	-9.4	15.255	38.3	25	C	0.5M(KNO_3) A=1,2-Diamino-N,N,N',N'-tetramethylethane R: $2CuA^{2+} + 2L^- = [Cu_2(L)_2A_2]^{2+}$	72Ad 72Be	
ZnA^{2+}	-3.77	5.20(<u>1</u>)	11.2	25	C	0.1M(KCl) A=3,3'-Diaminodipropyl amine	66Pa	
	-2.0	4.03(<u>1</u>)	11.7	25	C	μ=0.5M(KNO_3) A=4,7-Diazadecane-1,10-diamine	73Bc	
HgA^+	-8.5	9.394(<u>1</u>)	13.7	20	C	μ=0.1(KNO_3) A=CH_3	65Sa	
HgA_2	-1.24	4.09(<u>1</u>)	14.6	25	C	μ=0. A=Cl	65Pa	
	-1.21	3.77(<u>2</u>)	13.2	25	T	μ=0	65Pa	
BL_3	-10.25	(<u>1</u>)	-12.65	0	T	μ=0 ΔCp=3.9	72Me	b
	-10.12	(<u>1</u>)	-12.18	25	T	μ=0 ΔCp=6.7	72Me	b
	-9.92	(<u>1</u>)	-11.54	50	T	μ=0 ΔCp=9.4	72Me	b
	-9.65	(<u>1</u>)	-10.73	75	T	μ=0 ΔCp=12.2	72Me	b
	-9.31	(<u>1</u>)	-9.79	100	T	μ=0 ΔCp=15.0	72Me	b
	-8.90	(<u>1</u>)	-8.73	125	T	μ=0 ΔCp=17.7	72Me	b
	-8.42	(<u>1</u>)	-7.57	150	T	μ=0 ΔCp=20.5	72Me	b
	-7.88	(<u>1</u>)	-6.32	175	T	μ=0 ΔCp=23.3	72Me	b

Metal Ion	ΔH, kcal /mole	Log K	ΔS, cal /mole $^\circ$K	T, $^\circ$C	M	Conditions	Ref.	Re-marks
H34, cont.								
BL$_3$, cont.	-7.26	(1)	-4.98	200	T	μ=0	72Me	b,ΔCp= 26.0
	-6.58	(1)	-3.57	225	T	μ=0	72Me	b,ΔCp= 28.8
	-5.82	(1)	-2.09	250	T	μ=0	72Me	b,ΔCp= 31.5
	-5.00	(1)	-0.56	275	T	μ=0	72Me	b,ΔCp= 34.3
	-4.11	(1)	1.04	300	T	μ=0	72Me	b,ΔCp= 37.7
	-10.3	(1)	-12.0	25	T	μ=1M(KCl)	72Me	b
	-9.2	(2-1)	-9.1	25	T	μ=1M(KCl)	72Me	b
	-14.4	(3-1)	-17.1	25	T	μ=1M(KCl)	72Me	b
	-34	(4-2)	-55	25	T	μ=1M(KCl)	72Me	b,q: ±6
	-43	(5-3)	-58	25	T	μ=1M(KCl)	72Me	b,q: ±4
Al^{3+}	-9.0	(2-2)	61	25-125	T	μ=1M(KCl)	71Mc	b,q: ±2.6
	-23.4	(3-4)	111	25-125	T	μ=1M(KCl)	71Mc	b,q: ±3.4
	-201.6	(14-34)	961	25-125	T	μ=1M(KCl)	71Mc	b
Tl$^+$	0.37	0.82(1)	5.1	25	T	μ=0	53Ba	
	1.8	0.088(1)	6.4	25	C	μ=3.0(LiClO$_4$)	73Ki	q: ±0.2
	5.0	-0.953(2)	12.50	25	C	μ=3.0(LiClO$_4$)	73Ki	q: ±0.6
SiL$_4$	-1.0	5.6(25°)	22	15-65	T	μ=5.0M(NaCl) R: SiL$_4$ + L$^-$ = SiOL$_3^-$ + H$_2$O	73Pi	b,q: ±0.5
	-1.5	7.1(25°)	28	15-65	T	μ=5.0M(NaCl) R: SiL$_4$ + 2L$^-$ = SiO$_2$L$_2^{2-}$ + 2H$_2$O	73Pi	b,q: ±0.5
	-3	12.0(25°)	45	15-65	T	μ=5.0M(NaCl) R: 4SiL$_4$ + 2L$^-$ - Si$_4$O$_6^{4-}$ + 6H$_2$O	73Pi	b,q: ±1

$$\underline{I}$$

I1 IMIDAZOLE			$C_3H_4N_2$					L
[Fe^{3+}]	1.1	1.462(18°,1)	10.2	18-32	T	0.1M Phosphate Buffer. pH=7.4 Ferricytochrome C	69Sf	b
[Fe^{3+}]	-4.05	2.199(1)	-3.5	25	T	μ=0.2. Ferrimy-globin	64Ga	
Co^{2+}	-4.2	2.47(25°,1)	-2.6	10-50	T	μ=0.16(KNO$_3$)	66Sc	b
	-3.9	1.93(25°,2)	-4.7	10-50	T	μ=0.16(KNO$_3$)	66Sc	b

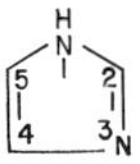

Metal Ion	ΔH, kcal /mole	Log K	ΔS, cal /mole °K	T, °C	M	Conditions	Ref.	Remarks
I1, cont.								
Co^{2+}, cont.	-3.5	1.45(25°,3)	-5.1	10-50	T	μ=0.16(KNO_3)	66Sc	b
	-3.7	1.00(25°,4)	-7.9	10-50	T	μ=0.16(KNO_3)	66Sc	b
	-2.9	0.54(25°,5)	-7.1	10-50	T	μ=0.16(KNO_3)	66Sc	b
	-3.6	-0.04(25°,6)	-12.0	10-50	T	μ=0.16(KNO_3)	66Sc	b
[Co]	-7.20	4.59	-3.2	25	T	μ=0. Aquocabalamin R: B-Co^+(H_2O) + L = B-Co^+ - L + H_2O B-Co^+(H_2O) = aquocobalamine	64Hb	
[Co]	-7.9	4.00(0°,1)	-10	-23-23	T	S: Toluene [Co]=Cobalt(II) protoporphyrin IX dimethyl ester	73Sr	b,q: ±0.6
[Co]	-6.3	4.09	-2.4	25	T	μ=0. Cobalamin Factor B. R: NC-Co^+(H_2O) + L = NC-Co^+ - L + H_2O NC-Co^+(H_2O) = cobalamine factor B	66Ha	
Ni^{2+}	-4	2.94(25°,1)	-4	0&25	T	μ=0.135	55Ma	a,b, f:4
	-5.2	3.09(25°,1)	-2.9	10-50	T	μ=0.16(KNO_3)	66Sc	b
	-5.3	3.10(25°,1)	-3.7	20-50	T	C_M=0.01 C_L=0.100	65Sd	b
	-4	2.41(25°,2)	-4	0&25	T	μ=0.135	55Ma	a,b, f:4
	-4.6	2.47(25°,2)	-4.1	10-50	T	μ=0.16(KNO_3)	66Sc	b
	-4	1.99(25°,3)	-4	0&25	T	μ=0.135	55Ma	a,b, f:4
	-4.3	2.00(25°,3)	-5.3	10-50	T	μ=0.16(KNO_3)	66Sc	b
	-4	1.3(25°,4)	-4	0&25	T	μ=0.135	55Ma	a,b, f:4
	-3.8	1.54(25°,4)	-5.4	10-50	T	μ=0.16(KNO_3)	66Sc	b
	-3.2	1.06(25°,5)	-5.9	10-50	T	μ=0.16(KNO_3)	66Sc	b
	-2.8	0.45(25°,6)	-7.2	10-50	T	μ=0.16(KNO_3)	66Sc	b
	-4.6	-0.49(25°,1-2)	-3.7	20-50	T	C_M=0.01 C_L=0.100	65Sd	b
	-4.2	1.98(25°,1-3)	-4.9	20-50	T	C_M=0.01 C_L=0.100	65Sd	b
	-3.7	1.53(25°,1-4)	-5.5	20-50	T	C_M=0.01 C_L=0.100	65Sd	b
	-3.6	1.01(25°,1-5)	-7.4	20-50	T	C_M=0.01 C_L=0.100	65Sd	b
	-18.4	11.4(1-6)	-10	25	C	μ=0.16(KNO_3)	64Bc	
	-2.7	0.39(25°,1,6)	-7.4	20-50	T	C_M=0.10 C_L=0.100	65Sd	b
Cu^{2+}	-6	4.20(25°,1)	-7	0&25	T	μ=0.135	55Ma 54Ea 57Nd	a,b, f:4
	-7.2	4.31(1)	-4.2	25	T	μ=0.16(KNO_3)	66Sc	
	-6	3.47(25°,2)	-7	0&25	T	μ=0.135	55Ma	a,b, f:4

Metal Ion	ΔH, kcal /mole	Log K	ΔS, cal /mole °K	T, °C	M	Conditions	Ref.	Re-marks
I1, cont.								
Cu^{2+}, cont.	-5.4	3.53(2)	-1.8	25	T	μ=0.16(KNO_3)	66Sc	
	-6	2.84(25°,3)	-7	0&25	T	μ=0.135	55Ma	a,b, f:4
	-4.6	2.92(3)	-2.1	25	T	μ=0.16(KNO_3)	66Sc	
	-6	2.0(25°,4)	-7	0&25	T	μ=0.135	55Ma	a,b, f:4
	-3.1	2.14(4)	-0.5	25	T	μ=0.16(KNO_3)	66Sc	
	-23.0	12.6(1-4)	-20	25	T	μ=0.16(KNO_3)	64Bc	
Ag^{+}	-4.54	3.516(1)	-0.48	0	T	μ=0.1. ΔCp=-47.7	66Da	
	-5.01	3.381(1)	-2.2	10	T	μ=0.1. ΔCp=-47.7	66Da	
	-5.49	3.243(1)	-4.1	20	T	μ=0.1. ΔCp=-47.7	66Da	
	-5.97	3.100(1)	-5.7	30	T	μ=0.1. ΔCp=-47.7	66Da	
	-6.68	2.955(1)	-7.6	40	T	μ=0.1. ΔCp=-47.7	66Da	
	-7.16	2.807(1)	-9.3	50	T	μ=0.1. ΔCp=-47.7	66Da	
	-7.3	3.05(1)	-11.0	25	C	μ=1(KNO_3)	64Bc	
	-6.0	4.361(2)	-2.1	0	T	μ=0.1. ΔCp=-334	66Da	
	-8.8	4.148(2)	-12.3	10	T	μ=0.1. ΔCp=-215	66Da	
	-10	3.893(2)	-17.4	20	T	μ=0.1. ΔCp=-71.6	66Da	
	-10	3.635(2)	-17.7	30	T	μ=0.1. ΔCp=-72	66Da	
	-8.8	3.409(2)	-12.9	40	T	μ=0.1. ΔCp=-215	66Da	
	-6.0	3.245(2)	-3.3	50	T	μ=0.1. ΔCp=-382	66Da	
	-8.4	3.83(2)	-11.2	25	C	μ=1(KNO_3)	64Bc	
	-15.6	6.89(1-2)	-21	25	C	μ=1(KNO_3)	64Bc	
Zn^{2+}	-16.2	9.24(1-4)	-12	25	C	μ=0.16(KNO_3)	64Bc	
Cd^{2+}	-7.0	3.03(25°,1)	-9	0-45	T	μ=0.1(KNO_3)	64Ac	b
	-5	2.8(1)	4.0	25	T	μ=0.15	53Ta	
	-5.6	2.11(25°,2)	-10	0-45	T	μ=0.1(KNO_3)	64Ac	b
	-4.3	1.34(25°,3)	-9	0-45	T	μ=0.1(KNO_3)	64Ac	b
	-2.8	0.79(25°,4)	-6	0-45	T	μ=0.1(KNO_3)	64Ac	b
	-10.8	7.48(25°,1-4)	-2.2	0-35	T	μ=0.15(KNO_3) S: 18.8% C_2H_5OH	54La	b
I2 IMIDAZOLE, N-acetyl-			$C_5H_6ON_2$			$C_3H_3N_2(CH_3CO)$		
CoA	-10.1	2.895(25°,1)	-20	20-50	T	S: Toluene A=α,β,γ,δ-Tetra-(p-methoxy-phenyl) porphyrin	73Wa	b,q: ±1
I3 IMIDAZOLE, 4(5)-acetic acid			$C_5H_6O_2N_2$			$C_3H_3N_2(CH_2CO_2H)$		L^-
Co^{2+}	-2.3	4.00(15°,1)	10.4	0-40	T	μ=0.25(KCl)	65Ad	b
	-3.4	3.03(15°,2)	2.1	0-40	T	μ=0.25(KCl)	65Ad	b
Ni^{2+}	-2.3	4.83(15°,1)	14.2	0-40	T	μ=0.25(KCl)	65Ad	b
	-3.4	3.71(15°,2)	5.2	0-40	T	μ=0.25(KCl)	65Ad	b
Cu^{2+}	-3.4	7.34(15°,1)	21.9	0-40	T	μ=0.25(KCl)	65Ad	b
	-2.3	5.81(15°,2)	18.8	0-40	T	μ=0.25(KCl)	65Ad	b
Zn^{2+}	-2.3	3.83(15°,1)	9.4	0-40	T	μ=0.25(KCl)	65Ad	b
	0.0	3.32(15°,2)	15.3	0-40	T	μ=0.25(KCl)	65Ad	b
	-3.4	2.63(15°,3)	5.2	0-40	T	μ=0.25(KCl)	65Ad	b

Metal Ion	ΔH, kcal /mole	Log K	ΔS, cal /mole °K	T, °C	M	Conditions	Ref.	Remarks
I4 IMIDAZOLE, 4(5)-(2'amino-ethyl)-			$C_5H_9N_3$			$C_3H_3N_2(CH_2CH_2NH_2)$		L
Ni^{2+}	-7.76	6.84(1)	5.3	25	C	μ=0.30(ClO_4^-)	67Hf	
	-7.00	5.04(2)	-0.4	25	C	μ=0.30(ClO_4^-)	67Hf	
	-6.05	3.15(3)	-5.9	25	C	μ=0.30(ClO_4^-)	67Hf	
Cu^{2+}	-10.29	9.56(1)	9.2	25	C	μ=0.30(ClO_4^-)	67Hf	
	-10.05	6.64(2)	-3.3	25	C	μ=0.30(ClO_4^-)	67Hf	
I5 IMIDAZOLE, 4(5)-aminomethyl-			$C_4H_7N_3$			$C_3H_3N_2(CH_2NH_2)$		L
Ni^{2+}	-8.69	5.85(1)	-2.4	25	C	μ=0.30(ClO_4^-)	67Hf	
	-8.67	4.82(2)	-7.0	25	C	μ=0.30(ClO_4^-)	67Hf	
	-8.65	3.12(3)	-14.8	25	C	μ=0.30(ClO_4^-)	67Hf	
Cu^{2+}	-11.44	8.72(1)	1.6	25	C	μ=0.30(ClO_4^-)	67Hf	
	-11.41	7.72(2)	-2.9	25	C	μ=0.30(ClO_4^-)	67Hf	
I6 IMIDAZOLE, 2-methoxy-			$C_4H_6ON_2$			$C_3H_3N_2(OCH_3)$		
Cu^{2+}	-6.84	4.27(1)	-3.4	25	C	μ=3.00($NaClO_4$)	68Wa	f:4
	-6.84	3.92(2)	-5.0	25	C	μ=3.00($NaClO_4$)	68Wa	f:4
	-6.84	3.13(3)	-8.6	25	C	μ=3.00($NaClO_4$)	68Wa	f:4
	-6.84	2.10(4)	-13.3	25	C	μ=3.00($NaClO_4$)	68Wa	f:4
I7 IMIDAZOLE, 4-methoxy-			$C_4H_6ON_2$			$C_3H_3N_2(OCH_3)$		
Cu^{2+}	-7.04	4.25(1)	-4.2	25	C	μ=3.0($NaClO_4$)	68Wa	f:4
	-7.04	3.90(2)	-5.8	25	C	μ=3.0($NaClO_4$)	68Wa	f:4
	-7.04	3.11(3)	-9.4	25	C	μ=3.0($NaClO_4$)	68Wa	f:4
	-7.04	2.08(4)	-14.1	25	C	μ=3.0($NaClO_4$)	68Wa	f:4
I8 IMIDAZOLE, 1-methyl			$C_4H_6N_2$			$C_3H_3N_2(CH_3)$		L
CoA	-11.4	3.375(25°,1)	-23	20-50	T	S: Toluene A=α,β,γ,δ-Tetra-(p-methoxy-phenyl) porphyrin	73Wa	b,q: ±1
[Co]	-13.8	(1)	--	25	C	S: CH_2Cl_2 [Co]=Methylcobaloxime	74Tc	
[Co]	-22.1	--	--	24	C	S: Dichloromethane [Co]=Methylcobaloxime R: $[Co]_2$ + 2L = 2[Co]L	73Cf	
[Co]	-10.7	4.40(0°,1)	-19	-23-23	T	S: Toluene [Co]=Cobalt(II) proto-porphyrin IX dimethyl ester	73Sr	b,q: ±0.6
Cu^{2+}	-6.7	3.46(29°,1)	6.3	2.5&29	T	μ=0.16	57Nd	a,b, q:±1
	-6.4	3.24(29°,2)	6.3	2.5&29	T	μ=0.16	57Nd	a,b, q:±3
	-5.3	2.99(29°,3)	3.6	2.5&29	T	μ=0.16	57Nd	a,b, q:±3

I8, cont.

Metal Ion	ΔH, kcal /mole	Log K	ΔS, cal /mole °K	T, °C	M	Conditions	Ref.	Remarks
CuA_2	-16.1	(1)	--	25	C	S: CCl_4 A-Hexafluoroacetyl acetone	74Tc	
Ag^+	-15.6	6.9(1-2)	-21	25	C	1M(KNO_3)	64Bc	
Zn^{2+}	-2.8	1.83(27°,1)	-0.97	2.5&27	T	μ=0.16	57Nd	a,b, q:±1 kcal
	-3.1	2.33(27°,2)	0.33	2.5&27	T	μ=0.16	57Nd	a,b, q:±3 kcal
	-3.1	2.80(27°,3)	2.5	2.5&27	T	μ=0.16	57Nd	a,b, q:±3 kcal
ZnA_2	-14.4	(1)	--	25	C	S: n-Hexane A=NSi$(CH_3)_3$	74Tc	

I9 IMIDAZOLE, 2-(2-pyridyl)- $C_8H_7N_3$

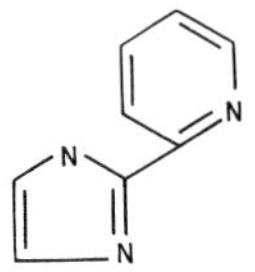

Metal Ion	ΔH, kcal /mole	Log K	ΔS, cal /mole °K	T, °C	M	Conditions	Ref.	Remarks
Fe^{2+}	-19.4	11.60(1-3)	-12.1	25	C	μ=0.1M(NO_3^-)	70Eb	
Co^{2+}	-7.73	5.26(1)	-1.6	25	C	μ=0.1M(NO_3^-)	70Eb	
	-6.28	4.79(2)	2.9	25	C	μ=0.1M(NO_3^-)	70Eb	
	-4.19	3.82(3)	0.9	25	C	μ=0.1M(NO_3^-)	70Eb	
Ni^{2+}	-9.25	6.39(1)	-1.9	25	C	μ=0.1M(NO_3^-)	70Eb	
	-8.62	6.22(2)	-1.5	25	C	μ=0.1M(NO_3^-)	70Eb	
	-5.16	5.19(3)	6.4	25	C	μ=0.1M(NO_3^-)	70Eb	
Cu^{2+}	-10.56	7.94(1)	0.8	25	C	μ=0.1M(NO_3^-)	70Eb	
	-6.59	5.70(2)	3.9	25	C	μ=0.1M(NO_3^-)	70Eb	
	-5.91	3.28(3)	4.9	25	C	μ=0.1M(NO_3^-)	70Eb	
Zn^{2+}	-6.37	4.39(1)	0.1	25	C	μ=0.1M(NO_3^-)	70Eb	
	-6.79	4.57(2)	-3.3	25	C	μ=0.1M(NO_3^-)	70Eb	
	-1.72	3.11(3)	8.3	25	C	μ=0.1M(NO_3^-)	70Eb	

I10 IMIDAZOLE, 4-(2-pyridyl)- $C_8H_7N_3$

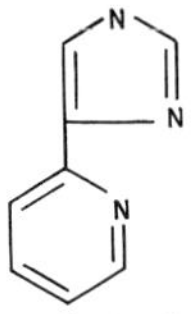

Metal Ion	ΔH, kcal /mole	Log K	ΔS, cal /mole °K	T, °C	M	Conditions	Ref.	Remarks
Fe^{2+}	-22.0	13.76(1-3)	-11.0	25	C	μ=0.1M(NO_3^-)	70Eb	
Ni^{2+}	-10.05	7.20(1)	-0.9	25	C	μ=0.1M(NO_3^-)	70Eb	
	-9.55	6.75(2)	-1.1	25	C	μ=0.1M(NO_3^-)	70Eb	
	-5.85	5.87(3)	7.2	25	C	μ=0.1M(NO_3^-)	70Eb	
Cu^{2+}	-11.71	8.76(1)	0.7	25	C	μ=0.1M(NO_3^-)	70Eb	
	-7.26	6.40(2)	4.8	25	C	μ=0.1M(NO_3^-)	70Eb	
	-4.54	3.25(3)	-0.4	25	C	μ=0.1M(NO_3^-)	70Eb	

I10, cont.

Metal Ion	ΔH, kcal /mole	Log K	ΔS, cal /mole °K	T, °C	M	Conditions	Ref.	Remarks
Zn^{2+}	-7.08	5.42(1)	1.0	25	C	μ=0.1M(NO_3^-)	70Eb	
	-5.80	4.82(2)	3.5	25	C	μ=0.1M(NO_3^-)	70Eb	
	-3.71	3.60(3)	7.8	25	C	μ=0.1M(NO_3^-)	70Eb	

I11 2-IMIDAZOLIDONE, N,N'-dimethyl- $C_5H_{10}ON_2$

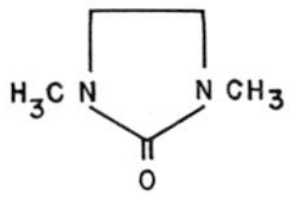

Metal Ion	ΔH, kcal /mole	Log K	ΔS, cal /mole °K	T, °C	M	Conditions	Ref.	Remarks
SbA_5	-27.7	--	--	25	C	S: CH_2Cl_2 A=Cl	71Oa	j

I12 INDENE C_9H_8

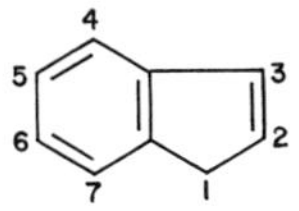

Metal Ion	ΔH, kcal /mole	Log K	ΔS, cal /mole °K	T, °C	M	Conditions	Ref.	Remarks
Li^+	5.0	--	20	-40-80	T	S: 1,2-Dimethoxy-ethane R: NaSL=NaL (Solvent separated = Contact ion pair)	72Gi	b
Na^+	3.5	--	15	-40-80	T	S: 1,2-Dimethoxy-ethane R: NaSL=NaL (Solvent separated = Contact ion pair)	72Gi	b

I13 INOSINE-5'-triphosphoric acid $C_{10}H_{15}O_{14}N_4P_3$

Metal Ion	ΔH, kcal /mole	Log K	ΔS, cal /mole °K	T, °C	M	Conditions	Ref.	Remarks
Mg^{2+}	1.5	3.76(25°,1)	22	25-45	T	μ=0.10M(KNO_3)	73Ke	b,g: Na,q: ±0.8
	4.48	3.91(1)	32.7	30	C	μ=0.2 (Tetramethyl-ammonium chloride) pH=8.50	73Sa	
Ca^{2+}	-1.1	3.41(25°,1)	12	25-45	T	μ=0.10M(KNO_3)	73Ke	b,g: Na,q: ±0.1
Mn^{2+}	-2.4	4.45(25°,1)	12	25-45	T	μ=0.10M(KNO_3)	73Ke	b,g: Na
Co^{2+}	-1.2	4.97(25°,1)	19	25-45	T	μ=0.10M(KNO_3)	73Ke	b,g: Na,q: ±0.1
Ni^{2+}	-2.0	5.06(25°,1)	19	25-45	T	μ=0.10M(KNO_3)	73Ke	b,g: Na
Cu^{2+}	-6.0	5.76(25°,1)	11	25-45	T	μ=0.10M(KNO_3)	73Ke	b,g: Na
Zn^{2+}	-1.7	4.57(25°,1)	15	25-45	T	μ=0.10M(KNO_3)	73Ke	b,g: Na,q: ±0.2

Metal Ion	ΔH, kcal /mole	Log K	ΔS, cal /mole °K	T, °C	M	Conditions	Ref.	Re-marks
I14 INSULIN, bovine								
[Zn]	-17.2	6.38	-29	25	C	μ=0.1 pH=2 [Zn]=Bovine zinc insulin. R: 2ZnL = $[ZnL]_2$ monomer=dimer	73Lc	
I15 IODATE ION			IO_3			IO_3^-		L^-
Na^+	4.58	1.73(1)	23.3	25	T	μ=0 S: Ethanol	72Kd	
Eu^{3+}	2.65	1.14(25°,1)	14.3	--	T	μ=0.10M($NaClO_4$)	73Cb	b,q: ±0.4,u
Ag^+	5.14	0.801(1)	20.3	35	T	μ=1.0($LiClO_4$)	56Ra	
	-5.20	1.79(1-2)	-8.50	35	T	μ=1.0($LiClO_4$)	56Ra	
I16 IODIDE ION			I			I^-		L^-
Li^+	13.1	11.32(1)	96	25	T	$C_M=10^{-4}$. S: 25 wt% Butanol. 75 wt% Hexane. D=2.85	61Rd	
	2.40	15.56(1)	79	25	T	$C_M=10^{-4}$. S: 12.65 wt% Butanol. 87.35 wt% Hexane. D=2.16	61Rd	
	-4.60	17.83(1)	66	25	T	$C_M=10^{-4}$. S: 7 wt% Butanol. 93 wt% Hexane	61Rd	
Na^+	1.00	3.96(30°,1)	21	25&35	T	S: 2-Methylpropan-1-ol	70Ec	a,b,q: ±0.15
	1.20	3.84(30°,1)	22	25&35	T	S: 2-Methylpropan-1-ol Pressure=506 bar $\Delta V=17 cm^3$/mole	70Ec	a,b,q: ±0.15
	1.51	3.72(30°,1)	22	25&35	T	S: 2-Methylpropan-1-ol Pressure=1013 bar $\Delta V=13 cm^3$/mole	70Ec	a,b,q: ±0.15
	1.70	3.55(30°,1)	22	25&35	T	S: 2-Methylpropan-1-ol Pressure=2026 bar $\Delta V=10 cm^3$/mole	70Ec	a,b,q: ±0.15
	1.89	3.45(30°,1)	22	25&35	T	S: 2-Methylpropan-1-ol Pressure=3040 bar $\Delta V=6 cm^3$/mole	70Ec	a,b,q: ±0.15
CrA_6^{3+}	7.6	-4.155(25°)	6.6	15-45	T	μ=4.2M(KI-HI) C_{H^+}=0.26M A=H_2O R: CuA_6^{3+} + L^- = CuA_5L^{2+} + A	68Se	b
CoA_6^{3+}	2.13	1.95(1)	16	25	T	μ=0. A=NH_3	55Na	
	1.63	1.24(1)	11	25	T	μ=0.054. A=NH_3	53Ea	
	1.22	0.938(1)	8	25	T	μ=0.054. A= Ethylenediamine	53Ea	
NiA_2	0.179	0.438(25°)	2.60	-36-50	T	S: Chloroform A=[(p-C_6H_5)(p-C_6H_5) MeP] R: NiA_2L_2(planar) = NiA_2L_2(tetrahedral)	70Pc	b
	0.869	0.081(25°)	3.29	-29-50	T	S: Chloroform	70Pc	b

Metal Ion	ΔH, kcal /mole	Log K	ΔS, cal /mole °K	T, °C	M	Conditions	Ref.	Re-marks
I16, cont.								
NiA_2, cont.						A=[(p-C_6H_5)(p-ClC_6H_4)MeP] R: NiA_2L_2(planar) = NiA_2L_2(tetrahedral)		
	0.296	0.484(25°)	3.21	-41-42	T	S: Chloroform	70Pc	b
						A=[p-C_6H_5)(p-MeC_6H_4)MeP] R: NiA_2L_2(planar) = NiA_2L_2(tetrahedral)		
	1.500	-0.242(25°)	3.92	-47-30	T	S: Chloroform	70Pc	b
						A=[(p-C_6H_5)(p-$CF_3C_6H_4$)MeP] R: NiA_2L_2(planar) = NiA_2L_2(tetrahedral)		
	1.530	-0.246(25°)	4.00	-50-45	T	S: Chloroform	70Pc	b
						A=[(p-ClC_6H_4)(p-ClC_6H_4)MeP] R: NiA_2L2(planar) = NiA_2L_2(tetrahedral)		
	0.0052	0.693(25°)	3.19	-36-40	T	S: Chloroform	70Pc	b
						A=[(p-MeC_6H_4)(p-MeC_6H_4)MeP] R: NiA_2L_2(planar) = NiA_2L_2(tetrahedral)		
$NiAL_2$	3.02	0.428(27°)	12.1	7-137		S: Nitrobenzene A=$(C_6H_5)_2P(CH_2)_3$-$P(C_6H_5)_2$. R: $NiAL_2$ (planar) = $NiAL_2$ (tetrahedral)	66Va	b
	4.51	0.585(27°)	17.7	-43-57		S: Chloroform A=$(C_6H_5)_2P(CH_2)_3$-$P(C_6H_5)_2$. R: $NiAL_2$ (planar) = $NiAL_2$ (tetrahedral)	66Va	b
$RhA_2B_2^+$	-2.62	0.934(70°)	-3.4	70&90	T	μ=2.0M A=Ethylenediamine B=Br. R: <u>trans</u> $RhA_2B_2^+$ + L^- = <u>trans</u> RhA_2BL^+ + B^-	66Bi	a,b,q: ±0.39
	0.9	0.781(85°)	6	85&96	T	μ=1.5M A=Ethylenediamine B=Cl. R: <u>trans</u> $RhA_2B_2^+$ + L = <u>trans</u> RhA_2BL^+ + B^-	66Bi	a,b,q: ±1.9
RhA_2BL^+	-6.18	0.757(50°)	-15.6	45-55	T	μ=0.6M A=Ethylenediamine B=Cl. R: <u>trans</u> RhA_2BL^+ + L^- = <u>trans</u> $RhA_2L_2^+$ + B^-	66Bi	b,q: ±0.4
	-3.38	0.508(50°)	-8.1	45-55	T	μ=0.5M A=Ethylenediamine B=Br. R: <u>trans</u> RhA_2BL^+ + L^- = <u>trans</u> $RhA_2L_2^+$ + B^-	66Bi	b,q: ±0.34
RhA_2BC	-7.1	3.699	1	50	T	μ=0.2M A=Ethylenediamine B=I . C=H_2O R: <u>trans</u> RhA_2BC^{2+} + L^- = <u>trans</u> RhA_2BL^+ + C	67Bb	
RhA_5B	-1.9	--	--	--	-	A=NH_3. B=H_2O	67Pa	q:

Metal Ion	ΔH, kcal /mole	Log K	ΔS, cal /mole °K	T, °C	M	Conditions	Ref.	Re-marks
I16, cont.								
RhA_5B, cont.						R: $RhA_5B^{3+} + L^- = RhA_5L^{2+} + B$		±0.3,u
PdA_2	-5.7	3.63(1)	-2.4	25	T	μ=0 A=o-Phenylenebisdi-methylarsine	67Ec 67Ed	
	-5.8	3.348(35°,1)	-3.2	22-55	T	μ=0.003($NaNO_3$) A=o-Phenylenebisdi-methylarsine	67Ed	b
$PdAB^+$	-3.7	1.950(27°)	-3.4	24-63	T	μ=0.5 A=Diethylenetriamine $B=Cl^-$. R: $PdAB^+ + L^- = PdAL^+ + B^-$	67Hd	b
	-2.5	1.480(27°)	-1.6	19-85	T	μ=0.5 A=Diethylenetriamine $B=Br^-$. R: $PdAB^+ + L^- = PdAL^+ + B^-$	67Hd	b
PtA_2	-3.8	5.68(23°,1)	13.2	21-55	T	μ=0. S: CH_3OH A=o-Phenylenebis-dimethylarsine	66Db	b
	-5.4	3.279(1)	-3.2	25	T	μ=0 A=o-Phenylenebis-dimethylarsine	67Ec 67Ed	
	-5.5	2.969(40°,1)	-4.1	25-55	T	μ=0.003($NaNO_3$) A=o-Phenylenebis-dimethylarsine	67Ed	b
PtA_3^{4+}	1.3	1.14(1)	9.6	25	T	μ=0. A=Ethylene-diamine	63Gb	
PtA_6^{2-}	-19	18.0	-19	25	T	μ=0.5. R: $PtA_6^{2-} + 6L^- = PtL_6^{2-} + 6A^-$ A=Cl	60Pf	
	-19	--	--	0-44.5	T	μ=0.5. A=Cl R: $PtA_6^{2-} + 6L^- = PtL_6^{2-} + 6A^-$	59Pa	
	-23	16.1	3	25	T	μ=0.5. R: $PtA_6^{2-} + 6L^- = PtL_6^{2-} + 6A^-$ A=Br	60Pf	
CuA^{2+}	1.19	2.53(1)	15.5	25	C	S: Methanol A=NN'-bis(2-amino-ethyl) propane-1,3-diamine	72Bd	
	-0.36	3.10(1)	12.9	25	C	S: Methanol A=NN'-bis(3-amino-propyl) propane-1,3-diamine	72Bd	
	1.57	2.83(1)	18.2	25	C	S: Methanol A=Triethylene-tetramine	72Bd	
CuA_2^{2+}	1.29	2.89(1)	17.5	25	C	S: Methanol A=NN-Dimethyl-ethylenediamine	72Bf	

Metal Ion	ΔH, kcal /mole	Log K	ΔS, cal /mole °K	T, °C	M	Conditions	Ref.	Re-marks
I16, cont.								
CuA_2^{2+}, cont.	0.64	3.05(1)	16.1	25	C	S: Methanol A=NN'-Dimethyl-ethylenediamine	72Bf	
	2.84	1.73(1)	17.4	25	C	S: Methanol A=Ethylenediamine	72Bd 72Bf	
	1.01	2.62(1)	15.3	25	C	S: Methanol A=N-Methyl-ethylenediamine	72Bf	
	0.51	2.52(1)	13.2	25	C	S: Methanol A=Trimethylene-diamine	72Bd	
Zn^{2+}	0	-2.93(1)	-13.4	25	C	μ=4.5	56Sa	n
	2.90	-1.76(1-2)	1.7	25	C	μ=4.5	56Sa	n
	13.20	-1.83(1-3)	35.9	25	C	μ=4.5	56Sa	n
	-18.30	-2.42(1-4)	-72.4	25	C	μ=4.5	56Sa	n
Cd^{2+}	-2.05	(1)	--	15&25	T	μ=0	59Aa 66Ga	a,b
	-2.32	1.94(1)	1.1	25	C	μ=0.25($NaClO_4$)	68Gc 54Ya	
	-2.43	1.88(1)	0.5	25	C	μ=0.50($NaClO_4$)	68Gc 60Ca	
	-2.45	1.88(1)	0.4	25	C	μ=1($NaClO_4$)	68Gc	
	-2.32	1.94(1)	1.2	25	C	μ=2($NaClO_4$)	68Gc	
	-2.26	2.08(1)	1.9	25	C	μ=3($NaClO_4$)	68Gc	
	-1.7	0.70(2)	-2.5	25	C	μ=0.25($NaClO_4$)	68Gc	
	-0.93	0.75(2)	0.3	25	C	μ=0.50($NaClO_4$)	68Gc	
	-0.48	0.78(2)	1.9	25	C	μ=1($NaClO_4$)	68Gc	
	-0.29	0.63(2)	1.9	25	C	μ=2($NaClO_4$)	68Gc	
	-0.20	0.70(2)	2.5	25	C	μ=3($NaClO_4$)	68Gc	
	-0.75	1.64(3)	5.0	25	C	μ=0.50($NaClO_4$)	68Gc	
	-1.38	1.68(3)	3.1	25	C	μ=1($NaClO_4$)	68Gc	
	-1.21	2.10(3)	5.6	25	C	μ=2($NaClO_4$)	68Gc	
	-0.73	2.13(3)	7.3	25	C	μ=3($NaClO_4$)	68Gc	
	-4.23	1.2(4)	-8.7	25	C	μ=0.50($NaClO_4$)	68Gc	
	-3.98	1.28(4)	-7.5	25	C	μ=1($NaClO_4$)	68Gc	
	-3.80	1.33(4)	-6.6	25	C	μ=2($NaClO_4$)	68Gc	
	-3.81	1.60(4)	-5.4	25	C	μ=3($NaClO_4$)	68Gc	
	10.8	(1-4)	--	25	C	--	52Ya	
Hg^{2+}	-17.6	13.5(1)	3	25	C	μ=0	61Ma	
	-18.9	13.53(1)	-5.3	8	C	μ=0.50	64Cc	
	-18.0	12.87(1)	-1.5	25	C	μ=0.50	64Cc	
	-17.3	12.44(1)	1.5	40	C	μ=0.50	64Cc	
	-16.6	13.5(1)	6	25	C	--	54Ya	
	-15.3	11.59(2)	-1.4	8	C	μ=0.50	64Cc	
	-16.2	11.69(2)	-4.2	25	C	μ=0.50	64Cc	
	-17.2	10.64(2)	-6.4	40	C	μ=0.50	64Cc	
	-34.0	24.97(1-2)	-6.7	8	C	μ=0.50	64Cc	
	-34.15	23.78(1-2)	-5.7	25	C	μ=0.50	64Cc	
	-34.5	23.08(1-2)	-4.9	40	C	μ=0.50	64Cc	
	-44.2	29.86(1-4)	-13.4	25	C	μ=0.5	60Ga	
HgA_2	-11.2	5.76(20°)	-11.7	10-30	T	μ=0 A=Cl R: $HgA_2 + L^- =$	72Bh	b

Metal Ion	ΔH, kcal /mole	Log K	ΔS, cal /mole °K	T, °C	M	Conditions	Ref.	Remarks
I16, cont.								
HgA_2, cont.	-23.4	10.62(20°)	-31.2	10-30	T	$HgAL + A^-$ μ=0 A=Cl R: $HgA_2 + 2L^- = HgL_2 + 2A^-$	72Bh	b
	1.1	0.902(20°)	7.9	10-30	T	μ=0 A=Cl R: $HgA_2 + HgL_2 = 2HgAL$	72Bh	b
HgAL	-12.3	4.86(20°)	-19.6	10-30	T	μ=0 A=Cl R: $HgLA + L^- = HgL_2 + A^-$	72Bh	b
HgL_2	-0.4	0.55(25°)	1.2	5-60	T	$\mu=10^{-3}$. R: $1/2HgL_2 + 1/2HgBr_2 = HgLBR$	64Ea	b,r
	-0.7	0.88(25°)	1.7	5-60	T	$\mu=10^{-3}$. R: $1/2HgL_2 + 1/2HgCl_2 = HgLCl$	64Ea	b,r
In^{3+}	-0.73	1.03(1)	2.3	25	C	μ=2.00M($NaClO_4$)	69Rd	
	0.81	1.25(2)	8.4	25	C	μ=2.00M($NaClO_4$)	69Rd	
Tl^+	-17.530	7.284(1)	-25.5	25	T	μ=0	34La	
Pb^{2+}	-0.300	(1)	--	25	C	μ=0	59Aa	
	-0.415	(1)	--	25	C	$\mu=6.5x10^{-3}$	59Aa	
	-11.8	1.68(25°,1)	-31	5-35	T	μ=3M($LiClO_4$ + LiI)	72Fh	b
	-1.00	2.29(1)	7	25	C	--	54Ya	
	-1.0	1.32(25°,2)	2	5-35	T	μ=3M($LiClO_4$ + LiI)	72Fh	b
	2.6	1.36(25°,3)	15	5-35	T	μ=3M($LiClO_4$ + LiI)	72Fh	b
	-1.6	0.40(25°,4)	-4	5-35	T	μ=3M($LiClO_4$ + LiI)	72Fh	b
	-59	6.201(25°,1-4)	-169	0-35	T	1M	49Kb	b
	15.6	(1-4)	--	--	C	--	52Ya	
PbL_2	3.60	0.041(25°)	13	5-25	T	μ=3M($LiClO_4$ + LiI) R: $2PbL_2 = PbL^+ + PbL_3^-$	72Fh	b
[N]	8.30	11.56(1)	81	25	T	$C_M \approx 10^{-4}$ iso $(C_5H_{11})_4N$ S: 25 w % Butanol 75 w % Hexane. D=2.85	61Rd	
[N]	5.40	13.13(1)	78	25	T	$C_M \approx 10^{-4}$ iso $(C_5H_{11})_4N$ S: 20 w % Butanol 80 w % Hexane. D=2.51	61Rd	
[N]	-1.50	15.53(1)	66	25	T	$C_M \approx 10^{-4}$ iso $(C_5H_{11})_4N$ S: 12.65 w % Butanol 87.35 w % Hexane. D=2.16	61Rd	
Bi^{3+}	-8.6	2.91(25°,1)	-15.5	6-45	T	μ=3	72Fe	b,q: ±0.8
	-2.6	3.65(25°,2)	7.9	6-45	T	μ=3	72Fe	b
	-0.1	3.34(25°,3)	15.0	6-45	T	μ=3	72Fe	b
	-4.1	2.46(25°,4)	-2.4	6-45	T	μ=3	72Fe	b
	-3.5	2.02(25°,5)	-2.7	6-45	T	μ=3	72Fe	b
	1.5	0.56(25°,6)	7.7	6-45	T	μ=3	72Fe	b
I_2	-3.215	2.69(46.65°,1)	1.1	25-63	T	μ=0. ΔCp=33	52Da	b
	-3.645	(31.69°,1)	--	25-63	T	μ=0. ΔCp=33	52Da	b

Metal Ion	ΔH, kcal /mole	Log K	ΔS, cal /mole °K	T, °C	M	Conditions	Ref.	Re-marks
I16, cont.								
I_2, cont.	-3.272	(44.01°,1)	--	25-63	T	μ=0. ΔCp=33	52Da	b
	-2.815	(56.35°,1)	--	25-63	T	μ=0. ΔCp=33	52Da	b
	5.10	2.85(25°,1)	4.1	1-39	T	μ=0.05(KCl)	51Aa	b
	-3.60	--	--	20-30	T	R: $I_2 + I^- = I_3^-$	55Ka	b
	-7.20	5.58(38.38°)	2.4	25-49.65	T	μ=0. R: $2I^- + 2I_2 = I_6^{2-}$	52Da	b
I17 IODINE			I_2			I_2		L
PbA_4	-32.57	--	--	--	C	S: $CHCl_3$ A=Phenyl R: $PbA_4 + 2L_2 = PbL_2A_2 + 2LA$	70Bg	u
PbA_3L	-14.16	--	--	--	C	S: $CHCl_3$ A=Phenyl R: $PbA_3L + L_2 = PbA_2L_2 + LA$	70Bg	u
[N]	-6.0	-2.60(1)	-32	25	C	C_5H_5N. S: Benzene	65Ga	
[O]	-3.3	0(1)	-11.0	25	T	$O(C_2H_4)_2O$. S: CCl_4	61Md	
[O]	-5.9	(1)	--	25	C	C_4H_8O. S: not stipulated	65Ga	
[S]	-4.6	0.836(1)	-11.7	25	T	$C_M=1.6x10^{-3}-6.0x10^{-3}$. $C_L=2.1x10^{-4}-6.9x10^{-4}$. $C_3H_6S_3$ (s-Trithiane). S: CCl_4	62Mc	
[S]	-6.63	1.55(25°,1)	15.2	5-25	T	trans-2,3-Butylene sulfide S: n-Heptane	62Ta	b
[S]	-7.46	2.05(25°,1)	15.6	5-25	T	Thiacyclobutane S: n-Heptane	62Ta	b
[S]	-6.57	1.89(25°,1)	13.4	5-25	T	Thiacyclobutane S: CCl_4	62Ta	b
[S]	-7.77	2.42(25°,1)	-15.0	5-25	T	Thiacyclopentane S: n-Heptane	62Ta	b
[S]	-8.6	(1)	--	25	C	Thiacyclo-pentane. S: not stipulated	65Ga	
[S]	-7.1	2.15(1)	-14.1	25	T	$C_M=1.6x10^{-3}-6.0x10^{-3}$. $C_L=2.1x10^{-4}-6.9x10^{-4}$. Thiacyclohexane S: CCl_4	62Mc	
[S]	-7.06	2.20(25°,1)	13.6	5-25	T	Thiacyclohexane S: n-Heptane	62Ta	b
[S]	-8.7	2.27(1)	-8.7	25	T	$C_M=1.6x10^{-3}-6.0x10^{-3}$. $C_L=2.1x10^{-4}-6.0x10^{-4}$. Thiacyclopentane S: CCl_4	62Mc	

Metal Ion	ΔH, kcal /mole	Log K	ΔS, cal /mole °K	T, °C	M	Conditions	Ref.	Re-marks
I17, cont.								
[S]	-8.90	2.30(25°,1)	19.4	5-25	T	Diethyl sulfide S: n Heptane	62Ta	b
[S]	-6.2	1.90(1)	-12.1	25	T	$S(C_2H_4)_2S$. 1,4-Dithiane. S: CCl_4	61Md	
[S]	-5.2	1.77(1)	-9.5	25	T	$O(C_2H_4)_2S$. S: CCl_4	57Mb	
[S]	-8.20	(1)	--	25	C	$(C_4H_9)_2S$. S: Octane	65Ga	
[S]	-6.70	(1)	--	25	C	$(C_4H_9)_2S$. S: Benzene	65Ga	
[S]	-2.30	(1)	--	25	C	$(C_4H_9)_2S$. S: Tetrahydrofuran	65Ga	
[Se]	-8.6	2.70(1)	-16.4	25	T	Dilute solution. Dimethyl Se. $Se(CH_3)_2$. S: CCl_4	60Ma	
[Se]	-7.6	2.13(1)	-15.6	25	T	$Se(C_2H_4)_2O$. S: CCl_4	61Md	
[Se]	-7.4	2.29(1)	-14.2	25	T	$Se(C_2H_4)_2S$. S: CCl_4	61Md	
[Se]	-7.0	2.46(1)	-12.3	25	T	$Se(C_2H_4)_2Se$. S: CCl_4	61Md	
[Se]	-9.8	3.17(1)	-18	25	T	$C_M=C_L=1.6x10^{-4}-2.0x10^{-3}$. Selenacyclohexane. SeC_5H_{10}. S: CCl_4	67Mg	
[Se]	-11.1	3.36(1)	-22	25	T	$C_M=1.5x10^{-3}-8x10^{-3}$ $C_L=1.0x10^{-4}-1.6x10^{-3}$. Selenacyclopentane. SeC_4H_8. S: CCl_4	67Mg	
[Se]	-7.6	1.82(1)	-17.1	25	T	Dilute solution. 4,4'-Dimethoxydiphenyl Se. S: CCl_4	60Ma	
[Se]	-6.9	1.60(1)	-16.4	25	T	Dilute solution. 4,4'-Dimethyldiphenyl Se. S: CCl_4	60Ma	
[Se]	-4.2	0.923(1)	-10.0	25	T	Dilute solution. 4,4'-Dichlorodiphenyl Se. S: CCl_4	60Ma	
I_2	-1.96	(1)	--	8-40	T	low μ. S: CCl_4	56Da	b
I18 ISOQUINOLINE			C_9H_7N			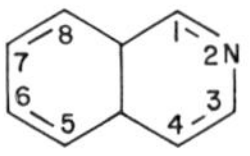		L
CoL_2A_2	-14.7	0.984(1-2)	-45.7	20	C	μ=0. S: Chloroform A=Cl	64Ka	
	-16.6	0.865(1-2)	-52.7	20	C	μ=0. S: Chloroform A=Br	64Ka	

Metal Ion	ΔH, kcal /mole	Log K	ΔS, cal /mole °K	T, °C	M	Conditions	Ref.	Remarks
I18, cont.								
CoL_2A_2, cont.	-17.4	0.36(1-2)	-57.7	20	C	μ=0. S: Chloroform A=I	64Ka	
NiA	-14.13	3.78(1-2)	-30.1	25	T	C_M≃C_L≃10^{-4} S: Benzene A=Diacetylbisbenzoyl hydrazine	58Sa	
I19 ISOSERINE			$C_3H_7O_3N$			$H_2NCH_2CH(OH)CO_2H$		L^-
Ni^{2+}	-3.09	4.19(1)	8.8	25	C	μ=0.16	70Lb	
	-3.73	3.66(2)	4.2	25	C	μ=0.16	70Lb	
Cu^{2+}	-5.78	7.31(1)	14.1	25	C	μ=0.16	70Lb	
	-6.16	7.06(2)	11.6	25	C	μ=0.16	70Lb	
I20 ISOTHIOCYANIC ACID, methyl ester			C_2H_3NS			$SCNCH_3$		L
I_2	3.3	-0.149(25°,1)	10.4	25-39.5	T	C_M=0.3x10^{-2} C_L=0.36. S: CCl_4	65Wb	b

L

Metal Ion	ΔH, kcal /mole	Log K	ΔS, cal /mole °K	T, °C	M	Conditions	Ref.	Remarks
L1 LEUCINE			$C_6H_{13}O_2N$			$(CH_3)_2CHCH_2CH(NH_2)CO_2H$		
Y^{3+}	7.986	4.26(25°,1)	46.28	25&35	T	μ=0.1(KCl)	73Sg	a,b,q: ±0.6
	12.609	3.90(25°,2)	60.13	25&35	T	μ=0.1(KCl)	73Sg	a,b
Ce^{3+}	9.666	4.69(25°,1)	53.88	25&35	T	μ=0.1(KCl)	73Sg	a,b,q: ±0.6
L2 LIMONENE			$C_{10}H_{16}$			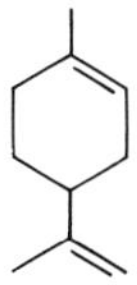		
PdA_2B_2	-6.8	--	--	25	C	S: CH_2Cl_2 C_L=3.8x10^{-3}M C_M=3.5x10^{-4}M A=Cl B=C_6H_5CN R: PdA_2B_2 + L = PdA_2L + 2B	72Pc	

M

Metal Ion	ΔH, kcal /mole	Log K	ΔS, cal /mole °K	T, °C	M	Conditions	Ref.	Remarks
M1 MALEIC ACID			$C_4H_4O_4$			$HO_2CCH:CHCO_2H$		L^{2-}
La^{3+}	3.06	3.44(1)	26.0	25	C	μ=0.10M($NaClO_4$)	73Cc	
Pr^{3+}	2.94	3.63(1)	26.5	25	C	μ=0.10M($NaClO_4$)	73Cc	
	2.58	2.814(1)	21.5	25	C	μ=1.0M($NaClO_4$)	73De	
	2.20	1.882(2)	16.0	25	C	μ=1.0M($NaClO_4$)	73De	
Nd^{3+}	2.89	3.66(1)	26.4	25	C	μ=0.10M($NaClO_4$)	73Cc	
Sm^{3+}	3.04	3.82(1)	27.6	25	C	μ=0.10M($NaClO_4$)	73Cc	

Metal Ion	ΔH, kcal /mole	Log K	ΔS, cal /mole °K	T, °C	M	Conditions	Ref.	Re-marks
M1, cont.								
Sm^{3+}, cont.	2.55	3.000(1)	22.3	25	C	μ=1.0M($NaClO_4$)	73De	
	1.69	1.906(2)	14.3	25	C	μ=1.0M($NaClO_4$)	73De	
Eu^{3+}	3.40	3.83(1)	28.9	25	C	μ=0.10M($NaClO_4$)	73Cc	
Gd^{3+}	3.57	3.79(1)	29.3	25	C	μ=0.10M($NaClO_4$)	73Cc	
	2.99	2.964(1)	23.6	25	C	μ=1.0M($NaClO_4$)	73De	
	1.75	1.819(2)	14.1	25	C	μ=1.0M($NaClO_4$)	73De	
Tb^{3+}	4.15	3.74(1)	31.0	25	C	μ=0.10M($NaClO_4$)	73Cc	
Dy^{3+}	4.16	3.74(1)	31.1	25	C	μ=0.10M($NaClO_4$)	73Cc	
Ho^{3+}	4.36	3.67(1)	31.4	25	C	μ=0.10M($NaClO_4$)	73Cc	
	3.69	2.886(1)	25.6	25	C	μ=1.0M($NaClO_4$)	73De	
	2.07	1.775(2)	15	25	C	μ=1.0M($NaClO_4$)	73De	
Er^{3+}	4.31	3.64(1)	31.1	25	C	μ=0.10M($NaClO_4$)	73Cc	
Tm^{3+}	4.38	3.62(1)	31.2	25	C	μ=0.10M($NaClO_4$)	73Cc	
Yb^{3+}	4.43	3.64(1)	31.5	25	C	μ=0.10M($NaClO_4$)	73Cc	
	3.90	2.805(1)	25.9	25	C	μ=1.0M($NaClO_4$)	73De	
	2.56	1.843(2)	16.9	25	C	μ=1.0M($NaClO_4$)	73De	
Lu^{3+}	4.41	3.59(1)	31.2	25	C	μ=0.10M($NaClO_4$)	73Cc	

M2 MALEIC ACID, anhydride $C_4H_2O_3$

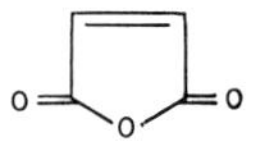

Metal Ion	ΔH, kcal /mole	Log K	ΔS, cal /mole °K	T, °C	M	Conditions	Ref.	Re-marks
NiAB	-12.8	4.52(27°)	-22.0	22-47	T	S: Tetrahydrofuran A=Dipyridyl B=Tetrahydrofuran R: NiAB + L = NiAL + B	71Ya	b

M3 MALONIC ACID $C_3H_4O_4$ $CH_2(CO_2H)_2$ L^{2-}

Metal Ion	ΔH, kcal /mole	Log K	ΔS, cal /mole °K	T, °C	M	Conditions	Ref.	Re-marks
Mg^{2+}	3.40	2.86(25°,1)	23.9	20-25	T	μ=0	52Ea	b
	3.20	2.86(1)	23.9	25	T	μ=0	54Cd	
	3.10	(1)	--	25-30	T	μ=0	52Ea	b
	3.20	(1)	--	30-35	T	μ=0	52Ea	b
Ca^{2+}	4.45	2.50(25°,1)	26.4	25-40	T	μ=0	70Ge	b
Sc^{3+}	3.41	5.87(1)	38.3	25	C	μ=0.1($NaClO_4$)	69Ge	
	2.93	4.25(2)	29.3	25	C	μ=0.1($NaClO_4$)	69Ge	
	1.46	2.89(3)	18.1	25	C	μ=0.1($NaClO_4$)	69Ge	
La^{3+}	4.8	4.98(25°,1)	39	25&35	T	μ=0	56Ga	a,b
	3.74	3.92(1)	30.5	25	C	μ=0.10M($NaClO_4$)	72Db	
	2.89	3.068(1)	23.7	25	C	μ=1.00M($NaClO_4$)	73Dc	q: ±0.17
	4.0	2.56(1)	25.0	25	C	μ=2.00M($NaClO_4$)	72Db	
	1.98	2.085(2)	16.2	25	C	μ=1.00M($NaClO_4$)	73Dc	q: ±0.41
	1.24	6.307	33.0	25	C	μ=1.00M($NaClO_4$) R: La^{3+} + HL^- = $LaHL^{2+}$	73Dc	q: ±0.17

Metal Ion	ΔH, kcal /mole	Log K	ΔS, cal /mole °K	T, °C	M	Conditions	Ref.	Re-marks
M3, cont.								
La^{3+}, cont.	4.25	9.180	56.2	25	C	μ=1.00M($NaClO_4$) R: $La^{3+} + L^{2-} + HL^- = LaHL_2$	73Dc	q: ±0.41
Ce^{3+}	3.61	4.12(1)	30.9	25	C	μ=0.10M($NaClO_4$)	72Db	
	2.92	3.259(1)	24.6	25	C	μ=1.00M($NaClO_4$)	73Dc	q: ±0.19
	1.62	1.980(2)	16.0	25	C	μ=1.00M($NaClO_4$)	73Dc	q: ±2.1
	1.27	6.377	33.5	25	C	μ=1.00M($NaClO_4$) R: $La^{3+} + HL^- = LaHL^{2+}$	73Dc	q: ±0.19
	4.78	9.303	58.6	25	C	μ=1.00M($NaClO_4$) R: $La^{3+} + L^{2-} + HL^- = LaHL_2$	73Dc	q: ±0.6
Pr^{3+}	3.73	4.20(1)	31.7	25	C	μ=0.10M($NaClO_4$)	72Db	
	3.04	3.294(1)	25.3	25	C	μ=1.00M($NaClO_4$)	73Dc	q: ±0.2
	2.10	2.313(2)	42.8	25	C	μ=1.00M($NaClO_4$)	73Dc	q: ±0.26
	1.27	6.465	33.9	25	C	μ=1.00M($NaClO_4$) R: $La^{3+} + HL^- = LaHL^{2+}$	73Dc	q: ±0.21
	4.54	9.28	57.4	25	C	μ=1.00M($NaClO_4$) R: $La^{3+} + L^{2-} + HL^- = LaHL_2$	73Dc	q: ±1.0
Nd^{3+}	3.95	4.24(1)	32.6	25	C	μ=0.10M($NaClO_4$)	72Db	
	3.13	3.380(1)	26.0	25	C	μ=1.00M($NaClO_4$)	73Dc	q: ±0.17
	1.91	2.540(2)	44.0	25	C	μ=1.00M($NaClO_4$)	73Dc	q: ±0.62
	1.20	6.482	33.7	25	C	μ=1.00M($NaClO_4$) R: $La^{3+} + HL^- = LaHL^{2+}$	73Dc	q: ±0.17
	4.59	9.443	58.6	25	C	μ=1.00M($NaClO_4$) R: $La^{3+} + L^{2-} + HL^- = LaHL_2$	73Dc	q: ±0.35
Sm^{3+}	3.29	4.53(1)	31.7	25	C	μ=0.10M($NaClO_4$)	72Db	
	2.99	3.672(1)	26.8	25	C	μ=1.00M($NaClO_4$)	73Dc	
	3.47	3.20(1)	26.3	25	C	μ=2.00M($NaClO_4$)	72Db	
	2.10	2.400(2)	18.2	25	C	μ=1.00M($NaClO_4$)	73Dc	q: ±0.36
	1.34	6.588	34.6	25	C	μ=1.00M($NaClO_4$) R: $La^{3+} + HL^- = LaHL^{2+}$	73Dc	q: ±0.17
	4.37	9.864	59.8	25	C	μ=1.00M($NaClO_4$) R: $La^{3+} + L^{2-} + HL^- = LaHL_2$	73Dc	q: ±0.33
Eu^{3+}	3.23	4.62(1)	31.9	25	C	μ=0.10M($NaClO_4$)	72Db	
	3.06	3.721(1)	27.2	25	C	μ=1.00M($NaClO_4$)	73Dc	
	1.87	3.02(2)	20.1	25	C	μ=0.10M($NaClO_4$)	72Db	
	1.72	2.516(2)	17.2	25	C	μ=1.00M($NaClO_4$)	73Dc	q: ±0.79

Metal Ion	ΔH, kcal /mole	Log K	ΔS, cal /mole °K	T, °C	M	Conditions	Ref.	Remarks
M3, cont.								
Eu^{3+}, cont.	1.41	6.482	34.4	25	C	μ=1.00M($NaClO_4$) R: La^{3+} + HL^- = $LaHL^{2+}$	73Dc	q: ±0.19
	4.02	9.986	59.3	25	C	μ=1.00M($NaClO_4$) R: La^{3+} + L^{2-} + HL^- = $LaHL_2$	73Dc	q: ±0.24
Gd^{3+}	5.1	5.36(25°,1)	42	25&35	T	μ=0	56Ga	a,b
	3.17	4.70(1)	32.1	25	C	μ=0.10M($NaClO_4$)	72Db	
	3.01	3.732(1)	27.2	25	C	μ=1.00M($NaClO_4$)	73Dc	
	1.90	3.00(2)	20.1	25	C	μ=0.10M($NaClO_4$)	72Db	
	1.79	2.50(2)	17.4	25	C	μ=1.00M($NaClO_4$)	73Dc	q: ±0.38
	1.55	6.50	34.9	25	C	μ=1.00M($NaClO_4$) R: La^{3+} + HL^- = $LaHL^{2+}$	73Dc	q: ±0.19
	4.32	9.78	59.3	25	C	μ=1.00M($NaClO_4$) R: La^{3+} + L^{2-} + HL^- = $LaHL_2$	73Dc	q: ±0.36
Tb^{3+}	3.14	4.74(1)	32.2	25	C	μ=0.10M($NaClO_4$)	72Db	
	3.014	3.816(1)	27.5	25	C	μ=1.00M($NaClO_4$)	73Dc	
	1.64	3.04(2)	19.4	25	C	μ=0.10M($NaClO_4$)	72Db	
	2.25	2.561(2)	19.4	25	C	μ=1.00M($NaClO_4$)	73Dc	
	1.82	6.36	35.1	25	C	μ=1.00M($NaClO_4$) R: La^{3+} + HL^- = $LaHL^{2+}$	73Dc	q: ±0.17
	4.11	10.09	60.0	25	C	μ=1.00M($NaClO_4$) R: La^{3+} + L^{2-} + HL^- = $LaHL_2$	73Dc	
Dy^{3+}	3.40	4.86(1)	33.6	25	C	μ=0.10M($NaClO_4$)	72Db	
	3.066	3.849(1)	28.0	25	C	μ=1.00M($NaClO_4$)	73Dc	
	2.00	3.06(2)	20.7	25	C	μ=0.10M($NaClO_4$)	72Db	
	2.27	2.488(2)	18.9	25	C	μ=1.00M($NaClO_4$)	73Dc	
	0.17	1.19(3)	5.7	25	C	μ=1.00M($NaClO_4$)	73Dc	q: ±2.4
	1.96	6.31	35.4	25	C	μ=1.00M($NaClO_4$) R: La^{3+} + HL^- = $LaHL^{2+}$	73Dc	q: ±0.29
	4.30	9.81	35.4	25	C	μ=1.00M($NaClO_4$) R: La^{3+} + L^{2-} + HL^- = $LaHL_2$	73Dc	q: ±0.41
Ho^{3+}	3.229	3.832(1)	28.4	25	C	μ=1.00M($NaClO_4$)	73Dc	
	2.06	2.54(2)	18.4	25	C	μ=1.00M($NaClO_4$)	73Dc	
	1.89	1.30(3)	12.9	25	C	μ=1.00M($NaClO_4$)	73Dc	q: ±0.81
	1.65	6.31	34.4	25	C	μ=1.00M($NaClO_4$) R: La^{3+} + HL^- = $LaHL^{2+}$	73Dc	q: ±0.29
	4.30	9.95	60.0	25	C	μ=1.00M($NaClO_4$) R: La^{3+} + L^{2-} + HL^- = $LaHL_2$	73Dc	q: ±0.29
Er^{3+}	3.11	4.78(1)	32.3	25	C	μ=0.10M($NaClO_4$)	72Db	

Metal Ion	ΔH, kcal /mole	Log K	ΔS, cal /mole °K	T, °C	M	Conditions	Ref.	Remarks
M3, cont.								
Er^{3+}, cont.	3.20	3.851(1)	28.4	25	C	μ=1.00M($NaClO_4$)	73Dc	
	3.27	3.32(1)	26.2	25	C	μ=2.00M($NaClO_4$)	72Db	
	1.69	3.05(2)	19.6	25	C	μ=0.10M($NaClO_4$)	72Db	
	2.17	2.54(2)	18.9	25	C	μ=1.00M($NaClO_4$)	73Dc	
	1.31	1.23(3)	10.0	25	C	μ=1.00M($NaClO_4$)	73Dc	q: ±2.4
	1.50	6.34	34.2	25	C	μ=1.00M($NaClO_4$) R: La^{3+} + HL^- = $LaHL^{2+}$	73Dc	q: ±0.35
	4.90	9.93	61.9	25	C	μ=1.00M($NaClO_4$) R: La^{3+} + L^{2-} + HL^- = $LaHL_2$	73Dc	q: ±0.48
Tm^{3+}	3.12	4.80(1)	32.4	25	C	μ=0.10M($NaClO_4$)	72Db	
	3.456	3.846(1)	29.2	25	C	μ=1.00M($NaClO_4$)	73Dc	
	1.90	2.98(2)	20.0	25	C	μ=0.10M($NaClO_4$)	72Db	
	2.089	2.57(2)	18.9	25	C	μ=1.00M($NaClO_4$)	73Dc	
	2.56	1.19(3)	14.1	25	C	μ=1.00M($NaClO_4$)	73Dc	
	1.39	6.27	33.5	25	C	μ=1.00M($NaClO_4$) R: La^{3+} + HL^- = $LaHL^{2-}$	73Dc	q: ±0.43
	4.06	9.99	59.3	25	C	μ=1.00M($NaClO_4$) R: La^{3+} + L^{2-} + HL^- = $LaHL_2$	73Dc	q: ±0.52
Yb^{3+}	3.16	4.80(1)	32.5	25	C	μ=0.10M($NaClO_4$)	72Db	
	3.403	3.870(1)	2.92	25	C	μ=1.00M($NaClO_4$)	73Dc	
	3.55	3.42(1)	27.5	25	C	μ=2.00M($NaClO_4$)	72Db	
	1.95	3.09(2)	20.6	25	C	μ=0.100M($NaClO_4$)	72Db	
	2.428	2.54(2)	19.8	25	C	μ=1.00M($NaClO_4$)	73Dc	
	2.08	1.37(3)	13.1	25	C	μ=1.00M($NaClO_4$)	73Dc	
	1.89	6.22	34.9	25	C	μ=1.00M($NaClO_4$) R: La^{3+} + HL^- = $LaHL^{2+}$	73Dc	q: ±0.24
	5.16	9.76	61.9	25	C	μ=1.00M($NaClO_4$) R: La^{3+} + L^{2-} + HL^- = $LaHL_2$	73Dc	q: ±0.38
Lu^{3+}	5.2	5.69(25°,1)	44	25&35	T	μ=0	56Ga	a,b
	3.30	4.74(1)	32.7	25	C	μ=0.10M($NaClO_4$)	72Db	
	3.487	3.879(1)	29.4	25	C	μ=1.00M($NaClO_4$)	73Dc	
	2.17	3.09(2)	21.4	25	C	μ=0.10M($NaClO_4$)	72Db	
	2.58	2.533(2)	20.3	25	C	μ=1.00M($NaClO_4$)	73Dc	
	2.34	1.47(3)	14.6	25	C	μ=1.00M($NaClO_4$)	73Dc	
	1.91	6.25	35.1	25	C	μ=1.00M($NaClO_4$) R: La^{3+} + HL^- = $LaHL^{2+}$	73Dc	q: ±0.36
	5.88	9.46	62.1	25	C	μ=1.00M($NaClO_4$) R: La^{3+} + L^{2-} + HL^- = $LaHL_2$	73Dc	q: ±0.93
Mn^{2+}	3.68	3.29(1)	27.4	25	C	μ=0	63Mb	
	3.53	3.29(1)	26.8	25	T	μ=0. ΔCp=92	61Na	
Co^{2+}	2.90	3.76(1)	26.9	25	C	μ=0	63Mb	
	2.57	3.76(1)	25.8	25	T	μ=0. ΔCp=68	61Na	

Metal Ion	ΔH, kcal /mole	Log K	ΔS, cal /mole °K	T, °C	M	Conditions	Ref.	Remarks
M3, cont.								
Ni^{2+}	1.88	4.10(1)	25.0	25	C	μ=0	66Mc 66Na 65Ac	
	1.77	4.10(1)	24.8	25	T	μ=0. ΔCp=46	61Na	
	3	1.398(25°,1)	16	12-305	T	μ=0. R: $Ni(H_2O)_6^{2+}$ + L = $[(H_2O)_5Ni(H_2O)L]$ (ion pair)	72Hc	b
Cu^{2+}	2.85	5.64(1)	35.4	25	C	μ=0	66Mc 66Na 65Ac	
	3.50	5.09(25°,1)	35.0	25-40	T	μ=0	70Ge	b
	6.58	0.90(25°)	26.2	25-40	T	μ=0 R: Cu^{2+} + HL^- = $CuHL^+$	70Ge	b
Zn^{2+}	3.13	3.83(1)	28.0	25	C	μ=0	66Mc 66Na 51Jb	
	2.98	3.81(1)	27.4	25	T	μ=0. ΔCp=86.5 Equations are given for ΔH as a function of T in 65Na	65Na 65Ac	
M4 MALONIC ACID, benzyl-			$C_{10}H_{10}O_4$			$C_6H_5CH_2CH(CO_2H)_2$		
Mg^{2+}	7.13	2.785(30°,1)	36.2	25-35	T	μ=0	72Pd	b
Cu^{2+}	0.47	5.430(1)	26.4	25	C	μ=0	73Pg	
M5 MANNITOL			$C_6H_{14}O_6$			$HOCH_2[CH(OH)]_4CH_2OH$		
BA_4^-	-8.05	3.041(25°,1)	-13	0-35	T	C_L=0-0.9M A=OH^-	67Cc	b,q: ±0.8
	-4.45	5.137(25°,1-2)	8.6	0-35	T	C_L=0-0.9M A=OH^-	67Cc	b,q: ±0.5
[B]	0.008	4.26(1)	19	0	T	0.1M(NaCl) [B]=$H_2BO_3^-$	57Wa	
	-0.023	4.18(1)	19	10	T	0.1M(NaCl) [B]=$H_2BO_3^-$	57Wa	
	-0.058	4.12(1)	18	20	T	0.1M(NaCl) [B]=$H_2BO_3^-$	57Wa	
	-0.098	3.91(1)	18	30	T	0.1M(NaCl) [B]=$H_2BO_3^-$	57Wa	
	-0.143	3.64(1)	16	40	T	0.1M(NaCl) [B]=$H_2BO_3^-$	57Wa	
	-0.193	3.26(1)	14	50	T	0.1M(NaCl) [B]=$H_2BO_3^-$	57Wa	
	-0.094	1.20(2)	5	0	T	0.1M(NaCl) [B]=$H_2BO_3^-$	57Wa	
	-0.060	1.01(2)	4	10	T	0.1M(NaCl) [B]=$H_2BO_3^-$	57Wa	
	-0.021	0.90(2)	5	20	T	0.1M(NaCl) [B]=$H_2BO_3^-$	57Wa	
	0.025	0.94(2)	4	30	T	0.1M(NaCl) [B]=$H_2BO_3^-$	57Wa	

Metal Ion	ΔH, kcal /mole	Log K	ΔS, cal /mole °K	T, °C	M	Conditions	Ref.	Remarks
[B], cont.	0.077	0.98(2)	5	40	T	0.1M(NaCl) [B]=$H_2BO_3^-$	57Wa	
	0.135	1.22(2)	6	50	T	0.1M(NaCl) [B]=$H_2BO_3^-$	57Wa	
M6 METHANE, amino-			CH_5N			CH_3NH_2		L
NiA	-15.9	3.74(1)	-36.4	25	T	$C_M=10^{-4}$ S: Benzene A=Diacetyl-bisbenzoyl hydrazine	60Sb	
Ag^+	-3	3.07(1)	3.4	25	C	μ=0	71He	q:±2
	-8	3.82(2)	-10.1	25	C	μ=0	71He	q:±2
	-14.4	7.48(1-2)	-18.1	2.5	T	μ=0	65Le	b,h: 2.5-45
	-11.72	6.89(1-2)	-7.8	25	C	μ=0	71He	
	-14.4	6.69(1-2)	-17.7	25	T	μ=0	65Le	b,h: 2.5-45
	-14.4	6.18(1-2)	-18.5	35	T	μ=0	65Le	b,h: 2.5-45
	-14.4	5.96(1-2)	-18.0	45	T	μ=0	65Le	b,h: 2.5-45
	-11.5	6.78(1-2)	-7.5	25	C	--	55Fa	
	-14.4	7.86(1-2)	-16.4	2.5	T	μ=0. S: 50% CH_3OH	65Le	b,h: 2.5-45
	-14.4	6.97(1-2)	-16.4	25	T	μ=0. S: 50% CH_3OH	65Le	b,h: 2.5-45
	-14.4	6.96(1-2)	-16.2	35	T	μ=0. S: 50% CH_3OH	65Le	b,h: 2.5-45
	-14.4	6.36(1-2)	-16.5	45	T	μ=0. S: 50% CH_3OH	65Le	b,h: 2.5-45
Cd^{2+}	-7.02	4.808(25°,1)	-1.5	25	T	μ=2	53Sa	
	-13.7	6.55(25°,1-2)	-16.0	25	T	μ=2	53Sb	
HgA_2	-6.8	2.40	-11.7	25	C	μ=0. R: HgA_2 + L = $HgAL^+$ + A^- A=Cl	66Pd	
	-1.0	2.21	6.7	25	C	μ=0. R: $HgAL^+$ + L = HgL_2^{2+} + A^- A=Cl	66Pd	
AlA_3	-30.02	(1)	--	25-28	C	S: n-Hexane A=CH_3	68Hc	
I_2	-7.1	2.724(20°,1)	-12.3	20&30	T	S: n-Heptane	60Ya	a,b
M7 METHANE, 1,1-diphosphonic acid			$CH_6O_6P_2$			$CH_2(PO_3H_2)_2$		
Li^+	0.70	2.48(1)	13.7	25	C	0.5M$(CH_3)_4$NCl	68Ca	
Na^+	1.6	1.13(1)	10.5	25	C	0.5M$(CH_3)_4$NCl	68Ca	
K^+	1.4	1.02(1)	9.6	25	C	0.5M$(CH_3)_4$NCl	68Ca	
Mg^{2+}	4.4	4.396(25°,1)	34	25-50	T	μ=0	62Ia	b
Ca^{2+}	0	4.765(25°,1)	22	25-50	T	μ=0	62Ia	b,q: ±2.0
	2.6	4.70(1)	30	25	C	0.5M$(CH_3)_4$NCl	68Ca	

Metal Ion	ΔH, kcal /mole	Log K	ΔS, cal /mole °K	T, °C	M	Conditions	Ref.	Re-marks
M8 METHANE, iodo-			CH_3I			CH_3I		L
I_2	-4.5	-2.38(25°,1)	-25.8	5-45	T	S: n-Heptane	57Sa	b
M9 METHANE, tetrachloro-			CCl_4			CCl_4		
I_2A	-0.20	0.188(25°)	0.55	-20-40	T	S: Hexane A=Hexane R: $I_2A + L = I_2L + A$	73Si	b,q: ±0.05
M10 METHANE, triphenyl-			$C_{19}H_{16}$			$(C_6H_5)_3CH$		
Na^+	0.6	4.41(1)	23	-60	T	S: Tetrahydrofuran	72Cb	
	6.2	5.07(1)	44	25	T	S: Tetrahydrofuran	72Cb	
NaL	8.2	--	28	-40-80	T	S: Tetrahydrofuran R: NaSL = NaL (Solvent separate = Contact ion pair)	72Gi	b
KL	6.7	--	23	-40-80	T	S: Tetrahydrofuran R: KSL = KL (Solvent separated = Contact ion pair)	72Gi	b
	5.3	--	16	-40-80	T	S: 1,2-Dimethoxy-ethane. R: KSL = KL (Solvent separated = Contact ion pair)	72Gi	b
RbL	3.4	--	13	-40-80	T	S: 1,2-Dimethoxy-ethane. R: RbSL = RbL (Solvent separated = Contact ion pair)	72Gi	b
M11 METHANETHIOL, (2-furyl)-			C_5H_6OS			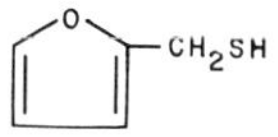		
La^{3+}	-4.37	8.703(35°,1-3)	25.65	25-45	T	S: 50% Ethanol μ=0.1M($NaClO_4$)	73Sj	b
Ce^{3+}	-4.39	9.518(35°,1-3)	29.32	25 45	T	S: 50% Ethanol μ=0.1M($NaClO_4$)	73Sj	b
UO_2^{2+}	-1.56	15.09(25°,1-2)	63.56	15-35	T	S: 50% Ethanol μ=0.1M($NaClO_4$)	73Se	b
M12 METHANOL			CH_4O			CH_3OH		L
VOA_2	-5.77	-0.26(1)	-20.6	25	C	μ=0. S: Nitro-benzene. A=Acetylacetone	65Ca	
SbA_5	-19.08	(1)	--	25	C	S: Ethylene Chloride. A=Cl	68Oa	
I_2	-1.90	0.67(25°,1)	-3.32	7-40	T	C_L=0.06 mole fraction. S: CCl_4	57Dc	b
	-2.1	(1)	--	--	C	S: CH_3OH	61Aa	i: 50Ha
	-1.7	(1)	--	--	C	S: CH_3OH	50Ha	

Metal Ion	ΔH, kcal /mole	Log K	ΔS, cal /mole °K	T, °C	M	Conditions	Ref.	Remarks
M12, cont.								
I_2, cont.	-1.9	(1)	--	--	C	S: CCl_4	60Bc	
M13 METHIONINE			$C_5H_{11}O_2NS$			$CH_3SCH_2CH_2CH(NH_2)CO_2H$		L^-
Mn^{2+}	-1.7	2.87(20°,1)	7	20-60	T	μ=0.1(KNO_3)	73Bf	b
	-0.2	2.05(20°,2)	9	20-60	T	μ=0.1(KNO_3)	73Bf	b
Ni^{2+}	-5.17	5.56(25°,1)	8.1	20-40	T	--	57Pa 60Pc	b
	-4.0	6.84(1)	17.9	25	T	S: 44.6% Dioxane	60Pd	
	-3.0	7.3(1)	23.3	25	T	S: 59.7% Dioxane	60Pd	
	-5.17	4.63(25°,2)	3.9	20-40	T	--	60Pc	b
	-6.0	5.91(2)	6.9	25	T	S: 44.6% Dioxane	60Pd	
	-4.0	6.55(2)	-16.6	25	T	S: 59.7% Dioxane	60Pd	
	-6.42	2.63(25°,3)	-9.5	20-40	T	--	60Pc	b
	-5.5	3.94(3)	-0.4	25	T	S: 44.6% Dioxane	60Pd	
M14 MONACTIN			$C_{41}H_{66}O_2$					
Na^+	-5.35	(1)	--	25	C	S: Methanol	73Za 71Fe	i:71Fe
	-6.00	2.6(1)	-8.1	25	C	S: Methanol H. Diebler, M. Eigen, G. Ilgenfritz, G. Maass & R. Winkler, Pure Appl. Chem. 20, 93 (1969)	73Za	
M15 MONENSIN			$C_{36}H_{62}O_{11}$					
Na^+	-3.9	6.0(1)	15	25	C	S: Methanol	71Li	
K^+	-3.87	4.477(1)	7.4	25	C	S: Methanol	71Fe	
	-3.7	4.6(1)	8.4	25	C	S: Methanol	71Li	
M16 MORPHOLINE			C_4H_9ON			$OCH_2CH_2NHCH_2CH_2$ (ring)		L
VOA_2	-9.34	2.48(1)	-20.0	25	C	μ=0. S: Nitro- benzene. A=Acetylacetone	65Ca	
NiA_2	-7.7	1.863(24.6°,1)	-17.5	8-39.5	T	S: Benzene. A=O,O'- Diethyldithiophosphate	72Mc	b
	-8.7	2.301(24.6°,1)	-18.8	8-39.5	T	S: Benzene. A=O,O'- Dimethyldithiophosphate	72Mc	b
M17 MUREXIDE			$C_8H_8O_6N_6$			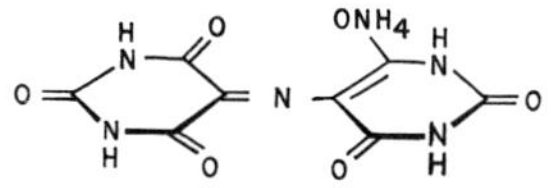		L^{3-}
Ni^{2+}	-1.6	3.218(20°)	9.4	12-32	T	μ=0.1M($NaClO_4$) R: Ni^{2+} + H_2L^- = NiH_2L^+	71Le	b,q: ±0.3
	-5.6	4.079(20°)	-0.3	12-32	T	S: 50 v % DMSO μ=0.1M($NaClO_4$)	71Le	b,q: ±0.5

Metal Ion	ΔH, kcal /mole	Log K	ΔS, cal /mole °K	T, °C	M	Conditions	Ref.	Remarks
M17, cont.								
Ni^{2+}, cont.						R: $Ni^{2+} + H_2L^- = NiH_2L^+$		
M18 MYOSIN B								
K^+	0	2.903(27°)	13.3	5&27	T	C_L=1.5-2.5% pH=7.7 Reaction for binding of K^+ to first set of sites (imidazole groups)	57La	a,b,q: ±3.1
	-10.3	1.699(27°)	-26.8	5&27	T	C_L=1.5-2.5% pH=7.7 Reaction for binding of K^+ to second set of sites (amino groups)	57La	a,b,q: ±2.5

N

N1 1-NAPHTHALDEHYDE, 2-hydroxy-6-nitro- $C_{11}H_7O_3N$

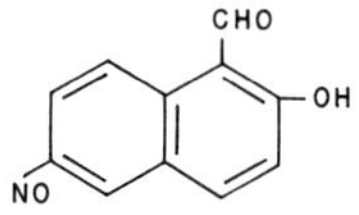

Metal Ion	ΔH, kcal /mole	Log K	ΔS, cal /mole °K	T, °C	M	Conditions	Ref.	Remarks
Cu^{2+}	-5.47	6.9556(25°,1)	13.28	15-50	T	--	68Rb	b
	-4.90	5.6069(25°,2)	9.10	15-50	T	--	68Rb	b

N2 NAPHTHALENE $C_{10}H_8$ L

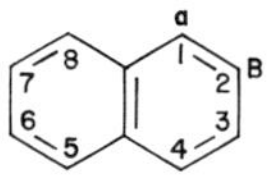

Metal Ion	ΔH, kcal /mole	Log K	ΔS, cal /mole °K	T, °C	M	Conditions	Ref.	Remarks
Li^+	0.5	-5.027(1)	26	-65	T	S: Tetrahydrofuran	68Nb	q: ±0.4
	2.8	-5.420(1)	35	15	T	S: Tetrahydrofuran	68Nb	q: ±0.4
Na^+	-4.7	(1)		-65-30	T	S: 2-Methyl-tetrahydrofuran	61Ab	b
	1.8	6.647(1)	30	-65	T	μ=0. S: Tetra-hydrofuran	66Cb	b
	1.8	(1)	30	-65	T	S: Tetrahydrofuran	68Nb	q: ±0.4
	8.2	6.66(1)	58	15	T	μ=0. S: Tetra-hydrofuran	66Cb	b
	8.2	(1)	58	15	T	S: Tetrahydrofuran	68Nb	
	9.2	0.208(-35°,1)	62	-70--35	T	S: Tetrahydrofuran	68He	b
	7.5	(1)	--	-65-30	T	S: Tetrahydrofuran	61Ab	q: ±1.5
	2.6	(1)	--	-65-30	T	S: Tetrahydropyran	61Ab	
NaL	-4.8	--	-20	-58-12	T	S: Tetrahydrofuran $C_L=10^{-5}-10^{-4}M$ R: Tight ion pair = Loose ion pair	67He	b,q: ±0.5
	-6.4	--	-33	-101--51	T	S: 50% Tetrahydro-furan. 50% Diethyl	67He	b,q: ±0.8

Metal Ion	ΔH, kcal /mole	Log K	ΔS, cal /mole °K	T, °C	M	Conditions	Ref.	Re-marks
N2, cont.								
NaL, cont.						ether. $C_L=10^{-5}-10^{-4}M$ R: Tight ion pair = Loose ion pair		
VAB_3	-0.170	(1)	--	10-26	T	S: Cyclohexane A=O B=Cl	71Ha	
WA_6	-0.170	(1)	--	-24-26	T	S: CCl_4 A=Cl	71Hd	
OsA_4	-0.400	(1)	--	0-31.5	T	S: CCl_4 A=O	71Hc	
PA_5	-0.070	(1)	--	-30-31	T	S: CCl_4 A=Cl	71Hb	
	-0.110	(1)	--	-30-31	T	S: Dichloromethane A=Cl	71Hb	
I_2	-1.65	(1)	--	1&45	T	C_L=0.07-1.76 S: CCl_4	60Dc	a,b
	-1.64	(1)	--	22&45	T	C_L=0.07-1.76 S: $CHCl_3$	60Dc	a,b
	-1.56	(1)	--	22&45	T	C_L=0.07-1.76 S: Cyclohexane	60Dc	a,b
	-1.63	(1)	--	22&45	T	C_L=0.07-1.76 S: n-Hexane	60Dc	a,b
	-1.80	(1)	--	--	T	S: n-Hexane	60Dc	
	-1.80	0.364(1)	-4.3	25	T	S: n-Hexane	54Ka	1
	-1.38	(1)	--	22&45	T	S: n-Heptane	60Dc	a,b
	-1.8	(1)	--	--	-	S: n-Heptane	60Ta	
N3 NAPHTHALENE, 2-methoxy-			$C_{11}H_{10}O$			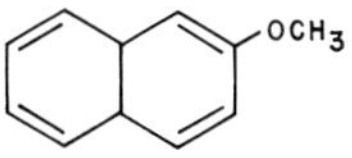		
Al_2A_6	-15.4	--	--	25	C	S: Benzene C_M=0.07M A=Br R: $0.5(Al_2A_6) + L = AlA_3L$	68Rc	
N4 NAPHTHALENE, 1-methyl-			$C_{11}H_{10}$			$C_{10}H_7(CH_3)$		L
I_2	-2.10	0.444(1)	-5.0	25	T	S: n-Hexane	54Ka	1
N5 NAPHTHALENE, 2-methyl-			$C_{11}H_{10}$			$C_{10}H_7(CH_3)$		L
I_2	-2.10	0.569(1)	-4.4	25	T	S: n-Hexane	54Ka	1
N6 NAPHTHALENE, octafluoro-			$C_{10}F_8$			$C_{10}H_7(F_8)$		F
SbA_5	0.110	(1)	--	-30-31	T	S: CCl_4 A=Cl	71Hb	
N7 2,7-NAPHTHALENEDISULFONIC ACID, 4,5-dihydroxy-			$C_{10}H_8O_8S_2$			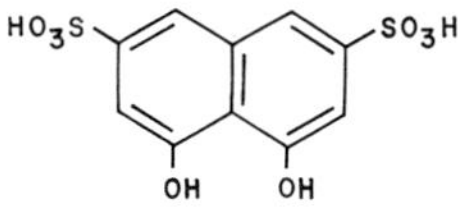		L^{2-}
ThA	10.7	--	--	25	C	μ=0.10(KNO_3) A=EDTA	70Kb	g:Na

Metal Ion	ΔH, kcal /mole	Log K	ΔS, cal /mole °K	T, °C	M	Conditions	Ref.	Remarks
N7, cont.								
ThA, cont.						R: ThA + H_2L = $ThAL^{2-}$ + $2H^+$		
	11.4	--	--	25	C	μ=0.10(KNO_3) A=1,2-Diaminocyclohexane-N,N,N',N'-tetraacetic acid R: ThA + H_2L = $ThAL^{2-}$ + $2H^+$	70Kb	g:Na

N8 2,7-NAPHTHALENEDISULFONIC ACID, $C_{16}H_{10}O_9N_2S_2$
4,5-dihydroxy-3-(phenyl)azo-

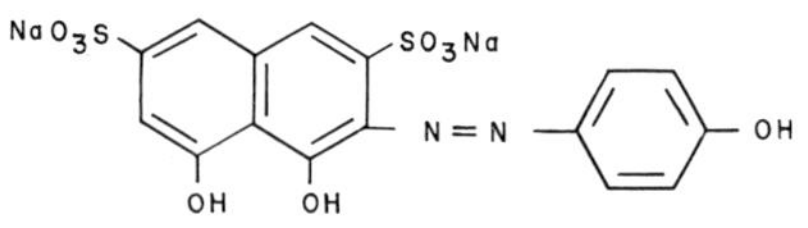

Metal Ion	ΔH, kcal /mole	Log K	ΔS, cal /mole °K	T, °C	M	Conditions	Ref.	Remarks
Cu^{2+}	-17.92	4.43(25°,1)	-39.82	15-30	T	μ=0.1M($NaClO_4$)	73Df	b

N9 1-NAPHTHOIC ACID, 2-hydroxy- $C_{11}H_8O_3$

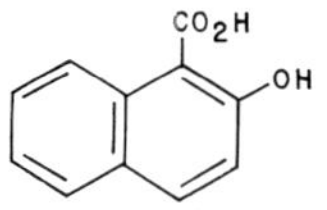

Metal Ion	ΔH, kcal /mole	Log K	ΔS, cal /mole °K	T, °C	M	Conditions	Ref.	Remarks
UO_2^{2+}	-2.58	3.52(30°,1)	7.59	20-50	T	μ=0.1	72Pb	b
Fe^{3+}	-7.26	7.83(30°,1-2)	11.90	20-50	T	μ=0.1	72Pb	b

N10 2-NAPHTHOIC ACID, 1-hydroxy- $C_{11}H_8O_3$

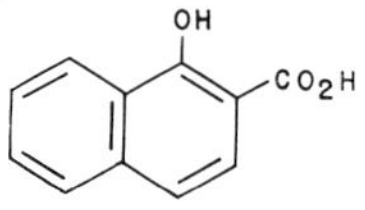

Metal Ion	ΔH, kcal /mole	Log K	ΔS, cal /mole °K	T, °C	M	Conditions	Ref.	Remarks
UO_2^{2+}	-1.41	3.42(30°,1)	11.01	20-60	T	μ=0.1 pH=4.6	71Wb	b
Fe^{3+}	-5.642	4.20(25°,1)	0.28	20-40	T	μ=0.1M S: 50% Ethanol	72Pa	b

N11 2-NAPHTHOIC ACID, 3-hydroxy- $C_{11}H_8O_3$ L^{2-}

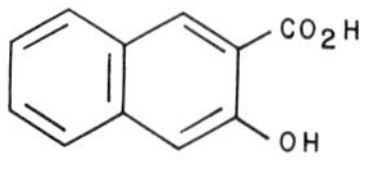

Metal Ion	ΔH, kcal /mole	Log K	ΔS, cal /mole °K	T, °C	M	Conditions	Ref.	Remarks
Fe^{3+}	4.09	4.983(30°,1)	35.5	10-60	T	μ=0.02 pH=2.6	69Gg	b
Al^{3+}	9.45	4.772(30°)	52.2	20-60	T	μ=0.02($NaClO_4$) R: Al^{3+} + HL^- = AlL^+ + H^+	67Gd	b

N12 NIGERICIN $C_{40}H_{67}O_{11}$ L^-

Metal Ion	ΔH, kcal /mole	Log K	ΔS, cal /mole °K	T, °C	M	Conditions	Ref.	Remarks
Na^+	1.6	3.9(1)	23	25	C	S: Methanol	71Li	q: ±0.7
K^+	-0.98	5.6(1)	22	25	C	S: Methanol	71Li	q: ±0.3

Metal Ion	ΔH, kcal /mole	Log K	ΔS, cal /mole °K	T, °C	M	Conditions	Ref.	Remarks
N13 NITRATE ION			NO_3			NO_3^-		L^-
Sr^{2+}	-0.03	0.59(1)	2.6	25	C	μ=0.5($LiNO_3$)	64Va	c:Sr $(NO_3)_2$
	-2.42	0.52(1)	-5.7	25	C	μ=0.5M(KNO_3)	64Va	c:Sr $(NO_3)_2$
Ba^{2+}	-1.90	0.94(1)	-2.1	25	C	μ=0.5($LiNO_3$)	64Va	c:Ba $(NO_3)_2$
	-3.22	0.93(1)	-6.5	25	C	μ=0.5M(KNO_3)	64Va	c:Ba $(NO_3)_2$
	-10.20	2.4(1-2)	-23.5	25	T	μ=0	34La	
Eu^{3+}	-0.57	0.30(1)	-0.54	25	T	μ=1.0	65Cf	
Cr^{3+}	4.5	-2.01(1)	5.9	25	T	μ=1.0. ΔH changes with NO_3^- conc., K decreases with increasing NO_3^- conc.	67Ac	c
UO_2^{++}	-4.3	1.5	-7.7	25	T	R: UO_2^{++} + $2L^-$ + 2TBP = $UO_2(NO_3)_2(TBP)_2$ where TBP = tributyl phosphate	60Na	
Co^{2+}	-1.7	-0.757(25°,1)	-9.3	15-35	T	μ=1.0(HNO_3, $HClO_4$)	69Me	b,q: ±1.2
Cu^{2+}	0.85	-0.824(1)	-1	25	C	3.0N($LiClO_4$ + LiCl) Values are for formation of inner sphere complexes	70Bf	
	0.48	-0.398(1)	0	25	C	3.0N($LiClO_4$ + LiCl) Values are for formation of outer sphere complexes	70Bf	
Cd^{2+}	-5.24	0.31(1)	-16.2	25	C	1M(KNO_3)	62Va 67Va	
	-5.7	0.22(25°,1)	-18	18&25	T	--	62Va	a,b
Tl^+	-0.65	0.33(1)	-1.0	25	T	μ=0	57Na	
	-9.970	1.31(1)	-27.4	25	T	μ=0	34La	
	-6.2	-0.48(1)	-23	25	C	3M($LiClO_4$ or $LiNO_3$)	65Ke	
Tl^{3+}	0	-0.90(1)	4.1	25	C	2.0M($LiClO_4$) 1.0M($HClO_4$)	67Md	
Pb^{2+}	-0.57	1.19(1)	3.5	25	T	μ=0	55Na	
	-1.31	0.518(1)	-2.0	25	C	μ=3M($LiClO_4$ & $LiNO_3$)	72Fg	
	-1.4	0.505(25°,1)	-2.4	2-65	T	μ=3.0M($LiClO_4$ & $LiNO_3$)	67Fa	b
	-1.4	0.506(25°,1)	-2.4	18-65	T	μ=3M($LiClO_4$)	67Fb 68Fa	b
	-0.93	0.58(25°,1)	-0.6	5-65	T	μ=3M Li[ClO_4, NO_3) S: 3M Ethanol D=70.1	71Gj	b,q: ±0.30
	-1.66	0.83(25°,1)	-1.8	5-65	T	μ=3M Li(ClO_4, NO_3) S: 6M Ethanol D=61.0	71Gj	b,q: ±0.27

Metal Ion	ΔH, kcal /mole	Log K	ΔS, cal /mole °K	T, °C	M	Conditions	Ref.	Remarks
N13, cont.								
Pb^{2+}, cont.	-1.50	0.98(25°,1)	-0.6	5-65	T	μ=3M Li(ClO_4, NO_3) S: 9M Ethanol D=51.2	71Gj	b,q: ±0.30
	-1.68	1.23(25°,1)	0.0	5-65	T	μ=3M Li(ClO_4, NO_3) S: 12M Ethanol D=41.5	71Gj	b,q: ±0.47
	-0.65	0.72(25°,1)	1.1	5-65	T	μ=3M Li(ClO_4, NO_3) S: 6M Methanol D=69.0	71Gj	b,q: ±0.28
	-1.31	0.83(25°,1)	-0.6	5-65	T	μ=3M Li(ClO_4, NO_3) S: 12M Methanol D=58.3	71Gj	b,q: ±0.22
	-2.13	1.15(25°,1)	-1.9	5-65	T	μ=3M Li(ClO_4, NO_3) S: 18M Methanol D=45.9	71Gj	b,q: ±0.20
	-2.93	1.49(25°,1)	-3.0	5-65	T	μ=3M Li(ClO_4, NO_3) S: 24.72M Methanol D=31.5	71Gj	b,q: ±0.47
	-0.29	-0.070(2)	-1.3	25	C	μ=3M($LiClO_4$ & $LiNO_3$)	72Fg	
	-1.3	-0.18(25°,2)	-5.2	2-65	T	μ=3.0M($LiClO_4$ & $LiNO_3$)	67Fa	b
	-1.3	-0.183(25°,2)	-5.2	18-65	T	μ=3M($LiClO_4$)	68Fa 67Fb	b
	-1.77	0.25(25°,2)	-4.5	5-65	T	μ=5M Li(ClO_4, NO_3) S: 3M Ethanol D=70.1	71Gj	b,q: ±0.5
	-0.04	0.21(25°,2)	1.0	5-65	T	μ=5M Li(ClO_4, NO_3) S: 6M Ethanol D=61.0	71Gj	b,q: ±0.6
	-0.2	0.28(25°,2)	0.5	5-65	T	μ=5M Li(ClO_4, NO_3) S: 9M Ethanol D=51.2	71Gj	b,q: ±0.7
	0.58	0.63(25°,2)	4.9	5-65	T	μ=5M Li(ClO_4, NO_3) S: 12M Ethanol D=41.5	71Gj	b,q: ±0.3
	-3.55	0.11(25°,2)	-11.1	5-65	T	μ=3M Li(ClO_4, NO_3) S: 6M Methanol D=69.0	71Gj	b,q: ±0.9
	-1.89	0.32(25°,2)	-4.8	5-65	T	μ=3M Li(ClO_4, NO_3) S: 12M Methanol D=58.3	71Gj	b,q: ±0.4
	0.23	0.59(25°,2)	3.4	5-65	T	μ=3M Li(ClO_4, NO_3) S: 18M Methanol D=45.9	71Gj	b,q: ±0.4
	-1.87	0.88(25°,2)	-2.3	5-65	T	μ=3M Li(ClO_4, NO_3) S: 24.72 Methanol D=31.5	71Gj	b,q: ±1.2
	1.3	0.022(25°,3)	3.5	2-65	T	μ=3.0M($LiClO_4$ & $LiNO_3$)	67Fa	b
	0.8	-0.194(25°,3)	1.7	2-65	T	μ=3M($LiClO_4$ & $LiNO_3$)	72Fg	b,q: ±0.5
	1.3	0.022(25°,3)	4.5	18-65	T	μ=3M($LiClO_4$)	68Fa 67Fb	b

Metal Ion	ΔH, kcal /mole	Log K	ΔS, cal /mole °K	T, °C	M	Conditions	Ref.	Re-marks
N13, cont.								
Pb^{2+}, cont.	-6.5	-0.553(25°,4)	-24	2-65	T	μ=3M($LiClO_4$ & $LiNO_3$)	72Fg	b,q: ±1.0
Bi^{3+}	-0.99	0.733(1)	0	25	T	μ=3	71Fb	
N14 NITRITE ION			NO_2			NO_2^-		L^-
[Fe^{3+}]	-10	2.4(1)	-22	25	-	Hemoglobin	60Gb	
NiA_2L_2	-2.3	--	-7.2	3-51	T	S: Chloroform C_M=0.015 A=Diethylethylene-diamine R: $NiA_2(ONO)_2$ = $NiA_2(NO_2)_2$ (Nitro = Nitrito equilibrium)	66Ge	b,q: ±0.6
Cd^{2+}	-2.09	1.76(1)	1.1	25	C	3.0M($NaClO_4$)	66Gb	
	-2.11	1.32(2)	-1.0	25	C	3.0M($NaClO_4$)	66Gb	
	-1.58	0.62(3)	-2.4	25	C	3.0M($NaClO_4$)	66Gb	
N15 NITROGEN			N_2			N_2		L
RuA_5	-22	(2-1)	--	25	C	0.1M(H_2SO_4) C_M=0.03-0.1 A=NH_3	70Fa	q:±2
RuA_5B	-9.4	4.690	-10	20-35	T	μ=0.10(HCl) A=NH_3 B=H_2O R: RuA_5B^{2+} + L = RuA_5L^{2+} + B K's were measured as a function of temperature by Kinetic method	70Ab	b,q: ±0.9
	-10.1	4.518	-13	25-45	T	μ=0.10(HCl) A=NH_3 B=H_2O R: RuA_5B^{2+} + L = RuA_5L^{2+} + B K's were measured as a function of temperature by Static method	70Ab	b,q: ±1.4
	-11.2	3.863	-20	25-45	T	μ=0.10(HCl) A=NH_3 B=H_2O R: RuA_5B^{2+} + RuA_5L^{2+} = $[RuA_5LRuA_5]^{4+}$ + B	70Ab	b,q: ±1.4
N16 NITROGEN MONOXIDE			NO			NO		L
CoA_5B	-8.8	4.0	-11	25	-	R: CoA_5B^{2+} + L = CoA_5L^{2+} + B A=NH_3 B=H_2O	60Gb	
N17 NITROXIDE, di-tert-butyl			$C_8H_{18}ON$			$ON[C(CH_3)_3]_2$		
SiA_4	-3.1	0.90(1)	-10	-53	T	S: $SiCl_4$. A=Cl	74Ce	q: ±0.3

Metal Ion	ΔH, kcal /mole	Log K	ΔS, cal /mole °K	T, °C	M	Conditions	Ref.	Remarks
N17, cont.								
GeA_4	-3.0	1.70(1)	-6.1	-53	T	S: $GeCl_4$. A=Cl	74Ce	
$SnAB_2C_5 \cdot D$	-2.2	--	--	-53	T	S: Toluene A=Cr. B=t-Butyl. C=CO D=Tetrahydrofuran R: $SnAB_2C_5 \cdot D$ + L = $SnAB_2C_5 \cdot L$ + D	74Ce	q: ±0.2
N18 NONACTIN			$C_{40}H_{64}O_2$					
Na^+	-6.55	3.27(1)	7.0	25	C	S: Ethanol	73Az	
	-2.65	2.71(1)	3.5	25	C	S: Methanol	73Za	i: 71Fe
	-3.39	(1)	--	25	C	S: Methanol	71Fe	
K^+	-12.48	5.26(1)	-17.8	25	C	S: Ethanol	73Za	
	-11.0	4.301(1)	-17	25	C	S: Methanol	71Fe	
	-10.42	4.49(1)	-14.4	25	C	S: Methanol	73Za	i: 71Fe
N19 NONANE, 1-amino-			$C_9H_{21}N$			$[CH_3(CH_2)_7CH_2NH_2]$		L
NiA	-16.9	6.10(1-2)	-28.8	25	T	$C_M=10^{-4}$. S: Benzene A=Diacetyl-bisbenzoyl hydrazine	60Sb	

O

Metal Ion	ΔH, kcal /mole	Log K	ΔS, cal /mole °K	T, °C	M	Conditions	Ref.	Remarks
O1 OCTANE, 1-amino			$C_8H_{19}N$			$CH_3(CH_2)_7NH_2$		L
NiA	-16.9	6.06(1-2)	-29.1	25	T	$C_M=10^{-4}$ S: Benzene A=Diacetyl-bisbenzoyl hydrazine	60Sb	
O2 OCTANE, 1,8-diamino-3,6-dithio-			$C_6H_{16}N_2S_2$			$H_2N(CH_2CH_2S)_2CH_2CH_2NH_2$		L
Ni^{2+}	-12	7.90(30°,1)	-3	0-50	T	μ=1(KCl or KNO_3)	54Gb	b
Cu^{2+}	-15.5	10.43(30°,1)	3	10-40	T	μ=0	59Ma	b
O3 OCTANE, 1,8-diamino-3-oxa-6-thio-			$C_6H_{16}ON_2S$			$H_2NCH_2CH_2OCH_2CH_2SCH_2CH_2NH_2$		L
Ni^{2+}	-8.2	6.17(30°,1)	1	10-40	T	μ=0	59Lb	b
Cu^{2+}	-13.1	8.86(30°,1)	-3	10-40	T	μ=0	59Lb	b
Ag^+	-14.4	8.12(30°,1)	-10	10-40	T	μ=0	59Lb	b
O4 OCTANE, 1,8-diamino-N,N,N',N'-tetraacetic acid			$C_{16}H_{28}O_8N_2$			$(CH_2CO_2H)_2N(CH_2)_8N-(CH_2CO_2H)_2$		L^{4-}
Mn^{2+}	0.5	9.01(1)	42.8	20	C	μ=0.1(KNO_3)	64Ab	
Co^{2+}	-4.76	12.91(1)	42.6	20	C	μ=0.1(KNO_3)	64Ab	
Ni^{2+}	-8.5	13.62(1)	33.2	20	C	μ=0.1(KNO_3)	64Ab	

Metal Ion	ΔH, kcal /mole	Log K	ΔS, cal /mole °K	T, °C	M	Conditions	Ref.	Remarks
04, cont.								
Cu^{2+}	-10.28	15.8(1)	37.2	20	C	μ=0.1(KNO_3)	64Ab	
Zn^{2+}	-4.36	12.66(1)	43.05	20	C	μ=0.1(KNO_3)	64Ab	
Cd^{2+}	-4.59	11.99(1)	39.2	20	C	μ=0.1(KNO_3)	64Ab	
Hg^{2+}	-19.74	21.83(1)	32.5	20	C	μ=0.1(KNO_3)	64Ab	
05 OCTANE, 3,6-dioxa-1,8-diamino-N,N',N',N'-tetraacetic acid			$C_{14}H_{24}O_{10}N_2$			$(CH_2CO_2H)_2N(CH_2CH_2O)_2$-$CH_2CH_2N(CH_2CO_2H)_2$		L^{4-}
Mg^{2+}	5.18	5.2(1)	41.5	20	C	μ=0.1(KNO_3)	64Ab	
	4.4	5.4(1)	40	25	C	μ=0.1(KNO_3)	65Wd	
	5.49	5.28(1)	42.6	25	C	μ=0.1(KCl or KNO_3)	65Bc	
Ca^{2+}	-8.38	10.97(1)	21.6	20	C	μ=0.1(KNO_3)	65Ab	
	-8.0	10.8(1)	23	25	C	μ=0.1(KNO_3)	65Wd	
	-7.94	10.89(1)	23.2	25	C	μ=0.1(KCl or KNO_3)	65Bc	
Sr^{2+}	-6.4	8.06(1)	15	25	C	μ=0.1(KNO_3)	65Wd	
	-5.74	8.43(1)	19.3	25	C	μ=0.1(KCl or KNO_3)	65Bc	
Ba^{2+}	-8.8	7.99(1)	7.1	25	C	μ=0.1(KNO_3)	65Wd	
	-8.99	8.30(1)	7.8	25	C	μ=0.1(KCl or KNO_3)	65Bc	
La^{3+}	-5.46	15.79(1)	53.6	20	C	μ=0.1(KNO_3)	64Ab	
UO_2^{2+}	2.4	9.49(25°)	51	20-40	T	μ=0.1M(KNO_3) R: UO_2 + HL = UO_2HL	69Sg	b
Mn^{2+}	-8.16	12.28(1)	21.5	20	C	μ=0.1(KNO_3)	64Ab	
	-8.8	12.2(1)	27	25	C	μ=0.1(KNO_3)	65Wd	
Fe^{2+}	-5.2	11.8(1)	37	25	C	μ=0.1(KNO_3)	65Wd	
Co^{2+}	-2.83	12.28(1)	46.5	20	C	μ=0.1(KNO_3)	64Ab	
	-3.4	12.2(1)	45	25	C	μ=0.1(KNO_3)	65Wd	
Ni^{2+}	-3.83	11.82(1)	41.0	20	C	μ=0.1(KNO_3)	64Ab	
	-5.0	13.6(1)	45	25	C	μ=0.1(KNO_3)	65Wd	
Pd^{2+}	-13.2	11.8(1)	9.1	20	C	μ=0.1(KNO_3)	64Ab	
Cu^{2+}	-11.0	17,71(1)	43.5	20	C	μ=0.1(KNO_3)	64Ab	
	-10.5	17.7(1)	46	25	C	μ=0.1(KNO_3)	65Wd	
Zn^{2+}	-4.28	12.91(1)	44.4	20	C	μ=0.1(KNO_3)	64Ab	
	-3.8	14.4(1)	53	25	C	μ=0.1(KNO_3)	65Wd	
	-5.02	12.86(1)	42.1	25	C	μ=0.1(KCl or KNO_3)	65Bc	
Cd^{2+}	-14.8	16.1(1)	23.2	20	C	μ=0.1(KNO_3)	64Ab	
	-14.1	16.6(1)	29	25	C	μ=0.1(KNO_3)	65Wd	
	-14.89	16.53(1)	25.7	25	C	μ=0.1(KCl or KNO_3)	65Bc	
Hg^{2+}	-23.7	23.2(1)	25.2	20	C	μ=0.1(KNO_3)	64Ab	
	-23.3	23.7(1)	31	25	C	μ=0.1(KNO_3)	65Wd	
Pb^{2+}	-12.5	14.6(1)	25	25	C	μ=0.1(KNO_3)	65Wd	
06 OCTANEDIOIC ACID, 3,6-oxy-			$C_6H_{10}O_6$			$HO_2CCH_2OCH_2CH_2OCH_2CO_2H$		L^{2-}
La^{3+}	1.61	4.354(1)	25.3	25	C	μ=1.00($NaClO_4$)	74Gc	
	-2.58	3.360(2)	11.3	25	C	μ=1.00($NaClO_4$)	74Gc	

Metal Ion	ΔH, kcal /mole	Log K	ΔS, cal /mole °K	T, °C	M	Conditions	Ref.	Re-marks
06, cont.								
Ce^{3+}	1.32	4.644(1)	25.6	25	C	μ=1.00($NaClO_4$)	74Gc	
	-2.10	3.243(2)	12.4	25	C	μ=1.00($NaClO_4$)	74Gc	
Pr^{3+}	1.03	4.806(1)	25.3	25	C	μ=1.00($NaClO_4$)	74Gc	
	-1.54	3.085(2)	9.1	25	C	μ=1.00($NaClO_4$)	74Gc	
Nd^{3+}	0.79	4.920(1)	25.1	25	C	μ=1.00($NaClO_4$)	74Gc	
	-0.65	3.035(2)	11.7	25	C	μ=1.00($NaClO_4$)	74Gc	
Sm^{3+}	0.38	5.074(1)	24.4	25	C	μ=1.00($NaClO_4$)	74Gc	
	1.52	2.909(2)	18.1	25	C	μ=1.00($NaClO_4$)	74Gc	
Gd^{3+}	0.85	4.894(1)	25.3	25	C	μ=1.00($NaClO_4$)	74Gc	
	2.07	2.978(2)	20.6	25	C	μ=1.00($NaClO_4$)	74Gc	
Tb^{3+}	1.74	4.753(1)	27.5	25	C	μ=1.00($NaClO_4$)	74Gc	
	1.62	3.145(2)	19.8	25	C	μ=1.00($NaClO_4$)	74Gc	
Dy^{3+}	2.59	4.668(1)	30.1	25	C	μ=1.00($NaClO_4$)	74Gc	
	1.66	3.258(2)	20.3	25	C	μ=1.00($NaClO_4$)	74Gc	
Ho^{3+}	3.10	(1)	31.5	25	C	μ=1.00($NaClO_4$)	74Gc	
	1.54	(2)	20.1	25	C	μ=1.00($NaClO_4$)	74Gc	
Er^{3+}	3.73	4.606(1)	33.7	25	C	μ=1.00($NaClO_4$)	74Gc	
	3.21	3.432(2)	26.3	25	C	μ=1.00($NaClO_4$)	74Gc	
Tm^{3+}	4.00	4.637(1)	34.6	25	C	μ=1.00($NaClO_4$)	74Gc	
	3.60	3.597(2)	28.5	25	C	μ=1.00($NaClO_4$)	74Gc	
Yb^{3+}	3.81	4.847(1)	34.9	25	C	μ=1.00($NaClO_4$)	74Gc	
	3.72	3.829(2)	30.1	25	C	μ=1.00($NaClO_4$)	74Gc	
Lu^{3+}	3.64	(1)	34.2	25	C	μ=1.00($NaClO_4$)	74Gc	
	3.39	(2)	29.8	25	C	μ=1.00($NaClO_4$)	74Gc	
07 OCTANEDIOIC ACID, 3,6-thio-			$C_6H_{10}O_4S_2$			$HO_2CCH_2SCH_2CH_2SCH_2CO_2H$		L^{2-}
Sm^{3+}	2.94	2.320(1)	20.5	25	C	μ=1.00($NaClO_4$)	74Gc	
	2.13	1.171(2)	12.5	25	C	μ=1.00($NaClO_4$)	74Gc	
Dy^{3+}	4.88	2.086(1)	25.8	25	C	μ=1.00($NaClO_4$)	74Gc	
	0.93	1.391(2)	9.6	25	C	μ=1.00($NaClO_4$)	74Gc	
Er^{3+}	4.68	2.009(1)	24.8	25	C	μ=1.00($NaClO_4$)	74Gc	
	3.21	1.137(2)	16.1	25	C	μ=1.00($NaClO_4$)	74Gc	
08 7-OCTENOIC ACID, 3-seleno-			$C_7H_{12}O_2Se$			CH_2:CH(CH_2)$_3$SeCH_2CO_2H		L^-
Ag^+	-9.13	4.62(25°,1)	-9.62	0-40	T	μ=0.20M pH=5.8	71Bd 71Bc	b,q: ±0.5
	-7.89	3.05(25°,2)	-12.82	0-40	T	μ=0.20M pH=5.8	71Bd 71Bc	b,q: ±1.7
	-3.58	1.80(25°)	-4.01	0-40	T	μ=0.20M R: AgL + Ag^+ = Ag_2L^+ pH=5.8	71Bd	b,q: ±1.2
09 OXALIC ACID			$C_2H_2O_4$			HO_2CCO_2H		L^{2-}
Ca^{2+}	-5.090	8.642	-22.5	25	T	μ=0. R: Ca^{2+} + L^{2-} + H_2O = CaL·H_2O	34La	

Metal Ion	ΔH, kcal /mole	Log K	ΔS, cal /mole °K	T, °C	M	Conditions	Ref.	Re-marks
09, cont.								
Er^{3+}	-7.13	8.42(30°,1-3)	15.1	18&30	T	μ=0.10 Signs on ΔG and ΔH are reverse those reported in article	69Ka	a,b,q: ±0.66
CrA_5B	-1.3	0.672(40°,1)	-0.9	40-60	T	μ=1.0M($NaClO_4$) A=NH_3 B=H_2O	73Na	b,q: ±1.0
	-4.3	0.146(40°)	-13.0	40-60	T	μ=1.0M($NaClO_4$) A=NH_3 B=H_2O R: CrA_5B^{3+} + HL^- = $CrA_5B^{3+}HL^-$	73Na	b,q: ±0.4
NpO_2^+	0.0	0.290(25°,1)	1.3	10-47	T	μ=0.5	53Ga	b
Pu^{3+}	-1.300	7.936(20°,1-2)	31.9	20&70	T	μ=1	57Ga 58Gb	b
	-1.200	8.252(20°,1-3)	33.7	20&70	T	μ=1	57Ga 58Gb	b
	-1.300	8.602(20°,1-4)	34.9	20&70	T	μ=1	57Ga 58Gb	b
Mn^{2+}	1.42 0.20	3.96(25°,1) 5.4(1-2)	22.9 23.8	0-45 25	T C	μ=0. ΔCp=65 C_M=1-2.2 C_L=0.25-0.8	61Mb 54Yb	b
Fe^{3+}	-0.3	9.68(1)	43	--	-	--	52Ua	r
Co^{2+}	0.59 0.80	4.79(25°,1) 6.7(1-2)	23.9 28.2	0-45 25	T C	μ=0. ΔCp=20 C_M=1-2.2 C_L=0.25-0.80	61Mb 54Yb	b
Ni^{2+}	0.15 1.23	5.17(25°,1) 7.2(1-2)	24.2 28.8	0-45 25	T C	μ=0. ΔCp=35 C_M=1-2.2 C_L=0.25-0.80	61Mb 54Yb	b
Cu^{2+}	-0.05	6.22(1)	28.3	25	C	μ=0	66Mc 67Mf	
	1.59	7.48(1-2)	29	25	C	C_M=1-2.2 C_L=0.25-0.80	54Yb	
Zn^{2+}	1.55	7.33(1-2)	28.4	25	C	C_M=1-2.2 C_L=0.25-0.80	54Yb	
SbO^+	59.54	12.30(27°)	142	25-30	T	R: SbO^+ + $2L^=$ + H_2O = SbL^-_2 + $2OH^-$	53Na	b
010 OXYGEN (aq)			O_2					
[Fe]	-6.4	--	--	20	C	[Fe]=Ox Haemoglobin pH=6.8 Buffer: 0.07M Phosphate R: 1 mole O_2 (solution) + [Fe] (solution) = O_2[Fe] (solution)	35Ra	q: ±0.70
	-10.5	--	--	20	C	[Fe]=Ox Haemoglobin pH=9.5. R: 1 mole O_2 (solution) + [Fe] (solution) = O_2[Fe] (solution)	35Ra	q: ±2.1

Metal Ion	ΔH, kcal /mole	Log K	ΔS, cal /mole °K	T, °C	M	Conditions	Ref.	Re-marks
010, cont.								
[Fe], cont.	-17.9	(1)	-33.0	0-30	T	0.001M THAM buffer pH=8.5. [Fe]=Sperm whale myoglobin Standard state for L = unit mole fraction in organic solvent	71Ka	b
CoA_2	-30.1	6.63(2-1)	-70	25	C	μ=0.02M A=Histidinate	72Ph	
	-15.03	4.728(20°,1)	-29	15-31.5	T	S: Pyridine L(gas) A=(acetyl acetone) ethylenediimine	70Aa	b
CoAB	-9.3	2.84(-65°,1)	-32	-80- -20	T	S: Toluene A=α,β,γ,δ-Tetra(p-methoxyphenyl) porphyrin B=Pyridine. Standard state of 1M O_2	73Wb	b,q: ±1.1
	-8.8	2.95(-65°,1)	-29	-80- -20	T	S: Toluene A=α,β,γ,δ-Tetra(p-methoxyphenyl) porphyrin B=4-Picoline. Standard state of 1M O_2	73Wb	b,q: ±0.9
	-9.2	2.81(-65°,1)	-31	-80- -20	T	S: Toluene A=α,β,γ,δ-Tetra(p-methoxyphenyl) porphyrin B=3,4-Lutidine Standard state of 1M O_2	73Wb	b
	-8.5	3.41(-65°,1)	-25	-80- -20	T	S: Toluene A=α,β,γ,δ-Tetra(p-methoxyphenyl) porphyrin. B=4-Dimethylamino pyridine Standard state of 1M O_2	73Wb	b,q: ±0.8
	-9.5	2.41(-65°,1)	-35	-80- -20	T	S: Toluene A=α,β,γ,δ-Tetra(p-methoxyphenyl) porphyrin B=γ-Collidine. Standard state of 1M O_2	73Wb	b,q: ±1.3
	-8.6	2.10(-65°,1)	-31	-80- -20	T	S: Toluene A=α,β,γ,δ-Tetra(p-methoxyphenyl) porphyrin. B=5-Chloro-N-methylimidazole Standard state of 1M O_2	73Wb	b
	-8.9	3.58(-65°,1)	-26	-80- -20	T	S: Toluene A=α,β,γ,δ-Tetra(p-methoxyphenyl) porphyrin. B=N-Methylimidazole Standard state of 1M O_2	73Wb	b,q: ±0.5

Metal Ion	ΔH, kcal /mole	Log K	ΔS, cal /mole °K	T, °C	M	Conditions	Ref.	Remarks
010, cont.								
CoAB, cont.	-8.2	2.87(-65°,1)	-26	-80- -20	T	S: Toluene A=α,β,γ,δ-Tetra(p-methoxyphenyl) porphyrin B=Piperidine. Standard state of 1M O_2	73Wb	b
	-8.5	-2.18(-65,1)	-49.5	-80- -20	T	S: Carbon Disulfide A=α,β,γ,δ-Tetra(p-methoxyphenyl) porphyrin B=Pyridine. Standard state of 1M O_2	73Wb	b,q: ±0.5
	-9.0	-2.10(-65,1)	-51.8	-80- -20	T	S: Carbon Disulfide A=α,β,γ,δ-Tetra(p-methoxyphenyl) porphyrin. B=3,4-Lutidine. Standard state of 1M O_2	73Wb	b,q: ±0.6
$CoAB_2$	-7.5	2.863(-65,1)	-23	-80- -20	T	S: Toluene A=α,β,γ,δ-Tetra(p-methoxyphenyl) porphyrin B=3,4-Lutidine	73Wb	b,q: ±1.0
	-6.7	2.322(-65°,1)	-21	-80- -20	T	S: Toluene A=α,β,γ,δ-Tetra(p-methoxyphenyl) porphyrin B=Piperidine	73Wb	b,q: ±1.0
CoA_2B_2	-29.4	10.84	-49	25	C	μ=1.0M(KCl) A=Ethylenediamine B=H_2O. R: $2CoA_2B_2^+$ + L = Co_2A_4LOH + H_3O^+	72Ph	
	-29.4	7.33	-65	25	-	μ=1.0M(KCl) A=Ethylenediamine B=H_2O. R: $2CoA_2B_2^+$ + L - $Co_2A_4LB_2$	72Ph	
[Co]A	-11.0	-2.27(-45°,1)	-59	-63.5- -31	T	S: Toluene [Co]=Cobalt(II) protoporphyrin IX dimethyl ester A=N,N-Dimethyl-formamide. L(gas)	73Sr	b,cc
	-9.2	-2.25(-57.5°,1)	-53	-63.5- -45	T	S: Toluene [Co]=Cobalt(II)	72Sj 73Sr	b,cc

Metal Ion	ΔH, kcal /mole	Log K	ΔS, cal /mole °K	T, °C	M	Conditions	Ref.	Remarks
010, cont.								
[Co]A, cont.						protoporphyrin IX dimethyl ester A=Pyridine. L(gas)		
	-9.8	-2.77(-45°,1)	-56	-63.5- -37.4	T	S: Toluene [Co]=Cobalt(II) protoporphyrin IX dimethyl ester A=4-tert-butyl-pyridine. L(gas)	73Sr 72Sj	b,cc
	-11.3	-1.84(-45°,1)	-58	-57.5- -31	T	S: Toluene [Co]=Cobalt(II) protoporphyrin IX dimethyl ester A=Imidazole. L(gas)	73Sr	b,cc
	-11.8	-2.04(-37.4,1)	-59	-45- -31		S: Toluene [Co]=Cobalt(II) protoporphyrin IX dimethyl ester A=1-Methylimidazole L(gas)	73Sr	b,cc
	-9.9	-2.05(-45°,1)	-53	-63.5- -31		S: Toluene [Co]=Cobalt(II) protoporphyrin IX dimethyl ester A=4-Aminopyridine L(gas)	73Sr	b,cc
	-9.0	-2.35(-45°,1)	-50	-63.5- -31		S: Toluene [Co]=Cobalt(II) protoporphyrin IX dimethyl ester A=Piperidine. L(gas)	73Sr	b,cc

P

P1 PENTAAMMINE COBALT(II)FUMARATE $C_4H_{18}O_4N_5Co$

Cu^{II}	-11.5	3.643(23°,1)	-22	5-40	T	μ=1.0(HClO$_4$)	73Hg	b,q: ±3.2

P2 PENTAAMMINE (PYRAZINE) RUTHENIUM(II) $C_4H_{19}N_7Ru$ $Ru(NH_3)_5$  L^{2+}

Ni^{2+}	-6	1.230(25°,1)	-15	5-30	T	μ=1.5M(LiClO$_4$)	74Pb	b,q: ±0.8
Cu^{2+}	-4.3	1.505(25°,1)	-7	5-30	T	μ=1.0M(LiClO$_4$)	74Pb	b,q: ±0.5
Zn^{2+}	-4.3	0.491(25°,1)	-12	5-30	T	μ=2.0M(LiClO$_4$)	74Pb	b,q: ±0.5

P3 PENTADECANE, 5,11-dithio- $C_{13}H_{28}S_2$ $C_4H_9S(CH_2)_5SC_4H_9$

SnA_4	-25.1	(1)	--	--	C	S: Benzene C_M=0.05-0.08M. A=Cl	65Gb	u

Metal Ion	ΔH, kcal /mole	Log K	ΔS, cal /mole °K	T, °C	M	Conditions	Ref.	Re-marks
P3, cont.								
TeA_4	-12.0	2.98(1)	-27.0	25	C	S: Benzene. A=Cl	73Ph	
	-5.0	(2)	--	25	C	S: Benzene. A=Cl	73Ph	
P4 PENTAETHYLENE GLYCOL, dimethyl ether			$C_{12}H_{26}O_6$			$CH_3OCH_2(CH_2OCH_2)_4CH_2OCH_3$		L
NaA	-9.2	2.653(25°)	-18.5	0-50	T	S: Tetrahydrofuran A=Fluorenyl ion R: NaA + L = NaLA	70Ce	b
P5 PENTANE, 1-amino-			$C_5H_{13}N$			$CH_3(CH_2)_4NH_2$		L
NiA	-17.1	5.96(1-2)	-30.0	25	T	$C_M=10^{-4}$ S: Benzene A=Diacetyl-bisbenzoyl hydrazine	60Sb	
Ag^+	-5	3.57(1)	-3.4	25	C	$\mu=0$	71He	q:±2
	-7	3.93(2)	-6.7	25	C	$\mu=0$	71He	q:±2
	-12.34	7.50(1-2)	-7.1	25	C	$\mu=0$	71He	
P6 PENTANE, 1-amino-5-hydroxy-3-thia-			$C_4H_{11}ONS$			$HOCH_2CH_2SCH_2CH_2NH_2$		L
Cu^{2+}	-5.1	5.26(30°,1)	7	10-30	T	$\mu=0$	59Lb	b
	-5.9	4.85(30°,2)	3	10-30	T	$\mu=0$	59Lb	b
Ag^+	-8.8	4.78(30°,1)	-7	10-40	T	$\mu=0$	59Lb	b
	-5.9	3.95(30°,2)	-2	10-40	T	$\mu=0$	59Lb	b
P7 PENTANE, 3-aza-1,5-diamino-N,N,N',N",N"-pentaacetic acid			$C_{14}H_{23}O_{10}N_3$			$(HO_2CCH_2)_2N(CH_2)_2N(CH_2CO_2H)-(CH_2)_2N(CH_2CO_2H)_2$		L^{5-}
Mg^{2+}	3.0	9.09(1)	52.4	20	C	0.1M(KNO_3)	65Ac	
	3.6	9.09(1)	54	25	C	$\mu=0.1(KNO_3)$	65Wd	
	2.6	9.10(1)	50	27	C	$\mu=0.1(KNO_3,HNO_3)$	68Cb	
Ca^{2+}	-5.95	10.9(1)	29.9	20	C	0.1M(KNO_3)	65Ac	
	-6.1	10.7(1)	29	25	C	$\mu=0.1(KNO_3)$	65Wd	
	-5.6	10.6(1)	30	27	C	$\mu=0.1(KNO_3,HNO_3)$	68Cb	
Sr^{2+}	-7.5	9.6(1)	19	25	C	$\mu=0.1(KNO_3)$	65Wd	
	-6.7	9.5(1)	21	27	C	$\mu=0.1(KNO_3,HNO_3)$	68Cb	
Ba^{2+}	-7.3	8.50(1)	14	25	C	$\mu=0.1(KNO_3)$	65Wd	
	-6.7	8.52(1)	21	27	C	$\mu=0.1(KNO_3,HNO_3)$	68Cb	
Y^{3+}	-5.2	22.05(1)	83.4	25	T	0.1M(KNO_3)	62Mh	
	-6.5	22.0(1)	79	27	C	0.1M(KNO_3)	68Cc	
La^{+3}	-4.7	19.6(1)	73.5	20	C	0.1M(KNO_3)	65Ac	
	-5.7	19.48(1)	70.0	25	T	0.1M(KNO_3)	62Mh	
	-5.2	19.4(1)	72	27	C	0.1M(KNO_3)	68Cc	
Ce^{3+}	-5.75	(1)	--	27	C	$\mu=0.10$	68Cd	
Pr^{3+}	-7.1	21.07(1)	72	25	T	0.1M(KNO_3)	62Mh	
	-6.45	21.0(1)	74.8	27	C	$\mu=0.10$	68Cd	

Metal Ion	ΔH, kcal /mole	Log K	ΔS, cal /mole °K	T, °C	M	Conditions	Ref.	Re-marks
P7, cont.								
Nd^{3+}	-5.8	21.60(1)	79.4	25	T	0.1M(KNO_3)	62Mh	
	-7.1	21.6(1)	75.0	27	C	μ=0.10	68Cd	
Sm^{3+}	-8.2	22.34(1)	74.7	25	T	0.1M(KNO_3)	62Mh	
	-7.9	22.3(1)	75.7	27	C	μ=0.10	68Cd	
Eu^{3+}	-8.1	22.39(1)	75.3	25	T	0.1M(KNO_3)	62Mh	
	-7.9	22.4(1)	76.0	27	C	μ=0.10	68Cd	
Gd^{3+}	-7.5	22.46(1)	77.6	25	T	0.1M(KNO_3)	62Mh	
	-7.8	22.4(1)	76.7	27	C	μ=0.10	68Cd	
Tb^{3+}	-7.7	22.71(1)	78.1	25	T	0.1M(KNO_3)	62Mh	
	-7.7	22.6(1)	78.0	27	C	μ=0.10	68Cd	
Dy^{3+}	-8.0	22.82(1)	77.6	25	T	0.1M(KNO_3)	62Mh	
	-7.9	22.8(1)	78.0	27	C	μ=0.10	68Cd	
Ho^{3+}	-7.6	22.78(1)	78.7	25	T	0.1M(KNO_3)	62Mh	
	-7.5	22.7(1)	79.0	27	C	μ=0.10	68Cd	
Er^{3+}	-7.3	22.74(1)	79.6	25	T	0.1M(KNO_3)	62Mh	
	-7.4	22.7(1)	79.3	27	C	μ=0.10	68Cd	
Tm^{3+}	-5.5	22.72(1)	85.5	25	T	0.1M(KNO_3)	62Mh	
	-6.6	22.7(1)	82.0	27	C	μ=0.10	68Cd	
Yb^{3+}	-5.5	22.62(1)	85.1	25	T	0.1M(KNO_3)	62Mh	
	-6.2	22.6(1)	82.7	27	C	μ=0.10	68Cd	
Lu^{3+}	-4.6	22.44(1)	87.2	25	T	0.1M(KNO_3)	62Mh	
	-5.1	22.4(1)	85.7	27	C	μ=0.10	68Cd	
Mn^{2+}	-7.18	15.60(1)	47.0	20	C	0.1M(KNO_3)	65Ac	
	-7.5	15.5(1)	46	25	C	μ=0.1(KNO_3)	65Wd	
Fe^{2+}	-7.7	16.4(1)	49	25	C	μ=0.1(KNO_3)	65Wd	
Co^{2+}	-9.41	19.26(1)	56.2	20	C	0.1M(KNO_3)	65Ac	
	-9.5	19.1(1)	56	25	C	μ=0.1(KNO_3)	65Wd	
Ni^{2+}	-11.7	20.2(1)	52.7	20	C	0.1M(KNO_3)	65Ac	
	-11.2	20.0(1)	54	25	C	μ=0.1(KNO_3)	65Wd	
Cu^{2+}	-13.6	21.5(1)	52.2	20	C	0.1M(KNO_3)	65Ac	
	-13.4	21.3(1)	53	25	C	μ=0.1(KNO_3)	65Wd	
Zn^{2+}	-8.8	18.54(1)	55.0	20	C	0.1M(KNO_3)	65Ac	
	-10.6	18.8(1)	50	25	C	μ=0.1(KNO_3)	65Wd	
Cd^{2+}	-12.35	19.3(1)	46.4	20	C	0.1M(KNO_3)	65Ac	
	-12.4	18.9(1)	45	25	C	μ=0.1(KNO_3)	65Wd	
Hg^{2+}	-23.7	26.7(1)	41.3	20	C	0.1M(KNO_3)	65Ac 62Mh	
	-23.6	26.9(1)	44	25	C	μ=0.1(KNO_3)	65Wd 66Mf	
Al^{3+}	8.4	18.51(1)	113.0	25	T	μ=0.1(KNO_3)	66Mf	
	7.4	18.6(1)	110	27	C	μ=0.1(KNO_3)	68Cc	
Pb^{+2}	-18.8	18.9(1)	21.8	25	C	0.1M(KNO)	65Ac	
	-18.8	18.5(1)	22	27	C	μ=0.1(KNO_3)	65Wd	

Metal Ion	ΔH, kcal /mole	Log K	ΔS, cal /mole °K	T, °C	M	Conditions	Ref.	Remarks
P8 PENTANE, 1,5-diamino-3-oxa-N,N,N',N'-tetraacetic acid			$C_{12}H_{20}O_5N_2$			$O[CH_2CH_2N(CH_2CO_2H)_2]_2$		L^{4-}
Mg^{2+}	3.6	8.4(1)	50	25	C	μ=0.1(KNO_3)	65Wd	
Ca^{2+}	-6.4	9.9(1)	24	25	C	μ=0.1(KNO_3)	65Wd	
Sr^{2+}	-8.1	8.6(1)	12	25	C	μ=0.1(KNO_3)	65Wd	
Ba^{2+}	-6.5	8.1(1)	15	25	C	μ=0.1(KNO_3)	65Wd	
Mn^{2+}	-5.6	13.0(1)	41	25	C	μ=0.1(KNO_3)	65Wd	
Fe^{2+}	-6.4	14.7(1)	46	25	C	μ=0.1(KNO_3)	65Wd	
Co^{2+}	-6.6	14.6(1)	45	25	C	μ=0.1(KNO_3)	65Wd	
Ni^{2+}	-5.1	14.6(1)	50	25	C	μ=0.1(KNO_3)	65Wd	
Cu^{2+}	-9.7	17.5(1)	48	25	C	μ=0.1(KNO_3)	65Wd	
Zn^{2+}	-7.6	15.2(1)	44	25	C	μ=0.1(KNO_3)	65Wd	
Cd^{2+}	-9.7	16.1(1)	41	25	C	μ=0.1(KNO_3)	65Wd	
Hg^{2+}	-19.5	23.0(1)	40	25	C	μ=0.1(KNO_3)	65Wd	
Pb^{2+}	-12.2	14.4(1)	25	25	C	μ=0.1(KNO_3)	65Wd	
P9 PENTANE, 1,5-diamino-N,N,N',N'-tetraacetic acid			$C_{13}H_{22}O_8N_2$			$(CH_2CO_2H)_2N(CH_2)_5N-(CH_2CO_2H)_2$		L^{4-}
Mn^{2+}	0.9	8.7(1)	43	20	C	μ=0.1(KNO_3)	64Ab	
Co^{2+}	-3.1	13.38(1)	50.7	20	C	μ=0.1(KNO_3)	64Ab	
Ni^{2+}	-6.7	13.9(1)	40.8	20	C	μ=0.1(KNO_3)	64Ab	
Cu^{2+}	-10.9	16.24(1)	37.2	20	C	μ=0.1(KNO_3)	64Ab	
Zn^{2+}	-2.7	12.67(1)	48.8	20	C	μ=0.1(KNO_3)	64Ab	
Cd^{2+}	-4.46	11.6(1)	37.5	20	C	μ=0.1(KNO_3)	64Ab	
P10 2,4-PENTANEDIONE			$C_5H_8O_2$			$CH_3COCH_2COCH_3$		L^-
Be^{2+}	-2.0	7.8(30°,1)	29	10-40	T	μ=0	55Ia	b
	-6.9	6.7(30°,2)	8.0	10-40	T	μ=0	55Ia	b
Mg^{2+}	-1.8	3.6(30°,1)	11	10-40	T	μ=0	55Ia	b
	-4.3	3.1(30°,2)	2.5	10-40	T	μ=0	55Ia	b
Ce^{3+}	0	5.3(30°,1)	24	10-40	T	μ=0	55Ia	b
	0.2	4.0(30°,2)	18	10-40	T	μ=0	55Ia	b
ZrA_2L_2	0.05	0.806(25°)	3.4	7.5-53.1	T	S: Benzene A=Trifluoroacetyl-acetone. R: ZrA_2L_2 + ZrL_4 = $2ZrAL_3$	66Pf	b,q: ±0.54
	-0.10	0.755(25°)	3.1	18-79.8	T	S: CCl_4 A=Trifluoroacetyl-acetone. R: ZrA_2L_2 + ZrL_4 = $2ZrAL_3$	66Pf	b,q: ±0.19
	0.02	0.843(25°)	3.9	7.5-63.9	T	S: Benzene A=Trifluoroacetyl-acetone. R: ZrA_2L_2 + ZrA_4 = $2ZrA_3L$	66Pf	b,q: ±0.26

Metal Ion	ΔH, kcal /mole	Log K	ΔS, cal /mole °K	T, °C	M	Conditions	Ref.	Remarks
P10, cont.								
ZrA_2L_2, cont.	-0.20	0.894(25°)	3.4	-18.5-51.6	T	S: CCl_4 A=Trifluoroacetyl acetone. R: ZrA_2L_2 + ZrA_4 = $2ZrA_3L$	66Pf	b,q: ±0.25
ZrA_3L	-0.05	0.806(25°)	3.5	7.5-77.8	T	S: Benzene A=Trifluoroacetyl-acetone. R: ZrA_3L + $ZrAL_3$ = $2ZrA_2L_2$	66Pf	b,q: ±0.18
	0.25	0.799(25°)	4.5	-18.5-63.4	T	S: CCl_4 A=Trifluoroacetyl-acetone. R: ZrA_3L + $ZrAL_3$ = $2ZrA_2L_2$	66Pf	b,q: ±0.18
VO^{++}	-1.7	8.65(25°)	34.0	25-40	T	R: VO^{++} + HL = VOL^+ + H^+	56Td	b
	-6.8	7.04(25°)	9.6	25-40	T	R: VOL^+ + HL = VOL_2 + H^+	56Td	b
UO_2^{2+}	-4.0	5.3(30°,2)	16	10-40	T	μ=0	55Ia	b
NpO_2^+	-6.3	4.08(25°,1)	-2.3	18-32	T	μ=0.1M	72Gh	b,q: ±1.0
	-3.2	2.92(25°,2)	2.6	18-32	T	μ=0.1M	72Gh	b,q: ±1.0
Mn^{2+}	-2.5	4.2(30°,1)	11	10-40	T	μ=0	55Ia	b
	-1.5	4.07(1)	14	25	C	μ=0.1($NaClO_4$)	68Gh	
	-0.6	4.24(1)	19	0	T	S: 1-Propanol-Water Mixture 1/D=0.0113	71Da	
	-0.4	4.18(1)	17.6	25	T	S: 1-Propanol-Water Mixture 1/D=0.0127	71Da	
	-0.5	4.26(1)	16.9	40	T	S: 1-Propanol-Water Mixture 1/D=0.0137	71Da	
	-0.6	4.48(1)	20.0	0	T	S: 1-Propanol-Water Mixture 1/D=0.0131	71Da	
	-0.1	4.40(1)	19.3	25	T	S: 1-Propanol-Water Mixture 1/D=0.0149	71Da	
	-0.2	4.47(1)	19.4	40	T	S: 1-Propanol-Water Mixture 1/D=0.0159	71Da	
	-0.3	4.80(1)	22.5	0	T	S: 1-Propanol-Water Mixture 1/D=0.0160	71Da	
	0.2	4.76(1)	22.3	25	T	S: 1-Propanol-Water Mixture 1/D=0.0183	71Da	
	0.1	4.82(1)	21.0	40	T	S: 1-Propanol-Water Mixture 1/D=0.0199	71Da	
	-0.3	5.04(1)	23.9	0	T	S: 1-Propanol-Water Mixture 1/D=0.0182	71Da	
	0.5	5.13(1)	24.2	25	T	S: 1-Propanol-Water Mixture 1/D=0.0213	71Da	
	0.4	5.10(1)	23.6	40	T	S: 1-Propanol-Water Mixture 1/D=0.0230	71Da	
	0.1	5.44(1)	26.3	0	T	S: 1-Propanol-Water Mixture 1/D=0.0212	71Da	
	0.7	5.50(1)	27.1	25	T	S: 1-Propanol-Water Mixture 1/D=0.0247	71Da	

Metal Ion	ΔH, kcal /mole	Log K	ΔS, cal /mole °K	T, °C	M	Conditions	Ref.	Remarks
P10, cont.								
Mn^{2+}, cont.	0.8	5.51(1)	26.9	40	T	S: 1-Propanol-Water Mixture 1/D=0.0269	71Da	
	0.4	5.84(1)	29.1	0	T	S: 1-Propanol-Water Mixture 1/D=0.0248	71Da	
	1.1	6.01(1)	31.0	25	T	S: 1-Propanol-Water Mixture 1/D=0.0294	71Da	
	1.3	6.00(1)	30.7	40	T	S: 1-Propanol-Water Mixture 1/D=0.0323	71Da	
	0.8	6.40(1)	33.2	0	T	S: 1-Propanol-Water Mixture 1/D=0.0297	71Da	
	1.9	6.60(1)	35.7	25	T	S: 1-Propanol-Water Mixture 1/D=0.0352	71Da	
	2.2	6.70(1)	37.0	40	T	S: 1-Propanol-Water Mixture 1/D=0.0392	71Da	
	-4.7	3.0(30°,2)	-1.5	10-40	T	μ=0	55Ia	b
	-1.8	3.12(2)	7.1	0	T	S: 1-Propanol-Water Mixture 1/D=0.0113	71Da	
	-1.4	3.00(2)	8.2	25	T	S: 1-Propanol-Water Mixture 1/D=0.0127	71Da	
	-1.3	3.00(2)	9.0	40	T	S: 1-Propanol-Water Mixture 1/D=0.0137	71Da	
	-1.6	3.36(2)	8.7	0	T	S: 1-Propanol-Water Mixture 1/D=0.0131	71Da	
	-1.3	3.22(2)	9.8	25	T	S: 1-Propanol-Water Mixture 1/D=0.0149	71Da	
	-1.1	3.14(2)	10.7	40	T	S: 1-Propanol-Water Mixture 1/D=0.0159	71Da	
	-1.5	3.60(2)	11.6	0	T	S: 1-Propanol-Water Mixture 1/D=0.0160	71Da	
	-1.1	3.52(2)	12.6	25	T	S: 1-Propanol-Water Mixture 1/D=0.0183	71Da	
	-0.4	3.49(2)	13.8	40	T	S: 1-Propanol-Water Mixture 1/D=0.0199	71Da	
	-1.2	3.84(2)	13.5	0	T	S: 1-Propanol-Water Mixture 1/D=0.0182	71Da	
	-0.4	3.74(2)	15.4	25	T	S: 1-Propanol-Water Mixture 1/D=0.0213	71Da	
	-0.2	3.77(2)	16.5	40	T	S: 1-Propanol-Water Mixture 1/D=0.0230	71Da	
	-0.8	4.08(2)	16.0	0	T	S: 1-Propanol-Water Mixture 1/D=0.0212	71Da	
	0.0	4.10(2)	18.3	25	T	S: 1-Propanol-Water Mixture 1/D=0.0247	71Da	
	0.6	4.12(2)	19.8	40	T	S: 1-Propanol-Water Mixture 1/D=0.0269	71Da	
	-0.6	4.48(2)	19.1	0	T	S: 1-Propanol-Water Mixture 1/D=0.0248	71Da	
	0.6	4.47(2)	22.7	25	T	S: 1-Propanol-Water Mixture 1/D=0.0294	71Da	
	1.4	4.54(2)	24.4	40	T	S: 1-Propanol-Water Mixture 1/D=0.0323	71Da	
	0.2	4.96(2)	23.4	0	T	S: 1-Propanol-Water Mixture 1/D=0.0297	71Da	
	0.6	5.06(2)	28.2	25	T	S: 1-Propanol-Water Mixture 1/D=0.0353	71Da	

Metal Ion	ΔH, kcal /mole	Log K	ΔS, cal /mole °K	T, °C	M	Conditions	Ref.	Remarks
P10, cont.								
Mn^{2+}, cont.	2.3	5.16(2)	30.8	40	T	S: 1-Propanol-Water Mixture 1/D=0.0392	71Da	
MnA_3	-1.170	5.094	19.4	25	T	R: $MnA_3 + 2H_2O + H^+ = MnA_2(H_2O)_2^+ + HL$. A=L	52Ca	
	3.030	3.92	-7.8	25	T	R: $MnA_3 + 2H_2O = MnA_2(H_2O)_2^+ + A^-$ A=L	52Ca	
Co^{2+}	-1.2	5.4(30°,1)	21	10-40	T	μ=0	55Ia	b
	-5.0	4.1(30°,2)	2.4	10-40	T	μ=0	55Ia	b
CoL_2AB	-3.1	0.394(29°)	-8.3	29-60	T	S: $CDCl_3$ C_M=0.1M A=Pyridine. $B=N_3^-$ R: CoL_2AB(trans) = CoL_2AB(cis)	73Hd	b,q: ±2.1
	-5.5	0.570(29°)	-20.8	29-60	T	S: $CDCl_3$ C_M=0.1M A=Pyridine. $B=NCO^-$ R: CoL_2AB(trans) = CoL_2AB(cis)	73Hd	b,q: ±3.3
	-3.3	-0.226(29°)	-12.0	29-60	T	S: $CDCl_3$ C_M=0.1M A=Pyridine. $B=NO_2^-$ R: CoL_2AB(trans) = CoL_2AB(cis)	73Hd	b,q: ±0.9
Ni^{2+}	-5.6	6.3(1)	8.4	0	T	--	68Gb	
	-4.5	6.1(1)	12.8	25	T	--	68Gb	
	-3.8	5.9(1)	14.7	40	T	--	68Gb	
	-6.7	5.9(30°,1)	12	10-40	T	μ=0	55Ia	b
	-4.6	5.69(25°,1)	11	15-40	T	μ=0	71Yb	b,q: ±0.5
	-3.4	5.72(1)	15	25	C	μ=0.1($NaClO_4$)	68Gh	
	-2.1	8.77(25°,1)	33	15-40	T	μ=0 S: 75 v % Dioxane	71Yb	b,q: ±0.5
	-5.2	6.6(1)	11.0	0	T	0.098 w fraction CH_3OH	68Gb	
	-4.0	6.3(1)	15.4	25	T	0.099 w fraction CH_3OH	68Gb	
	-3.2	6.2(1)	17.9	40	T	0.099 w fraction CH_3OH	68Gb	
	-4.6	6.8(1)	14.3	0	T	0.220 w fraction CH_3OH	68Gb	
	-3.1	6.7(1)	20.1	25	T	0.222 w fraction CH_3OH	68Gb	
	-2.0	6.5(1)	23.3	40	T	0.224 w fraction CH_3OH	68Gb	
	-4.2	7.0(1)	16.5	0	T	0.289 w fraction CH_3OH	68Gb	
	-2.5	6.8(1)	22.8	25	T	0.295 w fraction CH_3OH	68Gb	
	-1.3	6.6(1)	26.9	40	T	0.299 w fraction CH_3OH	68Gb	

Metal Ion	ΔH, kcal /mole	Log K	ΔS, cal /mole °K	T, °C	M	Conditions	Ref.	Re-marks
P10, cont.								
Ni^{2+}, cont.	-3.7	7.2(1)	19.4	0	T	0.374 w fraction CH_3OH	68Gb	
	-1.8	7.1(1)	26.5	25	T	0.383 w fraction CH_3OH	68Gb	
	-0.4	7.0(1)	30.7	40	T	0.387 w fraction CH_3OH	68Gb	
	-3.0	7.4(1)	23.1	0	T	0.474 w fraction CH_3OH	68Gb	
	-0.8	7.4(1)	31.2	25	T	0.485 w fraction CH_3OH	68Gb	
	1.0	7.4(1)	37.0	40	T	0.489 w fraction CH_3OH	68Gb	
	-2.2	7.8(1)	27.5	0	T	0.590 w fraction CH_3OH	68Gb	
	0.4	7.8(1)	37.2	25	T	0.610 w fraction CH_3OH	68Gb	
	2.5	8.0(1)	44.3	40	T	0.621 w fraction CH_3OH	68Gb	
	-5.8	6.3(1)	7.7	0	T	0 w % 1-propanol	68Gb	
	-5.3	6.0(1)	9.6	25	T	0 w % 1-propanol	68Gb	
	-5.0	5.8(1)	10.3	40	T	0 w % 1-propanol	68Gb	
	-5.5	6.6(1)	9.8	0	T	0.056 w fraction 1-propanol	68Gb	
	-4.9	6.2(1)	12.1	25	T	0.056 w fraction 1-propanol	68Gb	
	-4.6	6.0(1)	12.9	40	T	0.056 w fraction 1-propanol	68Gb	
	-5.1	6.9(1)	12.8	0	T	0.132 w fraction 1-propanol	68Gb	
	-4.1	6.6(1)	16.3	25	T	0.134 w fraction 1-propanol	68Gb	
	-3.7	6.4(1)	17.7	40	T	0.134 w fraction 1-propanol	68Gb	
	-4.6	7.1(1)	15.5	0	T	0.183 w fraction 1-propanol	68Gb	
	-3.4	6.9(1)	20.0	25	T	0.188 w fraction 1-propanol	68Gb	
	-3.0	6.8(1)	21.5	40	T	0.188 w fraction 1-propanol	68Gb	
	-4.1	7.4(1)	18.9	0	T	0.251 w fraction 1-propanol	68Gb	
	-2.7	7.3(1)	24.3	25	T	0.256 w fraction 1-propanol	68Gb	
	-2.0	7.1(1)	26.2	40	T	0.256 w fraction 1-propanol	68Gb	
	-3.4	7.8(1)	23.3	0	T	0.334 w fraction 1-propanol	68Gb	
	-1.5	7.8(1)	30.4	25	T	0.346 w fraction 1-propanol	68Gb	
	-0.8	7.6(1)	32.5	40	T	0.346 w fraction 1-propanol	68Gb	
	-2.3	8.3(1)	29.7	0	T	0.450 w fraction 1-propanol	68Gb	

Metal Ion	ΔH, kcal /mole	Log K	ΔS, cal /mole °K	T, °C	M	Conditions	Ref.	Remarks
P10, cont.								
Ni^{2+}, cont.	-0.1	8.4(1)	38.3	25	T	0.473 w fraction 1-propanol	68Gb	
	1.0	8.3(1)	41.1	40	T	0.473 w fraction 1-propanol	68Gb	
	-6.7	5.1(2)	-1.8	0	T	--	68Gb	
	-6.3	4.7(2)	0.3	25	T	--	68Gb	
	-5.7	4.5(2)	2.2	40	T	--	68Gb	
	-6.3	4.5(30°,2)	-0.4	10-40	T	μ=0	55Ia	b
	-5.3	4.47(25°,2)	3	15-40	T	μ=0	71Yb	b,q: $\pm$0.5
	-2.3	7.21(25°,2)	25	15-40	T	μ=0 S: 75 v % Dioxane	71Yb	b,q: $\pm$0.5
	-6.6	5.3(2)	-0.2	0	T	0.098 w fraction CH_3OH	68Gb	
	-5.9	4.8(2)	2.3	25	T	0.099 w fraction CH_3OH	68Gb	
	-5.2	4.6(2)	4.5	40	T	0.099 w fraction CH_3OH	68Gb	
	-6.2	5.4(2)	1.9	0	T	0.220 w fraction CH_3OH	68Gb	
	-5.4	5.1(2)	5.0	25	T	0.222 w fraction CH_3OH	68Gb	
	-4.5	4.8(2)	7.7	40	T	0.224 w fraction CH_3OH	68Gb	
	-6.0	5.5(2)	3.3	0	T	0.289 w fraction CH_3OH	68Gb	
	-5.0	5.1(2)	6.7	25	T	0.295 w fraction CH_3OH	68Gb	
	-4.2	5.0(2)	9.6	40	T	0.299 w fraction CH_3OH	68Gb	
	-5.7	5.6(2)	4.7	0	T	0.374 w fraction CH_3OH	68Gb	
	-4.7	5.3(2)	8.4	25	T	0.383 w fraction CH_3OH	68Gb	
	-3.5	5.2(2)	12.5	40	T	0.387 w fraction CH_3OH	68Gb	
	-5.3	5.8(2)	7.1	0	T	0.474 w fraction CH_3OH	68Gb	
	-4.0	5.5(2)	11.7	25	T	0.485 w fraction CH_3OH	68Gb	
	-2.6	5.5(2)	16.6	40	T	0.489 w fraction CH_3OH	68Gb	
	-4.8	6.0(2)	9.9	0	T	0.590 w fraction CH_3OH	68Gb	
	-3.3	5.8(2)	15.4	25	T	0.610 w fraction CH_3OH	68Gb	
	-1.6	5.7(2)	21.1	40	T	0.621 w fraction CH_3OH	68Gb	
	-6.3	5.0(2)	0.2	0	T	0 wt % 1-propanol	68Gb	
	-5.8	4.5(2)	1.3	25	T	0 wt % 1-propanol	68Gb	
	-5.7	4.4(2)	1.7	40	T	0 wt % 1-propanol	68Gb	
	-6.2	5.1(2)	1.0	0	T	0.056 w fraction 1-propanol	68Gb	

Metal Ion	ΔH, kcal /mole	Log K	ΔS, cal /mole °K	T, °C	M	Conditions	Ref.	Re-marks
P10, cont.								
Ni^{2+}, cont.	-5.5	4.8(2)	3.4	25	T	0.056 w fraction 1-propanol	68Gb	
	-5.4	4.6(2)	3.6	40	T	0.056 w fraction 1-propanol	68Gb	
	-5.7	5.4(2)	3.8	0	T	0.132 w fraction 1-propanol	68Gb	
	-4.9	5.1(2)	6.8	25	T	0.134 w fraction 1-propanol	68Gb	
	-4.7	4.9(2)	7.5	40	T	0.134 w fraction 1-propanol	68Gb	
	-5.4	5.6(2)	5.9	0	T	0.183 w fraction 1-propanol	68Gb	
	-4.3	5.4(2)	10.0	25	T	0.188 w fraction 1-propanol	68Gb	
	-4.2	5.2(2)	10.4	40	T	0.188 w fraction 1-propanol	68Gb	
	-5.0	5.9(2)	8.9	0	T	0.251 w fraction 1-propanol	68Gb	
	-3.7	5.6(2)	13.6	25	T	0.256 w fraction 1-propanol	68Gb	
	-3.3	5.5(2)	14.4	40	T	0.256 w fraction 1-propanol	68Gb	
	-4.5	6.2(2)	12.3	0	T	0.334 w fraction 1-propanol	68Gb	
	-2.8	6.2(2)	18.6	25	T	0.346 w fraction 1-propanol	68Gb	
	-2.4	6.0(2)	19.5	40	T	0.346 w fraction 1-propanol	68Gb	
	-3.6	6.7(2)	17.5	0	T	0.450 w fraction 1-propanol	68Gb	
	-1.5	6.7(2)	25.5	25	T	0.473 w fraction 1-propanol	68Gb	
	-0.8	6.5(2)	27.2	40	T	0.473 w fraction 1-propanol	68Gb	
	-6.7	2.2(30°,3)	-12	10-40	T	μ=0	55Ia	b
	-7.6	9.66(1-2)	19	25	C	μ=0.1($NaClO_4$)	68Gh	
NiL_2	45.89	6.0	178.8	30	C	S: Benzene C_M=0.007-0.070M R: $3NiL_2 = Ni_3L_6$	73Af	q
Pd^{2+}	-18	16.2(30°,1)	14	20-40	T	μ≈0.45	57De	b
	-18	10.9(30°,2)	-11	20-40	T	μ≈0.45	57De	b
Cu^{2+}	-4.7	8.2(30°,1)	22	10-40	T	μ=0	55Ia	b
	-7.1	8.29(25°,1)	14	15-40	T	μ=0.02[$N(CH_3)_4Cl$]	71Yb	b
	-4.8	8.16(1)	21	25	C	μ=0.1($NaClO_4$)	68Gh	
	-4.1	12.06(25°,1)	41	15-40	T	μ=0 S: 75 v % Dioxane	71Yb	b,q: ±0.5
	-6.6	6.7(30°,2)	9	10-40	T	μ=0	55Ia	b
	-7.7	6.70(25°,2)	5	15-40	T	μ=0.02[$N(CH_3)_4Cl$]	71Yb	b
	-4.2	10.43(25°,2)	34	15-40	T	μ=0 S: 75 v % Dioxane	71Yb	b,q: ±0.5
	-10.1	14.76(1-2)	34	25	C	μ=0.1($NaClO_4$)	68Gh	

Metal Ion	ΔH, kcal /mole	Log K	ΔS, cal /mole °K	T, °C	M	Conditions	Ref.	Remarks
P10, cont.								
CuL_2	4.0	--	13	0-60	T	S: Pyridine R: CuL_2 (low temperature species) = CuL_2 (high temperature species)	72Yb	b,q: 11
Zn^{2+}	-1.9	5.0(30°,1)	17	10-40	T	μ=0	55Ia	b
	-1.32	5.36(1)	20	25	C	μ=0.1	67Mc	
	-1.5	4.68(1)	16	25	C	μ=0.1($NaClO_4$)	68Gh	
	-1.5	5.20(1)	19.5	0	T	S: 1-Propanol-Water Mixture 1/D=0.0113	71Da	
	-1.0	5.06(1)	19.2	25	T	S: 1-Propanol-Water Mixture 1/D=0.0127	71Da	
	-0.7	5.02(1)	19.3	40	T	S: 1-Propanol-Water Mixture 1/D=0.0137	71Da	
	-1.2	5.36(1)	21.8	0	T	S: 1-Propanol-Water Mixture 1/D=0.0131	71Da	
	-0.7	5.28(1)	21.6	25	T	S: 1-Propanol-Water Mixture 1/D=0.0149	71Da	
	-0.3	5.30(1)	21.8	40	T	S: 1-Propanol-Water Mixture 1/D=0.0159	71Da	
	-0.7	5.68(1)	25.3	0	T	S: 1-Propanol-Water Mixture 1/D=0.0160	71Da	
	0.0	5.64(1)	25.7	25	T	S: 1-Propanol-Water Mixture 1/D=0.0183	71Da	
	0.7	5.72(1)	24.4	40	T	S: 1-Propanol-Water Mixture 1/D=0.0199	71Da	
	-0.4	5.92(1)	27.7	0	T	S: 1-Propanol-Water Mixture 1/D=0.0182	71Da	
	0.7	5.94(1)	29.5	25	T	S: 1-Propanol-Water Mixture 1/D=0.0213	71Da	
	1.4	6.00(1)	30.4	40	T	S: 1-Propanol-Water Mixture 1/D=0.0230	71Da	
	0.2	6.24(1)	31.4	0	T	S: 1-Propanol-Water Mixture 1/D=0.0212	71Da	
	1.5	6.30(1)	33.5	25	T	S: 1-Propanol-Water Mixture 1/D=0.0247	71Da	
	2.5	6.42(1)	35.2	40	T	S: 1-Propanol-Water Mixture 1/D=0.0269	71Da	
	1.0	6.64(1)	36.1	0	T	S: 1-Propanol-Water Mixture 1/D=0.0248	71Da	
	2.5	6.82(1)	39.2	25	T	S: 1-Propanol-Water Mixture 1/D=0.0294	71Da	
	3.8	6.98(1)	41.9	40	T	S: 1-Propanol-Water Mixture 1/D=0.0323	71Da	
	1.9	7.20(1)	42.1	0	T	S: 1-Propanol-Water Mixture 1/D=0.0297	71Da	
	3.9	7.40(1)	46.6	25	T	S: 1-Propanol-Water Mixture 1/D=0.0352	71Da	
	5.5	7.68(1)	50.6	40	T	S: 1-Propanol-Water Mixture 1/D=0.0392	71Da	
	-1.43	4.49(2)	16	25	C	μ=0.1	67Mc	
	1.0	4.64(2)	27.5	0	T	S: 1-Propanol-Water Mixture 1/D=0.0160	71Da	

Metal Ion	ΔH, kcal /mole	Log K	ΔS, cal /mole °K	T, °C	M	Conditions	Ref.	Remarks
P10, cont.								
Zn^{2+}, cont.	0.7	4.76(2)	23.2	25	T	S: 1-Propanol-Water Mixture 1/D=0.0183	71Da	
	0.5	4.75(2)	22.2	40	T	S: 1-Propanol-Water Mixture 1/D=0.0199	71Da	
	1.1	4.88(2)	26.1	0	T	S: 1-Propanol-Water Mixture 1/D=0.0182	71Da	
	0.5	4.98(2)	23.9	25	T	S: 1-Propanol-Water Mixture 1/D=0.0213	71Da	
	0.4	5.02(2)	23.1	40	T	S: 1-Propanol-Water Mixture 1/D=0.0230	71Da	
	1.0	5.20(2)	28.9	0	T	S: 1-Propanol-Water Mixture 1/D=0.0212	71Da	
	0.3	5.28(2)	24.9	25	T	S: 1-Propanol-Water Mixture 1/D=0.0247	71Da	
	0.3	5.30(2)	24.5	40	T	S: 1-Propanol-Water Mixture 1/D=0.0269	71Da	
	0.8	5.60(2)	30.1	0	T	S: 1-Propanol-Water Mixture 1/D=0.0248	71Da	
	0.0	5.72(2)	26.1	25	T	S: 1-Propanol-Water Mixture 1/D=0.0294	71Da	
	0.0	5.72(2)	26.1	40	T	S: 1-Propanol-Water Mixture 1/D=0.0323	71Da	
	0.6	6.16(2)	31.8	0	T	S: 1-Propanol-Water Mixture 1/D=0.0297	71Da	
	-0.3	6.16(2)	28.0	25	T	S: 1-Propanol-Water Mixture 1/D=0.0353	71Da	
	-0.1	6.28(2)	28.4	40	T	S: 1-Propanol-Water Mixture 1/D=0.0392	71Da	
	-3.4	7.92(1-2)	24	25	C	μ=0.1($NaClO_4$)	68Gh	
Cd^{2+}	-1.4	3.8(30°,1)	13	10-40	T	μ=0	55Ia	b
	-0.95	3.59(1)	13	25	C	μ=0.1	67Mc	
	-1.89	2.55(2)	5	25	C	μ=0.1	67Mc	
AlA_3	-3.894	--	--	25	C	S: 50 v % Dioxane (0.174 mole fraction dioxane) R: AlA_3 + 3NaL = AlL_3 + 3NaA A=Cl	66Hb	
GaA_3	0.32	0.413(25)	2.98	5.7-43	T	S: Benzene A=Dibenzoylmethane R: 2/3GaA_3 + 1/3GaL_3 = $GaLA_2$	69Pc	b,q: ±0.25
	-4.1	3.049(25°)	0.1	36-98	T	S: Benzene A=Hexafluoroacetyl-acetone. R: 2/3GaA_3 + 1/3GaL_3 = $GaLA_2$	69Pc	b,q: ±0.8
	0.04	0.393(25°)	1.91	5.7-43	T	S: Benzene A=Dibenzoylmethane R: 1/3GaA_3 + 2/3GaL_3 = $GaAL_2$	69Pc	b,q: ±0.24
	-4.5	3.086(25°)	-0.9	36-98	T	S: Benzene A=Hexafluoroacetyl-acetone. R: 1/3GaA_3 + 2/3GaL_3 = $GaAL_2$	69Pc	b,q: ±1

Metal Ion	ΔH, kcal /mole	Log K	ΔS, cal /mole °K	T, °C	M	Conditions	Ref.	Remarks
P11	2,4-PENTANEDIONE, 3-benzyl-		$C_{12}H_{14}O_2$			$CH_3C(O)CH(CH_2C_6H_5)C(O)CH_3$		
NiL_2	27.2	4.0	<u>108</u>	30	C	S: Benzene C_M=0.002-0.044M R: $3NiL_2 = Ni_3L_6$	73Af	
P12	2,4-PENTANEDIONE, 3-butyl-		$C_9H_{16}O_2$			$CH_3C(O)CH(C_4H_9)C(O)CH_3$		
NiL_2	48.8	2.9	<u>174</u>	30	C	S: Benzene C_M=0.006-0.045M R: $3NiL_2 = Ni_3L_6$	73Af	
P13	2,4-PENTANEDIONE, 1,1,1-trifluoro-		$C_5H_5O_2F_3$			$CF_3C(O)CH_2C(O)CH_3$		
VL_3	-1.4	0.785	-1.0	25	T	S: $CDCl_3$ R: VL_3(cis) = VL_3(trans)	71Gf	q: ±0.3
MnL_3	-1.1	0.643	-0.7	25	T	S: $CDCl_3$ R: MnL_3(cis) = MnL_3(trans)	71Gf	q: ±0.4
CoL_3	-0.34	0.650(25°)	1.8	79.1-115		S: Chloroform R: CoL_3(cis) = CoL_3(trans)	63Fa	b,q: ±0.71
RuL_3	-1.1	0.690	0.6	150	T	S: Sym-$C_2H_2Cl_4$ R: RuL_3(cis) = RuL_3(trans)	71Gf	q: ±0.6
AlL_3	-0.24	0.658(25°)	2.2	-25.5-3		S: Chloroform R: AlL_3(cis) = AlL_3(trans)	63Fa	b,q: ±0.19
	-0.47	0.682(25°)	1.5	-25.5-29.5		S: Chloroform R: AlL_3(cis) = AlL_3(trans)	63Fa	b,q: ±0.22
GaL_3	-0.47	0.672	1.5	25	T	S: $CDCl_3$ R: GaL_3(cis) = GaL_3(trans)	71Gf	q: ±0.22
P14	2,4-PENTANEDIONE, 1,1,1-trifluoro-, 4-imine, N-isopropyl-		$C_8H_{11}ONF_3$			$CF_3C(O):CHC(CH_3):NCH_2CH_2CH_3$		
Ni^{2+}	-0.002	0.602(25°)	2.7	17&107	T	S: $CDCl_3$ R: $NiL_2 + NiA_2 = 2NiAL$. Where A = N-isopropylacetyl-acetoneimine	73La	a,b,q: ±0
P15	1-PENTANESULPHONIC ACID, 5-amino-		$C_5H_{13}O_3NS$			$H_2N(CH_2)_5SO_3H$		L^-
Ag^+	-5.54	3.489(<u>1</u>)	-2.6	25	C	μ=0.5M(KNO_3)	72Va	
	-7.40	3.908(<u>2</u>)	-7.0	25	C	μ=0.5M(KNO_3)	72Va	

Metal Ion	ΔH, kcal /mole	Log K	ΔS, cal /mole °K	T, °C	M	Conditions	Ref.	Remarks
P16	PENTANOIC ACID, nitrile		C_5H_9N			$CH_3(CH_2)_3CN$		L
I_2	-2.70	-0.303(25°,1)	-10.3	--	T	S: CCl_4 C_L=0.2-1.0M C_M=2x10^{-3}. Data analyzed by method 1	72Ma	j,u,q: ±0.35
	-2.93	-0.303(25°,1)	-10.99	--	T	S: CCl_4 C_L=0.2-1.0M C_M=2x10^{-3}. Data analyzed by method 2	72Ma	j,u
	-3.0	0.009(25°,1)	-9.8	25-55	T	S: Heptane C_L=0.2-0.9M C_M=8x10^{-4}-1.6x10^{-3}	67Ma	b,j,q: ±0.3
	-4.25	0.009(25°,1)	-14.12	--	T	S: n-Heptane C_L=0.2-1.0M C_M=2x10^{-3}. Data analyzed by method 2	72Ma	j,u
P17	PENTANOIC ACID, 2-amino-		$C_5H_{11}O_2N$			$CH_3CH_2CH_2CH(NH_2)CO_2H$		
Mn^{2+}	-3.1	3.30(20°,1)	5	20-60	T	μ=0.1(KNO_3)	73Bf	b
	1.6	1.89(20°,2)	14	20-60	T	μ=0.1(KNO_3)	73Bf	b
Cu^{2+}	-5.12	8.07(1)	20	25	C	μ=0.2M(KCl)	73Ga	
	-6.88	6.75(2)	8	25	C	μ=0.2M(KCl)	73Ga	
P18	2-PENTANONE		$C_5H_{10}O$			$CH_3CH_2CH_2COCH_3$		L
SbA_5	-16.1	(1)	--	25	C	S: CCl_4 A=Cl	73Oa	
TeA_4	-5.6	(1)	--	25	C	S: Benzene A=Cl	72Pe	
P19	3-PENTANONE		$C_5H_{10}O$			$CH_3CH_2COCH_2CH_3$		L
SbA_5	-16.53	(1)	--	25	C	S: 1,2-Dichloroethane. A=Cl	65Oa	
P20	3-PENTANONE, 2,2-dimethyl-		$C_7H_{14}O$			$CH_3CH_2COC(CH_3)_3$		L
SbA_5	-14.58	(1)	--	25	C	S: 1,2-Dichloroethane. A=Cl	65Oa	
P21	3-PENTANONE, 2,4-dimethyl-		$C_7H_{14}O$			$(CH_3)_2CHCOCH(CH_3)_2$		L
SbA_5	-16.20	(1)	--	25	C	S: 1,2-Dichloroethane. A=Cl	65Oa	
I_2	-2.5	(1)	--	0-45	T	S: Cyclohexane C_L≈1M	66Bc	b,q: ±1
P22	3-PENTANONE, 2-methyl-		$C_6H_{12}O$			$CH_3CH_2COCH(CH_3)_2$		L
SbA_5	-16.16	(1)	--	25	C	S: 1,2-Dichloroethane. A=Cl	65Oa	
P23	3-PENTANONE, 2,2,4,4-tetramethyl-		$C_9H_{18}O$			$(CH_3)_3CCOC(CH_3)_3$		L
SbA_5	-10.30	(1)	--	25	C	S: 1,2-Dichloroethane. A=Cl	65Oa	
	-9.6	2.52(1)	-27.0	--	C	S: Ethylene chloride. A=Cl	67Of	r

Metal Ion	ΔH, kcal /mole	Log K	ΔS, cal /mole °K	T, °C	M	Conditions	Ref.	Re-marks
P24	3-PENTANONE, 2,2,4-trimethyl-		$C_8H_{16}O$			$(CH_3)_3CCOCH(CH_3)_2$		L
SbA_5	-11.34	(1)	--	25	C	S: 1,2-Dichloro-ethane. A=Cl	65Oa	
	-11.60	3.57(1)	-28.9	--	C	S: Ethylene chloride. A=Cl	67Of	
P25	1-PENTENE		C_5H_{10}			$CH_3CH_2CH_2CH:CH_2$		L
Ag^+	3.6	0.826(25°,1)	-8.1	0-40	T	S: Ethylene glycol	65Ci	b
P26	2-PENTENE(cis)		C_5H_{10}			$CH_3CH_2CH:CHCH_3$		L
Ag^+	3.6	0.748(25°,1)	-8.5	0-40	T	S: Ethylene glycol	65Ci	b
P27	2-PENTENE(trans)		C_5H_{10}			$CH_3CH_2CH:CHCH_3$		L
Ag^+	2.9	0.255(25°,1)	-8.7	0-40	T	S: Ethylene glycol	65Ci	b
P28	3-PENTEN-2-ONE, 4-amino-		C_5H_9ON			$CH_3C(NH_2):CHCOCH_3$		L^-
Co^{2+}	0.706	0.492(25°)	4.62	-33-72	T	S: 15 v % Tetra-methylsilane in $CHCl_3$	68Ea	b,t
P29	3-PENTEN-2-ONE, 4-amino-1,1,1-trifluoro-		$C_5H_6ONF_3$			$CH_3C(NH_2):CHCOCF_3$		L^-
Co^{2+}	∿0	0.307	1.41	-43-67	T	S: 15 v % Tetra-methylsilane in $CHCl_3$	68Ea	b,t
P30	3-PENTEN-2-ONE, 4-methyl-amino-		$C_6H_{11}ON$			$CH_3C(NHCH_3):CHCOCH_3$		L^-
Ni^{2+}	2.950	-1.30(25°)	3.98	-43-77	T	S: $CDCl_3$	68Ea	b,t
P31	3-PENTEN-2-ONE, 4-methyl-amino-1,1,1-trifluoro-		$C_6H_8ONF_3$			$CH_3C(NHCH_3):CHCOCF_3$		L^-
Ni^{2+}	2.760	1.129(25°)	4.11	17-102	T	S: $CDCl_3$	68Ea	b,t
P32	PERCHLORATE ION		ClO_4			ClO_4^-		L^-
Ce^{3+}	-11.8	1.9(1)	-31	25	T	$\mu \to 0$	55Ha	
	-17	0.079(26.5°,1)	-57	17.5-31.4	T	μ=1.14	56Sc	b
	-14	-0.387(26°,1)	-48	17.5-31.4	T	μ=5.11	56Sc	b
[Fe]	-3.5	--	--	25	C	Tris-cacodylate buffer (≈0.1M Tris) [Fe]=G. gouldii Hemerythrin=$HrFe_2O_3$ R: $HrFe_2O_3 + L^- = HrFe_2O_3L^-$	71Lb	
$[Fe^{2+}]$	-4.5	--	--	25	C	Tris-cacodylate buffer (≈0.1M Tris)	71Lb	q:Na

Metal Ion	ΔH, kcal /mole	Log K	ΔS, cal /mole °K	T, °C	M	Conditions	Ref.	Remarks
P32, cont.								
$[Fe^{2+}]$, cont.						[Fe]=G. gouldii Hemerythrin = $HrFe_2$ R: $HrFe_2 + L^- = HrFe_2L^-$		
PtA_4	-1.48	0.455(1)	-2.8	25	C	μ=1M(NaF) C_M=0.040-0.050M $A=NH_3$	73Ma	q: ±0.10
	-1.28	0.484(1)	-2.1	25	C	μ=1M(NaF) C_M=0.040-0.050M A=Ethylenediamine	73Ma	q: ±0.10
NA_4^+	0.22	0.25(1)	1.9	25	C	μ=0. $A=CH_3$	69Ia	
	0.19	0.18(1)	1.5	25	C	μ=0. $A=CH_3CH_2$	69Ia	
	2.5	0.03(1)	8.4	25	C	μ=0. $A=CH_3CH_2CH_2$	69Ia	
P33 PERIODATE ION			IH_2O_6			$H_2IO_6^{3-}$		L^{3-}
CuA_4^{2-}	-7.5	10.468(40°)	23.8	30-50	T	$1.3M(ClO_4^-)$ R: $CuA_4^{2-} + 2L^{3-} = Cu(IO_6)_2^{7-} + 4H_2O$ A=OH	53Lb	b
P34 PERSULFATE ION			O_8S_2			$S_2O_8^{2-}$		L^{2-}
Na^+	4.3	0.580(25°,1)	17.0	25&37	T	μ=0	71Cb	a,b
K^+	1.6	0.914(25°,1)	9.6	25&37	T	μ=0	71Cb	a,b
Rb^+	1.3	1.167(25°,1)	9.8	25&37	T	μ=0	71Cb	a,b
Cs^+	-1.0	1.415(25°,1)	3.2	25&37	T	μ=0	71Cb	a,b
P35 PERYLENE			$C_{20}H_{12}$			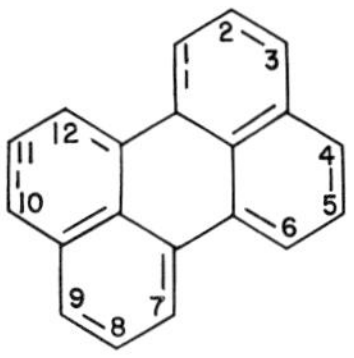		L
Li^+	0.0	5.194(1)	24	-65	T	S: Tetrahydrofuran	68Nb	q: ±0.4
	1.4	5.301(1)	29	15	T	S: Tetrahydrofuran	68Nb	q: ±0.4
Na^+	0	4.72(1)	21.5	-55	T	μ≃0. S: Dimethoxy ethane	66Cb	
	2.5	5.15(15°,1)	32.5	20	T	μ≃0. S: Dimethoxy ethane	66Cb	
	0.0	4.54(1)	21	-65	T	μ≃0. S: Tetra- hydrofuran	66Cb 68Nb	q: ±0.4
	2.2	4.75(1)	29	15	T	μ≃0. S: Tetra- hydrofuran	66Cb 68Nb	q: ±0.4

Metal Ion	ΔH, kcal /mole	Log K	ΔS, cal /mole °K	T, °C	M	Conditions	Ref.	Re-marks
P36 1,10-PHENANTHROLINE			$C_{12}H_8N_2$			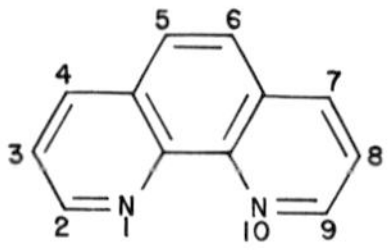		L
Mn^{2+}	-3.5	4.13(1)	6.8	20	C	μ=0.1($NaNO_3$)	63Ac	
	-3	3.8(25°,1)	8	-46-0	T	μ=0.2($NaClO_4$) S: Methanol	73Be	b,q
	-7.0	7.61(1-2)	10.9	20	C	μ=0.1($NaNO_3$)	63Ac	
	-9.0	10.3(1-3)	10.4	20	C	μ=0.1($NaNO_3$)	63Ac	
Fe^{2+}	-31.32	20.48(1-3)	-11.3	25	T	μ≈10^{-3}	64Lc	
	-33.0	21.3(1-3)	-15.4	20	C	μ=0.1($NaNO_3$)	63Ac	
Fe^{3+}	-9.9	13.79(1-3)	29	25	T	μ≈10^{-3}	64Lc	
Co^{2+}	-9.1	7.25(1)	2.1	20	C	μ=0.1($NaNO_3$)	63Ac	
	-15.8	13.95(1-2)	9.9	20	C	μ=0.1($NaNO_3$)	63Ac	
	-23.8	19.9(1-3)	9.9	20	C	μ=0.1($NaNO_3$)	63Ac	
Ni^{2+}	-11.2	8.8(1)	2.1	20	C	μ=0.1($NaNO_3$)	63Ac	
	-9.23	6.46(1)	-1.4	25	C	μ=0.1M(KNO_3) S: 50 w % Ethanol	72Bj	
	-9.16	5.98(2)	-3.3	25	C	μ=0.1M(KNO_3) S: 50 w % Ethanol	72Bj	
	-8.17	5.24(3)	-3.4	25	C	μ=0.1M(KNO_3) S: 50 w % Ethanol	72Bj	
	-20.5	17.1(1-2)	8.2	20	C	μ=0.1($NaNO_3$)	63Ac	
	-30.0	24.8(1-3)	11.2	20	C	μ=0.1($NaNO_3$)	63Ac	
NiA_2	-8.2	5.15(20°,1)	-4.6	10-30	T	S: $CHCl_3$. A=N-(p-Chlorophenyl)-β-mercaptocinnamamide	74Cc	b
	-8.0	5.03(20°,1)	-4.2	10-30	T	S: $CHCl_3$. A=N-(p-Ethoxyphenyl)-β-mercaptocinnamamide	74Cc	b
	-5.9	4.05(20°,1)	-1.5	10-30	T	S: $CHCl_3$. A=N-Ethyl-β-mercapto-cinnamamide	74Cc	b
	-8.8	5.46(20°,1)	-5.0	10-30	T	S: $CHCl_3$. A=Ethyl-β-mercaptothiocinnamate	74Cc	b
	-9.9	5.90(20°,1)	-7.1	10-30	T	S: $CHCl_3$. A=Ethyl-β-mercaptocinnamate	74Cc	b
	-4.8	3.39(20°,1)	-0.2	10-30	T	S: $CHCl_3$. A=N-Methyl-N-phenyl-β-mercaptocinnamamide	74Cc	b
	-7.7	4.99(20°,1)	-3.4	10-30	T	S: $CHCl_3$. A=N-(α-Naphtyl)-β-mercaptocinnamamide	74Cc	b
	-5.5	3.87(20°,1)	-0.9	10-30	T	S: $CHCl_3$. A=Mono-thiodibenzoylmethane	74Cc	b
	-8.0	5.10(20°,1)	-4.3	10-30	T	S: $CHCl_3$. A=N-Phenyl-β-mercapto-cinnamamide	74Cc	b
	-5.2	3.83(20°,1)	-0.1	10-30	T	S: $CHCl_3$ A=Thiobenzoylacetone	74Cc	b

Metal Ion	ΔH, kcal /mole	Log K	ΔS, cal /mole °K	T, °C	M	Conditions	Ref.	Remarks
P36, cont.								
Cu^{2+}	-11.03	9.14(1)	4.8	25	C	μ=0	70Ea	
	-11.7	9.25(1)	2.4	20	C	μ=0.1($NaNO_3$)	63Ac	
	-5.42	6.87(2)	13.2	25	C	μ=0	70Ea	
	-5.1	5.42(3)	1.1	25	C	μ=0	70Ea	
	-18.2	16.00(1-2)	11.2	20	C	μ=0.1($NaNO_3$)	63Ac	
	-26.4	21.35(1-3)	7.5	20	C	μ=0.1($NaNO_3$)	63Ac	
CuA_2	-20.8	(1)	--	30	C	S: Benzene A=1,1,1,5,5,5-Hexafluoropentane-2,4-dione	74Gb	
	-12.73	(1)	--	30	C	S: Benzene A=1,1,1-Trifluoropentane-2,4-dione	74Gb	
	-13.0	(1)	--	30	C	S: Benzene A=1,1,1-Trifluoro-4-phenylbutane-2,4-dione	74Gb	
	-14.9	(1)	--	30	C	S: Benzene A=1,1,1-Trifluoro-4-(2-thienyl) butane-2,4-dione	74Gb	
	-8.49	5.40(1)	-3.70	25	C	S: Chloroform A=4,4,4-Trifluoro-1-(2-thienyl)-1,3-butanedione	72Ka	
Zn^{2+}	-7.5	6.17(1)	3.1	25	C	μ=0	70Ea	
	-7.5	6.55(1)	4.4	20	C	μ=0.1($NaNO_3$)	63Ac	
	-6.67	5.97(1)	5.0	25	C	μ=0.1M(KNO_3) S: 50 w % Ethanol	72Bj	
	-4.76	5.91(2)	11.1	25	C	μ=0	70Ea	
	-7.26	5.50(2)	0.7	25	C	μ=0.1M(KNO_3) S: 50 w % Ethanol	72Bj	
	-2.9	5.25(3)	14.3	25	C	μ=0	70Ea	
	-4.70	4.97(3)	7.0	25	C	μ=0.1M(KNO_3) S: 50 w % Ethanol	72Bj	
	-15.0	12.35(1-2)	5.5	20	C	μ=0.1($NaNO_3$)	63Ac	
	-19.3	17.55(1-3)	14.3	20	C	μ=0.1($NaNO_3$)	63Ac	
ZnA_2	-11.08	2.50(1)	-26.83	25	C	S: Chloroform A=4,4,4-Trifluoro-1-(2-thienyl)-1,3-butanedione	72Ka	
Cd^{2+}	-6.3	5.78(1)	4.8	20	C	μ=0.1($NaNO_3$)	63Ac	
	-5.37	5.48(1)	7.0	25	C	μ=0.1M(KNO_3) S: 50 w % Ethanol	72Bj	
	-5.83	5.31(2)	4.8	25	C	μ=0.1M(KNO_3) S: 50 w % Ethanol	72Bj	
	-3.98	4.05(3)	5.2	25	C	μ=0.1M(KNO_3) S: 50 w % Ethanol	72Bj	
	-13.1	10.82(1-2)	4.8	20	C	μ=0.1($NaNO_3$)	63Ac	
	-16.1	14.92(1-3)	13.3	20	C	μ=0.1($NaNO_3$)	63Ac	
Tl^{3+}	-11.8	11.6(1)	-13	25	C	μ=1.0	62Kc	

Metal Ion	ΔH, kcal /mole	Log K	ΔS, cal /mole °K	T, °C	M	Conditions	Ref.	Re-marks
P36, cont.								
Tl^{3+}, cont.	-23.6	18.3(2)	-5	25	C	μ=1.0	62Kc	
	-8.8	5.4	-5	25	C	μ=1.0. R: Tl^{3+} + HL^+ = TlL^{3+} + H^+	62Kc	
	-11.0	10.6	12	25	C	μ=1.0. R: TlL^{3+} + HL^+ = $TlL_2{}^{3+}$ + H^+	62Kc	
P37 1,10-PHENANTHROLINE, 2,9-dimethyl-			$C_{14}H_{12}N_2$			$C_{12}H_6N_2(CH_3)_2$		
NiA_2	-11.2	(1)	--	30	C	S: Benzene A=Diethyldithio-phosphate	71Dc	q: ±1.2
CuA_2	-13.1	(1)	--	30	C	S: Benzene A=1,1,1,5,5,5-Hexa-fluoropentane-2,4-dione	74Gb	
	-7.17	3.531(1)	-7.41	30	C	S: Benzene A=1,1,1-Trifluoro-pentane-2,4-dione	74Gb	
	-8.13	3.672(1)	-10.0	30	C	S: Benzene A=1,1,1-Trifluoro-4-phenylbutane-2,4-dione	74Gb	
	-9.08	(1)	--	30	C	S: Benzene A=1,1,1-Trifluoro-4-(2-thienyl) butane-2,4-dione	74Gb	
Zn^{2+}	-1.32	3.79(1)	12.5	25	C	μ=0.1M(KNO_3) S: 50 w % Ethanol	72Ed	
	-4.36	2.58(2)	-2.4	25	C	μ=0.1M(KNO_3) S: 50 w % Ethanol	72Ed	
Cd^{2+}	-4.4	3.94(1)	3.2	25	C	μ=0.1M(KNO_3) S: 50 w % Ethanol	72Ed	
	-4.46	2.40(2)	-3.9	25	C	μ=0.1M(KNO_3) S: 50 w % Ethanol	72Ed	
P38 1,10-PHENANTHROLINE, 4,7-dimethyl-			$C_{14}H_{12}N_2$			$C_{12}H_6N_2(CH_3)_2$		
Ni^{2+}	-9.32	6.54(1)	-1.3	25	C	μ=0.1M(KNO_3) S: 50 w % Ethanol	72Ed	
	-11.40	5.59(2)	-12.7	25	C	μ=0.1M(KNO_3) S: 50 w % Ethanol	72Ed	
	-6.65	6.30(3)	6.7	25	C	μ=0.1M(KNO_3) S: 50 w % Ethanol	72Ed	
Zn^{2+}	-5.44	6.55(1)	11.7	25	C	μ=0.1M(KNO_3) S: 50 w % Ethanol	72Ed	
	-5.80	6.73(2)	11.3	25	C	μ=0.1M(KNO_3) S: 50 w % Ethanol	72Ed	
	-6.29	5.86(3)	5.8	25	C	μ=0.1M(KNO_3) S: 50 w % Ethanol	72Ed	
Cd^{2+}	-5.06	6.19(1)	11.4	25	C	μ=0.1M(KNO_3) S: 50 w % Ethanol	72Ed	

Metal Ion	ΔH, kcal /mole	Log K	ΔS, cal /mole °K	T, °C	M	Conditions	Ref.	Re-marks
P38, cont.								
Cd^{2+}, cont.	-4.74	5.92(2)	11.1	25	C	μ=0.1M(KNO_3) S: 50 w % Ethanol	72Ed	
	-5.46	4.76(3)	3.5	25	C	μ=0.1M(KNO_3) S: 50 w % Ethanol	72Ed	
P39	1,10-PHENANTHROLINE, 5,6-dimethyl-		$C_{14}H_{12}N_2$			$C_{12}H_6N_2(CH_3)_2$		L
Fe^{2+}	-30.75	21.95(25°,1-3)	-2.6	20-40	T	μ=0	68La	b
Fe^{3+}	-18.44	18.54(25°,1-3)	23.0	20-40	T	μ=0	68La	b
Ni^{2+}	-10.37	5.83(1)	-8.1	25	C	μ=0.1M(KNO_3) S: 50 w % Ethanol	72Ed	
	-11.11	6.30(2)	-8.4	25	C	μ=0.1M(KNO_3) S: 50 w % Ethanol	72Ed	
	-9.37	4.53(3)	-10.7	25	C	μ=0.1M(KNO_3) S: 50 w % Ethanol	72Ed	
Zn^{2+}	-7.31	5.71(1)	1.6	25	C	μ=0.1M(KNO_3) S: 50 w % Ethanol	72Ed	
	-9.31	6.19(2)	-2.9	25	C	μ=0.1M(KNO_3) S: 50 w % Ethanol	72Ed	
	-5.83	5.15(3)	4.0	25	C	μ=0.1M(KNO_3) S: 50 w % Ethanol	72Ed	
Cd^{2+}	-7.2	5.77(1)	2.2	25	C	μ=0.1M(KNO_3) S: 50 w % Ethanol	72Ed	
	-7.3	5.34(2)	0	25	C	μ=0.1M(KNO_3) S: 50 w % Ethanol	72Ed	
	-4.12	4.13(3)	5.1	25	C	μ=0.1M(KNO_3) S: 50 w % Ethanol	72Ed	
P40	1,10-PHENANTHROLINE, 6-methyl-		$C_{13}H_{10}N_2$			$C_{12}H_7N_2(CH_3)$		L
Fe^{2+}	-33.8	21.01(25°,1-3)	-17.2	15-45	T	μ=0	67La	b
Ni^{2+}	-10.09	6.19(1)	-5.5	25	C	μ=0.1M(KNO_3) S: 50 w % Ethanol	72Bj	
	-8.87	5.93(2)	-2.6	25	C	μ=0.1M(KNO_3) S: 50 w % Ethanol	72Bj	
	-11.00	4.93(3)	-14.4	25	C	μ=0.1M(KNO_3) S: 50 w % Ethanol	72Bj	
Zn^{2+}	-6.86	5.83(1)	3.6	25	C	μ=0.1M(KNO_3) S: 50 w % Ethanol	72Bj	
	-8.18	5.70(2)	-1.3	25	C	μ=0.1M(KNO_3) S: 50 w % Ethanol	72Bj	
	-5.83	5.01(3)	3.4	25	C	μ=0.1M(KNO_3) S: 50 w % Ethanol	72Bj	
Cd^{2+}	-6.26	5.32(1)	3.4	25	C	μ=0.1M(KNO_3) S: 50 w % Ethanol	72Bj	
	-5.47	5.27(2)	5.7	25	C	μ=0.1M(KNO_3) S: 50 w % Ethanol	72Bj	
	-4.84	4.43(3)	4.1	25	C	μ=0.1M(KNO_3) S: 50 w % Ethanol	72Bj	

Metal Ion	ΔH, kcal /mole	Log K	ΔS, cal /mole °K	T, °C	M	Conditions	Ref.	Re-marks
P41	1,10-PHENANTHROLINE, 6-nitro-		$C_{12}H_7O_2N_3$			$C_{12}H_7N_2(NO_2)$		L
Fe^{2+}	-25.2	15.63(<u>1</u>-<u>3</u>)	-13.0	25	T	μ=0	64Ld	
	-18.13	5.94	-35.0	25	T	μ=0. R: $Fe^{2+} + 3HL^+ = FeL_3^{2+} + 3H^+$	64Ld	
P42	1,10-PHENANTHROLINE, 5-phenyl-		$C_{18}H_{12}N_2$			$C_{12}H_7N_2(C_6H_5)$		
Ni^{2+}	-9.65	5.67(<u>1</u>)	-6.4	25	C	μ=0.1M(KNO_3) S: 50 w % Ethanol	72Bj	
	-8.81	4.90(<u>2</u>)	-7.1	25	C	μ=0.1M(KNO_3) S: 50 w % Ethanol	72Bj	
	-9.74	4.66(<u>3</u>)	-11.4	25	C	μ=0.1M(KNO_3) S: 50 w % Ethanol	72Bj	
Zn^{2+}	-6.84	5.51(<u>1</u>)	2.3	25	C	μ=0.1M(KNO_3) S: 50 w % Ethanol	72Bj	
	-5.26	5.39(<u>2</u>)	7.0	25	C	μ=0.1M(KNO_3) S: 50 w % Ethanol	72Bj	
	-4.59	4.08(<u>3</u>)	3.3	25	C	μ=0.1M(KNO_3) S: 50 w % Ethanol	72Bj	
Cd^{2+}	-5.50	5.12(<u>1</u>)	4.9	25	C	μ=0.1M(KNO_3) S: 50 w % Ethanol	72Bj	
	-5.00	4.77(<u>2</u>)	5.1	25	C	μ=0.1M(KNO_3) S: 50 w % Ethanol	72Bj	
P43	PHENOL		C_6H_6O			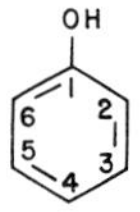		L^-
Fe^{3+}	-0.1	7.84(<u>1</u>)	36	25	T	μ=0.1	67Mi <u>69Da</u>	
[P]	-8.2	2.710(25°,<u>1</u>)	-15.2	20-70	T	[P]=$Et_2P(O)OEt$ S: CCl_4	68Aa	b
	-7.9	2.512(25°,<u>1</u>)	-15.1	20-70	T	[P]=$EtP(O)(OEt)_2$ S: CCl_4	68Aa	b
	-8.1	2.338(25°,<u>1</u>)	-16.5	20-70	T	[P]=$(EtO)_3P{=}O$ S: CCl_4	68Aa	b
	-7.8	2.227(25°,<u>1</u>)	-15.9	20-70	T	[P]= 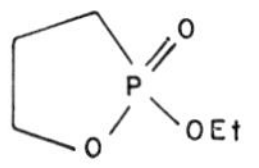 S: CCl_4	68Aa	b
	-7.6	1.981(25°,<u>1</u>)	-16.3	20-70	T	[P]= 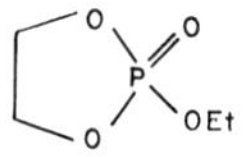 S: CCl_4	68Aa	b
	-7.6	2.487(25°,<u>1</u>)	-14.2	20-70	T	[P]= 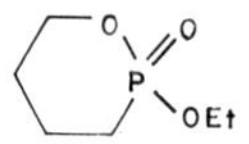	68Aa	b

Metal Ion	ΔH, kcal /mole	Log K	ΔS, cal /mole °K	T, °C	M	Conditions	Ref.	Re-marks
P43, cont.								
[P], cont.								
	-7.6	2.257(25°,1)	-15.0	20-70	T	S: CCl_4 [P]= 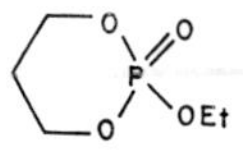	68Aa	b
	-8.3	2.520(25°,1)	-16.4	20-70	T	S: CCl_4 [P]= 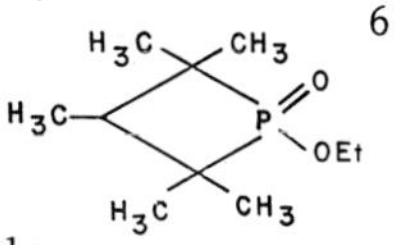	68Aa	b
	-9.6	2.797(25°,1)	-19.4	20-70	T	S: CCl_4 [P]=Ph_3P=O	68Aa	b
	-9.5	2.834(25°,1)	-19.2	20-70	T	S: CCl_4 [P]=$Ph_2(CH_3)P$=O	68Aa	b
	-9.6	2.959(25°,1)	-18.7	20-70	T	S: CCl_4 [P]=$Ph(CH_3)_2P$=O	68Aa	b
	-9.6	3.000(25°,1)	-18.5	20-70	T	S: CCl_4 [P]=Bu_3P=O	68Aa	b
	-10.5	3.245(25°,1)	-20.5	20-70	T	S: CCl_4 [P]= 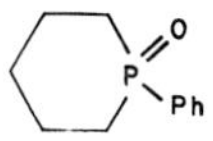	68Aa	b
	-12.8	3.332(25°,1)	-27.7	20-70	T	S: CCl_4 [P]= 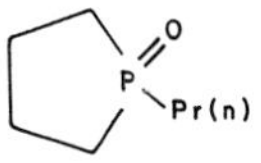	68Aa	b
	-8.9	2.281(25°,1)	-19.4	20-70	T	S: CCl_4 [P]= 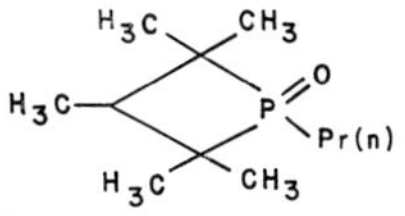	68Aa	b
	-7.5	1.806(20°,1)	-17.4	20&50	T	S: CCl_4 [P]=Pyridine S: CCl_4	60Aa	a,b
P44 PHENOL, 2-amino-			C_6H_7ON			$C_6H_5O(NH_2)$		
La^{3+}	7.1	3.54(25°,1)	55	15-35	T	μ=0.12M($NaClO_4$) S: 50% DMSO	69Bd	b
	-2.0	4.50(25°,1)	14	15-35	T	μ=0.12M($NaClO_4$) S: 50% Dioxane	69Bd	b
	-2.0	3.59(25°,1)	10	15-35	T	μ=0.12M($NaClO_4$) S: 50% Ethanol	69Bd	b
Nd^{3+}	6.9	3.72(25°,1)	52	15-35	T	μ=0.12M($NaClO_4$) S: 50% DMSO	69Bd	b
	-3.0	4.62(25°,1)	11	15-35	T	μ=0.12M($NaClO_4$) S: 50% Dioxane	69Bd	b
	-2.6	4.31(25°,1)	11	15-35	T	μ=0.12M($NaClO_4$) S: 50% Ethanol	69Bd	b

Metal Ion	ΔH, kcal /mole	Log K	ΔS, cal /mole °K	T, °C	M	Conditions	Ref.	Remarks
P44, cont.								
Gd^{3+}	6.1	3.80(25°,1)	52	15-35	T	μ=0.12M($NaClO_4$) S: 50% DMSO	69Bd	b
	-1.8	5.04(25°,1)	17	15-35	T	μ=0.12M($NaClO_4$) S: 50% Dioxane	69Bd	b
	-2.3	4.48(25°,1)	13	15-35	T	μ=0.12M($NaClO_4$) S: 50% Ethanol	69Bd	b
Tb^{3+}	4.4	3.82(25°,1)	44	15-35	T	μ=0.12M($NaClO_4$) S: 50% DMSO	69Bd	b
	-2.2	5.14(25°,1)	16	15-35	T	μ=0.12M($NaClO_4$) S: 50% Dioxane	69Bd	b
	-2.7	4.65(25°,1)	12	15-35	T	μ=0.12M($NaClO_4$) S: 50% Ethanol	69Bd	b
Dy^{3+}	3.4	3.98(25°,1)	41	15-35	T	μ=0.12M($NaClO_4$) S: 50% DMSO	69Bd	b
	-2.7	5.25(25°,1)	15	15-35	T	μ=0.12M($NaClO_4$) S: 50% Dioxane	69Bd	b
	-2.7	4.75(25°,1)	13	15-35	T	μ=0.12M($NaClO_4$) S: 50% Ethanol	69Bd	b
Ho^{3+}	4.8	4.23(25°,1)	48	15-35	T	μ=0.12M($NaClO_4$) S: 50% DMSO	69Bd	b
	-2.8	5.32(25°,1)	15	15-35	T	μ=0.12M($NaClO_4$) S: 50% Dioxane	69Bd	b
	-2.8	4.79(25°,1)	13	15-35	T	μ=0.12M($NaClO_4$) S: 50% Ethanol	69Bd	b
Er^{3+}	6.4	4.41(25°,1)	57	15-35	T	μ=0.12M($NaClO_4$) S: 50% DMSO	69Bd	b
	-2.5	5.52(25°,1)	17	15-35	T	μ=0.12M($NaClO_4$) S: 50% Dioxane	69Bd	b
	-3.0	4.89(25°,1)	13	15-35	T	μ=0.12M($NaClO_4$) S: 50% Ethanol	69Bd	b
Tm^{3+}	6.9	4.48(25°,1)	49	15-35	T	μ=0.12M($NaClO_4$) S: 50% DMSO	69Bd	b
	-2.3	5.65(25°,1)	18	15-35	T	μ=0.12M($NaClO_4$) S: 50% Dioxane	69Bd	b
	-3.2	5.01(25°,1)	12	15-35	T	μ=0.12M($NaClO_4$) S: 50% Ethanol	69Bd	b
Yb^{3+}	7.3	4.54(25°,1)	62	15-35	T	μ=0.12M($NaClO_4$) S: 50% DMSO	69Bd	b
	-2.4	5.82(25°,1)	19	15-35	T	μ=0.12M($NaClO_4$) S: 50% Dioxane	69Bd	b
	-3.2	5.12(25°,1)	13	15-35	T	μ=0.12M($NaClO_4$) S: 50% Ethanol	69Bd	b
P45 PHENOL, 2-chloro-			C_6H_5OCl			$C_6H_5O(Cl)$		L^-
Fe^{3+}	0.2	6.08(1)	28	25	T	μ=0.1	69Da	
P46 PHENOL, 3-chloro-			C_6H_5OCl			$C_6H_5O(Cl)$		L^-
Fe^{3+}	-0.2	6.89(1)	31	25	T	μ=0.1	69Da	

Metal Ion	ΔH, kcal /mole	Log K	ΔS, cal /mole °K	T, °C	M	Conditions	Ref.	Re-marks
P47	PHENOL, 4-chloro-		C_6H_5OCl			$C_6H_5O(Cl)$		L^-
Fe^{3+}	-0.2	7.26(1)	32	25	T	μ=0.1	67Mi 69Da	
P48	PHENOL, 2-cyano-		C_7H_5ON			$C_6H_5O(CN)$		L^-
Fe^{3+}	0.65	5.53(1)	27	25	T	μ=0.1	69Da	
P49	PHENOL, 3-cyano-		C_7H_5ON			$C_6H_5O(CN)$		L^-
Fe^{3+}	0.9	6.30(1)	31	25	T	μ=0.1	69Da	
P50	PHENOL, 4-cyano-		C_7H_5ON			$C_6H_5O(CN)$		L^-
Fe^{3+}	-0.2	5.79(1)	26	25	T	μ=0.1	69Da	
P51	PHENOL, 3-nitro-		$C_6H_5O_3N$			$C_6H_5O(NO_2)$		L^-
Fe^{3+}	1.2	6.2(1)	33	25	T	μ=0.1	67Mi 69Da	
P52	PHENOL, 4-nitro-		$C_6H_5O_3N$			$C_6H_5O(NO_2)$		L^-
Fe^{3+}	0.3	5.3(1)	25	25	T	μ=0.1	67Mi 69Da	
P53	PHENOL, 2(2-thiazolylazo)- 4-chlorophenol		$C_9H_8ON_3ClS$			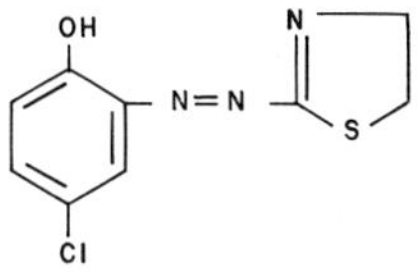		
La^{3+}	-14.58	6.63(25°,1)	-18.59	15-35	T	μ=0.1(KNO_3)	74Ka	b
Pr^{3+}	-12.35	7.48(25°,1)	-7.18	15-35	T	μ=0.1(KNO_3)	74Ka	b
Sm^{3+}	-13.97	7.77(25°,1)	-11.17	15-35	T	μ=0.1(KNO_3)	74Ka	b
Eu^{3+}	-11.78	7.82(25°,1)	-3.74	15-35	T	μ=0.1(KNO_3)	74Ka	b
Gd^{3+}	-10.12	7.94(25°,1)	2.43	15-35	T	μ=0.1(KNO_3)	74Ka	b
Dy^{3+}	-12.96	8.38(25°,1)	-5.24	15-35	T	μ=0.1(KNO_3)	74Ka	b
Ho^{3+}	-11.75	8.24(25°,1)	-1.70	15-35	T	μ=0.1(KNO_3)	74Ka	b
Er^{3+}	-12.15	8.18(25°,1)	-3.29	15-35	T	μ=0.1(KNO_3)	74Ka	b
Yb^{3+}	-9.11	8.43(25°,1)	7.98	15-35	T	μ=0.1(KNO_3)	74Ka	b
P54	PHENOL, 2(2-thiazolylazo)- 4-methoxy		$C_{10}H_{11}O_2N_3S$			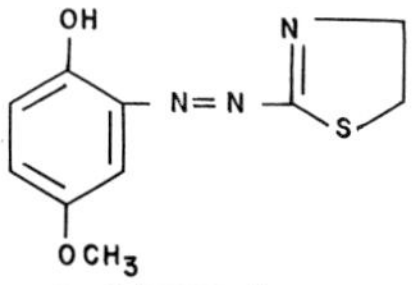		
Pr^{3+}	-6.31	8.52(25°,1)	17.81	15-35	T	μ=0.1(KNO_3)	74Ka	b
Nd^{3+}	-5.76	8.53(25°,1)	19.71	15-35	T	μ=0.1(KNO_3)	74Ka	b
Sm^{3+}	-8.43	8.98(25°,1)	12.81	15-35	T	μ=0.1(KNO_3)	74Ka	b
Eu^{3+}	-6.58	9.09(25°,1)	19.52	15-35	T	μ=0.1(KNO_3)	74Ka	b

Metal Ion	ΔH, kcal /mole	Log K	ΔS, cal /mole °K	T, °C	M	Conditions	Ref.	Remarks
P54, cont.								
Gd^{3+}	-6.32	8.96(25°,1)	19.81	15-35	T	μ=0.1(KNO_3)	74Ka	b
Dy^{3+}	-10.45	9.39(25°,1)	7.87	15-35	T	μ=0.1(KNO_3)	74Ka	b
Ho^{3+}	-8.39	9.63(25°,1)	15.91	15-35	T	μ=0.1(KNO_3)	74Ka	b
Er^{3+}	-7.29	9.49(25°,1)	18.97	15-35	T	μ=0.1(KNO_3)	74Ka	b
Yb^{3+}	-4.82	9.60(25°,1)	27.76	15-35	T	μ=0.1(KNO_3)	74Ka	b
P55 PHOSPHATE ION			O_4P			PO_4^{3-}		L^{3-}
Li^+	≈6	0.72(25°,1)	≈24	0&25	T	μ=0.2	56Sc	a,b
Na^+	≈6	0.60(25°,1)	≈24	0&25	T	μ=0.2	56Sc	a,b
K^+	≈6	0.49(25°,1)	≈24	0&25	T	μ=0.2	56Sc	a,b
Mg^{2+}	2.917	2.708(25°,1)	22.2	25	T	μ=0 ΔCp=60	54Cd	
	3.640	2.791(37°,1)	24.6	37	T	μ=0 ΔCp=62	54Cd	
	≈5	1.881(25°)	≈25	0&25	T	μ=0.2[n-$(C_3H_7)_4N$]$^+$ R: Mg^{2+} + HL^{2-} = MgHL	56Sb	a,b
Ca^{2+}	3.1	0.462(25°,1)	40	25&37	T	μ=0	68Ce	a,b,q: ±2.0
	3.3	2.739(25°,1)	23	25&37	T	μ=0 R: Ca^{2+} + HL^{2-} = CaHL	68Ce	a,b,q: ±0.6
	≈5	1.699(25°)	≈25	0&25	T	μ=0.2[n-$(C_3H_7)_4N$]$^+$ R: Cu^{2+} + HL^{2-} = CaHL	56Sb	a,b
	3.4	1.408(25°,1)	18	25&37	T	μ=0 R: Ca^{2+} + H_2L^- = CaH_2L^+	68Ce	a,b,q: ±1.7
Sr^{2+}	≈5	1.518(25°)	≈25	0&25	T	μ=0.2[n-$(C_3H_7)_4N$]$^+$ R: Sr^{2+} + HL^{2-} = SrHL	56Sb	a,b
Mn^{2+}	≈5	2.582(25°)	≈25	0&25	T	μ=0.2[n-$(C_3H_7)_4N$]$^+$ R: Mn^{2+} + HL^{2-} = MnHL	56Sb	a,b
Fe^{2+}	12.8	7.340(30,°1)	75	20&30	T	μ=0	64Lb	a,b,q: ±2
Fe^{3+}	3.77	10.88(25°)	37	25-50	T	μ=0.037 R: Fe^{3+} + HL^{2-} = $FeHL^+$	65La	b
P56 PHOSPHATE ION, diimidotri-			$P_3H_2O_8N_2$			$[P_3O_8(NH)_2]^{5-}$		L^{5-}
Ca^{2+}	-9.1	7.10(1)	1.9	25	T	μ=0	61Ia	
P57 PHOSPHATE ION, dithio-, dicyclohexyl ester			$C_{12}H_{22}O_2S_2P$			$(C_6H_{11}O)_2PS_2$		
CdL_2	-3.10	3.40	-2.63	30	C	S: Benzene R: $2CdL_2$ = Cd_2L_4	71Db	q: ±0.6
P58 PHOSPHATE ION, dithio-, diisobutyl ester			$C_8H_{18}O_2S_2P$			$[(CH_3)_2CHCH_2O]_2PS_2$		
ZnL_2	-4.13	0.85	-11.7	30	C	S: Benzene R: $2ZnL_2$ = Zn_2L_4	71Db	q: ±0.4
CdL_2	-4.09	3.70	-5.02	30	C	S: Benzene R: $2CdL_2$ = Cd_2L_4	71Db	q: ±0.7

Metal Ion	ΔH, kcal /mole	Log K	ΔS, cal /mole °K	T, °C	M	Conditions	Ref.	Remarks
P59 PHOSPHATE ION, dithio-, diisopropyl ester			$C_6H_{14}O_2S_2P$			$[(CH_3)_2CHO]_2PS_2$		
ZnL_2	-3.94	0.68	-11.5	30	C	S: Benzene R: $2ZnL_2 = Zn_2L_4$	71Db	
CdL_2	-4.78	3.80	-7.2	30	C	S: Benzene R: $2CdL_2 = Cd_2L_4$	71Db	q: ±0.7
P60 PHOSPHATE ION, dithio-, dipropyl ester			$C_6H_{14}O_2S_2P$			$(CH_3CH_2CH_2O)_2PS_2$		
HgL_2	-0.76	0.00	-2.63	30°	C	S: Benzene R: $2HgL_2 = Hg_2L_4$	71Db	q: ±0.2
P61 PHOSPHATE, hexa (meta)			$P_6O_{18}Na_6$			$(NaPO_3)_6$		L^-
Ba^{2+}	-11.16	10.32(35°,3)	11.0	35-45	T	C_M=3x10^{-4}-4x10^{-3} C_L=1x10^{-3}-1.5x10^{-2} $(NaPO_3)_n$	69Bh	b
P62 PHOSPHATE ION, imidodi-			P_2HO_6N			$[P_2O_6(NH)]^{4-}$		L^{4-}
Ca^{2+}	-5.7	6.07(1)	8.7	25	T	μ=0	61Ia	
P63 PHOSPHATE, long chain			P_n			P_n		L
Ca^{2+}	-0.7	7.28	31	25	T	μ=0. R: Ca + P_6 = CaP_6	60Ia	
	-1.2	7.23	29	25	T	μ=0. R: Ca + P_{14} = CaP_{14}	60Ia	
	1	7.54	38	25	T	μ=0. R: Ca + P_{60} = CaP_{60}	60Ia	
P64 PHOSPHATE, tri-n-butyl-			$C_{12}H_{27}O_4P$			$PO_4(CH_2CH_2CH_2CH_3)_3$		L
UO_2^{2+}	-4.3	1.5	-7.7	25	T	R: $UO_2^{2+} + 2NO_3^- + 2L = UO_2(NO_3)_2L_2$	60Na	
I_2	-2.94	1.290(25°,1)	-3.99	10-35	T	C_M=0.67x10^{-3} C_L=0.03-0.12 S: n-Heptane	60Td	b
P65 PHOSPHATE ION, tripoly-			$O_{10}H_5P_3$					L^{5-}
Be^{2+}	4.7	(1)	--	20	C	0.1M$(CH_3)_4NNO_3$	65Aa	
Mg^{2+}	4.34	7.05(1)	47	20	C	0.1M$(CH_3)_4NNO_3$	65Aa	
	-2.11	4.93(25°,1)	30	2-45	T	μ=0.10M(KNO_3)	73Kd	b
	1.4	3.33(25°)	19.9	2-45	T	μ=0.10M(KNO_3) R: $Mg^{2+} + HL^{4-} = MgHL^{2-}$	73Kd	b,q: ±0.4
Ca^{2+}	-3.2	6.90(1)	21	25	T	μ=0	60Ia	
	3.3	6.31(1)	40.0	20	C	0.1M$(CH_3)_4NNO_3$	65Aa	
	-6.6	4.80(25°,1)	9.5	2-45	T	μ=0.10M(KNO_3)	73Kd	b,q: ±3.0

Metal Ion	ΔH, kcal /mole	Log K	ΔS, cal /mole °K	T, °C	M	Conditions	Ref.	Re-marks
P65, cont.								
Ca^{2+}, cont.	-0.82	3.25(25°)	14.7	2-45	T	μ=0.10M(KNO_3) R: Ca^{2+} + HL^{4-} = $CaHL^{2-}$	73Kd	b
Sr^{2+}	3.16	5.46(1)	35.7	20	C	0.1M($(CH_3)_4NNO_3$)	65Aa	
	-5.96	4.00(25°,1)	5.0	2-45	T	μ=0.10M(KNO_3)	73Kd	b
	-1.4	2.86(25°)	8.4	2-45	T	μ=0.10M(KNO_3) R: Sr^{2+} + HL^{4-} = $SrHL^{2-}$	73Kd	b,q: ±0.3
Y^{3+}	-1.5	9.1(25°,1-2)	36	25-45	T	μ=0.1M(KNO_3)	74Ta	b,q: ±0.2
	-4.5	6.7(25°)	15	25-45	T	μ=0.1M(KNO_3) R: Y^{3+} + $2HL^{8-}$ = $YH_2L_2^{5-}$	74Ta	b
YA^-	-5.6	5.43(25°,1)	8	2-45	T	μ=0.10M(KNO_3) A=$ETAT^{2-}$	73Kf	b,g: Na,q: ±1.0
	-3.5	4.02(25°)	7	2-45	T	μ=0.10M(KNO_3) A=$EDTA^{2-}$ R: YA^- + HL^{4-} = $YAHL^{5-}$	73Kf	b,g: Na,q: ±1.0
La^{3+}	-2.5	8.0(25°,1-2)	28	25-45	T	μ=0.1M(KNO_3)	74Ta	q: ±0.2
	-4.7	6.1(25°)	12	25-45	T	μ=0.1M(KNO_3) R: La^{3+} + $2HL^{8-}$ = $LaH_2L_2^{5-}$	74Ta	
LaA^-	-3.4	3.90(25°,1)	6.0	2-45	T	μ=0.10M(KNO_3) A=$EDTA^{2-}$	73Kf	b,g: Na
	-2.9	2.54(25°)	2.1	2-45	T	μ=0.10M(KNO_3) A=$EDTA^{2-}$ R: LaA^- + HL^{4-} = $LaAHL^{5-}$	73Kf	b,g: Na,q: ±0.5
Ce^{3+}	-2.4	8.2(25°,1-2)	29	25-45	T	μ=0.1M(KNO_3)	74Ta	b,q: ±0.2
	-2.5	6.2(25°)	20	25-45	T	μ=0.1M(KNO_3) R: Ce^{3+} + $2HL^{8-}$ = $CeH_2L_2^{5-}$	74Ta	b,q: ±0.2
CeA^-	-2.9	4.25(25°,1)	10	2-45	T	μ=0.10M(KNO_3) A=$EDTA^{2-}$	73Kf	b,g: Na
	-3.4	2.90(25°)	2.0	2-45	T	μ=0.10M(KNO_3) A=$EDTA^{2-}$ R: CeA^- + HL^{4-} = $CeAHL^{5-}$	73Kf	b,g: Na,q: ±0.6
Pr^{3+}	-2.4	8.3(25°,1-2)	29	25-45	T	μ=0.1M(KNO_3)	74Ta	b,q: ±0.2
	-2.6	6.3(25°)	20	25-45	T	μ=0.1M(KNO_3) R: Pr^{3+} + $2HL^{8-}$ = $PrH_2L_2^{5-}$	74Ta	b,q: ±0.2
PrA^-	2.2	4.75(25°,1)	14	2-45	T	μ=0.10M(KNO_3) A=$EDTA^{2-}$	73Kf	b,g: Na

Metal Ion	ΔH, kcal /mole	Log K	ΔS, cal /mole °K	T, °C	M	Conditions	Ref.	Remarks
P65, cont.								
PrA^-, cont.	-3.8	3.10(25°)	1.5	2-45	T	μ=0.10M(KNO_3) $A=EDTA^{2-}$ R: $PrA^- + HL^{4-} = PrAHL^{5-}$	73Kf	b,g: Na,q: ±1.0
Nd^{3+}	-4.5	8.5(25°,<u>1</u>-<u>2</u>)	23	25-45	T	μ=0.1M(KNO_3)	74Ta	b
	-2.6	6.4(25°)	20	25-45	T	μ=0.1M(KNO_3) R: $Nd^{3+} + 2HL^{8-} = NdH_2L_2{}^{5-}$	74Ta	b,q: ±0.2
NdA^-	-2.5	4.90(25°,<u>1</u>)	14	2-45	T	μ=0.10M(KNO_3) $A=EDTA^{2-}$	73Kf	b,g: Na,q: ±0.4
	-4.1	3.20(25°)	1	2-45	T	μ=0.10M(KNO_3) $A=EDTA^{2-}$ R: $NdA^- + HL^{4-} = NdAHL^{5-}$	73Kf	b,g: Na,q: ±0.9
Sm^{3+}	-6.8	8.7(25°,<u>1</u>-<u>2</u>)	17	25-45	T	μ=0.1M(KNO_3)	74Ta	b
	-2.6	6.5(25°)	21	25-45	T	μ=0.1M(KNO_3) R: $Sm^{3+} + 2HL^{8-} = SmH_2L_2{}^{5-}$	74Ta	b,q: ±0.2
SmA^-	-2.0	5.10(25°,<u>1</u>)	16	2-45	T	μ=0.10M(KNO_3) $A=EDTA^{2-}$	73Kf	b,g: Na,q: ±0.7
	-3.6	3.35(25°)	3	2-45	T	μ=0.10M(KNO_3) $A=EDTA^{2-}$ R: $SmA^- + HL^{4-} = SmAHL^{5-}$	73Kf	b,g: Na,q: ±0.7
Eu^{3+}	-6.9	8.8(25°,<u>1</u>-<u>2</u>)	17	25-45	T	μ=0.1M(KNO_3)	74Ta	b
	-2.6	6.6(25°)	21	25-45	T	μ=0.1M(KNO_3) R: $Eu^{3+} + 2HL^{8-} = EuH_2L_2{}^{5-}$	74Ta	b,q: ±0.2
EuA^-	-2.3	5.24(25°,<u>1</u>)	16	2-45	T	μ=0.10M(KNO_3) $A=EDTA^{2-}$	73Kf	b,g: Na,q: ±0.7
	-2.8	3.44(25°)	6	2-45	T	μ=0.10M(KNO_3) $A=EDTA^{2-}$ R: $EuA^- + HL^{4-} = EuAHL^{5-}$	73Kf	b,g: Na,q: ±0.7
Gd^{3+}	-6.9	8.9(25°,<u>1</u>-<u>2</u>)	17	25-45	T	μ=0.1M(KNO_3)	74Ta	b
	-2.6	6.6(25°)	21	25-45	T	μ=0.1M(KNO_3) R: $Gd^{3+} + 2HL^{8-} = GdH_2L_2$	74Ta	b,q: ±0.2
GdA^-	-2.1	5.31(25°,<u>1</u>)	17	2-45	T	μ=0.10M(KNO_3) $A=EDTA^{2-}$	73Kf	b,g: Na,q: ±0.5
	-2.3	3.58(25°)	9	2-45	T	μ=0.10M(KNO_3) $A=EDTA^{2-}$ R: $GdA^- + HL^{4-} = GdAHL^{5-}$	73Kf	b,g: Na,q: ±0.2
Tb^{3+}	-6.8	9.1(25°,<u>1</u>-<u>2</u>)	18	25-45	T	μ=0.1M(KNO_3)	74Ta	b

Metal Ion	ΔH, kcal /mole	Log K	ΔS, cal /mole °K	T, °C	M	Conditions	Ref.	Remarks
P65, cont.								
Tb^{3+}, cont.	-4.7	6.8(25°)	15	25-45	T	μ=0.1M(KNO_3) R: Tb^{3+} + $2HL^{8-}$ = $TbH_2L_2^{5-}$	74Ta	b
TbA^-	-4.9	5.54(25°,1)	9	2-45	T	μ=0.10M(KNO_3) A=$EDTA^{2-}$	73Kf	b,g: Na,q: ±0.3
	-3.1	3.95(25°)	8	2-45	T	μ=0.10M(KNO_3) A=$EDTA^{2-}$ R: TbA^- + HL^{4-} = $TbAHL^{5-}$	73Kf	b,g: Na,q: ±0.4
Dy^{3+}	-4.7	9.2(25°,1-2)	26	25-45	T	μ=0.1M(KNO_3)	74Ta	b
	-4.7	6.9(25°)	15	25-45	T	μ=0.1M(KNO_3) R: Dy^{3+} + $2HL^{8-}$ = $DyH_2L_2^{5-}$	74Ta	b
DyA^-	-2.8	5.71(25°,1)	17	2-45	T	μ=0.10M(KNO_3) A=$EDTA^{2-}$	73Kf	b,g: Na,q: ±0.3
	-3.2	4.18(25°)	8	2-45	T	μ=0.10M(KNO_3) A=$EDTA^{2-}$ R: DyA^- + HL^{4-} = $DyAHL^{5-}$	73Kf	b,g: Na,q: ±0.6
Ho^{3+}	-6.8	9.3(25°,1-2)	26	25-45	T	μ=0.1M(KNO_3)	74Ta	b
	-6.8	7.1(25°)	9	25-45	T	μ=0.1M(KNO_3) R: Ho^{3+} + $2HL^{8-}$ = $HoH_2L_2^{5-}$	74Ta	b
HoA^-	-2.4	6.50(25°,1)	21	2-45	T	μ=0.10M(KNO_3) A=$EDTA^{2-}$	73Kf	b,g: Na,q: ±0.4
	-1.8	4.43(25°)	14	2-45	T	μ=0.10M(KNO_3) A=$EDTA^{2-}$ R: HoA^- + HL^{4-} = $HoAHL^{5-}$	73Kf	b,g: Na,q: ±0.4
Er^{3+}	-6.7	9.5(25°,1-2)	20	25-45	T	μ=0.1M(KNO_3)	74Ta	b
	-4.6	7.2(25°)	17	25-45	T	μ=0.1M(KNO_3) R: Er^{3+} + $2HL^{8-}$ = $ErH_2L_2^{5-}$	74Ta	b
ErA^-	-2.2	7.01(25°,1)	25	2-45	T	μ=0.10M(KNO_3) A=$EDTA^{2-}$	73Kf	b,g: Na,q: ±0.3
	-3.3	4.95(25°)	11	2-45	T	μ=0.10M(KNO_3) A=$EDTA^{2-}$ R: ErA^- + HL^{4-} = $ErAHL^{5-}$	73Kf	b,g: Na,q: ±2.0
Tm^{3+}	-6.7	9.7(25°,1-2)	21	25-45	T	μ=0.1M(KNO_3)	74Ta	b
	-4.6	7.3(25°)	18	25-45	T	μ=0.1M(KNO_3) R: Tm^{3+} + $2HL^{8-}$ = $TmH_2L_2^{5-}$	74Ta	b
TmA^-	-3.9	7.64(25°,1)	22	2-45	T	μ=0.10M(KNO_3) A=$EDTA^{2-}$	73Kf	b,g: Na,q: ±1.0

Metal Ion	ΔH, kcal /mole	Log K	ΔS, cal /mole °K	T, °C	M	Conditions	Ref.	Re-marks
P65, cont.								
TmA^-, cont.	-4.7	5.10(25°)	6	2-45	T	μ=0.10M(KNO_3) A=$EDTA^{2-}$ R: TmA^- + HL^{4-} = $TmAHL^{5-}$	73Kf	b,g: Na,q: ±0.9
Mn^{2+}	2.8	8.04(1)	46.4	20	C	0.1M$(CH_3)_4NNO_3$	65Aa	
	-6.0	6.20(25°,1)	8.4	2-45	T	μ=0.10M(KNO_3)	73Kd	b
	-0.01	4.30(25°)	19.6	2-45	T	μ=0.10M(KNO_3) R: Mn^{2+} + HL^{4-} = $MnHL^{2-}$	73Kd	b,q: ±0.01
Co^{2+}	4.51	7.95(1)	51.7	20	C	0.1M$(CH_3)_4NNO_3$	65Aa	
	-7.0	6.95(25°,1)	9.7	2-45	T	μ=0.10M(KNO_3)	73Kd	b,q: ±2.2
	-0.20	4.05(25°)	17.8	2-45	T	μ=0.10M(KNO_3) R: Co^{2+} + HL^{4-} = $CoHL^{2-}$	73Kd	b,q: ±0.01
Ni^{2+}	5.0	7.8(1)	52.7	20	C	0.1M$(CH_3)_4NNO_3$	65Aa	
	-1.6	7.07(25°,1)	25	2-45	T	μ=0.10M(KNO_3)	73Kd	b,q: ±0.1
	0.03	3.95(25°)	18.2	2-45	T	μ=0.10M(KNO_3) R: Ni^{2+} + HL^{4-} = $NiHL^{2-}$	73Kd	b,q: ±0.01
Cu^{2+}	4.9	9.3(1)	59.2	20	C	0.1M$(CH_3)_4NNO_3$	65Aa	
	-8.2	8.20(25°,1)	10	2-45	T	μ=0.10M(KNO_3)	73Kd	b,q: ±0.5
	-2.8	5.20(25°)	14.3	2-45	T	μ=0.10M(KNO_3) R: Cu^{2+} + HL^{4-} = $CuHL^{2-}$	73Kd	b,q: ±0.5
Zn^{2+}	6.32	8.35(1)	59.8	20	C	0.1M$(CH_3)_4NNO_3$	65Aa	
	-4.4	6.83(25°,1)	16	2-45	T	μ=0.10M(KNO_3)	73Kd	b,q: ±1.9
	-1.4	3.75(25°)	12.4	2-45	T	μ=0.10M(KNO_3) R: Zn^{2+} + HL^{4-} = $ZnHL^{2-}$	73Kd	b,q: ±0.1
Cd^{2+}	2.7	8.1(1)	46.2	20	C	0.1M$(CH_3)_4NNO_3$	65Aa	
P66	PHOSPHINE, bis(dimethyamine) fluoro-		$C_4H_{12}N_2FP$			$[(CH_3)_2N]_2PF$		
CoA_2	-6.0	-0.252(30°,1)	-21	24-46	T	S: CH_2Cl_2 C_M=0.37M C_L=0.9M A=Cl	72Nc	b
	-6.0	-0.420(28°,1)	-22	9-46	T	S: CH_2Cl_2 C_M=0.37M C_L=1.0M A=Cl	72Nc	b
	-5.4	0.577(30°,1)	-20	23-43	T	S: CH_2Cl_2 C_M=0.37M C_L=1.25M A=Cl	72Nc	b
	-4.3	0.471(22°,1)	-13	8-41	T	S: CH_2Cl_2 C_M=0.37M C_L=1.0M A=Br	72Nc	b

Metal Ion	ΔH, kcal /mole	Log K	ΔS, cal /mole °K	T, °C	M	Conditions	Ref.	Remarks
P67	**PHOSPHINE, chlorodiethyl, sulfide**		$C_4H_{10}ClSP$			$(CH_3CH_2)_2P(Cl)S$		
I_2	-3.6	0.462(20°,1)	-10	10-40	T	S: n-Heptane Results from UV data	71La	b,q: ±0.4
	-4.4	0.491(20°,1)	-13	10-40	T	S: n-Heptane Results from Visible data	71La	b
P68	**PHOSPHINE, cyclohexldiethyl-**		$C_8H_{21}P$			$C_6H_{11}P(C_2H_5)_2$		L
CoL_2A_2	-4.1	0.301(10°,1)	-13	-30-20	T	S: 1,2-Dichloro-ethane. A=NCS	67Bd	b
P69	**PHOSPHINE, diethylphenyl-**		$C_8H_{15}P$			$(C_2H_5)_2PC_6H_5$		L
CoL_2A_2	-18.5	3.556(10°,1)	-49	-30-20	T	S: 1,2-Dichloro-ethane. A=NCS	67Bd	b
NiL_2A_2	-18.6	1.970(1)	-54	20	T	S: $C_2H_4Cl_2$. A=CN	69Rc	
	-16.1	3.120(1)	-41	20	T	S: C_2H_5OH. A=CN	69Rc	
	-17.0	(1)	--	20	C	S: C_2H_5OH. A=CN	69Rc	
P70	**PHOSPHINE, diphenylethyl-**		$C_{14}H_{15}P$			$(C_6H_5)_2PC_2H_5$		L
CoL_2A_2	-13.9	1.00(10°,1)	-45	-30-20	T	S: 1,2-Dichloro-ethane. A=NCS	67Bd	b
NiL_2A_2	-8.9	0.380(1)	-29	20	T	S: $C_2H_4Cl_2$. A=CN	69Rc	
	-6.8	1.508(1)	-16	20	T	S: C_2H_5OH. A=CN	69Rc	
Ga_2A_6	-27.5	--	--	25	C	S: Benzene C_M=0.03-0.04M A=Cl R: $1/2Ga_2A_6 + L = GaA_3L$	73Rd	
P71	**PHOSPHINE, dipropylphenyl-**		$C_{12}H_{19}P$			$(CH_3CH_2CH_2)_2PC_6H_5$		
Al_2A_6	-27.7	--	--	25	C	S: Benzene C_M=0.03-0.04M A=Br R: $1/2Al_2A_6 + L = AlA_3L$	73Rd	
P72	**PHOSPHINE, tribenzyl-**		$C_{21}H_{21}P$			$P(CH_3C_6H_4)_3$		
Ni^o	-13	1.204(30°,4)	-37	10-70	T	S: Benzene	70Gi	b,q: ±1.5
P73	**PHOSPHINE, tributyl-**		$C_{12}H_{27}P$			$P(CH_2CH_2CH_2CH_3)_3$		L
NiA	-14.8	5.75(1-2)	-20.6	25	T	$C_M=10^{-4}$. S: Benzene A=Diacetylbisbenzoyl hydrazine	60Sd	
NiL_2A_2	-11.6	0.447(0°,1)	-38	20	T	S: C_2H_5OH. A=CN	69Rc	b
	-10.8	0.681(0°,1)	-36	20	T	S: n-Hexane. A=CN	69Rc	b
HgA_2	-29.4	(1)	--	30	C	S: Benzene $C_M=1-5x10^{-3}M$ A=Cl	73Fa	
	-29.2	(1)	--	30	C	S: Benzene	73Fa	

Metal Ion	ΔH, kcal /mole	Log K	ΔS, cal /mole °K	T, °C	M	Conditions	Ref.	Remarks
P73, cont.								
HgA_2, cont.						C_M=1-5x10^{-3}M. A=Br		
	-23.4	(1)	--	30	C	S: Benzene C_M=1-5x10^{-3}M. A=I	73Fa	
	-12.9	(2)	--	30	C	S: Benzene C_M=1-5x10^{-3}M. A=Cl	73Fa	
	-13.9	(2)	--	30	C	S: Benzene C_M=1-5x10^{-3}M. A=Br	73Fa	
	-16.5	(2)	--	30	C	S: Benzene C_M=1-5x10^{-3}M. A=I	73Fa	
BA_3C	-19.7	--	--	25	C	S: Benzene C_M=0.03-0.04M. A=F C=Diethyl ether R: BA_3C + L = BA_3L + C	73Rd	
Al_2A_6	-34.2	--	--	25	C	S: Benzene C_M=0.03-0.04M. A=Br R: 1/2Al_2A_6 + L = AlA_3L	73Rd	
Ga_2A_6	-35.0	--	--	25	C	S: Benzene C_M=0.03-0.04M. A=Cl R: 1/2Ga_2A_6 + L = GaA_3L	73Rd	
P74 PHOSPHINE, triethyl-			$C_6H_{15}P$			$P(CH_2CH_3)_3$		L
CoL_2A_2	-19.0	3.285(10°,1)	-52	-30-20	T	S: 1,2-Dichloro-ethane. A=NCS	67Bc	b
Ni^o	-16	1.921(25°,4)	-45	-50-70	T	S: Benzene	74Tb	b,q: ±2
NiA	-13.9	6.18(1-2)	-18.3	25	T	C_M=10^{-4}. S: Benzene A=Diacetylbisbenzoyl hydrazine	60Sd	
NiL_2A_2	-13.4	0.279(0°,1)	-44	20	T	S: $C_2H_4Cl_2$. A=CN	69Rc	
	-10.7	1.127(0°,1)	-31	20	T	S: C_2H_5OH. A=CN	69Rc	
AlA_3	-22.12	(1)	--	25-28	C	S: n-Hexane. A=CH_3	68Hc	
P75 PHOSPHINE, trihexyl-			$C_{18}H_{39}P$			$P[CH_3(CH_2)_5]_3$		L
Al_2A_6	-33.2	--	--	25	C	S: Benzene C_M=0.03-0.04M. A=Br R: 1/2Al_2A_6 + L = AlA_3L	73Rd	
P76 PHOSPHINE, trimethyl-			C_3H_9P			$P(CH_3)_3$		
AlA_3	-21.02	(1)	--	25-28	C	S: n-Hexane. A=CH_3	68Hc	
P77 PHOSPHINE, (trimethylol-propane) ester			$C_6H_{15}O_3P$			$CH_3CH_2C(CH_2OH)_2CH_2OPH_2$		L
[Co]	-19.7	--	--	24	C	[Co]=methylcobal oxime.	73Cf	

Metal Ion	ΔH, kcal /mole	Log K	ΔS, cal /mole °K	T, °C	M	Conditions	Ref.	Remarks
P77, cont.								
[Co], cont.						S: Dichloromethane R: $[Co]_2$ + 2L = 2[Co]L		
P78 PHOSPHINE, trimethyl-, oxide			C_3H_9OP			$(CH_3)_3PO$		L
NbA_5	-4.2	-0.29	-19	-48	T	S: CH_3CN. A=OCH_3 R: $[NbA_5]_2$ + 2L = $2NbA_5L$	71Hi	q: ±0.4
TaA_5	-3.4	1.84	-7	-48	T	S: CH_3CN. A=OCH_3 R: $[TaA_5]_2$ + 2L = $2TaA_5L$	71Hi	q: ±0.3
P79 PHOSPHINE, trioctyl-			$C_{24}H_{51}P$			$[CH_3(CH_2)_6CH_2]_3P$		L
Ga_2A_6	-36.3	--	--	25	C	S: Benzene C_M=0.03-0.04M. A=Cl R: $1/2Ga_2A_6$ + L = GaA_3L	73Rd	
P80 PHOSPHINE, trioctyl-, sulfide			$C_{24}H_{51}SP$			$[CH_3(CH_2)_6CH_2]_3PS$		
I_2	-11.4	3.929(20°,1)	-21	10-40	T	S: n-Heptane Results from UV data	71La	b
	-11.5	3.944(20°,1)	-21	10-40	T	S: n-Heptane Results from Visible data	71La	b
P81 PHOSPHINE, triphenyl-			$C_{18}H_{15}P$			$(C_6H_5)_3P$		L
RuA_2L_2	-4.2	2.569(25°,1)	-2.2	25-49	T	μ=0 S: Benzene	74Ja	b,q: ±0.3
	-3.1	1.337(25°,1)	-4.1	25-49	T	μ=0 S: N,N-Dimethyl-acetamide	74Ja	b,q: ±0.3
$RhAB_2$	-27.3	--	--	21-24	C	S: Dichloromethane A=Cl B=1,5-Cyclooctadiene R: $[RhAB]_2$ + 2L = 2RhABL	73Pe	
	-29.1	--	--	21-24	C	S: Dichloromethane A=Br B=1,5-Cyclooctadiene R: $[RhAB]_2$ + 2L = 2RhABL	73Pe	
	-30.4	--	--	21-24	C	S: Dichloromethane A=I B=1,5-Cyclooctadiene R: $[RhAB]_2$ + 2L = 2RhABL	73Pe	
PdA_2B_2	-39.0	--	--	21-24	C	S: Dichloromethane A=Cl. B=Benzonitrile R: PdA_2B_2 + 2L = PdA_2L_2 + 2B	73Pe	

Metal Ion	ΔH, kcal /mole	Log K	ΔS, cal /mole °K	T, °C	M	Conditions	Ref.	Remarks
P81, cont.								
AgA	-11.3	(2)	--	25	C	S: CH_2Cl_2. A=Hexafluoroacetylacetone	73Pf	
AgAB	-13.3	--	--	25	C	S: CH_2Cl_2. A=Hexafluoroacetylacetone B=Cyclohexene R: [AgAB] + L = [AgAL] + B	73Pf	
	-18.0	--	--	25	C	S: CH_2Cl_2. A=Hexafluoroacetylacetone B=Cyclopentene R: [AgAB] + L = [AgAL] + B	73Pf	
	-14.3	--	--	25	C	S: CH_2Cl_2. A=Hexafluoroacetylacetone B=Cycloheptene R: [AgAB] + L = [AgAL] + B	73Pf	
	-9.88	--	--	25	C	S: CH_2Cl_2. A=Hexafluoroacetylacetone B=Cyclooctene R: [AgAB] + L = [AgAL] + B	73Pf	
	-11.7	--	--	25	C	S: CH_2Cl_2. A=Hexafluoroacetylacetone B=1,5-Cyclooctadiene R: [AgAB] + L = [AgAL] + B	73Pf	
	-10.7	--	--	25	C	S: CH_2Cl_2. A=Hexafluoroacetylacetone B=1,3,5,7-Cyclooctatetraene. R: [AgAB] + L = [AgAL] + B	73Pf	
ZnA	-7.7	1.429(25°,1)	-19.4	15-35	T	S: Benzene A=α,β,γ,δ-Tetraphenylporphine	73Vd	b,q: ±0.6
HgA_2	-17.9	(1)	--	30	C	S: Benzene C_M=1-5x10^{-3}M. A=Cl	73Fa	
	-18.4	(1)	--	30	C	S: Benzene C_M=1-5x10^{-3}M. A=Br	73Fa	
	-15.5	(1)	--	30	C	S: Benzene C_M=1-5x10^{-3}M. A=I	73Fa	
	-7.89	3.712(2)	-9.1	30	C	S: Benzene C_M=1-5x10^{-3}M. A=Cl	73Fa	q: ±0.86
	-10.0	3.550(2)	-17	30	C	S: Benzene C_M=1-5x10^{-3}M. A=Br	73Fa	q: ±1.2
	-9.56	3.204(2)	-17	30	C	S: Benzene C_M=1-5x10^{-3}M. A=I	73Fa	
AlA_3	-17.63	(1)	--	25-28	C	S: n-Hexane. A=CH_3	68Hc	
Al_2A_6	-21.6	--	--	25	C	S: Benzene C_M=0.03-0.04M. A=Br	73Rd	

Metal Ion	ΔH, kcal /mole	Log K	ΔS, cal /mole °K	T, °C	M	Conditions	Ref.	Remarks
P81, cont.								
Al_2A_6, cont.						R: $1/2Al_2A_6 + L = AlA_3L$		
	-22.6	--	--	25	C	S: Benzene C_M=0.03-0.04M. A=Br R: $Al_2A_6 + L = L(AlA_3)_2$	73Rd	
Ga_2A_6	-24.4	--	--	25	C	S: Benzene C_M=0.03-0.04M. A=Cl R: $1/2Ga_2A_6 + L = GaA_3L$	73Rd	
	-25.7	--	--	25	C	S: Benzene C_M=0.03-0.04M. A=Cl R: $Ga_2A_6 + L = L(GaA_3)_2$	73Rd	
P82 PHOSPHINE, triphenyl-, oxide-			$C_{18}H_{15}OP$			$(C_6H_5)_3PO$		L
VOA_2	-5.91	1.50(1)	-13.0	25	C	μ→0. S: Nitro-benzene. A=Acetylacetonate	65Ca	
P83 PHOSPHINE, tripropyl-			$C_9H_{21}P$			$(C_3H_7)_3P$		L
CoL_2A_2	-17.9	2.505(10°,1)	-52	-30-20	T	S: 1,2-Dichloro-ethane. A=NCS	67Bd	b
NiA	-13.8	5.75(1-2)	-19.9	25	T	$C_M=10^{-4}$. S: Benzene A=Diacetylbisbenzoyl hydrazine	60Sd	
NiL_2A_2	-12.0	0.914(0°,1)	-40	20	T	S: C_2H_5OH. A=CN	69Rc	b
	-11.6	0.643(0°,1)	-39	20	T	S: n-Hexane. A=CN	69Rc	b
P84 PHOSPHINE, tris(dimethyl-amino)-, oxide			$C_3H_{18}ON_3P$			$[N(CH_3)_2]_3PO$		L
NbA_5	-1.6	-0.78	-11	-48	T	S: $C_6H_5CH_3$. A=OCH_3 R: $[NbA_5]_2 + 2L = 2NbA_5L$	71Hi	q: ±2
	-1.4	-0.51	-9	-60	T	S: CH_3CN. A=OCH_3 R: $[NbA_5]_2 + 2L = 2NbA_5L$	71Hi	q: ±0.3
P85 PHOSPHORIC ACID, triamide, hexamethyl-			$C_6H_{18}ON_3P$			$[(CH_3)_2N]_3PO$		L
VOA_2	-5.93	2.02(1)	-10.7	25	C	μ→0. S: Nitro-benzene. A=Acetylacetonate	65Ca	
$SnAB_3$	-10.1	2.584(1)	21.9	26	C	S: Isooctane A=Cl. B=CH_3	66Bg	
P86· PHOSPHORIC ACID, tri(4-chloro-phenyl) ester			$C_{18}H_{12}O_3Cl_3P$			$[C_6H_4O(Cl)]_3P$		

Metal Ion	ΔH, kcal /mole	Log K	ΔS, cal /mole °K	T, °C	M	Conditions	Ref.	Remarks
P86, cont.								
Ni^{o}	-21	9.699(25°,4)	-26	-50-70	T	S: Benzene	74Tb	b,q: ±7
P87	PHOSPHORIC ACID, triethyl-ester		$C_6H_{15}O_4P$			$(C_2H_5O)_3PO$		
I_2	-3.2	0.792(20°,1)	-7.9	20&50	T	S: CCl_4	62Ga	a,b
P88	PHOSPHORIC ACID, triisopropyl ester		$C_9H_{21}O_3P$			$P(C_3H_7O)_3$		
Ni^{o}	-23.5	4.569(25°,4)	-58	-50-70	T	S: Benzene	74Tb	b,q: ±1.5
P89	PHOSPHOROUS ACID, tri(4-chloro-2-methylphenyl) ester		$C_{21}H_{18}O_3Cl_3P$			$[C_6H_3O(Cl)(CH_3)]_3P$		
Ni^{o}	-12.5	2.155(25°,4)	-32	-50-70	T	S: Benzene	74Tb	b,q: ±1
P90	PHOSPHOROUS ACID, tri(2,4-dimethylphenyl) ester		$C_{24}H_{27}O_3P$			$[C_6H_3O(CH_3)_2]_3P$		
Ni^{o}	-12	1.509(25°,4)	-33	-50-70	T	S: Benzene	74Tb	b,q: ±2
P91	PHOSPHOROUS ACID, trimethyl ester		C_3H_9OP			$(CH_3O)_3P$		L
CoAB	-4.7	--	--	-90-50	T	S: CH_2Cl_2. A=CH_3 B=N,N'-Ethylenebis (acetylacetoniminato) R: CoAB + L = CoABL	73Gd	b,q: ±1
[Co]	-19.2	--	--	24	C	S: Dichloromethane [Co]=Methyl cobaloxime R: $[Co]_2$ + 2L = 2[Co]L	73Cf	
CuA_2	-6.0	(1)	--	25	C	S: Methylene chloride. A=Hexafluoroacetylacetonate	74Da	
P92	PHOSPHOROUS ACID, triphenyl ester		$C_{18}H_{15}O_3P$			$(C_6H_5O)_3P$		L
RhAB	-23.8	--	--	25	C	S: CCl_4. A=Acetylacetonate. B=1,5-Cyclooctadiene R: RhAB + 2L = $RhAL_2$ + B	73Pd	
	-28.3	--	--	25	C	S: CCl_4. A=Acetylacetonate. B=1,3,5,7-Cyclooctatetraene R: RhAB + 2L = $RhAL_2$ + B	73Pd	
	-21.4	--	--	25	C	S: CH_2Cl_2. A=Acetylacetonate	73Pd	

Metal Ion	ΔH, kcal /mole	Log K	ΔS, cal /mole °K	T, °C	M	Conditions	Ref.	Re-marks
P92, cont.								
RhAB, cont.						B=1,5-Cyclooctadiene R: RhAB + 2L = $RhAL_2$ + B		
	-25.8	--	--	25	C	S: CH_2Cl_2. A=Acetylacetonate B=1,3,5,7-Cyclo-octatetraene. R: RhAB + 2L = $RhAL_2$ + B	73Pd	
	-28.2	--	--	25	C	S: CH_2Cl_2. A=Cl B=1,5-Cyclooctadiene R: $[RbAB]_2$ + 2L = $[Rb_2A_2BL_2]$ + B	73Pd	
	-28.2	--	--	21-24	C	S: Dichloromethane A=Cl B=1,5-Cyclooctadiene R: $[RhAB]_2$ + 2L = $Rh_2A_2BL_2$ + B	73Pe	
	-27.6	--	--	25	C	S: CH_2Cl_2. A=Cl B=1,3,5,7-Cyclo-octatetraene R: $[RhAB]_2$ + 2L = $[Rh_2A_2BP_2]$ + B	73Pd	
	-31.8	--	--	25	C	S: CH_2Cl_2. A=Cl B=<u>cis</u>-Cyclooctene R: $[RhAB]_2$ + 2L = $[Rh_2A_2BL_2]$ + B	73Pd	
	-28.5	--	--	25	C	S: CH_2Cl_2. A=Cl B=Dicyclopentadiene R: $[RhAB]_2$ + 2L = $[Rh_2A_2BL_2]$ + B	73Pd	
	-28.1	--	--	25	C	S: CH_2Cl_2. A=Cl B=Norbornadiene R: $[RbAB]_2$ + 2L = $[Rb_2A_2BL_2]$ + B	73Pd	
	-30.1	--	--	21-24	C	S: Dichloromethane A=Br. B=1,5-Cyclo-octadiene. R: $[RhAB]_2$ + 2L = $Rh_2A_2BL_2$ + B	73Pe	
	-30.2	--	--	21-24	C	S: Dichloromethane A=I. B=1,5-Cyclo-octadiene. R: $[RhAB]_2$ + 2L = $Rh_2A_2BL_2$ + B	73Pe	
$RhAL_2$	-23.0	--	--	25	C	S: CH_2Cl_2. A=Cl R: $[RhAL_2]_2$ + 2L = $2[RhAL_3]$	73Pd	
	-23.6	--	--	21-24	C	S: Dichloromethane A=Cl. R: $[RhAL_2]_2$ + 2L = $2RhAL_3$	73Pe	
	-23.4	--	--	21-24	C	S: Dichloromethane A=Br. R: $[RhAL_2]_2$ + 2L = $2RhAL_3$	73Pe	
	-23.0	--	--	21-24	C	S: Dichloromethane A=I. R: $[RhAL_2]_2$ + 2L = $2RhAL_3$	73Pe	

Metal Ion	ΔH, kcal /mole	Log K	ΔS, cal /mole °K	T, °C	M	Conditions	Ref.	Remarks
P92, cont.								
$Rh_2A_2BL_2$	-33.5	--	--	25	C	S: CH_2Cl_2. A=Cl B=Benzoquinone R: $[Rh_2A_2BL_2]$ + 2L = $[RhAL_2]_2$ + B	/3Pd	
	-23.0	--	--	25	C	S: CH_2Cl_2. A=Cl B=1,5-Cyclooctadiene R: $[Rh_2A_2BL_2]$ + 2L = $[RhAL_2]_2$ + B	73Pd	
	-23.5	--	--	21-24	C	S: Cichloromethane A=Cl. B=1,5-Cyclo-octadiene. R: $Rh_2A_2BL_2$ + 2L = $[RhAL_2]_2$ + B	73Pe	
	-25.2	--	--	25	C	S: CH_2Cl_2. A=Cl B=1,3,5,7-Cyclo-octatetraene R: $[Rh_2A_2BL_2]$ + 2L = $[RhAL_2]_2$ + B	73Pd	
	-30.3	--	--	25	C	S: CH_2Cl_2. A=Cl B=<u>cis</u>-Cyclooctene R: $[Rh_2A_2BL_2]$ + 2L = $[RhAL_2]_2$ + B	73Pd	
	-26.3	--	--	25	C	S: CH_2Cl_2. A=Cl B=Dicyclopentadiene R: $[Rh_2A_2BL_2]$ + 2L = $[RhAL_2]_2$ + B	73Pd	
	-22.6	--	--	25	C	S: CH_2Cl_2. A=Cl R: Norbornadiene R: $[Rh_2A_2BL_2]$ + 2L = $[RhAL_2]_2$ + B	73Pd	
	-25.0	--	--	21-24	C	S: Dichloromethane A=Br. B=1,5-Cyclo-octadiene. R: $Rh_2A_2BL_2$ + 2L = $[RhAL_2]_2$ + B	73Pe	
	-24.6	--	--	21-24	C	S: Dichloromethane A=I. B=1,5-Cyclo-octadiene. R: $Rh_2A_2BL_2$ + 2L = $[RhAL_2]_2$ + B	73Pe	
PdA_2B	-21.2	--	--	21-24	C	S: Dichloromethane A=Cl. B=1,5-Cyclo-octadiene. R: PdA_2B + 2L = PdA_2L_2 + B	73Pe	
P93 PHOSPHOROUS ACID, tri(2-tolyl) ester $C_{21}H_{21}O_3P$ $(C_7H_7O)_3P$								
Ni^0	-23	9.222(25°,<u>4</u>)	-35	-50-70	T	S: Benzene	74Tb	b,q: ±3
P94 PHOSPHORIC ACID, thiono-, triethyl ester $C_6H_{15}O_3SP$ $(C_2H_5O)_3PS$								
I_2	-5.9	0.602(20°,<u>1</u>)	-18.7	20&50	T	S: CS_2	62Ga	a,b
	-5.9	1.000(20°,<u>1</u>)	-16.7	20&50	T	S: CCl_4	62Ga	a,b

Metal Ion	ΔH, kcal /mole	Log K	ΔS, cal /mole °K	T, °C	M	Conditions	Ref.	Remarks
P95 PHTHALIC ACID			$C_8H_6O_4$			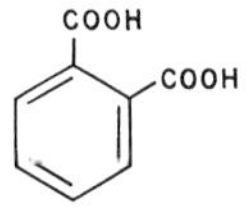		L^{2-}
Ca^{2+}	7.11	2.41(25°,1)	34.8	25-45	T	μ=0	70Ge	b
ThA	5.62	3.10(1)	33.0	25	C	μ=0.10(KNO_3) A=EDTA	70Kb	g:k
	4.95	2.63(1)	28.7	25	C	μ=0.10(KNO_3) A=1,2-Diaminocyclo-hexane-N,N,N',N'-tetraacetic acid	70Kb	g:k
	5.69	--	--	25	C	μ=0.10(KNO_3) A=EDTA R: ThA + HL = $ThAL^-$ + H^+	70Kb	g:k
	5.02	--	--	25	C	μ=0.10(KNO_3) A=1,2-Diaminocyclo-hexane-N,N,N',N'-tetraacetic acid R: ThA + HL = $ThAL^-$ + H^+	70Kb	g:k
Mn^{2+}	2.54	(1)	--	25	C	μ=0	71Wc	
	2.20	2.741(1)	19.9	25	T	μ=0 ΔCp=63	62Da	
	2.13	(1)	--	25	C	μ=0.024M	71Wc	
Co^{2+}	1.87	2.829(1)	19.2	25	T	μ=0 ΔCp=51	62Da	
Ni^{2+}	1.88	(1)	--	25	C	μ=0	71Wc	
	1.765	2.951(1)	68	25	T	μ=0 ΔCp=68	62Da	
	1.42	(1)	--	25	C	μ=0.028M	71Wc	
Cu^{2+}	2.01	4.038(1)	25.2	25	C	μ=0	71Wc	
	9.49	3.14(25°,1)	46.2	25-45	T	μ=0	70Ge	b
	1.56	(1)	--	25	C	μ=0.034M	71Wc	
	5.53	0.99(25°)	23.1	25-45	T	μ=0 R: Cu^{2+} + HL^- = $CuHL^+$	70Ge	b
Zn^{2+}	3.16	2.90(1)	23.8	25	T	μ=0 ΔCp=85.7 Equations of ΔH as a function of T are given	65Na	
P96 PHTHALOCYANINE, 4,4',4",4'''-tetraoctadecyl-sulfonamido			$C_{104}H_{162}O_8N_{12}S_4$			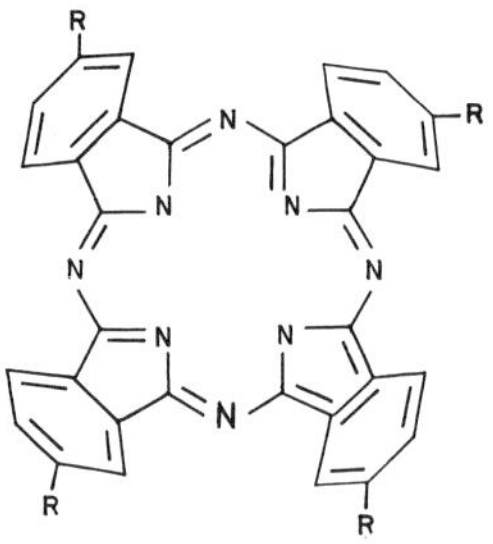		
[Cu]	-10.0	4.114	-15.0	25	C	S: Benzene [Cu]=4,4',4",4'''-tetraoctadecyl-sulfonamido copper(II) phthalocyanine. R: 2[Cu] = $[Cu]_2$ (dimerization)	73Gc	q: ±1.0

Metal Ion	ΔH, kcal /mole	Log K	ΔS, cal /mole °K	T, °C	M	Conditions	Ref.	Remarks
P97	PHTHALOCYANINE, 4,4',4",4"'-tetrasulfo-		$C_{32}H_{18}O_{12}N_8S_4$, see P96, R=SO_3H					L^{2-}
CoL_2	-14	5.280(58°)	-18	38-58	T	μ=0 R: $2CoL_2$ = $(CoL_2)_2$	70Sa	q: ±0.9
CuL	-0.76	0.42(25°)	-0.62	25-58	T	R: 2CuL = $(CuL)_2$	73Bj	b
	-1.08	0.28(25°)	-2.29	25-58	T	S: 20% Ethanol R: 2CuL = $(CuL)_2$	73Bj	b
	-0.98	0.35(25°)	-1.65	25-58	T	S: 20% Ethanol 10^{-2}M(NaCl) R: 2CuL = $(CuL)_2$	73Bj	b
	-0.67	0.37(25°)	-0.57	25-58	T	S: 200g/l Formamide in H_2O R: 2CuL = $(CuL)_2$	73Bj	b
	-0.19	0.28(25°)	0.62	25-58	T	S: Formamide 1M(CH_3COOH) R: 2CuL = $(CuL)_2$	73Bj	b
	-0.93	0.32(25°)	-1.72	25-58	T	S: 20% Methanol R: 2CuL = $(CuL)_2$	73Bj	b
	-0.57	0.35(25°)	-0.36	25-58	T	S: 100g/l Thiourea in H_2O R: 2CuL = $(CuL)_2$	73Bj	b
	-0.52	0.33(25°)	-0.22	25-58	T	S: 250g/l Urea in H_2O R: 2CuL = $(CuL)_2$	73Bj	b
	-0.33	0.32(25°)	0.35	25-58	T	S: 500g/l Urea in H_2O R: 2CuL = $(CuL)_2$	73Bj	b
	-0.36	0.30(25°)	0.19	25-58	T	S: 250g/l Urea + 100 g/l Thiourea in H_2O R: 2CuL = $(CuL)_2$	73Bj	b
P98	PIPERZAINE		$C_4H_{10}N_2$			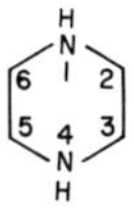		
Ag^+	-6.41	3.33(1)	-6.2	25	C	μ=0.1M(KNO_3)	73He	
	-7.779	3.42(1)	-10.5	25	C	μ=0.1M(KNO_3) S: 50 w % Ethanol	73Ed	
	-4.04	2.71(2)	-1.2	25	C	μ=0.1M(KNO_3)	73He	
	-4.403	2.87(2)	-1.6	25	C	μ=0.1M(KNO_3) S: 50 w % Ethanol	73Ed	
P99	PIPERAZINE, N-(2-aminomethyl)-		$C_6H_{15}N_3$			$CH_2CH_2NHCH_2CH_2N(CH_2CH_2NH_2)$		L
Cu^{2+}	-4.6	5.51(20°,1)	10	10-40	T	μ→0	61Pa	b
	-1.9	3.77(20°,2)	11	10-40	T	μ→0	61Pa	b
P100	PIPERAZINE, 1-(p-methoxy-phenyl)-2-methyl-		$C_{12}H_{18}ON_2$			$C_4H_8N_2(C_6H_4OCH_3)(CH_3)$		
Ag^+	-6.415	3.23(1)	-6.7	25	C	μ=0.1M(KNO_3) S: 52 w % Ethanol	73Ed	

Metal Ion	ΔH, kcal /mole	Log K	ΔS, cal /mole °K	T, °C	M	Conditions	Ref.	Re-marks
P100, cont.								
Ag^+, cont.	-6.035	1.80(2)	-12.0	25	C	μ=0.1M(KNO_3) S: 52 w % Ethanol	73Ed	
P101	PIPERAZINE, 1-methyl-		$C_5H_{12}N_2$			$C_4H_9N_2(CH_3)$		
Ag^+	-5.58	2.93(1)	-5.3	25	C	μ=0.1M(KNO_3)	73He	
	-3.97	2.33(2)	-2.6	25	C	μ=0.1M(KNO_3)	73He	
P102	PIPERAZINE, 2-methyl-		$C_5H_{12}N_2$			$C_4H_9N_2(CH_3)$		
Ag^+	-6.28	3.44(1)	-5.3	25	C	μ=0.1M(KNO_3)	73He	
	-7.758	3.53(1)	-9.9	25	C	μ=0.1M(KNO_3) S: 52 w % Ethanol	73Ed	
	-3.81	2.74(2)	-0.2	25	C	μ=0.1M(KNO_3)	73He	
	-4.522	3.00(2)	-1.4	25	C	μ=0.1M(KNO_3) S: 52 w % Ethanol	73Ed	
P103	PIPERAZINE, 2-methyl-1-(m-tolyl)-		$C_{12}H_{18}N_2$			$C_4H_8N_2(C_6H_4CH_3)(CH_3)$		
Ag^+	-6.249	3.06(1)	-6.9	25	C	μ=0.1M(KNO_3) S: 52 w % Ethanol	73Ed	
	-6.238	2.11(2)	-11.3	25	C	μ=0.1M(KNO_3) S: 52 w % Ethanol	73Ed	
P104	PIPERAZINE, 2-methyl-1-(p-tolyl)-		$C_{12}H_{18}N_2$			$C_4H_8N_2(C_6H_4CH_3)(CH_3)$		
Ag^+	-6.152	3.18(1)	-6.1	25	C	μ=0.1M(KNO_3) S: 52 w % Ethanol	73Ed	
	-6.343	2.10(2)	-11.6	25	C	μ=0.1M(KNO_3) S: 52 w % Ethanol	73Ed	
P105	PIPERAZINE, 1-phenyl-		$C_{10}H_{14}N_2$			$C_4H_9N_2(C_6H_5)$		
Ag^+	-5.35	2.97(1)	-4.6	25	C	μ=0.1M(KNO_3)	73He	
	-5.18	2.77(2)	-4.4	25	C	μ=0.1M(KNO_3)	73He	
[Ni]	-2.42	-1.07(34°,1)	-12.6	4-24	T	[Ni]=Nickel meso-porphyrin IX dimethyl ester S: Mixtures of piperidine and tetrahydrofuran	64Ba	b
CuA_2	-7.5	0.591(25°,1)	-22	0-65	T	S: Methylcyclo-hexane-Piperidine A=(di-n-Butyldi-thiocarbamate)	71Cf	b
[Cu]	-2.30	-0.916(34°,1)	-11.6	11-37	T	[Cu]=Copper meso-porphyrin IX dimethyl ester S: Mixtures of piperidine and tetrahydrofuran	64Ba	b

Metal Ion	ΔH, kcal /mole	Log K	ΔS, cal /mole °K	T, °C	M	Conditions	Ref.	Re-marks
P105, cont.								
I_2	-10.3	3.973(20°,1)	-16.1	10-30	T	$C_M=10^{-5}$ $C_L=10^{-4}-10^{-3}$ S: n-Heptane	60Ya	b
P106 PIPERIDINE		$C_5H_{11}N$				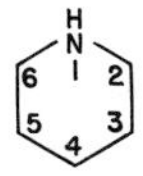		L
VOA_2	-10.06	3.00(1)	-20.0	25	C	$\mu\to 0$ S: Nitrobenzene A=Acetylacetonate	65Ca	
CoA	-6.8	3.386(25°,1)	-7	20-50	T	S: Toluene A=α,β,γ,δ-Tetra-(p-methoxyphenyl) porphyrin	73Wa	b
CoAL	-1.7	0.602(-65,1)	-6	-80- -20	T	S: Toluene A=α,β, γ,δ-Tetra(p-methoxy-phenyl), porphyrin	73Wb	b,q: ±1.0
CoA_2	-13.8	2.64(25°,1)	-34.2	25	C	S: Acetone A=Cl	73Lb	
[Co]	-10.4	4.61(0°,1)	-17	-23-23	T	S: Toluene [Co]=Cobalt (II) protoporphyrin IX dimethyl ester	73Sr	b,q: ±0.6
NiA	-13.88	5.70(1-2)	-20.5	25	T	$C_L=10^{-4}$ S: Benzene A=Diacetylbisbenzoyl hydrazine	58Sa	
NiA_2	-8.9	2.601(24.6°,1)	-18.0	8-39.5	T	S: Benzene A=0,0'-Diethyldithio-phosphate	72Mc	b
	-7.6	3.025(24.6°,1)	-11.8	8-39.5	T	S: Benzene A=0,0'-Dimethyl-dithiophosphate	72Mc	b
P107 PIPERIDINE, 2-(2-aminoethyl)-		$C_7H_{16}N_2$				$C_5H_{10}N(CH_2CH_2NH_2)$		L
Ni^{2+}	-1	5.28(30°,1)	20.8	10-40	T	$\mu=0$	63Ha	b
	4	4.42(30°,2)	33.4	20-40	T	$\mu=0$	63Ha	b
Cu^{2+}	-8	8.60(30°,1)	10.2	10-40	T	$\mu=0$	63Ha	b
	-8	5.86(30°,2)	0.4	10-40	T	$\mu=0$ Log K_2 values vary slightly with $\bar{n}$	63Ha	b
P108 PIPERIDINE, 2-aminomethyl-		$C_6H_{14}N_2$				$C_5H_{10}N(CH_2NH_2)$		L
Co^{2+}	-3	4.81(30°,1)	12.1	10-40	T	$\mu=0$	63Ha	b
	-7	3.96(30°,2)	-5.0	10-40	T	$\mu=0$	63Ha	b
Ni^{2+}	-7	6.23(30°,1)	5.4	10-40	T	$\mu=0$	63Ha	b
	-7	4.98(30°,2)	-0.7	10-40	T	$\mu=0$	63Ha	b
	-6	2.90(30°,3)	-6.5	10-40	T	$\mu=0$	63Ha	b
Cu^{2+}	-12	9.60(30°,1)	4.3	10-40	T	$\mu=0$	63Ha	b
	-11	7.96(30°,2)	0.1	10-40	T	$\mu=0$	63Ha	b

Metal Ion	ΔH, kcal /mole	Log K	ΔS, cal /mole °K	T, °C	M	Conditions	Ref.	Remarks
P109	PIPERIDINE, 2-methyl-		$C_6H_{13}N$			$C_5H_{10}N(CH_3)$		L
VOA_2	-7.69	1.28(1)	-19.9	25	C	μ→0 S: Nitrobenzene A=Acetylacetonate	65Ca	
P110	PIPERIDINE, 2,2,6,6-tetramethyl-		$C_9H_{19}N$			$C_5H_7N(CH_3)_4$		
CuA_2	-11.7	(1)	--	25	C	S: Cyclohexane	72Lc	
BF_3	-26.5	(1)	--	25	C	S: 1,2-Dichloro-ethane	72Lc	
P111	2-PIPERIDONE, N-methyl-		$C_6H_{11}ON$			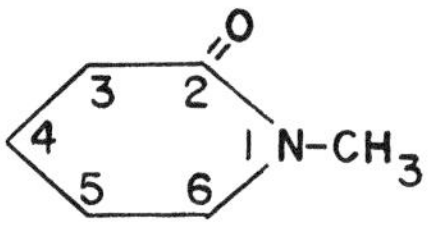		L
I_2	-6.0	1.47(1)	-13.4	25	T	S: n-Heptane	66Kc	
P112	PLUMBATE(II) ION, tetrabromo		Br_4Pb			$PbBr_4^{2-}$		L^{-2}
K^+	-5.0	0.0(25°,1)	-17	25&35	T	4M(Li,K,ClO_4,Cl)	63Mc	a,b
P113	POLYACRYLIC ACID		$(C_3H_4O_2)_n$			$[-CH_2CH(CO_2H)-]_n$		L^-_n
Cu^{2+}	3.7	3.5(25°)	-3.9	10-40	T	1M(KCl). L=Linear polyacrylic acid. R: $Cu^{2+} + 2HL = CuL_2 + 2H^+$	55Lb	b
	0.5	5.9(25°,1-2)	28	10-40	T	1M(KCl). L=Linear polyacrylic acid.	55Lb	b
	0.8	3.4(25°)	-13	3-40	T	1M(KCl). L=Cross-linked polyacrylic acid. R: $Cu^{2+} + 2HL = CuL_2 + 2H^+$	55Lb	b
	0.4	7.0(25°,1-2)	33	3-40	T	1M(KCl). L=Cross-linked poly-acrylic acid	55Lb	b
P114	POLYADENYLIC ACID		$[C_{10}H_{14}O_7N_5P]_n$					L
Mg^{2+}						R: Mg^{2+} + L (Na^+ form) = MgL + iNa^+ θ=fraction of nucleotide sites with bound Mg^{2+}		
	-3.9	--	--	10-40	T	C_{Na^+} = 0.01M θ=0	71Kd	b,q: ±1.5
	-3.4	--	--	10-40	T	C_{Na^+} = 0.029M θ=0	71Kd	b,q: ±1.5
	-1.8	--	--	10-40	T	C_{Na^+} = 0.06M θ=0	71Kd	b,q: ±1.5
	-0.6	--	--	10-40	T	C_{Na^+} = 0.01M θ=0.25	71Kd	b,q: ±1.5
	-0.3	--	--	10-40	T	C_{Na^+} = 0.029M θ=0.25	71Kd	b,q: ±1.5

Metal Ion	ΔH, kcal /mole	Log K	ΔS, cal /mole °K	T, °C	M	Conditions	Ref.	Remarks
P114, cont.								
Mg^{2+}, cont.	0.7	--	--	10-40	T	C_{Na^+} = 0.06M θ=0.25	71Kd	b,q: ±1.5
	2.5	--	--	10-40	T	C_{Na^+} = 0.01M θ=0.50	71Kd	b,q: ±1.5
	3.0	--	--	10-40	T	C_{Na^+} = 0.029M θ=0.50	71Kd	b,q: ±1.5
	2.0	--	--	10-40	T	C_{Na^+} = 0.06M θ=0.50	71Kd	b,q: ±1.5
P115 POLY (ADENYLIC + URIDYLIC) ACID								L
Mg^{2+}						R: Mg^{2+} + L (Na^+ form) = MgL + iNa^+ θ=fraction of nucleotide sites with bound Mg^{2+}		
	2.4	--	--	10-40	T	C_{Na^+} = 0.01M θ=0	71Kd	b,q: ±1.5
	2	--	--	10-40	T	C_{Na^+} = 0.029M θ=0	71Kd	b,q: ±1.5
	-0.7	--	--	10-40	T	C_{Na^+} = 0.06M θ=0	71Kd	b,q: ±1.5
	2.2	--	--	10-40	T	C_{Na^+} = 0.01M θ=0.25	71Kd	b,q: ±1.5
	1	--	--	10-40	T	C_{Na^+} = 0.029M θ=0.25	71Kd	b,q: ±1.5
	0.5	--	--	10-40	T	C_{Na^+} = 0.06M θ=0.25	71Kd	b,q: ±1.5
	2.1	--	--	10-40	T	C_{Na^+} = 0.01M θ=0.50	71Kd	b,q: ±1.5
	5	--	--	10-40	T	C_{Na^+} = 0.029M θ=0.50	71Kd	b,q: ±1.5
	1	--	--	10-40	T	C_{Na^+} = 0.06M θ=0.50	71Kd	b,q: ±1.5
P116 POLY (ADENYLIC + 2 URIDYLIC) ACID								L
Mg^{2+}						R: Mg^{2+} + L (Na^+ form) = MgL + iNa^+ θ=fraction of nucleotide sites with bound Mg^{2+}		
	3.1	--	--	10-40	T	C_{Na^+} = 0.01M θ=0	71Kd	b,q: ±1.5
	-1	--	--	10-40	T	C_{Na^+} = 0.029M θ=0	71Kd	b,q: ±1.5
	0.7	--	--	10-40	T	C_{Na^+} = 0.06M θ=0	71Kd	b,q: ±1.5
	2.3	--	--	10-40	T	C_{Na^+} = 0.01M θ=0.25	71Kd	b,q: ±1.5
	0.7	--	--	10-40	T	C_{Na^+} = 0.029M θ=0.25	71Kd	b,q:

Metal Ion	ΔH, kcal /mole	Log K	ΔS, cal /mole °K	T, °C	M	Conditions	Ref.	Remarks
P116, cont.								
Mg^{2+}, cont.	0.8	--	--	10-40	T	C_{Na^+} = 0.06M θ=0.25	71Kd	b,q: ±1.5
	1.3	--	--	10-40	T	C_{Na^+} = 0.01M θ=0.50	71Kd	b,q: ±1.5
	2	--	--	10-40	T	C_{Na^+} = 0.029M θ=0.50	71Kd	b,q: ±1.5
	1	--	--	10-40	T	C_{Na^+} = 0.06M θ=0.50	71Kd	b,q: ±1.5
P117 POLYURIDYLIC ACID			$(C_9H_{13}O_9N_2P)_n$					
Mg^{2+}						R: Mg^{2+} + L (Na^+ form) = MgL + iNa^+ θ=fraction of nucleotide sites with bound Mg^{2+}		
	3.4	--	--	10-40	T	C_{Na^+} = 0.01M θ=0	71Kd	b,q: ±1.5
	1.9	--	--	10-40	T	C_{Na^+} = 0.029M θ=0	71Kd	b,q: ±1.5
	2.2	--	--	10-40	T	C_{Na^+} = 0.01M θ=0.25	71Kd	b,q: ±1.5
	1.9	--	--	10-40	T	C_{Na^+} = 0.029M θ=0.25	71Kd	b,q: ±1.5
	1.0	--	--	10-40	T	C_{Na^+} = 0.01M θ=0.50	71Kd	b,q: ±1.5
	1.8	--	--	10-40	T	C_{Na^+} = 0.029M θ=0.50	71Kd	b,q: ±1.5
P118 PROLINE			$C_5H_9O_2N$			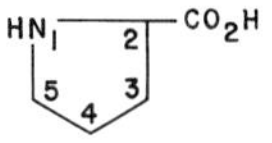		
Y^{3+}	6.304	5.40(25°,1)	45.85	25&35	T	μ=0.1(KCl)	73Sg	a,b,q: ±0.6
	10.087	4.81(25°,1)	55.83	25&35	T	μ=0.1(KCl)	73Sg	a,b,q: ±0.6
Ce^{3+}	2.942	6.00(25°,1)	37.36	25&35	T	μ=0.1(KCl)	73Sg	a,b,q: ±0.6
CoA	-4.7	4.26(25°,1)	3.8	15-55	T	μ=0.1M(KNO_3) A=Methyliminodiacetate	73Ib	b,q: ±0.3
	-5.5	3.85(25°,1)	-0.8	15-70	T	μ=0.1M(KNO_3) A=Nitrilotriacetate	71Ia	b,q: ±0.4
Ni^{2+}	-5.7	5.95(1)	8.1	25	C	μ=0.1M	73Ia	
	-6.1	4.95(2)	2.0	25	C	μ=0.1M	73Ia	
NiA	-6.1	5.18(25°,1)	3.4	15-55	T	μ=0.1M(KNO_3) A=Methyliminodiacetate	73Ib	b,q: ±0.7
	-6.5	4.99(25°,1)	0.9	15-70	T	μ=0.1M(KNO_3) A=Nitrilotriacetate	71Ia	b,q: ±0.5

Metal Ion	ΔH, kcal /mole	Log K	ΔS, cal /mole °K	T, °C	M	Conditions	Ref.	Re-marks
P118, cont.								
Cu^{2+}	-7.4	8.84(1)	15.3	25	C	μ=0.1M	73Ia	
	-7.7	7.44(2)	8.4	25	C	μ=0.1M	73Ia	
	-14.3	16.4(1-2)	27.2	25	C	μ=0.1M	71Be	
CuA	-8.0	6.68(25°,1)	2.1	15-55	T	μ=0.1M(KNO_3) A=Methylimino-diacetate	73Ib	b,q: ±0.8
	-10.9	6.24(25°,1)	-12.0	15-70	T	μ=0.1M(KNO_3) A=Nitrilotriacetate	71Ia	b
ZnA	-5.7	3.98(25°,1)	-0.8	15-70	T	μ=0.1M(KNO_3) A=Nitrilotriacetate	71Ia	b
CdA	-4.2	3.95(25°,1)	4.0	15-55	T	μ=0.1M(KNO_3) A=Methylimino-diacetate	73Ib	b,q: ±0.3
	-3.9	3.05(25°,1)	0.9	15-70	T	μ=0.1M(KNO_3) A=Nitrilotriacetate	71Ia	b
P119 PROLINE (L)			$C_5H_9O_2N$					
Cu^{2+}	-14.2	16.4(2)	27.5	25	C	μ=0.1M	71Be	
P120 PROLINE, N-glycyl-(L)			$C_7H_{12}O_3N_2$			$C_5H_8O_2N(COCH_2NH_2)$		
Cu^{2+}	-6.3	6.34(1)	7.7	25	C	μ=0.1M(KNO_3)	72Bl	
	-6.8	5.01(2)	0.0	25	C	μ=0.1M(KNO_3)	72Bl	
P121 PROLINE, 4-hydroxy-			$C_5H_9O_3N$			$C_5H_8O_2N(OH)$		
Y^{3+}	1.681	4.52(1)	26.32	25&35	T	μ=0.1(KCl)	73Sg	a,b,q: ±0.6
	3.362	4.40(2)	31.41	25&35	T	μ=0.1(KCl)	73Sg	a,b,q: ±0.6
Ce^{3+}	12.609	4.90(1)	64.71	25&35	T	μ=0.1(KCl)	73Sg	a,b
UO_2^{2+}	-6.94	7.023(25°,1)	8.86	25&45	T	μ=0.05M(KCl)	73Sh	a,b
	-6.94	6.817(25°,2)	7.92	25&45	T	μ=0.05M(KCl)	73Sh	a,b
	-4.37	13.891(25°,1-2)	49.02	25&45	T	μ=0	73Sh	a,b
Co^{2+}	-3.3	4.83(1)	11.0	25	C	μ=0.1M	73Ia	q: ±0.2
	-3.36	4.05(2)	7.2	25	C	μ=0.1M	73Ia	
Ni^{2+}	-5.0	5.94(1)	10.4	25	C	μ=0.1M	73Ia	q: ±0.2
	-5.48	5.01(2)	4.6	25	C	μ=0.1M	73Ia	
Cu^{2+}	-6.9	8.38(1)	15.2	25	C	μ=0.1M	73Ia	
	-7.8	7.04(2)	6.0	25	C	μ=0.1M	73Ia	
Zn^{2+}	-2.6	5.03(1)	14.4	25	C	μ=0.1M	73Ia	q: ±0.2
	-2.9	4.35(2)	10.3	25	C	μ=0.1M	73Ia	
P122 PROPANE, 1-amino-			C_3H_9N			$CH_3CH_2CH_2NH_2$		L
NiA	-16.9	5.89(1)	-29.7	25	T	C_M≈10^{-4}. S: Benzene A=Dibenzoyl hydrazine	60Sb	

Metal Ion	ΔH, kcal /mole	Log K	ΔS, cal /mole °K	T, °C	M	Conditions	Ref.	Re-marks
P122, cont.								
Ag^+	-3	3.45(1)	3.4	25	C	μ=0	71He	q:±2
	-9	3.99(2)	-13.4	25	C	μ=0	71He	q:±2
	-12.72	7.44(1-2)	-8.6	25	C	μ=0	71He	
P123	PROPANE, 2-amino-		C_3H_9N			$CH_3CH(NH_2)CH_3$		L
NiA	-17.3	5.20(1)	-34.2	25	T	C_M≈10^{-4} S: Benzene A=Dibenzoyl hydrazine	60Sb	
P124	PROPANE, 2-amino-1,3-dihydroxy-2-hydroxymethyl-		$C_4H_{11}O_3N$			$(HOCH_2)_3CNH_2$		L
Ag^+	-2.4	3.503(1)	7.6	0	T	μ=0.05 ΔCp=-358	66Da	
	-5.2	3.362(1)	-2.9	10	T	μ=0.05 ΔCp=-239	66Da	
	-6.9	3.251(1)	-8.6	20	T	μ=0.05 ΔCp=-95.6	66Da	
	-6.9	3.040(1)	-9.3	30	T	μ=0.05 ΔCp=47.8	66Da	
	-5.7	2.909(1)	-5.0	40	T	μ=0.05 ΔCp=215	66Da	
	-2.9	2.796(1)	4.1	50	T	μ=0.05 ΔCp=382	66Da	
	2	2.793(1)	17.9	60	T	μ=0.05 ΔCp=550	66Da	
	-10	4.042(2)	-18.6	0	T	μ=0.05 ΔCp=71.7	66Da	
	-9.2	3.766(2)	-16.0	10	T	μ=0.05 ΔCp=71.7	66Da	
	-8.6	3.529(2)	-13	20	T	μ=0.05 ΔCp=71.7	66Da	
	-7.9	3.328(2)	-10	30	T	μ=0.05 ΔCp=71.7	66Da	
	-6.9	3.158(2)	-7.6	40	T	μ=0.05 ΔCp=95.6	66Da	
	-6.0	3.017(2)	-5.0	50	T	μ=0.05 ΔCp=95.6	66Da	
	-5.2	2.902(2)	-2.4	60	T	μ=0.05 ΔCp=95.6	66Da	
P125	PROPANE, 1-amino-2-methyl-		$C_4H_{11}N$			$(CH_3)_2CHCH_2NH_2$		L
NiA	-16.6	4.32(1-2)	-35.9	25	T	C_M=10^{-4} S: Benzene A=Diactylbisbenzoyl hydrazine	60Sb	
P126	PROPANE, 2-amino-2-methyl-		$C_4H_{11}N$			$(CH_3)_3CNH_2$		L
NiA	-16.1	3.39(1-2)	-38.5	25	T	C_M=10^{-4} S: Benzene A=Diactylbisbenzoyl hydrazine	60Sb	
ZnA_2	-9.8	2.580(25°,1)	-21	17.8-194	T	S: Toluene A=$[S_2CN(CH_3)_2]$	65Ch	b
P127	PROPANE, 2,2-bis(hydroxymethyl)-1,3-dihydroxy-		$C_5H_{12}O_4$			$CH_2OHC(CH_2OH)_2CH_2OH$		L
$H_5TeO_6^-$	-4.3	0.49	-12	25	T	μ=0.1 R: $H_5TeO_6^-$ + L = $HTeO_4L^-$ + $2H_2O$	62Ea	
P128	PROPANE, 1,2-diamino		$C_3H_{10}N$			$CH_3CH(NH_2)CH_2NH_2$		L
Ni^{2+}	-8.24	7.30(1)	5.7	25	C	μ=0	73Pk	
	-8.24	6.28(2)	1.1	25	C	μ=0	73Pk	
	-10.3	4.98(3)	-11.7	25	C	μ=0	73Pk	

Metal Ion	ΔH, kcal /mole	Log K	ΔS, cal /mole °K	T, °C	M	Conditions	Ref.	Remarks
P128, cont.								
Cu^{2+}	-11.95	10.322(1)	7.1	25	C	$\mu\to 0$	67Pc	
	-13.0	10.54(1)	4.6	25	T	$\mu\to 0$	62Na	
	-14.1	10.59	1.1	15-40	T	μ=0.3M($NaClO_4$)	72Md	b
	-11.90	8.881(2)	0.9	25	C	$\mu\to 0$	67Pc	
	-13.2	9.06(2)	-2.7	25	T	$\mu\to 0$	62Na	
	-13.8	9.15(2)	-4.3	15-40	T	μ=0.3M($NaClO_4$)	72Md	b
	-23.9	19.20(1-2)	8.0	25	T	$\mu\to 0$	67Pc	
Hg^{2+}	-33.8	23.40(1-2)	-6	25	T	0.1M(KNO_3)	61Rc	
P129	PROPANE, 1,3-diamino-		$C_3H_{10}N_2$			$H_2NCH_2CH_2CH_2NH_2$		L
Ni^{2+}	-10.2	6.18(30°,1)	-5	10-40	T	μ=0	58Ba	b
	-8.8	6.39(25°,1)	-0.3	0-50	T	μ=0.15	55Cc	b
	-7.24	6.91(1)	7.3	25	C	μ=0.30(ClO_4^-)	67Hf	
	-7.8	6.39(1)	3.0	25	C	1M(KNO_3)	55Pb	
	-8.2	4.28(30°,2)	-8	10-40	T	μ=0	58Ba	b
	-8.2	4.38(25°,2)	7.4	0-50	T	μ=0.15	55Cc	b
	-7.35	5.34(2)	-0.2	25	C	μ=0.30(ClO_4^-)	67Hf	
	-7.2	4.39(2)	-4.1	25	C	1M(KNO_3)	55Pb	
	-6	1.5(25°,3)	-13	0-50	T	μ=0.15	55Cc	b
	-7.12	3.01(3)	-10.1	25	C	μ=0.30(ClO_4^-)	67Hf	
	-6.3	1.23(3)	-15.5	25	C	1M(KNO_3)	55Pb	
	-21.3	12.01(1-3)	-16.6	25	C	1M(KNO_3)	55Pb	
Cu^{2+}	-13.9	9.45(30°,1)	-3	10-40	T	μ=0	58Ba	b
	-12.5	9.68(25°,1)	2.3	0-50	T	μ=0.15	55Cc	b
	-11.02	10.16(1)	9.6	25	C	μ=0.30(ClO_4^-)	67Hf	
	-14	9.62(30°,1)	-3	0-30	T	1M(KNO_3)	56Ha	b
	-12.3	7.12(25°,2)	-8.7	0-50	T	μ=0.15	55Cc	b
	-11.02	7.14(2)	-4.2	25	C	μ=0.30(ClO_4^-)	67Hf	
	-13	7.00(30°,2)	-12	0-30	T	1M(KNO_3)	56Ha	b
	-22.8	17.17(1-2)	2.0	25	C	1M(KNO_3)	55Pb	
	-1.08	2.26(25°)	6.7	-75&25	T	S: 50 v % Methanol R: Cu + CuL_2 = 2CuL	73Ya	a,b,q: ±0.54
Ag^+	-14.6	5.56(30°,1)	-23	10-40	T	μ=0	58Ba	b
Cd^{2+}	-5.3	4.62(25°,1)	3.4	0-50	T	μ=0.15	55Cc	b
	-4.4	3.05(25°,2)	-0.6	0-50	T	μ=0.15	55Cc	b
P130	PROPANE, 1,3-diamino-N-(2-aminoethyl)-		$C_5H_{15}N_3$			$NH_2CH_2CH_2NHCH_2CH_2CH_2NH_2$		
Co^{2+}	-6.7	8.27(25°,1)	15	15-40	T	μ=0.1M(KCl)	73Sm	b
Ni^{2+}	-9.5	11.01(25°,1)	19	15-40	T	μ=0.1M(KCl)	73Sm	b
Cu^{2+}	-16.3	16.30(25°,1)	20	15-40	T	μ=0.1M(KCl)	73Sm	b
Zn^{2+}	-4.9	8.62(25°,1)	23	15-40	T	μ=0.1M(KCl)	73Sm	b
P131	PROPANE, 1,3-diamino-N,N'-bis(2-aminoethyl)-		$C_7H_{20}N_4$			$H_2NCH_2CH_2NHCH_2CH_2CH_2NHCH_2CH_2NH_2$		
Ni^{2+}	-17.9	16.4(1)	15.1	25	C	μ=0.5M(KCl)	72Fa	
	-4.1	3.66(2)	3.0	25	C	μ=0.5M(KCl)	72Fa	
Cu^{2+}	-27.7	23.9(1)	16.4	25	C	μ=0.5M(KCl)	72Fa	

Metal Ion	ΔH, kcal /mole	Log K	ΔS, cal /mole °K	T, °C	M	Conditions	Ref.	Remarks
P131, cont.								
Zn^{2+}	-11.9	12.8(1)	18.8	25	C	μ=0.5M(KCl)	72Fa	
P132	PROPANE, 1,3-diamino, 2,2-dimethyl-		$C_5H_{14}N_2$			$H_2NCH_2C(CH_3)_2CH_2NH_2$		L
Co^{2+}	-7	4.88(1)	0	30	T	1M(KNO_3)	56Ha	
	-6	3.07(2)	-6	30	T	1M(KNO_3)	56Ha	
Ni^{2+}	-8	6.59(1)	3	30	T	1M(KNO_3)	56Ha	
	-7	4.41(2)	-3	30	T	1M(KNO_3)	56Ha	
Cu^{2+}	-12	9.94(1)	7	30	T	1M(KNO_3)	56Ha	
	-12	7.45(2)	-7	30	T	1M(KNO_3)	56Ha	
Zn^{2+}	-5	5.21(30°,1)	8	0-30	T	1M(KNO_3)	56Ha	b
	-5	5.20(30°,2)	8	0-30	T	1M(KNO_3)	56Ha	b
P133	PROPANE, 1,2-diamino-2-methyl		$C_4H_{12}N_2$			$CH_3C(NH_2)(CH_3)CH_2NH_2$		L
Ni^{2+}	-7.05	6.47(1)	6.0	25	C	μ=0. Data refers to reactions involving paramagnetic bis-diamine complex.	73Pk	
	-8.39	(2)	--	25	C	μ=0. Data refers to reactions involving paramagnetic bis-diamine complex.	73Pk	
	-7.41	(2)	--	25	C	μ=0. Data refers to mixed diamagnetic-paramagnetic complex.	73Pk	
	-3.01	(3)	--	25	C	μ=0. Data refers to reactions involving paramagnetic bis-diamine complex.	73Pk	q: ±0.41
	-3.99	(3)	--	25	C	μ=0. Data refers to mixed diamagnetic-paramagnetic complex.	73Pk	q: ±0.41
NiL_2	-2.34	0.079	-7.49	25	T	μ=0. R: NiL_2^{2+} (diamagnetic) = NiL_2^{2+} (paramagnetic)	73Pk	
Cu^{2+}	-11.27	9.765(1)	6.9	25	C	μ→0	67Pc	
	-12.26	8.819(2)	-0.7	25	C	μ→0	67Pc	
	-23.5	18.59(1-2)	6.2	25	C	μ→0	67Pc	
P134	PROPANE, 1,2-diamino-N,N,N',N'-tetraacetic acid		$C_{11}H_{18}O_8N_2$			$(CH_2CO_2H)_2NCH_2CH_2CH_2N(CH_2CO_2H)_2$		L
Mg^{2+}	9.09	6.21(1)	59.0	20	C	0.1M(KNO_3)	64Ab	
	1.00	9.603(1)	47.3	20	C	μ=1M(HCl). C_L=0.01	69Sh	g:k
Ca^{2+}	-1.74	7.28(1)	27.4	20	C	0.1M(KNO_3)	64Ab	
	-3.02	11.250(1)	41.2	20	C	μ=1M(HCl). C_L=0.01	69Sh	g:k
Sr^{2+}	-0.55	9.237(1)	40.4	20	C	μ=1M(HCl). C_L=0.01	69Sh	g:k
Ba^{2+}	-1.94	8.231(1)	31.0	20	C	μ=1M(HCl). C_L=0.01	69Sh	g:k

Metal Ion	ΔH, kcal /mole	Log K	ΔS, cal /mole °K	T, °C	M	Conditions	Ref.	Remarks
P134, cont.								
La^{3+}	3.76	11.23(1)	64.2	20	C	0.1M(KNO_3)	64Ab	
	3.78	11.23(1)	63.7	20	C	μ=0.1(KNO_3)	71Af	
Ce^{3+}	3.60	11.70(1)	65.8	20	C	μ=0.1(KNO_3)	71Af	
Pr^{3+}	3.92	11.99(1)	68.1	20	C	μ=0.1(KNO_3)	71Af	
Nd^{3+}	4.2	12.34(1)	69.5	20	C	μ=0.1(KNO_3)	71Af	
Sm^{3+}	5.52	13.14(1)	78.9	20	C	μ=0.1($KClO_4$)	71Af	
	5.65	13.14(1)	79.4	20	C	μ=0.1(KNO_3)	71Af	
Eu^{3+}	5.81	13.54(1)	81.7	20	C	μ=0.1($KClO_4$)	71Af	
	5.95	13.54(1)	82.2	20	C	μ=0.1(KNO_3)	71Af	
Gd^{3+}	4.95	13.70(1)	79.2	20	C	μ=0.1(KNO_3)	71Af	
Dy^{3+}	4.56	14.65(1)	82.6	20	C	μ=0.1(KNO_3)	71Af	
Er^{3+}	4.21	14.80(1)	80.7	20	C	μ=0.1(KNO_3)	71Af	
Yb^{3+}	3.65	15.42(1)	80.3	20	C	μ=0.1(KNO_3)	71Af	
Lu^{3+}	3.53	15.54(1)	83.4	20	C	μ=0.1(KNO_3)	71Af	
Mn^{2+}	-0.72	9.99(1)	52.9	20	C	0.1M(KNO_3)	64Ab	
Co^{2+}	-2.6	15.55(1)	62.2	20	C	0.1M(KNO_3)	64Ab	
Ni^{2+}	-6.66	18.15(1)	60.3	20	C	0.1M(KNO_3)	64Ab	
Cu^{2+}	-7.74	18.92(1)	60.1	20	C	0.1M(KNO_3)	64Ab	
Zn^{2+}	-2.27	15.26(1)	61.8	20	C	0.1M(KNO_3)	64Ab	
Cd^{2+}	-5.44	13.90(1)	45.0	20	C	0.1M(KNO_3)	64Ab	
Hg^{2+}	-18.9	19.9(1)	26.6	20	C	0.1M(KNO_3)	64Ab	
Pb^{2+}	-6.4	13.78(1)	40.8	20	C	0.1M(KNO_3)	64Ab	
P135	PROPANE, 1,3-diethoxy-		$C_7H_{16}O_2$			$C_2H_5O(CH_2)_3OC_2H_5$		
TiA_4	-23.2	3.74(1)	-60.7	25	C	S: Benzene C_M=0.05-0.08M. A=Cl	68Ge 70Gh	
SnA_4	-20.9	3.81(1)	-52.7	25	C	S: Benzene C_M=0.05-0.08M. A=Cl	68Ge 70Gh	
P136	PROPANE, 2,2-diphosphonic acid		$C_3H_{10}O_6P_2$			$CH_3C(PO_3H_2)_2CH_3$		L^{4-}
Li^+	-1.32	3.83(1)	13.1	25	C	0.5M$(CH_3)_4NCl$	68Ca	
Na^+	0.3	2.08(1)	10.7	25	C	0.5M$(CH_3)_4NCl$	68Ca	
K^+	0.0	1.60(1)	8.0	25	C	0.5M$(CH_3)_4NCl$	68Ca	
Ca^{2+}	-1.02	6.34(1)	25.6	25	C	0.5M$(CH_3)_4NCl$	68Ca	
P137	PROPANE, 1,2-epoxy		C_3H_6O			OCH_2CHCH_3 (ring)		
I_2	-3.8	0.81(25°,1)	-9.0	1-25	T	S: n-Heptane	60Ta 60Bc	b

Metal Ion	ΔH, kcal /mole	Log K	ΔS, cal /mole °K	T, °C	M	Conditions	Ref.	Re-marks
P138	PROPANE, 1,3-epoxy-		C_3H_6O			$CH_2CH_2CH_2$ (ring through O)		L
I_2	-6.4	1.41(25°,1)	-15.0	2.5-25	T	S: n-Heptane	60Bc 60Ta	b
P139	1,2-PROPANEDIOL		$C_3H_8O_2$			$CH_3CH(OH)CH_2OH$		L
BA_4^-	-3.02	0.607(25°,1)	-7.3	0-35	T	C_L=0-0.9M. A=OH^-	67Cc	b,q: ±0.3
	-7.22	0.585(25°,1-2)	-22	0-35	T	C_L=0-0.9M. A=OH^-	67Cc	b,q: ±0.7
$H_5TeO_6^-$	-5.0	1.52	-10	25	T	μ=0.1. R: $H_5TeO_6^-$ + L = $HTeO_4L^-$ + $2H_2O$	62Ea	
P140	1,3-PROPANEDIOL		$C_3H_8O_2$			$HOCH_2CH_2CH_2OH$		L
BA_4^-	-4.55	0.104(25°,1)	-15	0-35	T	C_L=0-0.9M. A=OH^-	67Cc	b,q: ±0.5
	-6.18	-0.959(25°,1-2)	-16	0-35	T	C_L=0-0.9M. A=OH^-		b,q: ±0.6
$H_5TeO_6^-$	-4.8	0.51	-14	25	T	μ=0.1. R: $H_5TeO_6^-$ + L = $HTeO_4L^-$ + $2H_2O$	62Ea	
P141	1,2-PROPANEDIOL, carbonate		$C_4H_6O_3$			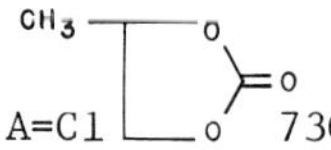		
SbA_5	-13.9	(1)	--	40	C	S: CCl_4. A=Cl	73Oa	
P142	1,3-PROPANEDIOL, 2,2-diethyl-		$C_7H_{16}O_2$			$HOCH_2C(CH_2CH_3)_2CH_2OH$		L
$H_5TeO_6^-$	-4.7	0.23	-15	25	T	μ=0.1. R: $H_5TeO_6^-$ + L = $HTeO_4L^-$ + $2H_2O$	62Ea	
P143	1,3-PROPANEDIOL, 2-hydroxy-methyl-2-methyl-		$C_5H_{12}O_3$			$OHCH_2C(CH_2OH)(CH_3)CH_2OH$		L
$H_5TeO_6^-$	-3.4	0.21	-12	25	T	μ=0.1. R: $H_5TeO_6^-$ + L = $HTeO_4L^-$ + $2H_2O$	62Ea	
P144	1,2-PROPANEDIOL, 3-methoxy-		$C_4H_{10}O_3$			$CH_3OCH_2CH(OH)CH_2OH$		L
$H_5TeO_6^-$	-3.8	1.48	-6	25	T	μ=0.1. R: $H_5TeO_6^-$ + L = $HTeO_4L^-$ + $2H_2O$	62Ea	
P145	1,3-PROPANEDIONE, 1,3-diphenyl-		$C_{15}H_{12}O_2$			$C_6H_5COCH_2COC_6H_5$		L
LiA	-1.896	5.95(30°,1)	21.0	25.5	C	S: 75 v % Dioxane C_L=0.02. A=NO_3	68Mb	
Ni^{2+}	-3.4	9.93(25°,1)	34	15-40	T	μ=0 S: 75 v % Dioxane	71Rb	b,q: ±0.5
	-4.1	8.73(25°,2)	26	15-40	T	μ=0 S: 75 v % Dioxane	71Rb	b,q: ±0.5
Cu^{2+}	-9.7	13.32(25°,1)	28	15-40	T	μ=0 S: 75 v % Dioxane	71Rb	b
	-9.2	11.79(25°,2)	23	15-40	T	μ=0 S: 75 v % Dioxane	71Rb	b

Metal Ion	ΔH, kcal /mole	Log K	ΔS, cal /mole °K	T, °C	M	Conditions	Ref.	Re-marks
P145, cont.								
TlA_2B	-0.967	6.90(30°)	28.4	25.5	C	S: 75 v % Dioxane C_L=0.02 R: TlA_2B + L = TlA_2L + B A=CH_3. B=ClO_4	68Mb	
	-1.360	7.07(30°)	27.9	25.5	C	S: 75 v % Dioxane C_L=0.02 R: TlA_2B + L = TlA_2L + B A=CH_3CH_2. B=ClO_4	68Mb	
	-1.237	6.96(30°)	27.8	25.5	C	S: 75 v % Dioxane C_L=0.02 R: TlA_2B + L = TlA_2L + B A=CH_3CH_2. B=NO_3	68Mb	b
	-1.668	7.51(30°)	38.9	25.5	C	S: 75 v % Dioxane C_L=0.02 R: TlA_2B + L = TlA_2L + B A=$CH_3CH_2CH_2$. B=NO_3	68Mb	b
	-1.976	8.43(30°)	31.7	25.5	C	S: 75 v % Dioxane C_L=0.02 R: TlA_2B + L = TlA_2L + B A=C_6H_5. B=ClO_4	68Mb	b
P146	1-PROPANESULPHONIC ACID, 3-amino-		$C_3H_9O_3NS$			$H_2N(CH_2)_3SO_3H$		L^-
Ag^+	-5.12	3.167(1)	-2.7	25	C	μ=0.5M(KNO_3)	72Va	
	-7.01	3.584(2)	-7.1	25	C	μ=0.5M(KNO_3)	72Va	
P147	PROPANOIC ACID		$C_3H_6O_2$			$CH_3CH_2CO_2H$		L^-
Y^{3+}	3.88	1.61(1)	20.4	25	C	μ=2.0	65Cd	
	2.02	1.20(2)	12.2	25	C	μ=2.0	65Cd	
La^{3+}	2.47	1.53(1)	15.3	25	C	μ=2.0	65Cd	
	1.66	0.89(2)	9.6	25	C	μ=2.0	65Cd	
Ce^{3+}	2.16	1.67(1)	14.9	25	C	μ=2.0	65Cd	
	1.77	1.00(2)	10.5	25	C	μ=2.0	65Cd	
Pr^{3+}	1.89	1.78(1)	14.5	25	C	μ=2.0	65Cd	
	1.68	1.08(2)	10.6	25	C	μ=2.0	65Cd	
Nd^{3+}	1.78	1.93(1)	14.8	25	C	μ=2.0	65Cd	
	1.47	1.15(2)	10.2	25	C	μ=2.0	65Cd	
Sm^{3+}	1.56	2.02(1)	14.5	25	C	μ=2.0	65Cd	
	1.33	1.22(2)	10.0	25	C	μ=2.0	65Cd	
Eu^{3+}	1.76	1.98(1)	15.0	25	C	μ=2.0	65Cd	
	2.32	1.927(1)	16.5	25	C	μ=2.0($NaClO_4$)	71Ai	
	0.99	1.30(2)	9.3	25	C	μ=2.0	65Cd	
	1.82	1.314(2)	12.2	25	C	μ=2.0($NaClO_4$)	71Ai	
Gd^{3+}	2.22	1.84(1)	15.9	25	C	μ=2.0	65Cd	

Metal Ion	ΔH, kcal /mole	Log K	ΔS, cal /mole °K	T, °C	M	Conditions	Ref.	Re-marks
P147, cont.								
Gd^{3+}, cont.	0.86	1.33(2)	8.9	25	C	μ=2.0	65Cd	
Tb^{3+}	3.03	1.73(1)	18.1	25	C	μ=2.0	65Cd	
	0.65	1.37(2)	8.5	25	C	μ=2.0	65Cd	
Dy^{3+}	3.67	1.63(1)	19.8	25	C	μ=2.0	65Cd	
	0.83	1.34(2)	8.9	25	C	μ=2.0	65Cd	
Ho^{3+}	3.90	1.62(1)	20.5	25	C	μ=2.0	65Cd	
	1.63	1.23(2)	11.1	25	C	μ=2.0	65Cd	
Er^{3+}	3.96	1.60(1)	20.6	25	C	μ=2.0	65Cd	
	2.32	1.12(2)	12.9	25	C	μ=2.0	65Cd	
Tm^{3+}	3.98	1.61(1)	20.7	25	C	μ=2.0	65Cd	
	2.83	1.05(2)	14.2	25	C	μ=2.0	65Cd	
Yb^{3+}	3.77	1.63(1)	20.1	25	C	μ=2.0	65Cd	
	2.73	1.07(2)	14.0	25	C	μ=2.0	65Cd	
Lu^{3+}	3.82	1.66(1)	20.4	25	C	μ=2.0	65Cd	
	2.50	1.12(2)	13.5	25	C	μ=2.0	65Cd	
P148	PROPANOIC ACID, amide, N,N-dimethyl-		$C_5H_{11}ON$			$CH_3CH_2CON(CH_3)_2$		L
I_2	-4.0	0.59(25°,1)	-10.7	16-43	T	$C_M=4.87x10^3$ $C_L=4.99x10^3$ S: CCl_4	62Dc	b
P149	PROPANOIC ACID, chloride		C_3H_5OCl			CH_3CH_2COCl		L
SbA_5	-3.3	1.15(1)	-12	25	C	S: Ethylene chloride. A=Cl	67Ob	
P150	PROPANOIC ACID, ethyl ester		$C_5H_{10}O_2$			$CH_3CH_2CO_2C_2H_5$		L
SbA_5	-16.82	(1)	--	25	C	Excess $SbCl_5$ C_L low, variable, ∿0.02. S: 1,2-Dichloroethane A=Cl	64Oa	
	-16.82	5.09(1)	-33.1	25	C	Excess $SbCl_5$ C_L low, variable, ∿0.02. S: Ethylene chloride. A=Cl	67Od	
P151	PROPANOIC ACID, nitrile		C_3H_5N			CH_3CH_2CN		
I_2	-2.31	-0.296(25°)	-9.2	--	T	S: CCl_4 C_L=0.2-1.0M $C_M=2x10^{-3}$. Data analyzed by method 1	72Ma	j,u,q: ±0.45
	-2.94	-0.296(25°)	-11.26	--	T	S: CCl_4 C_L=0.2-1.0M $C_M=2x10^{-3}$. Data analyzed by method 2	72Ma	j,u
	-3.93	-0.021(25°)	-13.32	--	T	S: n-Heptane	72Ma	j,u

Metal Ion	ΔH, kcal /mole	Log K	ΔS, cal /mole °K	T, °C	M	Conditions	Ref.	Remarks
P151, cont.								
I_2, cont.						C_L=0.2-1.0M C_M=2x10^{-3}. Data analyzed by method 2.		
P152	PROPANOIC ACID, 2-amino-2-methyl-		$C_4H_9O_2N$			$(CH_3)_2C(NH_2)CO_2H$		L^-
Cu^{2+}	-5.4	8.55(20°,1)	21	25	C	μ=0	64Ia	
	-5.7	7.05(20°,2)	13	25	C	μ=0	64Ia	
P153	PROPANOIC ACID, 2,3-diamino-, methyl ester (dl)		$C_4H_{10}O_2N_2$			$H_2NCH_2CH(NH_2)CO_2CH_3$		
Cu^{2+}	-11.2	8.986(25°,1)	3.6	25-50	T	μ=0.1M	71Hg	b
	-11.2	7.763(25°,2)	-2.1	25-50	T	μ=0.1M	71Hg	b
P154	PROPANOIC ACID, 2,2-dimethyl-, ethyl ester		$C_7H_{14}O_2$			$(CH_3)_3CCO_2CH_2CH_3$		L
SnA_4	-5.6	0.623(30°,1)	15.6	30-40	T	S: Benzene. A=Cl	64La	b
SbA_5	-12.93	(1)	--	25	C	Excess $SbCl_5$ C_L low, variable, ∿0.02. S: 1,2-Dichloroethane A=Cl	64Oa	
	-13.03	3.362(1)	-34.7	25	C	Excess $SbCl_5$ C_L low, variable, ∿0.02. S: Ethylene chloride. A=Cl	67Oe	
	-13.03	1.97(1)	-34.7	25	C	Excess $SbCl_5$ C_L low, variable, ∿0.02. S: Ethylene chloride. A=Cl	67Od	
P155	PROPANOIC ACID, 2,2-dimethyl-, methyl ester		$C_6H_{12}O_2$			$CH_3C(CH_3)_2CO_2CH_3$		L
SbA_5	-12.83	2.89(1)	-36.2	25	C	Excess $SbCl_5$. low μ. C_L low, variable, ∿0.02 S: Ethylene chloride. A=Cl	67Oe	
P156	PROPANOIC ACID, 2,2-dimethyl-, nitrile		C_5H_9N			$CH_3C(CH_3)_2CN$		L
I_2	-2.10	-0.226(25°)	-8.2	--	T	S: CCl_4 C_L=0.2-1.0M C_M=2x10^{-3}	72Ma	j,u
P157	PROPANOIC ACID, 2,2-dimethyl-, 2-propyl ester		$C_8H_{16}O_2$			$CH_3C(CH_3)_2CO_2CH(CH_3)_2$		L
SbA_5	-13.5	(1)	--	25	C	Excess $SbCl_5$. low μ. C_L low, variable, ∿0.02	67Oe	

Metal Ion	ΔH, kcal /mole	Log K	ΔS, cal /mole °K	T, °C	M	Conditions	Ref.	Remarks
P157, cont.								
SbA_5, cont.						S: Ethylene chloride. A=Cl		
P158	PROPANOIC ACID, 2,3-diphosphoric acid		$C_3H_8O_{10}P_2$			$H_2O_3POCH_2CH(OPO_3H_2)CO_2H$		
[Fe]	-10.8	4.301(1)	-19	5	C	[Fe]=Horse Oxy-hemoglobin C_{Cl^-} = 0 pH(initial)=7.32	72He	q: ±0.8
	-11.2	5.342(1)	-16	5	C	[Fe]=Horse Oxy-hemoglobin C_{Cl^-} = 0.005 pH(initial)=6.50	72He	q: ±1.0
	-9.1	3.602(1)	-14	25	C	[Fe]=Horse Oxy-hemoglobin C_{Cl^-} = 0 pH(initial)=6.93	72He	q: ±0.6
	-8.5	2.602(1)	-16	25	C	[Fe]=Horse Oxy- C_{Cl^-} = 0.10 pH(initial)=7.05	72He	q: ±0.8
P159	PROPANOIC ACID, 2-hydroxy-		$C_3H_6O_3$			$CH_3CH(OH)CO_2H$		L^-
Ca^{2+}	10.91	1.55(25°,1)	43.7	25-45	T	μ=0	70Gd	b
La^{3+}	-1.58	2.27(1)	5.1	25	C	μ=2.0($NaClO_4$)	66Cc	
	-1.1	1.68(2)	4.0	25	C	μ=2.0($NaClO_4$)	66Cc	
	-2.3	1.11(3)	-2.6	25	C	μ=2.0($NaClO_4$)	66Cc	
Ce^{3+}	-1.69	2.29(1)	4.8	25	C	μ=2.0($NaClO_4$)	66Cc	
	-0.83	2.33(1)	7.9	25	T	μ=2($NaClO_4$)	61Ca	
	0.6	1.79(2)	10.1	25	C	μ=2.0($NaClO_4$)	66Cc	
	-5.3	1.01(3)	-13.0	25	C	μ=2.0($NaClO_4$)	66Cc	
Nd^{3+}	-3.23	2.47(1)	0.5	25	C	μ=2.0($NaClO_4$)	66Cc	
	1.5	1.89(2)	13.7	25	C	μ=2.0($NaClO_4$)	66Cc	
Sm^{3+}	-2.33	2.57(1)	4.0	25	C	μ=2.0($NaClO_4$)	66Cc	
	0.45	2.01(2)	10.7	25	C	μ=2.0($NaClO_4$)	66Cc	
	-7.2	1.36(3)	-17.8	25	C	μ=2.0($NaClO_4$)	66Cc	
Eu^{3+}	-1.94	2.54(1)	5.1	25	C	μ=2.0($NaClO_4$)	66Cc	
	-1.03	2.479(1)	7.89	25	C	μ=2.0($NaClO_4$)	71Ai	
	0.8	2.03(2)	12.1	25	C	μ=2.0($NaClO_4$)	66Cc	
	-1.03	2.08(2)	5.98	25	C	μ=2.0($NaClO_4$)	71Ai	
	-4.4	1.36(3)	-8.6	25	C	μ=2.0($NaClO_4$)	66Cc	
	-3.44	1.26(3)	-5.74	25	C	μ=2.0($NaClO_4$)	71Ai	
Gd^{3+}	-1.97	2.60(1)	5.3	25	C	μ=2.0($NaClO_4$)	66Cc	
	-0.54	2.53(1)	9.8	25	T	μ=2($NaClO_4$)	61Ca	
	-0.2	1.89(2)	7.9	25	C	μ=2.0($NaClO_4$)	66Cc	
	-2.8	1.68(3)	-1.7	25	C	μ=2.0($NaClO_4$)	66Cc	
Tb^{3+}	-1.64	2.66(1)	6.6	25	C	μ=2.0($NaClO_4$)	66Cc	
Dy^{3+}	-2.16	2.71(1)	5.2	25	C	μ=2.0($NaClO_4$)	66Cc	
Er^{3+}	-1.91	2.85(1)	6.6	25	C	μ=2.0($NaClO_4$)	66Cc	

Metal Ion	ΔH, kcal /mole	Log K	ΔS, cal /mole °K	T, °C	M	Conditions	Ref.	Re-marks
P159, cont.								
Er^{3+}, cont.	1.0	2.12(2)	12.9	25	C	μ=2.0($NaClO_4$)	66Cc	
	-4.2	1.95(3)	-5.1	25	C	μ=2.0($NaClO_4$)	66Cc	
Yb^{3+}	-2.19	2.89(1)	5.9	25	C	μ=2.0($NaClO_4$)	66Cc	
	-1.4	2.29(2)	5.9	25	C	μ=2.0($NaClO_4$)	66Cc	
	1.4	1.90(3)	3.9	25	C	μ=2.0($NaClO_4$)	66Cc	
Cu^{2+}	8.90	2.36(25°,1)	40.6	25-45	T	μ=0	70Gd	b
	-2.77	1.54(25°,2)	-2.2	25-45	T	μ=0	70Gd	b
In^{3+}	3.2	3.21(35°,1)	25	25-45	T	μ=0.2M($NaClO_4$)	73Sc	b,q: ±0.5
	2.4	2.66(35°,2)	20	25-45	T	μ=0.2M($NaClO_4$)	73Sc	b,q: ±0.5
P160	PROPANOIC ACID, 3-hydroxy-		$C_3H_6O_3$			$HOCH_2CH_2CO_2H$		
Y^{3+}	4.6	1.431(1)	20.5	25	C	μ=2.00M($NaClO_4$)	69Ja	q: ±0.3
La^{3+}	2.57	1.556(1)		25	C	μ=2.00M($NaClO_4$)	69Ja	
Ce^{3+}	2.55	1.568(1)		25	C	μ=2.00M($NaClO_4$)	69Ja	
Pr^{3+}	1.88	1.623(1)		25	C	μ=2.00M($NaClO_4$)	69Ja	q: ±0.27
Nd^{3+}	1.50	1.663(1)		25	C	μ=2.00M($NaClO_4$)	69Ja	q: ±0.30
Sm^{3+}	2.47	1.748(1)		25	C	μ=2.00M($NaClO_4$)	69Ja	
Gd^{3+}	3.80	1.613(1)		25	C	μ=2.00M($NaClO_4$)	69Ja	q: ±0.34
Tb^{3+}	4.36	1.544(1)	21.5	25	C	μ=2.00M($NaClO_4$)	69Ja	q: ±0.33
Dy^{3+}	4.21	1.447(1)	20.8	25	C	μ=2.00M($NaClO_4$)	69Ja	q: ±0.46
Ho^{3+}	4.65	1.477(1)	22.5	25	C	μ=2.00M($NaClO_4$)	69Ja	
Er^{3+}	4.90	1.322(1)	22.9	25	C	μ=2.00M($NaClO_4$)	69Ja	
Tm^{3+}	5.25	1.447(1)	24.1	25	C	μ=2.00M($NaClO_4$)	69Ja	
Yb^{3+}	4.46	1.505(1)	21.9	25	C	μ=2.00M($NaClO_4$)	69Ja	q: ±0.33
Lu^{3+}	4.52	1.398(1)	21.7	25	C	μ=2.00M($NaClO_4$)	69Ja	q: ±0.42
In^{3+}	10.5	7.20(35°,1-2)	67	25-45	T	μ=0	72Sa	b,q: ±1.5
	8.7	6.98(35°,1-2)	60	25-45	T	μ=0.10M($NaClO_4$)	72Sa	b,q: ±1.5
	3.0	3.80(35°,1)	27	25-45	T	μ=0.20M($NaClO_4$)	73Sc	b
P161	PROPANOIC ACID, 2-hydroxy-2-methyl-		$C_4H_8O_3$			$(CH_3)_2C(OH)CO_2H$		L^-
Y^{3+}	-1.23	2.82(1)	8.8	25	C	μ=2.0($NaClO_4$)	66Cc	

Metal Ion	ΔH, kcal /mole	Log K	ΔS, cal /mole °K	T, °C	M	Conditions	Ref.	Remarks
P161, cont.								
Y^{3+}, cont.	-2.9	2.63(2)	2.3	25	C	μ=2.0($NaClO_4$)	66Cc	
	-0.4	1.80(3)	6.8	25	C	μ=2.0($NaClO_4$)	66Cc	
La^{3+}	-1.96	2.28(1)	3.9	25	C	μ=2.0($NaClO_4$)	66Cc	
	-0.4	1.69(2)	6.4	25	C	μ=2.0($NaClO_4$)	66Cc	
	-3.5	1.20(3)	-6.3	25	C	μ=2.0($NaClO_4$)	66Cc	
Ce^{3+}	-1.48	2.41(1)	6.0	25	C	μ=2.0($NaClO_4$)	66Cc	
	-0.83	2.43(1)	8.4	25	T	μ=2($NaClO_4$)	61Ca	
	-0.4	1.93(2)	7.3	25	C	μ=2.0($NaClO_4$)	66Cc	
	-5.7	0.894(3)	-14.9	25	C	μ=2.0($NaClO_4$)	66Cc	
Nd^{3+}	-1.60	2.58(1)	6.5	25	C	μ=2.0($NaClO_4$)	66Cc	
	-1.7	1.98(2)	3.5	25	C	μ=2.0($NaClO_4$)	66Cc	
	-0.7	1.27(3)	3.4	25	C	μ=2.0($NaClO_4$)	66Cc	
Eu^{3+}	-1.51	2.74(1)	7.5	25	C	μ=2.0($NaClO_4$)	66Cc	
	-1.8	2.16(2)	3.8	25	C	μ=2.0($NaClO_4$)	66Cc	
	-1.6	1.69(3)	2.4	25	C	μ=2.0($NaClO_4$)	66Cc	
Gd^{3+}	-0.89	2.82(1)	9.9	25	T	μ=2($NaClO_4$)	61Ca	
Tb^{3+}	-1.47	2.84(1)	8.1	25	T	μ=2.0($NaClO_4$)	66Cc	
	-3.2	2.39(2)	0.14	25	T	μ=2.0($NaClO_4$)	66Cc	
	1.0	1.81(3)	11.5	25	C	μ=2.0($NaClO_4$)	66Cc	
Er^{3+}	-0.98	3.13(1)	11.0	25	C	μ=2.0($NaClO_4$)	66Cc	
	-4.3	2.53(2)	-2.8	25	C	μ=2.0($NaClO_4$)	66Cc	
	1.6	2.24(3)	15.7	25	C	μ=2.0($NaClO_4$)	66Cc	
Yb^{3+}	-0.7	3.57(1)	14	25	C	μ=0.10M($NaClO_4$)	71Bf	
	-1.11	3.04(1)	10.2	25	C	μ=2.0($NaClO_4$)	66Cc	
	-2.8	4.24(1)	10	25	C	μ=0.10M($NaClO_4$) S: 40 v % DMSO	71Bf	
	-4.0	5.34(1)	11	25	C	μ=0.10M($NaClO_4$) S: DMSO	71Bf	
	-10.8	9.66(1)	8	25	C	μ=0.10M($NaClO_4$) S: Ethanol	71Bf	
	-1.6	4.89(1)	17	25	C	μ=0.10M($NaClO_4$) S: Ethylene Glycol	71Bf	
	-1.6	2.92(2)	8	25	C	μ=0.10M($NaClO_4$)	71Bf	
	-5.2	2.84(2)	-4.5	25	C	μ=2.0($NaClO_4$)	66Cc	
	-4.9	3.81(2)	1	25	C	μ=0.10M($NaClO_4$) S: 40 v % DMSO	71Bf	
	-4.1	4.54(2)	7	25	C	μ=0.10M($NaClO_4$) S: DMSO	71Bf	
	-3.8	8.47(2)	26	25	C	μ=0.10M($NaClO_4$) S: Ethanol	71Bf	
	1.0	2.14(3)	13.1	25	C	μ=2.0($NaClO_4$)	66Cc	
Fe^{3+}	-2.95	3.24(30°,1)	5.08	30&50	T	$C_L \approx C_M \approx 1x10^{-3}M$	71Sb	a,b
Cu^{2+}	-0.2	2.99(1)	13	25	C	μ=0.10M($NaClO_4$)	71Bf	
	-3.7	5.99(1)	15	25	C	μ=0.10M($NaClO_4$) S: DMSO	71Bf	
	-2.2	4.67(1)	14	25	C	μ=0.01M($NaClO_4$) S: Ethylene Glycol	71Bf	
	-0.8	1.90(2)	6	25	C	μ=0.10M($NaClO_4$)	71Bf	

Metal Ion	ΔH, kcal /mole	Log K	ΔS, cal /mole °K	T, °C	M	Conditions	Ref.	Remarks
P161, cont.								
Cu^{2+}, cont.	-5.6	4.76(2)	3	25	C	μ=0.10M($NaClO_4$) S: DMSO	71Bf	
	-1.8	3.50(2)	10	25	C	μ=0.01M($NaClO_4$) S: Ethylene Glycol	71Bf	
P162	PROPANOIC ACID, 2-hydroxy-2-phenyl-		$C_9H_{10}O_3$			$C_6H_5C(OH)(CH_3)CO_2H$		L
Fe^{3+}	-2.70	3.45(30°,1)	6.86	30&50	T	$C_L≈C_M≈1x10^{-3}M$	71Sb	a,b
P163	PROPANOIC ACID, 2-mercapto-		$C_3H_6O_2S$			$CH_3CH(SH)CO_2H$		L^-
Y^{3+}	1.34	1.699(1)	12.0	--	C	μ=2.00($NaClO_4$)	69Cc	u:25°
La^{3+}	2.64	1.230(1)	13.6	--	C	μ=2.00($NaClO_4$)	69Cc	u:25°
Ce^{3+}	2.31	1.362(1)	14.1	--	C	μ=2.00($NaClO_4$)	69Cc	u:25°
Pr^{3+}	1.42	1.892(1)	13.4	--	C	μ=2.00($NaClO_4$)	69Cc	u:25°
Nd^{3+}	1.34	1.929(1)	13.1	--	C	μ=2.00($NaClO_4$)	69Cc	u:25°
Sm^{3+}	0.98	2.064(1)	12.3	--	C	μ=2.00($NaClO_4$)	69Cc	u:25°
Eu^{3+}	1.21	2.000(1)	13.2	--	C	μ=2.00($NaClO_4$)	69Cc	u:25°
Gd^{3+}	1.77	1.748(1)	13.3	--	C	μ=2.00($NaClO_4$)	69Cc	u:25°
Tb^{3+}	1.57	1.653(1)	11.6	--	C	μ=2.00($NaClO_4$)	69Cc	u:25°
Dy^{3+}	1.73	1.580(1)	12.7	--	C	μ=2.00($NaClO_4$)	69Cc	u:25°
Ho^{3+}	1.87	1.531(1)	13.3	--	C	μ=2.00($NaClO_4$)	69Cc	u:25°
Er^{3+}	2.45	1.531(1)	15.1	--	C	μ=2.00($NaClO_4$)	69Cc	u:25°
Tm^{3+}	2.56	1.477(1)	14.8	--	C	μ=2.00($NaClO_4$)	69Cc	u:25°
Yb^{3+}	3.19	1.431(1)	17.2	--	C	μ=2.00($NaClO_4$)	69Cc	u:25°
Lu^{3+}	1.66	1.505(1)	12.4	--	C	μ=2.00($NaClO_4$)	69Cc	u:25°
Ni^{2+}	-9.00	13.68(30°,1-2)	33.0	20-40	T	μ=0.1M($NaClO_4$)	72Sf	b
In^{3+}	-5.8	12.15(35°,1)	37	25-45	T	μ=0.2M($NaClO_4$)	73Sc	b,q: ±0.5
	-6.5	10.56(35°,2)	27	25-45	T	μ=0.2M($NaClO_4$)	73Sc	b,q: ±0.5
	-7.4	6.37(35°,3)	5	25-45	T	μ=0.2M($NaClO_4$)	73Sc	b,q: ±0.5
P164	PROPANOIC ACID, 3-mercapto-		$C_3H_6O_2S$			$HSCH_2CH_2CO_2H$		L^-
Y^{3+}	5.04	1.505(1)	23.7	--	C	μ=2.00($NaClO_4$)	69Cc	u:25°
La^{3+}	2.64	1.544(1)	15.8	--	C	μ=2.00($NaClO_4$)	69Cc	u:25°
Ce^{3+}	2.39	1.568(1)	15.3	--	C	μ=2.00($NaClO_4$)	69Cc	u:25°
Pr^{3+}	2.59	1.568(1)	15.6	--	C	μ=2.00($NaClO_4$)	69Cc	u:25°
Nd^{3+}	2.45	1.740(1)	16.2	--	C	μ=2.00($NaClO_4$)	69Cc	u:25°
Sm^{3+}	1.73	1.740(1)	13.7	--	C	μ=2.00($NaClO_4$)	69Cc	u:25°
Eu^{3+}	2.20	1.643(1)	15.2	--	C	μ=2.00($NaClO_4$)	69Cc	u:25°

Metal Ion	ΔH, kcal /mole	Log K	ΔS, cal /mole °K	T, °C	M	Conditions	Ref.	Re-marks
P164, cont.								
Gd^{3+}	2.59	1.708(1)	16.5	--	C	μ=2.00($NaClO_4$)	69Cc	u:25°
Tb^{3+}	2.93	1.623(1)	17.2	--	C	μ=2.00($NaClO_4$)	69Cc	u:25°
Dy^{3+}	3.54	1.602(1)	19.2	--	C	μ=2.00($NaClO_4$)	69Cc	u:25°
Ho^{3+}	3.23	1.462(1)	17.5	--	C	μ=2.00($NaClO_4$)	69Cc	u:25°
Er^{3+}	4.77	1.415(1)	22.9	--	C	μ=2.00($NaClO_4$)	69Cc	u:25°
Tm^{3+}	4.71	1.398(1)	22.1	--	C	μ=2.00($NaClO_4$)	69Cc	u:25°
Yb^{3+}	3.98	1.431(1)	20.1	--	C	μ=2.00($NaClO_4$)	69Cc	u:25°
Lu^{3+}	4.87	1.491(1)	22.8	--	C	μ=2.00($NaClO_4$)	69Cc	u:25°
UO_2^{2+}	-10.1	15.05(30°,1-2)	35.5	20-40	T	μ=0.1M(KNO_3)	69Sd	b
Ni^{2+}	-7.03	10.74(30°,1-2)	25.87	20-40	T	μ=0.1M(KNO_3)	69Sc 69Se	b
In^{3+}	-6.0	11.60(35°,1)	34	25-45	T	μ=0.20M($NaClO_4$)	73Sc	b
	-16.4	26.40(35°,1-3)	67	25-45	T	μ=0	72Sa	b,q: ±1.5
	-16.70	25.40(35°,1-3)	62	25-45	T	μ=0.10M($NaClO_4$)	72Sa	b,q: ±1.5
Tl^{+}	-3.34	2.850(30°,1)	2.01	20-40	T	μ=0.1M(KNO_3)	68Sd	b
Tl^{3+}	-27.8	18.75(30°,1-2)	-6.2	20-40	T	μ=0.1M(KNO_3)	69Sb	b
P165	PROPANOIC ACID, 2-methyl-		$C_4H_8O_2$			$(CH_3)_2CHCO_2H$		L^-
Y^{3+}	5.36	1.64(1)	25.5	25	C	μ=2.0	65Cd	
	3.22	1.15(2)	16.0	25	C	μ=2.0	65Cd	
La^{3+}	3.47	1.57(1)	18.8	25	C	μ=2.0	65Cd	
	2.50	0.90(2)	12.5	25	C	μ=2.0	65Cd	
Ce^{3+}	3.33	1.62(1)	18.6	25	C	μ=2.0	65Cd	
	2.55	1.10(2)	13.6	25	C	μ=2.0	65Cd	
Pr^{3+}	3.02	1.80(1)	18.4	25	C	μ=2.0	65Cd	
	2.51	1.11(2)	13.5	25	C	μ=2.0	65Cd	
Nd^{3+}	2.84	1.91(1)	18.3	25	C	μ=2.0	65Cd	
	2.36	1.18(2)	13.3	25	C	μ=2.0	65Cd	
Sm^{3+}	2.66	2.00(1)	18.1	25	C	μ=2.0	65Cd	
	2.08	1.25(2)	12.7	25	C	μ=2.0	65Cd	
Eu^{3+}	2.91	1.98(1)	18.8	25	C	μ=2.0	65Cd	
	1.85	1.31(2)	12.2	25	C	μ=2.0	65Cd	
Gd^{3+}	3.45	1.86(1)	20.1	25	C	μ=2.0	65Cd	
	1.66	1.34(2)	11.7	25	C	μ=2.0	65Cd	
Tb^{3+}	4.38	1.73(1)	22.6	25	C	μ=2.0	65Cd	
	1.46	1.39(2)	11.3	25	C	μ=2.0	65Cd	
Dy^{3+}	5.04	1.56(1)	24.5	25	C	μ=2.0	65Cd	
	1.79	1.32(2)	12.0	25	C	μ=2.0	65Cd	
Ho^{3+}	5.31	1.63(1)	25.3	25	C	μ=2.0	65Cd	
	2.55	1.21(2)	14.1	25	C	μ=2.0	65Cd	

Metal Ion	ΔH, kcal /mole	Log K	ΔS, cal /mole °K	T, °C	M	Conditions	Ref.	Remarks
P165, cont.								
Er^{3+}	5.49	1.61(1)	25.8	25	C	μ=2.0	65Cd	
	3.43	1.11(2)	16.6	25	C	μ=2.0	65Cd	
Tm^{3+}	5.39	1.61(1)	25.5	25	C	μ=2.0	65Cd	
	4.10	1.06(2)	18.6	25	C	μ=2.0	65Cd	
Yb^{3+}	5.5	2.09(1)	28	25	C	μ=0.10M($NaClO_4$)	71Bf	
	5.35	1.62(1)	25.4	25	C	μ=2.0	65Cd	
	3.6	3.04(1)	26	25	C	μ=0.10M($NaClO_4$) S: 40 v % DMSO	71Bf	
	-5.6	8.04(1)	18	25	C	μ=0.10M($NaClO_4$) S: Ethanol	71Bf	
	3.7	3.63(1)	29	25	C	μ=0.10M($NaClO_4$) S: Ethylene Glycol	71Bf	
	-1.1	3.21(1)	11	25	C	μ=0.10M($NaClO_4$) S: Formamide	71Bf	
	0.8	1.60(2)	10	25	C	μ=0.10M($NaClO_4$)	71Bf	
	3.97	1.05(2)	18.1	25	C	μ=2.0	65Cd	
	1.0	2.54(2)	15	25	C	μ=0.10M($NaClO_4$) S: 40 v % DMSO	71Bf	
	-6.9	6.81(2)	8	25	C	μ=0.10M($NaClO_4$) S: Ethanol	71Bf	
	3.0	3.05(2)	24	25	C	μ=0.10M($NaClO_4$) S: Ethylene Glycol	71Bf	
Lu^{3+}	5.35	1.65(1)	25.5	25	C	μ=2.0	65Cd	
	3.65	1.12(2)	17.4	25	C	μ=2.0	65Cd	
Cu^{2+}	-1.2	6.56(1)	26	25	C	μ=0.10M($NaClO_4$) S: DMSO	71Bf	
	1.7	4.22(1)	25	25	C	μ=0.10M($NaClO_4$) S: Ethylene Glycol	71Bf	
	0.6	2.84(1)	15	25	C	μ=0.10M($NaClO_4$) S: Formamide	71Bf	
	-8.2	6.23(2)	1	25	C	μ=0.10M($NaClO_4$) S: DMSO	71Bf	
	-1.3	4.89(2)	18	25	C	μ=0.10M($NaClO_4$) S: Ethylene Glycol	71Bf	
	-0.8	2.34(2)	8	25	C	μ=0.10M($NaClO_4$) S: Formamide	71Bf	
P166	PROPANOIC ACID, 2-methyl-, ethyl ester		$C_6H_{12}O_2$			$CH_3CH(CH_3)CO_2C_2H_5$		L
SbA_5	-16.44	(1)	--	25	C	Excess $SbCl_5$. C_L low, variable, ∿0.02. S: 1,2-Dichloroethane. A=Cl	64Oa	
	-16.37	4.55(1)	-34.1	25	C	Excess $SbCl_5$. C_L low, variable, ∿0.02. S: 1,2-Dichloroethane. A=Cl	67Od	
P167	PROPANOIC ACID, 2-oxo-		$C_3H_4O_3$			$CH_3C(O)CO_2H$		
Eu^{3+}	-1.17	1.973(1)	5.26	25	C	μ=2.0($NaClO_4$)	71Ai	

Metal Ion	ΔH, kcal /mole	Log K	ΔS, cal /mole °K	T, °C	M	Conditions	Ref.	Re-marks
P167, cont.								
Eu^{3+}, cont.	-2.41	1.344(2)	-2.39	25	C	μ=2.0($NaClO_4$)	71Ai	
P168 2-PROPANOL			C_3H_8O			$CH_3CH(OH)CH_3$		
CeA_2B_4	-8.18	0.210(25°,1)	-26.51	25-35	T	μ=1.08M(HNO_3) $C_M=4x10^{-3}M$ $C_L=5x10^{-2}M$	71Tc	b
P169 2-PROPANOL, 1,3-diamino-			$C_3H_{10}ON_2$			$H_2NCH_2CH(OH)CH_2NH_2$		L
Ni^{2+}	-5.8	5.42(30°,1)	6	10-40	T	μ=0	58Ba	b
	-5.0	4.16(30°,2)	2	10-40	T	μ=0	58Ba	b
Cu^{2+}	-5.2	3.57(30°)	-1	10-40	T	μ=0. R: Cu^{2+} + HL^+ = CuL^{2+} + H^+	58Bb	b
Ag^+	-13.2	5.31(30°,1)	-19	10-40	T	μ=0	58Ba	b
P170 2-PROPANONE			C_3H_6O			CH_3COCH_3		L
CoA_2B_2	3.2	-3.917(27°)	-6.6	-80-60	T	S: Acetone A=N-Methyl thiourea B=Cl. R: CoA_2B_2 + S = $CoASB_2$ + A	71Ea	b,q: ±0.5
	3.5	-3.84(27°)	-5.8	-80-60	T	S: Acetone A=Thiourea. B=Cl R: CoA_2B_2 + S = $CoASB_2$ + A	71Ea	b,q: ±0.4
	2.3	-3.686(27°)	-8.9	-80-60	T	S: Acetone A=N,N'-Dimethyl thiourea. B=Br R: CoA_2B_2 + S = $CoASB_2$ + A	71Ea	b
	3.9	-3.889(27°)	-4.5	-80-60	T	S: Acetone A=N-Methyl thiourea B=Br. R: CoA_2B_2 + S = $CoASB_2$ + A	71Ea	b
	4.0	-3.55(27°)	-2.5	-80-60	T	S: Acetone A=Thiourea. B=Br R: CuA_2B_2 + S = $CoASB_2$ + A	71Ea	b,q: ±0.4
	3.3	-3.436(27°)	-5.1	-80-60	T	S: Acetone A=N,N'-Dimethyl thiourea. B=I R: CoA_2B_2 + S = $CoASB_2$ + A	71Ea	b,q: ±0.9
	6.1	-3.41(27°)	4.9	-80-60	T	S: Acetone A=Thiourea. B=I R: CoA_2B_2 + S = $CoASB_2$ + A	71Ea	b,q: ±0.8
	-8.36	-2.74	--	-84-30	T	S: Acetone C_M=0.2-0.4M. A=Cl R: CoL_2A_2 + 2L = CoL_4A_2	72Bg	b,1

Metal Ion	ΔH, kcal /mole	Log K	ΔS, cal /mole °K	T, °C	M	Conditions	Ref.	Remarks
P170, cont.								
CoA_2B_2, cont.	3.82	-0.770	--	-84-30	T	S: Acetone C_M=0.2-0.4M. A=Cl R: $2CoL_2A_2$ + 2L = CoL_6^{2+} + CoA_4^{2-}	72Bg	b,1
	2.7	-3.532(27°)	-2.2	-80-60	T	S: Acetone A=N,N'-Dimethyl thiourea. B=ClO_4 R: CoA_4B_2 + S = CoA_3SB_2 + A	71Ea	b,j,q: ±0.4
	2.9	-2.499(27°)	-1.7	-80-60	T	S: Acetone A=N-Methyl thiourea B=ClO_4. R: CoA_4B_2 + S = CoA_3SB_2 + A	71Ea	b,j
	2.9	-3.272(27°)	0.1	-80-60	T	S: Acetone A=Thiourea. B=ClO_4 R: CoA_4B_2 + S = CoA_3SB_2 + A	71Ea	b,j,q: ±0.4
	4.6	-3.842(27°)	2.6	-80-60	T	S: Acetone A=Thiourea. B=NO_3 R: CoA_4B_2 + S = CoA_3SB_2 + A	71Ea	b,j
$SiAB_3$	-3.25	1.36(30°,<u>1</u>)	-4.52	10-45	T	S: CCl_4 X_L=5x10^{-2} X_M=5x10^{-4} A=OH. B=CH_3	69Lb	b
	-3.58	1.36(30°,<u>1</u>)	-5.61	20-60	T	S: CCl_4 X_L=5x10^{-2} X_M=5x10^{-4} A=OH. B=C_2H_5	69Lb	b
	-3.78	1.54(30°,<u>1</u>)	-5.45	0-50	T	S: CCl_4 X_L=5x10^{-2} X_M=4x10^{-4} A=OH. B=Phenyl	69Lb	b
$SnAB_3$	-2.61	0.740(<u>1</u>)	-<u>5.4</u>	25	C	S: Benzene A=C_4H_9. B=Cl^-	71Ge	
SnA_2B_2	-1.13	0.290(<u>1</u>)	-<u>2.4</u>	25	C	S: Benzene A=C_4H_9. B=Cl	71Ge	q: ±0.22
SnA_3B	-5.7	-0.377(<u>1</u>)	<u>21</u>	26	C	S: CCl_4. A=CH_3. B=Cl	66Bg	
	-6.0	-0.41(<u>1</u>)	<u>22</u>	26	C	S: Isooctane A=CH_3. B=Cl	66Bg	
SbA_5	-17.03	(<u>1</u>)	--	25	C	S: 1,2-Dichloro-ethane. A=Cl	63Oa	
TeA_4	-5.4	(<u>1</u>)	--	25	C	S: Benzene. A=Cl	72Pe	
I_2	-2.5	(<u>1</u>)	--	0-45	T	S: Cyclohexane C_L≃1M	66Bc	b,q: ±1
	-5.8	1.19(20°,<u>1</u>)	-14	-15-35	T	S: Freon	65Wc	b,k
	-3.8	0.094(25°,<u>1</u>)	-12	5-25	T	S: Heptane	71Gk	b
I_2L	-21	2.91(20°)	-57	-15-35	T	S: Freon R: $2I_2L$ = $(I_2L)_2$	65Wc	b,k b,k

Metal Ion	ΔH, kcal /mole	Log K	ΔS, cal /mole °K	T, °C	M	Conditions	Ref.	Remarks
P171	2-PROPANONE, 3-(p-methoxy-benzoyl)-		$C_{11}H_{12}O_3$			$(CH_3OC_6H_4CO)CH_2COCH_3$		
Ni^{2+}	-2.7	9.23(25°,1)	33	15-40	T	μ=0 S: 70 v % Dioxane	71Yb	b,q: ±0.5
	3.4	7.79(25°,2)	24	15-40	T	μ=0 S: 70 v % Dioxane	71Yb	b,q: ±0.5
Cu^{2+}	-9.2	12.93(25°,1)	28	15-40	T	μ=0 S: 75 v % Dioxane	71Yb	b
	-8.5	11.24(25°,2)	23	15-40	T	μ=0 S: 70 v % Dioxane	71Yb	b
P172	1-PROPANONE, 2-methyl-1-phenyl-		$C_{10}H_{12}O$			$C_6H_5C(O)CH(CH_3)_2$		L
I_2	-2.8	-0.14(25°,1)	-10	5-25	T	S: Heptane	71Te	b
P173	1-PROPANONE, 1-phenyl-		$C_9H_{10}O$			$C_6H_5C(O)C_2H_5$		L
I_2	-2.9	-0.10(25°,1)	-10	5-25	T	S: Heptane	71Te	b
P174	2-PROPANONE, 3-thenoyl-1,1,1-trifluoro-		$C_8H_5O_2F_3S$			F_3CCOCH_2CO— 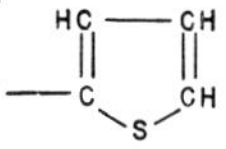		
	-4.93	1.18(50°)	-9.8	30-70	T	S: Benzene pH=1.6(HCl) A=Dibenzyl sulphoxide R: UO_2^{2+} + 2HL + A = UO_2L_2A + $2H^+$	70Sb	b
	-4.58	0.75(50°)	-10.7	30-70	T	S: Benzene pH=1.6(HCl) A=Diphenyl sulphoxide R: UO_2^{2+} + 2HL + A = UO_2L_2A + $2H^+$	70Sb	b
	-5.21	-1.54(50°)	-9.1	30-70	T	S: Benzene pH=1.6(HCl) A=Tributyl phosphate R: UO_2^{2+} + 2HL + A = UO_2L_2A + $2H^+$	70Sb	b
	-3.77	2.56(50°)	0.2	30-70	T	S: Benzene pH=1.6(HCl) A=Trioctylphosphine oxide. R: UO_2^{2+} + 2HL + A = UO_2L_2A + $2H^+$	70Sb	b
P175	PROPENAL		C_3H_4O			CH_2:CHCHO		L
NiAB	-19.0	3.62(27°)	-47.0	27-42	T	S: Tetrahydrofuran A=Dipyridyl B=Tetrahydrofuran R: NiAB + L = NiAL + B	71Ya	b
P176	PROPENE		C_3H_6			$CH_3CH:CH_2$		L

Metal Ion	ΔH, kcal /mole	Log K	ΔS, cal /mole °K	T, °C	M	Conditions	Ref.	Re-marks
P176, cont.								
$RhAB_2$	1.4	-1.108(25°)	-0.5	0&25	T	S: Toluene A=2,4-Pentanedionato B=Ethylene R: L + $RhAB_2$ = RhABL + B	67Cd	a,b,q: ±0.9
PdA_4^{2-}	0	0.90(20°)	4	5-25	T	μ=4.0. R: PdA_4^{2-} + L = PdA_3L^- + A^- A=Cl	65Pb	b
	0	0.66(20°)	3	5-25	T	μ=4.0. R: PdA_4^{2-} + L + H_2O = $PdA_2L(OH)_2$ + $2A^-$. A=Cl	65Pb	b
Cu^+	-10.11	5.018(25°,1)	-10.95	20-50	T	μ=0	70Oa	b
Ag^+	-3.5	0.88(25°,1)	-7.5	0-40	T	S: Ethylene glycol	65Ci	b
AgAB	-0.1	-0.366(25°)	-1.5	--	-	A = NO_3. B=Ethylene R: L + AgAB = AgAL + B	67Cd	i: 65Cg,u
P177 PROPENE, 3-bromo-			C_3H_5Br			$BrCH_2CH:CH_2$		L
I_2A	-1.10	0.745(25°)	-0.29	-20-40	T	S: Hexane A=Hexane R: I_2A + L = I_2L + A	73Si	b,q: ±0.10
P178 PROPENE, 3-chloro-			C_3H_5Cl			$ClCH_2CH:CH_2$		L
CuA_3^{2-}	-17.05	0(1)	-56.3	25	T	C_{Cl}→0. A=Cl	65Ta	
	-7.1	0.046	-24.0	25	T	C_{Cl}→0. R: CuA_3^{2-} + L = CuA_2L^- + A^-. A=Cl	65Ta	
P179 PROPENE, 2-methyl-			C_4H_8			$(CH_3)_2C:CH_2$		L
$RhAB_2$	3.9	-3.456(25°)	-2.6	0&25	T	S: Toluene A=2,4-Pentonedionato B=Ethylene R: L + $RhAB_2$ = RhABL + B	67Cd	a,b,q: ±0.7
Cu^+	-7.05	4.495(25°,1)	-3.05	20-50	T	μ=0	70Oa	b
Ag^+	-5.08	1.773(25°,1)	-8.95	20-80	T	μ=0.1M($AgNO_3$ + KNO_3)	70Pb	b
P180 PROPENE, 3-sulphonate			$C_3H_5O_3S$			$CH_2:CHCH_2SO_3^-$		L^-
PtA_4	-6.1	3.606(25°)	-4.1	25-45	T	1.90M(NaCl) 0.1M(HCl) A=Cl R: PtA_4^{2-} + L^- = PtA_3L^{2-} + A^-	68Mc	b,g:k
P181 PROPENOIC ACID, amide			C_3H_5ON			$CH_2:CHCONH_2$		L
NiAB	-17.8	3.31(27°)	-44.2	22-46	T	S: Tetrahydrofuran A=Dipyridyl B=Tetrahydrofuran R: NiAB + L = NiAL + B	71Ya	b

Metal Ion	ΔH, kcal /mole	Log K	ΔS, cal /mole °K	T, °C	M	Conditions	Ref.	Remarks
P182	PROPENOIC ACID, nitrile		C_3H_3N			CH_2:CHCN		L
NiAB	-16.0	2.82(27°)	-40.5	20-42	T	S: Tetrahydrofuran A=Dipyridyl B=Tetrahydrofuran R: NiAB + L = NiAL + B	71Ya	b
P183	PROPENOIC ACID, 2-methyl-, nitrile		C_4H_5N			CH_2:C(CH_3)CN		L
NiAB	-14.1	2.37(27°)	-36.2	22-43	T	S: Tetrahydrofuran A=Dipyridyl B=Tetrahydrofuran R: NiAB + L = NiAL + B	71Ya	b
P184	2-PROPEN-1-ol		C_3H_6O			CH_2:$CHCH_2OH$		L
PtA_4	-8.1	4.212(25°)	-7.8	25-45	T	1.90M(NaCl) 0.1M(HCl) A=Cl R: PtA_4^{2-} + L = PtA_3L^- + A^-	68Mc	b,q: ±0.5
	-8.1	3.860(44.5°)	-7.6	30-60	T	μ=2.0M(NaCl + HCl) A=Cl. R: PtA_4^{2-} + L = $PtLA_3^-$ + A^-	67Hb	b,q: ±0.5
CuA_2^-	-1.95	0.834(1)	1.83	25	T	$C_{Cl}\to 0$. A=Cl	65Ta	
CuA_3^{2-}	-6.02	1.333	-14.15	25	T	$C_{Cl}\to 0$. R: CuA_3^- + L = CuA_2L^- + A^- A=Cl	65Ta	
P185	PROPYNE, 3-acetoxy-		$C_5H_6O_2$			CH_3COOCH_2C:CH		L
Ag^+	-2.7	0.146(25°,1)	-8.5	10-40	T	μ=1M($NaNO_3$) S: 52 v % Ethanol	73St	b,q: ±0.5
P186	PROPYNE, 3-chloro-		C_3H_3Cl			$ClCH_2C$:CH		L
Ag^+	-4.9	0.049(25°,1)	-16.0	10-40	T	μ=1M($NaNO_3$) S: 52 v % Ethanol	73St	b,q: ±0.8
	-6.8	0.00(25°,1)	-22.8	10-40	T	μ=1M($NaNO_3$) S: 75 v % Ethanol	73St	b,q: ±0.4
P187	PROPYNE, 3-hydroxy-		C_3H_4O			$HOCH_2C$:CH		L
Ag^+	-2.4	0.633(25°,1)	-5	10-40	T	μ=1M($NaNO_3$) S: 52 v % Ethanol	73St	b,q: ±0.4
	-4.3	0.518(25°,1)	-12.1	10-40	T	μ=1M($NaNO_3$) S: 75 v % Ethanol	73St	b,q: ±0.4
P188	PROPYNE, 3-methoxy-		C_4H_6O			CH_3OCH_2C:CH		L
Ag^+	-2.4	0.431(25°,1)	-6.4	10-40	T	μ=1M($NaNO_3$) S: 52 v % Ethanol	73St	b,q: ±0.5

Metal Ion	ΔH, kcal /mole	Log K	ΔS, cal /mole °K	T, °C	M	Conditions	Ref.	Remarks
P189	PURINE, 6-amino-9-cyclohexyl-		$C_{11}H_{14}N_5$			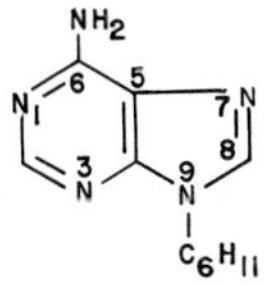		L
I_2	-9.5	2.250(25°,1)	20.9	15-35	T	μ=0. S: CCl_4	74La	b
P190	PYRAN, tetrahydro-		$C_5H_{10}O$			$CH_2CH_2CH_2CH_2CH_2O$ (ring)		L
I_2	-4.9	0.122(25°,1)	-10.7	5.5-25	T	S: n-Heptane	60Ta 60Bc	b
P191	PYRAZINE		$C_4H_4N_2$					L
Ni^{2+}	-3.2	1.013(25°,1)	-6	10-35	T	μ=1.0M($LiClO_4$) pH=4	72Mb	b,q: ±0.5
Ag^+	-4.07	1.38(1)	-7.34	25	C	μ=0.1M(KNO_3)	73Bg	
	-4.03	1.03(2)	-8.76	25	C	μ=0.1M(KNO_3)	73Bg	
P192	PYRAZINE, 2-methyl-		$C_5H_6N_2$			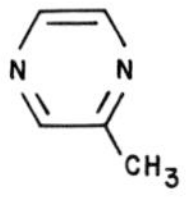		
I_2	-6.12	1.478(1)	--	30-50	T	S: Cyclohexane K has been corrected for interation between I_2 and solvent	73Ic	b
	-6.71	1.549(1)	--	30-50	T	S: n-Heptane K has been corrected for interation between I_2 and solvent	73Ic	b
P193	PYRAZOLE		$C_3H_4N_2$			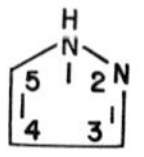		L
Cd^{2+}	-3.94	1.50(25°,1)	-6.3	0-45	T	0.1M(KNO_3). C_M= 5x10^{-4}[$Cd(NO_3)_2$] C_L=0.05-2.0M	63Ad	b
	-8.10	2.18(25°,1-2)	-20.1	0-45	T	0.1M(KNO_3). C_M= 5x10^{-4}[$Cd(NO_3)_2$] C_L=0.05-2.0M	63Ad	b
	-12.14	2.32(25°,1-3)	-30.1	0-45	T	0.1M(KNO_3). C_M= 5x10^{-4}[$Cd(NO_3)_2$] C_L=0.05-2.0M	63Ad	b
P194	PYRIDAZINE		$C_4H_4N_2$			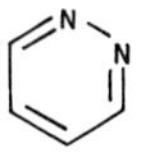		

Metal Ion	ΔH, kcal /mole	Log K	ΔS, cal /mole °K	T, °C	M	Conditions	Ref.	Re-marks
P194, cont.								
Ag^+	-7.39	1.48(1)	-18.0	25	C	μ=0.1M(KNO_3)	73Bg	
	-0.67	1.34(2)	3.9	25	C	μ=0.1M(KNO_3)	73Bg	
P195 PYRIDINE			C_5H_5N					L
Be^{2+}	-4.48	2.37(40°,1)	-3.2	35&45	T	--	67Ra	a,b
MgAL	-2.28	-1.024(30°)	-12.0	--	T	S: Piperidine C_L=12.36M A=Etioporphyrin II R: MgAL + L = $MgAL_2$	66Sd	u
	-2.55	-1.125(30°)	-13.7	--	T	S: Pyridine-2,6-lutidine. C_L=12.36M A=Desoxophyllorythrin monomethyl ester R: MgAL + L = $MgAL_2$	66Sd	u
	-3.89	-0.613(30°)	-15.6	--	T	S: Pyridine-2,6-lutidine. C_L=12.36M A=Deuteroporphyrin IX dimethyl ester R: MgAL + L = $MgAL_2$	66Sd	u
	-3.90	-0.548(30°)	-15.4	--	T	S: Pyridine-2,6-lutidine. C_L=12.36M A=Dichlorodeutero-porphin IX dimethyl ester. R: MgAL + L = $MgAL_2$	66Sd	u
	-2.92	-1.168(30°)	-15.0	--	T	S: Pyridine-2,6-lutidine. C_L=12.36M A=Etioporphyrin II R: MgAL + L = $MgAL_2$	66Sd	u
	-3.77	-0.836(30°)	-16.3	--	T	S: Pyridine-2,6-lutidine. C_L=12.36M A=Mesoporphyrin IX dicholesteryl ester R: MgAL + L = $MgAL_2$	66Sd	u
	-3.98	-1.154(30°)	-18.4	--	T	S: Pyridine-2,6-lutidine. C_L=12.36M A=Mesoporphyrin IX dimethyl ester R: MgAL + L = $MgAL_2$	66Sd	u
	-4.25	-0.606(30°)	-16.8	--	T	S: Pyridine-2,6-lutidine. C_L=12.36M A=Protoporphyrin IX dimethyl ester R: MgAL + L = $MgAL_2$	66Sd	u
	-2.37	-0.779(30°)	-11.4	--	T	S: Pyridine-2,6-lutidine. C_L=12.36M A=Tetraphenylporphin R: MgAL + L = $MgAL_2$	66Sd	u

Metal Ion	ΔH, kcal /mole	Log K	ΔS, cal /mole °K	T, °C	M	Conditions	Ref.	Re-marks
P195, cont.								
MgAL, cont.	-2.46	-1.024(30°)	-12.6	--	T	S: Pyrrolidine C_L=12.36M A=Etioporphyrin II R: MgAL + L = $MgAL_2$	66Sd	u
[Mg]	-2.60	-0.079(30°,2)	-9.0	20-40	T	S: Benzene [Mg]=Magnesium Tetraphenylchlorin	52Ma	b
	-2.60	-0.239(30°,2)	-9.8	20-40	T	S: Benzene [Mg]=Magnesium Tetraphenylporphine	52Ma	b
VOA_2	-7.35	1.76(1)	-16.6	25	C	μ→0. S: Nitrobenzene A=Acetylacetonate	65Ca	
NbA_5	-2.4	-1.4	-17	-54	T	S: $C_6H_5CH_3$. A=OCH_3 R: $[NbA_5]_2$ + 2L = $2NbA_5L$	71Hi	q: ±0.4
	-2.6	-0.68	-15	-48	T	S: CH_3CN. A=OCH_3 R: $[NbA_5]_2$ + 2L = $2NbA_5L$	71Hi	q: ±0.4
TaA_5	-3.4	-1.3	-21	-54	T	S: $C_6H_5CH_3$. A=OCH_3 R: $[TaA_5]_2$ + 2L = $2TaA_5L$	71Hi	q: ±0.6
Mn^{2+}	-2.39	1.86(1)	0.5	25	C	1M($NaClO_4$)	63Ae	f:4
	-2.39	1.59(2)	-1.5	25	C	1M($NaClO_4$)	63Ae	f:4
	-2.39	0.90(3)	-3.9	25	C	1M($NaClO_4$)	63Ae	f:4
	-2.39	0.60(4)	-5.3	25	C	1M($NaClO_4$)	63Ae	f:4
[Fe]	-2.4	2.56(25°,1-2)	4	18-40	T	[Fe]=Iron(II) deuteroporphyrin dimethyl ester S: Benzene	70Ch	b,q: ±0.2
	-3.5	1.76(25°,1-2)	-4	18-40	T	[Fe]=Iron(II) deuteroporphyrin dimethyl ester S: Benzene	71Cd	b,q: ±0.2
	-5.2	2.42(25°,1-2)	-6	18-40	T	[Fe]=Iron(II) mesoporphyrin dimethyl ester S: Benzene	70Ch	b,q: ±1.4
	-3.3	1.32(25°,1-2)	-5	18-40	T	[Fe]=Iron(II) protoporphyrin dimethyl ester S: Benzene	70Ch 71Cd	b,q: ±0.6
	-7.5	1.32(25°,1-2)	32	18-40	T	[Fe]=Iron(II) diacetyldeuter-porphyrin dimethyl ester. S: Benzene	70Ch	b,q: ±0.9
	-14.2	2.64(25°,1-2)	-35	18-40	T	[Fe]=Iron(II) deuteroporphyrin dimethyl ester S: CCl_4	71Cd	b,q:

Metal Ion	ΔH, kcal /mole	Log K	ΔS, cal /mole °K	T, °C	M	Conditions	Ref.	Re-marks
P195, cont.								
[Fe], cont.	-16.0	0.44(25°,1-2)	-50	18-40	T	[Fe]=Iron(II) protoporphyrin dimethyl ester. S: CCl_4	71Cd	b
	-10.8	0.22(25°,1-2)	-35	18-40	T	[Fe]=Iron(II) protoporphyrin dimethyl ester S: $CHCl_3$	71Cd	b,q: ±1.5
	2.35	0.88(20°)	12.1	5-30	T	pH=11.2 Protohaemin R: $(Haemin)_2$ + 2L = $(HaeminL)_2$ + $2H_2O$	69Md	b
CoA	-10.4	(1)	--	25	C	S: CH_2Cl_2. A=N,N'-bis(salicylidene)-o-phenylenediamino	74Tc	
	-8.5	2.686(25°,1)	-16	20-50	T	S: Toluene A=α,β,γ,δ-Tetra (p-methoxyphenyl) porphyrin. S: Toluene	73Wa	b,q: ±1
CoA_2	-5.1	3.15(25°,1)	-2.7	25	C	S: Acetone. A=Cl	73Lb	
	-10.7	5.4(1-2)	-11.4	25	C	S: Acetone $C_M=2x10^{-3}M$ $C_L=2x10^{-2}M$. A=Cl	71Lg	
	-9.8	5.4(1-2)	-8.4	25	C	S: Acetone $C_M=2x10^{-3}M$ $C_L=2x10^{-2}M$. A=Br	71Lg	
	-9.7	5.0(1-2)	-9.7	25	C	S: Acetonitrile $C_M=2x10^{-3}M$ $C_L=2x10^{-2}M$. A=Cl	71Lg	
	-9.8	4.8(1-2)	-11.1	25	C	S: Acetonitrile $C_M=2x10^{-3}M$ $C_L=2x10^{-2}M$. A=Br	71Lg	
	-6.3	3.4(1-2)	-5.7	25	C	S: n-Butyl alcohol $C_M=2x10^{-3}M$ $C_L=8x10^{-2}M$. A=Cl	71Lh	
	-5.3	3.2(1-2)	-3.4	25	C	S: n-Butyl alcohol $C_M=2x10^{-3}M$ $C_L=8x10^{-2}M$. A=Br	71Lh	
	-7.0	4.2(1-2)	-4.4	25	C	S: t-Butyl alcohol $C_M=2x10^{-3}M$ $C_L=8x10^{-2}M$. A=Cl	71Lh	
	-5.9	4.5(1-2)	1.0	25	C	S: t-Butyl alcohol $C_M=2x10^{-3}M$ $C_L=8x10^{-2}M$. A=Br	71Lh	
	-13.3	5.8(1-2)	-18.1	25	C	S: Cyclohexanone $C_M=2x10^{-3}M$ $C_L=2x10^{-2}M$. A=Cl	71Lg	
	-14.5	5.8(1-2)	-22.2	25	C	S: Cyclohexanone $C_M=2x10^{-3}M$ $C_L=2x10^{-2}M$. A=Br	71Lg	
	-4.89	4.17(1-2)	2.65	25	C	S: Dimethyl-formamide. A=Cl	72Ae	
	-7.68	3.22(1-2)	-11.04	25	C	S: Ethanol. A=Cl	72Af	

Metal Ion	ΔH, kcal /mole	Log K	ΔS, cal /mole °K	T, °C	M	Conditions	Ref.	Remarks
P195, cont.								
CoL_2A_2	-15.2	1.10(1-2)	-46.8	20	C	$\mu\to0$. S: $CHCl_3$ A=Cl	62Ka 63Ka	s
	-15.6	0.95(1-2)	-48.8	20	C	$\mu\to0$. S: $CHCl_3$ A=Br	62Ka 63Ka	s
	-16.6	0.36(1-2)	-54.9	20	C	$\mu\to0$. S: $CHCl_3$ A=I	62Ka 63Ka	s
	-13.7	1.36(1-2)	-40.5	20	C	$\mu\to0$. S: $CHCl_3$ A=NCO	62Ka 63Ka	s
	-16.6	4.92(1-2)	-34.1	20	C	$\mu\to0$. S: $CHCl_3$ A=NCS	62Ka 63Ka	s
	-11.2	-0.146(25°,1-2)	-38.2	27-57	T	μ=0.1(Et_4NClO_4) S: Nitrobenzene A=Br	72Fb	b
	-13.4	1.40(40°,1-2)	-36.7	32-60	T	S: Pyridine. A=Cl	61Ka	b
	-13.4	-0.672(25°,1-2)	-48.0	32-63	T	μ=0.1(Et_4NClO_4) S: Pyridine. A=Br	72Fb	b,q: ±1.3
CoA_6	-2.6	2.724	1.8	-30	T	S: Methanol A=CH_3OH. R: CoA_6^{2+} + L = CoA_5L^{2+} + A	73Pj	
[Co]	-20.8	--	--	24	C	[Co]=Methylcobal-oxime S: Dichloromethane R: $[Co]_2$ + 2L = 2[Co]L	73Cf	
	-6.9	4.03(0°,1)	-6	-23-23	T	[Co]=Cobalt(II) protoporphyrin IX dimethyl ester S: Toluene	73Sr	b,q: ±0.6
Ni^{2+}	-2.65	2.13(1)	0.8	25	C	1M($NaClO_4$)	63Ae	f:4
	-2.65	1.66(2)	-1.3	25	C	1M($NaClO_4$)	63Ae	f:4
	-2.65	1.12(3)	-3.8	25	C	1M($NaClO_4$)	63Ae	f:4
	-2.65	0.65(4)	-5.9	25	C	1M($NaClO_4$)	63Ae	f:4
NiA	-14.95	3.78(1-2)	-32.9	25	C	$C_M \approx C_L \approx 10^{-4}$ S: Benzene A=Diacetylbisbenzoyl hydrazine	58Sa	
	-6.7	1.82(25°,1)	-14	10-45	T	S: Benzene A= 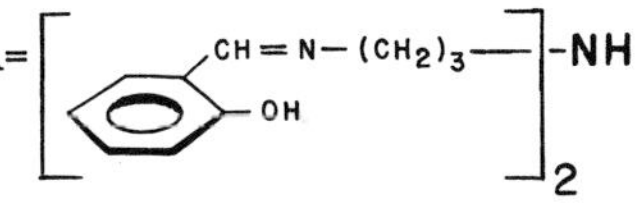	66Sa	b
	-5.5	1.35(25°,1)	-12	10-45	T	S: Benzene A= 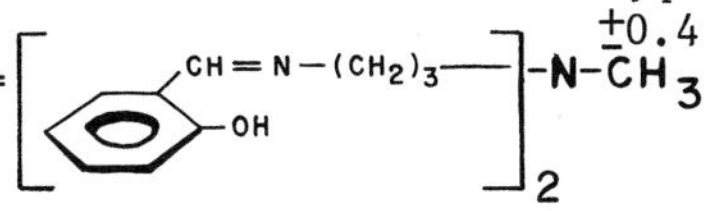	66Sa	b,q: ±0.4
	-7.6	2.25(25°,1)	-15	10-45	T	S: Benzene A= 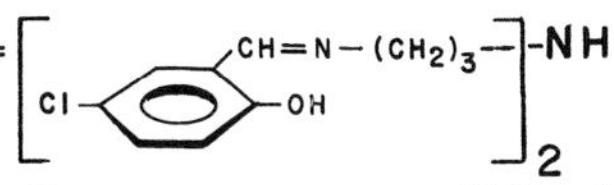	66Sa	b
	-7.9	2.33(25°,1)	-16	10-45	T	S: Benzene	66Sa	b

Metal Ion	ΔH, kcal/mole	Log K	ΔS, cal/mole $^\circ$K	T, $^\circ$C	M	Conditions	Ref.	Remarks
P195, cont.								
NiA, cont.						$A=\left[Br-\bigcirc\begin{smallmatrix}CH=N-(CH_2)_3-\\-OH\end{smallmatrix}\right]_2-NH$		
	−8.1	1.91(25°,1)	−18	10−45	T	S: Benzene	66Sa	b
						$A=\left[Br-\bigcirc\begin{smallmatrix}CH=N-(CH_2)_3-\\-OH\end{smallmatrix}\right]_2-N-CH_3$		
	−7.6	1.66(25°,1)	−18	10−45	T	S: Benzene	66Sa	b
						$A=\left[C_2H_5-\bigcirc\begin{smallmatrix}CH=N-(CH_2)_3-\\-OH\end{smallmatrix}\right]_2-NH$		
	−7.9	2.44(25°,1)	−15	10−45	T	S: Benzene	66Sa	b
						$A=\left[\bigcirc\bigcirc\begin{smallmatrix}CH=N-(CH_2)_3-\\-OH\end{smallmatrix}\right]_2-NH$		
	−7.1	2.46(25°,1)	−12	10−45	T	S: Benzene	66Sa	b
						$A=\left[\bigcirc\bigcirc\begin{smallmatrix}CH=N-(CH_2)_3-\\-OH\end{smallmatrix}\right]_2-NH$		
NiA₂	−17.4	1.18(1)	−52.1	30	C	S: Benzene A=Diethyldithiophosphato	71Dc	q: ±1.7
	−4.6	0.43(25°,1)	−13	−−	T	S: Benzene A=Diethyldithiophosphate. NiA₂ = bis(oo'-diethyldithiophosphate) nickel(II)	64Cb	b,u
	−7.0	1.46(25°,1)	−17	−−	T	S: Benzene A=Diethyldithiophosphate. NiA₂ = bis(oo'diethyldithiophosphato) nickel(II)	64Cb	b,u
	−7.0	1.462(25°,1)	−17	−−	T	S: Benzene A=oo'-Diethyldithiophosphate	64Ca	u
	−17.0	1.18(1)	−50.4	30	C	S: Benzene A=Dipropyldithiophosphato	71Dc	q: ±1.7
	−6.1	2.953(1)	−6.9	25	C	S: Benzene A=2,2,6,6-Tetramethyl-3,5-heptanedione	73Ce	
	0.96	1.34(2)	9.6	30	C	S: Benzene A=Diethyldithiophosphato	71Dc	q: ±1.7
	−4.6	0.430(25°,2)	−13	−−	T	S: Benzene	64Ca	u

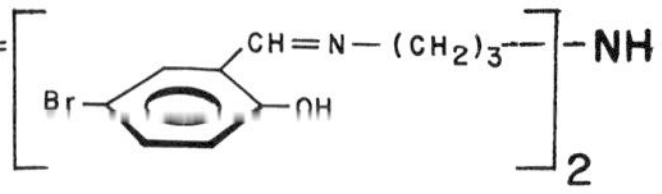

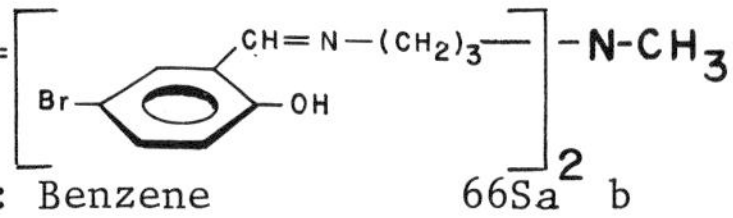

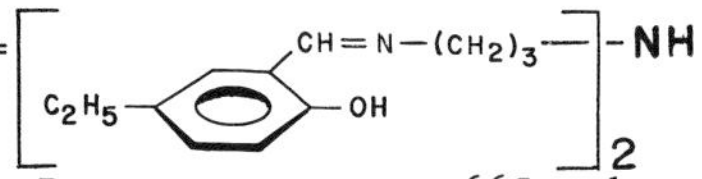

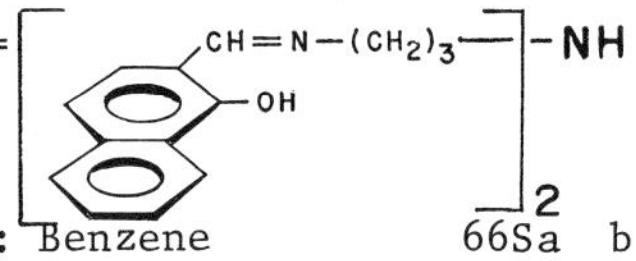

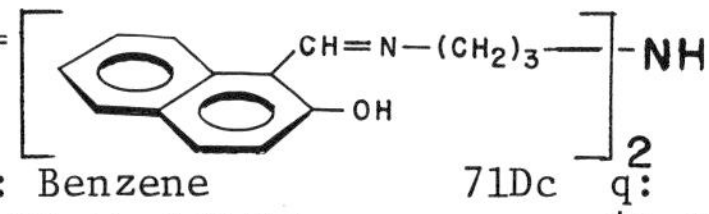

Metal Ion	ΔH, kcal /mole	Log K	ΔS, cal /mole °K	T, °C	M	Conditions	Ref.	Remarks
P195, cont.								
NiA_2, cont.						A=oo'-Diethyldithio- phosphate		
	0.49	1.43(2)	8.1	30	C	S: Benzene A=Dipropyldithio- phosphato	71Dc	q: ±1.7
	-13.0	4.301(2)	-23.9	25	C	S: Benzene A=2,2,6.6-Tetra- methyl-3,5-heptanedione	73Ce	
	-10.0	0.978(1-2)	-28.7	25	C	S: Benzene A=N-Benzyl-5-chloro- 1'-phenylsalicyl- aldehyde	73Da	q: ±1.0
	-8.60	5.5(1-2)	-3.2	30	C	S: Benzene C_M=0.002-0.044M A=3-Benzyl-pentane- 2,4-dione	73Af	
	-14.8	1.179(1-2)	-44.2	25	C	S: Benzene A=N-Butyl-5-chloro- 1'-phenylsalicyl- aldehyde	73Da	q: ±1.2
	-14.6	0.914(1-2)	-44.2	25	C	S: Benzene A=N-Butyl-5-chloro- 1'-phenylsalicyl- aldehyde	73Da	q: ±1.7
	-16.0	2.989(1-2)	-39.0	30	C	S: Benzene A=N-Butyl-5-chloro- salicyl-aldehyde	73Da	
	-4.06	4.8(1-2)	8.6	30	C	S: Benzene C_M=0.006-0.045M A=3-Butyl-pentane- 2,4-dione	73Af	
	-15.0	2.270(1-2)	-40.2	25	C	S: Benzene A=N-Butylsalicyl- aldehyde	73Da	
	-14.3	2.114(1-2)	-39.2	30	C	S: Benzene A=N-Butylsalicyl- aldehyde	73Ma	
	-11.2	0.929(1-2)	-33.0	25	C	S: Benzene A=N-(2,4-Dimethyl- benzene)-5-chloro- 1'-phenylsalicyl- aldehyde	73Da	q: ±1.2
	-10.0	1.531(1-2)	-26.3	30	C	S: Benzene A=N-(2,5-Dimethyl- benzene) salicyl- aldehyde	73Da	q: ±1.0
	-16	7.00(1-2)	-20.8	30	C	S: Benzene A=2,6-Dimethyl- heptane-3,5-dione	74Ga	
	-17.9	3.228(1-2)	-44.4	30	C	S: Benzene A=Ethylxanthate	71Dc	
	-11	3.204(28°,1-2)	-23	--	T	S: Benzene A=Ethylxanthate	64Ca	u

Metal Ion	ΔH, kcal /mole	Log K	ΔS, cal /mole °K	T, °C	M	Conditions	Ref.	Remarks
P195, cont.								
NiA_2, cont.	-8.6	0.949(1-2)	-23.9	25	C	S: Benzene A=N-Hexyl-5-chloro-1'-phenylsalicyl-aldehyde	73Da	q: ±1.0
	-15.3	3.041(1-2)	-37.0	30	C	S: Benzene A=N-Isopropy-5-chlorosalicyl-aldehyde	73Da	
	-14.6	2.176(1-2)	-38.2	30	C	S: Benzene A=N-Isopropylsalicyl-aldehyde	73Da	
	-10.0	1.204(1-2)	-27.5	30	C	S: Benzene A=N-(2-Methylbenzene) salicylaldehyde	73Da	q: ±1.0
	-16.5	3.371(1-2)	-38.7	30	C	S: Benzene A=N-Methyl-5-chloro-salicyl-aldehyde	73Da	
	-14.1	2.845(1-2)	-33.5	30	C	S: Benzene A=N-Methylsalicyl-aldehyde	73Da	
	0.72	6.7(1-2)	33	30	C	S: Benzene C_M=0.007-0.070M A=Pentane-2,4-dione	73Af	
	1.7	6.8(1-2)	36.7	30	C	S: Benzene C_M=0.006-0.062M A=1-Phenyl-butane-1,3-dione	73Af	
	-13.6	2.623(1-2)	-33.0	25	C	S: Benzene A=N-Phenyl-5-chloro-1'-phenylsalicyl-aldehyde	73Da	q: ±1.2
	-9.56	3.580(1-2)	-15.0	30	C	S: Benzene A=N-Phenyl-5-chloro-salicylaldehyde	73Da	
	9.08	3.00(1-2)	-16.2	30	C	S: Benzene A=N-Phenylsalicyl-aldehyde	73Da	q: ±1.0
	-14.8	2.00(1-2)	-39.4	30	C	S: Benzene A=N-Propylsalicyl-aldehyde	73Da	q: ±1.2
	-15.5	7.0(1-2)	-19.1	30	C	S: Benzene C_M=0.0004-0.003 A=2,2,6,6-Tetra-methylheptane-3,5-dione	73Af	
	-11.7	1.85(20°,1-2)	-31.9	10-30	T	S: $CHCl_3$. A=Ethyl-β-mercaptocinnamate	74Cc	b
	-9.9	1.56(20°,1-2)	-27.0	10-30	T	S: $CHCl_3$. A=Ethyl-β-mercaptothio-cinnamate	74Cc	b
	-9.5	0.60(20°,1-2)	-30.0	10-30	T	S: $CHCl_3$. A=N-(p-chlorophenyl)-β-mercaptocinnamamide	74Cc	b

Metal Ion	ΔH, kcal /mole	Log K	ΔS, cal /mole °K	T, °C	M	Conditions	Ref.	Remarks
P195, cont.								
NiA_2, cont.	-8.8	0.35(20°,1-2)	-28.4	10-30	T	S: $CHCl_3$. A=N-Phenyl-β-mercapto-cinnamamide	74Cc	b
	-8.2	0.26(20°,1-2)	-27.4	10-30	T	S: $CHCl_3$. A=N-(α-naphtyl)-β-mercapto-cinnamamide	74Cc	b
	-7.0	0.03(20°,1-2)	-24.0	10-30	T	S: $CHCl_3$. A=N-(p-ethoxyphenyl)-β-mercaptocinnamamide	74Cc	b
	-4.8	-0.29(20°,1-2)	-17.8	10-30	T	S: $CHCl_3$. A=Mono-thiodibenzoylmethane	74Cc	b
	-4.9	-0.46(20°,1-2)	-19.0	10-30	T	S: $CHCl_3$. A=Thio-benzoylacetone	74Cc	b
	-4.6	-0.66(20°,1-2)	-18.7	10-30	T	S: $CHCl_3$. A=N-Ethyl-β-mercapto-cinnamamide	74Cc	b
	-3.3	-0.93(20°,1-2)	-15.5	10-30	T	S: $CHCl_3$. A=N-methyl-N-phenyl-β-mercaptocinnamamide	74Cc	b
	-19.92	(1-2)	--	25	C	S: Cyclohexane A=2,2,6,6-Tetra-methyl-3,5-heptane-dione	73Ce	
	-11	(1-2)	-39	40-70	T	S: Pyridine A=N-Butyl-2-hydroxy-acetophenimine	70Ka	b,q: ±1
	-10	(1-2)	-37	40-70	T	S: Pyridine A=N-Butyl-2-hydroxy-butyrophenimine	70Ka	b,q: ±1
	-8	(1-2)	-33	40-70	T	S: Pyridine A=N-Butyl-2-hydroxy-propiophenimine	70Ka	b,q: ±1
	-5	(1-2)	-27	40-70	T	S: Pyridine A= 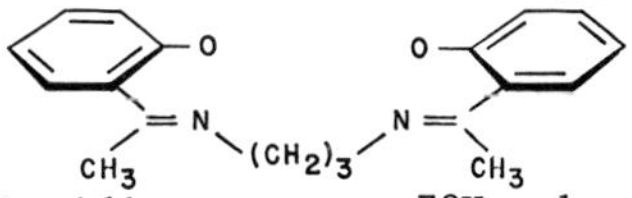	70Ka	b,q: ±1
	-4	(1-2)	-25	40-70	T	S: Pyridine A= 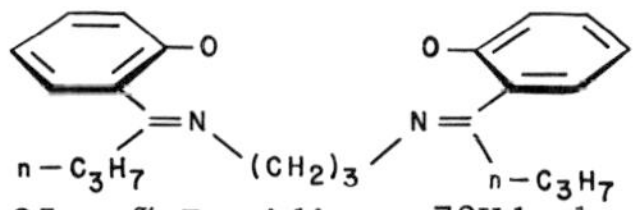	70Ka	b,q:
	-10.40	-1.553(25°,1-2)	-42.0	-10-20	T	S: 25 v % Pyridine, 75 v % CH_2Cl_2 A=N,N'-di-n-butyl dithiocarbamate	72Vd	b
	-6.70	-1.060(25°,1-2)	-27.3	-10-20	T	S: 25 v % Pyridine, 75 v % CH_2Cl_2 A=N,N'-diethyl dithiocarbamate	72Vd	b
	-8.25	-1.268(25°,1-2)	-33.5	-10-20	T	S: 25 v % Pyridine, 75 v % CH_2Cl_2 A=N,N'-di-n-propyl dithiocarbamate	72Vd	b

Metal Ion	ΔH, kcal /mole	Log K	ΔS, cal /mole °K	T, °C	M	Conditions	Ref.	Remarks
P195, cont.								
NiA_2, cont.	-16.8	1.89(25°,1-2)	-47.7	25	C	S: Nitrobenzene A=Diethyldithio phosphate. K determined in Benzene	70Cb	
	-16.8	(1-2)	--	25	C	S: Nitrobenzene A=Diethyldithiophosphate	70Cc	
	-5	--	-15	27-80	T	S: Pyridine A= 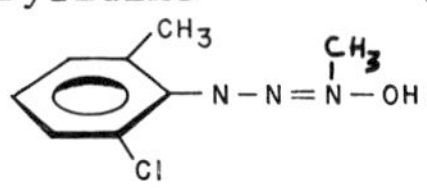 R: NiA_2(planar, S=0) + nL = NiA_2L_n (paramagnetic, S=1)	72Za	b,q: ±1
	-6	--	-19	25-70	T	S: Pyridine A= 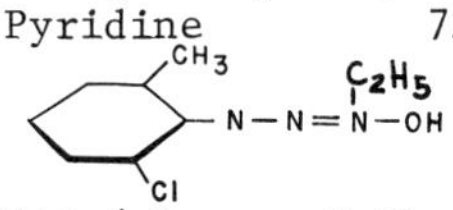 R: NiA_2(planar, S=0) + nL = NiA_2L_n (paramagnetic, S=1)	72Za	b,q: ±1
	-2.39	4.7(2-1)	13.4	30	C	S: Benzene C_M=0.002-0.044M A=3-Benzyl-pentane-2,4-dione	73Af	
	7.17	4.0(2-1)	42.1	30	C	S: Benzene C_M=0.006-0.045M A=3-Butyl-pentane-2,4-dione	73Af	
	-12	7.00(2-1)	-7.2	30	C	S: Benzene A=2,6-Dimethyl-heptane-3,5-dione	74Ga	q: ±2.4
	20.6	8.0(2-1)	104.5	30	C	S: Benzene C_M=0.007-0.070M A=Pentane-2,4-dione	73Af	q
	23.4	6.7(2-1)	107.8	30	C	S: Benzene C_M=0.006-0.062M A=1-Phenyl-butane-1,3-dione	73Af	q
	-12.0	6.3(2-1)	-10.5	30	C	S: Benzene C_M=0.0004-0.003 A=2,2,6,6-Tetramethylheptane-3,5-dione	73Af	
NiL_2A_2	-23.8	4.49(1-2)	-60.7	20	C	μ→0. S: $CHCl_3$ A=I	64Ka	
Ni_2A_4L	-14.8	6.06	-20.6	30	C	S: Benzene. A=3-Benzyl-pentane-2,4-dione. R: Ni_2A_4L + 3L = $2NiA_2L_2$	73Af	
	-15.3	5.68	-24.4	30	C	S: Benzene A=3-Butyl-pentane-	73Af	q: ±2.4

Metal Ion	ΔH, kcal /mole	Log K	ΔS, cal /mole °K	T, °C	M	Conditions	Ref.	Remarks
P195, cont.								
Ni_2A_4L, cont.						2,4-dione R: $Ni_2A_4L + 3L = 2NiA_2L_2$		
	-20	7.00	-34.4	30	C	S: Benzene A=2,6-Dimethyl-heptane-3,5-dione R: $Ni_2A_4L + 3L = 2NiA_2L_2$	74Ga	q: ±2.4
	-19.1	5.41	-38.2	30	C	S: Benzene A=Pentane-2,4-dione R: $Ni_2A_4L + 3L = 2NiA_2L_2$	73Af	q: ±2.4
	-19.6	6.99	-32.5	30	C	S: Benzene A=1-Phenyl-butane-1,3-dione. R: $Ni_2A_4L + 3L = 2NiA_2L_2$	73Af	q: ±2.4
	-19.1	7.10	-30.6	30	C	S: Benzene A=2,2,6,6-Tetra-methylheptane-3,5-dione. R: $Ni_2A_4L + 3L = 2NiA_2L_2$	73Af	q: ±2.4
Ni_3A_6	-61.7	6.10	-175	30	C	S: Benzene A=3-Benzyl-pentane-2,4-dione. R: $Ni_3A_6 + 3L = 3Ni_2A_4L$	73Af	q: ±7.2
	-76.0	5.58	-225	30	C	S: Benzene A=3-Butyl-pentane-2,4-dione. R: $Ni_3A_6 + 3L = 3Ni_2A_4L$	73Af	q: ±7.2
	-30.1	11.99	44.4	30	C	S: Benzene A=Pentane-2,4-dione R: $Ni_3A_6 + 3L = 3Ni_2A_4L$	73Af	q: ±7.2
	-37.3	8.06	-86.0	30	C	S: Benzene A=1-Phenyl-butane-1,3-dione. R: $Ni_3A_6 + 3L = 3Ni_2A_4L$	73Af	q: ±7.2
	-8.6	4.79	-6.2	30	C	S: Benzene A=2,6-Dimethylheptane-3,5-dione R: $2/3Ni_3A_6 + L = Ni_2A_4L$	74Ga	q: ±2.4
[Ni]	-6.3	3.15(25°,<u>1</u>)	-7	18-40	T	[Ni]=Nickel(II) mesotetraphenyl porphine. S: Benzene	72Ce	b
	-4.2	3.00(25°,<u>1</u>)	0	18-40	T	[Ni]=Nickel(II) mesotetraphenyl porphine. S: $CHCH_3$	72Ce	b,q:
$RhAB_2$	-19.2	--	--	21-24	C	S: Dichloromethane A=Cl. B=1,5-Cyclo-octadiene. R: $[RhAB]_2 + 2L = 2RhABL$	73Pe	

Metal Ion	ΔH, kcal /mole	Log K	ΔS, cal /mole °K	T, °C	M	Conditions	Ref.	Re-marks
P195, cont.								
$RhAB_2$, cont.	-19.7	--	--	21-24	C	S: Dichloromethane A=Br. B=1,5-Cyclo-octadiene. R: $[RhAB]_2 + 2L = 2RhABL$	73Pe	
	-18.4	--	--	21-24	C	S: Dichloromethane A=I. B=1,5-Cyclo-octadiene. R: $[RhAB]_2 + 2L = 2RhABL$	73Pe	
PdA_2B	-21.1	--	--	25	C	S: CH_2Cl_2. A=Cl B=Ethylene R: $(PdA_2B)_2 + 2L = 2PdA_2BL$	74Pa	
	-16.2	--	--	25	C	S: CH_2Cl_2. A=Cl B=cis-Butene R: $(PdA_2B)_2 + 2L = 2PdA_2BL$	74Pa	
	-21.5	--	--	25	C	S: CH_2Cl_2. A=Cl B=Cyclopentene R: $(PdA_2B)_2 + 2L = 2PdA_2BL$	74Pa	
	-27.8	--	--	25	C	S: CH_2Cl_2. A=Cl B=Cyclohexene R: $(PdA_2B)_2 + 2L = 2PdA_2BL$	74Pa	
	-16.2	--	--	25	C	S: CH_2Cl_2. A=Cl B=Cycloheptene R: $(PdA_2B)_2 + 2L = 2PdA_2BL$	74Pa	
	-16.6	--	--	25	C	S: CH_2Cl_2. A=Cl B=cis-Cyclooctene R: $(PdA_2B)_2 + 2L = 2PdA_2BL$	74Pa	
	-22.7	--	--	25	C	S: CH_2Cl_2. A=Cl B=Styrene R: $(PdA_2B)_2 + 2L = 2PdA_2BL$	74Pa	
PdA_2BL	-13.8	--	--	25	C	S: CH_2Cl_2. A=Cl B=Ethylene R: $PdA_2BL + L = PdA_2L_2 + B$	74Pa	
	-12.4	--	--	25	C	S: CH_2Cl_2. A=Cl B=cis-Butene R: $PdA_2BL + L = PdA_2L_2 + B$	74Pa	
	-14.3	--	--	25	C	S: CH_2Cl_2. A=Cl B=Cyclopentene R: $PdA_2BL + L = PdA_2L_2 + B$	74Pa	
	-13.9	--	--	25	C	S: CH_2Cl_2. A=Cl B=Cyclohexene R: $PdA_2BL + L = PdA_2L_2 + B$	74Pa	

Metal Ion	ΔH, kcal /mole	Log K	ΔS, cal /mole °K	T, °C	M	Conditions	Ref.	Re-marks
P195, cont.								
PdA_2BL, cont.	-12.7	--	--	25	C	S: CH_2Cl_2. A=Cl B=Cycloheptene R: $PdA_2BL + L = PdA_2L_2 + B$	74Pa	
	-11.6	--	--	25	C	S: CH_2Cl_2. A=Cl B=cis-Cyclooctene R: $PdA_2BL + L = PdA_2L_2 + B$	74Pa	
	-14.0	--	--	25	C	S: CH_2Cl_2. A=Cl B=Styrene R: $PdA_2BL + L = PdA_2L_2 + B$	74Pa	
PdA_2B_2	-26.0	--	--	21-24	C	S: Dichloromethane A=Cl. B=Benzonitrile R: $PdA_2B_2 + 2L = PdA_2L_2 + 2B$	73Pe	
	-26.0	--	--	25	C	S: CH_2Cl_2. A=Cl B=C_6H_5CN. R: $PdA_2B_2 + L = PdA_2L + 2B$ C_L=8.3x10^{-3}M C_M=1x10^{-4}M	72Pd	
Cu^{2+}	-4.02	2.50(1)	-2.0	25	C	μ=0	68Ib	
	-4.38	(1)	--	--	-	C_L=0.2M C_M=3x10^{-2}M C_{HNO_3}=0.01M	68Wb	f:2.98 u,y
	-2.95	2.46(1)	1.3	25	C	1M($NaClO_4$)	63Ae	f:4
	-4.4	2.58(1)	-2.7	25	C	1.0M(KNO_3)	60La	f:4
	-2.95	1.95(2)	-1.0	25	C	1M($NaClO_4$)	63Ae	f:4
	-4.4	1.85(2)	-6.0	25	C	1.0M(KNO_3)	60La	f:4
	-2.95	1.27(3)	-4.1	25	C	1M($NaClO_4$)	63Ae	f:4
	-4.4	1.41(3)	-8.0	25	C	1.0M(KNO_3)	60La	f:4
	-2.95	0.84(4)	-6.0	25	C	1M($NaClO_4$)	63Ae	f:4
	-4.4	0.82(4)	-10.6	25	C	1.0M(KNO_3)	60La	f:4
	>1.1	0.31(5)	2	25	C	1.0M(KNO_3)	66Cb	
	-8.86	4.30(1-2)	-10.0	25	C	μ=0	68Ib	
	-16.1	5.16(1-3)	-30.6	25	C	μ=0	68Ib	
	-21.5	6.04(1-4)	-41	25	C	μ=0	68Ib	
CuA_2	-6.45	0.672(1)	-18.2	30	C	S: Benzene A=Acetylacetone	71Gg	q: ±0.65
	-6.69	0.977(1)	-17.4	30	C	S: Benzene A=Benzoylacetone	71Gg	q: ±0.67
	-6.4	1.312(1)	-15.5	25	C	S: Benzene A=t-Butylaceto-acetate	69La	
	-4.8	-0.420(27°,1)	-16	7-47	T	S: Benzene A=Diethyldithio-carbamate	72Fc	b,q: ±0.5
	-2.94	0.298(1)	-8.4	30	C	S: Benzene A=3-Ethylacetyl-acetone	71Gg	q: ±0.29
	-3.18	0.302(1)	-9.1	30	C	S: Benzene A=3-Methylacetyl-acetone	71Gg	q: ±0.31

Metal Ion	ΔH, kcal /mole	Log K	ΔS, cal /mole °K	T, °C	M	Conditions	Ref.	Re-marks
P195, cont.								
CuA_2, cont.	-6.31	1.176(1)	-15.5	30	C	S: Benzene. A=N-Nitroso N Phenyl hydroxylamine	71Gh	q: ±0.62
	-6.05	0.89(1)	-16.2	25	C	μ≃0.005 S: Benzene. A=2,4-Pentanedionate	63Ma 69Ga	
	-3.31	(1)	--	25	T	μ≃0.005 S: Benzene. A=2,4-Pentanedionate	63Ma	
	-3.15	0.655(1)	-7.4	30	C	S: Benzene. A=3-Phenylacetylacetone	71Gg	q: ±0.31
	-4.78	0.832(1)	-12.0	30	C	S: Benzene. A=3-o-Tolylacetylacetone	71Gg	q: ±0.48
	-5.62	0.488(1)	-16.2	30	C	S: Benzene. A=3-p-Tolyacetylacetone	71Gg	q: ±0.58
	-7.38	2.846(1)	-11.5	30	C	S: Benzene. A=1,1,1-Trifluoropentane-2,4-dione	74Gb	
	-7.17	2.736(1)	-11.0	30	C	S: Benzene. A=1,1,1-Trifluoro-4-phenylbutane-2,4-dione	74Gb	
	-7.84	3.037(1)	-12.0	30	C	S: Benzene. A=1,1,1-Trifluoro-4-(2-thienyl) butane-2,4-dione	74Gb	
	-9.08	(1)	--	30	C	S: Benzene. A=1,1,1,5,5,5-Hexafluoro-pentane-2,4-dione	74Gb	
	-6.0	1.276(1)	-14.3	25	C	S: CCl_4. A=t-Butyl-acetoacetate. Slight ppt.	69La	
	-6.1	1.640(1)	-13.0	25	C	S: CCl_4. A=Ethylacetoacetate	70Pa	q: ±0.5
	-11.8	(1)	--	25	C	S: CCl_4. A=Hexa-fluoroacetylacetone	70Pa	
	-4.6	-0.620(25°,1)	-18	-3-57	T	S: Chloroform A=Diethyldithio-carbamate	74Hc	b
	-7.1	1.576(1)	-16.5	25	C	S: Cyclohexane A=t-Butylaceto-acetate	69La	
	-13.4	(1)	--	25	C	S: Cyclohexane A=Hexafluoroacetyl-acetone	70Pa	
	-5	-0.398(25°,1)	-19	10-30	T	S: Methylcyclo-hexane-Pyridine A=(di-n-Butyldithio-carbamate)	71Cf	b,q: ±2
	-10.6	(1)	--	25	C	S: Methylene chloride. A=Hexa-fluoroacetylacetonate	74Da	
	-4.8	-0.444(25°,1)	-18	-3-57	T	S: Toluene. A=Diethyldithio-carbamate	74Hc	b

Metal Ion	ΔH, kcal /mole	Log K	ΔS, cal /mole °K	T, °C	M	Conditions	Ref.	Re-marks
P195, cont.								
CuA_2, cont.	-4.8	0.681(1)	-13.0	25	C	S: $CHCl_3$. A=t-Butylacetoacetate	69La	
	-9.44	2.892(2)	-17.9	30	C	S: Benzene. A=1,1 1,5,5,5-Hexafluoro-pentane-2,4-dione	74Gb	
	-12.1	2.756(2)	-28.0	25	C	S: CCl_4. A=Hexa-fluoroacetylacetone	70Pa	
Ag^+	-4.6	2.04(1)	-6.2	25	C	μ=0	68Ib	
	-4.77	2.24(1)	-5.7	25	C	μ=0	63Ba	
	-4.83	2.00(1)	-7.0	25	C	0.5M(KNO_3)	66Pc	
	-4.83	2.00(1)	-7.0	25	C	μ=0.5M(KNO_3)	66Pb	
	4.73	2.063(1)	6.42	25	C	μ=0.5M(KNO_3)	72Ec	
	-4.4	2.08(1)	-5.4	25	C	μ=0.5M(KNO_3) Thermodynamic quantities were evaluated by two different math-matical treatment of the data.	72Bi	
	-6.76	1.95(2)	-13.8	25	C	μ=0	63Ba	
	-6.51	2.11(2)	-12.2	25	C	0.5M(KNO_3)	66Pc	
	-6.51	2.11(2)	12.2	25	C	μ=0.5M(KNO_3)	66Pb	
	-4.31	2.04(1)	-5.1	25	C	μ=0.5M(KNO_3) Thermodynamic quantities were evaluated by two different math-matical treatment of the data.	72Bi	
	6.45	2.123(2)	11.92	25	C	μ=0.5M(KNO_3)	72Ec	
	-6.3	2.05(2)	-11.6	25	C	μ=0.5M(KNO_3) Thermodynamic quantities were evaluated by two different math-matical treatment of the data.	72Bi	
	-6.58	2.08(2)	-12.5	25	C	μ=0.5M(KNO_3) Thermodynamic quantities were evaluated by two different math-matical treatment of the data.	72Bi	
	-11.25	4.09(1-2)	-19.0	25	C	μ=0	68Ib	
	-11.53	4.19(1-2)	-19.5	25	C	μ=0	63Ba	
	-11.34	4.11(1-2)	-19.2	25	C	0.5M(KNO_3)	66Pc	
Zn^{2+}	-2.58	2.08(1)	0.9	25	C	1M($NaClO_4$)	63Ae	f:4
	-2.58	1.69(2)	-0.9	25	C	1M($NaClO_4$)	63Ae	f:4
	-2.58	1.03(3)	-4.0	25	C	1M($NaClO_4$)	63Ae	f:4
	-2.58	0.64(4)	-5.7	25	C	1M($NaClO_4$)	63Ae	f:4
ZnA	-9.2	3.780(25°,1)	-14.0	15-35	T	S: Benzene	73Vd	b,i:

Metal Ion	ΔH, kcal /mole	Log K	ΔS, cal /mole °K	T, °C	M	Conditions	Ref.	Remarks
P195, cont.								
ZnA, cont.						A=α,β,γ,δ-Tetra-phenylporphine		52Ma
ZnA_2	-8.08	2.550(1)	-15.0	30	C	S: Benzene A=Acetylacetone C_M=(9.9-18.3)x10^{-3}M	73Db	q: ±0.8
	-7.62	3.015(1)	-11.5	30	C	S: Benzene A=Benzoylacetone C_M=(2.9-14.3)x10^{-3}M	73Db	q: ±0.8
	-9.03	3.291(1)	-13.1	30	C	S: Benzene A=Dibenzoylmethane C_M=(2.8-12.3)x10^{-3}M	73Db	q: ±0.9
	-10.1	3.95(1)	-15	30	C	S: Benzene A=Diisobutyldithio-phosphato	71Db	
	-10.2	3.53(1)	-17	30	C	S: Benzene A=Diisopropyldithio-phosphato	71Db	
	-3.68	2.705(1)	0.24	30	C	S: Benzene A=2,2,6,6-Tetra-methylheptane-3,5-dione. C_M=(10.7-18.9)x 10^{-3}M	73Db	a: ±0.86
	-5.2	0.34(2)	-16.0	30	C	S: Benzene A=Diisobutyldithio-phosphato	71Db	q: ±0.7
	-6.0	0.83(2)	-15.8	30	C	S: Benzene A=Diisopropyldithio-phosphato	71Db	q: ±0.7
	-7.6	1.906(25°,1)	-17	18.5-65	T	S: Toluene A=[$S_2CN(CH_3)_2$]	65Ch	b
[Zn]	-8.8	3.59(25°,1)	-13	18-40	T	S: Benzene [Zn]=Zinc(II) Meso-tetraphenylporphine	72Ce	b,q: ±1.2
	-9.00	3.722(30°,1)	-13	20-40	T	S: Benzene [Zn]=Zinc Tetra-phenylchlorin	52Ma	b
	-9.20	3.570(30°,1)	-14	20-40	T	S: Benzene [Zn]=Zinc Tetra-phenylporphine	52Ma	b
	-4.0	2.78(25°,1)	0	18-40	T	S: Chloroform [Zn]=Zinc(II) Meso-tetraphenylporphine	72Ce	b,q: ±0.4
Cd^{2+}	-4.58	1.330(25°,1)	-9.25	18-35	T	μ=0.1	65Cc	b
	-12.26	1.813(25°,1-2)	-32.84	18-35	T	μ=0.1	65Cc	b
	-12.53	1.886(25°,1-3)	-33.41	18-35	T	μ=0.1	65Cc	b
	-11.90	1.477(25°,1-4)	-33.17	18-35	T	μ=0.1	65Cc	b
CdA_2	-9.6	0.58(2)	-29.4	30	C	S: Benzene. A=Dicyclohexyldithio-phosphato	71Db	q: ±1.2
	-10.8	0.41(2)	-33.7	30	C	S: Benzene	71Db	q:±1

Metal Ion	ΔH, kcal /mole	Log K	ΔS, cal /mole °K	T, °C	M	Conditions	Ref.	Remarks
P195, cont.								
CdA_2, cont.						A=Diisobutyldithio-phosphato		
Cd_2A_4	-6.45	2.46(1-2)	-10.0	30	C	S: Benzene A=Dicyclohexyldithio-phosphato	71Db	q: ±0.7
	-6.17	2.70(1-2)	-7.9	30	C	S: Benzene A=Diisobutyldithio-phosphato	71Db	q: ±0.7
[Cd]	-8.70	3.430(30°,1)	-13	20-40	T	[Cd]=Cadmium Tetra-phenylporphine S: Benzene	52Ma	b
HgA_2	-15.2	2.336(1)	-39.7	30	C	S: Benzene. A=Cl C_M=1-5x10^{-3}M	73Fa	
	-16.2	2.354(1)	-43.0	30	C	S: Benzene. A=Br C_M=1-5x10^{-3}M	73Fa	
	-15.3	2.120(1)	-40.9	30	C	S: Benzene. A=I C_M=1-5x10^{-3}M	73Fa	
	-4.32	0.322(1)	-13.0	25	C	S: C_6H_6. A=C_6F_5	73Pl	
	-5.40	1.303(1)	-11.4	25	C	S: CCl_4. A=C_6F_5	73Pl	
	≃-2.9	≃-0.52(1)	≃-12	25	C	S: CCl_4. A=C_6H_5	73Pl	
[Hg]	-4.90	1.083(30°,1)	-11	20-40	T	[Hg]=Mercury Tetra-phenylporphine S: Benzene	52Ma	b
BA_3	-25	(1)	--	25	C	S: Nitrobenzene A=F	55Bb	
	-15.3	(1)	--	25	C	S: Nitrobenzene A=CH_3	56Ba	
BA_3C	-12.9	(1)	--	25	C	S: Benzene. A=F C=$(CH_3CH_2)_2O$	68Rd	
	-12.2	(1)	--	25	C	S: Nitrobenzene A=F. C=Tetra-hydropyran	55Bb	
	-14.1	(1)	--	25	C	S: Nitrobenzene A=F. C=$(C_2H_5)_2O$	55Bb	
	-15.9	(1)	--	25	C	S: Nitrobenzene A=F. C=n-Butyl ether	55Bb	
	-25.0	--	--	25	C	S: Nitrobenzene R: BA_3C + L = BA_3L + C. A=F C=$C_5H_5NO_2$	56Bc	
	-30.8	--	--	25	C	S: Nitrobenzene R: BA_3C+ L = BA_3L + C. A=Cl C=$C_5H_5NO_2$	56Bc	
	-32.0	--	--	25	C	S: Nitrobenzene R: BA_3C + L = BA_3L + C. A=Br C=$C_5H_5NO_2$	56Bc	
AlA_3	-27.56	(1)	--	25-28	C	S: n-Hexane. A=CH_3	68Hc	

Metal Ion	ΔH, kcal /mole	Log K	ΔS, cal /mole °K	T, °C	M	Conditions	Ref.	Remarks
P195, cont.								
Al_2A_6	-31.4	--	--	25	C	S: Benzene C_M=0.07M. A=Br R: 0.5(Al_2A_6) + L = AlA_3L	68Rc	
$SiAB_3$	-5.44	1.90(30 ,1)	-9.29	10-45	T	S: CCl_4. X_L=1x10^{-2} X_M=5x10^{-4} A=OH. B=CH_3	69Lb	b
	-5.65	1.76(30 ,1)	-10.59	20-60	T	S: CCl_4. X_L=1x10^{-2} X_M=5x10^{-4} A=OH. B=C_2H_5	69Lb	b
	-5.98	2.28(30 ,1)	-9.31	0-50	T	S: CCl_4. X_L=1x10^{-2} X_M=4x10^{-4} A=OH. B=Phenyl	69Lb	b
SnA_3B	-6.5	0.267(1)	-20.6	26	C	S: CCl_4. A=CH_3 B=Cl	66Bg	
PA_5	-22.0	(1)	--	30	C	S: Nitrobenzene A=Cl	63Hb	
PA_4B	-27.0	(1)	--	30	C	S: Nitrobenzene A=Cl. B=F	63Hb	
PA_3B_2	-35.3	(1)	--	30	C	S: Nitrobenzene A=Cl. B=F	63Hb	
SbA_5	-34.0	(1)	--	25	C	S: $C_2H_4Cl_2$. A=Cl	72Lb	
	-28.3	(1)	--	30	C	S: Nitrobenzene A=Cl	63Hb	
Br_2	-10.3	2.436(1)	-23.5	25	C	S: 1,2-Dichloro- ethane	73Tb	
I_2	-6.0	2.346(1)	-9.4	25	C	S: Benzene	66Ad	
	-8.0	1.99(1)	-18.1	22	C	low C. S: CCl_4	64Sd	b
	-7.9	2.017(1)	-17.3	25	C	S: CCl_4	73Bh	
	-7.47	2.02(25°,1)	-15.8	22-43	T	S: CCl_4	62Ka	
	-7.47	2.021(25°,1)	-15.8	--	T	S: CCl_4 C_L<3x10^{-2}M C_M<10^{-4}M	69Ma	u
	-7.82	1.890(25°,1)	-17.5	--	T	S: Chloroform C_L<3x10^{-2}M C_M<10^{-4}M	69Ma	u
	-8.4	2.107(1)	-18.4	25	C	S: Cyclohexane	73Bh	
	-7.8	(1)	--	25	C	S: 1,2-Dichloro- ethane	73Tb	
	-7.77	2.276(25°,1)	-15.6	--	T	S: 1,2-Dichloro- ethane C_L<3x10^{-2}M. C_M<10^{-4}M	69Ma	u
	-8.59	2.179(25°,1)	-18.9	--	T	S: Dichloromethane C_L<3x10^{-2}M. C_M<10^{-4}M	69Ma	u
	-7.8	2.5(1)	-15.5	17	T	S: Heptane	54Ra	
	-7.8	2.46(1)	-15.2	22	C	low C. S: n-Heptane	64Sd	
	-8.16	2.196(25°,1)	-17.3	--	T	S: n-Heptane C_L<3x10^{-2}M. C_M<10^{-4}M	69Ma	u
	-7.9	(1)	--	--	C	S: Pyridine	61Aa	i: 50Ha

Metal Ion	ΔH, kcal /mole	Log K	ΔS, cal /mole °K	T, °C	M	Conditions	Ref.	Re-marks
P195, cont.								
I_2, cont.	-7.95	(1)	--	--	C	S: Pyridine	50Ha	
P196	PYRIDINE, 3-acetyl-		C_7H_7ON			$C_5H_4N(OCCH_3)$		L
CuA_2	-1.76	0.99(1)	-1.38	25	C	μ≃0.005. S: Benzene A=2,4-Pentanedione	63Ma	
	-1.73	(1)	--	25	T	μ≃0.005. S: Benzene A=2,4-Pentanedione	63Ma	
P197	PYRIDINE, 4-acetyl-		C_7H_7ON			$C_5H_4N(OCCH_3)$		L
NiA_2	-18.79	(1-2)	--	25	C	S: Benzene. A=2,2,6,6-Tetramethyl-3,5-heptanedione	73Ce	
P198	PYRIDINE, 2-aldehyde-2'-pyridylhydrazone-		$C_{11}H_{10}N_4$			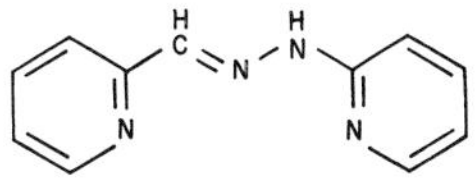		L
Fe^{2+}	-25.3	16.57(25°,1-2)	-9	5-60	T	μ=0	68Gg	b
Zn^{2+}	-11.0	5.82(25°,1)	-10	5-60	T	μ=0	68Gg	b
	-13.9	11.08(25°,2)	4	5-60	T	μ=0	68Gg	b
P199	PYRIDINE, 2-aldoxime-		$C_6H_6N_2$			$C_5H_4N(CH{:}NOH)$		L
Fe^{2+}	-20	24.85(1-3)	47	25	T	μ=0.045	62Hb	
P200	PYRIDINE, 2-amino-		$C_5H_6N_2$			$C_5H_4N(NH_2)$		L
Be^{+2}	-6.82	4.31(40°,1)	4.3	35&45	T	--	67Rb	a,b,p
P201	PYRIDINE, 4-amino-		$C_5H_6N_2$			$C_5H_4N(NH_2)$		
$[Fe^{2+}]$	-16.1	5.64(25°,1-2)	-28	18-40	T	[Fe]=Iron(II) protoporphyrin dimethyl ester S: Benzene	70Ch	b,q: ±2.2
$[Co^{2+}]$	-8.1	4.48(0°,1)	-9	-23-23	T	[Co]=Cobalt(II) protoporphyrin IX dimethyl ester S: Toluene	73Sr	b,q: ±0.6
$[Ni^{2+}]$	-7.6	2.57(25°,1)	-14	18-40	T	[Ni]=Nickel(II) mesotetraphenyl-porphine. S: $CHCl_3$	72Ce	b,q: ±0.7
	-6.6	3.08(25°,1)	-8	18-40	T	[Ni]=Nickel(II) mesotetraphenyl-porphine. S: Benzene	72Ce	b,q: ±1.4
$[Zn^{2+}]$	-5.1	3.52(25°,1)	-1	18-40	T	[Zn]=Zinc(II) mesotetraphenyl-porphine. S: $CHCl_3$	72Ce	b
	-13.9	4.62(25°,1)	-26	18-40	T	[Zn]=Zinc(II) mesotetraphenyl-porphine. S: Benzene	72Ce	b

Metal Ion	ΔH, kcal /mole	Log K	ΔS, cal /mole °K	T, °C	M	Conditions	Ref.	Remarks
P202	PYRIDINE, N-(4-aminobutane)-		$C_9H_{15}N$			$C_5H_5N[(CH_2)_4NH_2]$		L^+
Ag^+	-5.43	3.240(1)	-3.4	25	C	μ=0.5M(KNO_3)	72Va	
	-8.04	3.365(2)	-11.5	25	C	μ=0.5M(KNO_3)	72Va	
P203	PYRIDINE, N-(2-aminoethane)-		$C_7H_{11}N$			$C_5H_5N[(CH_2)_2NH_2]$		L^+
Ag^+	-5.56	2.096(1)	-9.1	25	C	μ=0.5M(KNO_3)	72Va	
	-5.54	2.361(2)	-7.7	25	C	μ=0.5M(KNO_3)	72Va	
P204	PYRIDINE, 2-(2'-aminoethyl)-		$C_7H_{10}N_2$			$C_5H_4N(CH_2CH_2NH_2)$		L
Ni^{2+}	-7.14	5.14(30 ,1)	0	10-40	T	μ=0	59Gb	b
	-4.06	3.20(30 ,1)	1.3	10-40	T	μ=0	59Gb	b
	-6.92	5.42(1)	1.6	25	C	μ=0.30(ClO_4^-)	67Hf	
	-7.10	5.37(1)	0.8	25	C	μ=0.5M(KNO_3)	71Gd	
	-7.00	3.39(2)	-8.0	25	C	μ=0.30(ClO_4^-)	67Hf	
	-6.10	3.37(2)	-5.0	25	C	μ=0.5M(KNO_3)	71Gd	
	-6.99	2.35(3)	-12.7	25	C	μ=0.30(ClO_4^-)	67Hf	
Cu^{2+}	-8.61	7.48(30°,1)	5	10-40	T	μ=0	59Gb	b
	-8.84	7.54(1)	4.9	25	C	μ=0.30(ClO_4^-)	67Hf	
	-9.38	7.64(1)	3.6	25	C	μ=0.5M(KNO_3)	71Gd	
	-8.85	5.64(2)	-3.9	25	C	μ=0.30(ClO_4^-)	67Hf	
	-8.28	5.59(2)	-2.2	25	C	μ=0.5M(KNO_3)	71Gd	
P205	PYRIDINE, 2-aminomethyl-		$C_6H_8N_2$			$C_5H_4N(CH_2NH_2)$		L
Co^{2+}	-6.77	5.41(30°,1)	2.4	10-40	T	μ=0	59Gb	b
	-7.42	5.54(1)	0.5	25	C	μ=0.5M(KNO_3)	71Gd	
	-16.68	10.39(1)	-8.39	25	C	μ=0.1(KNO_3) S: 50 w % Dioxane	66Wa	
	-7.13	4.52(30°,2)	-2.8	10-40	T	μ=0	59Gb	b
	-7.47	4.79(2)	-3.1	25	C	μ=0.5M(KNO_3)	71Gd	
	-6.14	3.33(30°,3)	-5.0	10-40	T	μ=0	59Gb	b
	-6.85	3.50(3)	-7.0	25	C	μ=0.5M(KNO_3)	71Gd	
Ni^{2+}	-8.35	7.09(30°,1)	4.9	10-40	T	μ=0	59Gb	b
	-8.82	7.18(1)	4.3	25	C	μ=0.30(ClO_4^-)	67Hf	
	-9.91	7.11(1)	-0.5	25	C	μ=0.5M(KNO_3)	71Gd	
	-20.52	13.15(1)	-8.63	25	C	μ=0.1(KNO_3) S: 50 w % Dioxane	66Wa	
	-9.25	6.08(30°,2)	-2.7	10-40	T	μ=0	59Gb	b
	-8.54	5.96(2)	-1.4	25	C	μ=0.30(ClO_4^-)	67Hf	
	-9.43	6.41(2)	-2.2	25	C	μ=0.5M(KNO_3)	71Gd	
	-8.43	4.95(30°,3)	-5.1	10-40	T	μ=0	59Gb	b
	-8.53	4.41(3)	-8.4	25	C	μ=0.30(ClO_4^-)	67Hf	
	-9.67	5.14(3)	-8.8	25	C	μ=0.5M(KNO_3)	71Gd	
Cu^{2+}	-9.61	9.45(30°,1)	11.5	10-40	T	μ=0	59Gb	b
	-9.92	9.34(1)	9.5	25	C	μ=0.30(ClO_4^-)	67Hf	
	-21.57	15.69(1)	-0.55	25	C	μ=0.1(KNO_3) S: 50 w % Dioxane	66Wa	
	-8.98	7.80(30°,2)	6.1	10-40	T	μ=0	59Gb	b
	-9.96	7.92(2)	2.8	25	C	μ=0.30(ClO_4^-)	64Hf	
Zn^{2+}	-6.81	5.17(30°,1)	1.2	10-40	T	μ=0	59Gb	b
	-6.48	5.37(1)	2.9	25	C	μ=0.5M(KNO_3)	71Gd	
	-12.44	9.84(1)	3.22	25	C	μ=0.1(KNO_3)	66Wa	

Metal Ion	ΔH, kcal /mole	Log K	ΔS, cal /mole °K	T, °C	M	Conditions	Ref.	Remarks
P205, cont.								
Zn^{2+},						S: 50 w % Dioxane		
cont.	-7.28	4.21(30°,2)	-4.8	10-40	T	μ=0	59Gb	b
	-6.43	4.47(2)	-1.3	25	C	μ=0.5M(KNO_3)	71Gd	
	-5.85	3.07(30°,3)	-5.2	10-40	T	μ=0	59Gb	b
	-6.90	2.80(3)	-8.9	25	C	μ=0.5M(KNO_3)	71Gd	
Cd^{2+}	-5.88	4.59(30°,1)	1.6	10-40	T	μ=0	59Gb	b
	-5.81	4.76(1)	-2.3	25	C	μ=0.5M(KNO_3)	71Gd	
	-12.59	9.68(1)	2.05	25	C	μ=0.1(KNO_3)	66Wa	
						S: 50 w % Dioxane		
	-6.14	3.82(30°,2)	-2.8	10-40	T	μ=0	59Gb	b
	-5.71	3.94(2)	-1.3	25	C	μ=0.5M(KNO_3)	71Gd	
	-6.19	2.59(3)	-8.9	25	C	μ=0.5M(KNO_3)	71Gd	
P206	PYRIDINE, 2-aminomethyl-6-methyl-		$C_7H_{10}N_2$			$C_5H_3N(CH_2NH_2)(CH_3)$		L
Co^{2+}	-10.55	7.00(1)	-3.35	25	C	μ=0.1(KNO_3)	66Wa	
						S: 50 w % Dioxane		
Ni^{2+}	-5.86	4.71(30°,1)	2.2	10-40	T	μ=0	60Wa	b
	-9.02	8.43(1)	8.32	25	C	μ=0.1(KNO_3)	66Wa	
						S: 50 w % Dioxane		
	-4.99	3.02(30°,2)	-2.6	10-40	T	μ=0	60Wa	b
Cu^{2+}	-6.86	6.81(30°,1)	8.5	10-40	T	μ=0	60Wa	b
	-15.28	12.69(1)	6.82	25	C	μ=0.1(KNO_3)	66Wa	
						S: 50 v % Dioxane		
	-7.23	5.65(30°,2)	2.0	10-40	T	μ=0	60Wa	b
Ag^+	-9.34	3.97(30°,1)	-12.7	10-40	T	μ=0	60Wa	b
	-1.07	3.45(30°,2)	12.2	10-40	T	μ=0	60Wa	b
Zn^{2+}	-5.34	7.80(1)	17.77	25	C	μ=0.1(KNO_3)	66Wa	
						S: 50 w % Dioxane		
Cd^{2+}	-4.52	4.01(30°,1)	3.4	10-40	T	μ=0	60Wa	b
	-5.97	7.07(1)	12.32	25	C	μ=0.1(KNO_3)	66Wa	
						S: 50 w % Dioxane		
	0.52	2.69(30°,2)	14.1	10-40	T	μ=0	60Wa	b
P207	PYRIDINE, N-(5-aminopentane)-		$C_{10}H_{17}N$			$C_5H_5N[(CH_2)_5NH_2]$		L^+
Ag^+	-5.81	3.489(1)	-3.5	25	C	μ=0.5M(KNO_3)	72Va	
	-7.71	3.732(2)	-8.8	25	C	μ=0.5M(KNO_3)	72Va	
P208	PYRIDINE, N-(3-aminopropane)-		$C_8H_{13}N$			$C_5H_5N[(CH_2)_3NH_2]$		L^+
Ag^+	-5.19	2.910(1)	-4.0	25	C	μ=0.5M(KNO_3)	72Va	
	-7.70	3.101(2)	-11.7	25	C	μ=0.5M(KNO_3)	72Va	
P209	PYRIDINE, 2-(azo-p-dimethyl-aniline)-		$C_{13}H_{14}N_4$					
Co^{2+}	-6.0	(1)	5	15-50	T	μ=0.1M($NaNO_3$)	72Ca	b
Ni^{2+}	-6.5	(1)	3	15-50	T	μ=0.1M($NaNO_3$)	72Ca	b

Metal Ion	ΔH, kcal /mole	Log K	ΔS, cal /mole °K	T, °C	M	Conditions	Ref.	Remarks
P209, cont.								
ZnA_4^{2+}	-1.75	2.360(25°)	4.8	25&35	T	μ=0.15($NaNO_3$) A=H_2O. R: ZnA_4^{2+} + L = ZnA_2L^{2+} + 2A	53Kb	a,b
P210	PYRIDINE, 4-benzoyl-		$C_{12}H_9ON$			$C_5H_4N(COC_6H_5)$		L
CuA_2	-1.42	1.23(1)	0.87	25	C	μ≈0.005. S: Benzene A=2,4-Pentanedione	63Ma	
	-1.30	(1)	--	--	T	μ≈0.005. S: Benzene A=2,4-Pentanedione	63Ma	
	-5.76	2.376(1)	-8.13	30	C	S: Benzene A=1,1,1-Trifluoro-pentane-2,4-dione	74Gb	
P211	PYRIDINE, 2-benzoyl-, oxime		$C_{12}H_{10}ON_2$			$C_5H_4N[C(NOH)(C_6H_5)]$		L
Ni^{2+}	-2.31	9.21(30°,1)	33.47	20-40	T	μ=0 S: 40 w % Acetone	65Sc	b
	-0.42	7.05(30°,2)	29.95	20-40	T	μ=0 S: 40 w % Acetone	65Sc	b
Zn^{2+}	-6.30	8.40(30°,1)	17.12	20-40	T	μ=0 S: 40 w % Acetone	65Sc	b
	-0.63	6.95(30°,2)	28.83	20-40	T	μ=0 S: 40 w % Acetone	65Sc	b
Cd^{2+}	-12.39	6.95(30°,1)	-8.80	20-40	T	μ=0 S: 40 w % Acetone	65Sc	b
	-4.20	6.44(30°,2)	15.14	20-40	T	μ=0 S: 40 w % Acetone	65Sc	b
Pb^{2+}	-6.93	7.52(30°,1)	11.20	20-40	T	μ=0 S: 40 w % Acetone	65Sc	b
	-3.99	7.11(30°,2)	18.78	20-40	T	μ=0 S: 40 w % Acetone	65Sc	b
P212	PYRIDINE, 4-benzyl-		$C_{12}H_{11}N$			$C_5H_4N(CH_2C_6H_5)$		L
CuA_2	-7.43	3.075(1)	-10.5	30	C	S: Benzene. A=1,1,1-Trifluoropentane-2,4-dione	74Gb	
P213	PYRIDINE, 2-[benzyl(2-dimethylaminoethyl)amino]-		$C_{16}H_{21}N_3$			$C_5H_4N[N(CH_2CH_2N(CH_3)_2)(CH_2C_6H_5)]$		L
Be^{2+}	-9.9	5.56(25°,1)	-7	0-45	T	μ=0.12M(KCl) C_M≈2x10^{-3}M C_L≈8x10^{-3}M	69Cb	b
	-1.0	4.55(25°,2)	17.	0-45	T	μ=0.12M(KCl) C_M≈2x10^{-3}M C_L≈8x10^{-3}M	69Cb	b
Co^{2+}	7.5	6 34(25°,1-2)	54.0	0&25	T	0.06M(KCl)	62Aa	a,b
Ni^{2+}	16.4	6.94(25°,1-2)	86.8	0&25	T	0.06M(KCl)	62Aa	a,b
Cu^{2+}	-2.5	8.80(25°,1-2)	32.0	0&25	T	0.06M(KCl)	62Aa	a,b

Metal Ion	ΔH, kcal /mole	Log K	ΔS, cal /mole °K	T, °C	M	Conditions	Ref.	Remarks
P214	PYRIDINE, 1-benzyl-2-formamido-		$C_{13}H_{13}ON_2$			$C_5H_4N(CH_2C_6H_5)(NHCOH)$		
Fe(III)A_5	-2.84	3.302(25°,1)	5.57	--	T	μ=0.05. pH=8. A=CN	72Hb	u
P215	PYRIDINE, 3-bromo-		C_5H_4NBr			$C_5H_4N(Br)$		L
CoA_2	-4.27	2.646(1)	-2.2	25	C	S: Acetone C_M=0.002M. A=Cl	73Dg	
	-1.62	0.66(1)	2.42	25	C	μ≃0.005. S: Benzene A=2,4-Pentanedione	63Ma	
	-2.17	(1)	--	25	T	μ≃0.005. S: Benzene A=2,4-Pentanedione	63Ma	
Br_2	-8.0	1.243(1)	-21.2	25	C	S: 1,2-Dichloro-ethane	73Tb	
P216	PYRIDINE, 3-chloro-		C_5H_4NCl			$C_5H_4N(Cl)$		L
CoA_2	-4.95	3.196(1)	-2.0	25	C	S: Acetone C_M=0.002M. A=Cl	73Dg	
P217	PYRIDINE, 4-bromo-1-oxide-		C_5H_4ONBr			$C_5H_4N(Br)(\rightarrow O)$		
VAB_2	-6.07	(1)	--	25	T	S: CH_2Cl_2. A=O B=2,4-Pentanedione	69Pd	
P218	PYRIDINE, 4-butanoic acid		$C_{10}H_{13}O_2N$			$C_5H_4N[(CH_2)_4CO_2H]$		
[Fe]	-6.7	3.52(25°,1-2)	-7	18-40	T	[Fe]=Iron(II) deuteroporphyrin dimethyl ester S: Benzene	70Ch	b,q: ±1.0
	7.0	3.66(25°,1-2)	42	18-40	T	[Fe]=Iron(II) diacetyldeutero-porphyrin dimethyl ester. S: Benzene	70Ch	b
	-8.9	3.15(25°,1-2)	-15	18-40	T	[Fe]=Iron(II) mesoporphyrin dimethyl ester S: Benzene	70Ch	b,q: ±1.0
	-10.1	2.93(25°,1-2)	-20	18-40	T	[Fe]=Iron(II) protoporphyrin dimethyl ester S: Benzene	70Ch	b
P219	PYRIDINE, 4-tert-butyl-		$C_9H_{13}N$			$C_5H_4N[C(CH_3)_3]$		
[Co]	-9.0	4.40(0°,1)	-13	-23-23	T	[Co]=Cobalt(II) protoporphyrin IX dimethyl ester S: Toluene	73Sr	b,q: ±0.6
CuA_2	-7.8	1.836(1)	-17.9	25	C	S: Cyclohexane A=t-Butylaceto-acetate	69La	
	-7.9	1.876(25°,1)	-17.9	25-40	C	S: Cyclohexane A=t-Butylaceto-acetate	69Ga	b

Metal Ion	ΔH, kcal /mole	Log K	ΔS, cal /mole °K	T, °C	M	Conditions	Ref.	Re-marks
P220	PYRIDINE, 4-carbomethoxy-1-oxide-		$C_7H_7O_3N$			$C_5H_4N(CO_2CH_3)(\rightarrow O)$		
VAD$_2$	3.60	(1)		25	T	S: CH_2Cl_2. A=O B=2,4-Pentanedione	69Pd	
	-0.85	(1)	--	25	C	S: $CHCl_3$. A=O B=2,4-Pentanedione	69Pd	
P221	PYRIDINE, 4-chloro-		C_5H_4NCl			$C_5H_4N(Cl)$		L
CuA$_2$	-0.66	0.73(1)	1.12	25	C	μ≈0.005. S: Benzene A=2,4-Pentanedione	63Ma	
	-2.11	(1)	--	25	C	μ≈0.005. S: Benzene A=2,4-Pentanedione	63Ma	
P222	PYRIDINE-2-[4-chloro-α-(2-dimethylaminoethyl) benzyl]-		$C_{16}H_{19}N_2Cl$			$C_5H_4N[CH(CH_2CH_2N(CH_3)_2)(C_6H_4Cl)]$		L
Cu^{2+}	-24.8	9.26(25°,1-2)	-52	0&25	T	0.06M(KCl)	62Aa	a,b
P223	PYRIDINE, 4-chloro-1-oxide		C_5H_4ONCl			$C_5H_4N(Cl)(\rightarrow O)$		L
VAB$_2$	-5.72	(1)	--	25	C	S: CH_2Cl_2. A=O B=2,4-Pentanedione	69Pd	
	-2.18	(1)	--	25	C	S: $CHCl_3$. A=O B=2,4-Pentanedione	69Pd	
I_2	-4.65	1.564(25°,1)	-8.5	18-32	T	S: CCl_4	68Ga	b
P224	PYRIDINE, 4-cyano-1-oxide		$C_6H_4ON_2$			$C_5H_4N(CN)(\rightarrow O)$		L
VAB$_2$	-3.09	(1)	--	25	T	S: CH_2Cl_2. A=O B=2,4-Pentanedione	69Pd	
P225	PYRIDINE, 1,3-dimethyl-		$C_7H_{10}N$			$C_5H_4N(CH_3)_2$		L^+
[Fe]	-3.1	5.8(20°)	15.8	5-30	T	pH-11.2 Protohaemin R: $(Haemin)_2 + 2L = (HaeminL)_2 + H_2O$	69Md	b w:I
P226	PYRIDINE, 2,3-dimethyl-		C_7H_9N			$C_5H_3N(CH_3)_2$		L
[Fe]	0	6.2(20°)	28.3	5-30	T	pH=11.2 Protohaemin R: $(Haemin)_2 + 2HL^+ + 2OH^- = (HaeminL)_2 + 2H_2O$	69Md	b
NiA$_2$	-5.46	-2.386(35°,1-2)	-29	25-85	T	S: L A=Salicylaldoximate	71Ma	b
P227	PYRIDINE, 2,4-dimethyl-		C_7H_9N			$C_5H_3N(CH_3)_2$		L
NiA	-13.64	1.02(1-2)	-41.2	25	T	$C_M \approx C_L \approx 10^{-4}$ S: Benzene A=Diacetyl-bisbenzoyl hydrazine	58Sa	

Metal Ion	ΔH, kcal /mole	Log K	ΔS, cal /mole °K	T, °C	M	Conditions	Ref.	Remarks
P227, cont.								
NiA_2	-5.55	-1.851(35°,1-2)	-26	25-95	T	S: L A=Salicylaldoximate	71Ma	b
Br_2	-11.8	3.090(1)	-25.7	25	C	S: 1,2-Dichloro-ethane	73Tb	
P228	PYRIDINE, 2,5-dimethyl-		C_7H_9N			$C_5H_3N(CH_3)_2$		L
NiA	-12.58	0.54(25°,1-2)	-39.7	25	T	$C_M \simeq C_L \simeq 10^{-4}$ S: Benzene A=Diacetyl-bisben-zoyl hydrazine	58Sa	
NiA_2	-4.26	-1.955(35°,1-2)	-23	25-95	T	S: L A=Salicylaldoximate	71Ma	b
P229	PYRIDINE, 2,6-dimethyl-		C_7H_9N			$C_5H_3N(CH_3)_2$		L
VOA_2	-7.36	0.78(1)	-21.2	25	C	$\mu \to 0$ S: Nitrobenzene A=Acetylacetonate	65Ca	
[Fe]	3.1	2.04(20°)	20.0	5-30	T	pH=11.2 Protohaemin R: $(Haemin)_2 + 2L = (HaeminL)_2 + 2H_2O$	69Md	b
CoA_2	-7.24	4.73(1-2)	-2.65	25	C	S: Acetone. A=Cl	72Ae	
	-10.98	4.10(1-2)	-18.09	25	C	S: Acetonitrile A=Cl	72Ae	
	-4.32	2.99(1-2)	-0.84	25	C	S: n-Butanol. A=Cl	72Af	
	-11.03	4.83(1-2)	-14.93	25	C	S: Cyclohexanone A=Cl	72Ae	
	-3.72	3.46(1-2)	3.36	25	C	S: Dimethyl-formamide. A=Cl	72Ae	
	-4.86	2.59(1-2)	-5.13	25	C	S: Ethanol. A=Cl	72Af	
	-4.13	3.20(1-2)	0.77	25	C	S: 2-Methyl propan-2-ol. A=Cl	72Af	
	-0.16	1.54(1-2)	6.51	25	C	S: Ethylene Glycol A=Cl. R: $CoA_2 + 2L + 2S = [CoL_2S_2]A_2$	72Af	
NiA	-13.64	1.05(1-2)	-41.0	25	T	$C_M \simeq C_L \simeq 10^{-4}$ S: Benzene A=Diacetyl-bisben-zoyl hydrazine	58Sa	
NiA_2	-7.28	-1.114(35°,1-2)	-28	25-95	T	S: L A=Salicylaldoximate	71Ma	b
BA_3	-4.74	(1)		25	C	S: Nitrobenzene $C_L \simeq 0.18M$ $C_M \simeq 0.30M$. A=F	56Bb	
AlA_3	-19.87	(1)	--	25-28	C	S: n-Hexane. $A=CH_3$	68Hc	
Br_2	-9.9	2.666(1)	-21.1	25	C	S: 1,2-Dichloro-ethane	73Tb	

Metal Ion	ΔH, kcal /mole	Log K	ΔS, cal /mole °K	T, °C	M	Conditions	Ref.	Remarks
P230	PYRIDINE, 3,4-dimethyl-		C_7H_9N			$C_5H_3N(CH_3)_2$		L
VOA_2	-7.31	2.12(1)	-14.8	25	C	μ→0. S: Nitrobenzene A=Acetylacetonate	65Ca	
CoA	-9.0	3.035(25°,1)	-16	20-50	T	S: Toluene A=α,β,γ,δ-Tetra(p-methoxyphenyl) porphyrin	73Wa	b,q: ±1
CoAL	-1.8	-0.097(-65,1)	-10	-80--20	T	S: Toluene A=α,β, γ,δ-Tetra(p-methoxyphenyl), porphyrin	73Wb	b,q: ±1.0
CoL_2A_2	-16.8	4.255(1-2)	-38.0	20	C	S: $CHCl_3$. A=NCS	66Ca	s
CuA_2	-5.2	-0.013(25°,1)	-17	-3-57	T	S: Benzene A=Diethyldithiocarbamate	74Hc	b
P231	PYRIDINE, 3,5-dimethyl-		C_7H_9N			$C_5H_3N(CH_3)_2$		L
VOA_2	-6.76	2.01(1)	-13.5	25	C	μ→0 S: Nitrobenzene A=Acetylacetonate	65Ca	
CoL_2A_2	-14.6	3.70(1-2)	-33.0	20	C	S: $CHCl_3$. A=NCS	66Ca	s
NiA_2	-19.32	(1-2)	--	25	C	S: Benzene. A=2,2, 6,6-Tetramethyl-3, 5-heptanedione	73Ce	
Br_2	-11.7	3.021(1)	-25.4	25	C	S: 1,2-Dichloroethane	73Tb	
I_2	-6.0	(1)	--	25	C	S: 1,2-Dichloroethane	73Tb	
P232	PYRIDINE, 4-(dimethylamino)-		$C_7H_{10}N_2$			$C_5H_4N[N(CH_3)_2]$		L
NiA_2	-21.12	(1-2)	--	25	C	S: Benzene. A=2,2, 6,6-Tetramethyl-3,5-heptanedione	73Ce	
P233	PYRIDINE, 2-[α-(2-dimethylaminoethyl)benzyl]-		$C_{16}H_{20}N_2$			$C_5H_4N[CH(CH_2CH_2N(CH_3)_2)(C_6H_5)]$		L
Cu^{2+}	-16.4	10.04(25°,1-2)	-9.2	0&25	T	0.06M(KCl)	62Aa	a,b
P234	PYRIDINE, 4-(dimethylamino)-, 1-oxide-		$C_7H_{10}ON_2$			$C_5H_4N[N(CH_3)_2](\rightarrow O)$		
VAB_2	-10.5	(1)	--	25	T	S: CH_2Cl_2. A=O B=2,4-Pentanedione	69Pd	
P235	PYRIDINE, 2,3-dimethyl-,1-oxide-		C_7H_9ON			$C_5H_4N(CH_3)_2(\rightarrow O)$		
VAB_2	-4.55	(1)	--	25	C	S: CH_2Cl_2. A=O B=2,4-Pentanedione	69Pd	
P236	PYRIDINE, 2,4-dimethyl-, 1-oxide-		C_7H_9ON			$C_5H_3N(CH_3)_2(\rightarrow O)$		

Metal Ion	ΔH, kcal /mole	Log K	ΔS, cal /mole °K	T, °C	M	Conditions	Ref.	Remarks
P236, cont.								
VAB_2	-5.37	(1)	--	25	C	S: CH_2Cl_2. A=O B=2,4-Pentanedione	69Pd	
P237	PYRIDINE, 2,6-dimethyl-, 1-oxide		C_7H_9ON			$C_5H_3N(CH_3)_2(\rightarrow O)$		
VAB_2	-0.87	(1)	--	25	C	S: CH_2Cl_2. A=O B=2,4-Pentanedione	69Pd	
I_2	-6.69 -5.6	2.18(25°,1) (1)	-12.5 --	10-40 --	T -	S: CCl_4 S: CCl_4	57Ka 60Bc	b
P238	PYRIDINE, 3,4-dimethyl-, 1-oxide		C_7H_9ON			$C_5H_3N(CH_3)_2(\rightarrow O)$		L
VAB_2	-7.10	(1)	--	25	C	S: CH_2Cl_2. A=O B=2,4-Pentanedione	69Pd	
P239	PYRIDINE, 3,5-dimethyl-, 1-oxide-		C_7H_9ON			$C_5H_3N(CH_3)_2(\rightarrow O)$		L
VAB_2	-6.72	(1)	--	25	C	S: CH_2Cl_2. A=O B=2,4-Pentanedione	69Pd	
P240	PYRIDINE, 2-ethyl-		C_7H_9N			$C_5H_4N(CH_2CH_3)$		
BA_3	-9.0	(1)	--	25	C	S: Nitrobenzene A=CH_3	56Ba	
Br_2	-10.5	2.686(1)	-22.9	25	C	S: 1,2-Dichloro- ethane	73Tb	
I_2	-7.8	(1)	--	25	C	S: 1,2-Dichloro- ethane	73Tb	
P241	PYRIDINE, 3-ethyl-		C_7H_9N			$C_5H_4N(CH_2CH_3)$		L
CoL_2A_2	-13.2 -15.3	2.31(1-2) 4.34(1-2)	-43.1 -32.3	20 20	C C	S: $CHCl_3$. A=Cl S: $CHCl_3$. A=NCS	66Ca 66Ca	s s
CuA_2	-7.6	1.643(1)	-17.9	25	C	S: Cyclohexane A=t-Butylaceto- acetate	69La	
Br_2	-11.1	2.700(1)	-25.2	25	C	S: 1,2-Dichloro- ethane	73Tb	
P242	PYRIDINE, 4-ethyl-		C_7H_9N			$C_5H_4N(CH_2CH_3)$		L
CoL_2A_2	-16.0 -16.5	1.053(1-2) 4.89(1-2)	-49.7 -33.8	20 20	C C	S: $CHCl_3$. A=Cl S: $CHCl_3$. A=NCS	66Ca 66Ca	s s
CuA_2	-7.9	1.785(1)	-18.2	25	C	S: Cyclohexane A=t-Butylaceto- acetate	69La 69Ga	
SbA_5	-30.7	(1)	--	30	C	S: Nitrobenzene	63Hb	
Br_2	-11.3	2.867(1)	-24.8	25	C	S: 1,2-Dichloro- ethane	73Tb	

Metal Ion	ΔH, kcal /mole	Log K	ΔS, cal /mole °K	T, °C	M	Conditions	Ref.	Re-marks
P242, cont.								
I_2	-9.0	(1)	--	25	C	S: 1,2-Dichloro-ethane	73Tb	
P243	PYRIDINE, 2-ethyl-, 1-oxide-		C_7H_9ON			$C_5H_4N(CH_2CH_3)(\rightarrow O)$		
VAB_2	-3.65	(1)	--	25	C	S: CH_2Cl_2. A=O B=2,4-Pentanedione	69Pd	
P244	PYRIDINE, 2-(2-hydroxy-ethyl)-		C_7H_9ON			$C_5H_4N(CH_2CH_2OH)$		L
Cu^{2+}	-13.82	18.54(1-2)	38.5	25	C	$\mu=1.5\times10^{-4}$ S: 50 v % 1,4-Dioxane	64Le	
	-13.65	18.54(25°,1-2)	39.1	0-25	T	$\mu=4.5\times10^{-3}$ S: 50 v % 1,4-Dioxane	64Le	b
P245	PYRIDINE, 2-(2-hydroxy-ethyl)-6-methyl-		$C_8H_{11}ON$			$C_5H_3N(CH_2CH_2OH)(CH_3)$		L
Cu^{2+}	-12.43	18.12(1-2)	41.2	25	C	$\mu=1.5\times10^{-4}$ S: 50 v % 1,4-Dioxane	64Le	
	-12.20	18.12(25°,1-2)	42.0	0-25	T	$\mu=4.5\times10^{-3}$ S: 50 v % 1,4-Dioxane	64Le	b
P246	PYRIDINE, 2-hydroxymethyl-		C_6H_7ON			$C_5H_4N(CH_2OH)$		L
Cu^{2+}	-23.81	19.15(1-2)	7.8	25	C	$\mu=7\times10^{-5}$ S: 50 v % 1,4-Dioxane	64Le	
	-23.32	19.15(1-2)	9.4	25	C	$\mu=1.5\times10^{-4}$ S: 50 v % 1,4-Dioxane	64Le	
	-14.59	19.15(25°,1-2)	38.7	0-25	T	$\mu=4.5\times10^{-3}$ S: 50 v % 1,4-Dioxane	64Le	b
P247	PYRIDINE, 2-hydroxymethyl-6-methyl-		C_7H_9ON			$C_5H_3N(CH_2OH)(CH_3)$		L
Cu^{2+}	-14.52	18.35(25°,1-2)	35.3	0-25	T	$\mu=4.5\times10^{-3}$ S: 50 v % 1,4-Dioxane	64Le	b
P248	PYRIDINE, 2-(2-propyl)-		$C_8H_{11}N$			$C_5H_4N[CH(CH_3)_2]$		
BA_3	-7.6	(1)	--	25	C	S: Nitrobenzene A=CH_3	56Ba	
P249	PYRIDINE, 4-(2-propyl)-		$C_8H_{11}N$			$C_5H_4N[CH(CH_3)_2]$		L
SbA_5	-30.1	(1)	--	30	C	S: Nitrobenzene A=Cl	63Hb	

Metal Ion	ΔH, kcal /mole	Log K	ΔS, cal /mole °K	T, °C	M	Conditions	Ref.	Remarks
P250	PYRIDINE, 4-methoxy-, 1-oxide-		$C_6H_7O_2N$			$C_5H_4N(OCH_3)(\rightarrow O)$		
VAB_2	-8.03	(1)	--	25	C	S: CH_2Cl_2. A=O B=2,4-Pentanedione	69Pd	
	-4.75	(1)	--	25	C	S: $CHCl_3$. A=O B=2,4-Pentanedione	69Pd	
I_2	-7.2	3.118(25°,1)	-9.8	18-32	T	S: CCl_4	68Ga	b
P251	PYRIDINE, 1-methyl-		C_5H_6N			$C_5H_5N(H)$		L^+
[Fe]	-3.9	5.7(20°)	12.7	5-30	T	pH=11.2 Protohaemin R: $(Haemin)_2 + 2L = (HaeminL)_2 + 2H_2O$	69Md	b w:I
P252	PYRIDINE, 2-methyl-		C_6H_7N			$C_5H_4N(CH_3)$		L
Be^{+2}	-4.98	3.52(35°,1)	-0.03	35&45	T	--	67Rb	a,b
CoA_2	-8.03	4.83(1-2)	-4.87	25	C	S: Acetone. A=Cl	72Ae	
	-9.42	4.19(1-2)	-12.42	25	C	S: Acetonitrile A=Cl	72Ae	
	-5.62	3.75(1-2)	-1.71	25	C	S: n-Butanol. A=Cl	72Af	
	-12.20	5.14(1-2)	-17.43	25	C	S: Cyclohexanone A=Cl	72Ae	
	-2.72	3.79(1-2)	8.19	25	C	S: Dimethyl-foramamide. A=Cl	72Ae	
	-6.80	2.69(1-2)	-10.50	25	C	S: Ethanol. A=Cl	72Af	
	-6.01	3.81(1-2)	-2.75	25	C	S: 2-Methyl propan-2-ol. A=Cl	72Af	
	-0.61	1.76	5.97	25	C	S: Ethylene Glycol A=Cl. R: $CoA_2 + 2L + 2S = [CoL_2S_2]A_2$	72Af	
NiA	-12.83	0.48(1-2)	-40.9	25	T	$C_M=C_L=10^{-4}$ S: Benzene A=Diacetyl-bisbenzoyl hydrazine	58Sa	
NiA_2	-5.46	-2.558(35°,1-2)	-29	25-95	T	S: L A=Salicylaldoximate	71Ma	b
CuA_2	-3.1	-1.301(25°,1)	-17	-3-57	T	S: Benzene A=Diethyldithio-carbamate	74Hc	b
	-6.3	0.964(1)	-16.7	25	C	S: Cyclohexane A=t-Butylaceto-acetate	69La	
	-5.42	0.633(1)	-15.1	30	C	S: Benzene. A=N-Nitroso-N-Phenyl-hydroxylamine	71Gh	q: ±0.55
Ag^+	5.81	2.333(1)	8.81	25	C	μ=0.5M(KNO_3)	72Ec	
	4.37	2.333(2)	3.88	25	C	μ=0.5M(KNO_3)	72Ec	
HgA_2	-18.3	1.839(1)	-49.7	30	C	S: Benzene. A=Cl $C_M=1\text{-}5\times10^{-3}M$	73Fa	

Metal Ion	ΔH, kcal /mole	Log K	ΔS, cal /mole °K	T, °C	M	Conditions	Ref.	Re-marks
P252, cont.								
HgA_2, cont.	-17.3	1.778(1)	-49.0	30	C	S: Benzene. A=Br C_M=1-5x10^{-3}M	73Fa	
	-15.0	1.415(1)	-43.0	30	C	S: Benzene. A=I C_M=1-5x10^{-3}M	73Fa	
BA_3	-10.0	(1)	--	25	C	S: Nitrobenzene A=CH_3	56Ba	
	18.8	-2.398(70°,1)	43.8	-50-100	T	Equimolar amounts of L and BA_3 A=C_3H_5	71Bj	b
	13.1	1.217(70°,1)	43.7	-50-100	T	Equimolar amounts of L and BA_3 A=C_3H_7	71Bj	b
AlA_3	-26.13	(1)	--	25-28	C	S: n-Hexane. A=CH_3	68Hc	
$SiAB_3$	-5.06	1.85(30°,1)	-8.22	10-45	T	S: CCl_4 X_L=1x10^{-2} X_M=5x10^{-4} A=OH. B=CH_3	69Lb	b
	-5.97	1.76(30°,1)	-11.65	20-60	T	S: CCl_4 X_L=1x10^{-2} X_M=5x10^{-4} A=OH. B=C_2H_5	69Lb	b
	-6.10	2.31(30°,1)	-9.57	0-50	T	S: CCl_4 X_L=1x10^{-2} X_M=4x10^{-4} A=OH. B=Phenyl	69Lb	b
SbA_5	-26.2	(1)	--	30	C	S: Nitrobenzene A=Cl	63Hb	
P253	PYRIDINE, 3-methyl-		C_6H_7N			$C_5H_4N(CH_3)$		L
Be^{+2}	-4.08	2.94(35°,1)	-0.18	35&45	T	--	67Rb	a,b
VOA_2	-8.10	1.81(1)	-18.9	25	C	μ→0. S: Nitrobenzene A=Acetylacetonate	65Ca	
CoA_2	-10.42	4.90(1-2)	-12.55	25	C	S: Acetone. A=Cl	72Ae	
	-9.78	4.62(1-2)	-11.68	25	C	S: Acetonitrile A=Cl	72Ae	
	-6.66	4.10(1-2)	-3.62	25	C	S: n-Butanol. A=Cl	72Af	
	-13.27	5.64(1-2)	-18.72	25	C	S: Cyclohexanone A=Cl	72Ae	
	-4.54	4.06(1-2)	3.36	25	C	S: Dimethylformamide. A=Cl	72Ae	
	-7.59	2.85(1-2)	-12.42	25	C	S: Ethanol. A=Cl	72Af	
	-7.50	4.47(1-2)	-4.73	25	C	S: 2-Methyl propan-2-ol. A=Cl	72Af	
	-1.83	2.57(1-2)	5.60	25	C	S: Ethylene Glycol A=Cl. R: CoA_2 + 2L + 2S → $[CoL_2S_2]A_2$	72Af	
CoL_2A_2	-13.4	0.348(1-2)	-44.1	20	C	S: $CHCl_3$. A=Cl	66Ca	s
	-10.4	0.785(1-2)	-31.7	20	C	S: $CHCl_3$. A=NCO	66Ca	s

Metal Ion	ΔH, kcal /mole	Log K	ΔS, cal /mole °K	T, °C	M	Conditions	Ref.	Re-marks
P253, cont.								
CoL_2A_2, cont.	-15.2	4.204(1-2)	-32.6	20	C	S: $CHCl_3$. A=NCS	66Ca	s
NiA	-15.17	3.90(25°,1-2)	-33.1	25	T	$C_M \approx C_L \approx 10^{-4}$ S: Benzene A=Diacetyl-bisbenzoyl hydrazine	58Sa	
NiA_2	-3.8	-0.44(20°,1-2)	-15.4	10-30	T	S: $CHCl_3$. A=N-Ethyl-β-mercapto-cinnamamide	74Cc	b
	-12.4	2.05(20°,1-2)	-33.0	10-30	T	S: $CHCl_3$. A=Ethyl-β-mercaptocinnamate	74Cc	b
	-11.5	1.92(20°,1-2)	-30.2	10-30	T	S: $CHCl_3$. A=Ethyl-β-mercaptothio-cinnamate	74Cc	b
	-4.6	-0.35(20°,1-2)	-17.2	10-30	T	S: $CHCl_3$. A=Mono-thiodibenzoylmethane	74Cc	b
	-9.5	0.37(20°,1-2)	-30.9	10-30	T	S: $CHCl_3$. A=N-Phenyl-β-mercapto-cinnamamide	74Cc	b
NiL_2A_2	-21.7	3.91(1-2)	-55.8	20	C	$\mu \approx 0$. S: $CHCl_3$. A=I	65Nc	
CuA_2	-7.5	1.618(1)	-17.7	25	C	S: Cyclohexane A=t-Butylaceto-acetate	69La	
	-5.2	-0.310(25°,1)	-19	-3-57	T	S: Benzene A=Diethyldithio-carbamate	74Hc	b
	-8.04	0.68(1)	-23.9	25	C	$\mu \approx 0.005$ S: Benzene A=2,4-Pentanedione	63Ma	
	-3.38	(1)	--	25	T	$\mu \approx 0.005$ S: Benzene A=2,4-Pentanedione	63Ma	
	-6.48	1.364(1)	-15.1	30	C	S: Benzene. A=N-Nitroso-N-Phenyl hydroxylamine	71Gh	q: ±0.65
Ag^+	5.17	2.253(1)	7.0	25	C	μ=0.5M(KNO_3)	72Ec	
	6.76	2.233(2)	12.5	25	C	μ=0.5M(KNO_3)	72Ec	
BA_3	-15.6	(1)		25	C	S: Nitrobenzene A=CH_3	56Ba	
Br_2	-10.9	2.644(1)	-24.7	25	C	S: 1,2-Dichloro-ethane	73Tb	
I_2	-8.7	(1)	--	25	C	S: 1,2-Dichloro-ethane	73Tb	
P254 PYRIDINE, 4-methyl-			C_6H_7N			$C_5H_4N(CH_3)$		L
Be^{+2}	-4.98	3.53(35°,1)	0	35&45	T	--	67Rb	a,b
VOA_2	-8.80	2.14(1)	-19.7	25	C	$\mu \approx 0$ S: Nitrobenzene A=Acetylacetonate	65Ca	

P254, cont.

Metal Ion	ΔH, kcal /mole	Log K	ΔS, cal /mole °K	T, °C	M	Conditions	Ref.	Remarks
[Fe]	2.35	1.74(20°)	16.0	5-30	T	pH=11.2 Protohaemin R: $(Haemin)_2$ + 2L = $(HaeminL)_2$ + $2H_2O$	69Md	b
	0.8	2.42(25°,1-2)	13	18-40	T	[Fe]=Iron(II) deuteroporphyrin dimethyl ester S: Benzene	70Ch 71Cd	b,q: ±0.6
	-2.6	1.10(25°,1-2)	-4	18-40	T	[Fe]=Iron(II) diacetyldeutero-porphyrin dimethyl ester. S: Benzene	70Ch	b,q: ±0.7
	-8.7	3.74(25°,1-2)	-12	18-40	T	[Fe]=Iron(II) mesoporphyrin dimethyl ester S: Benzene	70Ch	b,q: ±1.2
	-14.0	6.60(25°,1-2)	-16	18-40	T	[Fe]=Iron(II) protoporphyrin dimethyl ester S: Benzene	70Ch 71Cd	b
	-7.5	2.78(25°,1-2)	-12	18-40	T	[Fe]=Iron(II) deuteroporphyrin dimethyl ester C: CCl_4	71Cd	b,q: ±1.9
	-24.6	2.78(25°,1-2)	-70	18-40	T	[Fe]=Iron(II) protoporphyrin dimethyl ester S: CCl_4	71Cd	b,q: ±2.6
	-12.2	1.10(25°,1-2)	-36	18-40	T	[Fe]=Iron(II) protoporphyrin dimethyl ester S: $CHCl_3$	71Cd	b,q: ±1.0
CoA	-7.3	2.900(25°,1)	-11	20-50	T	S: Toluene A=α,β,γ,δ-Tetra(p-methoxyphenyl) porphyrin	73Wa	b,q: ±1
CoA_2	-11.17	5.14(1-2)	-13.96	25	C	S: Acetone. A=Cl	73Ah	
	-10.91	4.97(1-2)	-13.86	25	C	S: Acetonitrile A=Cl	73Ah	
	-8.32	4.22(1)	-8.62	25	C	S: n-Butyl alcohol A=Cl	73Ai	q: ±0.72
	-9.42	4.51(1)	-10.97	25	C	S: t-Butyl alcohol A=Cl	73Ai	q: ±0.74
	-15.14	5.70(1-2)	-24.73	25	C	S: Cyclohexanone A=Cl	73Ah	
	-6.46	4.12(1-2)	-2.82	25	C	S: Dimethyl-formamide. A=Cl	73Ah	
	-2.04	2.62(1)	5.13	25	C	S: Ethylene glycol A=Cl	73Ai	q: ±0.14
	-8.03	2.88(1)	-13.76	25	C	S: Ethyl alcohol A=Cl	73Ai	q: ±0.50
CoL_2A_2	-15.7	1.049(1-2)	-48.8	20	C	S: $CHCl_3$. A=Cl	66Ca	s

Metal Ion	ΔH, kcal /mole	Log K	ΔS, cal /mole °K	T, °C	M	Conditions	Ref.	Remarks
P254, cont.								
CoL_2A_2, cont.	-13.3	1.182(1-2)	-40.0	20	C	S: $CHCl_3$. A=NCO	66Ca	s
	-16.7	4.892(1-2)	-34.2	20	C	S: $CHCl_3$. A=NCS	66Ca	s
NiA	-15.65	4.45(25°,1-2)	-32.3	25	T	$C_M \simeq C_L \simeq 10^{-4}$ S: Benzene A=Diacetyl-bisbenzoyl hydrazine	58Sa	
	-10.12	2.062(1)	-24.52	25	C	S: Toluene A=bis(salicylidene-γ-iminopropyl) methylamine	73Kg	
	-6.73	1.082(1)	-17.63	25	C	S: CH_2Cl_2. A=bis-(salicylidene-γ-iminopropyl) methylamine	73Kg	
NiA_2	-19.8	1.48(1)	-59.3	30	C	S: Benzene A=Diethyldithiophosphato	71Dc	q: ±2
	1.7	1.88(2)	14.1	30	C	S: Benzene A=Diethyldithiophosphato	71Dc	q: ±1.7
	-20.52	(1-2)	--	25	C	S: Benzene. A=2,2,6,6-Tetramethyl-3,5-heptanedione	73Ce	
	-12.6	2.10(20°,1-2)	-33.3	10-30	T	S: $CHCl_3$. A=Ethyl-β-mercaptocinnamate	74Cc	b
	-4.6	-0.28(20°,1-2)	-17.4	10-30	T	S: $CHCl_3$. A=N-Ethyl-β-mercapto cinnamamide	74Cc	b
	-12.1	2.07(20°,1-2)	-31.6	10-30	T	S: $CHCl_3$. A=Ethyl-β-mercaptothiocinnamate	74Cc	b
	-6.0	-0.18(25°,1-2)	-21.3	10-30	T	S: $CHCl_3$. A=Monothiodibenzoylmethane	74Cc	b
	-10.2	0.58(20°,1-2)	-32.0	10-30	T	S: $CHCl_3$. A=N-Phenyl-β-mercapto-cinnamamide	74Cc	b
Cu^{2+}	-4.70	(1)	--	--	-	C_L=0.2M C_M=3x10^{-2}M C_{HNO_3}=0.01M	68Wb	f: 3.44, u,y
CuA_2	-5.7	-0.208(25°,1)	-20	-3-57	T	S: Benzene A=Diethyldithiocarbamate	74Hc	b
	-11.5	0.99(1)	-34.2	25	C	μ≃0.005 S: Benzene A=2,4-Pentanedione	63Ma 69Ga	
	-2.71	(1)	--	25	T	μ≃0.005 S: Benzene A=2,4-Pentanedione	63Ma	
	-6.79	1.412(1)	-16.0	30	C	S: Benzene. A=N-Nitro-N-Phenylhydroxylamine	71Gh	q: ±0.67

Metal Ion	ΔH, kcal /mole	Log K	ΔS, cal /mole °K	T, °C	M	Conditions	Ref.	Remarks
P254, cont.								
CuA_2, cont.	-7.6	1.735(1)	-17.6	25	C	S: Cyclohexane A=t-Butylaceto-acetate	69La	
	-7.34	3.175(1)	-11.71	30	C	S: Benzene. A=1,1 1-Trifluoropentane-2,4-dione	74Gb	
Ag^+	6.09	2.183(1)	10.44	25	C	μ=0.5M(KNO_3)	72Vc	
	6.71	2.464(2)	11.23	25	C	μ=0.5M(KNO_3)	72Vc	
ZnA_2	-9.6	2.336(25°,1)	-21	25&46	T	S: Toluene A=[$S_2CN(CH_3)_2$]	65Ch	a,b
HgA_2	-17.6	2.412(1)	-47.1	30	C	S: Benzene. A=Cl C_M=1-5x10^{-3}M	73Fa	
	-17.4	2.579(1)	-45.6	30	C	S: Benzene. A=Br C_M=1-5x10^{-3}M	73Fa	
	-16.2	2.425(1)	-42.5	30	C	S: Benzene. A=I C_M=1-5x10^{-3}M	73Fa	
BA_3	-15.9	(1)	--	25	C	S: Nitrobenzene A=CH_3	56Ba	
SbA_5	-31.0	(1)	--	30	C	S: Nitrobenzene A=Cl	63Hb	
Br_2	-11.3	2.867(1)	-24.8	25	C	S: 1,2-Dichloro-ethane	73Tb	
P255	PYRIDINE, 4-(methylacetate)-		$C_8H_9O_2N$			$C_5H_4N(CH_2CO_2CH_3)$		
[Ni]	1.5	3.15(25°,1)	6	18-40	T	[Ni]=Nickel(II) mesotetraphenyl-porphine. S: $CHCl_3$	72Ce	b,q: ±0.7
[Zn]	-4.7	3.22(25°,1)	-1	18-40	T	[Zn]=Zinc(II) mesotetraphenyl-porphine. S: $CHCl_3$	72Ce	b,q: ±0.5
P256	PYRIDINE, 2-(methylamino-ethyl)-		$C_8H_{12}N_2$			$C_5H_4N(CH_2CH_2NHCH_3)$		
Ni^{2+}	-6.30	4.65(1)	0.2	25	C	μ=0.5M(KNO_3)	71Gd	
P257	PYRIDINE, 2-(methylamino-methyl)-		$C_7H_{10}N_2$			$C_5H_4N(CH_2NHCH_3)$		L
Co^{2+}	-6.59	5.10(30°,1)	1.6	10-40	T	μ=0	59Gb	b
	-6.76	5.52(1)	1.2	25	C	μ=0.5M(KNO_3)	71Gd	
	-16.88	(1)	--	25	C	μ=0.1(KNO_3) S: 50 w % Dioxane	66Wa	
	-5.61	3.84(30°,2)	-0.92	10-40	T	μ=0	59Gb	b
	-5.98	3.98(2)	-1.8	25	C	μ=0.5M(KNO_3)	71Gd	
	-2.65	2.63(30°,3)	0.33	10-40	T	μ=0	59Gb	b
Ni^{2+}	-9.04	6.66(30°,1)	0.7	10-40	T	μ=0	59Gb	b
	-8.50	6.91(1)	3.2	25	C	μ=0.5M(KNO_3)	71Gd	
	-16.91	(1)	--	25	C	μ=0.1(KNO_3) S: 50 w % Dioxane	66Wa	

Metal Ion	ΔH, kcal /mole	Log K	ΔS, cal /mole °K	T, °C	M	Conditions	Ref.	Remarks
P257, cont.								
Ni^{2+} cont.	-8.44	5.30(30°,2)	-3.6	10-40	T	μ=0	59Gb	b
	-8.50	5.53(2)	-3.2	25	C	μ=0.5M(KNO_3)	71Gd	
	-3.76	2.79(30°,3)	0.4	10-40	T	μ=0	59Gb	b
Cu^{2+}	-7.67	9.27(30°,1)	17.1	10-40	T	μ=0	59Gb	b
	-10.69	9.07(1)	5.8	25	C	μ=0.5M(KNO_3)	71Gd	
	-19.24	14.60(1)	2.28	25	C	μ=0.1(KNO_3) S: 50 w % Dioxane	66Wa	
	-9.84	6.55(30°,2)	-2.5	10-40	T	μ=0	59Gb	b
	-8.13	6.71(2)	3.5	25	C	μ=0.5M(KNO_3)	71Gd	
Zn^{2+}	-5.26	4.73(30°,1)	4.3	10-40	T	μ=0	59Gb	b
	-5.78	4.96(1)	3.3	25	C	μ=0.5M(KNO_3)	71Gd	
	-11.00	(1)	--	25	C	μ=0.1(KNO_3) S: 50 w % Dioxane	66Wa	
	-4.38	3.62(2)	1.9	25	C	μ=0.5M(KNO_3)	71Gd	
Cd^{2+}	-4.89	4.30(30°,1)	3.5	10-40	T	μ=0	59Gb	b
	-5.22	4.49(1)	2.9	25	C	μ=0.5M(KNO_3)	71Gd	
	-10.02	8.54(1)	5.48	25	C	μ=0.1(KNO_3) S: 50 w % Dioxane	66Wa	
	-4.79	3.35(2)	0.2	25	C	μ=0.5M(KNO_3)	71Gd	
P258	PYRIDINE, 2-(methylamino-methyl)-6-methyl-		$C_8H_{12}N_2$			$C_5H_3N(CH_2NHCH_3)(CH_3)$		L
Co^{2+}	-3.86	3.41(30°,1)	3	10-40	T	μ=0	61Rb	b
	-4.37	3.57(1)	1.7	25	C	μ=0.5M(KNO_3)	71Gd	
	-11.44	7.87(1)	-2.35	25	C	μ=0.1(KNO_3) S: 50 w % Dioxane	66Wa	
	-2.37	1.41(2)	-1.5	25	C	μ=0.5M(KNO_3)	71Gd	
Ni^{2+}	-7.10	4.58(30°,1)	-3	10-40	T	μ=0	61Ra	b
	-5.97	4.74(1)	1.7	25	C	μ=0.5M(KNO_3)	71Gd	
	-7.52	7.57(1)	9.41	25	C	μ=0.1(KNO_3) S: 50 w % Dioxane	66Wa	
	-4.64	2.52(2)	-4.0	25	C	μ=0.5M(KNO_3)	71Gd	
Cu^{2+}	-6.98	6.48(30°,1)	7	10-40	T	μ=0	61Rb	b
	-6.96	6.88(1)	8.2	25	C	μ=0.5M(KNO_3)	71Gd	
	-12.01	12.26(1)	15.80	25	C	μ=0.1(KNO_3) S: 50 w % Dioxane	66Wa	
	-4.37	4.94(30°,2)	8	10-40	T	μ=0	61Rb	b
	-6.57	5.44(2)	2.9	25	C	μ=0.5M(KNO_3)	71Gd	
Zn^{2+}	-3.92	7.76(1)	22.36	25	C	μ=0.1(KNO_3) S: 50 w % Dioxane	66Wa	
P259	PYRIDINE, 2-methyl-, 1-oxide		C_6H_7ON			$C_5H_4N(\rightarrow O)(CH_3)$		L
VAB_2	-4.06	(1)	--	25	C	S: CH_2Cl_2. A=0 B=2,4-Pentanedione	69Pd	
I_2	-6.02	2.05(25°,1)	-10.8	9-42	T	S: CCl_4	57Ka	b
	-6.0	(1)	--	--	-	S: CCl_4	60Bc	
P260	PYRIDINE, 3-methyl-,1-oxide		C_6H_7ON			$C_5H_4N(\rightarrow O)(CH_3)$		L

Metal Ion	ΔH, kcal /mole	Log K	ΔS, cal /mole °K	T, °C	M	Conditions	Ref.	Remarks
P260, cont.								
VAB_2	-5.57	(1)	--	25	C	S: CH_2Cl_2. A=O B-2,4-Pentanedione	69Pd	
I_2	-6.09	2.13(25°,1)	-10.7	11-38	T	S: CCl_4	57Ka	b
	-6.1	(1)	--	--	-	S: CCl_4	60Bc	
P261	PYRIDINE, 4-methyl-,1-oxide		C_6H_7ON			$C_5H_4N(\rightarrow O)(CH_3)$		L
VAB_2	-7.22	(1)	--	25	C	S: CH_2Cl_2. A=O	69Pd	
	-3.50	(1)	--	25	C	S: $CHCl_3$. A=O B=2,4-Pentanedione	69Pd	q: ±0.20
SbA_5	-36.3	(1)	--	25	C	S: $C_2H_4Cl_2$. A=Cl	72Lb	
I_2	-6.15	2.14(25°,1)	-10.8	12-38	T	S: CCl_4	57Ka	b
	-6.3	2.312(25°,1)	-10.6	25&32	T	S: CCl_4	68Ga	a,b
	-6.2	(1)	--	--	-	S: CCl_4	60Bc	
P262	PYRIDINE, 4-nitro-,1-oxide		$C_5H_4O_3N_2$			$C_5H_4N(NO_2)(\rightarrow O)$		L
VAB_2	-2.24	(1)	--	25	C	S: CH_2Cl_2. A=O B=2,4-Pentanedione	69Pd	
P263	PYRIDINE, 1-oxide		C_5H_5ON			$C_5H_5N(\rightarrow O)$		L
VAB_2	-6.21	(1)	--	25	C	S: CH_2Cl_2. A=O B=2,4-Pentanedione	69Pd	
	-2.57	(1)	--	25	C	S: $CHCl_3$. A=O B=2,4-Pentanedione	69Pd	
NiA_2	-5.68	2.456(1)	-7.8	25	C	S: Benzene. A=2,2,6,6-Tetramethyl-3,5-heptanedione	73Ce	
	-9.3	0.699(2)	-28.0	25	C	S: Benzene. A=2,2,6,6-Tetramethyl-3,5-heptanedione	73Ce	
I_2	-5.75	1.891(23°,1)	-10.67	7-33	T	S: CCl_4	68Ga	b
	-5.85	1.892(25°,1)	-11.0	7-33	T	$C_L=10^{-4}-10^{-3}$ S: CCl_4	65Kc	b
	-5.59	1.85(25°,1)	-10.3	10-38	T	S: CCl_4	57Ka	b
	-5.6	(1)	--	--	-	S: CCl_4	60Bc	
P264	PYRIDINE, 4-propyl-		$C_8H_{11}N$			$C_5H_4N(CH_2CH_2CH_3)$		L
CoL_2A_2	-15.8	1.07(1-2)	-49.0	20	C	S: $CHCl_3$. A=Cl	66Ca	s
	-16.4	4.90(1-2)	-33.6	20	C	S: $CHCl_3$. A=NCS	66Ca	s
P265	PYRIDINE, 1,2,4,6-tetramethyl-		$C_9H_{14}N$			$C_5H_2N(CH_3)_4$		L^+
[Fe]	0	6.4(20°)	29.3	5-30	T	pH=11.2 Protohaemin R: $(Haemin)_2 + 2L = (HaeminL)_2 + 2H_2O$	69Md	b w:I
P266	PYRIDINE, 1,2,4-trimethyl-		$C_8H_{12}N$			$C_5H_3N(CH_3)_3$		L^+

Metal Ion	ΔH, kcal /mole	Log K	ΔS, cal /mole °K	T, °C	M	Conditions	Ref.	Remarks
P266, cont.								
[Fe]	0	6.8(20°)	31.2	5-30	T	pH=11.2 Protohaemin R: $(Haemin)_2 + 2L = (HaeminL)_2 + 2H_2O$	69Md	b w:I
P267	PYRIDINE, 2,4,6-trimethyl-		$C_8H_{11}N$			$C_5H_2N(CH_3)_3$		L
CuA_2	-3.54	0.415(1)	-9.8	30	C	S: Benzene. A=N-Nitroso-N-phenylhydroxylamine	71Gh	q:
BA_3	-5.68	(1)	--	25	C	S: Nitrobenzene $C_L \simeq 0.20M$ $C_M \simeq 0.25M$. A=F	56Bb	
Br_2	-11.4	2.725(1)	-25.9	25	C	S: 1,2-Dichloroethane	73Tb	
I_2	-7.4	(1)	--	25	C	S: 1,2-Dichloroethane	73Tb	
IA_3B	-5.0	0.95(25°,1)	-12.6	-30-40	T	C_L=0-0.48 S: Cyclopentane A=F. $B=CH_3$	65Ld 65Lc	b,1
P268	PYRIDINE, 2,4,6-trimethyl, 1-oxide-		$C_8H_{11}ON$			$C_5H_2N(CH_3)_3(\rightarrow O)$		L
VAB_2	-2.06	(1)	--	25	C	S: CH_2Cl_2. A=O B=2,4-Pentanedione	69Pd	
P269	PYRIDINE, 4-vinyl-		C_7H_7N			$C_5H_4N(CH:CH_2)$		
[Fe]	-1.6	2.86(25°,1-2)	8	18-40	T	[Fe]=Iron(II) deuteroporphrin dimethyl ester S: Benzene	70Ch	b,q: ±0.9
	-3.6	2.64(25°,1-2)	0	18-40	T	[Fe]=Iron(II) diacetyldeuteroporphyrin dimethyl ester. S: Benzene	70Ch	b,q: ±0.9
	-10.2	3.66(25°,1-2)	-16	18-40	T	[Fe]=Iron(II) mesoporphyrin dimethyl ester S: Benzene	70Ch	b,q: ±1.0
	-7.4	3.74(25°,1-2)	-8	18-40	T	[Fe]=Iron(II) protoporphyrin dimethyl ester S: Benzene	70Ch	b,q: ±1.1
P270	4-PYRIDINECARBOXALDEHYDE		C_6H_5ON			$C_5H_4N(COH)$		
[Ni]	6.0	2.27(25°,1)	31	18-40	T	[Ni]=Nickel(II) mesotetraphenylporphine. S: $CHCl_3$	72Ce	b,q: ±1.0
Cu^{2+}	-4.42	(1)	--	--	-	C_L=0.2M $C_M=3x10^{-2}M$	68Wb	f: 3.06,

Metal Ion	ΔH, kcal /mole	Log K	ΔS, cal /mole °K	T, °C	M	Conditions	Ref.	Remarks
P270, cont.								
Cu^{2+}, cont.						C_{HNO_3}=0.01M		u,y
P271	4-PYRIDINECARBOXYLIC ACID, amide		$C_6H_6ON_2$			$C_5H_4N(CONH_2)$		
Cu^{2+}	-3.08	(1)	--	--	-	C_L=0.2M C_M=3x10^{-2}M C_{HNO_3}=0.01M	68Wb	f: 2.20, u,y
P272	4-PYRIDINECARBOXYLIC ACID, butyl ester		$C_{10}H_{13}O_2N$			$C_5H_4N(CO_2HCH_2CH_2CH_2CH_3)$		L
CuA_2	-6.02	2.258(1)	-9.56	30	C	S: Benzene. A=1,1,1-Trifluoropentane-2,4-dione	74Gb	
P273	4-PYRIDINECARBOXYLIC ACID, methyl ester		$C_7H_7O_2N$			$C_5H_4N(CO_2CH_3)$		
NiA_2	-19.75	(1-2)	--	25	C	S: Cyclohexane A=2,2,6,6-Tetramethyl-3,5-heptanedione	73Ce	
Cu^{2+}	-3.02	(1)	--	--	-	C_L=0.2M C_M=3x10^{-2}M C_{HNO_3}=0.01M	68Wb	f: 2.20, u,y
P274	3-PYRIDINECARBOXYLIC ACID, nitrile		$C_6H_4N_2$			$C_5H_4N(CN)$		L
CuA_2	-2.50	0.79(1)	-4.77	25	C	μ≃0.005 S: Benzene A=2,4-Pentandione	63Ma	
	-2.92	(1)	--	25	T	μ≃0.005 S: Benzene A=2,4-Pentanedione	63Ma	
P275	4-PYRIDINECARBOXYLIC ACID, nitrile		$C_6H_4N_2$			$C_5H_4N(CN)$		L
[Fe]	-3.9	3.8(20°)	4.0	5-30	T	pH=11.2 R: $(Haemin)_2$ + 2L = $(HaeminL)_2$ + $2H_2O$	69Md	b
	-6.9	3.52(25°,1-2)	-7	18-40	T	[Fe]=Iron(II) deuteroporphyrin dimethyl ester S: Benzene	70Ch 71Cd	b,q: ±1.1
	7.5	1.25(25°,1-2)	32	18-40	T	[Fe]=Iron(II) diacetyldeuteroporphyrin dimethyl ester. S: Benzene	70Ch	b
	-11.7	2.78(25°,1-2)	-24	18-40	T	[Fe]=Iron(II) mesoporphyrin dimethyl ester S: Benzene	70Ch	b,q: ±1.1

Metal Ion	ΔH, kcal /mole	Log K	ΔS, cal /mole °K	T, °C	M	Conditions	Ref.	Remarks
P275, cont.								
[Fe], cont.	-9.0	3.00(25°,1-2)	-16	18-40	T	[Fe]=Iron(II) protoporphyrin dimethyl ester S: Benzene	70Ch 71Cd	b,q: ±1.0
	-17.4	3.08(25°,1-2)	-44	18-40	T	[Fe]=Iron(II) deuteroporphyrin dimethyl ester S: CCl_4	71Cd	b,q: ±2.3
	-25.6	3.44(25°,1-2)	-70	18-40	T	[Fe]=Iron(II) protoporphyrin dimethyl ester S: CCl_4	71Cd	b,q: ±3.0
	-12.0	1.54(25°,1-2)	-33	18-40	T	[Fe]=Iron(II) deuteroporphyrin dimethyl ester S: $CHCl_3$	71Cd	b
	-19.3	2.71(25°,1-2)	-52	18-40	T	[Fe]=Iron(II) protoporphyrin dimethyl ester S: $CHCl_3$	71Cd	b,q: ±3.6
[Co]	-10.0	3.88(0°,1)	-19	-23-23	T	[Co]=Cobalt(II) protoporphyrin IX dimethyl ester S: Toluene	73Sr	b,q: ±0.6
NiA_2	-8.2	-0.88(20°,1-2)	-24.7	10-30	T	S: $CHCl_3$. A=Ethyl-β-mercaptocinnamate	74Cc	b
	-7.7	0.52(20°,1-2)	-23.5	10-30	T	S: $CHCl_3$. A=Ethyl-β-mercaptothio-cinnamate	74Cc	b
	-8.0	0.06(20°,1-2)	-26.4	10-30	T	S: $CHCl_3$. A=N-Phenyl-β-mercapto-cinnamamide	74Cc	b
[Ni]	-4.3	2.86(25°,1)	-1	18-40	T	[Ni]=Nickel(II) mesotetraphenyl-porphine. S: Benzene	72Ce	b
	7.0	2.56(25°,1)	35	18-40	T	[Ni]=Nickel(II) mesotetraphenyl-porphine. S: $CHCl_3$	72Ce	b,q: ±1.5
Cu^{2+}	-2.49	(1)	--	--	-	C_L=0.2M C_M=3x10^{-2}M C_{HNO_3}=0.01M	68Wb	f:1.34 u,y
CuA_2	-4.37	0.716(1)	-11.2	30	C	S: Benzene A=N-nitroso-N-Phenylhydroxylamine	71Gh	q: ±0.43
	-3.66	1.892(1)	-3.35	30	C	S: Benzene A=1,1,1-Trifluoro-pentane-2,4-dione	74Gb	
[Zn]	-4.3	2.20(25°,1)	-1	18-40	T	[Zn]=Zinc(II) mesotetraphenyl-porphine. S: Benzene	72Ce	b,q: ±0.5

Metal Ion	ΔH, kcal /mole	Log K	ΔS, cal /mole °K	T, °C	M	Conditions	Ref.	Remarks
P275, cont.								
[Zn], cont.	-4.8	2.05(25°,1)	-1	18-40	T	[Zn]=Zinc(II) mesotetraphenyl-porphin . S: $CHCl_3$	72Ce	b,q: +0.4
P276	2,6-PYRIDINEDICARBOXYLIC ACID		$C_7H_5O_4N$			$C_5H_3N(CO_2H)_2$		L^{2-}
Sc^{3+}	-3.99	11.2(1)	37.9	25	C	μ=0.50($NaClO_4$)	69Ge	
	-6.73	7.64(2)	12.3	25	C	μ=0.50($NaClO_4$)	69Ge	
Y^{3+}	-1.438	8.44(1)	33.8	25	C	μ=0.5	63Gc	
	-5.311	15.66(1-2)	53.9	25	C	μ=0.5	63Gc	
	-12.236	21.18(1-3)	55.9	25	C	μ=0.5	63Gc	
La^{3+}	-3.125	7.99(1)	25.8	25	C	μ=0.5	63Gc	
	-6.401	13.71(1-2)	41.3	25	C	μ=0.5	63Gc	
	-8.631	17.96(1-3)	53.2	25	C	μ=0.5	63Gc	
Ce^{3+}	-3.547	8.30(1)	26.1	25	C	μ=0.5	63Gc	
	-7.140	14.33(1-2)	41.6	25	C	μ=0.5	63Gc	
	-10.296	18.68(1-3)	50.9	25	C	μ=0.5	63Gc	
Pr^{3+}	-3.913	8.58(1)	26.1	25	C	μ=0.5	63Gc	
	-7.869	15.00(1-2)	42.3	25	C	μ=0.5	63Gc	
	-11.370	19.79(1-3)	52.4	25	C	μ=0.5	63Gc	
Nd^{3+}	-4.012	8.73(1)	26.5	25	C	μ=0.5	63Gc	
	-8.109	15.39(1-2)	43.2	25	C	μ=0.5	63Gc	
	-11.883	20.41(1-3)	53.5	25	C	μ=0.5	63Gc	
Sm^{3+}	-4.283	8.80(1)	25.9	25	C	μ=0.5	63Gc	
	-8.971	15.77(1-2)	42.1	25	C	μ=0.5	63Gc	
	-13.265	21.08(1-3)	52.0	25	C	μ=0.5	63Gc	
Eu^{3+}	-4.073	8.79(1)	26.6	25	C	μ=0.5	63Gc	
	-9.123	15.86(1-2)	42.0	25	C	μ=0.5	63Gc	
	-13.732	21.32(1-3)	51.5	25	C	μ=0.5	63Gc	
Gd^{3+}	-3.582	8.70(1)	27.8	25	C	μ=0.5	63Gc	
	-8.731	15.96(1-2)	43.7	25	C	μ=0.5	63Gc	
	-13.619	21.67(1-3)	53.5	25	C	μ=0.5	63Gc	
Tb^{3+}	-2.689	8.65(1)	30.5	25	C	μ=0.5	63Gc	
	-8.007	16.01(1-2)	46.4	25	C	μ=0.5	63Gc	
	-13.811	21.86(1-3)	53.7	25	C	μ=0.5	63Gc	
Dy^{3+}	-2.169	8.67(1)	32.4	25	C	μ=0.5	63Gc	
	-7.158	16.10(1-2)	49.7	25	C	μ=0.5	63Gc	
	-13.820	21.97(1-3)	54.2	25	C	μ=0.5	63Gc	
Ho^{3+}	-1.946	8.68(1)	33.2	25	C	μ=0.5	63Gc	
	-6.166	15.99(1-2)	52.4	25	C	μ=0.5	63Gc	
	-13.557	21.91(1-3)	54.8	25	C	μ=0.5	63Gc	
Er^{3+}	-1.850	8.75(1)	38.8	25	C	μ=0.5	63Gc	
	-5.752	16.32(1-2)	55.4	25	C	μ=0.5	63Gc	
	-13.406	21.97(1-3)	55.5	25	C	μ=0.5	63Gc	
Tm^{3+}	-1.834	8.82(1)	34.2	25	C	μ=0.5	63Gc	
	-5.471	16.48(1-2)	57.0	25	C	μ=0.5	63Gc	

Metal Ion	ΔH, kcal /mole	Log K	ΔS, cal /mole °K	T, °C	M	Conditions	Ref.	Remarks
P276, cont.								
Tm^{3+}, cont.	-12.870	21.89(1-3)	57.0	25	C	μ=0.5	63Gc	
Yb^{3+}	-1.925	8.83(1)	34.0	25	C	μ=0.5	63Gc	
	-5.784	16.54(1-2)	56.3	25	C	μ=0.5	63Gc	
	-12.907	21.58(1-3)	55.4	25	C	μ=0.5	63Gc	
Lu^{3+}	-2.191	9.01(1)	33.9	25	C	μ=0.5	63Gc	
	-5.995	16.73(1-2)	56.4	25	C	μ=0.5	63Gc	
	-12.658	21.34(1-3)	55.2	25	C	μ=0.5	63Gc	
P277	PYRIMIDINE		$C_4H_4N_2$					
Ag^+	-4.21	1.61(1)	-6.74	25	C	μ=0.1M(KNO_3)	73Bg	
	-4.14	1.37(2)	-7.61	25	C	μ=0.1M(KNO_3)	73Bg	
P278	PYRIMIDINE, 2-amino-N-(dimethylaminoethyl)-N-(4-methoxybenzyl)-		$C_{16}H_{22}ON_4$			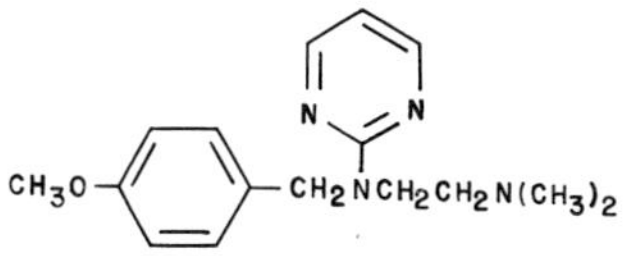		L
Co^{2+}	-4.4	6.30(25°,1-2)	14	0&25	T	0.06M(KCl)	62Aa	a,b
Ni^{2+}	16.2	8.32(25°,1-2)	92.4	0&25	T	0.06M(KCl)	62Aa	a,b
P279	PYROPHOSPHATE ION		O_7P_2			$P_2O_7{}^{4-}$		L^{4-}
Li^+	1.03	3.702(1)	20.4	25	C	μ=0	73Va	q: ±0.09
	0.39	(1)	--	5	C	μ=0.50[$(CH_3)_4NNO_3$]	73Va	g: $(CH_3)_4$-N^+,q: ±0.04
	0.40	(1)	--	15	C	μ=0.50[$(CH_3)_4NNO_3$]	73Va	g: $(CH_3)_4$-N^+,q: ±0.04
	0.48	(1)	--	25	C	μ=0.50[$(CH_3)_4NNO_3$]	73Va	g: $(CH_3)_4$-N^+,q: ±0.04
	0.29	(1)	--	5	C	μ=0.75[$(CH_3)_4NNO_3$]	73Va	g: $(CH_3)_4$-N^+,q: ±0.04
	0.32	(1)	--	15	C	μ=0.75[$(CH_3)_4NNO_3$]	73Va	g: $(CH_3)_4$-N^+,q: ±0.04
	0.37	(1)	--	25	C	μ=0.75$(CH_3)_4$	73Va	g: $(CH_3)_4$-N^+,q: ±0.04

Metal Ion	ΔH, kcal /mole	Log K	ΔS, cal /mole °K	T, °C	M	Conditions	Ref.	Remarks
P279, cont.								
Li^+, cont.	0.24	(1)	--	5	C	μ=1.0[$(CH_3)_4NNO_3$]	73Va	g: $(CH_3)_4$-N^+,q: ±0.04
	0.24	(1)	--	15	C	μ=1.0[$(CH_3)_4NNO_3$]	73Va	g: $(CH_3)_4$-N^+,q: ±0.04
	0.26	2.390(1)	12.0	25	C	μ=1.0[$(CH_3)_4NNO_3$]	73Va	g: $(CH_3)_4$-N^+,q: ±0.04
	0.29	2.001	10.1	25	C	μ=0 R: $Li^+ + HL^{3-} = LiHL^{2-}$	73Va	q: ±0.08
	-0.18	--	--	25	C	μ=0.75[$(CH_3)_4NNO_3$] R: $Li^+ + HL^{3-} = LiHL^{2-}$	73Va	g: $(CH_3)_4$-N^+,q: ±0.06
	-0.21	1.026(1)	4.0	25	C	μ=1.0 [$(CH_3)_4NNO_3$] R: $Li^+ + HL^{3-} = LiHL^{2-}$	73Va	g: $(CH_3)_4$-N^+,q: ±0.06
	-0.30	(1)	--	25	C	μ=1.50[$(CH_3)_4NNO_3$] R: $Li^+ + HL^{3-} = LiHL^{2-}$	73Va	g: $(CH_3)_4$-N^+,q: ±0.06
	-0.30	(1)	--	25	C	μ=1.75[$(CH_3)_4NNO_3$] R: $Li^+ + HL^{3-} = LiHL^{2-}$	73Va	g: $(CH_3)_4$-N^+,q: ±0.06
Na^+	1.37	2.309(1)	15.2	25	C	μ=0	73Va	q: $(CH_3)_4$-N^+,q: ±0.13
	0.700	(1)	--	25	C	μ=0.50[$(CH_3)_4NNO_3$]	73Va	g: $(CH_3)_4$-N^+,q: ±0.06
	0.53	(1)	--	25	C	μ=0.75[$(CH_3)_4NNO_3$]	73Va	g: $(CH_3)_4$-N^+,q: ±0.06
	0.46	0.997(1)	6.1	25	C	μ=1.00[$(CH_3)_4NNO_3$]	73Va	g: $(CH_3)_4$-N^+,q: ±0.06
	0.53	--	--	25	C	R: $4Na^+ + H_2L^{2-} = Na_4L_4 + 2H^+$	67Wc	
K^+	1.74	2.111(1)	15.5	25	C	μ=0	73Va	g: $(CH_3)_4$-N^+,q: ±0.15

Metal Ion	ΔH, kcal /mole	Log K	ΔS, cal /mole °K	T, °C	M	Conditions	Ref.	Re-marks
P279, cont.								
K^+, cont.	0.91	(1)	--	15	C	μ=0.5M[$(CH_3)_4NNO_3$]	73Va	g: $(CH_3)_4N^+$,q: ±0.1
	1.00	(1)	--	25	C	μ=0.5M[$(CH_3)_4NNO_3$]	73Va	g: $(CH_3)_4N^+$,q: ±0.1
	1.11	(1)	--	35	C	μ=0.5M[$(CH_3)_4NNO_3$]	73Va	g: $(CH_3)_4N^+$,q: ±0.1
	0.61	(1)	--	15	C	μ=1.00M[$(CH_3)_4NNO_3$]	73Va	g: $(CH_3)_4N^+$,q: ±0.1
	0.70	0.799(1)	6.0	25	C	μ=1.00M[$(CH_3)_4NNO_3$]	73Va	g: $(CH_3)_4N^+$,q: ±0.1
	0.83	(1)	--	35	C	μ=1.00M[$(CH_3)_4NNO_3$]	73Va	g: $(CH_3)_4N^+$,q: ±0.1
	0.42	(1)	--	15	C	μ=1.5M[$(CH_3)_4NNO_3$]	73Va	g: $(CH_3)_4N^+$,q: ±0.1
	0.43	(1)	--	25	C	μ=1.5M[$(CH_3)_4NNO_3$]	73Va	g: $(CH_3)_4N^+$,q: ±0.1
	0.53	(1)	--	35	C	μ=1.5M[$(CH_3)_4NNO_3$]	73Va	g: $(CH_3)_4N^+$,q: ±0.1
Mg^{2+}	2.9	5.7(1)	35.8	25	T	--	57Va	
Ca^{2+}	4.6	5.6(1)	46	25	T	μ=0	60Ia	
Ni^{2+}	4.21	5.82(1)	40.8	25	C	C_L=0.025-0.18	56Ya	
	-2.21	1.37(2)	-1.2	25	C	C_L=0.025-0.18	56Ya	
	2.00	7.19(1-2)	39.6	25	C	C_L=0.025-0.18	56Ya	
Cu^{2+}	-0.67	9.00(1-2)	39.1	25	C	C_L=0.025-0.184	56Ya	
Zn^{2+}	2.64	6.50(1-2)	38.6	25	C	C_L=0.025-0.184	56Ya	
Pb^{2+}	-1.01	6.24(1-2)	25.2	25	C	C_L=0.025-0.184	56Ya	

P280 PYRROLIDINE C_4H_9N L

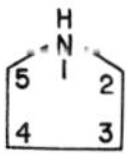

Metal Ion	ΔH, kcal /mole	Log K	ΔS, cal /mole °K	T, °C	M	Conditions	Ref.	Remarks
P280, cont.								
VOA_2	-10.47	(1)	--	25	C	μ→0 S: Nitrobenzene A=Acetylacetonate	65Ca	
NiA	-14.94	6.84(1-2)	-18.8	25	T	$C_M \simeq C_L \simeq 10^{-4}$ S: Benzene A=Diactyl-bisbenzoyl hydrazine	58Sa	
NiA_2	-8	3.196(24.6°,1)	-12	8-39.5	T	S: Benzene. A=O,O'-Diethyldithiophosphate	72Mc	b,q: ±1
	-4.9	3.472(24.6°,1)	-0.5	8-39.5	T	S: Benzene. A=O,O'-Dimethyldithiophosphate	72Mc	b
P281 2-PYRROLIDONE, N-methyl-		C_5H_9ON				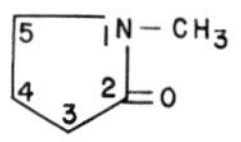		L
I_2	-5.6	1.28(1)	-13	25	T	S: n-Heptane	66Kc	
			Q					
Q1 QUINOLINE		C_9H_7N				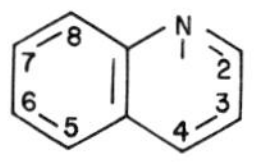		L
VOA_2	-4.83	0.26(1)	-15.0	25	C	μ→0 S: Nitrobenzene A=Acetylacetonate	65Ca	
CoA	0	1.201(25 ,1)	-6	20-50	T	S: Toluene A=α,β,γ,δ-Tetra-(p-methoxyphenyl) porphyrin	73Wa	b,q: ±1
CoA_2	-2.4	2.42(25°,1)	3.7	25	C	S: Acetone. A=Cl	73Lb	q: ±0.3
	-7.9	4.3(1-2)	-6.7	25	C	S: Acetone $C_M=2x10^{-3}M$ $C_L=4x10^{-2}M$. A=Cl	71Lg	
	-7.4	4.8(1-2)	-3.0	25	C	S: Acetone $C_M=2x10^{-3}M$ $C_L=4x10^{-2}M$. A=Br	71Lg	
	-5.5	4.1(1-2)	0.3	25	C	S: Acetonitrile $C_M=2x10^{-3}M$ $C_L=4x10^{-2}M$. A=Cl	71Lg	
	-3.8	3.7(1-2)	4.4	25	C	S: Acetonitrile $C_M=2x10^{-3}M$ $C_L=4x10^{-2}M$. A=Br	71Lg	
	-1.6	3.5(1-2)	10.7	25	C	S: n-Butyl alcohol $C_M=2x10^{-3}M$ $C_L=0.1M$. A=Cl	71Lh	
	-1.4	3.4(1-2)	11.1	25	C	S: n-Butyl alcohol		

Metal Ion	ΔH, kcal /mole	Log K	ΔS, cal /mole °K	T, °C	M	Conditions	Ref.	Re-marks
Q1, cont.								
CoA_2, cont.	-11.5	4.6(1-2)	-17.5	25	C	$C_M=2x10^{-3}M$ $C_L=0.1M$. A=Br S: Cyclohexanone	71Lg	
	-11.1	4.8(1-2)	-15.1	25	C	$C_M=2x10^{-3}M$ $C_L=4x10^{-2}M$. A=Cl S: Cyclohexanone $C_M=2x10^{-3}M$ $C_L=4x10^{-2}M$. A=Br	71Lg	
NiA	-15.80	2.08(25°,1-2)	-43.5	25	T	$C_M \approx C_L \approx 10^{-4}$ S: Benzene A=Diacetyl-bisbenzoyl hydrazine	58Sa	
Cu^{2+}	-17.73	16.66(25°,1)	17	5-35	T	μ=0.5($LiClO_4$) S: 50 w % C_2H_5OH R: $Cu^{2+} + A^{2-} + L = CuAL$. A=2-Carboxy-phenyl-azo-β-naphthol	58Ja	b
	-26.12	19.21(25°,1)	0.00	5-35	T	μ=0.5($LiClO_4$) S: 50 w % C_2H_5OH R: $Cu^{2+} + A^{2-} + L = CuAL$. A=2-Carboxy-phenyl-azo-β-naphthylamine	58Ja	b
	-39.85	22.52(25°,1)	-31	5-35	T	μ=0.5($LiClO_4$) S: 50 w % C_2H_5OH R: $Cu^{2+} + A^{2-} + L = CuAL$. A=o,o Dihydroxy-azobenzene	58Ja	b
CuA_2	-6.93	2.064(1)	-13.38	30	C	S: Benzene. A=1,1,1-Trifluoropentane-2,4-dione	74Gb	
Al_2A_6	-30.1	--	--	25	C	S: Benzene C_M=0.07M. A=Br R: $0.5(Al_2A_6) + 2L = AlA_3L$	68Rc	
$SiAB_3$	-5.34	1.80(30°,1)	-9.41	10-45	T	S: CCl_4 $X_L=1x10^{-2}$ $X_M=5x10^{-4}$ A=OH. B=CH_3	69Lb	b
	-5.53	1.73(30°,1)	-10.33	20-60	T	S: CCl_4 $X_L=1x10^{-2}$ $X_M=5x10^{-4}$ A=OH. B=C_2H_5	69Lb	b
	-5.91	2.27(30°,1)	-9.11	0-50	T	S: CCl_4 $X_L=1x10^{-2}$ $X_M=4x10^{-4}$ A=OH. B=Phenyl	69Lb	b
I_2	-5.6	2.477(1)	-7.4	25	C	S: Benzene	66Ad	

Metal Ion	ΔH, kcal /mole	Log K	ΔS, cal /mole °K	T, °C	M	Conditions	Ref.	Remarks
Q2 QUINOLINE, 1,2-dimethyl-			$C_{11}H_{12}N$			$C_9H_6N(CH_3)_2$		L^+
[Fe]	-13.3	9.0(25)	-4.4	5-30	T	pH=11.2 Protohaemin R: $(Haemin)_2$ + 2L = $(HaeminL)_2$ + $2H_2O$	69Md	b w:I
Q3 QUINOLINE, 8-hydroxy-			C_9H_7ON			$C_9H_6N(OH)$		L^-
Mn^{2+}	-3.5	7.30(1)	22	25	C	μ=0.1($NaClO_4$) S: 50 v % Dioxane	68Gh	
	-10.5	13.44(2)	27	25	C	μ=0.1($NaClO_4$) S: 50 v % Dioxane	68Gh	
Co^{2+}	-7.2	9.65(1)	20	25	C	μ=0.1($NaClO_4$) S: 50 v % Dioxane	68Gh	
	-4.9	(1)	--	25	C	ΔH value cal. by two different methods S: 50 v % Dioxane	59Fa	
	-6.4	(1)	--	25	C	ΔH value cal. by two different methods S: 50 v % Dioxane	59Fa	
	-15.4	18.05(1-2)	31	25	C	μ=0.1($NaClO_4$) S: 50 v % Dioxane	68Gh	
	-21.6	19.6(1-2)	21	25	C	ΔH value cal. by two different methods S: 50 v % Dioxane	59Fa	
	-20.4	19.6(1-2)	21	25	C	ΔH value cal. by two different methods S: 50 v % Dioxane	59Fa	
	-11.1	19.6(25°,1-2)	52	0-40	T	S: 50 v % Dioxane	59Fa	b
Ni^{2+}	-9.3	10.50(1)	17	25	C	μ=0.1($NaClO_4$) S: 50 v % Dioxane	68Gh	
	-6.6	(1)	--	25	C	ΔH value cal. by two different methods S: 50 v % Dioxane	59Fa	
	-8.5	(1)	--	25	C	ΔH value cal. by two different methods S: 50 v % Dioxane	59Fa	
	-19.3	20.27(1-2)	28	25	C	μ=0.1($NaClO_4$) S: 50 v % Dioxane	68Gh	
	-21.3	21.5(1-2)	29	--	-	ΔH value cal. by two different methods S: 50 v % Dioxane	59Fa	
	-20.8	21.5(1-2)	29	--	-	ΔH value cal. by two different methods S: 50 v % Dioxane	59Fa	
	-19.0	21.5(25°,1-2)	35	0-40	T	S: 50 v % Dioxane	59Fa	b

Metal Ion	ΔH, kcal /mole	Log K	ΔS, cal /mole °K	T, °C	M	Conditions	Ref.	Remarks
Q3, cont.								
Cu^{2+}	-10.2	13.29(1)	27	25	C	μ=0.1($NaClO_4$) S: 50 v % Dioxane	68Gh	
	-10.5	(1)	--	25	C	S: 50 v % Dioxane	59Fa	
	-19.45	17.67(25°,1)	16	5-35	T	μ=0.5($LiClO_4$) S: 50 w % C_2H_5OH R: Cu^{2+} + A^{2-} + L = CuAL. A=2-Carboxylphenyl-azo-β-naphthol	58Ja	b
	27.08	20.22(25°,1)	2	5-35	T	μ=0.5($LiClO_4$) S: 50 w % C_2H_5OH R: Cu^{2+} + A^{2-} + L = CuAL. A=2-Carboxyphenyl-azo-β-naphthylamine	58Ja	b
	-41.76	23.51(25°,1)	-33	5-35	T	μ=0.5($LiClO_4$) S: 50 w % C_2H_5OH R: Cu^{2+} + A^{2-} + L = CuAL. A=o,o Dihydroxy-azo-benzene	58Ja	b
	-19.6	25.90(1-2)	53	25	C	μ=0.1($NaClO_4$) S: 50 v % Dioxane	68Gh	
	-20.6	25.8(1-2)	48	25	C	ΔH value cal. by two different methods S: 50 v % Dioxane	59Fa	
	-20.8	25.8(1-2)	48	25	C	ΔH value cal. by two different methods S: 50 v % Dioxane	59Fa	
	-19.2	25.8(25°,1-2)	53	0-40	T	S: 50 v % Dioxane	59Fa	b
Zn^{2+}	-5.9	9.45(1)	23	25	C	μ=0.1($NaClO_4$) S: 50 v % Dioxane	68Gh	
	-9.6	18.15(1-2)	51	25	C	μ=0.1($NaClO_4$) S: 50 v % Dioxane	68Gh	
Cd^{2+}	-5.5	8.22(1)	19	25	C	μ=0.1($NaClO_4$) S: 50 v % Dioxane	68Gh	
	-11.6	15.22(1-2)	31	25	C	μ=0.1($NaClO_4$) S: 50 v % Dioxane	68Gh	
Pb^{2+}	-6.6	10.03(1)	24	25	C	μ=0.1($NaClO_4$) S: 50 v % Dioxane	68Gh	
	-15.1	17.34(1-2)	29	25	C	μ=0.1($NaClO_4$) S: 50 v % Dioxane	68Gh	

Q4 QUINOLINE, 8-hydroxy-2-methyl- $C_{10}H_9ON$ $C_9H_5N(OH)(CH_3)$ L^-

Metal Ion	ΔH, kcal /mole	Log K	ΔS, cal /mole °K	T, °C	M	Conditions	Ref.	Remarks
Mn^{2+}	-3.3	6.81(1)	20	25	C	μ=0.1($NaClO_4$) S: 50 v % Dioxane	68Gh	
	-6.6	14.00(25°,1-2)	42	0-40	T	μ=0.005	53Fa 54Ja	b
	-6.3	13.10(1-2)	39	25	C	μ=0.1($NaClO_4$) S: 50 v % Dioxane	68Gh	

Metal Ion	ΔH, kcal /mole	Log K	ΔS, cal /mole °K	T, °C	M	Conditions	Ref.	Re-marks
Q4, cont.								
Co^{2+}	-4.2	8.59(1)	25	25	C	μ=0.1($NaClO_4$) S: 50 v % Dioxane	68Gh	
	-4.6	(1)	--	25	C	S: 50 v % Dioxane	59Fa	
	-10.3	18.44(25°,1-2)	50	0-40	T	μ=0.005	53Fa 54Ja	b
	-13.8	17.38(1-2)	33	25	C	μ=0.1($NaClO_4$) S: 50 v % Dioxane	68Gh	
	-11.6	18.50(1-2)	45	25	C	S: 50 v % Dioxane	59Fa	
Ni^{2+}	-5.0	8.96(1)	24	25	C	μ=0.1($NaClO_4$) S: 50 v % Dioxane	68Gh	
	-6.5	(1)	--	25	C	S: 50 v % Dioxane	59Fa	
	-10.7	17.85(25°,1-2)	46	0-40	T	μ=0.005	53Fa 54Ja	b
	-12.1	16.94(1-2)	37	25	C	μ=0.1($NaClO_4$) S: 50 v % Dioxane	68Gh	
	-10.3	17.8(1-2)	47	25	C	ΔH value cal. two different ways. S: 50 v % Dioxane	59Fa	
	-10.6	17.8(1-2)	47	25	C	ΔH value cal. two different ways. S: 50 v % Dioxane	59Fa	
Cu^{2+}	-8.0	11.92(1)	28	25	C	μ=0.1($NaClO_4$) S: 50 v % Dioxane	68Gh	
	-8.7	(1)	--	25	T	ΔH value cal. two different ways. S: 50 v % Dioxane	59Fa	
	-6.5	(1)	--	25	T	ΔH value cal. two different ways. S: 50 v % Dioxane	59Fa	
	-17.3	23.85(25°,1-2)	50	0-40	T	μ=0.005	53Fa 54Ja	b
	-15.5	22.82(1-2)	53	25	C	μ=0.1($NaClO_4$) S: 50 v % Dioxane	68Gh	
	-15.5	23.8(1-2)	52	25	T	S: 50 v % Dioxane	59Fa	
Zn^{2+}	-3.8	9.06(1)	29	25	C	μ=0.1($NaClO_4$) S: 50 v % Dioxane	68Gh	
	-10.3	18.72(25°,1-2)	51	0-40	T	μ=0.005	53Fa 54Ja	b
	-11.1	17.90(1-2)	45	25	C	μ=0.1($NaClO_4$) S: 50 v % Dioxane	68Gh	
Pb^{2+}	-6.3	9.97(1)	25	25	C	μ=0.1($NaClO_4$) S: 50 v % Dioxane	68Gh	
	-13.7	17.18(1-2)	33	25	C	μ=0.1($NaClO_4$) S: 50 v % Dioxane	68Gh	
Q5 QUINOLINE, 8-hydroxy-4-methyl-		$C_{10}H_9ON$				$C_9H_5N(OH)(CH_3)$		L^-
Mn^{2+}	-4.1	7.74(1)	22	25	C	μ=0.1($NaClO_4$) S: 50 v % Dioxane	68Gh	
	-11.6	15.50(25°,1-2)	32	0-40	T	μ=0.005	53Fa 54Ja	b
	-6.4	14.81(1-2)	46	25	C	μ=0.1($NaClO_4$) S: 50 v % Dioxane	68Gh	

Metal Ion	ΔH, kcal /mole	Log K	ΔS, cal /mole °K	T, °C	M	Conditions	Ref.	Remarks
Q5, cont.								
Co^{2+}	-6.8	9.95(1)	23	25	C	μ=0.1($NaClO_4$) S: 50 v % Dioxane	68Gh	
	-7.1	(1)	--	25	C	S: 50 v % Dioxane	59Fa	
	-20.3	20.00(25°,1-2)	23	0-40	T	μ=0.005	53Fa 54Ja	b
	-17.8	18.92(1-2)	27	25	C	μ=0.1($NaClO_4$) S: 50 v % Dioxane	68Gh	
	-25.0	20.0(1-2)	8	25	C	S: 50 v % Dioxane	59Fa	
Ni^{2+}	-9.4	10.56(1)	17	25	C	μ=0.1($NaClO_4$) S: 50 v % Dioxane	68Gh	
	-6.8	(1)	--	25	C	S: 50 v % Dioxane	59Fa	
	-26.0	22.28(25°,1-2)	15	0-40	T	μ=0.005	53Fa 54Ja	b
	-18.3	20.47(1-2)	32	25	C	μ=0.1($NaClO_4$) S: 50 v % Dioxane	68Gh	
	-25.8	22.3(1-2)	15	25	C	S: 50 v % Dioxane	59Fa	
Cu^{2+}	-10.8	14.04(1)	28	25	C	μ=0.1($NaClO_4$) S: 50 v % Dioxane	68Gh	
	-9	(1)	--	25	C	S: 50 v % Dioxane	59Fa	
	-34.1	27.55(25°,1-2)	12	0-40	T	μ=0.005	53Fa 54Ja	b
	-21.6	26.96(1-2)	51	25	C	μ=0.1($NaClO_4$) S: 50 v % Dioxane	68Gh	
	-24.4	27.6(1-2)	44	25	C	S: 50 v % Dioxane	59Fa	
Zn^{2+}	-5.6	9.76(1)	26	25	C	μ=0.1($NaClO_4$) S: 50 v % Dioxane	68Gh	
	-19.6	20-25(25°,1-2)	26	0-40	T	μ=0.05	53Fa 54Ja	b
	-12.8	18.96(1-2)	44	25	C	μ=0.1($NaClO_4$) S: 50 v % Dioxane	68Gh	
Pb^{2+}	-6.8	10.46(1)	25	25	C	μ=0.1($NaClO_4$) S: 50 v % Dioxane	68Gh	
	-15.4	18.55(1-2)	33	25	C	μ=0.1($NaClO_4$) S: 50 v % Dioxane	68Gh	

Q6 QUINOLINE, 8-hydroxy-5-sulfo- $C_9H_8O_4N$

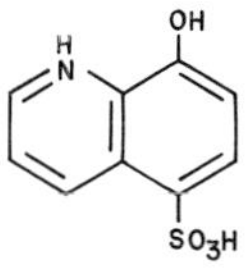

Metal Ion	ΔH, kcal /mole	Log K	ΔS, cal /mole °K	T, °C	M	Conditions	Ref.	Remarks
VO^{2+}	-1.7	11.79(25°,1)	45	25-35	T	μ=0.10(KNO_3)	66Mg	b,q: ±0.5

Q7 QUINOLINE, 8-mercapto- C_9H_7NS $C_9H_6N(SH)$ L^-

Metal Ion	ΔH, kcal /mole	Log K	ΔS, cal /mole °K	T, °C	M	Conditions	Ref.	Remarks
Mn^{2+}	-3.5	6.74(1)	19	25	C	μ=0.1($NaClO_4$) S: 50 v % Dioxane	68Gh	
Co^{2+}	-15.4	7.9(1)	-15	25	C	μ=0.1($NaClO_4$) S: 50 v % Dioxane	68Gh	
Ni^{2+}	-11.4	11.0(1)	12	25	C	μ=0.1($NaClO_4$) S: 50 v % Dioxane	68Gh	

Metal Ion	ΔH, kcal /mole	Log K	ΔS, cal /mole °K	T, °C	M	Conditions	Ref.	Re-marks
Q7, cont.								
Cu^{2+}	-14.7	12.7(1)	9	25	C	μ=0.1($NaClO_4$) S: 50 v % Dioxane	68Gh	
Zn^{2+}	-7.2	11.0(1)	26	25	C	μ=0.1($NaClO_4$) S: 50 v % Dioxane	68Gh	
Pb^{2+}	-10.1	11.85(1)	20	25	C	μ=0.1($NaClO_4$) S: 50 v % Dioxane	68Gh	
Q8 QUINOLINE, 2-methyl-8-thiol-			$C_{10}H_9NS$			$C_9H_5N(CH_3)(SH)$		L^-
Co^{2+}	-5.7	9.6(1)	25	25	C	μ=0.1($NaClO_4$) S: 50 v % Dioxane	68Gh	
Ni^{2+}	-8.1	9.2(1)	15	25	C	μ=0.1($NaClO_4$) S: 50 v % Dioxane	68Gh	
Cu^{2+}	-20.5	11.7(1)	-16	25	C	μ=0.1($NaClO_4$) S: 50 v % Dioxane	68Gh	
Zn^{2+}	-8.1	11.1(1)	24	25	C	μ=0.1($NaClO_4$) S: 50 v % Dioxane	68Gh	
Q9 5-QUINOLINE SULFONIC ACID, 7-bromo-8-hydroxy-			$C_9H_6O_4NBrS$			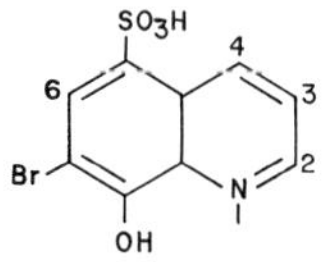		
Ga^{3+}	7.78	13.6(20°,1-3)	87.8	20-45	T	μ=0.2(NaAc & HAc) pH=4	72Ah	b,n,v: ±0.2
Q10 7-QUINOLINE SULFONIC ACID, 5-bromo-8-hydroxy-			$C_9H_6O_4NBrS$			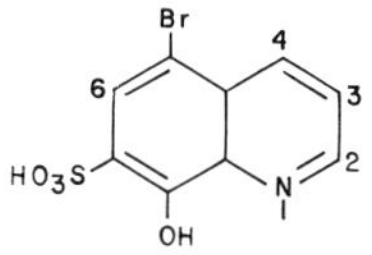		
UO_2^{2+}	3.5	8.623(25°,1-2)	51.1	--	T	--	72Ba	u
Q11 5-QUINOLINE SULPHONIC ACID, 7-chloro-8-hydroxy-			$C_9H_6O_4NClS$			$C_9H_4N(Cl)(OH)(SO_3H)$		
UO_2^{2+}	-8.70	9.43(30°,1-2)	12.85	20-45	T	μ=0.1($NaClO_4$)	71Ah	b,n,v: ±0.35
Cu^{2+}	-14.00	11.6(20°,1-2)	6.805	20-45	T	μ=0.1($NaClO_4$)	70Ad	b,n,v: ±0.30
Ga^{3+}	1.44	13.5(20°,1-3)	65.4	20-50	T	μ=0.2(NaAc & HAc) pH=4	72Ah	b,n,v: ±0.2
Q12 7-QUINOLINE SULFONIC ACID, 5-chloro-8-hydroxy-			$C_9H_6O_4NClS$			$C_9H_4N(Cl)(OH)(SO_3H)$		
UO_2^{2+}	-4.1	8.50(35°,1-2)	52.3	25-45	T	μ=0.02	70Bb	
	3.2	8.556(25°,1-2)	49.4	--	T	--	72Ba	u
Cu^{2+}	-3.66	9.2(25°,1-2)	30.0	25-45	T	μ=0.02 Value of ΔS was reported as 52.9.	69Bc	b

Metal Ion	ΔH, kcal /mole	Log K	ΔS, cal /mole °K	T, °C	M	Conditions	Ref.	Remarks
Q13	5-QUINOLINE SULFONIC ACID, 8-hydroxy-		$C_9H_7O_4NS$			$C_9H_5N(OH)(SO_3H)$		L^{2-}
Mg^{2+}	-4.0	5.01(1)	10	0	T	μ=0	57Ua	
	-2.2	4.79(1)	15	25	T	μ=0	57Ua	
	0.7	4.74(1)	23	50	T	μ=0	57Ua	
Y^{3+}	-5.64	22.39	83.2	20	C	μ=0.1M(KNO_3) R: Y^{3+} + $EDTA^{4-}$ + L^{2-} = $Y(EDTA)L^{3-}$	71Ga	x: 62Ma
	-19.1	17.16(25°,1-3)	15	25-45	T	μ=0.2M($NaClO_4$)	73Pb	b
YA	-5.05	4.31(1)	2.5	20	C	μ=0.1M(KNO_3) A=EDTA	71Ga	
La^{3+}	-2.4	5.63(1)	18	25	T	μ=0	58Fa	
	-4.2	4.5(2)	6	25	T	μ=0	58Fa	
	-4.5	3.7(3)	2	25	T	μ=0	58Fa	
	-5.94	18.87	66.0	20	C	μ=0.1M(KNO_3) R: La^{3+} + $EDTA^{4-}$ + L^{2-} = $La(EDTA)L^{3-}$	71Ga	x: 62Ma
LaA	-3.01	3.36(1)	5.1	20	C	μ=0.1M(KNO_3) A=EDTA	71Ga	
Ce^{3+}	-3.7	6.05(1)	15	25	T	μ=0	58Fa	
	-4.2	5.0(2)	9	25	T	μ=0	58Fa	
	-4.6	3.9(3)	3	25	T	μ=0	58Fa	
	-7.42	19.77	65.1	20	C	μ=0.10M(KNO_3) R: Ce^{3+} + $EDTA^{4-}$ + L^{2-} = $Ce(EDTA)L^{3-}$	71Ga	x: 62Ma
CeA	-4.48	3.80(1)	2.1	20	C	μ=0.1M(KNO_3) A=EDTA	71Ga	
Pr^{3+}	-3.4	6.17(1)	17	25	T	μ=0	58Fa	
	-2.8	5.20(2)	14	25	T	μ=0	58Fa	
	-4.8	4.3(3)	4	25	T	μ=0	58Fa	
	-8.03	20.41	66.0	20	C	μ=0.1M(KNO_3) R: Pr^{3+} + $EDTA^{4-}$ + L^{2-} = $Pr(EDTA)L^{3-}$	71Ga	x: 62Ma
PrA	-4.83	4.01(1)	1.9	20	C	μ=0.1M(KNO_3) A=EDTA	71Ga	
Nd^{3+}	-3.0	6.3(1)	19	25	T	μ=0	58Fa	
	-2.8	5.3(2)	15	25	T	μ=0	58Fa	
	-2.8	4.4(3)	11	25	T	μ=0	58Fa	
	-8.97	20.68	64.0	20	C	μ=0.1M(KNO_3) R: Nd^{3+} + $EDTA^{4-}$ + L^{2-} = $Nd(EDTA)L^{3-}$	71Ga	x: 62Ma
NdA	-5.35	4.07(1)	0.4	20	C	μ=0.1M(KNO_3) A=EDTA	71Ga	
Sm^{3+}	-2.9	6.58(1)	20	25	T	μ=0	58Fa	
	-3.6	5.70(2)	14	25	T	μ=0	58Fa	
	-3.8	4.76(3)	9	25	T	μ=0	58Fa	
	-9.05	21.50	67.4	20	C	μ=0.1M(KNO_3) R: Sm^{3+} + $EDTA^{4-}$ + L^{2-} = $Sm(EDTA)L^{3-}$	71Ga	x: 62Ma

Metal Ion	ΔH, kcal /mole	Log K	ΔS, cal /mole °K	T, °C	M	Conditions	Ref.	Re-marks
Q13, cont.								
SmA	-5.70	4.35(1)	0.4	20	C	μ=0.1M(KNO_3) A=EDTA	71Ga	
Eu^{3+}	-8.33	21.78	71.2	20	C	μ=0.1M(KNO_3) R: Eu^{3+} + $EDTA^{4-}$ + L^{2-} = $Eu(EDTA)L^{3-}$	71Ga	x: 62Ma
EuA	-5.77	4.41(1)	0.5	20	C	μ=0.1M(KNO_3) A=EDTA	71Ga	
Gd^{3+}	-3.5	6.64(1)	19	25	T	μ=0	58Fa	
	-4.3	5.73(2)	12	25	T	μ=0	58Fa	
	-4.7	4.9(3)	7	25	T	μ=0	58Fa	
	-7.83	21.75	72.7	20	C	μ=0.1M(KNO_3) R: Gd^{3+} + $EDTA^{4-}$ + L^{2-} = $Gd(EDTA)L^{3-}$	71Ga	x: 62Ma
GdA	-6.10	4.36(1)	-0.9	20	C	μ=0.1M(KNO_3) A=EDTA	71Ga	
Tb^{3+}	-6.62	22.43	80.0	20	C	μ=0.1M(KNO_3) R: Tb^{3+} + $EDTA^{4-}$ + L^{2-} = $Tb(EDTA)L^{3-}$	71Ga	x: 62Ma
TbA	-5.51	4.50(1)	1.8	20	C	μ=0.1M(KNO_3) A=EDTA	71Ga	
Dy^{3+}	-6.71	22.87	81.7	20	C	μ=0.1M(KNO_3) R: Dy^{3+} + $EDTA^{4-}$ + L^{2-} = $Dy(EDTA)L^{3-}$	71Ga	x: 62Ma
DyA	-5.50	4.56(1)	2.1	20	C	μ=0.1M(KNO_3) A=EDTA	71Ga	
Ho^{3+}	-7.03	23.40	83.1	20	C	μ=0.1M(KNO_3) R: Ho^{3+} + $EDTA^{4-}$ + L^{2-} = $Ho(EDTA)L^{3-}$	71Ga	x: 62Ma
HoA	-5.67	4.65(1)	2.0	20	C	μ=0.1M(KNO_3) A=EDTA	71Ga	
Er^{3+}	-5.4	7.16(1)	15	25	T	μ=0	58Fa	
	-5.7	6.18(2)	9	25	T	μ=0	58Fa	
	-5.6	5.22(3)	5	25	T	μ=0	58Fa	
	-7.68	23.52	81.4	20	C	μ=0.1M(KNO_3) R: Er^{3+} + $EDTA^{4-}$ + L^{2-} = $Er(EDTA)L^{3-}$	71Ga	x: 62Ma
ErA	-5.97	4.66(1)	1.0	20	C	μ=0.1M(KNO_3) A=EDTA	71Ga	
Tm^{3+}	-7.91	24.07	83.1	20	C	μ=0.1M(KNO_3) R: Tm^{3+} + $EDTA^{4-}$ + L^{2-} = $Tm(EDTA)L^{3-}$	71Ga	x: 62Ma
TmA	-6.04	4.76(1)	1.1	20	C	μ=0.1M(KNO_3) A=EDTA	71Ga	
Yb^{3+}	-8.64	24.35	81.9	20	C	μ=0.1M(KNO_3) R: Yb^{3+} + $EDTA^{4-}$ + L^{2-} = $Yb(EDTA)L^{3-}$	71Ga	x: 62Ma

Metal Ion	ΔH, kcal /mole	Log K	ΔS, cal /mole °K	T, °C	M	Conditions	Ref.	Remarks
Q13, cont.								
YbA	-6.33	4.82(1)	0.5	20	C	μ=0.1M(KNO_3) A=EDTA	71Ga	
Lu^{3+}	-8.81	24.65	82.7	20	C	μ=0.1M(KNO_3) R: Lu^{3+} + $EDTA^{4-}$ + L^{2-} = $Lu(EDTA)L^{3-}$	71Ga	x: 62Ma
LuA	-6.30	4.82(1)	0.5	20	C	μ=0.1M(KNO_3) A=EDTA	71Ga	
ThA	-4.42	6.71(1)	15.8	25	C	μ=0.10(KNO_3) A=1,2-Diaminocyclo-hexane-N,N,N',N'-tetraacetic acid	70Kb	
	-1.62	6.99(1)	26.5	25	C	μ=0.10(KNO_3) A=EDTA	70Kb	
	4.78	--	--	25	C	μ=0.10(KNO_3) A=1,2-Diaminocyclo-hexane-N,N,N',N'-tetraacetic acid R: ThA + H_2L = $ThAL^{2-}$ + $2H^+$	70Kb	
	7.60	--	--	25	C	μ=0.10(KNO_3) A=EDTA. R: ThA + H_2L =$ThAL^{2-}$ + $2H^+$	70Kb	
Mn^{2+}	-3.2	7.05(1)	22	25	C	μ=0.1($NaClO_4$) S: 50 v % Dioxane	68Gh	
	-6.7	13.18(2)	46	25	C	μ=0.1($NaClO_4$) S: 50 v % Dioxane	68Gh	
Co^{2+}	-6.3	7.38(1)	22	25	C	μ=0.1($NaClO_4$) S: 50 v % Dioxane	68Gh	
	-14.5	17.55(1-2)	32	25	C	μ=0.1($NaClO_4$) S: 50 v % Dioxane	68Gh	
Ni^{2+}	-6.4	9.11(1)	20	25	C	μ=0.1($NaClO_4$)	68Gh	
	-7.3	10.22(1)	22	25	C	μ=0.1($NaClO_4$) S: 50 v % Dioxane	68Gh	
	-14.8	17.34(1-2)	30	25	C	μ=0.1($NaClO_4$)	68Gh	
	-15.9	19.25(1-2)	35	25	C	μ=0.1($NaClO_4$) S: 50 v % Dioxane	68Gh	
	-25.6	23.23(1-3)	20	25	C	μ=0.1($NaClO_4$)	68Gh	
	-25.4	25.55(1-3)	32	25	C	μ=0.1($NaClO_4$) S: 50 v % Dioxane	68Ch	
Cu^{2+}	-8.1	12.45(1)	30	25	C	μ=0.1($NaClO_4$)	68Gh	
	-9.0	13.16(1)	30	25	C	μ=0.1($NaClO_4$) S: 50 v % Dioxane	68Gh	
	-17.6	22.99(1-2)	46	25	C	μ=0.1($NaClO_4$)	68Gh	
	-18.6	25.55(1-2)	54	25	C	μ=0.1($NaClO_4$) S: 50 v % Dioxane	68Gh	
Zn^{2+}	-5.1	7.95(1)	19	25	C	μ=0.1($NaClO_4$)	68Gh	
	-5.2	9.23(1)	25	25	C	μ=0.1($NaClO_4$) S: 50 v % Dioxane	68Gh	
	-9.6	14.97(1-2)	36	25	C	μ=0.1($NaClO_4$)	68Gh	

Metal Ion	ΔH, kcal /mole	Log K	ΔS, cal /mole °K	T, °C	M	Conditions	Ref.	Remarks
Q13, cont.								
Zn^{2+}, cont.	-10.3	17.56(1-2)	46	25	C	μ=0.1($NaClO_4$) S: 50 v % Dioxane	68Gh	
Q14 7-QUINOLINE SULFONIC ACID, 8-hydroxy			$C_9H_7O_4NS$			$C_9H_5N(OH)(SO_3H)$		
UO_2^{2+}	-5.3	9.04(35°,1-2)	59	25-45	T	μ=0.02. pH=5.8	70Bc	b
	4.6	8.929(25°,1-2)	56.5	--	T	--	72Ba	u
Pd^{2+}	5.50	11.20(25°,1-2)	70.08	25-45	T	μ=0	70Ba	b
Cu^{2+}	-5.3	9.8(25°,1-2)	27.5	25-45	T	μ=0.02	69Bb	b
Tl^{3+}	-8.445	6.550(32°,1-3)	-20.388	23.2-38.8	T	$C_L<1x10^{-4}M$ $C_M<2x10^{-4}M$	71Ca	b
Q15 5-QUINOLINE SULFONIC ACID, 8-hydroxy-7-iodo-			$C_9H_6O_4NIS$			$C_9H_4N(OH)(I)(SO_3H)$		L^{2-}
Mg^{2+}	-2.1	4.21(1)	12	0	T	μ=0	57Ua	
	-0.3	4.12(1)	18	25	T	μ=0	57Ua	
	2.3	4.17(1)	26	50	T	μ=0	57Ua	
Y^{3+}	-5.7	15.15(25°,1-3)	50	25-45	T	μ=0.2M($NaClO_4$)	73Sd	b
Fe^{3+}	2.07	3.50(30°,1)	22.9	20-40	T	--	72Pj	b
	2.35	4.88(30°,2)	30.2	20-40	T	--	72Pj	b
	1.51	4.61(30°,3)	26.4	20-40	T	--	72Pj	b
Tl^{3+}	-9.435	7.149(32°,1-3)	-15.143	23.2-38.8	T	$C_L<1x10^{-4}M$ $C_M<2x10^{-4}M$	71Ca	b
Q16 7-QUINOLINE SULFONIC ACID, 8-hydroxy-5-iodo-			$C_9H_6O_4NIS$			$C_9H_4N(OH)(I)(SO_3H)$		L^{2-}
UO_2^{2+}	4.00	8.80(1-2)	53.98	25	T	μ=0.05	72Ba	
	3.98	8.85(1-2)	53.93	30	T	μ=0.05	72Ba	
	4.02	8.89(1-2)	53.94	35	T	μ=0.05	72Ba	
	4.01	8.94(1-2)	53.96	40	T	μ=0.05	72Ba	
	3.99	8.98(1-2)	53.96	45	T	μ=0.05	72Ba	
Cu^{2+}	4.57	9.71(35°,1-2)	59.1	25-45	T	μ=0.02. pH=3	68Ba	
Q17 5-QUINOLINE SULFONIC ACID, 8-hydroxy-7-nitro-			$C_9H_6O_6N_2S$			$C_9H_4N(OH)(NO_2)(SO_3H)$		L^{2-}
Mg^{2+}	-1.5	3.32(1)	9	0	T	μ=0	57Ua	
	0.7	3.28(1)	17	25	T	μ=0	57Ua	
	3.9	3.41(1)	26	50	T	μ=0	57Ua	
Q18 5-QUINOLINE SULFONIC ACID, 8-hydroxy-7-(4-nitrophenyl)-			$C_{15}H_{10}O_6N_2S$			$C_9H_4N(OH)(C_6H_4NO_2)(SO_3H)$		L^{2-}
Mg^{2+}	-1.0	4.02(1)	14	0	T	μ=0	57Ua	
	-0.7	3.96(1)	16	25	T	μ=0	57Ua	
	-0.4	3.93(1)	18	50	T	μ=0	57Ua	
Ca^{2+}	-0.7	3.27(1)	12	0	T	μ=0	57Ua	
	-0.1	3.23(1)	14	25	T	μ=0	57Ua	
Ba^{2+}	0.9	2.51(1)	15	25	T	μ=0	57Ua	

Metal Ion	ΔH, kcal /mole	Log K	ΔS, cal /mole °K	T, °C	M	Conditions	Ref.	Re-marks
Q18, cont.								
Ba^{2+}, cont.	0.7	2.56(1)	18	50	T	μ=0	57Ua	
Mn^{2+}	-2.1	5.31(1)	17	25	T	μ=0	57Ua	
	0	5.25(1)	24	50	T	μ=0	57Ua	
Co^{2+}	-6.2	6.49(1)	9	0	T	μ=0	57Ua	
	-3.8	6.14(1)	16	25	T	μ=0	57Ua	
	-0.2	6.02(1)	25	50	T	μ=0	57Ua	
Ni^{2+}	-4.2	6.91(1)	18	25	T	μ=0	57Ua	
	-1.3	6.75(1)	26	50	T	μ=0	57Ua	
Cu^{2+}	-5.9	9.17(1)	20	0	T	μ=0	57Ua	
	-5.8	8.77(1)	21	25	T	μ=0	57Ua	
	-5.5	8.44(1)	22	50	T	μ=0	57Ua	
Zn^{2+}	-5.4	6.43(1)	10	0	T	μ=0	57Ua	
	-3.8	6.11(1)	15	25	T	μ=0	57Ua	
	-1.4	5.96(1)	23	50	T	μ=0	57Ua	
Cd^{2+}	-4.1	5.94(1)	12	0	T	μ=0	57Ua	
	-3.2	5.69(1)	15	25	T	μ=0	57Ua	
Q19	5-QUINOLINE SULFONIC ACID, 8-hydroxy-7-phenylazo		$C_{15}H_{11}O_4N_3S$			$C_9H_4N(OH)(C_6H_5N{\equiv}N)(SO_3H)$		L^{2-}
Mg^{2+}	-2.2	4.04(1)	10	0	T	μ=0	57Ua	
	-0.8	3.94(1)	15	25	T	μ=0	57Ua	
	1.3	3.95(1)	22	50	T	μ=0	57Ua	
Ba^{2+}	0.70	2.16(1)	12	0	T	μ=0	57Ua	
	1.2	2.22(1)	14	25	T	μ=0	57Ua	
	1.9	2.29(1)	17	50	T	μ=0	57Ua	
Mn^{2+}	-3.9	5.57(1)	11	0	T	μ=0	57Ua	
	-2.2	5.36(1)	17	25	T	μ=0	57Ua	
Ni^{2+}	-6.0	7.57(1)	13	0	T	μ=0	57Ua	
	-4.9	7.21(1)	17	25	T	μ=0	57Ua	
	-3.1	6.97(1)	22	50	T	μ=0	57Ua	
Cu^{2+}	-7.2	9.50(1)	17	0	T	μ=0	57Ua	
	-6.6	9.03(1)	19	25	T	μ=0	57Ua	
	-5.3	8.96(1)	23	50	T	μ=0	57Ua	
Zn^{2+}	-5.7	6.86(1)	10	0	T	μ=0	57Ua	
	-4.0	6.53(1)	17	25	T	μ=0	57Ua	
	-1.5	6.36(1)	24	50	T	μ=0	57Ua	

R

R1 RIBOFLAVIN $C_{17}H_{20}O_6N_4$ L

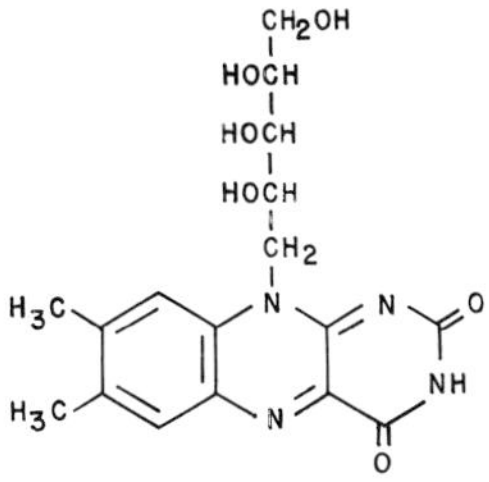

Metal Ion	ΔH, kcal /mole	Log K	ΔS, cal /mole °K	T, °C	M	Conditions	Ref.	Re-marks
R1, cont.								
Co^{2+}	5	4.3(25°,2-1)	52	10-40	T	C_M=0.01. C_L=0.05	59Ha	b,p
Ni^{2+}	3	4.2(25°,2-1)	46	10-40	T	C_M=0.01. C_L=0.05	59Ha	b,p
Cu^{2+}	-3	5.9(25°,2-1)	44	10-40	T	C_M=0.01. C_L=0.05	59Ha	b,p
Zn^{2+}	1	4.9(25°,2-1)	47	10-40	T	C_M=0.01. C_L=0.05	59Ha	b,p
R2 RIBONUCLEASE A			$C_{587}H_{909}N_{171}O_{197}S_{12}$					
[P]	-15.3	4.556(1)	-30.6	25	C	[P]=3'-Cytidine monophosphate μ=0.05(NaAc). pH=5.5	71Bk	
	-15.8	4.580(1)	-29.5	25	C	[P]=3'-Cytidine monophosphate μ=0.05(KCl). pH=5.5	71Bk	
	-13.2	4.591(1)	-23.5	30	C	[P]=3'-Cytidine monophosphate μ=0.15(NaAc). pH=5.5	71Bk	
	-9.2	3.724(1)	-14.0	25	C	[P]=3'-Cytidine monophosphate μ=0.50(NaAc). pH=5.5	71Bk	
	-9.8	3.800(1)	-15.4	25	C	[P]=3'-Cytidine monophosphate μ=0.50(KCl). pH=5.5	71Bk	
	-7.5	3.176(1)	-10.7	25	C	[P]=3'-Cytidine monophosphate μ=1.20(NaAc). pH=5.5	71Bk	
	-6.1	3.398(1)	-5.0	30	C	[P]=3'-Cytidine monophosphate μ=1.20(NaAc). pH=5.5	71Bk	
	-7.0	3.505(1)	-11.3	25	C	[P]=3'-Cytidine monophosphate μ=1.20(KCl). pH=5.5	71Bk	
	-6.0	3.322(1)	-4.9	30	C	[P]=3'-Cytidine monophosphate μ=3.00(NaAc). pH=5.5	71Bk	
S1 SALICYLALDIMINE, N-n-butyl-			$C_{11}H_{15}ON$			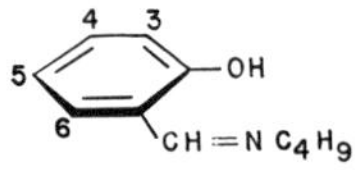		
Co^{2+}	-12.0	10.00(25°,1-2)	5	--	T	μ=0.1 S: 80 v % Methanol	72Kc	b,u
Cu^{2+}	-13.9	12.51(25°,1-2)	10	--	T	μ=0.1 S: 80 v % Methanol	72Kc	b,u
S2 SALICYLALDIMINE, N-tert-butyl-			$C_{11}H_{15}ON$			$C_7H_6ON(C_4H_9)$		
NiA_2	1.3	-0.208(25°)	3.5	17-107	T	S: C_2Cl_4. R: NiL_2 + NiA_2 = 2NiAL A=N-diethylmethyl salicylaldimine	73La	b,q: ±0.02
	-0.67	1.114(25°)	2.9	17-107	T	S: C_2Cl_4. R: NiL_2 + NiA_2 = 2NiAL	73La	b,q: ±0.07

Metal Ion	ΔH, kcal /mole	Log K	ΔS, cal /mole °K	T, °C	M	Conditions	Ref.	Remarks
S2, cont.								
NiA_2, cont.						A=N-isopropylacetyl-acetoneimine		
	6.0	1.114(25°)	25	--	T	S: C_2Cl_4. R: NiL_2 + NiA_2 = 2NiAL A=N-isopropylacetyl-acetoneimine	71Lf	b,u
	3.6	-0.796(25°)	9.1	17-107	T	S: C_2Cl_4. R: NiL_2 + NiA_2 = 2NiAL A=N-isopropyl-3-methoxysalicylaldimine	73La	b,q: ±0.05
	1.2	0.079(25°)	4.5	17-107	T	S: C_2Cl_4. R: NiL_2 + NiA_2 = 2NiAL A=N-isopropyl 4-methyl-salicylaldimine	73La	b,q: ±0.05
	-0.53	0.914(25°)	2.4	17-107	T	S: $CDCl_3$. R: NiL_2 + NiA_2 = 2NiAL where A=N-Isopropyl-benzoyl-acetoneimine	73La	b,q: ±0.05
	1.3	0.158(25°)	5.0	17-107	T	S: $CDCl_3$. R: NiL_2 + NiA_2 = 2NiAL where A=N-isopropyl-salicylaldimine	73La	b,q: ±0.05
	6.7	-0.886	18.4	17-107	T	S: $CDCl_3$. R: NiL_2 + NiA_2 = 2NiAL where A=N-phenyl-salicylaldimine	73La	b,q: ±0.1
S3 SALICYLALDIMINE, N-n-butyl-3-methoxy-			$C_{12}H_{17}O_2N$			$C_7H_5ON(C_4H_9)(OCH_3)$		
Co^{2+}	-12.0	10.45(25°,1-2)	7	--	T	μ=0.1 S: 80 v % Methanol	72Kc	b,u
Cu^{2+}	-14.0	13.47(25°,1-2)	14	--	T	μ=0.1 S: 80 v % Methanol	72Kc	b,u
S4 SALICYLALDIMINE, N-t-butyl-5-methyl-			$C_{12}H_{17}ON$			$C_7H_5ON(C_4H_9)(CH_3)$		
Ni^{2+}	0.2	0.556(25°)	3.1	17-107	T	S: C_2Cl_4. R: NiL_2 + NiA_2 = 2NiAL A=N-isopropylacetyl-acetoneimine	73La	b,q: ±0.7
S5 SALICYLALDIMINE, 3-chloro-N-isopropyl-			$C_{10}H_{12}ONCl$			$C_7H_5ON(Cl)[CH(CH_3)_2]$		
NiL_2	3.0	-0.194(120°)	7	80-170	T	S: Bibenzyl	64Sb 64Sa	b,t
S6 SALICYLALDIMINE, 5-chloro-N-n-propyl-			$C_{10}H_{12}ONCl$			$C_7H_5ON(Cl)[CH(CH_3)_2]$		
NiL_2	5.2	-1.501(120°)	6	80-170	T	S: Bibenzyl	64Sa 64Sb	b,t

Metal Ion	ΔH, kcal /mole	Log K	ΔS, cal /mole °K	T, °C	M	Conditions	Ref.	Remarks
S7 SALICYLALDIMINE, 5-chloro-N-isopropyl-			$C_{10}H_{12}ONCl$			$C_7H_5ON(Cl)[CH(CH_3)_2]$		
NiL_2	2.0	-0.089(120°)	4	80-170	T	S: Bibenzyl	64Sa 64Sb	b,t
S8 SALICYLALDIMINE, N-isobutyl-			$C_{11}H_{15}ON$			$C_7H_6ON[CH_2CH(CH_3)_2]$		
Co^{2+}	-11.5	10.08(25°,<u>1</u>-<u>2</u>)	7	--	T	μ=0.1 S: 80 v % Methanol	72Kc	b,u
Cu^{2+}	-13.2	13.13(25°,<u>1</u>-<u>2</u>)	17	--	T	μ=0.1 S: 80 v % Methanol	72Kc	b,u
S9 SALICYLALDIMINE, N-isobutyl-3-methoxy-			$C_{12}H_{17}O_2N$			$C_7H_5ON[CH_2CH(CH_3)_2](OCH_3)$		
Co^{2+}	-12.5	11.34(25°,<u>1</u>-<u>2</u>)	9	--	T	μ=0.1 S: 80 v % Methanol	72Kc	b,u
Cu^{2+}	-12.8	14.45(25°,<u>1</u>-<u>2</u>)	22	--	T	μ=0.1 S: 80 v % Methanol	72Kc	b,u
S10 SALICYLALDIMINE, N-isopentane			$C_{12}H_{17}ON$			$C_7H_6ON[CH(C_2H_5)_2]$		
NiL_2	2.6	-0.086(25°)	8.4	17-107	T	S: C_2Cl_4. R: NiL_2 + NiA_2 = 2NiAL A=N-Isopropyl-acetylacetoneimine	73La	b,q: ±0.38
	-0.2	0.556(25°)	1.7	17-107	T	S: $CDCl_3$. R: NiL_2 + NiA_2 = 2NiAL A=N-Isopropyl-benzoylacetoneimine	73La	b,q: ±0.72
S11 SALICYLALDIMINE, N-isopropyl-			$C_{10}H_{12}ON$			$C_7H_6ON[CH(CH_3)_2]$		
NiA_2	3.1	-0.097	10	17-107	T	S: C_2Cl_4. R: NiL_2 + NiA_2 = 2NiAL A=N-isopropyl-acetylacetoneimine	73La	b,q: ±0.72
	4.5	-0.276(25°)		--	T	S: C_2Cl_4. R: NiL_2 + NiA_2 = 2NiAL A=N-isopropyl-acetylacetoneimine	71Lf	
	4.3	0.121(25°)	15	17-107	T	S: C_2Cl_4. R: NiL_2 + NiA_2 = 2NiAL A=N-n-propyl-salicylaldimine	73La	b,q: ±2.2
NiL_2	3.2	0.289(120°)	10	80-170	T	S: Bibenzyl	64Sa 64Sb	b,t
S12 SALICYLALDIMINE, 5-methyl-N-isopropyl-			$C_{11}H_{15}ON$			$C_7H_5ON(CH_3)[CH(CH_3)_2]$		
NiL_2	2.5	-0.111(120°)	6	80-170	T	S: Bibenzyl R: NiL_2(planar) = NiL_2(tetrahedral)	64Sa 64Sb	b,t

Metal Ion	ΔH, kcal /mole	Log K	ΔS, cal /mole °K	T, °C	M	Conditions	Ref.	Remarks
S13	SALICYLALDIMINE, 5-methyl-N-n-propyl-		$C_{11}H_{15}ON$			$C_7H_5ON(CH_3)[O_3H_7]$		
NiL_2	4.6	-1.557(120°)	4	80-170	T	S: Bibenzyl	64Sa 64Sb	b,t
S14	SALICYLALDIMINE, N-phenyl-		$C_{13}H_{11}ON$			$C_7H_6ON(C_6H_5)$		
NiL_2	9.1	-1.060(25°)	25.6	17-107	T	S: $CDCl_3$. R: NiL_2 + NiA_2 = 2NiAL A=N-isopropyl-acetylacetoneimine	73La	b
S15	SALICYLALDIMINE, N-n-propyl-		$C_{10}H_{13}ON$			$C_7H_6ON(C_3H_7)$		
NiL_2	4.6	-1.612(120°)	4	80-170	T	S: Bibenzyl	64Sa 64Sb	b,t
S16	SARCOSINE		$C_3H_7O_2N$			$CH_3NHCH_2CO_2H$		L^-
CoA	-4.2	3.13(25°,1)	0.5	15-70	T	μ=0.1M(KNO_3) A=Nitrilotriacetate	71Id	b
NiA	-5.0	4.23(25°,1)	2.8	15-70	T	μ=0.1M(KNO_3) A=Nitrilotriacetate	71Id	b
Cu^{2+}	-4.6	8.16(20°,1)	22	25	C	μ=0	64Ia	b
	-5.4	6.89(20°,2)	13	25	C	μ=0	64Ia	b
CuA	-12.0	5.15(25°,1)	-15.9	15-70	T	μ=0.1M(KNO_3) A=Nitrilotriacetate	71Id	b
ZnA	-5.0	3.22(25°,1)	-1.9	15-70	T	μ=0.1M(KNO_3) A=Nitrilotriacetate	71Id	b
CdA	-2.8	2.64(25°,1)	2.9	15-70	T	μ=0.1M(KNO_3) A=Nitrilotriacetate	71Id	b
S17	SARCOSINE, N-glycyl-		$C_5H_{10}O_3N_2$			$H_2NCH_2CON(CH_3)CH_2CO_2H$		
Cu^{2+}	-6.0	6.41(1)	9.2	25	C	μ=0.1M(KNO_3)	72B1	
	-5.4	5.01(2)	4.8	25	C	μ=0.1M(KNO_3)	72B1	
S18	SELENATE ION		O_4Se			SeO_4^{2-}		L^{2-}
Mn^{2+}	3.51	2.430(25°,1)	22.9	0-45	T	μ=0	70Gf	b,q: ±0.3
Co^{2+}	2.91	2.696(25°,1)	22.2	0-45	T	μ=0	70Gf	b
Ni^{2+}	3.51	2.669(25°,1)	22.9	0-45	T	μ=0	70Gf	b,q: ±0.4
S19	SELENITE ION		HO_3Se			$HSeO_3^-$		L^-
CrA_3^{3+}	5.59	-0.021	18.6	25	C	μ=3.0($NaClO_4$) A=Ethylenediamine R: CrA_3^{3+} + L^{2-} = CrA_3L^+	73Ua	
	-3.01	-0.126	10.6	25	C	μ=3.0($NaClO_4$) A=Ethylenediamine R: CrA_3L^+ + L^{2-} = $CrA_3L_2^-$	73Ua	

Metal Ion	ΔH, kcal /mole	Log K	ΔS, cal /mole °K	T, °C	M	Conditions	Ref.	Remarks
S19, cont.								
Fe^{3+}	4.04	3.45(25°)	29.4	20-40	T	0.5-1.0M($HClO_4$) K: Fe^{3+} + HL = FeL^{2+} + H^+	65Ha	b
CoA_3	1.4	0.556(1)	7	25	C	μ=3.0M($NaClO_4$ + Na_2L) A=Ethylenediamine	73Mc	q: ±0.2
	0.4	0.146(2)	2	25	C	μ=3.0M($NaClO_4$ + Na_2L) A=Ethylenediamine	73Mc	q: ±0.2
	0.2	-0.15(3)	0	25	C	μ=3.0M($NaClO_4$ + Na_2L) A=Ethylenediamine	73Mc	q: ±0.2
	-2.0	-0.52(4)	-7	25	C	μ=3.0M($NaClO_4$ + Na_2L) A=Ethylenediamine	73Mc	q: ±0.6
S20 SELENOCYANATE ION			CNSe			$SeCN^-$		L^-
Fe^{3+}	-5.6	2.1(1)	-9	25	-	Hemoglobin	60Gb	
Fe^{3+}	-11.7	3.2(1)	-24.7	25	-	Chironomus Hemoglobin	60Gb	
PdA_2	-4.5	3.125(1)	-0.9	25	T	μ=0. A=o-Phenylenebisdimethylarsine	67Ec 67Ed	
	-4.6	2.873(40°,1)	-1.7	25-55	T	μ=0.003($NaNO_3$) A=o-Phenylenebisdimethylarsine	67Ed	b
PtA_2	-4.3	2.701(35°,1)	-1.5	25-55	T	μ=3x10^{-3}($NaNO_3$) A=o-Phenylenebisdimethylarsine	67Ec	b,q: ±0.4
	-4.2	2.934(1)	-0.7	25	T	μ=0. A=o-Phenylenebisdimethylarsine	67Ec	q: ±0.4
Hg^{2+}	-46.5	29.6(1-4)	-20.8	25	T	μ=0	56Ta	
Pb^{2+}	-35.6	(1-6)	--	20-30	T	--	59Gd	b
S21 SEMICARBAZIDE			CH_5ON_3			$H_2NCONHNH_2$		L
Cu^{2+}	-5.80	4.000(1)	-1	30	C	μ=0.1(HNO_3 + KNO_3)	71Ae	
	-6.20	2.940(2)	-7	30	C	μ=0.1(HNO_3 + KNO_3)	71Ae	
Ag^+	-2.2	1.979(1)	2	30	C	μ=0.1M(KNO_3)	73Ad	q: ±0.5
	-10.3	0.733(2)	-31	30	C	μ=0.1M(KNO_3)	73Ad	q: ±0.1
Hg^{2+}	≃-16	11.6(1-2)	0.3	30	C	μ=0.1	69Gc	
	≃-18	15.2(1-3)	10.2	30	C	μ=0.1	69Gc	
Cd^{2+}	-2.8	1.246(1)	-4	30	C	μ=0.1M(KNO_3)	73Ad	q: ±0.5
	-2.2	1.613(2)	0	30	C	μ=0.1M(KNO_3)	73Ad	q: ±0.4
Pb^{2+}	-3.5	2.126(1)	-2	30	C	μ=0.1M(KNO_3)	73Ad	

Metal Ion	ΔH, kcal /mole	Log K	ΔS, cal /mole °K	T, °C	M	Conditions	Ref.	Remarks
S21, cont.								
Pb^{2+}, cont.	-3.4	0.733(2)	-8	30	C	μ=0.1M(KNO_3)	73Ad	
S22 SEMICARBAZIDE, thiono			CH_5N_3S			$H_2NCSNHNH_2$		L
Cu^{2+}	-9.80	6.114(1)	-4.3	30	C	μ=0.1(HNO_3 + KNO_3)	71Ae	
	-8.00	5.477(2)	-1.4	30	C	μ=0.1(HNO_3 + KNO_3)	71Ae	
Ag^+	-22.0	10.70(1-2)	-24	30	C	μ=0.03M(KNO_3)	73Ad	
	-27.0	12.68(1-3)	-32	30	C	μ=0.03M(KNO_3)	73Ad	
	-18.0	12.9(25°,1-3)	-1.7	20-50	T	μ=0.80($NaNO_3$)	60Tb	b
Hg^{2+}	-35.3	22.5(1-2)	-13.7	30	C	μ=0.1	69Gc	
	-39.9	24.8(1-3)	-18.1	30	C	μ=0.1	69Gc	
	-41	26.1(25°,1-4)	-18.2	20-50	T	μ=0.80($NaNO_3$)	60Tb	b
Cd^{2+}	-4.3	2.27(1)	-4	30	C	μ=0.03M(KNO_3)	73Ad	q: ±0.5
	-4.6	2.20(2)	-5	30	C	μ=0.03M(KNO_3)	73Ad	q: ±0.5
Pd^{2+}	-4.2	1.90(1)	-5	30	C	μ=0.03M(KNO_3)	73Ad	
S23 SEMICARBAZIDE, thiono, 1,1-diacetic acid-			$C_5H_9O_4N_3S$			$H_2NCSNHN(CH_2CO_2H)_2$		L^{2-}
Mg^{2+}	-1.4	0.7(1)	-1	30	C	μ=0.1(KNO_3)	67Gc	
Sr^{2+}	-0.1	2.2(1)	10	30	C	μ=0.1(KNO_3)	67Gc	
Mn^{2+}	7.2	2.2(1)	33	30	C	μ=0.1(KNO_3)	67Gc	
Co^{2+}	2.6	5.5(1)	33	30	C	μ=0.1(KNO_3)	67Gc	
Ni^{2+}	-0.8	5.9(1)	24	30	C	μ=0.1(KNO_3)	67Gc	
Cu^{2+}	-2.6	8.1(1)	28	30	C	μ=0.1(KNO_3)	67Gc	
Zn^{2+}	3.1	5.8(1)	37	30	C	μ=0.1(KNO_3)	67Gc	
S24 SEMICARBAZIDE, 1,1-diacetic acid-			$C_5H_9O_5N_3$			$H_2NCONHN(CH_2CO_2H)_2$		L^{2-}
Mg^{2+}	9.8	1.4(1)	39	30	C	μ=0.1(KNO_3)	67Gc	
Sr^{2+}	-0.7	2.3(1)	9	30	C	μ=0.1(KNO_3)	67Gc	
Mn^{2+}	3.2	2.6(1)	22	30	C	μ=0.1(KNO_3)	67Gc	
Co^{2+}	0.7	5.9(1)	29	30	C	μ=0.1(KNO_3)	67Gc	
Ni^{2+}	-1.1	6.6(1)	26	30	C	μ=0.1(KNO_3)	67Gc	
Cu^{2+}	-0.5	8.4(1)	37	30	C	μ=0.1(KNO_3)	67Gc	
Zn^{2+}	-0.4	6.6(1)	29	30	C	μ=0.1(KNO_3)	67Gc	
S25 SEMICARBAZIDE, seleno			CH_5N_3Se			$NH_2CSeNHNH_2$		L
Cu^{2+}	-11.0	5.544(1)	-11	30	C	μ=0.1(HNO_3 + KNO_3)	71Ae	
	-13.7	5.279(2)	-21	30	C	μ=0.1(HNO_3 + KNO_3)	71Ae	
Ag^+	-28.2	13.85(1-2)	-31	30	C	μ=0.01M(KNO_3)	73Ad	
	-30.8	15.83(1-3)	-30	30	C	μ=0.01M(KNO_3)	73Ad	

Metal Ion	ΔH, kcal /mole	Log K	ΔS, cal /mole °K	T, °C	M	Conditions	Ref.	Re- marks
S25, cont.								
Hg^{2+}	≃-44	30.4(1-3)	≃-7	30	C	μ=0.1	69Gc	
	≃-61	32.4(1-4)	≃-33	30	C	μ=0.1	69Gc	
Cd^{2+}	-3.1	2.93(1)	3	30	C	μ=0.01M(KNO_3)	73Ad	q: ±0.4
	-8.3	2.20(1)	-17	30	C	μ=0.01M(KNO_3)	73Ad	
Pd^{2+}	-4.8	2.27(1)	-6	30	C	μ=0.01M(KNO_3)	73Ad	q: ±0.2
S26 SERINE			$C_3H_7O_3N$			$HOCH_2CH(NH_2)CO_2H$		L^-
Mn^{2+}	-2.1	3.91(20°,1)	11	20-60	T	μ=0.1(KNO_3)	73Bf	b
	0.3	2.40(20°,2)	11	20-60	T	μ=0.1(KNO_3)	73Bf	b
	-0.8	4.00(15°,1-2)	15	15&40	T	0.2M(KNO_3)	68Ra	a,b
Fe^{2+}	-1.50	6.45(15°,1-2)	24	15&40	T	0.2M(KNO_3)	68Ra	a,b
Co^{2+}	-2.7	4.38(1)	11	25	C	μ=0.05M(KCl)	72Gb 72Ga	
	-3.16	4.38(25°,1)	10	20-35	T	μ=0.05M(KCl)	72Gb 72Ga	b
	-2.2	3.62(2)	9	25	C	μ=0.05M(KCl)	72Gb 72Ga	
	-3.04	3.62(25°,2)	6	20-35	T	μ=0.05M(KCl)	72Gb 72Ga	b
	-4.0	7.66(25°,1-2)	22	15-40	T	0.2M(KNO_3)	68Ra	b
Ni^{2+}	-3.8	5.43(1)	12	25	C	μ=0.05M(KCl)	72Gb 72Ga	
	-3.4	5.43(25°,1)	14	20-35	T	μ=0.05M(KCl)	72Gb 72Ga	b
	-3.76	5.45(1)	12.3	25	C	μ=0.16	70Lb	
	-5.22	5.44(25°,1)	7.4	10-30	T	--	57Pa 60Pc	b
	-4.6	4.53(2)	6	25	C	μ=0.05M(KCl)	72Gb 72Ga	
	-4.6	4.53(25°,2)	5	20-35	T	μ=0.05M(KCl)	72Gb 72Ga	b
	-4.27	4.53(2)	6.4	25	C	μ=0.16	70Lb	
	-5.67	4.50(25°,2)	1.6	10-30	T	--	57Pa 60Pc	b
	-5.28	3.54(3)	-1.5	25	C	μ=0.16	70Lb	
	-6.27	3.03(25°,3)	-7.2	10-30	T	--	57Pa 60Pc	b
	-8.0	10.1(1-2)	19.0	22.0	C	0.1M(KNO_3)	67Sc	
	-7.8	9.76(25°,1-2)	19	15-40	T	0.2M(KNO_3)	68Ra	b
Cu^{2+}	-6.8	7.93(1)	13	25	C	μ=0.05M(KCl)	70Gc 71Gc	
	-5.5	7.93(1)	18	25	C	μ=0.05M(KCl)	72Gb 72Ga	
	-4.3	7.93(25°,1)	22	20-35	T	μ=0.05M(KCl)	72Gb 72Ga	b
	-5.5	7.92(1)	17.8	25	C	μ=0.1M(KNO_3)	72Ia	
	-5.51	7.85(1)	17.5	25	C	μ=0.16	70Lb	
	-5.5	6.74(2)	16	25	C	μ=0.05M(KCl)	70Gc 71Gc	

Metal Ion	ΔH, kcal /mole	Log K	ΔS, cal /mole °K	T, °C	M	Conditions	Ref.	Re-marks
S26, cont.								
Cu^{2+}, cont.	-6.3	6.55(2)	9	25	C	μ=0.05M(KCl)	72Gb 72Ga	
	-6.9	6.55(25°,2)	7	20-35	T	μ=0.05M(KCl)	72Gb 72Ga	b
	-6.1	6.65(2)	10.0	25	C	μ=0.1M(KNO_3)	72Ia	
	-6.13	6.65(2)	9.9	25	C	μ=0.16	70Lb	
	-13.0	(1-2)	--	24	C	μ=0.01M($HClO_4$)	72Sg	
	-14.1	14.6(1-2)	19	21.9	C	0.1M(KNO_3)	67Sc	
	-9.6	14.40(25°,1-2)	34	16-40	T	0.2M(KNO_3)	68Ra	b
	0.38	1.50(25°)	8.1	-75&25	T	S: 50 v % Methanol R: Cu + CuL_2 = 2CuL	73Ya	a,b,q: ±0.35
Zn^{2+}	-2.3	4.65(1)	13	25	C	μ=0.05M(KCl)	72Gb 72Ga	
	-3.0	4.65(25°,1)	11	20-35	T	μ=0.05M(KCl)	72Gb 72Ga	b
	-2.6	4.03(2)	10	25	C	μ=0.05M(KCl)	72Gb 72Ga	
	-3.1	4.03(25°,2)	9	20-35	T	μ=0.05M(KCl)	72Gb 72Ga	b
	-6.7	(1-2)	--	22.0	C	0.1M(KNO_3)	67Sc	
	-4.3	8.38(25°,1-2)	24.0	15-40	T	0.2M(KNO_3)	68Ra	b
S27 SILICIC ACID			SiH_4O_4			$Si(OH)_4$		
Fe^{3+}	3.73	-0.602(25°)	9.83	18-32	T	C_M=5x10^{-4}M C_L=5x10^{-2}M pH=0.5-3.0 R: Fe^{3+} + L = $FeOSi(OH)_3^{2+}$ + H^+	73Ob	b
S28 SOLOCHROME, Violet R								L^{3-}
Cr^{3+}	-6	17.05(85°,2)	60	75-100	T	μ=0	62Cf	b,q: ±2.0
S29 STYRENE			C_8H_8			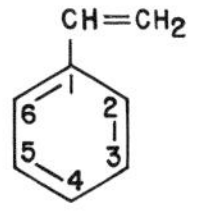		L
PdAB	-6.64	-0.175(25°)	22.9	0-39	T	S: $CDCl_3$ A=6-Acetoxynorbornenyl. B=Benzoyltrifluoroacetonate R: PdAB + L = PdCBL C=5-Acetoxynortricyclenyl	73Ba	b,q: ±0.46
Ag^+	-2.11	1.25(25°,1)	-1.35	0-40	T	μ=0	65Fa	b
S30 STYRENE, 4-chloro-			C_8H_7Cl			$C_8H_7(Cl)$		L
Ag^+	-2.19	1.03(25°,1)	-2.62	0-40	T	μ=0	65Fa	b

Metal Ion	ΔH, kcal /mole	Log K	ΔS, cal /mole °K	T, °C	M	Conditions	Ref.	Re-marks
S31	STYRENE, 4-dimethylamine		$C_{10}H_{13}N$			$C_8H_7[N(CH_3)_2]$		L
PdAB	-10.3	0.788(25°)	30.1	0-39	T	S: $CDCl_3$. A=6-Acetoxynorbornenyl B=Benzoyltrifluoro-acetonate R: PdAB + L = PdCBL C=5-Acetoxynortri-cyclenyl	73Ba	b
S32	STYRENE, 4-fluoro-		C_8H_7F			$C_8H_7(F)$		L
PdAB	-6.21	-0.263(25°)	22	0-39	T	S: $CDCl_3$. A=6-Acetoxynorbornenyl B=Benzoyltrifluoro-acetonate R: PdAB + L = PdCBL C=5-Acetoxynortri-cyclenyl	73Ba	b,q: ±0.46
S33	STYRENE, 4-methoxy-		$C_9H_{10}O$			$C_8H_7(OCH_3)$		L
PdAB	-7.89	0.175(25°)	25.1	0-39	T	S: $CDCl_3$. A=6-Acetoxynorbornenyl B=Benzoyltrifluoro-acetonate R: PdAB + L = PdCBL C=5-Acetoxynortri-cyclenyl	73Ba	b
S34	STYRENE, 4-methyl-		C_9H_{10}			$C_8H_7(CH_3)$		L
Ag^+	-1.88	1.38(25°,1)	0.03	0-40	T	μ=0	65Fa	b
S35	STYRENE, 4-nitro-		$C_8H_7O_2N$			$C_8H_7(NO_2)$		L
PdAB	-2.80	-0.578(25°)	12	0-39	T	S: $CDCl_3$. A=6-Acetoxynorbornenyl B=Benzoyltrifluoro-acetonate R: PdAB + L = PdCBL C=5-Acetoxynortri-cyclenyl	73Ba	b,q: ±0.46
S36	SUCCINIC ACID		$C_4H_6O_4$			$HO_2CCH_2CH_2CO_2H$		L^{2-}
Mn^{2+}	3.02	2.27(1)	20.5	25	C	μ=0	67Mf	
	2.95	2.27(25°,1)	20.3	0-45	T	μ=0.2. ΔCp=62	61Mc	b
Co^{2+}	3.15	2.22(1)	20.7	25	C	μ=0	67Mf	
	2.81	2.22(25°,1)	19.6	0-45	T	μ=0.2. ΔCp=72	61Mc	b
Ni^{2+}	2.46	2.35(1)	19.0	25	C	μ=0	67Mf	
	2.23	2.35(25°,1)	18.2	0-45	T	μ=0.2. ΔCp=42	61Mc	b
Cu^{2+}	4.56	3.24(1)	30.1	25	C	μ=0	67Mf	
Zn^{2+}	4.39	2.47(1)	26.0	25	C	μ=0	67Mf	

Metal Ion	ΔH, kcal /mole	Log K	ΔS, cal /mole °K	T, °C	M	Conditions	Ref.	Remarks
S37 SUCCINIC ACID, 2-hydroxy-			$C_4H_6O_5$			$HO_2CCH_2CH(OH)CO_2H$		
In^{3+}	4.2	4.60(35°,1)	35	25-45	T	μ=0.20M($NaClO_4$)	73Sc	b
	10.41	8.56(35°,1-2)	72.9	25-45	T	μ=0	72Sb	b
S38 SUCCINIC ACID, 2-mercapto-			$C_4H_6O_4S$			$HO_2CCH_2CH(SH)CO_2H$		L^{3-}
Mn^{2+}	-2	4.87(45°,1)	16	35-55	T	μ=0.2M($NaClO_4$)	69Rb	b,q: ±2
Fe^{3+}	5.1	3.446(25°)	33	10-25	T	μ=1.0M($NaClO_4$) R: $Fe^{3+} + HL^- = FeL^+ + H^+$	73Eb	b,q: ±1.5
	5.1	0.406(25°)	20	10-25	T	μ=1.0M($NaClO_4$) R: $Fe^{3+} + H_2L = FeL^+ + 2H^+$	73Eb	b,q: ±1.4
FeA^{2+}	-4.8	6.230(25°)	12	10-25	T	μ=1.0M($NaClO_4$) A=OH^-. R: $FeA^{2+} + HL^- = FeL^+ + HA$	73Eb	b,q: ±1.4
	-5.0	3.643(25°)	-2.5	10-25	T	μ=1.0M($NaClO_4$) A=OH^-. R: $FeA^{2+} + H_2L = FeL^+ + H^+ + HA$	73Eb	b,q: ±1.5
Ni^{2+}	-7	7.38(45°,1)	11	35-55	T	μ=0.2M($NaClO_4$)	69Rb	b,q: ±2
	-6	6.23(45°,2)	10	35-55	T	μ=0.2M($NaClO_4$)	69Rb	b,q: ±2
	-8.77	14.15(1-2)	35.36	25	T	0.1M(KNO_3)	68Sb	
Ag^+	-7.5	6.95(1)	7.0	30	T	μ=0.1M(KNO_3)	68Sa	b
Zn^{2+}	-5.3	15.60(30°,1-2)	53.8	25-35	T	μ=0.1M(KNO_3)	68Sc	b
In^{3+}	-6.1	14.46(35°,1)	45	25-45	T	μ=0.20M($NaClO_4$)	73Sc	
	-12.22	26.76(35°,1-2)	82.8	25-45	T	μ=0	72Sb	b
Tl^+	-2	3.86(45°,1)	11	35-55	T	μ=0.2M($NaClO_4$)	69Rb	b,q: ±2
S39 SULFATE ION			O_4S			SO_4^{2-}		L^{2-}
Li^+	≃0	(1)	--	25	C	μ=0	62Ac	
	≃0	(1)	--	25	C	μ=0.025	62Ac	
Na^+	-0.49	0.65(1)	-1.3	25	C	μ=0	69Ia	
	1.120	(1)	--	25	C	μ=0	62Ac	
	0.940	(1)	--	25	C	μ=0.025	62Ac	
K^+	1.01	0.75(1)	0.0	25	C	μ=0	69Ia	
Mg^{2+}	4.08	2.129(1)	24.7	0	T	μ=0. ΔCp=-7.8	67Mb	
	0.51	2.23(1)	12.0	25	C	μ=0	69Ib	
	1.55	(1)	--	25	C	μ=0	73Hb	
	1.27	2.31(1)	14.7	25	C	μ=0	70La	i: 59Lc
	4.01	2.399(1)	24.4	25	T	μ=0. ΔCp=2.8	67Mb	
	5.70	2.34(1)	31.0	25	T	μ=0	54Cd	
	2.04	2.130(25°,1)	16.6	0-40	T	μ=0	73Kc	b
	4.22	2.631(1)	25.1	50	T	μ=0. ΔCp=13.4	67Mb	
	4.68	2.846(1)	26.5	75	T	μ=0. ΔCp=24	67Mb	

Metal Ion	ΔH, kcal /mole	Log K	ΔS, cal /mole °K	T, °C	M	Conditions	Ref.	Remarks
S39, cont.								
Mg^{2+}, cont.	5.41	3.057(1)	28.5	100	T	μ=0. ΔCp=35	67Mb	
	6.41	3.274(1)	31.1	125	T	μ=0. ΔCp=45	67Mb	
	7.67	3.501(1)	34.1	150	T	μ=0. ΔCp=56	67Mb	
	9.20	3.743(1)	37.6	175	T	μ=0. ΔCp=66	67Mb	
	11.0	4.002(1)	41.5	200	T	μ=0. ΔCp=77	67Mb	
	15.4	4.58(1)	50.3	250	T	μ=0. ΔCp=98	67Mb	
	20.8	5.23(1)	60.2	300	T	μ=0. ΔCp=119	67Mb	
	27.3	5.96(1)	71.0	350	T	μ=0. ΔCp=140	67Mb	
	30.2	6.27(1)	75.6	370	T	μ=0. ΔCp=149	67Mb	
Ca^{+}	0.81	2.43(1)	13.7	25	C	μ=0	69Ib	
	1.50	2.31(1)	15.6	25	C	μ=0	70La	i: 59Lc
	1.53	2.309(1)	15.8	25	T	μ=0	73Ac	q: ±1.0
	1.65	2.31(1)	16.1	25	T	μ=0	53Ba	
	2.46	2.498(1)	18.9	40	T	μ=0. ΔCp=62.1	73Ac	q: ±0.5
	3.11	2.644(1)	20.8	50	T	μ=0	73Ac	q: ±1.0
Sc^{3+}	6.31	4.04(1)	39.6	25	C	μ=0	69Ib	
	3.96	1.66(2)	21.0	25	C	μ=0	69Ib	
Y^{3+}	3.61	3.34(1)	27.4	25	C	μ=0	69Ib	
	3.28	3.468(1)	26.9	25	C	μ=0	69Fa	
	4.04	1.23(1)	19.2	25	C	μ=2.0	67Cb	
	0.73	2.00(2)	11.6	25	C	μ=0	69Ib	
	1.5	0.45(2)	7.1	25	C	μ=2.0	67Cb	
La^{3+}	3.24	3.50(1)	26.9	25	C	μ=0	69Ib 52Jb	
	3.50	3.620(1)	28.3	25	C	μ=0	69Fa	
	2.5	1.4(1)	15	25	T	μ=1.0($NaClO_4$)	53Nb	
	3.72	1.29(1)	18.4	25	C	μ=2.0	67Cb	
	1.26	1.85(2)	12.7	25	C	μ=0	69Ib	
Ce^{3+}	3.46	3.48(1)	27.6	25	C	μ=0	69Ib	
	3.78	3.585(1)	29.1	25	C	μ=0	69Fa	
	4.69	1.23(1)	31.2	25	T	μ=0	53Nb	
	3.64	1.23(1)	17.8	25	T	μ=1.0($NaClO_4$)	53Nb	
	4.27	1.24(1)	20.0	25	C	μ=2.0	67Cb	
	4.20	1.32(1)	20.0	25	T	μ=2.0	67Cb	
	1.60	1.75(2)	13.4	25	C	μ=0	69Ib	
Pr^{3+}	3.49	3.58(1)	28.2	25	C	μ=0	69Ib	
	3.92	3.620(1)	29.7	25	C	μ=0	69Fa	
	3.94	1.27(1)	19.0	25	C	μ=2.0	67Cb	
	1.15	1.86(2)	12.4	25	C	μ=0	69Ib	
Nb^{3+}	3.62	3.43(1)	27.8	25	C	μ=0	69Ib	
	4.15	3.638(1)	30.6	25	C	μ=0	69Fa	
	4.18	1.25(1)	19.8	25	C	μ=2.0	67Cb	
	1.61	1.74(2)	13.4	25	C	μ=0	69Ib	
Pm^{3+}	3.9	1.32(1)	19	25	C	μ=2.0	67Cb	
Sm^{3+}	3.70	3.52(1)	28.6	25	C	μ=0	69Ib	

Metal Ion	ΔH, kcal /mole	Log K	ΔS, cal /mole °K	T, °C	M	Conditions	Ref.	Re-marks
S39, cont.								
Sm^{3+}, cont.	4.34	3.658(1)	31.3	25	C	μ=0	69Fa	
	4.27	1.30(1)	19.7	25	C	μ=2.0	67Cb	
	2.1	1.67(2)	14.7	25	C	μ=0	69Ib	
Eu^{3+}	3.64	3.54(1)	28.4	25	C	μ=0	69Ib	
	4.13	3.658(1)	30.6	25	C	μ=0	69Fa	
	5.4	3.671(1)	35	25	T	μ=0. ΔCp=46	72Ha	
	6.0	3.850(1)	37	38.5	T	μ=0. ΔCp=49	72Ha	
	6.7	4.040(1)	39	51.6	T	μ=0. ΔCp=53	72Ha	
	7.4	4.225(1)	41	65.1	T	μ=0. ΔCp=62	72Ha	
	4.1	2.757(1)	26	25	T	μ=0.045M. ΔCp=59	72Ha	
	4.9	2.900(1)	29	38.5	T	μ=0.045M. ΔCp=66	72Ha	
	5.8	3.053(1)	32	51.6	T	μ=0.045M. ΔCp=74	72Ha	
	6.9	3.227(1)	35	65.1	T	μ=0.045M. ΔCp=82	72Ha	
	3.88	1.37(1)	19.3	25	C	μ=2.0	67Cb	
	1.51	1.78(2)	13.2	25	C	μ=0	69Ib	
	2.4	0.71(2)	11.3	25	C	μ=2.0	67Cb	
Gd^{3+}	3.59	3.48(1)	28.0	25	C	μ=0	69Ib	
	4.10	3.658(1)	30.5	25	C	μ=0	69Fa	
	3.95	1.33(1)	19.3	25	C	μ=2.0	67Cb	
	1.50	1.73(2)	12.9	25	C	μ=0	69Ib	
	1.7	0.42(2)	7.6	25	C	μ=2.0	67Cb	
Tb^{3+}	3.60	3.47(1)	28.0	25	C	μ=0	69Ib	
	4.02	3.635(1)	30.1	25	C	μ=0	69Fa	
	4.24	1.28(1)	20.1	25	C	μ=2.0	67Cb	
	0.77	1.90(2)	11.3	25	C	μ=0	69Ib	
	1.6	0.62(2)	8.2	25	C	μ=2.0	67Cb	
Dy^{3+}	3.69	3.43(1)	28.1	25	C	μ=0	69Ib	
	3.58	3.611(1)	28.5	25	C	μ=0	69Fa	
	4.41	1.23(1)	20.4	25	C	μ=2.0	67Cb	
	1.37	1.75(2)	12.5	25	C	μ=0	69Ib	
	1.3	0.49(2)	6.6	25	C	μ=2.0	67Cb	
Ho^{3+}	3.66	3.38(1)	27.8	25	C	μ=0	69Ib	
	3.54	3.585(1)	28.3	25	C	μ=0	69Fa	
	4.22	1.24(1)	19.8	25	C	μ=2.0	67Cb	
	1.60	1.60(2)	12.7	25	C	μ=0	69Ib	
	1.7	0.53(2)	8.1	25	C	μ=2.0	67Cb	
Er^{3+}	3.61	3.41(1)	27.7	25	C	μ=0	69Ib	
	3.39	3.585(1)	27.8	25	C	μ=0	69Fa	
	4.21	1.22(1)	19.7	25	C	μ=2.0	67Cb	
	1.37	1.78(2)	12.7	25	C	μ=0	69Ib	
	1.5	0.48(2)	7.2	25	C	μ=2.0	67Cb	
Tm^{3+}	3.60	3.41(1)	27.7	25	C	μ=0	69Ib	
	3.15	3.585(1)	27.0	25	C	μ=0	69Fa	
	4.22	1.15(1)	19.4	25	C	μ=2.0	67Cb	
	0.86	1.80(2)	10.7	25	C	μ=0	69Ib	
	1.0	0.45(2)	5.4	25	C	μ=2.0	67Cb	
Yb^{3+}	3.60	3.33(1)	27.3	25	C	μ=0	69Ib	
	2.90	3.585(1)	26.1	25	C	μ=0	69Fa	
	4.14	1.15(1)	19.2	25	C	μ=2.0	67Cb	

Metal Ion	ΔH, kcal /mole	Log K	ΔS, cal /mole °K	T, °C	M	Conditions	Ref.	Remarks
S39, cont.								
Yb^{3+}, cont.	0.97	1.72(2)	11.1	25	C	μ=0	69Ib	
	1.2	0.45(2)	6.1	25	C	μ=2.0	67Cb	
Lu^{3+}	3.29	3.49(1)	27.4	25	C	μ=0	69Ib	
	3.54	3.585(1)	28.3	25	C	μ=0	69Fa	
	4.25	1.08(1)	19.2	25	C	μ=2.0	67Cb	
	1.04	1.8(2)	11.8	25	C	μ=0	69Ib	
	1.5	0.53(2)	7.4	25	C	μ=2.0	67Cb	
Th^{4+}	-0.54	2.220	8.4	25	C	μ=2.0($KClO_4$) R: $Th^{4+} + HL^- = ThL^{2+} + H^+$	59Za	
	-0.89	1.336	3.2	25	C	μ=2.0($KClO_4$) R: $ThL^{2+} + HL^- = ThL_2 + H^+$	59Za	
VO^{2-}	4.12	2.48(1)	25.2	25	C	μ=0	71Ba	
Cr^{3+}	7.2	1.50(48.2°,1)	29.3	48.2 & 82	T	μ=1	62Fb 64Ha	a,b
	7.2	0.76(56°,1)	25.3	56&65	T	μ=2	62Fb	a,b
CrA_3	2.2	0.146(1)	8.0	25	C	3N($LiClO_4$ + Li_2SO_4) A=Ethylenediamine	72Mg	
	1.5	-0.143(2)	4.5	25	C	3N($LiClO_4$ + Li_2SO_4) A=Ethylenediamine	72Mg	q: ±0.2
	-3.7	-0.155(3)	-13	25	C	3N($LiClO_4$ + Li_2SO_4) A=Ethylenediamine	72Mg	q: ±0.4
$HCrO_4^-$	-4.68	1.276(26°)	-9.8	26&38	T	μ=2.0M($NaClO_4$) R: $HCrO_4^- + 2H^+ + SO_4^= = HCrO_3SO_4^- + H_2O$	71Yc	b
U^{4+}	-3.2	2.5(1)	0.7	25	T	μ=2.0	55Da	
	-2.3	3.9(1-2)	10	25	T	μ=2.0	55Da	
UO_2^{2+}	4.98	2.72(1)	29.1	25	C	μ=0	71Ba	
	2.35	2.73(1)	20.4	25	T	μ→0	60Lc	
	5.10	3.262(35°,1)	31.5	25-50	T	μ=0	67Wb	q: ±0.43
	9.20	3.32(1)	43.7	50	T	μ→0	60Lc	
	22.9	4.45(1)	81.8	100	T	μ→0	60Lc	
	36.6	6.51(1)	116.3	150	T	μ→0	60Lc	
	50.3	8.87(1)	147.0	200	T	μ→0	60Lc	
	4.357	1.806(1)	22.87	25	C	μ=1.0($NaClO_4$)	71Ac	
	2.3	1.9(1)	16	25	T	μ=2.0($NaClO_4$)	54Db	
	0.93	1.48(2)	9.9	25	T	μ→0	60Lc	
	1.88	1.101(35°,2)	11.1	25-50	T	μ=0	67Wb	b,q: ±0.48
	1.11	1.52(2)	10.4	50	T	μ→0	60Lc	
	1.46	1.66(2)	11.5	100	T	μ→0	60Lc	
	1.82	1.77(2)	12.4	150	T	μ→0	60Lc	
	2.17	1.88(2)	13.2	200	T	μ→0	60Lc	
	4.034	0.950(2)	17.88	25	C	μ=1.0($NaClO_4$)	71Ac	
	1.4	2.8(2)	18	25	T	μ=2.0($NaClO_4$)	54Db	

Metal Ion	ΔH, kcal /mole	Log K	ΔS, cal /mole °K	T, °C	M	Conditions	Ref.	Remarks
S39, cont.								
Np^{4+}	4.0	3.51(25°,1)	29.5	10.2-35.3	T	μ=2.0($HClO_4$)	54Sa	b
	6.22	2.49(25°,2)	41.0	10.2-35.3	T	μ=2.0($HClO_4$)	54Sa	b
	-1.17	2.43(25°)	7.2	10.2-35.3	T	μ=2.0($HClO_4$) R: $Np^{4+} + HL^- = NpL^{2+} + H^+$	54Sa	b
	3.64	1.41(25°)	18.7	10.2-35.3	T	μ=2.0($HClO_4$) R: $NpL^{2+} + HL^- = NpL_2 + H^+$	54Sa	b
Am^{3+}	4.4	1.5(1)	21	25	C	μ=2.0	67Cb	
Cm^{3+}	4.1	1.3(1)	20	25	C	μ=2.0	67Cb	
Cf^{3+}	4.5	1.4(1)	21	25	C	μ=2.0	67Cb	
Mn^{2+}	0.61	2.86(1)	15.2	25	C	μ=0	69Ib	
	2.17	(1)	--	25	C	μ=0	73Hb	
	3.37	2.25(25°,1)	22.6	0-40	T	μ=0	59Na	b
	3.2	2.1(25°,1)	21	20-30	T	μ=0	67Af	b
Fe^{3+}	4.2	3.854(1)	32	20	T	μ=0	60Kb	
	0.56	2.20(1)	12.0	25	C	μ=0	69Ib	
	6.2	4.2(1)	39	25	T	μ=0.5($NaClO_4$ + $HClO_4$)	63Wa	
	4.2	1.8(1)	23	20	T	μ=1.0	60Kb	
Co^{2+}	0.50	2.69(1)	13.9	25	C	μ=0	69Ib 53Tb	
	1.47	(1)	--	25	C	μ=0	73Hb	i: 59Lc
	1.74	2.35(25°,1)	16.6	0-40	T	μ=0	59Na 66Gd	b
CoA_3	1.65	0.556(1)	8.0	25	C	3N($NaClO_4$ + Na_2SO_4) A=Ethylenediamine	72Mg	q: ±0.1
	2.05	0.204(1)	7.8	25	C	3N(Na_2SO_4 + $NaClO_4$) A=Propylenediamine	72Mh	
	0.8	0.146(2)	3.5	25	C	3N($NaClO_4$ + Na_2SO_4) A=Ethylenediamine	72Mg	q: ±0.2
	1.2	-0.081(3)	3.6	25	C	3N(Na_2SO_4 + $NaClO_4$) A=Propylenediamine	72Mh	
	0.4	-0.398(3)	≃0.5	25	C	3N($NaClO_4$ + Na_2SO_4) A=Ethlenediamine	72Mg	q: ±0.2
	2.7	0.041(2)	9.2	25	C	3N(Na_2SO_4 + $NaClO_4$) A=Propylenediamine	72Mh	
CoA_6^{3+}	0.40	3.32(1)	16.6	25	T	μ=0. A=NH_3	56Pa	
Ni^{2+}	0.41	2.81(1)	14.1	25	C	μ=0	69Ib	
	1.52	2.30(1)	15.7	25	C	μ=0	70La	i: 59Lc
	1.27	2.273(25°,1)	14.7	0-40	T	μ=0	73Kc	b
	3.31	2.32(25°,1)	21.7	0-45	T	μ=0	59Na	b
RhA_3	1.8	0.146(1)	6.7	25	C	3N($NaClO_4$ + Na_2SO_4) A=Ethylenediamine	72Mg	q: ±0.2

Metal Ion	ΔH, kcal /mole	Log K	ΔS, cal /mole °K	T, °C	M	Conditions	Ref.	Remarks
S39, cont.								
RhA_3, cont.	1.0	0.114(2)	3.9	25	C	3N($NaClO_4$ + Na_2SO_4) A=Ethylenediamine	72Mg	q: ±0.3
Cu^{2+}	1.22	2.26(1)	14.6	25	C	μ=0	69Ib	
	1.72	2.36(1)	16.6	25	C	μ=0	70La	i: 59Lc
	2.43	(1)	--	25	C	μ=0	73Hb	
	9.600	2.398(25°,1)	13.0	0&25	T	μ=0	72Pg	a,b
Ag^+	1.5	1.3(1)	11	25	C	μ=0	65Hc 34La 33La 60Sg	
Zn^{2+}	0.63	2.49(1)	13.5	25	C	μ=0	69Ib	
	1.36	2.32(1)	15.2	25	C	μ=0	70La	i: 59Lc
	1.59	(1)	--	25	C	μ=0	73Hb	
	4.01	2.38(25°,1)	24.4	0-45	T	μ=0	58Nb	b
Cd^{2+}	0.98	2.55(1)	15.0	25	C	μ=0	69Ib	
	2.15	2.31(1)	17.8	25	C	μ=0	70La	i: 59Lc
Al^{3+}	2.29	3.01(1)	21.4	25	C	μ=0	69Ib	
	0.78	1.89(2)	11.3	25	C	μ=0	69Ib	
Ga^{3+}	4.71	2.77(1)	28.4	25	C	μ=0	69Ib	
	-1.73	2.29(2)	4.8	25	C	μ=0	69Ib	
In^{3+}	6.95	3.04(1)	37.2	25	C	μ=0	69Ib	
	-1.75	1.96(2)	3.1	25	C	μ=0	69Ib	
Tl^+	-0.22	1.36(1)	5.5	25	T	μ=0	53Ba	
	-5.7	-0.484(1)	-23	25	C	μ=3.0M[Li(ClO_4, SO_4)]	69Sj	p
	4.4	-0.484(25°,1)	12	25-45	T	μ=3.0M[Na(ClO_4, SO_4]	69Sj	b,p
Tl^{3+}	-2.7	2.26(1)	1.1	25	C	2.0M($LiClO_4$) 1.0M($HClO_4$)	67Md	
S40 SULFIDE, benzylmethyl-			$C_8H_{10}S$			$CH_3SCH_2C_6H_5$		
PtA_2L_2	3.4	0.120(20°)	12	30-60	T	S: Chloroform A=Cl	73Re	b,bb
	3.2	-0.409(21°)	9	30-60	T	S: Dichloromethane A=Cl	73Re	b,bb
S41 SULFIDE, bis(dimethylthiocarbamyl)-			$C_6H_{12}N_2S_3$			$(CH_3)_2NC(S)SC(S)N(CH_3)_2$		
I_2	-7.30	2.636(20°,1)	-13.0	10-30	T	S: CCl_4	69Gd	b,q: ±0.66
S42 SULFIDE, diallyl-			$C_6H_{10}S$			$(CH_2:CHCH_2)_2S$		
I_2	-4.6	2.699(1)	-3.0	25	C	S: Octane	66Ad	

Metal Ion	ΔH, kcal /mole	Log K	ΔS, cal /mole °K	T, °C	M	Conditions	Ref.	Remarks
S43 SULFIDE, dibenzyl-			$C_{14}H_{14}S$			$(C_6H_5CH_2)_2S$		
PtA_2L_2	6.7	-0.796(40°)	18	30-60	T	S: Chloroform A=Cl	73Re	b,bb
	4.8	-0.319(36°)	14	30-60	T	S: Chloroform A=Br	73Re	b,bb
	2.0	0.531(33°)	9	30-60	T	S: Chloroform A=I	73Re	b,bb
SnA_4	-3.8	(<u>1</u>)	--	25	C	S: Benzene C_M=0.05-0.08M A=Cl	65Gb 68Ge	
I_2	-4.8	2.788(<u>1</u>)	-3.3	25	C	S: Octane	66Ad	
S44 SULFIDE, dibutyl-			$C_8H_{18}S$			$(C_4H_9)_2S$		
Al_2A_6	-17.6	--	--	25	C	S: Benzene C_M=0.07M. A=Br R: $0.5(Al_2A_6) + L = AlA_3L$	68Rc	
GaA_3	-19.0	4.03(<u>1</u>)	-45.3	25	C	S: Benzene. A=Cl	70Gh	
SnA_4	-12.6	(<u>1</u>)	--	25	C	S: Benzene C_M=0.05-0.08M A=Cl	68Ge 65Gb 70Gh	
	-11.6	(<u>2</u>)	--	25	C	S: Benzene C_M=0.05-0.08M A=Cl	68Ge 65Gb	
TeA_4	-11.2	2.24(<u>1</u>)	-27.3	25	C	S: Benzene. A=Cl	72Pe 73Ph	
I_2	-7.6	2.585(<u>1</u>)	-13.6	25	C	S: Octane	66Ad	
S45 SULFIDE, dicyclohexyl-			$C_{12}H_{22}S$			$(C_6H_{11})_2S$		L
Al_2A_6	-16.9	--	--	25	C	S: Benzene C_M=0.07M. A=Br R: $0.5(Al_2A_6) + L = AlA_3L$	68Rc	
I_2	-10.5	2.638(<u>1</u>)	-23.1	25	C	S: Octane	66Ad	
S46 SULFIDE, diethyl-			$C_4H_{10}S$			$(C_2H_5)_2S$		
TiA_4	-11.5	(<u>1</u>)	--	25	C	S: Benzene C_M=0.05-0.08M A=Cl	68Ge	
	-11.5	(2)	--	25	C	S: Benzene C_M=0.05-0.08M A=Cl	68Ge	
AlA_3	-16.75	(<u>1</u>)	--	28.5	C	S: Hexane. A=CH_3	67Hc	
I_2	-8.51	3.064(30°,<u>1</u>)	-14.0	20-40	T	S: CCl_4 K's determined from spectral data taken in visible region	61Ga	b

Metal Ion	ΔH, kcal /mole	Log K	ΔS, cal /mole °K	T, °C	M	Conditions	Ref.	Remarks
S46, cont.								
I_2, cont.	-8.05	3.212(30°,1)	-11.7	20-40	T	S: CCl_4 K's determined from spectral data taken in ultraviolet region	61Ga	b
	-7.8	2.322(25°,1)	-15.9	15-35	T	S: Cyclohexane	63Va	b
	-7.82	2.322(20°,1)	-15.9	0-30	T	C_M=5x10^{-5} C_L=0.001-0.05 S: CH_2Cl_2	61Ta	b
	-8.98	2.961(30°,1)	-15.95	20-40	T	S: n-Heptane K's determined from spectral data taken in visible region	61Ga	b
	-8.05	3.158(30°,1)	-12.0	20-40	T	S: n-Heptane K's determined from spectral data taken in ultraviolet region	61Ga	b
S47 SULFIDE, diethyl-,2,2'-diamino-			$C_4H_{12}N_2S$			$(NH_2CH_2CH_2)_2S$		L
Co^{2+}	-7	5.09(30°,1)	0	0-50	T	μ=0.1(KCl or KNO_3)	54Gb	b
	-8	3.92(30°,2)	-9	0-50	T	μ=0.1(KCl or KNO_3)	54Gb	b
	-15	9.01(30°,1-2)	-9	0-50	T	μ=0.1(KCl or KNO_3)	54Gb	b
Ni^{2+}	-10	7.27(30°,1)	0	0-50	T	μ=0.1(KCl or KNO_3)	54Gb	b
	-12	6.10(30°,2)	-11	0-50	T	μ=0.1(KCl or KNO_3)	54Gb	b
	-22	13.37(30°,1-2)	-11	0-50	T	μ=0.1(KCl or KNO_3)	54Gb	b
S48 SULFIDE, diethyl-, 2,2'-diamino-N-N,N',N'-tetraacetic acid			$C_{12}H_{20}O_8N_2S$			$(CH_2CO_2H)_2NCH_2CH_2SCH_2CH_2N-(CH_2CO_2H)_2$		L^{4-}
Mg^{2+}	4.13	4.61(1)	35.17	20	C	μ=0.1(KNO_3)	64Ab	
Ca^{2+}	-2.5	6.21(1)	23.2	20	C	μ=0.1(KNO_3)	64Ab	
La^{3+}	-0.2	12.8(1)	57.86	20	C	μ=0.1(KNO_3)	64Ab	
Mn^{2+}	-1.53	10.07(1)	41.86	20	C	μ=0.1(KNO_3)	64Ab	
Co^{2+}	-4.63	13.99(1)	48.2	20	C	μ=0.1(KNO_3)	64Ab	
Ni^{2+}	-7.7	15.7(1)	45.54	20	C	μ=0.1(KNO_3)	64Ab	
Cu^{2+}	-9.13	16.57(1)	44.65	20	C	μ=0.1(KNO_3)	64Ab	
Zn^{2+}	-3.7	13.44(1)	48.85	20	C	μ=0.1(KNO_3)	64Ab	
Cd^{2+}	-8.2	14.38(1)	37.8	20	C	μ=0.1(KNO_3)	64Ab	
Hg^{2+}	-22.8	23.09(1)	31.55	20	C	μ=0.1(KNO_3)	64Ab	
Pb^{2+}	-13.0	13.86(1)	19.07	20	C	μ=0.1(KNO_3)	64Ab	
S49 SULFIDE, diethyl-, 2,2'-dihydroxy-			$C_4H_{10}O_2S$			$(CH_2CH_2OH)_2S$		
Hg^{2+}	8.80	3.00(30°,1)	42.8	30-80	T	μ=1.5M($HClO_4$) C_M=0.01M	71Me	b,q: ±1.0

Metal Ion	ΔH, kcal /mole	Log K	ΔS, cal /mole °K	T, °C	M	Conditions	Ref.	Re-marks
S49, cont.								
Hg^{2+}, cont.	-22.3	5.08(30°,2)	-50.3	30-80	T	μ=1.5M($HClO_4$) C_M=0.01M	71Me	b,q: ±2.0
	-1.6	1.37(30°,3)	1.0	30-80	T	μ=1.5M($HClO_4$) C_M=0.01M	71Me	b,q: ±2.0
S50 SULFIDE, diheptyl-			$C_{14}H_{30}S$			$(C_7H_{15})_2S$		
GaA_3	-19.2	3.88(1)	-46.6	25	C	S: Benzene. A=Cl	70Gh	
TeA_4	-11.4	2.38(1)	-27.4	25	C	S: Benzene. A=Cl	73Ph	
	-11.4	(1)	--	25	C	S: Benzene. A=Cl	72Pe	
S51 SULFIDE, diisoheptyl (sec)-			$C_{14}H_{30}S$			$(C_7H_{15})_2S$		L
Al_2A_6	-17.1	--	--	25	C	S: Benzene C_M=0.07M. A=Br R: 0.5(Al_2A_6) + L = AlA_3L	68Rc	
S52 SULFIDE, diisooctyl (sec)-			$C_{16}H_{34}S$			$(C_8H_{17})_2S$		L
Al_2A_6	-16.1	--	--	25	C	S: Benzene C_M=0.07M. A=Br R: 0.5(Al_2A_6) + L = AlA_3L	68Rc	
SnA_4	-0.4	(1)	--	25	C	S: Benzene C_M=0.05-0.08M A=Cl	68Ge 65Gb	
S53 SULFIDE, diisopentyl-			$C_{10}H_{22}S$			$(C_5H_{11})_2S$		L
TeA_4	-11.6	2.18(1)	-28.9	25	C	S: Benzene. A=Cl	73Ph	
S54 SULFIDE, diisopropyl-			$C_6H_{14}S$			$(CH_3)_2CHSCH(CH_3)_2$		L
Al_2A_6	-18.2	--	--	25	C	S: Benzene C_M=0.07M. A=Br R: 0.5(Al_2A_6) + L = AlA_3L	68Rc	
S55 SULFIDE, dimethyl-			C_2H_6S			CH_3SCH_3		L
PtA_2L_2	1.9	0.806(29°)	10	30-60	T	S: Chloroform A=Cl	73Re	b,q: ±0.2 bb
	2.3	0.274(28°)	9	30-60	T	S: Dichloromethane A=Cl	73Re	b,bb
AlA_3	-16.69	(1)	--	28.5	C	S: Hexane. A=CH_3	67Hc	
S56 SULFIDE, N,N-dimethylthionocarbamyl methyl-			$C_4H_9NS_2$			$(CH_3)_2NC(S)SCH_3$		
I_2	-7.53	2.260(20°,1)	-15.3	10-30	T	S: CCl_4	69Gd	b,q: ±0.36

Metal Ion	ΔH, kcal /mole	Log K	ΔS, cal /mole °K	T, °C	M	Conditions	Ref.	Re-marks
S57 SULFIDE, dinonyl			$C_{18}H_{38}S$			$(C_9H_{19})_2S$		L
TeA_4	-10.1	2.10(1)	-24.3	25	C	S: Benzene. A=Cl	73Ph	
S58 SULFIDE, dioctyl-			$C_{16}H_{34}S$			$(C_8H_{17})_2S$		L
Al_2A_6	-17.1	--	--	25	C	S: Benzene C_M=0.07M. A=Br R: $0.5(Al_2A_6) + L = AlA_3L$	68Rc	
SnA_4	-11.9	(1)	--	25	C	S: Benzene C_M=0.05-0.08M A=Cl	68Ge 65Gb	
	-11.0	(2)	--	25	C	S: Benzene C_M=0.05-0.08M A=Cl	68Ge 65Gb	u
I_2	-7.4	2.873(1)	-11.6	25	C	S: Octane	66Ad	
S59 SULFIDE, dipentyl-			$C_{10}H_{22}S$			$(C_5H_{11})_2S$		L
Al_2A_6	-18.1	--	--	25	C	S: Benzene C_M=0.07M. A=Br R: $0.5(Al_2A_6) + L = AlA_3L$	68Rc	
TeA_4	-11.6	2.18(1)	-28.9	25	C	S: Benzene. A=Cl	73Ph	
I_2	-7.0	2.556(1)	-11.7	25	C	S: Octane	66Ad	
S60 SULFIDE, diphenyl-			$C_{12}H_{10}S$			$(C_6H_5)_2S$		L
Al_2A_6	-10.2	--	--	25	C	S: Benzene C_M=0.07M. A=Br R: $0.5(Al_2A_6) + L = AlA_3L$	68Rc	
I_2	-5.7	0.591(25°,1)	-16.6	17-35	T	S: Cyclohexane C_M=5x10^{-4}M	74Sa	b
	-0.3	(1)	--	25	C	S: Octane	66Ad	
S61 SULFIDE, diphenyl-, 2,2'-dimethyl-			$C_{14}H_{14}S$			$(C_6H_4CH_3)_2S$		L
I_2	-8.8	0.763(25°,1)	-26.2	17-35	T	S: Cyclohexane C_M=5x10^{-4}M	74Sa	b
S62 SULFIDE, diphenyl-, 4,4'-dimethyl-			$C_{14}H_{14}S$			$(C_6H_4CH_3)_2S$		L
I_2	-8.7	0.785(25°,1)	-25.0	17-35	T	S: Cyclohexane C_M=5x10^{-4}M	74Sa	b
S63 SULFIDE, diphenyl-, 4-methoxy-			$C_{13}H_{12}OS$			$(C_6H_5)S(C_6H_4OCH_3)$		L
I_2	-8.1	0.799(25°,1)	-23.6	17-35	T	S: Cyclohexane C_M=5x10^{-4}M	74Sa	b
S64 SULFIDE, diphenyl-, 4-methyl-			$C_{13}H_{12}S$			$(C_6H_5)S(C_6H_4CH_3)$		L

Metal Ion	ΔH, kcal /mole	Log K	ΔS, cal /mole °K	T, °C	M	Conditions	Ref.	Re-marks
S64, cont.								
I_2	-6.8	0.740(25°,1)	-19.3	17-35	T	S: Cyclohexane C_M=5x10^{-4}M	74Sa	b
S65 SULFIDE, diphenyl-, 2,2,2',2'-tetramethyl-		$C_{16}H_{20}S$				$[C_6H_4(CH_3)_2]_2S$		L
I_2	-10.2	0.568(25°,1)	-32.2	17-35	T	S: Cyclohexane C_M=5x10^{-4}M	74Sa	b
S66 SULFIDE, dipropyl-			$C_6H_{14}S$			$(C_3H_7)_2S$		L
Al_2A_6	-17.7	--	--	25	C	S: Benzene C_M=0.07M. A=Br R: 0.5(Al_2A_6) + L = AlA_3L	68Rc	
GaA_3	-19.6	3.81(1)	-48.3	25	C	S: Benzene. A=Cl	70Gh	
SnA_4	-12.9	(1)	--	25	C	S: Benzene C_M=0.05-0.08M A=Cl	68Ge 65Gb	
	-11.8	(2)	--	25	C	S: Benzene C_M=0.05-0.08M A=Cl	68Ge 65Gb	
TeA_4	-13.3	1.88(1)	-35.3	25	C	S: Benzene A=Cl	72Pe 73Ph	
I_2	-8.1	2.468(1)	-15.8	25	C	S: Octane	66Ad	
S67 SULFIDE, ethyl phenyl			$C_8H_{10}S$			$C_6H_5SC_2H_5$		L
Al_2A_6	-14.3	--	--	25	C	S: Benzene C_M=0.07M. A=Br R: 0.5(Al_2A_6) + L = AlA_3L	68Rc	
SnA_4	-2.0	(1)	--	25	C	S: Benzene C_M=0.05-0.08M A=Cl	65Gb 68Ge	
S68 SULFIDE, α-naphthyl phenyl-			$C_{16}H_{12}S$			$C_{10}H_7SC_6H_5$		L
Al_2A_6	-9.8	--	--	25	C	S: Benzene C_M=0.07M. A=Br R: 0.5(Al_2A_6) + L = AlA_3L	68Rc	
S69 SULFIDE, pentamethylene-			$C_5H_{10}S$			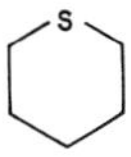		L
AlA_3	-16.99	(1)	--	28.5	C	S: Hexane. A=CH_3	67Hc	
S70 SULFIDE, propylene			C_3H_6S			$H_2CSCHCH_3$		L
AlA_3	-15.24	(1)	--	28.5	C	S: Hexane. A=CH_3	67Hc	

Metal Ion	ΔH, kcal /mole	Log K	ΔS, cal /mole °K	T, °C	M	Conditions	Ref.	Re- marks
S71	SULFIDE, trimethylene-		C_3H_6S			$(CH_2)_3S$		L
AlA$_3$	-16.04	(1)	--	28.5	C	S: Hexane. A=CH$_3$	67Hc	
S72	SULFITE ION		O_3S			SO_3^{2-}		L^{2-}
CrA$_3$	3.99	0.079(1)	13.7	25	C	μ=3.0(NaClO$_4$) A=Ethylenediamine	73Ua	
	0.69	0.042(2)	2.5	25	C	μ=3.0(NaClO$_4$) A=Ethylenediamine	73Ua	q: ±0.05
	-0.79	0.042(3)	-2.5	25	C	μ=3.0(NaClO$_4$) A=Ethylenediamine	73Ua	q: ±0.12
CoA$_3$	1.9	0.61(1)	9	25	C	μ=3.0M(NaClO$_4$ + Na$_2$L) A=Ethylenediamine	73Mc	q: ±0.1
	1.0	0.18(2)	4	25	C	μ=3.0M(NaClO$_4$+Na$_2$L) A=Ethylenediamine	73Mc	q: ±0.3
	0.5	-0.046(3)	2	25	C	μ=3.0M(NaClO$_4$ + Na$_2$L) A=Ethylenediamine	73Mc	q: ±0.3
	-2.5	-0.40(4)	-10	25	C	μ=3.0M(NaClO$_4$ + Na$_2$L) A=Ethylenediamine	73Mc	q: ±0.7
Ag$^+$	-4.95	8.326(20.15°, 1-2)	21.2	2-50	T	0.1-0.6M(Na$_2$SO$_3$) Done under 1 atm. of N$_2$	51Ja	b
	-5.41	8.332(25°,1-2)	20	20-60	T	0.1-1.4M(Na$_2$SO$_3$)	56Kb	b
S73	SULFITE, diethyl-		$C_4H_{10}O_3S$			$(CH_3CH_2)_2SO_3$		L
I$_2$	-2.2	-0.503(32.6°,1)	-9.5	23.5- 44.1	T	S: CCl$_4$	63Da	b
S74	SULFONE, tetramethylene-		$C_4H_8O_2S$			$\underline{CH_2CH_2CH_2CH_2SO_2}$		L
I$_2$	-2.2	-0.172(32.1°,1)	-8.0	25-45.5	T	S: CCl$_4$	63Da	b
S75	SULFOXIDE, bis(dimethylamino)-		$C_4H_{12}ON_2S$			$[(CH_3)_2N]_2SO$		L
AlA$_3$	-24.56	--	--	28	C	S: Hexane. A=CH$_3$ R: 1/2[AlA$_3$]$_2$ + L = AlLA$_3$	68Hd	
S76	SULFOXIDE, dibenzyl-		$C_{14}H_{14}OS$			$C_6H_5CH_2SOCH_2C_6H_5$		L
UA$_2$B$_2$	-3.84	2.994(40°)	1.4	30-60	T	S: Benzene. A=O B=Benzoyltrifluoro- acetone R: UA$_2$B$_2$·nH$_2$O + L = UA$_2$B$_2$L + nH$_2$O	73Ss	b
	-4.9	3.316(40°)	-0.47	30-60	T	S: Benzene. A=O B=Thenoyltrifluoro- acetone R: UA$_2$B$_2$·nH$_2$O + L = UA$_2$B$_2$L + nH$_2$O	73Ss	b

Metal Ion	ΔH, kcal /mole	Log K	ΔS, cal /mole °K	T, °C	M	Conditions	Ref.	Re-marks
S77 SULFOXIDE, didecyl-			$C_{20}H_{42}OS$			$(C_{10}H_{21})_2SO$		L
UA_2B_2	-3.07	2.785(40°)	2.9	30-60	T	S: Benzene. A=O B=Benzoylacetone R: $UA_2B_2 \cdot nH_2O + L = UA_2B_2L + nH_2O$	73Ss	b
	-4.58	3.581(40°)	1.8	30-60	T	S: Benzene. A=O B=Benzoyltrifluoro-acetone R: $UA_2B_2 \cdot nH_2O + L = UA_2B_2L + nH_2O$	73Ss	b
	-3.05	3.797(40°)	7.6	30-60	T	S: Benzene. A=O B=Thenoyltrifluoro-acetone R: $UA_2B_2 \cdot nH_2O + L = UA_2B_2L + nH_2O$	73Ss	b
S78 SULFOXIDE, dihexyl-			$C_{12}H_{26}OS$			$(C_6H_{11})_2SO$		L
UA_2B_2	-5.03	2.841(40°)	-3.1	30-60	T	S: Benzene. A=O B=Benzoylacetone R: $UA_2B_2 \cdot nH_2O + L = UA_2B_2L + nH_2O$	73Ss	b
	-3.81	3.678(40°)	4.7	30-60	T	S: Benzene. A=O B=Benzoyltrifluoro-acetone R: $UA_2B_2 \cdot nH_2O + L = UA_2B_2L + nH_2O$	73Ss	b
	-3.52	3.986(40°)	7.0	30-60	T	S: Benzene. A=O B=Thenoyltrifluoro-acetone R: $UA_2B_2 \cdot nH_2O + L = UA_2B_2L + nH_2O$	73Ss	b
TeA_4	-10.7	(1)	--	25	C	S: Benzene. A=Cl	72Pe	
S79 SULFOXIDE, diisopentyl			$C_{10}H_{22}OS$			$(C_5H_{11})_2SO$		L
UA_2B_2	-2.60	2.715(40°)	4.1	30-60	T	S: Benzene. A=O B=Benzoylacetone R: $UA_2B_2 \cdot nH_2O + L = UA_2B_2L + nH_2O$	73Ss	b
	-4.46	3.623(40°)	2.3	30-60	T	S: Benzene. A=O B=Benzoyltrifluoro-acetone R: $UA_2B_2 \cdot nH_2O + L = UA_2B_2L + nH_2O$	73Ss	b
	-3.81	3.832(40°)	5.4	30-60	T	S: Benzene. A=O B=Thenoyltrifluoro-acetone R: $UA_2B_2 \cdot nH_2O + L = UA_2B_2L + nH_2O$	73Ss	b
S80 SULFOXIDE, dimethyl-			C_2H_6OS			$(CH_3)_2SO$		L
NiA_2	-4.50	1.760(1)	-7.0	25	C	S: Benzene. A=2,2,6,6-Tetramethyl-3,5-heptanedione	73Ce	

Metal Ion	ΔH, kcal /mole	Log K	ΔS, cal /mole °K	T, °C	M	Conditions	Ref.	Remarks
S80, cont.								
NiA_2, cont.	-0.61	(2)	--	25	C	S: Benzene. A=2,2, 6,6 Tetramethyl-3,5-heptanedione	73Ce	
CuA_2	-8.5	(1)	--	25	C	S: CCl_4. A=Hexafluoroacetyl-acetone	70Pa	
	-7.2	3.549(1)	-7.9	25	C	S: o-Dichloro-benzene A=Hexafluoroacetyl-acetone	72Nd	q: ±1.3
	-6.3	(1)	--	25	C	S: Methylene chloride. A=Hexa-fluoroacetylacetonate	74Da	
CuA_2B	-0.5	--	--	25	C	S: CCl_4. A=Hexa-fluoroacetylacetone B=Dimethyl acetate R: $CuA_2B + L = CuA_2L + B$	72Nd	
	-2.6	--	--	25	C	S: CCl_4. A=Hexa-fluoroacetylacetone B=Ethylacetate R: $CuA_2B + L = CuA_2L + B$	72Nd	
	-0.6	--	--	25	C	S: $o\text{-}Cl_2C_6H_4$ A=Hexafluoroacetyl-acetone B=Dimethyl acetate R: $CuA_2B + L = CuA_2L + B$	72Nd	
	-2.5	--	--	25	C	S: $o\text{-}Cl_2C_6H_4$ A=Hexafluoroacetyl-acetone B=Ethyl acetate R: $CuA_2B + L = CuA_2L + B$	72Nd	
AlA_3	-28.64	--	--	28	C	S: 5% Benzene in Hexane. R: $1/2 [AlA_3]_2 + L = AlLA_3$ A=CH_3	68Hd	
$SnAB_3$	-8.2	0.959(1)	-23	26	C	S: CCl_4. A=Cl B=CH_3	66Bg	
I_2	-4.4	0.964(35.4°,1)	-9.8	23.4-46.1	T	S: CCl_4	63Da	b
S81 SULFOXIDE, dimethylamino, methyl-		C_3H_9ONS				$CH_3SON(CH_3)_2$		L
AlA_3	-26.27	--	--	28	C	S: Hexane R: $1/2[AlA_3]_2 + L = AlLA_3$. A=$CH_3$	68Hd	

Metal Ion	ΔH, kcal /mole	Log K	ΔS, cal /mole °K	T, °C	M	Conditions	Ref.	Remarks
S82 SULFOXIDE, diphenyl-			$C_{12}H_{10}OS$			$(C_6H_5)_2SO$		L
UA_2B	-4.29	2.520(40°)	-2.2	30-60	T	S: Benzene. A=O B=Benzoyltrifluoro-acetone R: $UA_2B_2 \cdot nH_2O + L = UA_2B_2L + nH_2O$	73Ss	b
UA_2B_2	-4.6	2.876(40°)	-1.5	30-60	T	S: Benzene. A=O B=Thenoyltrifluoro-acetone R: $UA_2B_2 \cdot nH_2O + L = UA_2B_2L + nH_2O$	73Ss	b
TeA_4	-11.1	(1)	--	25	C	S: Benzene. A=Cl	72Pe	
S83 SULFOXIDE, phenyl propyl-			$C_9H_{12}OS$			$CH_3CH_2CH_2SOC_6H_5$		L
TeA_4	-11.2	(1)	--	25	C	S: Benzene. A=Cl	72Pe	
S84 SULFOXIDE, tetramethylene			C_4H_8OS			$CH_2CH_2CH_2CH_2S{:}O$ (ring)		L
AlA_3	-28.25	--	--	28	C	S: 5% Benzene in Hexane. A=CH_3 R: $1/2[AlA_3]_2 + L = AlLA_3$	68Hd	
I_2	-4.4	1.070(35°,1)	-9.5	24.8-45	T	S: CCl_4	63Da	b
S85 SULFUR DIOXIDE			O_2S			SO_2		L
$RhABC_2$	-7.9	1.875(30°,1)	-17	10-60	T	S: Chlorobenzene A=Cl. B=CO. C=Ph_3P	71Vb	b
	-9.4	(1)	--	10-60	T	S: Chlorobenzene A=I. B=CO. C=Ph_3P	71Vb	b
$IrABC_2$	-10.5	3.100(30°,1)	-21	10-60	T	S: Chlorobenzene A=Cl. B=CO. C=Ph_3P	71Vb	b
	-25	(1)	--	10-60	T	S: Chlorobenzene A=I. B=CO. C=Ph_3P	71Vb	b

T

Metal Ion	ΔH, kcal /mole	Log K	ΔS, cal /mole °K	T, °C	M	Conditions	Ref.	Remarks
T1 TARTARIC ACID			$C_4H_6O_6$			$HO_2CCH(OH)CH(OH)CO_2H$		L^{2-}
Ca^{2+}	-3.51	1.84(1)	-3.4	25	-	--	53Ka	
La^{3+}	0.0	4.604(25°,1)	22.7	15-35	T	μ=0	72Dc	b
	10.7	2.986(25°,2)	47.8	15-35	T	μ=0	72Dc	b
	1.34	2.48(25°)	15.8	15-35	T	μ=0. R: $La^{3+} + HL^- = LaHL^{2+}$	72Dc	b
ZnA_3^+	-10.3	6.070(1)	6.79	25	C	C_M=0.01M C_L=0.01M. A=OH^-	73Kh	
	-9.1	6.070(1)	6.79	15-35	T	C_M=0.01M C_L=0.01M. A=OH^-	73Kh	b
T2 TELLURATE ION			O_6H_2Te			$H_2TeO_6^{4-}$		L^{4-}
CuA_4^-	-20	11.947(50°)	-7.1	40-60	T	1.2M(ClO_4^-) R: $CuA_4^- + 2L^{4-}$ =	53Lb	b

Metal Ion	ΔH, kcal /mole	Log K	ΔS, cal /mole °K	T, °C	M	Conditions	Ref.	Remarks
T2, cont.								
CuA_4^-, cont.						$Cu(HTeO_6)_2^{7-}$ + $2OH^-$ + $2H_2O$. $A=OH^-$		
T3 TELLURITE ION			O_3Te			TeO_3^{2-}		L^{2-}
CrA_3^{3+}	4.90	-0.046(1)	16.2	25	C	μ=3.0($NaClO_4$) A=Ethylenediamine	73Ua	
	-3.49	-0.126(2)	-12.3	25	C	μ=3.0($NaClO_4$) A=Ethylenediamine	73Ua	
CoA_3^{3+}	1.1	0.301(1)	5	25	C	μ=3.0M($NaClO_4$ + Na_2L) A=Ethylenediamine	73Mc	q: ±0.1
	0.5	0.041(2)	2	25	C	μ=3.0M($NaClO_4$ + Na_2L) A=Ethylenediamine	73Mc	q: ±0.2
	-1.6	-0.30(3)	-6	25	C	μ=3.0M($NaClO_4$ + Na_2L) A=Ethylenediamine	73Mc	q: ±0.3
T4 TELLURIUM TETRACHLORIDE			Cl_4Te			$TeCl_4$		L
Al_2A_6	-26.6	--	--	25	C	S: Benzene. A=Br R: 0.5[Al_2A_6] + L = [AlA_3L]	72Pe	
Ga_2A_6	-10.7	--	--	25	C	S: Benzene. A=Cl R: 0.5[Ga_2A_6] + L = [GaA_3L]	72Pe	
	-16.8	--	--	25	C	S: Benzene. A=Br R: 0.5[Ga_2A_6] + L = [GaA_3L]	72Pe	
T5 TETRACYCLINE, (hydrochloride)			$C_{22}H_{24}O_8N_2$			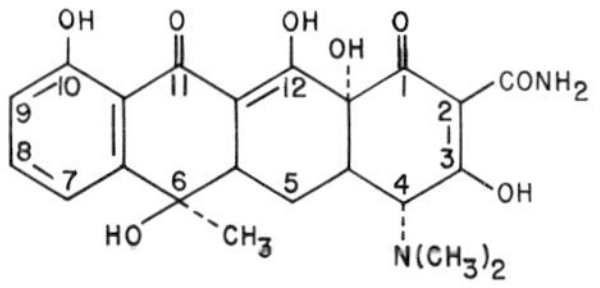		HL^-
Cu^{2+}	-7.6	7.80(1)	10.2	25	C	μ=0.01	66Bb	
	-1.80	5.34(2)	18.4	25	C	μ=0.01	66Bb	
T6 TETRACYCLINE, 7-chloro (hydrochloride)			$C_{22}H_{23}O_8N_2Cl$			$C_{22}H_{23}N_2O_8(Cl)$		HL^-
Cu^{2+}	-11.7	7.32(1)	-4.0	25	C	μ=0.01	66Bb	
	-1.80	5.09(2)	17.3	25	C	μ=0.01	66Bb	
T7 TETRACYCLINE, 7-chloro-6,6 dimethyl-(hydrochloride)			$C_{23}H_{25}O_7N_2$			$C_{22}H_{22}N_2O_7(Cl)(CH_3)_2$		HL^-
Cu^{2+}	-12.3	7.85(1)	-5.4	25	C	μ=0.01	66Bb	
	-2.60	5.46(2)	16.3	25	C	μ=0.01	66Bb	
T8 TETRACYCLINE, 7-chloro-4-epi-(hydrochloride)			$C_{22}H_{23}O_8N_2Cl$			$C_{22}H_{24}N_2O_8(Cl)$		HL^-

Metal Ion	ΔH, kcal /mole	Log K	ΔS, cal /mole °K	T, °C	M	Conditions	Ref.	Remarks
T8, cont.								
Cu^{2+}	-11.5	7.63($\underline{1}$)	-3.7	25	C	μ=0.01	66Bb	
	-2.00	5.07($\underline{2}$)	16.5	25	C	μ=0.01	66Bb	
T9 TETRADECANE, 5,10-dithio-			$C_{12}H_{26}S_2$			$CH_3(CH_2)_3S(CH_2)_4S(CH_2)_3CH_3$		L
TiA_4	-22.9	3.37($\underline{1}$)	-61.4	25	C	S: Benzene. A=Cl	70Gh	
SnA_4	-20.5	($\underline{1}$)	--	--	C	S: Benzene C_M=0.05-0.08M A=Cl	65Gb	u
TeA_4	-12.0	3.34($\underline{1}$)	-25.0	25	C	S: Benzene. A=Cl	73Ph	
	-4.4	($\underline{2}$)	--	25	C	S: Benzene. A=Cl	73Ph	
T10 TETRADECANE, 3,6,9,12-tetra-aza-1,14-diamino-			$C_{10}H_{24}N_6$			$H_2N(CH_2)_2N(CH_2)_2N(CH_2)_2N$-$(CH_2)_2N(CH_2)_2NH_2$		L
Cu^{2+}	-19.62	25.13(25°,$\underline{1}$)	$\underline{49.3}$	30-40	T	0.1M(KCl)	57Ja	b
T11 TETRAETHYLENE GLYCOL, dimethyl ether			$C_{10}H_{22}O_5$			$CH_3OCH_2(CH_2OCH_2)_3CH_2OCH_3$		L
NaA	-7.1	2.230(25°)	-14	0-50	T	S: Tetrahydrofuran A=Fluorenyl ion R: NaA + L = NaLA	70Ce	b
T12 TETRAPHENYLBORIDE ION			$C_{24}H_{20}B$			$(C_6H_5)_4B^-$		L^-
Na^+	1.8	4.27(25°,$\underline{1}$)	$\underline{25.6}$	-70-25	T	S: Dimethosyethane	65Cb	b,q: ±0.3
	2.0	4.266(20°,$\underline{1}$)	26	10-40	T	S: Dimethoxyethane	73Sk	b
	1.3	4.06(25°,$\underline{1}$)	$\underline{22.9}$	-70-25	T	S: Tetrahydrofuran	65Cb	b,q: ±0.3
NaA^+	2.5	4.222(20°,$\underline{1}$)	28	10-40	T	S: Dimethoxyethane A=Dibenzo-18-crown-6	73Sk	b
NAB_3^+	2.3	4.45(25°,$\underline{1}$)	$\underline{28.1}$	-70-25	T	S: Dimethoxyethane A=Butyl. B=Amyl	65Cb	b,q: ±0.3
	1.3	4.25(25°,$\underline{1}$)	$\underline{23.8}$	-70-25	T	S: Tetrahydrofuran A=Butyl. B=Amyl	65Cb	b,q: ±0.3
Cs^+	1.55	5.80(25°,$\underline{1}$)	$\underline{31.7}$	-70-25	T	S: Tetrahydrofuran	65Cb	b,q: ±0.3
	3.5	4.55(25°,$\underline{1}$)	$\underline{32.6}$	-70-25	T	S: Dimethoxyethane	65Cb	b,q: ±0.3
T13 2-THIAZOLIDINETHIONE			$C_3H_5NS_2$			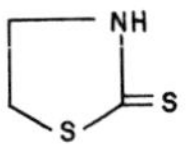		
I_2	-10.0	3.164(25°,$\underline{1}$)	$\underline{-19.1}$	0-45	T	S: Chloroform	66Bc	b,q: ±1
T14 THIOCYANIC ACID			CHNS			HSCN		L^-
Nd^{3+}	-5.47	0.81($\underline{1}$)	-14.7	25	T	μ=0.1($NaClO_4$)	65Ce	

Metal Ion	ΔH, kcal /mole	Log K	ΔS, cal /mole °K	T, °C	M	Conditions	Ref.	Re- marks
T14, cont.								
V^{2+}	-5.2	1.431(25°,1)	-11	11-45	T	μ=0.84($LiClO_4$)	68Ma	b,q: ±0.7
V^{3+}	-3.5	2.068(25°,1)	-2.2	5-37	T	μ=1.0($HClO_4$)	67Ba	b,q: ±0.3
	-3.6	2.0(25°,1)	-2.9	25-50	T	μ=2.6	51Fa	b
VO^{2+}	0.4	0.92(25°,1)	5.5	25-50	T	μ=2.6	51Fa	b
$CrA_2B_2^-$	1.2	0.653(25°,1)	7.3	25-55	T	μ=1.00F($NaNO_3$) A=C_2O_4. B=H_2O R: cis-$CrA_2B_2^-$ + L = cis-CrA_2LB^{2-}+ B	72Ag	b
CrA_6^{3+}	-2.13	3.09	6.9	25	T	μ=0. R: CrA_6^{3+} + L^- = CrA_5L^{2+} + A A=H_2O	55Pa 57Pb	
	2.76	3.09	21.7	95	T	μ=0. R: CrA_6^{3+} + L^- = $CrLA_5^{2+}$ + A A=H_2O	55Pa	
	-1.4	1.87	0.8	25	T	1M($NaClO_4$) R: CrA_6^{3+} + L^- = $CrLA_5^{2+}$ + A. A=H_2O	54Pb	
CrA_5B	-5	2.233	-6	45	T	μ=0.106M A=NH_3. B=H_2O R: CrA_5B^{3+} + L^- = CrA_5L^{2+} + B	67De	
U^{4+}	-5.7	1.9(1)	-10	25	T	μ=2.0	55Da	
	-3.8	2.86(1-2)	-0.3	25	T	μ=2.0	55Da	
UO_2^{2+}	-0.770	0.748(1)	0.84	25	C	μ=1.0($NaClO_4$)	71Ac	
	1.15	1.0(25°,1)	8.5	20-40	T	μ=0 S: 20 w % CH_3OH	57Ba	b
	-0.136	-0.032(2)	-4.73	25	C	μ=1.0($NaClO_4$)	71Ac	
	0.069	0.460(3)	4.3	25	C	μ=1.0($NaClO_4$)	71Ac	
Am^{3+}	-4.36	0.51(1)	-12.3	25	T	μ=0.1($NaClO_4$)	65Ce	
	1.6	0.34(25°,1)	7	17-45	T	μ=1.0($NaClO_4$)	72Hd	b,q: ±0.7
	-6	0.14(25°,1-3)	-20	17-45	T	μ=1.0($NaClO_4$)	72Hd	b,q: ±4
	-8	-0.20(25°)	-27	17-45	T	μ=1.0($NaClO_4$) R: AmL^{2+} + $2L^-$ = AmL_3	72Hd	b,q: ±5
Mn^{2+}	-0.92	1.23(1)	2.5	25	C	μ=0	67Na	
	-3.29	1.15(25°,1)	-5.8	25-45	T	μ=0	71Dd	b
Fe^{3+}	-1.5	3.029(1)	8.7	25	T	μ=0	56La 55La	
	-0.52	3.1(1)	12	25	T	μ=0	61Ya 60Gb	
	-1.6	2.937(1)	8	25	T	μ=0	53Bb	
	-2.2	2.17(1)	2	25	T	0.1M(Li^+, H^+, NO_3^-, SCN^-) at [H^+] = 0.1M	65Mb	
	-1.5	2.143(1)	4.8	25	T	μ=0.5	56La	

Metal Ion	ΔH, kcal /mole	Log K	ΔS, cal /mole °K	T, °C	M	Conditions	Ref.	Remarks
T14, cont.								
Fe^{3+}, cont.	-2.4	2.02(1)	1	25	T	0.6M(Li^+, H^+, NO_3^-, SCN^-) at $[H^+]$ = 0.1M	65Mb	
	-1.5	1.91(1)	4	25	T	1.1M(Li^+, H^+, NO_3^-, SCN^-) at $[H^+]$ = 0.1M	65Mb	
	-1.6	2.057(1)	4.1	25	T	μ=1.28	53Bb	
	-2.6	2.08(1)	1	25	T	3.1M(Li^+, H^+, NO_3^-, SCN^-) at $[H^+]$ = 0.1M	65Mb	
	-2.2	2.26(1)	3	25	T	6.1M(Li^+, H^+, NO_3^-, SCN^-) at $[H^+]$ = 0.1M	65Mb	
	-0.3	1.31(2)	5.0	25	T	μ=0.5	56La	
	-5.0	3.60(1-2)	-0.3	25	T	0.61M(Li^+, H^+, NO_3^-, SCN^-) at $[H^+]$ = 0.1M	65Mb	
	-5.2	2.99(1-2)	-3	25	T	1.1M(Li^+, H^+, NO_3^-, SCN^-) at $[H^+]$ = 0.1M	65Mb	
	-5.8	3.93(1-2)	-2	25	T	6.1M(Li^+, H^+, NO_3^-, SCN^-) at $[H^+]$ = 0.1M	65Mb	
FeA_5^{2-}	-8.8	2.41(25°,1)	-18.7	18.3-34.6	T	μ=1.00M. A=CN	72Ee	b,q: ±0.6
[Fe]	-11.7	3.37(1)	-23.9	--	-	Chironomus hemoglobin	60Gb	
	-8.8	2.20(1)	-19.7	--	-	Hemoglobin	60Gb	
$[Fe^{3+}]$	-14.4	3.602	-32.5	25	C	[Fe]=G. gouldii Hemerythrin = $HrFe_2$ Tris-cacodylate buffer (≈0.1M Tris) R: $HrFe_2(H_2O)_m + L = HrFe_2L + mH_2O$	71Lb	g:Na
$[Fe^{3+}]$	-4.99	2.83(1)	-4.06	20	T	6.0. Human Methaemoglobin A	68Ac	d: 0.05
	-6.23	2.76(1)	-8.60	20	T	6.5	68Ac	
	-7.47	2.66(1)	-13.3	20	T	7.0	68Ac	
	-6.37	2.53(1)	-10.2	20	T	7.5	68Ac	
	-4.01	2.43(1)	-2.6	20	T	8.0	68Ac	
	-2.59	2.38(1)	-2.1	20	T	8.5	68Ac	
$[Fe^{3+}]$	-4.20	2.94(1)	-0.89	20	T	6.0. Human Methaemoglobin C	68Ac	d: 0.05
	-5.56	2.93(1)	-5.56	20	T	6.5	68Ac	
	-6.22	2.87(1)	-8.08	20	T	7.0	68Ac	
	-7.80	2.75(1)	-14.0	20	T	7.5	68Ac	
	-6.22	2.53(1)	-9.65	20	T	8.0	68Ac	
	-3.00	2.31(1)	0.34	20	T	8.5	68Ac	
	-1.62	2.24(1)	4.74	20	T	9.0	68Ac	
Co^{2+}	-1.63	1.72(1)	2.2	25	C	μ=0	67Na	
	-4.45	1.87(25°,1)	-6.4	25-45	T	μ=0	71Dd	b,q: ±0.46
Ni^{2+}	-2.26	1.76(1)	0.5	25	C	μ=0	67Na 61Mg	
	-5.15	1.98(25°,1)	-8.2	25-45	T	μ=0	71Dd	b,q: ±0.52

Metal Ion	ΔH, kcal /mole	Log K	ΔS, cal /mole °K	T, °C	M	Conditions	Ref.	Remarks
T14, cont.								
PdA_2	-4.3	2.336(40°,1)	-3	25-55	T	μ=0.003($NaNO_3$) A=o-Phenylenebis-dimethylarsine	67Ed	b
	-4.2	2.60(1)	-2.2	25	T	μ=0. A=o-Phenylene-bisdimethylarsine	67Ec 67Ed	
$PdAB^{2+}$	-4.7	2.230(27°)	-5.4	10.6-61.4	T	μ=0.5 A=Diethylenetriamine B=Br. R: $PdAB^+ + L^- = PdAL^+ + B^-$	67Hd	b
	-4.4	2.021(27°)	4.4	20-82	T	μ=0.5. A=1,1,7,7-Tetraethyldi-ethylenetriamine B=Br. R: $PdAB^+ + L^- = PdAL^+ + B^-$	67Hd	b
PtA_2	-4.2	2.292(1)	-3.5	25	T	μ=0. A=o-Phenylene-bisdimethylarsine	67Ec	q: ±0.4
	-4.3	2.037(40°,1)	-4.3	25-55	T	μ=3x10^{-3}($NaNO_3$) A=o-Phenylenebisdi-methylarsine	67Ec	b,q: ±0.4
	-2.3	0.199(1)	-8.2	25	C	μ=1M(NaF) C_M=0.040-0.050M A=Ethylenediamine	73Ma	q: ±0.3
	-2.25	3.66(23°,1)	9.3	23-56	T	μ=0. S: CH_3OH A=o-Phenylene-bisdimethyl arsine	66Db	b
	-1.3	4.170(35°,1)	14.6	25-55	T	μ=0. S: Methanol A=o-Phenylene-bisdimethylarsine	67Ec	b,q: ±0.4
PtA_4	-3.0	0.041(1)	-9.9	25	C	μ=1M(NaF) C_M=0.040-0.050M A=NH_3	73Ma	q: ±0.6
Cu^{2+}	-3.00	2.33(1)	0.6	26	C	μ=0	67Na	
CuA^{2+}	0.82	2.73(1)	15.2	25	C	S: Methanol A=NN'-bis(2-amino-ethyl) propane-1,3-diamine	72Bd	
	-0.73	3.44(1)	13.3	25	C	S: Methanol A=NN'-bis(3-amino-propyl) propane-1,3-diamine	72Bd	
	0.70	3.22(1)	17.0	25	C	S: Methanol A=Triethylene-tetraamine	72Bd	
CuA_2^{2+}	-0.93	2.76(1)	9.5	25	C	S: Methanol A=NN-Dimethyl-ethylenediamine	72Bf	
	-0.83	3.93(1)	15.2	25	C	S: Methanol A=NN'-Dimethyl-ethylenediamine	72Bf	
	1.78	2.07(1)	15.4	25	C	S: Methanol	72Bd	

Metal Ion	ΔH, kcal /mole	Log K	ΔS, cal /mole °K	T, °C	M	Conditions	Ref.	Remarks
T14, cont.								
CuA_2^{2+}, cont.	0.57	2.79(1)	14.7	25	C	A=Ethylenediamine S: Methanol	72Bf	
	-0.02	2.70(1)	12.2	25	C	A=N-Methyl-ethylenediamine S: Methanol	72Bd	
						A=Trimethylene-diamine		
Ag^+	-28	11.18(25°,1-4)	-43	0-45	T	0.15	49Ja 56Gb	b
Zn^{2+}	0.22	1.85(1)	9.2	25	C	μ=0	67Na	
	-2.67	1.50(25°,1)		25-45	T	μ=0	71Dd	b,q: ±0.55
	-1.38	0.708(1)	-1.4	25	C	μ=1.0($NaClO_4$)	71Ad	
	-0.43	0.334(2)	0	25	C	μ=1.0($NaClO_4$)	71Ad	
	-0.19	0.135(3)	0	25	C	μ=1.0($NaClO_4$)	71Ad	
	-1.84	0.329(4)	-4.5	25	C	μ=1.0($NaClO_4$)	71Ad	
	-5.7	2.014(20°,1-4)	-10.2	20&40	T	--	66Gb 58Gc	a,b
Cd^{2+}	-0.70	2.51(1)	9.2	25	C	μ=0	67Na 57Gb	
	-3.43	2.15(25°,1)		25-45	T	μ=0	71Dd	b,q: ±0.49
	-2.24	1.43(1)	-1.0	25	C	μ=0.25($NaClO_4$)	68Gd	
	-2.24	1.36(1)	-1.3	25	C	μ=0.5($NaClO_4$)	68Gd	
	-2.29	1.32(1)	-1.6	25	C	μ=1($NaClO_4$)	68Gd	
	-2.22	1.34(1)	-1.3	25	C	μ=2($NaClO_4$)	68Gd	
	-1.94	1.41(1)	-0.1	25	C	μ=3($NaClO_4$)	68Gd	
	-1.94	1.42(1)	0.0	25	C	3.0M($NaClO_4$)	68Gc	
	-2.67	0.67(2)	-6.0	25	C	μ=0.25($NaClO_4$)	68Gd	
	-2.45	0.68(2)	-5.0	25	C	μ=0.5($NaClO_4$)	68Gd	
	-1.93	0.67(2)	-3.4	25	C	μ=1($NaClO_4$)	68Gd	
	-1.85	0.70(2)	-3.0	25	C	μ=2($NaClO_4$)	68Gd	
	-1.73	0.84(2)	-2.0	25	C	μ=3($NaClO_4$)	68Gd	
	-1.73	0.83(2)	-2.0	25	C	3.0M($NaClO_4$)	68Gc	
	-2.2	0.04(3)	-7.2	25	C	μ=1($NaClO_4$)	68Gd	
	-2.2	0.20(3)	-6.5	25	C	μ=2($NaClO_4$)	68Gd	
	-1.57	0.23(3)	-4.2	25	C	μ=3($NaClO_4$)	68Gd	
	-1.157	0.23(3)	-4.2	25	C	3.0M($NaClO_4$)	68Gc	
	-1.04	0.00(4)	-3.5	25	C	μ=3($NaClO_4$)	68Gd	
	-1.04	0.00(4)	-3.5	25	C	3.0M($NaClO_4$)	68Gc	
Hg^{2+}	-11.88	9.079(1)	1.7	25	C	μ=1.0($NaClO_4$)	71Ad	
	-12.05	7.778(2)	-4.8	25	C	μ=1.0($NaClO_4$)	71Ad	
	-4.88	2.842(3)	-3.3	25	C	μ=1.0($NaClO_4$)	71Ad	
	-5.02	1.973(4)	-7.9	25	C	μ=1.0($NaClO_4$)	71Ad	
	-30	17.3(1-2)	-20	25	T	μ=0	62Tb 66Gb	
	-29	19.9(1-3)	-6	25	T	μ=0	62Tb	
	-36	21.7(1-4)	-20	25	T	μ=0	62Tb	
	-35	21.3(1-4)	-20	25	T	μ=0	56Ta	
	-35.0	21.23(25°,1-4)	-20.5	20-30	T	μ≃0.3	56Ta	b

Metal Ion	ΔH, kcal /mole	Log K	ΔS, cal /mole °K	T, °C	M	Conditions	Ref.	Re-marks
Tl4, cont.								
HgA^+	-11.2	6.039(1)	-10.6	20	C	μ=0.1(KNO_3) A=CH_3	65Sa	
In^{3+}	-1.66	2.56(1)	6.1	25	C	μ=2.00M($NaClO_4$)	69Rd	
	-3.8	0.98(2)	-8.4	25	C	μ=2.00M($NaClO_4$)	69Rd	
	2.4	1.06(3)	12.7	25	C	μ=2.00M($NaClO_4$)	69Rd	
Tl^+	-2.96	0.80(1)	-6.4	25	T	μ=0	53Ba	
	-2.89	0.556(25°,1)	-7.2	10-60	T	μ=0($LiClO_4$)	72Fd	b,q: ±0.41
	-1.82	0.243(25°,1)	-4.9	10-60	T	μ=0.50($LiClO_4$)	72Fd	b,q: ±0.56
	-2.25	0.152(25°,1)	-7.1	10-60	T	μ=1.0($LiClO_4$)	72Fd	b,q: ±0.40
	-2.72	0.117(25°,1)	-8.8	10-60	T	μ=2.0($LiClO_4$)	72Fd	b,q: ±0.47
	-1.28	0.083(25°,1)	-3.9	10-60	T	μ=3.0($LiClO_4$)	72Fd	b,q: ±0.62
	-2.70	0.130(25°,1)	-8.6	10-60	T	μ=4.0($LiClO_4$)	72Fd	b,q: ±1.00
	0.21	0.255(25°)	2.2	10-60	T	μ=0. R: Tl^++ Cl^-+ L = TlClL	72Fd	b,q: ±0.91
	2.20	-0.155(25°)	6.6	10-60	T	μ=0.50($LiClO_4$) R: Tl^++ Cl^-+ L = TlClL	72Fd	b,q: ±0.70
	0.01	-0.022(25°)	0.5	10-60	T	μ=1.0($LiClO_4$) R: Tl^++ Cl^-+ L = TlClL	72Fd	b,q: ±2.40
	0.08	-0.180(25°)	-0.3	10-60	T	μ=2.0($LiClO_4$) R: Tl^++ Cl^-+ L = TlClL	72Fd	b,q: ±0.68
	-2.04	-0.201(25°)	-7.9	10-60	T	μ=3.0($LiClO_4$) R: Tl^++ Cl^-+ L = TlClL	72Fd	b,q: ±0.98
	-0.9	-0.222(25°)	-4.0	10-60	T	μ=4.0($LiClO_4$) R: Tl^++ Cl^-+ L = TlClL	72Fd	b,q: ±1.0
Pb^{2+}	0.30	1.09(1)	6.0	25	C	μ=0	67Na 57Gc	
	-4.1	0.418(25°,1)	-12	15-65	T	μ=3M($LiClO_4$)	69Fb	b,q: ±1.5
	1.1	0.293(25°,2)	5	15-65	T	μ=3M($LiClO_4$)	69Fb	b,q: ±2.4
	-0.9	0.205(25°,3)	-2	15-65	T	μ=3M($LiClO_4$)	69Fb	b,q: ±1.4
	0.5	-0.337(25°,4)	0	15-65	T	μ=3M($LiClO_4$)	69Fb	b,q: ±1.8
	-6.2	0.396(25°,5)	-18	15-65	T	μ=3M($LiClO_4$)	69Fb	b,q: ±1.7
Bi^{3+}	8.1	1.28(25°,1)	34	15-55	T	μ=3M	71Fa	b
	1.9	1.39(25°,2)	15	15-55	T	μ=3M	71Fa	b,q: ±1
	1.0	1.07(25°,3)	7	15-55	T	μ=3M	71Fa	b,q: ±2

Metal Ion	ΔH, kcal /mole	Log K	ΔS, cal /mole °K	T, °C	M	Conditions	Ref.	Remarks
T14, cont.								
Bi^{3+}, cont.	-4	1.46(25°,4)	-10	15-55	T	μ=3M	71Fa	b,q: ±4
I_2	-8.1	1.94(1)	-18.4	7.6-16.5	T	μ=1 to 5. ΔH and ΔG were extrapolated to 25°C	62Lb	b
T15 THIOCYANIC ACID, allyl ester			C_4H_5NS			$CH_2:CHCH_2SCN$		L
I_2	-3.6	0.015(20°,1)	-12	9-43.6	T	S: CCl_4 $C_M=1.2x10^{-3}M$	65Pc	b,q: ±0.8
T16 THIOCYANIC ACID, ethyl ester			C_3H_5NS			C_2H_5SCN		L
I_2	-3.9	0.045(20°,1)	-13	10.5-44.2	T	S: Heptane $C_M=8.0x10^{-4}M$	65Pc	b,q: ±0.8
T17 THIOCYANIC ACID, methyl ester			C_2H_3NS			CH_3SCN		L
I_2	2.2	0.176(25°,1)	8.2	25-41.7	T	S: CCl_4 $C_M=0.5x10^{-2}$ $C_L=0.35$	65Wb	b
T18 THIODIGLYCOLIC ACID			$C_4H_6O_4S$			$S(CH_2CO_2H)_2$		$L^=$
Ce^{3+}	3.06	2.654(1)	22.4	25	C	μ=1.00M($NaClO_4$)	73Dd	
	1.79	1.832(2)	14.4	25	C	μ=1.00M($NaClO_4$)	73Dd	
	2.03	5.359	31.3	25	C	μ=1.00M($NaClO_4$) R: $M^{3+} + L^{2-} + H^+ = MHL^{2+}$	73Dd	q: ±0.3
	1.72	2.295	16.5	25	C	μ=1.00M($NaClO_4$) R: $MHL^{2+} + L^{2-} = MHL_2$	73Dd	q: ±0.6
T19 THIOPHENE			C_4H_4S			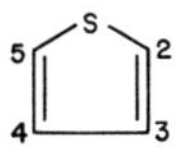		L
Al_2A_6	-18.4	--	--	25	C	S: Benzene $C_M=0.07M$. A=Br R: $0.5(Al_2A_6) + L = AlA_3L$	68Rc	
I_2	-0.4	(1)	--	25	C	S: Octane	66Ad	
T20 THIOPHENE, 2,5-dimethyl-			C_6H_8S			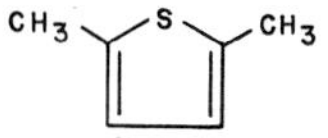		L
Al_2A_6	-18.9	--	--	25	C	S: Benzene $C_M=0.07M$. A=Br R: $0.5(Al_2A_6) + L = AlA_3L$	68Rc	

Metal Ion	ΔH, kcal /mole	Log K	ΔS, cal /mole °K	T, °C	M	Conditions	Ref.	Remarks
T21	THIOPHENE, tetrahydro-		C_4H_8S			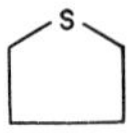		L
TiA_4	-12.0	(1)	--	25	C	S: Benzene C_M=0.05-0.08M A=Cl	68Ge	
[Co]	-15.3	--	--	24	C	[Co]=Methyl cobaloxime S: Dichloromethane R: $[Co]_2 + 2L = 2[Co]L$	73Cf	
CuA_2	-4.7	(1)	--	25	C	S: Methylene chloride. A=Hexafluoroacetylacetonate	74Da	
HgA_2	-15.9	1.699(1)	-44.9	30	C	S: Benzene. A=Cl $C_M=1-5x10^{-3}M$	73Fa	
	-16.3	1.826(1)	-45.4	30	C	S: Benzene. A=Br $C_M=1-5x10^{-3}M$	73Fa	
	-13.9	1.580(1)	-38.7	30	C	S: Benzene. A=I $C_M=1-5x10^{-3}M$	73Fa	
AlA_3	-16.95	(1)	--	28.5	C	S: Hexane. A=CH_3	67Hc	
I_2	-10.0	(1)	--	25	C	S: Octane	66Ad	
T22	THIOSULFATE ION		O_3S_2			$S_2O_3^{2-}$		L^{2-}
Ba^{2+}	2.6	2.21(25°,1)	18.8	25&35	T	μ=0	49Da 52Ua	a,b
CrA_3	1.74	0.543(1)	8.4	25	C	μ=3.0($NaClO_4$) A=Ethylenediamine	73Ua	q: ±0.06
	0.79	0.145(2)	3.3	25	C	μ=3.0($NaClO_4$) A=Ethylenediamine	73Ua	q: ±0.06
	0.29	0.042(3)	1.2	25	C	μ=3.0($NaClO_4$) A=Ethylenediamine	73Ua	q: ±0.12
Fe^{3+}	8.9	2.10(25°,1)	47.2	1-30	T	μ=0	54Pa	b
	9.8	2.10(25°,1)	42.5	1-30	T	μ=0.47	54Pa	b
CoA_3	0.8	0.72(1)	6.0	25	C	μ=3.0M($NaClO_4$+NaL) A=Ethylenediamine	73Mb	q: ±0.1
	0.4	0.23(2)	2.5	25	C	μ=3.0M($NaClO_4$+NaL) A=Ethylenediamine	73Mb	q: ±0.2
	0.2	0.00(3)	1	25	C	μ=3.0M($NaClO_4$+NaL) A=Ethylenediamine	73Mb	q: ±0.1
	-1.4	-0.30(4)	-6	25	C	μ=3.0M($NaClO_4$+NaL) A=Ethylenediamine	73Mb	q: ±0.4
CoA_6^{3+}	1.1	3.2(25°,1)	18	15-35	T	μ=0. A=NH_3	55Gb	b
PdA_2	-2.8	1.724(40°,1)	-1	25-55	T	μ=0.02($NaNO_3$) A=o-Phenylene-bisdimethylarsine	67Ed	b
	-2.6	2.305(1)	2.5	25	T	μ=0. A=o-Phenylene-bisdimethylarsine	67Ed	

Metal Ion	ΔH, kcal /mole	Log K	ΔS, cal /mole °K	T, °C	M	Conditions	Ref.	Remarks
T22, cont.								
CuA_2	1.5	0.097(1)	5.5	25	C	3N($NaClO_4$+$Na_2S_2O_3$) A=Ethylenediamine Values are for formation of inner-sphere complexes	71Bi	
	0.8	0.230(1)	3.5	25	C	3N($NaClO_4$+$Na_2S_2O_3$) A=Ethylenediamine Values are for formation of outer-sphere complexes	71Bi	
Ag^+	-7.7	(1)	--	5-45	T	μ=0.01	31Ca	b
	-13.1	8.82(20°,1)	-4.3	20-70	T		55Ca	b
	-21.96	14.36(1-2)	-14.2	2	T	μ=0	65Le 59Sb	h: 2-45°
	-21.96	13.25(1-2)	-13.1	25	T	μ=0	65Le	h: 2-45°
	-21.96	12.68(1-2)	-13.3	35	T	μ=0	65Le	h: 2-45°
	-21.96	12.21(1-2)	-13.2	45	T	μ=0	65Le	h: 2-45°
	-19	12.99(25°,1-2)	-7.7	0-45	T	μ≃0.15	49Ja	b
	-19.05	13.45(1-2)	-2.3	25	T	--	57Cb	
	-17.6	13.46(20°,1-2)	1.5	20-70	T		55Ca	b
	-19.05	14.54(25°,1-2)	-2.3	25-70	T		57Ca	b
	-21.96	15.40(1-2)	-9.4	2	T	μ=0 S: 50 % CH_3OH	65Le	h: 2-45°
	-21.96	14.19(1-2)	-8.8	25	T	μ=0 S: 50 % CH_3OH	65Le	h: 2-45°
	-21.96	13.71(1-2)	-8.6	35	T	μ=0 S: 50 % CH_3OH	65Le	h: 2-45°
	-21.96	13.16(1-2)	-8.8	45	T	μ=0 S: 50 % CH_3OH	65Le	h: 2-45°
Zn^{2+}	3.1	3.2(25°,1)	22	15-35	T	μ=0	55Gb	b
	3.5	1.24(1)	17	25	C	μ=1	57Ya	
Cd^{2+}	1.3	3.96(25°,1)	25	15-35	T	μ=0	55Gb	b
	0	2.71(1)	12	25	C	μ=1	57Ya	
	-1.5	2.51(2)	6	25	C	μ=1	57Ya	
	-1.9	1.03(3)	-1	25	C	μ=1	57Ya	
	-1.5	5.20(1-2)	18	25	C	μ=1	57Ya	
	-3.4	6.28(1-3)	17	25	C	μ=1	57Ya	
HgA^+	-11.7	10.885	9.9	20	C	μ=0.1(KNO_3) A=CH_3	65Sa	
T23 THREONINE			$C_4H_9O_3N$			$CH_3CH(OH)CH(NH_2)CO_2H$		L^-
UO_2^{2-}	-4.99	6.348(25°,1)	12.31	25&45	T	μ=0.05M(KCl)	73Sh	a,b
	-4.99	6.486(25°,2)	11.41	25&45	T	μ=0.05M(KCl)	73Sh	a,b
	-7.37	12.682(25°,1-2)	33.33	25&45	T	μ=0	73Sh	a,b
Mn^{2+}	-0.8	3.98(15°,1-2)	15	15&40	T	0.2M(KNO_3)	68Ra	a,b
Fe^{2+}	-2.0	6.62(15°,1-2)	23	15&40	T	0.2M(KNO_3)	68Ra	a,b

Metal Ion	ΔH, kcal /mole	Log K	ΔS, cal /mole °K	T, °C	M	Conditions	Ref.	Remarks
T23, cont.								
Co^{2+}	-1.1	4.38(1)	16	25	C	μ=0.05M(KCl)	72Gb 72Ga	
	-1.8	4.38(25°,1)	14	20-35	T	μ=0.05M(KCl)	72Gb 72Ga	b
	-4.4	3.63(2)	2	25	C	μ=0.05M(KCl)	72Gb 72Ga	
	-4.1	3.63(25°,2)	3	20-35	T	μ=0.05M(KCl)	72Gb 72Ga	b
	-4.3	7.84(25°,1-2)	22	15-40	T	$0.2M(KNO_3)$	68Ra	b
Ni^{2+}	-3.2	5.42(1)	14	25	C	μ=0.05M(KCl)	72Gb 72Ga	
	-3.4	5.42(25°,1)	13	20-35	T	μ=0.05M(KCl)	72Gb 72Ga	b
	-3.81	5.46(1)	12.2	25	C	μ=0.16	70Lb	
	-5.3	4.53(2)	3	25	C	μ=0.05M(KCl)	72Gb 72Ga	
	-4.3	4.53(25°,2)	7	20-35	T	μ=0.05M(KCl)	72Gb 72Ga	b
	-4.36	4.51(2)	6.0	25	C	μ=0.16	70Lb	
	-5.20	3.45(3)	-1.6	25	C	μ=0.16	70Lb	
	-8.1	9.94(25°,1-2)	19	15-40	T	$0.2M(KNO_3)$	68Ra	b
Cu^{2+}	-5.3	8.44(20°,1)	21	25	C	μ=0	64Ia	
	-5.5	8.02(1)	18	25	C	μ=0.05M(KCl)	72Gb 72Ga 70Gc 71Gc	
	-6.1	8.02(25°,1)	16	20-35	T	μ=0.05M(KCl)	72Gb 72Ga	b
	-5.56	7.95(1)	17.7	25	C	μ=0.16	70Lb	
	-6.1	6.96(20°,2)	11	25	C	μ=0	64Ia	
	-6.1	6.70(2)	10	25	C	μ=0.05M(KCl)	72Gb 72Ga 70Gc 71Gc	
	-6.7	6.70(25°,2)	8	20-35	T	μ=0.05M(KCl)	72Gb 72Ga	b
	-6.09	6.74(2)	10.4	25	C	μ=0.16	70Lb	
	-9.9	14.69(25°,1-2)	34	15-40	T	$0.2M(KNO_3)$	68Ra	b
Zn^{2+}	-2.5	4.67(1)	13	25	C	μ=0.05M(KCl)	72Gb 72Ga	
	-2.9	4.67(25°,1)	12	20-35	T	μ=0.05M(KCl)	72Gb 72Ga	b
	-2.8	3.99(2)	9	25	C	μ=0.05M(KCl)	72Gb 72Ga	
	-2.6	3.99(25°,2)	9	20-35	T	μ=0.05M(KCl)	72Gb 72Ga	b
	-4.5	8.51(25°,1-2)	24	15-40	T	$0.2M(KNO_3)$	68Ra	b

T24 TOLUENE C_7H_8 L

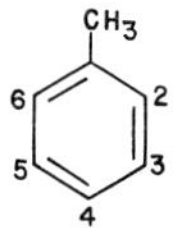

Metal Ion	ΔH, kcal /mole	Log K	ΔS, cal /mole °K	T, °C	M	Conditions	Ref.	Remarks
T24, cont.								
Ag^+	-2.72	0.076(1)	-8.77	25	T	μ=0.5 S: Equimolar H_2O-CH_3OH solution	560a	b
AgA	3.4	0.599(20°,1)	14.3	14&20	T	S: CH_3OH. A=ClO_4^-	64Tb	a,b
I_2	-2.1	(1)	--	--	C	S: Toluene	61Aa	i: 50Ha
	-1.45	(1)	--	--	C	S: Toluene	50Ha	
	-1.800	0.350(1)	-4.4	25	T	S: n-Hexane	54Ka	1
	-1.8	(1)	--	--	-	S: Cyclohexane	60Bc	
T25	TOLUENE, 2-amino-		C_7H_9N			$C_6H_4CH_3(NH_2)$		L
CoA_2	-3.36	3.548(1-2)	5.00	25	C	S: Acetone C_M=1x10^{-3}M C_L=0.2M. A=Cl	71Zc	q: ±0.6
T26	TOLUENE, 3-amino-		C_7H_9N			$C_6H_4CH_3(NH_3)$		
CoA_2	-9.77	4.039(1-2)	-14.28	25	C	S: Acetone C_M=1x10^{-3}M C_L=0.2M. A=Cl	71Zc	
T27	TOLUENE, 4-amino-		C_7H_9N			$C_6H_4CH_3(NH_2)$		
CoA_2	-9.10	3.768(1-2)	-13.29	25	C	S: Acetone C_M=1x10^{-3}M C_L=0.2M. A=Cl	71Zc	
T28	TOLUENE, 2-amino-N,N-dimethyl-		$C_9H_{13}N$			$C_6H_4CH_3[N(CH_3)_2]$		L
I_2	-2.3	0.38(17°,1)	-6.2	17&32	T	S: n-Heptane C_L=0.05-0.13 C_M=3x10^{-5}	60Tc	a,b
T29	TOLUENE, 4-amino-N,N-dimethyl-		$C_9H_{13}N$			$C_6H_4CH_3[N(CH_2)_2]$		L
I_2	-8.3	1.62(21.3°,1)	-20.8	12-33	T	S: n-Heptane C_L=0.005-0.03 C_M=3x10^{-5}	60Tc	b
T30	TOLUENE, 4-chloro-		C_7H_7Cl			$C_6H_4CH_3(Cl)$		L
IA	-1.67	0.55(25°,1)	-3.09	1.6-25	T	S: CCl_4 C_M=10^{-4} to 10^{-3} C_L=0.1-1. A=Cl	550a	b,k
	-1.75	0.55(25°,1)	-3.35	25-45.8	T	S: CCl_4 C_M=10^{-4} to 10^{-3} C_L=0.1-1. A=Cl	550a	b,k
T31	TOLUENE, (ℓ)α-hydroxy-α-(aminomethyl)-3,4-dihydroxy-		$C_8H_{11}O_3N$			$(OH)_2C_6H_3CH(OH)CH_2NH_2$		L
Co^{2+}	-19.0	7.36(25°,1-2)	-30	0&25	T	0.06M(KCl)	62Aa	a,b
Ni^{2+}	-22.4	8.00(25°,1-2)	-38.4	0&25	T	0.06M(KCl)	62Aa	a,b

Metal Ion	ΔH, kcal /mole	Log K	ΔS, cal /mole °K	T, °C	M	Conditions	Ref.	Re-marks
T31, cont.								
Cu^{2+}	-30.3	16.32(25°,1-2)	-26.8	0&25	T	0.06M(KCl)	62Aa	a,b
T32	TOLUENE, (d)-α-hydroxy-3,4-dihydroxy-α-(methylaminomethyl)-	$C_9H_{13}O_3N$				$(OH)_2C_6H_3CH(OH)CH_2NHCH_3$		L
Co^{2+}	6.9	10.06(25°,1-2)	69.2	0&25	T	0.06M(KCl)	62Aa	a,b
Ni^{2+}	3.9	9.26(25°,1-2)	55.6	0&25	T	0.06M(KCl)	62Aa	a,b
Cu^{2+}	-27.7	18.44(25°,1-2)	-8.4	0&25	T	0.06M(KCl)	62Aa	a,b
T33	TOLUENE, (ℓ)-α-hydroxy-3,4-dihydroxy-α-(methylamino-methyl)-	$C_9H_{13}O_3N$				$(OH)_2C_6H_3CH(OH)CH_2NHCH_3$		L
Co^{2+}	-20.2	8.94(25°,1-2)	-26.8	0&25	T	0.06M(KCl)	62Aa	a,b
Ni^{2+}	-7.4	8.40(25°,1-2)	13.6	0&25	T	0.06M(KCl)	62Aa	a,b
Cu^{2+}	-19.9	17.72(25°,1-2)	14.4	0&25	T	0.06M(KCl)	62Aa	a,b
T34	TOLUENE, α-hydroxy-α-[1-(methylamino)ethyl]-	$C_{10}H_{15}ON$				$C_6H_5CH(OH)CH(CH_3)NHCH_3$		L
Cd^{2+}	-10.8	6.49(1)	-8	25	T	μ=0.1(KNO_3)	64Ad	
T35	TOLUENE, 2-hydroxy-α(2-thiazolylazo)-	$C_{10}H_9ON_3S$				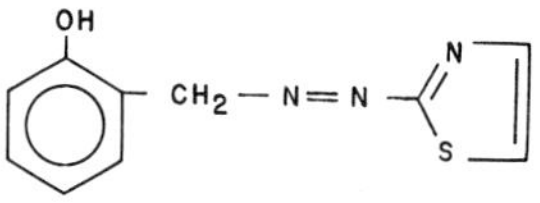		
La^{3+}	-4.25	8.17(25°,1)	22.65	15-35	T	μ=0.1M(KNO_3)	73Ka	b
Pr^{3+}	-3.65	8.91(25°,1)	27.82	15-35	T	μ=0.1M(KNO_3)	73Ka	b
Sm^{3+}	-5.87	9.41(25°,1)	22.20	15-35	T	μ=0.1M(KNO_3)	73Ka	b
Gd^{3+}	-6.07	9.61(25°,1)	22.51	15-35	T	μ=0.1M(KNO_3)	73Ka	b
Dy^{3+}	-4.05	9.71(25°,1)	30.02	15-35	T	μ=0.1M(KNO_3)	73Ka	b
Ho^{3+}	-6.28	9.61(25°,1)	22.32	15-35	T	μ=0.1M(KNO_3)	73Ka	b
Er^{3+}	-4.46	9.56(25°,1)	28.51	15-35	T	μ=0.1M(KNO_3)	73Ka	b
Ni^{2+}	-15.30	11.70(25°,1)	2.45	15-35	T	μ=0.1M(KNO_3)	73Ka	b
Zn^{2+}	-2.00	7.85(25°,1)	29.19	15-35	T	μ=0.1M(KNO_3)	73Ka	b
Cd^{2+}	-3.50	7.72(25°,1)	23.55	15-35	T	μ=0.1M(KNO_3)	73Ka	b
Hg^{2+}	-3.50	16.07(25°,1)	61.75	15-35	T	μ=0.1M(KNO_3)	73Ka	b
T37	TRANSFER RIBONUCLEIC ACID, yeast phenylalanine							L
Mg^{2+}	0.0	--	--	25	C	C_M=1x10^{-3}M Approximately 20 moles of Mg^{2+} bound per mole of L	72Rc	q: ±0.1
T38	TRIDECANE, 5,9-dithio-	$C_{11}H_{24}S_2$				$CH_3(CH_2)_3S(CH_2)_3S(CH_2)_3CH_3$		
TiA_4	-21.8	3.96(1)	-54.3	25	C	S: Benzene. A=Cl	70Gh 68Ge	

Metal Ion	ΔH, kcal /mole	Log K	ΔS, cal /mole °K	T, °C	M	Conditions	Ref.	Re- marks
T38, cont.								
TiA_4, cont.	-20.5	3.88(1)	-51.0	25	C	S: Benzene. A=Br	70Gh	
	-20.6	(1)	--	25	C	S: Benzene C_M-0.05-0.08M A=Br	70Gh 68Ge 65Gb	
SnA_4	-25.3	3.00(1)	-70.8	25	C	S: Benzene C_M=0.05-0.08M A=Cl	68Ge 70Gh	
	-20.1	3.08(1)	-53.3	25	C	S: Benzene. A=Br	70Gh 68Ge	
TeA_4	-13.6	3.44(1)	-29.9	25	C	S: Benzene. A=Cl	73Ph	
	-2.4	(2)	--	25	C	S: Benzene. A=Cl	73Ph	
T39 TRIDECANE, 1,5,9,13-tetraaza-			$C_9H_{24}N_4$			$NH_2(CH_2)_3NH(CH_2)_3NH(CH_2)_3NH_2$		
Ni^{2+}	-13.2	10.48(1)	3.7	25	C	0.1M($NaNO_3$)	72Bb	
Cu^{2+}	-19.45	17.05(1)	12.8	25	C	0.1M($NaNO_3$)	72Bb	
Zn^{2+}	-7.35	9.317(1)	18.0	25	C	0.1M($NaNO_3$)	72Bb	
T40 TRIGLYCINE			$C_6H_{11}O_4N_3$			$H_2NCH_2CONHCH_2CONHCH_2CO_2H$		L^-
Mg^{2+}	2	1.44(1)	14	30	T	μ=0.09(KCl)	57Mb	
Ni^{2+}	-5.0	3.69(1)	0.6	25	C	μ=0.1M(KNO_3)	71Ld	
	-4.2	3.04(2)	-0.5	25	C	μ=0.1M(KNO_3)	71Ld	
Cu^{2+}	-5.9	(1)	--	24	C	μ=0.01M($HClO_4$)	72Sg	
	-6.3	5.04(1)	1.9	25	C	μ=0.10($NaClO_4$)	68Bc	
T41 2,7,12-TRIMETHYL-3,7,11,17-tetra-azabicyclo [11.3.1] heptadeca-1(17),2,11,13,15-pentaene			$C_{16}H_{24}N_4$			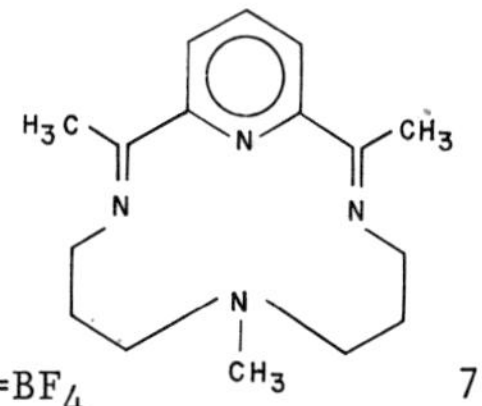		
NiA_2L	-4.58	--	-15.3	--	T	$A=BF_4$ R: $NiLA_2$(diamagnetic) = $NiLA_2$(paramagnetic)	71Re	u
	-4.06	--	-9.1	--	T	S: N,N-Dimethyl-formamide. $A=ClO_4$ R: $NiLA_2$(diamagnetic) = $NiLA_2$(paramagnetic)	71Re	u
T42 TRINACTIN								
Na^+	-7.29	3.2(1)	-9.8	25	C	S: Methanol	73Za	
T43 TRYPTOPHAN			$C_{11}H_{12}O_2N_2$			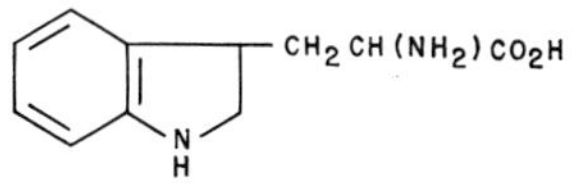		
Mn^{2+}	-1.3	2.88(20°,1)	9	20-60	T	μ=0.1(KNO_3)	73Bf	b
	-1.08	2.840(1)	9.4	25	C	μ=3.0($NaClO_4$)	70Wb	
	-1.19	2.312(2)	6.6	25	C	μ=3.0($NaClO_4$)	70Wb	

Metal Ion	ΔH, kcal /mole	Log K	ΔS, cal /mole °K	T, °C	M	Conditions	Ref.	Re-marks
T43, cont.								
Fe^{2+}	-2.22	3.923(1)	10.5	25	C	μ=3.0($NaClO_4$)	70Wb	
	-2.28	3.471(2)	8.2	25	C	μ=3.0($NaClO_4$)	70Wb	
Co^{2+}	-2.78	4.581(1)	11.6	25	C	μ=3.0($NaClO_4$)	70Wb	
	-2.86	4.321(2)	10.2	25	C	μ=3.0($NaClO_4$)	70Wb	
	-2.80	3.344(3)	5.9	25	C	μ=3.0($NaClO_4$)	70Wb	
CoA	-2.6	3.86(25°)	9.0	15-55	T	μ=0.1M(KNO_3) A=Methyliminodiacetate R: NiA + HL = $NiAL^-$ + H^+	71Ic	b,q: ±0.3
Ni^{2+}	-5.72	5.757(1)	7.2	25	C	μ=3.0($NaClO_4$)	70Wb	
	-6.10	5.227(2)	3.4	25	C	μ=3.0($NaClO_4$)	70Wb	
	-4.99	4.474(3)	3.8	25	C	μ=3.0($NaClO_4$)	70Wb	
NiA	-5.3	4.80(25°)	4.2	15-55	T	μ=0.1M(KNO_3) A=Methyliminodiacetate R: NiA + HL = $NiAL^-$ + H^+	71Ic	b,q: ±1.0
Cu^{2+}	-7.67	8.711(1)	14.1	25	C	μ=3.0($NaClO_4$)	70Wb	
	-7.74	7.953(2)	10.4	25	C	μ=3.0($NaClO_4$)	70Wb	
CuA	-8.4	6.38(25°)	0.9	15-55	T	μ=0.1M(KNO_3) A=Methyliminodiacetate R: CuA + HL = $CuAL^-$ + H^+	71Ic	b,q: ±1.6
Zn^{2+}	-4.10	5.013(1)	9.2	25	C	μ=3.0($NaClO_4$)	70Wb	
	-4.13	4.766(2)	7.9	25	C	μ=3.0($NaClO_4$)	70Wb	
CdA	-3.2	3.66(25°)	6.0	15-55	T	μ=0.1M(KNO_3) A=Methyliminodiocetate R: NiA + HL = $NiAL^-$ + H^+	71Ic	b,q: ±0.8

T44 2,2',2"-TRIPYRIDINE $C_{15}H_{11}N_3$

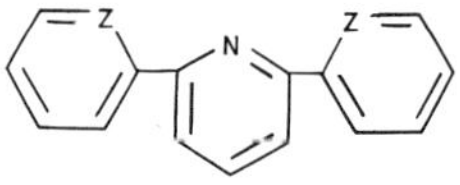

Metal Ion	ΔH, kcal /mole	Log K	ΔS, cal /mole °K	T, °C	M	Conditions	Ref.	Re-marks
Mn^{2+}	-1	5.0(25°,1)	19	-46-0	T	μ=0.2($NaClO_4$) S: Methanol	73Be	b,q
CuA_2	-18.6	(1)	--	30	C	S: Benzene. A=1,1,1,5,5,5-Hexafluoropentane-2,4-dione	74Gb	
	-8.36	1.898(1)	-18.9	30	C	S: Benzene. A=1,1,1-Trifluoropentane-2,4-dione	74Gb	
	-10.0	2.021(1)	-23.7	30	C	S: Benzene. A=1,1,1-Trifluoro-4-phenylbutane-2,4-dione	74Gb	
	-9.56	2.468(1)	-20.2	30	C	S: Benzene. A=1,1,1-Trifluoro-4-(2-thienyl) butane-2,4-dione	74Gb	

Metal Ion	ΔH, kcal /mole	Log K	ΔS, cal /mole °K	T, °C	M	Conditions	Ref.	Remarks
T45 TYROSINE			$C_9H_{11}O_3N$			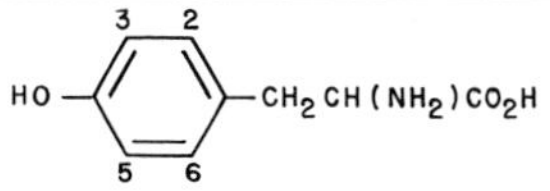		
Co^{2+}	-0.6	3.93(25°,1)	20	27	C	μ=0.05M(KCl)	71Gb	
	-0.2	3.93(25°,1)	19	20-35	T	μ=0.05M(KCl)	71Gb	b
	-1.7	3.45(25°,2)	11	27	C	μ=0.05M(KCl)	71Gb	
	-2.7	3.45(25°,2)	6	20-35	T	μ=0.05M(KCl)	71Gb	b
Ni^{2+}	-2.0	5.02(25°,1)	16	27	C	μ=0.05M(KCl)	71Gb	
	-1.7	5.02(25°,1)	17	20-35	T	μ=0.05M(KCl)	71Gb	b
	-2.8	4.25(25°,2)	10	27	C	μ=0.05M(KCl)	71Gb	
	-3.1	4.25(25°,2)	9	20-35	T	μ=0.05M(KCl)	71Gb	b
Cu^{2+}	-5.9	7.75(25°,1)	16	27	C	μ=0.05M(KCl)	71Gb	
	-4.0	7.75(25°,1)	21	20-35	T	μ=0.05M(KCl)	71Gb	b
	-6.6	6.82(25°,2)	10	27	C	μ=0.05M(KCl)	71Gb	
	-6.0	6.82(25°,2)	11	20-35	T	μ=0.05M(KCl)	71Gb	b
Zn^{2+}	-1.4	4.16(25°,1)	14	27	C	μ=0.05M(KCl)	71Gb	
	1.5	4.16(25°,1)	24	20-35	T	μ=0.05M(KCl)	71Gb	b
	-2.6	4.11(25°,2)	6	27	C	μ=0.05M(KCl)	71Gb	
	-3.5	4.11(25°,2)	7	20-35	T	μ=0.05M(KCl)	71Gb	b
U1 UNADENYLYLATED GLUTAMINE SYNTHETASE from Escherichia coli								
Mn^{2+}	≈3	6.27	≈38	37	C	pH=7.2. Values are per mole of Mn^{2+} bound to ligand	72Hg	
U2 UREA, 1,3-di-tert-butyl-2-thio-			$C_9H_{20}N_2S$			$(CH_3)_3CNHC(S)NHC(CH_3)_3$		
I_2	-18.6	4.611(25°,1)	-39.8	0-35	T	S: Chloroform C_L=4.3x10^{-5}-3.9x10^{-4}M	66Be	b
U3 UREA, 1,3-diethyl-2-thio-			$C_5H_{12}N_2S$			$CH_3CH_2NHC(S)NHCH_2CH_3$		
Ag^+	-1.3	6.009(1)	23.1	25	C	μ=0.5M(KNO_3)	73Ec	q: ±0.2
	-20.19	4.306(2)	-48.0	25	C	μ=0.5M(KNO_3)	73Ec	
	-0.29	3.035(3)	12.9	25	C	μ=0.5M(KNO_3)	73Ec	
	-9.04	0.817(4)	-26.55	25	C	μ=0.5M(KNO_3)	73Ec	
I_2	-14.4	4.191(25°,1)	-28.2	0-35	T	S: Chloroform C_L=3.6x10^{-5}-3.7x10^{-4}M	66Be	b
U4 UREA, 1,3-diisopropyl-2-thio-			$C_7H_{16}N_2S$			$(CH_3)_2CHNHC(S)NHCH(CH_3)_2$		
I_2	-11.3	4.216(25°,1)	-17.6	0-35	T	S: Chloroform C_L=3.7x10^{-5}-3.7x10^{-4}M	66Be	b
U5 UREA, 1,3-dimethyl-			$C_3H_8ON_2$			$CH_3NHCONHCH_3$		
SbA_5	-29.6	(1)	--	25	C	S: CH_2Cl_2. A=Cl	71Oa	j
U6 UREA, 1,3-dimethyl-2-thio-			$C_3H_8N_2S$			$CH_3NHC(S)NHCH_2$		
Ag^+	0	6.089(1)	27.85	25	C	μ=0.5M(KNO_3)	73Ec	
	-21.80	4.072(2)	-54.45	25	C	μ=0.5M(KNO_3)	73Ec	

Metal Ion	ΔH, kcal /mole	Log K	ΔS, cal /mole °K	T, °C	M	Conditions	Ref.	Re- marks
U6, cont.								
Ag^+, cont.	-0.35	2.598(3)	10.7	25	C	μ=0.5M(KNO_3)	73Ec	
	-9.46	1.116(4)	-26.6	25	C	μ=0.5M(KNO_3)	73Ec	
U7 UREA, 1,3-diphenyl-2-thio-			$C_{13}H_{12}N_2S$			$C_6H_5NHC(S)NHC_6H_5$		
I_2	-15.2	2.794(29°,1)	-37.5	25-32	T	S: Carbon Tetrachloride	72Aa	b
U8 UREA, 1-ethylidene-2-thio-			$C_3H_6N_2S$			$NH_2C(S)NHCH{:}CH_2$		
I_2	-8.1	2.747(29°,1)	-14.4	25-32	T	S: Carbon Tetrachloride	72Aa	b
U9 UREA, 1-methyl-2-thio-			$C_2H_6N_2S$			$NH_2C(S)NHCH_3$		
Ag^+	-1.6	6.710(1)	25.3	25	C	μ=0.5M(KNO_3)	73Ec	q: ±0.3
	-20.76	3.916(2)	-51.7	25	C	μ=0.5M(KNO_3)	73Ec	
	-0.41	2.434(3)	9.8	25	C	μ=0.5M(KNO_3)	73Ec	
	-9.64	1.021(4)	-27.7	25	C	μ=0.5M(KNO_3)	73Ec	
U10 UREA, seleno			CH_4N_2Se			NH_2CSeNH_2		L
Hg^{2+}	-30.5	24.0(1-2)	9.2	30	C	μ=0.1	69Gc	
	-37.4	30.2(1-3)	-14.8	30	C	μ=0.1	69Gc	
	-54.1	32.9(1-4)	-27.7	30	C	μ=0.1	69Gc	
U11 UREA, 1,1,3,3-tetramethyl-			$C_5H_{12}ON_2$			$(CH_3)_2NCON(CH_3)_2$		L
ZnA	-6.5	1.858(25°,1)	-13.3	15-35	T	S: Benzene A=α,β,γ,δ-Tetra-phenylporphine	73Vd	b
SbA_5	-29.64	(1)	--	25	C	C_T<.01. S: Ethylene chloride. A=Cl	64Ob	
I_2	-4.35	0.813(25°,1)	-10.9	24-43	T	S: CCl_4 $C_M=1.6x10^{-3}$ $C_L=3.8x10^{-2}$	64Ma	b
	-3.3	0.663(20°,1)	-8	10-30	T	S: Dichloromethane	68Lc	b,q: ±0.6
	-5.0	1.279(20°,1)	-11	10-30	T	S: n-Heptane	68Lc	b,q: ±0.4
U12 UREA, 1,1,3,3-tetramethyl-2-thio-			$C_5H_{12}N_2S$			$(CH_3)_2NC(S)N(CH_3)_2$		L
ZnA	-5.6	1.079(25°,1)	-13.8	15-35	T	S: Benzene A=α,β,γ,δ-Tetra-phenylporphine	73Vd	b
I_2	-10.5	3.903(25°,1)	-17.4	25-40	T	S: CCl_4 $C_M=10^{-3}$ $C_L=2x10^{-4}$	64Na	b
	-9.5	4.132(25°,1)	-12.3	0-35	T	S: Chloroform $C_L=3.7x10^{-5}-3.7x10^{-4}M$	66Be	b
	-9.0	4.690(20°,1)	-9	10-30	T	S: Dichloromethane	68Lc	b,q: ±1.6

Metal Ion	ΔH, kcal /mole	Log K	ΔS, cal /mole °K	T, °C	M	Conditions	Ref.	Remarks
U12, cont.								
I_2	-9.9	4.114(20°,1)	-15	10-30	T	S: n-Heptane	68Lc	b,q: ±1.2
U13 UREA, 2-thio-			CH_4N_2S			$H_2NC(S)NH_2$		L
PdA_2	-4.6	0.996(1)	-10.7	25	T	μ=0. A=o-Phenylene-bisdimethylarsine	67Ec 67Ed	
	-4.6	0.845(40°,1)	-10.7	25-55	T	μ=0.0021($NaNO_3$) A=o-Phenylenebis-dimethylarsine	67Ed	b
	-3.6	1.322(40°,1)	-5.4	25-55	T	μ=7x10^{-4} S: Methanol A=o-Phenylenebis-dimethylarsine	67Ec	b,q: ±0.4
PtA_2	-4.8	0.676(1)	-12.9	25	T	μ=0. A=o-Phenylene-bisdimethylarsine	67Ec	q: ±0.4
	-4.6	1.30(23°,1)	-9.5	23-54	T	μ=0. S: CH_3OH A=o-Phenylenebis-dimethylarsine	66Db	b
	-4.8	0.574(35°,1)	-12.9	25-55	T	μ=7x10^{-4}($NaNO_3$) A=o-Phenylenebis-dimethylarsine	67Ec	b,q: ±0.4
Ag^+	-2.6	7.050(1)	23.5	25	C	μ=0.5M(KNO_3)	73Ec	q: ±0.3
	0	7.043(25°,1)	32	10-30	T	μ=0.5M(KNO_3)	69Bg	b
	-19.33	3.575(2)	-48.5	25	C	μ=0.5M(KNO_3)	73Ec	
	-16.5	3.571(25°,2)	-38.5	10-30	T	μ=0.5M(KNO_3)	69Bg	b,q: ±4.0
	-1.13	2.233(3)	6.5	25	C	μ=0.5M(KNO_3)	73Ec	
	0	2.101(25°,3)	9.45	10-30	T	μ=0.5M(KNO_3)	69Bg	b
	-8.93	0.781(4)	-26.4	25	C	μ=0.5M(KNO_3)	73Ec	
	-7.5	0.778(25°,4)	-21.62	10-30	T	μ=0.5M(KNO_3)	69Bg	b,q: ±2.0
	-21	10.613(25°,1-2)	-21.8	0-45	T	μ≃0.15(AgCl)	49Ja	b
	-23	12.716(25°,1-3)	-18.8	0-45	T	μ≃0.15(AgBr)	49Ja	b
	-20.0	13.1(25°,1-3)	-7.38	20-50	T	μ=0.80($NaNO_3$)	60Tb	b
Hg^{2+}	-33	21.26(1-2)	-12.9	30	C	μ=0.1	69Gc	
	-34.6	24.22(1-3)	-3.3	30	C	μ=0.1	69Gc	
	-46.4	25.88(1-4)	-34.6	30	C	μ=0.1	69Gc	
	-44	26.3(25°,1-4)	-27.5	20-50	T	μ=0.80($NaNO_3$)	60Tb	b
	-46	26.50(25°,1-4)	-32.9	20-30	T	1M($NaNO_3$)	56Tb	b
HgA_2	-1.50	1.97(1)	4.0	25	C	μ=0. A=CN	68Ia	
	-2.4	2.03(1)	1.3	25	C	μ=0. A=CN S: 20.0 wt % C_2H_5OH	68Ia	
	-4.51	1.94(1)	-6.3	25	C	μ=0. A=CN S: 40.0 wt % C_2H_5OH	68Ia	
	-5.93	1.86(1)	-11.1	25	C	μ=0. A=CN S: 60.0 wt % C_2H_5OH	68Ia	
	-5.96	2.11(1)	-10.3	25	C	μ=0. A=CN S: 80.0 wt % C_2H_5OH	68Ia	
	-6.16	2.28(1)	-10.3	25	C	μ=0. A=CN S: 92.3 wt % C_2H_5OH	68Ia	

Metal Ion	ΔH, kcal /mole	Log K	ΔS, cal /mole °K	T, °C	M	Conditions	Ref.	Remarks
U13, cont.								
HgA_2	-1.4	2.08(1)	4.8	25	C	μ=0. S: 20 w % Formamide. A=CN	71If	q: ±0.1
	-1.7	1.95(1)	3.2	25	C	μ=0. S: 40 w % Formamide. A=CN	71If	
	-1.7	2.01(1)	3.5	25	C	μ=0. S: 60 w % Formamide. A=CN	71If	
	-1.6	2.02(1)	3.8	25	C	μ=0. S: 80 w % Formamide. A=CN	71If	
	-1.5	2.02(1)	4.0	25	C	μ=0. S: 100 w % Formamide. A=CN	71If	
	-7.9	0.58(2)	-23.8	25	C	μ=0. A=CN	68Ia	
	-8.6	0.77(2)	-25.5	25	C	μ=0. A=CN S: 20.0 wt % C_2H_5OH	68Ia	
	-8.2	1.04(2)	-22.7	25	C	μ=0. A=CN S: 40.0 wt % C_2H_5OH	68Ia	
	-5.21	1.26(2)	-11.5	25	C	μ=0. A=CN S: 60.0 w % C_2H_5OH	68Ia	
	-5.0	1.30(2)	-11.1	25	C	μ=0. A=CN S: 80.0 w % C_2H_5OH	68Ia	
	-4.8	1.34(2)	-10.1	25	C	μ=0. A=CN S: 92.3 w % C_2H_5OH	68Ia	
	-9.1	0.56(2)	-28	25	C	μ=0. S: 20 w % Formamide. A=CN	71If	
	-9.2	0.57(2)	-28	25	C	μ=0. S: 40 w % Formamide. A=CN	71If	
	-8.7	0.60(2)	-26	25	C	μ=0. S: 60 w % Formamide. A=CN	71If	
	-7.8	0.64(2)	-23	25	C	μ=0. S: 80 w % Formamide. A=CN	71If	
	-7.4	0.64(2)	-22	25	C	S: 100 w % Formamide. A=CN	71If	
I_2	-9.6	3.643(25.5°,1)	-14.7	0-35	T	S: Chloroform	66Be	b
	-9.6	3.929(20°,1)	-14.8	4-22	T	C_M=2.5x10^{-5} C_L=10^{-4}-4x10^{-4} S: CH_2Cl_2	62La	b
U14 UREA, 1,1,3-trimethyl-		$C_4H_{10}ON_2$				$H_3CNHC(O)N(CH_3)_2$		
SbA_5	-29.3	(1)	--	25	C	S: CH_2Cl_2. A=Cl	71Oa	j
U15 URIDINE-5'-diphosphoric acid		$C_9H_{14}O_{12}N_2P_2$						L^{3-}
Mg^{2+}	3.22	3.43(1)	26.4	30	C	μ=0.2(Tetramethyl-ammonium chloride) pH=8.50	73Sa	
U16 URIDINE-5'-phosphoric acid		$C_9H_{13}O_9N_2P$						
Mg^{2+}	1.82	1.69(1)	13.7	30	C	μ=0.2(Tetramethyl-ammonium chloride) pH=8.50	73Sa	

Metal Ion	ΔH, kcal /mole	Log K	ΔS, cal /mole °K	T, °C	M	Conditions	Ref.	Remarks
U17 URIDINE-5'-triphosphoric acid			$C_9H_{15}O_{15}N_2P_3$					
Mg^{2+}	4.40	4.13(1)	33.4	30	C	μ=0.2(Tetramethylammonium chloride) pH-8.50	73Sa	
V								
V1 VALINE			$C_5H_{11}O_2N$			$(CH_3)_2CHCH(NH_2)CO_2H$		L^-
Y^{3+}	-4.626	4.79(25°,1)	6.28	25&35	T	μ=0.1(KCl)	73Sg	a,b,q: ±0.6
	8.826	4.27(25°,2)	49.14	25&35	T	μ=0.1(KCl)	73Sg	a,b,q: ±0.6
Ce^{3+}	8.406	5.02(25°,1)	51.16	25&35	T	μ=0.1(KCl)	73Sg	a,b,q: ±0.6
Ni^{2+}	-4.33	5.45(25°,1)	10.4	10-30	T	--	60Pb 57Pa	b
	-3.75	6.87(1)	16.6	25	T	S: 44.6 w % Dioxane	60Pd	
	-5.0	7.38(1)	17.1	25	T	S: 59.7 w % Dioxane	60Pd	
	-7.17	8.40(1)	14.4	25	T	S: 69 w % Dioxane	60Pd	
	-5.42	4.27(25°,2)	1.4	10-30	T	--	60Pb 57Pa	b
	-3.25	5.56(2)	14.4	25	T	S: 44.6 w % Dioxane	60Pd	
	-8.25	6.31(2)	1.3	25	T	S: 59.7 w % Dioxane	60Pd	
	-8.50	7.04(2)	3.7	25	T	S: 69 w % Dioxane	60Pd	
	-4.50	2.48(25°,3)	-3.8	10-30	T	--	60Pb 57Pa	b
	-3.75	3.58(3)	3.7	25	T	S: 44.6 w % Dioxane	60Pd	
	-5.17	4.30(3)	2.3	25	T	S: 69 w % Dioxane	60Pd	
	-14.5	9.53(1-2)	-5.5	25	T	μ=0.15	56Lb	
Cu^{2+}	-5.56	8.11(1)	18.5	25	C	μ=0.1M(KNO_3)	72Ia	
	-11.6	15.05(2)	29.9	25	C	μ=0.1M (±)Valine (DL)	71Be	
	-11.7	15.0(2)	29.6	25	C	μ=0.1M (±)Valine (D)	71Be	
	-11.7	15.0(2)	29.9	25	C	μ=0.1M (-)Valine (L)	71Be	
	-20.5	14.76(1-2)	0	25	T	μ=0.15	56Lb	
	-11.65	14.84	28.8	25	C	μ=0.1M(KNO_3) R: Cu + L + Serine = [CuL(Serine)]	72Ia	
V2 VALINOMYCIN			$C_{54}H_{90}O_{18}N_6$					
K^+	-8.9	6.079(1)	-2.16	25	C	S: Ethanol $C_L=10^{-4}$. $C_M=10^{-5}$	71Mf	
	-5.72	(1)	--	25	C	S: Methanol	71Fe	
W								
W1 WATER			H_2O			H_2O		
Be^{2+}	5.0	-3.30(25°)	1.4	0-60	T	μ=1M(NaCl) R: $2Be^{2+}$ + L = $Be_2(OH)^{3+}$ + H^+	67Mh	b

Metal Ion	ΔH, kcal /mole	Log K	ΔS, cal /mole °K	T, °C	M	Conditions	Ref.	Remarks
W1, cont.								
Be^{2+}, cont.	4.43	-3.212	0.2	25	C	3M($NaClO_4$) R: $2Be^{2+}$ + L = Be_2OH^{3+} + H^+	62Cd	
	16.0	-8.94(25°)	15.3	0-60	T	μ=1M(NaCl) R: $3Be^{2+}$ + 3L = $Be_3(OH)_3^{3+}$ + $3H^+$	67Mh	b
	15.18	-8.664	11.3	25	C	3M($NaClO_4$) R: $3Be^{2+}$ + 3L = $Be_3(OH)_3^{3+}$ + $3H^+$	62Cd	
	45.3	-25.51(25°)	35.2	0-60	T	μ=1M(NaCl) R: $5Be^{2+}$ + 7L = $Be_5(OH)_7^{3+}$ + $7H^+$	67Mh	b
MgA^{2-}	-3.4	-2.34	-22	25	T	μ=0. A=ATP^{4-} R: MgA^{2-} + L = $Mg(ADP)^-$ + HPO_4^{2-} + H^+	63Ga	i: 59Bc
ScA_6^{3+}	6.4	-5.1	-0.7	10	T	μ=1($NaClO_4$) R: ScA_6^{3+} + L = ScA_5OH^{2+} + H^+ A=H_2O. ΔCp=190	54Kb	
	9.3	-4.8	9	25	T	μ=1($NaClO_4$) R: ScA_6^{3+} + L = ScA_5OH^{2+} + H^+ A=H_2O. ΔCp=200	54Kb	
	12	-4.4	19	40	T	μ=1($NaClO_4$) R: ScA_6^{3+} + L = ScA_5OH^{2+} + H^+ A=H_2O. ΔCp=210	54Kb	
	16	-6.6	27	10	T	μ=1($NaClO_4$) R: $2ScA_6^{3+}$ + 2L = $[ScA_5OH]_2^{4+}$ A=H_2O. ΔCp=-44	54Kb	
	15	-6.0	25	25	T	μ=1($NaClO_4$) R: $2ScA_6^{3+}$ + 2L = $[ScA_5OH]_2^{4+}$ A=H_2O. ΔCp=-46	54Kb	
	15	-5.5	22	40	T	μ=1($NaClO_4$) R: $2ScA_6^{3+}$ + 2L = $[ScA_5OH]_2^{4+}$ A=H_2O. ΔCp=-49	54Kb	
Ce^{4+}	16.1	0.806(24.5°)	57.7	7.6-30.8	T	μ=1.0($NaClO_4$) R: Ce^{4+} + L = $CeOH^{3+}$ + H^+	71Eb	b,q: ±1.3
	15.5	2.27(25°)	55.4	5-35	T	μ=2.0. R: Ce^{4+} + L = $CeOH^{3+}$ + H^+	51Ha	b
CeA	14.2	-1.681(25°)	39.9	2.4-25	T	μ=1.0($NaClO_4$) A=H_3AsA_3. R: CeA^{4+} + H_2O = $CeOH^{3+}$ + A + H^+	71Eb	b,q: ±2.6
	-2.2	-2.164(25°)	-17.3	2.4-2.5	T	μ=1.0($NaClO_4$) A=$H_2AsO_3^-$. R: CeA^{3+} + H_2O = $CeOH^{3+}$ + HA	71Eb	b,q: ±1.9

Metal Ion	ΔH, kcal /mole	Log K	ΔS, cal /mole °K	T, °C	M	Conditions	Ref.	Remarks
W1, cont.								
CeA, cont.	7.5	-0.921(24.5°)	20.9	7.6-30.8	T	μ=1.0($NaClO_4$) A=OH^-. R: CeA^{3+} + H_2O = $Ce(A)_2^{2+}$ + H^+	71Eb	b,q: ±1.0
Th^{4+}	5.9	-4.10	0.9	25	T	μ=1M($NaClO_4$) R: Th^{4+} + L = $ThOH^{3+}$ + H^+. ΔCp=224	65Ba	
	13.9	-7.84	11.0	25	T	μ=1M($NaClO_4$) R: Th^{4+} + 2L = $Th(OH)_2^{2+}$ + $2H^+$ ΔCp=304	65Ba	
	14.8	-4.62	28.4	25	T	μ=1M($NaClO_4$) R: $2Th^{4+}$ + $2H_2O$ = $Th_2(OH)_2^{6+}$ + $2H^+$ ΔCp=1	65Ba	
	57.7	-18.99	106.6	25	T	μ=1M($NaClO_4$) R: $4Th^{4+}$ + $8H_2O$ = $Th_4(OH)_8^{8+}$ + $8H^+$ ΔCp=108	65Ba	
	108.5	-36.73	195.7	25	T	μ=1M($NaClO_4$) R: $6Th^{4+}$ + $15H_2O$ = $Th_6(OH)_{15}^{9+}$ + $15H^+$ ΔCp=239	65Ba	
V^{3+}	11.1	-2.9(25°)	20	25-50	T	μ=1. R: V^{3+} + L = VOH^{2+} + H^+	50Fa	b
V^{5+}	92	--	--	1	T	R: $HV_{10}O_{28}^{5-}$ + 2L = $10VO_3^-$ + $5H^+$	61Sb	
$MoAB^{2-}$	-6.0	1.60	-12	25	T	C_M=3.7x10^{-4} 5M(HCl). R: $2MoAB^{2-}$ + L = $(MoA)_2O^{4-}$ + $2H^+$ + $2B^-$. A=OCl_4, B=Cl	62Ha	
MoO_4^{2-}	55.1	--	--	25	C	μ=1M(LiCl) R: $7MoO_4^{2-}$ + 4L = Mo_7O_{24} + $8OH^-$	74Kb	
$Mo_7O_{24}^{6-}$	27.8	--	--	25	C	μ=1M(LiCl) R: $Mo_7O_{24}^{6-}$ + 2L = $H_2Mo_7O_{24}^{4-}$ + $2OH^-$	74Kb	
	56.7	--	--	25	C	μ=2M(NaCl) R: $Mo_7O_{24}^{6-}$ + 4L = $7MoO_4^{2-}$ + $8H^+$	74Kb	
	55.1	--	--	25	C	μ=3M($NaClO_4$) R: $Mo_7O_{24}^{6-}$ + 4L = $7MoO_4^{2-}$ + $8H^+$	74Kb	
	56	-57.7	-76	25	C	3M($NaClO_4$) R: $Mo_7O_{24}^{6-}$ + 4L = $7MoO_4^{2-}$ + $8H^+$	70Ac	
$H_2Mo_7O_{24}^{4-}$	123.2	--	--	25	C	μ=1M(LiCl) R: $8H_2Mo_7O_{24}^{4-}$ + L = $7HMo_8O_{26}^{3-}$ + $11OH^-$	74Kb	

Metal Ion	ΔH, kcal /mole	Log K	ΔS, cal /mole °K	T, °C	M	Conditions	Ref.	Re-marks
W1, cont.								
$HMo_8O_{26}^{3-}$	39.3	--	--	25	C	μ=1M(LiCl) R: $HMo_8O_{26}^{3}$ + 3L = $H_4Mo_8O_{26}$ + $3OH^-$	74Kb	
$H_4Mo_8O_{26}$	98.8	--	--	25	C	μ=3M($NaClO_4$) R: $7H_4Mo_8O_{26}$ + 10L = $8H_2Mo_7O_{24}$ + $32H^+$	74Kb	
$W_6O_{21}^{6-}$	57.1	-52.5	-49	25	C	3M($NaClO_4$) R: $W_6O_{21}^{6-}$ + 3L = $6WO_4^{2-}$ + $6H^+$	70Ac 69Aa	
$HW_6O_{21}^{5-}$	64.5	65.2	82	25	C	μ≈0.3. R: $HW_6O_{21}^{5-}$ + 3L = $6WO_4^{2-}$ + $7H^+$	62Dd	
	47	65(25°)	140	17-35	T	μ≈0.3. R: $HW_6O_{21}^{5-}$ + 3L = $6WO_4^{2-}$ + $7H^+$	62Dd	b
	62.5	-60.8	-68	25	C	3M($NaClO_4$) R: $HW_6O_{21}^{5-}$ + 3L = $6WO_4^{2-}$ + $7H^+$	70Ac 69Aa	
$W_{12}O_{41}^{10-}$	1.9	-1.69	-1.3	25	C	μ=3M($NaClO_4$) R: $W_{12}O_{41}^{10-}$ + L = $2HW_6O_{21}^{5-}$	69Aa	
	126.9	-123.2	-138	25	C	3M($NaClO_4$) R: $W_{12}O_{41}^{10-}$ + 7L = $12WO_4^{2-}$ + $14H^+$	70Ac	
U^{4+}	11.7	-0.678	36	25	T	μ=0. R: U^{4+} + L = UOH^{3+} + H^+	55Kc	
	11.0	-0.66	33	25	T	μ=0. R: U^{4+} + L = UOH^{3+} + H^+	55Ba	
	10.7	-1.168	41	25	T	μ=0.19. R: U^{4+} + L = UOH^{3+} + H^+	55Ba	
	11.2	-1.466	31	25	T	μ=0.5. R: U^{4+} + L = UOH^{3+} + H^+	55Kc	
UO_2^{2+}	20.8	-5.9	43	25	T	μ=0.0347 R: UO_2^{2+} + L = UO_2OH^+ + H^+	57Ha	
	11.0	-5.7(25°)	11	25&94	T	μ=0.5. R: UO_2^{2+} + L = UO_2OH^+ + H^+	62Ba	a,b
	6.7	-6.2	-6	25	T	μ=0.0347 R: $2UO_2^{2+}$ + L = $UO_2UO_3^{2+}$ + $2H^+$	57Ha	
	10.2	-5.92(25°)	7.1	25&94	T	μ=0.5. R: $2UO_2^{2+}$ + 2L = $(UO_2)_2(OH)_2^{2+}$ + $2H^+$	62Ba	a,b
	9.5	-6.02	4.3	25	C	3M($NaClO_4$) R: $2UO_2^{2+}$ + 2L = $(UO_2)_2(OH)_2^{2+}$ + $2H^+$	68Ag	
	9.5	-6.01	4.3	25	C	3M($NaClO_4$) R: $2UO_2^{2+}$ + 2L = $(UO_2)_2(OH)_2^{2+}$ + $2H^+$	70Ac	i: 62Sa
	9.442	-6.030	4.09	25	C	3.0M($NaClO_4$) R: $2UO_2^{2+}$ + 2L = $(UO_2)_2(OH)_2^{2+}$ + $2H^+$	62Sa	

Metal Ion	ΔH, kcal /mole	Log K	ΔS, cal /mole °K	T, °C	M	Conditions	Ref.	Remarks
W1, cont.								
UO_2^{2+}, cont.	18	-13.20	0	25	C	3.0M($NaClO_4$) R: $3UO_2^{2+}$ + 4L = $(UO_2)_3(OH)_4$ + $4H^+$	62Sa	
	25.1	-16.2(25°)	10	25&94	T	μ=0.5. R: $3UO_2^{2+}$ + 5L = $(UO_2)_3(OH)_5^+$ + $5H^+$	62Ba	a,b
	24.4	-16.54	6	25	C	3M($NaClO_4$) R: $3UO_2^{2+}$ + 5L = $(UO_2)_3(OH)_5^+$ + $5H^+$	68Ae	
	24.4	-16.6	6	25	C	3M($NaClO_4$) R: $3UO_2^{2+}$ + 5L = $(UO_2)_3(OH)_5^+$ + $5H^+$	70Ac	i: 62Sa
	25	-16.55	8	25	C	3.0M($NaClO_4$) R: $3UO_2^{2+}$ + 5L = $(UO_2)_3(OH)_5$ + $5H^+$	62Sa	
	24	-19.42	-8	25	C	3.0M($NaClO_4$) R: $4UO_2^{2+}$ + 6L = $(UO_2)_4(OH)_6$ + $6H^+$	62Sa	
Pu^{4+}	8.5	-1.733(25°)	2.0	15.4& 25	T	μ=2.0($HClO_4$) R: Pu^{4+} + L = $PuOH^{3+}$ + H^+	60Ra	a,b
MnA_6^{3+}	4.8	-0.056(25°)	15.7	1-35	T	μ=4.0. R: MnA_6^{3+} + L = MnA_5OH^- + H^+ A=H_2O	65Wb 65Wd	b
Fe^+	12.0	-6.74(25°)	9.5	20-40	T	μ=0.505-2.19 R: Fe^{2+} + L = $FeOH^+$ + H^+	63Bc	b
[Fe]	24.5	--	--	25	C	[Fe]= G. gouldii Hemerythrin = $HrFe_2$ Tris-cacodylate buffer (≃0.1M Tris) R: $HrFe_2$ + mL + H_2 = $HrFe_2L_m$ + $2H^+$	71Lb	
Fe^{3+}	10.4	-2.17	25.0	25	T	μ=0. R: Fe^{3+} + L = $Fe(OH)^{2+}$ + H^+	57Ma	
	9.9	-2.32(35.3°)	21.5	18.2& 35.5	T	μ=0.15. R: Fe^{3+} + L = $FeOH^{2+}$ + H^+	60Rb	a,b
	18.2	-1.959(25°)	52.1	20&25	T	$C_{ClO_4^-}$ = 0.015M C_M=0.005M R: Fe^{3+} + L = $FeOH^{2+}$ + H^+	44La	a
	10.2	-2.78	21	25	T	μ=1. R: Fe^{3+} + L = $Fe(OH)^{2+}$ + H^+	57Ma	
	11.0	-3.046	23	25	C	3M($NaClO_4$). R: Fe^{3+} + L = $FeOH^{2+}$ + H^+	68Af	i: 62Sb
	11.0	-3.08	23	25	C	3M($NaClO_4$). R: Fe^{3+} + L = $FeOH^{2+}$ + H^+	70Ac	i: 62Sb
	13.5	-2.86	32	25	T	μ=0. R: $2Fe^{2+}$ + 2L = $Fe_2(OH)_2^{4+}$ + $2H^+$	57Ma	
	12.2	-2.71	28	25	T	μ=1. R: $2Fe^{2+}$ + 2L = $Fe_2(OH)_2^{4+}$ + $2H^+$	57Ma	

W1, cont.

Metal Ion	ΔH, kcal /mole	Log K	ΔS, cal /mole °K	T, °C	M	Conditions	Ref.	Re-marks
Fe^{3+}, cont.	10.0	-2.959	20	25	C	3M($NaClO_4$) R: $2Fe^{3+} + 2L = Fe_2(OH)_2^{4+} + 2H^+$	68Af	i: 62Sb
	10.0	-2.93	20.0	25	C	3M($NaClO_4$) R: $2Fe^{3+} + 2L = Fe_2(OH)_2^{4+} + 2H^+$	70Ac	i: 62Sb
	9.8	-2.914(25°)	19.5	15-51	T	μ=3($NaClO_4$) R: $2Fe^{3+} + 2L = Fe_2(OH)_2^{4+} + 2H^+$	55Mb	b
	9.8	--	--	--	T	3M($NaClO_4$) C_M=0.04 R: $2Fe^{3+} + 2L = Fe_2(OH)_2^{4+} + 2H^+$	54Ma	
	14.3	-5.770	22	25	C	3M($NaClO_4$) R: $3Fe^{3+} + 4L = Fe_3(OH)_4^{5+} + 4H^+$	68Af	i: 62Sb
	14.3	-5.79	22	25	C	3M($NaClO_4$) R: $3Fe^{3+} + 4L = Fe_3(OH)_4^{3+} + 4H^+$	70Ac	i: 62Sb
FeA^-	10.0	-9.3(25°)	-9	13.7-42.4	T	μ=0.1(KCl) $FeA^- + L = FeOHA^{2-} + H^+$. A=Cyclohexane diaminetetraacetate	63Gd	b
	10	-7.55(25°)	-2	13-42.4	T	μ=1(KCl). R: $FeA^- + L = FeOHL^{2-} + H^+$ A=EDTA	63Gd	b
	4.7	-12.2(25°)	-40	13-42.4	T	μ=1(KCl) R: $2FeA^{2-} + 2L = (FeOHL)_2^{4-} + 2H^+$ A=EDTA	63Gd	b
FeA_5B	12.3	-3.52	25.2	25	T	μ=0.1(NaCl) R: $FeA_5B + L = FeA_5L^{3-} + B^-$. A=CN. B=$NO_2$	66Se	
Co^{2+}	8.2	9.82(25°)	72.4	15-40	T	μ=0.327($NaClO_4$) R: $Co^{2+} + H_2O = CoOH^+ + H^+$	62Bb 63Be	b,q: ±2
CoA_2	-2.85	1.76(1)	-1.5	25	C	S: 1-Butanol A=ClO_4	68Hb	
	-4.70	3.08(1-2)	-1.7	25	C	S: 1-Butanol A=ClO_4	68Hb	
	-5.35	4.10(1-3)	0.8	25	C	S: 1-Butanol A=ClO_4	68Hb	
	-5.60	4.64(1-4)	2.4	25	C	S: 1-Butanol A=ClO_4	68Hb	
	-5.8	5.6(1-5)	6	25	C	S: 1-Butanol A=ClO_4	68Hb	
Co^{3+}	10	-1.8	25	25	T	μ=0.1 R: $Co^{3+} + L = CoOH^{2+} + H^+$	56Sd	
CoA_2B_2	8.65	-4.747(80°)	2.8	75-90	T	μ=0.01. R: $CoA_2B_2 + L = CoA_2BL + B$	70Fc	b

Metal Ion	ΔH, kcal /mole	Log K	ΔS, cal /mole °K	T, °C	M	Conditions	Ref.	Remarks
W1, cont.								
CoA_2B_2, cont.						A=Dimethylglyoximate B=Thiocyanate		
Co(III) AB	-4.4	-0.271(25°,1)	-16	-60-50	T	S: CD_3OD A=Corrin ring B=Methyl	68Fc	b
	-4.5	-0.088(25°,1)	-15.5	-60-50	T	S: CD_3OD A=Corrin ring B=Vinyl	68Fc	b
Ni^{2+}	7.76	-9.76(25°)	-18.6	25-50	T	0.03-0.1M($NaClO_4$) R: Ni^{2+} + L = $NiOH^+$ + H^+	63Bd	b
	43.0	-28.48	14.0	25	C	3M(NaCl) R: $4Ni^{2+}$ + 4L = $Ni_4(OH)_4{}^{4+}$ + $4H^+$	68Ad	
	43.0	-28.52	14	25	C	3M($NaClO_4$) R: $4Ni^{2+}$ + 4L = $Ni_4(OH)_4{}^{4+}$ + $4H^+$	70Ac	
NiA^{2+}	-4.8	(1-2)	--	12-58	T	C_M=0.1M. A=1,4,8,11-Tetraazaundecane	72If	b
NiA_2	-3.05	1.80(1)	-2.0	25	C	Variable μ S: 1-Butanol. A=ClO_4	68Hb	
	-4.80	3.23(1-2)	-1.3	25	C	Variable μ S: 1-Butanol. A=ClO_4	68Hb	
	-5.25	4.36(1-3)	2.3	25	C	Variable μ S: 1-Butanol. A=ClO_4	68Hb	
	-5.45	5.20(1-4)	5.5	25	C	Variable μ S: 1-Butanol. A=ClO_4	68Hb	
	-5.7	5.8(1-5)	8	25	C	Variable μ S: 1-Butanol. A=ClO_4	68Hb	
RhA_2BC	4.1	-2.70(50°)	0.1	55-81	T	μ=0.2M. A=Ethylenediamine. B=OH. C=I R: trans-RhA_2BC^+ + L = trans-RhA_2BL^{2+} + C^-	72Pi	b,q: ±0.2
RhA_3B_3	-4.8	-4.25(25°)	-35.6	25-85	T	0.1F($HClO_4$) R: RhA_3B_3 + 2L = $RhAB_3L_2{}^{2+}$ + $2A^-$ A=Cl. B=H_2O	66Ba	b
RhA_5B	-1.3	-2.249	-14.3	50	T	μ=0.2M. A=NH_3. B=Cl R: RhA_5B^{2+} + L = RhA_5L^{3+} + B^-	67Pb 67Pa	q: ±0.6
	-0.7	-2.159	-11.8	50	T	μ=0.2M. A=NH_3. B=Br R: RhA_5B^{2+} + L = RhA_5L^{3+} + B^-	67Pb 67Pa	q: ±0.4
	1.8	-2.682	-6.4	50	T	μ=0.2M. A=NH_3. B=I R: RhA_5B^{2+} + L = RhA_5L^{3+} + B^-	67Pb 67Pa	q: ±0.3
$PtAB_2$	4	-2.66(25°)		15-35	T	μ=0.318M A=Ethylenediamine	73Cd	b,q: ±1

Metal Ion	ΔH, kcal /mole	Log K	ΔS, cal /mole °K	T, °C	M	Conditions	Ref.	Remarks
W1, cont.								
$PtAB_2$, cont.						B=Cl. R: $PtAB_2$ + L = $PtALB^+$ + B^-		
$PtALB^+$	-1	-3.84(25°)		15-35	T	μ=0.318M A=Ethylenediamine B=Cl. R: $PtALB^+$ + L = $PtAL_2^{2+}$ + B^-	73Cd	b,q: ±2
PtA_4^{2-}	5	<u>-1.8</u>	9	25	T	μ→0. R: PtA_4^{2-} + L = PtA_3L^- + A^-. A=Cl	55Gc	
	4.4	1.89(25°)	6.0	15-35	T	0.5M($HClO_4$) R: PtA_4^{2-} + L = PtA_3L^- + A^-. A=Cl	67Dd	b
$PtAB_3^-$	2.1	-1.8	-1	25	T	μ→0. R: $PtAB_3^-$ + L = $PtAB_2L$ + A^- A=NH_3. B=Cl	58Ea	
Cu^{2+}	11 to 12	-8	≃0	25	C	3M($NaClO_4$). ΔH values set assuming ΔS=0. R: Cu^{2+} + L = $CuOH^+$ + H^+	68Ae	
	15.8	-10.6	4.4	25	C	3M($NaClO_4$) R: $2Cu^{2+}$ + 2L = $Cu_2(OH)_2^{2+}$ + $2H^+$	68Ae	
	15.8	-10.63	4.4	25	C	3M($NaClO_4$) R: $2Cu^{2+}$ + 2L = $Cu_2(OH)_2^{2+}$ + $2H^+$	70Ac	
CuAB	1	-1.39(25°)	-5	0.3-42.5	T	μ=0.1M(KNO_3). A=OH B=N,N'-Dihydroxy-ethylethylenediamine R: $(CuAB)_2^{2+}$ + 2L = $2CuABL^+$	59Ge	b,q: ±7
	-1.0	-3.81(25°)	-21	0.3-42.5	T	μ=0.1M(KNO_3). A=OH B=N,N'-Dimethyl-ethylenediamine R: $(CuAB)_2^{2+}$ + 2L = $2CuABL^+$	59Ge	b,q: ±0.4
	-4	-2.20(25°)	-23	0.3-42.5	T	μ=0.1M(KNO_3). A=OH B=N-Hydroxyethyl-ethylenediamine R: $(CuAB)_2^{2+}$ + 2L = $2CuABL^+$	59Ge	b,q: ±3
	-3	-3.88(25°)	-28	0.3-42.5	T	μ=0.1M(KNO_3). A=OH B=N,N,N',N'-Tetra-methylethylenediamine R: $(CuAB)_2^{2+}$ + 2L = $2CuABL^+$	59Ge	b,q: ±3
Zn^{2+}	13.4	-8.96(25°)	<u>4</u>	15-42	T	μ=0. R: Zn^{2+} + L = $ZnOH^+$ + H^+	62Pb	b
Cd^{2+}	13.1	-10.2	-2.7	25	C	3M($LiClO_4$). R: Cd^{2+} + L = $CdOH^+$ + H^+	67Ae 70Ac	

Metal Ion	ΔH, kcal /mole	Log K	ΔS, cal /mole °K	T, °C	M	Conditions	Ref.	Remarks
W1, cont.								
Cd^{2+}, cont.	10.9	-9.1	-5.1	25	C	3M($LiClO_4$) R: $2Cd^{2+} + L = Cd_2OH^{3+} + H^+$	67Ae 70Ac	
	40.6	-31.8	-9.4	25	C	3M($LiClO_4$) R: $4Cd^{2+} + 4H_2O = Cd_4(OH)_4^{4+} + 4H^+$	67Ae 70Ac	
CdA_2^{2+}	2.04	-1.53(1)	-0.2	25	C	S: 56.17 w % C_2H_5OH. A=H_2O	59Va	
CdA_3^{2+}	2.04	-1.67(1)	-0.8	25	C	S: 56.17 w % C_2H_5OH. A=H_2O	59Va	
CdA_4^{2+}	2.04	-1.70(1)	-0.9	25	C	S: 56.17 w % C_2H_5OH. A=H_2O	59Va	
Hg^{2+}	7.23	-3.56	7.95	25	C	3M($NaClO_4$) R: $Hg^{2+} + L = HgOH^+ + H^+$	67Ad	
	7.2	-3.59	7.9	25	C	3M($NaClO_4$) R: $Hg^{2+} + L = HgOH^+ + H^+$	70Ac	
	10.790	-6.22	8.4	20	T	μ=0. R: $Hg^{2+} + 2L = Hg(OH)_2 + 2H^+$	58Aa	
	9.84	-6.21	4.59	25	C	3M($NaClO_4$) R: $Hg^{2+} + 2L = Hg(OH)_2 + 2H^+$	67Ad	
	9.8	-6.23	4.6	25	C	3M($NaClO_4$) R: $Hg^{2+} + 2L = Hg(OH)_2 + 2H^+$	70Ac	
	3.06	-2.70	-2.08	25	C	3M($NaClO_4$) R: $2Hg^{2+} + L = Hg_2(OH)^{3+} + H^+$	67Ad	
	3.1	-2.71	-2.1	25	C	3M($NaClO_4$) R: $2Hg^{2+} + L = Hg_2OH^{3+} + H^+$	70Ac	
	12.7	-5.20	18.8	25	C	3M($NaClO_4$) R: $2Hg^{2+} + 2L = Hg_2(OH)_2^{2+} + 2H^+$	70Ac	
	12.70	-5.20	18.8	25	C	3M($NaClO_4$) R: $2Hg^{2+} + 2L = Hg_2(OH)_2^{2+} + 2H^+$	67Ad	
Al^{3+}	18.3	-7.45(25°)	27	25-125	T	μ=1M(KCl) R: $2Al^{3+} + 2L = Al_2(OH)_2^{4+} + 2H^+$ ΔCp=12	71Tc	b,q: ±2.6
	30.9	-13.36(25°)	42	25-125	T	μ=1M(KCl) R: $3Al^{3+} + 4L = Al_3(OH)_4^{5+} + 4H^+$ ΔCp=90	71Td	b,q: ±3.4
	262.8	-110.45(25°)	377	25-125	T	μ=1M(KCl) R: $14Al^{3+} + 34L = Al_{14}(OH)_{34}^{8+} + 34H^+$ ΔCp=733	71Td	b

Metal Ion	ΔH, kcal /mole	Log K	ΔS, cal /mole °K	T, °C	M	Conditions	Ref.	Remarks
W1, cont.								
In^{3+}	4.85	-4.42	-4.0	25	C	3.0M($NaClO_4$) R: In^{3+} + L = $In(OH)^{2+}$ + H^+	61Sc	
	14	-8.32	9	25	C	3.0M($NaClO_4$) R: In^{3+} + 2L = $In(OH)_2^+$ + $2H^+$	61Sc	
	n(10.18)	0.3($10^{-4.69n}$)	-2.4+ n(12.7)	25	C	3.0M($NaClO_4$) R: (n+1)In^{3+} + 2nL = $In[In(OH)]_n^{3+n}$ + $2nH^+$	61Sc	
Tl^{3+}	24	-1.41	74	25	C	μ=3.0M($LiClO_4$) R: Tl^{3+} + L = $Tl(OH)^{2+}$ + H^+	73Ki	
	21	-1.180	70	25	C	μ=3M($LiClO_4$, $HClO_4$) R: Tl^{3+} + HL = TlL^{2+} + H^+	67Ya	q: ±6
	26	-1.15	82	25	C	μ=3.0M($LiClO_4$). A=OH. R: $Tl(A)^{2+}$ + L = $Tl(A)_2^+$ + H^+	73Ki	q: ±3
	21	-1.415	65	25	C	μ=3M($LiClO_4$, $HClO_4$) A=OH. R: TlA^{2+} + HL = TlA_2^+ + H^+	67Fb	q: ±24
GeA_6^{2-}	9.5	<u>-5.28</u>(25°)	8	0-25	T	R: $GeA_6^=$ + 2L = $[GeA_4LOH]^-$ + HA + A^-. A=F	64Rb	a,b
	-0.59	-3.72	-19	25	T	μ=0.4(HCl + NaCl) R: GeA_6^{2-} + L = GeA_5L^- + A^-. A=F	64Ra	
SnA_5	-24.3	(<u>1</u>)	--	25	C	S: Ethylene Chloride. A=Cl	67Oc	
Pb^{2+}	26.5	-22.87	-15.8	25	C	3M($NaClO_4$) R: $3Pb^{2+}$ + 4L = $Pb_3(OH)_4^{2+}$ + $4H^+$	62Cc	
	20.07	-19.25	-20.8	25	C	3M($NaClO_4$) R: $4Pb^{2+}$ + 4L = $Pb_4(OH)_4^{4+}$ + $4H^+$	62Cc	
	49.44	-42.12	-27.0	25	C	3M($NaClO_4$) R: $6Pb^{2+}$ + 8L = $Pb_6(OH)_8^{4+}$ + $8H^+$	62Cc	
W2 WATER, deuterated			D_2O			D_2O		L
Pu^{4+}	19.4	-1.94(25°)	<u>11.5</u>	15.4& 25	T	μ=2.0($HClO_4$) S: D_2O. R: Pu^{4+} + L = $PuOD^{3+}$ + D^+	60Ra	a,b
X1 XANTHOGENIC ACID, <u>iso</u>-butyl			$C_5H_{10}OS_2$			C_4H_9OCSSH		L^-
Ni^{2+}	1.37	(<u>1</u>-<u>2</u>)	--	30	C	μ=0	74Cb	g:K, q: ±0.11

Metal Ion	ΔH, kcal /mole	Log K	ΔS, cal /mole °K	T, °C	M	Conditions	Ref.	Remarks
X2 XANTHOGENIC ACID, n-butyl			$C_5H_{10}OS_2$			C_4H_9OCSSH		L^-
Ni^{2+}	0.84	(1-2)	--	30	C	μ=0	74Cb	g:K
X3 XANTHOGENIC ACID, ethyl-			$C_3H_6OS_2$			C_2H_5OCSSH		L^-
Ni^{2+}	1.15	(1-2)	--	30	C	μ=0	74Cb	g:K
X4 XANTHOGENIC ACID, methyl-			$C_2H_4OS_2$			CH_3OCSSH		L^-
Ni^{2+}	2.54	(1-2)	--	30	C	μ=0	74Cb	g:K
X5 XANTHOGENIC ACID, propyl-			$C_4H_8OS_2$			C_3H_7OCSSH		L^-
Ni^{2+}	1.17	(1-2)	--	30	C	μ=0	74Cb	g:K
X6 XENON			Xe					
[Fe]	-9.0	2.12(25°,1)	-20	20-30	T	C_L=10% Solution of L. [Fe]=Cyanomet-myoglobin	70Ed	b,j
	-7.2	2.12(25°,1)	-14	20-30	T	C_L=10% Solution of L. [Fe]=Metmyoglobin	70Ed	b,j
	-5.1	1.98(25°,1)	-8.0	20-30	T	C_L=10% Solution of L. [Fe]=Myoglobin	70Ed	b,j

EMPIRICAL FORMULA INDEX

EMPIRICAL FORMULA INDEX

EMPIRICAL FORMULA INDEX

EMPIRICAL FORMULA INDEX

ELEMENT INDEX

ELEMENT INDEX

SYNONYM INDEX

SYNONYM INDEX

SYNONYM INDEX

REFERENCE INDEX

1974

74Ca R. D. Cannon and J. Gardiner, *Inorg. Chem.*, *13*, 390 (1974).

74Cb K. J. Cavell, J. O. Hill and R. J. Magee, *Thermochimica Acta*, *8*, 447 (1974).

74Cc M. Chikuma, A. Yokoyama and H. Tanaka, *J. Inor. Nucl. Chem.*, *36*, 1243 (1974).

74Cd J. J. Christensen, D. J. Eatough and R. M. Izatt, *Chem. Reviews*, *74*, 351 (1974).

74Ce A. H. Cohen and B. M. Hoffman, *Inorg. Chem.*, *13*, 1484 (1974).

74Da R. S. Drago, J. A. Nusz and R. C. Courtright, *J. Amer. Chem. Soc.*, *96*, 2082 (1974).

74Ga D. P. Graddon and T. T. Nyein, *Aust. J. Chem.*, *27*, 407 (1974).

74Gb D. P. Graddon and W. K. Ong, *Aust. J. Chem.*, *27*, 741 (1974).

74Gc I. Grenthe and G. Gardhammar, *Acta Chem. Scand.*, *28*, 125 (1974).

74Ha G. I. H. Hanania and S. A. Israelian, *J. Solution Chem.*, *3*, 57 (1974).

74Hb G. R. Hedwig and H. K. J. Powell, *J. C. S. Dalton*, 47 (1974).

74Hc F. G. Herring and R. L. Tapping, *J. Phys. Chem.*, *78*, 316 (1974).

74Ja B. R. James and L. D. Markham, *Inorg. Chem.*, *13*, 97 (1974).

74Ka F. Kai and Y. Sadakane, *J. Inorg. Nucl. Chem.*, *36*, 1404 (1974).

74Kb N. Kiba and T. Takeuchi, *J. Inorg. Nucl. Chem.*, *36*, 847 (1974).

74La L. N. Lin, S. D. Christian and J. D. Childs, *J. Amer. Chem. Soc.*, *96*, 2727 (1974).

74Ma M. H. Mihailov, V. Ts. Mihailova, and V. A. Khalkin, *J. Inorg. Nucl. Chem.*, *36*, 145 (1974).

74Na I. Nagypal, A. Gergely and E. Farkas, *J. Inorg. Nucl. Chem.*, *36*, 699 (1974).

74Nb R. M. Nikolic and I. J. Gal, *J. C. S. Dalton*, 985 (1974).

74Pa W. Partenheimer and B. Durham, *J. Amer. Chem. Soc.*, *96*, 3800 (1974).

74Pb M. S. Pereira and J. M. Malin, *Inorg. Chem.*, *13*, 386 (1974).

74Pc C. Pillot, J. P. Pascault and J. Gole, *Bull. Soc. Chim. France*, No. 3-4, 357 (1974).

74Sa S. Santini, G. Reichenback, and U. Mazzucato, *J. C. S. Perkin II*, 494 (1974).

74Ta M. M. Taqui Khan and P. Rabindra Reddy, *J. Inorg. Nucl. Chem.*, *36*, 607 (1974).

74Tb C. A. Tolman, W. C. Seidel and L. W. Gosser, *J. Amer. Chem. Soc.*, *96*, 53 (1974).

74Tc B. S. Tovrog and R. S. Drago, *J. Amer. Chem. Soc.*, *96*, 2743 (1974).

1973

73Aa R. Abu-Eittah and G. Arafa, *Z. Anorg. Allg. Chem.*, *399*, 244 (1973).

1973 (continued)

73Ab R. D. Adams and F. A. Cotton, *Inorg. Chim. Acta*, *7*, 153 (1973).

73Ac R. G. Ainsworth, *Trans. Faraday Soc.*, *69*, 1028 (1973).

73Ad S. O. Ajayi and D. R. Goddard, *J. C. S. Dalton*, 1751 (1973).

73Ae P. E. M. Allen and R. M. Lough, *Trans. Faraday Soc.*, *I*, *69* (*5*), 849 (1973).

73Af L. T. Ang and D. P. Graddon, *Aust. J. Chem.*, *26*, 1901 (1973).

73Ag A. C. I. Anusiem and R. Lumry, *J. Amer. Chem. Soc.*, *95*, 904 (1973).

73Ah E. P. Artyukhova and V. I. Dulova, *Russ. J. Inorg. Chem.* (*English Trans.*), *18*, 385 (1973). (Russian Page No. 735.)

73Ai E. P. Artyukhova and V. I. Dulova, *Russ. J. Phys. Chem.* (*English Trans.*), *47*, 317 (1973). (Russian Page No. 559.)

73Aj J. L. Ault and H. J. Harries, *J. C. S. Dalton*, 1095 (1973).

73Ba E. M. Ban, R. P. Hughes and J. Powell, *J. C. S. Chem. Comm.*, 591 (1973).

73Bb J. L. Banyasz and J. E. Stuehr, *J. Amer. Chem. Soc.*, *95*, 7226 (1973).

73Bc R. Barbucci, L. Fabbrizzi, P. Paoletti and A. Vacca, *J. C. S. Dalton*, 1763 (1973).

73Bd J. K. Beattie and N. Sutin, *J. Amer. Chem. Soc.*, *95*, 2052 (1973).

73Be D. J. Benton and P. Moore, *J. C. S. Dalton*, 399 (1973).

73Bf L. P. Berezina, V. G. Samoilenko and A. I. Pozigun, *Russ. J. Inorg. Chem.* (*English Trans.*), *18*, 205 (1973). (Russian Page No. 393.)

73Bg G. Berthon and O. Enea, *Thermochimica Acta*, *6*, 57 (1973).

73Bh G. L. Bertrand, D. E. Oyler, U. G. Eichelbaum and L. G. Hepler, *Thermochimica Acta*, *7*, 87 (1973).

73Bi J. R. Blackborow and J. C. Lockhart, *J. Chem. Thermodyn.* *5*, 603 (1973).

73Bj R. J. Blagrove and L. C. Gruen, *Aust. J. Chem.*, *26*, 225 (1973).

73Ca R. Cetina, J. Gomez-Lara and R. Contreras, *J. Inorg. Nucl. Chem.*, *35*, 4217 (1973).

73Cb G. R. Choppin and S. L. Bertha, *J. Inor. Nucl. Chem.*, *35*, 1309 (1973).

73Cc G. R. Choppin, A. Dadgar and R. Stampfli, *J. Inorg. Nucl. Chem.*, *35*, 875 (1973).

73Cd R. F. Coley and D. S. Martin, *Inorg. Chim. Acta*, *7*, 573 (1973).

73Ce M. J. Collins and H. F. Henneike, *Inorg. Chem.*, *12*, 2983 (1973).

73Cf R. L. Courtright, R. S. Drago, J. A. Nusz and M. S. Nozari, *Inorg. Chem.* *12*, 2809 (1973).

73Cg R. E. Cramer and R. L. Harris, *Inorg. Chem.*, *12*, 2575 (1973).

73Da D. R. Dakternieks and D. P. Graddon, *Aust. J. Chem.*, *26*, 2379 (1973).

73Db D. R. Dakternieks and D. P. Graddon, *Aust. J. Chem.*, *26*, 2537 (1973).

73Dc I. Dellien, *Acta Chem. Scand.*, *27*, 733 (1973).

1973 (continued)

73Dd I. Dellien, I. Grenthe and G. Hessler, *Acta Chem. Scand.*, *27*, 2431 (1973).

73De I. Dellien and L. Malmsten, *Acta Chem. Scand.*, *27*, 2877 (1973).

73Df S. C. Dhupar, K. C. Srivastava and S. K. Banerji, *J. Indian Chem. Soc.*, *20*, 19 (1973).

73Dg V. I. Dulova and T. A. Zhuravel, *Russ. J. Phys. Chem.*, *(English Trans.)*, *47*, 1510 (1973). (Russian Page No. 2685.)

73Ea B. L. Edgar, D. J. Duffy, M. C. Palazzotto and L. H. Pignolet, *J. Amer. Chem. Soc.*, *95*, 1125 (1973).

73Eb K. J. Ellis and A. McAuley, *J. C. S. Dalton*, 1533 (1973).

73Ec O. Enea and G. Berthon, *Thermochimica Acta*, *6*, 47 (1973).

73Ed O. Enea, K. Houngbossa and G. Berthon, *Thermochimica Acta*, *6*, 309 (1973).

73Fa Y. Farhang and D. P. Graddon, *Aust. J. Chem.*, *26*, 983 (1973).

73Ga A. Gergely and I. Sovago, *J. Inorg. Nucl. Chem.*, *35*, 4355 (1973).

73Gb I. P. Goldshtein, E. N. Guryanova, M. E. Peisakhova and R. R. Shifrina, *J. Gen. Chem. USSR*, *(English Trans.)*, *43*, 2332 (1973). (Russian Page No. 2347.)

73Gc R. C. Graham, G. H. Henderson, E. M. Eyring and E. M. Woolley, *J. Chem. Eng. Data*, *18*, 277 (1973).

73Gd R. J. Guschl and T. L. Brown, *Inorg. Chem.*, *12*, 2815 (1973).

73Ha G. R. Hedwig and H. K. J. Powell, *J. C. S. Dalton*, 793 (1973).

73Hb G. R. Hedwig and H. K. J. Powell, *J. C. S. Dalton*, 798 (1973).

73Hc G. R. Hedwig and H. K. J. Powell, *J. C. S. Dalton*, 1942 (1973).

73Hd D. R. Herrington and L. J. Boucher, *Inorg. Chem.*, *12*, 2378 (1973).

73He K. Houngbossa, G. Berthon, and O. Enea, *Thermochimica Acta*, *6*, 215 (1973).

73Hf D. H. Huchital and R. J. Hodges, *Inorg. Chem.*, *12*, 998 (1973).

73Hg J. K. Hurst and R. H. Lane, *J. Amer. Chem. Soc.*, *95*, 1703 (1973).

73Ia T. P. I., E. J. Burke, J. L. Meyer, and G. H. Nancollas, *Thermochimica Acta*, *5*, 463 (1973).

73Ib J. Israeli and R. Volpe, *Bull. Soc. Chim. France*, No. 1, 43 (1973).

73Ic P. B. Iyer and M. R. Padhye, *Indian J. Chem.*, *11*, 166 (1973).

73Ka F. Kai, Y. Sadakand, H. Yokoi and H. Aburada, *J. Inorg. Nucl. Chem.*, *35*, 2128 (1973).

73Kb R. C. Kapoor and B. S. Aggarwal, *Indian J. Chem.*, *11*, 71 (1973).

73Kc S. Katayama, *Bull. Chem. Soc. Japan*, *46*, 106 (1973).

73Kd M. M. T. Khan and P. R. Reddy, *J. Inorg. Nucl. Chem.*, *35*, 179 (1973).

73Ke M. M. T. Khan and P. R. Reddy, *J. Inorg. Nucl. Chem.*, *35*, 2813 (1973).

73Kf M. M. T. Khan and P. R. Reddy, *J. Inorg. Nucl. Chem.*, *35*, 2821 (1973).

73Kg N. K. Kildahl and R. S. Drago, *J. Amer. Chem. Soc.*, *95*, 6245 (1973).

1973 (continued)

73Kh Ts. B. Konunova and L. S. Kachkar, *Russ. J. Inorg. Chem. (English Trans.)*, *18*, 805 (1973). (Russian Page No. 1527.)

73Ki F. Ya. Kulba, E. A. Kopylov and Yu. B. Yakovlev, *Russ. J. Inorg. Chem. (English Trans.)*, *18*, 38 (1973). (Russian Page No. 76.)

73La J. C. Lockhart and W. J. Mossop, *J. C. S. Dalton*, 19 (1973).

73Lb V. A. Logachev, V. I. Dulova, and N. R. Molchanova, *Russ. J. Inorg. Chem. (English Trans.)*, *18*, 1361 (1973). (Russian Page No. 2565.)

73Lc R. S. Lord, F. Gubensek, and J. A. Rupley, *Biochemistry*, *12*, 4385 (1973).

73Ma V. E. Mironov and N. N. Knyazeva, *Russ. J. Phys. Chem. (English Trans.)*, *47*, 736 (1973). (Russian Page No. 1301.)

73Mb V. E. Mironov, G. K. Ragulin, N. P. Kolobov, V. M. Fadeev and Yu. B. Solovev, *Russ. J. Phys. Chem. (English Trans.)*, *47*, 456 (1973). (Russian Page No. 802.)

73Mc V. E. Mironov, G. K. Ragulin, N. P. Kolobov, V. M. Fadeev and Yu. B. Solovev, *Russ. J. Phys. Chem. (English Trans.)*, *47*, 458 (1973). (Russian Page No. 806.)

73Md V. E. Mironov, G. K. Ragulin, Yu. B. Solovev, V. M. Fadeev and N. P. Kolobov, *Russ. J. Phys. Chem. (English Trans.)*, *47*, 301 (1973). (Russian Page No. 530.)

73Me M. I. Mittal, R. S. Saxena and A. V. Pandey, *J. Inorg. Nucl. Chem.*, *35*, 1691 (1973).

73Na O. Nor and A. G. Sykes, *J. C. S. Dalton*, 1232 (1973).

73Oa G. Olofsson and I. Olofsson, *J. Amer. Chem. Soc.*, *95*, 7231 (1973).

73Ob L. L. Olson, *J. Inorg. Nucl. Chem.*, *35*, 1977 (1973).

73Oc H. Ots, *Acta Chem. Scand.*, *27*, 2344 (1973).

73Pa V. A. Palkin, T. A. Kuzina and N. N. Kuzmina, *Russ. J. Inorg. Chem. (English Trans.)*, *18*, 406 (1973).

73Pb S. P. Pande and K. N. Munshi, *Indian J. Chem.*, *11*, 1322 (1973).

73Pc A. N. Pant, R. N. Soni, and S. L. Gupta, *J. Inorg. Nucl. Chem.*, *35*, 1390 (1973).

73Pd W. Partenheimer and E. F. Hoy, *J. Amer. Chem. Soc.*, *95*, 2840 (1973).

73Pe W. Partenheimer and E. F. Hoy, *Inorg. Chem.*, *12*, 2805 (1973).

73Pf W. Partenheimer and E. H. Johnson, *Inorg. Chem.*, *12*, 1274 (1973).

73Pg S. Parthasarathy and S. Ambujavalli, *Thermochimica Acta*, *7*, 225 (1973).

73Ph M. E. Peisakhova, I. P. Goldshtein, E. N. Guryanova, and E. S. Shcherbakova, *J. General Chem. U.S.S.R. (English Trans.)*, *43*, 157 (1973). (Russian Page No. 159.)

73Pi I. Z. Pevzner, N. I. Eremin, N. N. Knyazeva, Ya. B. Rozen and V. E. Mironov, *Russ. J. Inorg. Chem. (English Trans.)*, *18*, 596 (1973). (Russian Page No. 1129.)

73Pj K. Plotkin, J. Copes, and J. R. Vriesenga, *Inorg. Chem.*, *12*, 1494 (1973).

1973 (continued)

73Pk H. K. J. Powell and N. F. Curtis, *Aust. J. Chem.*, *26*, 977 (1973).

73Pl W. H. Puhl and H. F. Henneike, *J. Phys. Chem.*, *77*, 558 (1973).

73Ra D. A. Redfield and J. H. Nelson, *Inorg. Chem.*, *12*, 15 (1973).

73Rb G. Reinhard, R. Dreyer and R. Munze, *Z. Phys. Chemie, Leipzig*, *254*, 226 (1973).

73Rc F. H. Reynolds, R. K. Burkhard and D. D. Mueller, *Biochem.*, *12*, 359 (1973).

73Rd I. P. Romm, E. M. Sadykova, E. N. Guryanova and I. D. Kolli, *J. Gen. Chem. U.S.S.R. (English Trans.)*, *43*, 727 (1973). (Russian Page No. 728.)

73Re R. Roulet and C. Barbey, *Helvetica Chimica Acta*, *56*, 2179 (1973).

73Sa J. C. Sari and J. P. Belaich, *J. Amer. Chem. Soc.*, *95*, 7491 (1973).

73Sb R. Sarin and K. N. Munshi, *J. Indian Chem. Soc.*, *1*, 307 (1973).

73Sc R. Sarin and K. N. Munshi, *J. Inorg. Nucl. Chem.*, *35*, 201 (1973).

73Sd R. S. Saxena and U. S. Chaturvedi, *Monatsheftle Chemie*, *104*, 1208 (1973).

73Se R. S. Saxena and S. S. Sheelwant, *J. Inorg. Nucl. Chem.*, *35*, 941 (1973).

73Sf J. J. Schaer, B. Torre and D. Janjic, *Helvetica Chimica Acta*, *56*, 2101 (1973).

73Sg B. S. Sekhon and S. L. Chopra, *Thermochimica Acta*, *7*, 151 (1973).

73Sh B. S. Sekhon and S. L. Chopra, *Thermochimica Acta*, *7*, 311 (1973).

73Si G. B. Sergeev, V. Y. Smirnov and M. S. Sytilin, *Russ. J. Phys. Chem. (English Trans.)*, *47*, 396 (1973). (Russian Page No. 704.)

73Sj R. S. Sexena and S. S. Sheelwant, *J. Inorg. Nucl. Chem.*, *35*, 3963 (1973).

73Sk E. Shchori and J. Jagur-Grodzinski, *Israel J. Chem.*, *11*, 243 (1973).

73Sl H. K. Sinha and B. Prasad, *J. Indian Chem. Soc.*, *1*, 177 (1973).

73Sm S. K. Srivastava and H. B. Mathur, *Indian J. Chem.*, *11*, 936 (1973).

73Sn S. K. Srivastava, E. V. Raju and H. B. Mathur, *J. Inorg. Nucl. Chem.*, *35*, 253 (1973).

73So R. Stampfli and G. R. Choppin, *J. Inorg. Nucl. Chem.*, *34*, 205 (1972).

73Sp G. R. Stevenson and A. E. Alegria, *J. Phys. Chem.*, *77*, 3100 (1973).

73Sq G. R. Stevenson and L. Echegoyen, *J. Phys. Chem.*, *77*, 2339 (1973).

73Sr D. V. Stynes, H. C. Stynes, B. R. James, and J. A. Ibers, *J. Amer. Chem. Soc.*, *95*, 1796 (1973).

73Ss M. S. Subramanian and S. A. Pai, *Aust. J. Chem.*, *26*, 77 (1973).

73St T. G. Sukhova, O. N. Temkin, L. I. Kuleshova and R. M. Flid, *Russ. J. Inorg. Chem. (English Trans.)*, *18*, 172 (1973). (Russian Page No. 325.)

73Ta P. H. Tedesco, V. B. DeRumi, J. A. Gonzalez Quintana, *J. Inorg. Nucl. Chem.*, *35*, 285 (1973).

73Tb B. Tilquin and L. Lamberts, *Z. Phys. Chemie N. F.*, *84*, 84 (1973).

1973 (continued)

73Ua I. E. Umova, B. I. Lobov, Yu. I. Rutkovskii and V. E. Mironov, *Russ. J. Inorg. Chem. (English Trans.)*, *18*, 430 (1973). (Russian Page No. 836.)

73Va V. P. Vasilev and S. A. Aleksandrova, *Russ. J. Inorg. Chem. (English Trans.)*, *18*, 1089 (1973). (Russian Page No. 2055.)

73Vb V. N. Vasileva, V. P. Vasilev, T. K. Korbut, and L. I. Bukoyazova, *Russ. J. Inorg. Chem. (English Trans.)*, *18*, 974 (1973). (Russian Page No. 1843.)

73Vc V. P. Vasilev and G. A. Zaitseva, *Russ. J. Inorg. Chem. (English Trans.)*, *18*, 70 (1973). (Russian Page No. 139.)

73Vd G. C. Vogel and L. A. Searby, *Inorg. Chem.*, *12*, 936 (1973).

73Wa F. A. Walker, *J. Amer. Chem. Soc.*, *95*, 1150 (1973).

73Wb F. A. Walker, *J. Amer. Chem. Soc.*, *95*, 1154 (1973).

73Ya H. Yokoi, M. Otagiri and T. Isobe, *Bull. Chem. Soc. Japan*, *46*, 442 (1973).

73Za Ch. U. Zust, P. U. Fruh and W. Simon, *Helv. Chim. Acta*, *56*, 495 (1973).

1972

72Aa R. Abu-Eittah and A. El-Kourashy, *J. Phys. Chem.*, *76*, 2405 (1972).

72Ab P. E. M. Allen, A. E. Byers and R. M. Lough, *J. C. S. Dalton Trans.*, No. 4, 479 (1972).

72Ac R. D. Allendoerfer and R. J. Papez, *J. Phys. Chem.*, *76*, 1012 (1972).

72Ad E. Arenare, P. Paoletti, A. Dei and A. Vacca, *J. C. S. Dalton Trans.*, No. 6, 736 (1972).

72Ae E. P. Artyukhova and V. I. Dulova, *Russ. J. Inorg. Chem. (English Trans.)*, *17*, 456 (1972). (Russian Page No. 873.)

72Af E. P. Artyukhova and V. I. Dulova, *Russ. J. Phys. Chem. (English Trans.)*, *46*, 295 (1972). (Russian Page No. 510.)

72Ag K. R. Ashley and S. Kulprathipanja, *Inorg. Chem.*, *11*, 444 (1972).

72Ah B. K. Avinashi and S. K. Banerji, *J. Indian Chem. Soc.*, *49*, 693 (1972).

72Ba K. Balachandran and S. K. Banerji, *J. Indian Chem. Soc.*, *49*, 543 (1972).

72Bb R. Barbucci, L. Fabbrizzi and P. Paoletti, *J. C. S. Dalton*, 745 (1972).

72Bd R. Barbucci, L. Fabbrizzi and P. Paoletti, *J. C. S. Dalton*, 1099 (1972).

72Be R. Barbucci, L. Fabbrizzi, P. Paoletti and A. Vacca, *J. C. S. Dalton Trans.*, No. 6, 740 (1972).

72Bf R. Barbucci, P. Paoletti and L. Fabbrizzi, *J. C. S. Dalton*, 2593 (1972).

72Bg G. Beech and K. Miller, *J. C. S. Dalton*, 801 (1972).

72Bh V. I. Belevantsev, B. I. Peshchevitskii, and Zh. O. Badmaeva, *Russ. J. Inorg. Chem. (English Trans.)*, *17*, 1522 (1972). (Russian Page No. 2897.)

72Bi G. Berthon and O. Enea, *Thermochimica Acta*, *5*, 107 (1972).

72Bj G. Berthon, O. Enea and Y. Bokra, *Thermochimica Acta*, *4*, 441 (1972).

1972 (continued)

72Bk O. K. Borggaard, *Acta Chem. Scand.*, *26*, 393 (1972).

72Bl A. P. Brunetti, E. J. Burke, M. C. Lim, and G. H. Nancollas, *J. Solution Chem.*, *1*, 153 (1972).

72Bm J. G. Bullitt, F. A. Cotton and T. J. Marks, *Inorg. Chem.*, *11*, 671 (1972).

72Bn M. J. Burkhart and R. C. Thompson, *J. Amer. Chem. Soc.*, *94*, 2999 (1972).

72Ca E. F. Caldin, M. W. Grant, B. B. Hasinoff, *Trans. Faraday Soc.*, *68*, 2247 (1972).

72Cb L. L. Chan and J. Smid, *J. Phys. Chem.*, *76*, 695 (1972).

72Cc P. B. Chock, *Proc. Nat. Acad. Sci., USA*, *69*, 1939 (1972).

72Cd G. R. Choppin and G. Degischer, *J. Inorg. Nucl. Chem.*, *34*, 3473 (1972).

72Ce S. J. Cole, G. C. Curthoys, E. A. Magnusson and J. N. Phillips, *Inorg. Chem.*, *11*, 1024 (1972).

72Da A. Dadgar and G. R. Choppin, *J. Inorg. Nucl. Chem.*, *34*, 1297 (1972).

72Db G. Degischer and G. R. Choppin, *J. Inorg. Nucl. Chem.*, *34*, 2823 (1972).

72Dc H. S. Dunsmore and D. Midgley, *J. C. S. Dalton*, 1138 (1972).

72Ea S. S. Eaton, J. R. Hutchison, R. H. Holm, and E. L. Muetterties, *J. Amer. Chem. Soc.*, *94*, 6411 (1972).

72Eb L. Elegant, A. Pagliardini, G. Torri and M. Azzaro, *Bull. Soc. Chim. (France)*, No. 11, 4422 (1972).

72Ec O. Enea and G. Berthon, *Comptes Rendus*, *274*, Series C, 1968 (1972).

72Ed O. Enea, G. Berthon and Y. Bokra, *Thermochimica Acta*, *4*, 449 (1972).

72Ee J. H. Espenson and S. G. Wolenuk, *Inorg. Chem.*, *11*, 2034 (1972).

72Fa L. Fabbrizzi, R. Barbucci and P. Paoletti, *J. C. S. Dalton*, 1529 (1972).

72Fb R. D. Farina and J. H. Swinehart, *Inorg. Chem.*, *11*, 645 (1972).

72Fc J. B. Farmer, F. G. Herring and R. L. Tapping, *Can. J. Chem.*, *50*, 2079 (1972).

72Fd V. A. Fedorov, I. D. Isaev, A. M. Robov, A. V. Vertiprakhov and V. E. Mironov, *Russ. J. Inorg. Chem. (English Trans.)*, *17*, 495 (1972). (Russian Page No. 951.)

72Fe V. A. Fedorov, T. N. Kalosh, L. I. Shmyd'ko, and V. E. Mironov, *Russ. J. Inorg. Chem. (English Trans.)*, *17*, 1086 (1972). (Russian Page No. 2089.)

72Ff V. A. Fedorov, L. I. Kiprin and V. E. Mironov, *Russ. J. Inorg. Chem. (English Trans.)*, *17*, 641 (1972). (Russian Page No. 1233.)

72Fg V. A. Fedorov, A. M. Robov, T. I. Grigor, and V. E. Mironov, *Russ. J. Inorg. Chem. (English Trans.)*, *17*, 990 (1972). (Russian Page No. 1909.)

72Fh V. A. Fedorov, N. P. Samsonova and V. E. Mironov, *Russ. J. Inorg. Chem. (English Trans.)*, *17*, 674 (1972). (Russian Page No. 1301.)

72Ga A. Gergely, J. Mojzes and Z. Kassai-Bazsa, *J. Inorg. Nucl. Chem.*, *34*, 1277 (1972).

1972 (continued)

72Gb A. Gergely, J. Mojzes and Z. Kassai-Bazsa, *Magyar Kemiai Folyoirat*, *78*, 76 (1972).

72Gc A. Gergely and I. Sovago, *Acta Chim. Acad. Sci. Hung.*, *74*, 273 (1972).

72Gd A. Gergely and I. Sovago, *Magyar Kemiai Folyoirat*, *78*, 274 (1972).

72Ge I. Grenthe and G. Gardhammar, *Acta Chem. Scand.*, *26*, 3207 (1972).

72Gf I. Grenthe and H. Ots, *Acta Chem. Scand*, *26*, 1229 (1972).

72Gg B. W. Griffin and J. A. Peterson, *Biochem*, *11*, 4740 (1972).

72Gh J. Gross and C. Keller, *J. Inorg. Nucl. Chem.*, *34*, 725 (1972).

72Gi J. B. Grutzner, J. M. Lawlor and L. M. Jackman, *J. Amer. Chem. Soc.*, *94*, 2306 (1972).

72Ha C. F. Hale and F. H. Spedding, *J. Phys. Chem.*, *76*, 1887 (1972).

72Hb V. Hankonyi, V. Ondrusek, V. Karas-Gasparec and Z. Binenfeld, *Z. Phys. Chemie, Leipzig*, *251*, 280 (1972).

72Hc S. Harada, K. Amidaiji and T. Yasunaga, *Bull. Chem. Soc. Japan*, *45*, 1752 (1972).

72Hd H. D. Harmon and J. R. Peterson, *J. Inorg. Nucl. Chem.*, *34*, 1381 (1972).

72He B. Hedlund, C. Danielson and R. Lovrien, *Biochemistry*, *11*, 4660 (1972).

72Hf J. B. Hunt, P. D. Ross and A. Ginsburg, *Biochemistry*, *11*, 3716 (1972).

72Hg J. B. Hunt, P. D. Ross and A. Ginsburg, *Biochemistry*, *11*, 3716 (1972).

72Ia T. P. I. and G. H. Nancollas, *Inorg. Chem.*, *11*, 2414 (1972).

72Ib Y. Ishino, T. Ogura, K. Noda, T. Hirashima and O. Manabe, *Bull. Chem. Soc. Japan*, *45*, 150 (1972).

72Ic J. Israeli and R. Volpe, *Bull. Soc. Chim. France*, No. 4, 1277 (1972).

72Id J. Israeli and R. Volpe, *Inorg. Chim. Acta*, *6*, 5 (1972).

72Ie M. Itoh, *Bull. Chem. Soc. Japan*, *45*, 1947 (1972).

72If K. J. Ivin, R. Jamison and J. J. McGarvey, *J. Amer. Chem. Soc.*, *94*, 1763 (1972).

72Ig R. M. Izatt, H. D. Johnson and J. J. Christensen, *J. C. S. Dalton*, 1152 (1972).

72Ja R. F. Jameson and M. F. Wilson, *J. C. S. Dalton*, 2614 (1972).

72Jb R. F. Jameson and M. F. Wilson, *J. C. S. Dalton*, 2617 (1972).

72Ka E. F. Kassierer and A. S. Kertes, *J. Inorg. Nucl. Chem.*, *34*, 3209 (1972).

72Kb A. S. Kertes and E. F. Kassierer, *Inorg. Chem.*, *11*, 2108 (1972).

72Kc V. A. Kogan, N. I. Dorokhova, O. A. Osipov, and S. G. Kochin, *Russ. J. Phys. Chem. (English Trans.)*, *46*, 120 (1972). (Russian Page No. 205.)

72Kd B. S. Krumgal'z, V. V. Subbotina and K. M. Mishchenko, *Russ. J. Phys. Chem. (English Trans.)*, *46*, 165 (1972). (Russian Page No. 272.)

72La T. H. Leong and L. A. Dunn, *J. Phys. Chem.*, *76*, 2294 (1972).

1972 (continued)

72Lb Y. Y. Lim and R. S. Drago, *Inorg. Chem.*, *11*, 202 (1972).

72Lc Y. Y. Lim and R. S. Drago, *Inorg. Chem.*, *11*, 1334 (1972).

72Ma J. A. Maguire, J. J. Banewicz, R. C. T. Hung and K. L. Wright, *Inorg. Chem.*, *11*, 3059 (1972).

72Mb J. M. Malin and R. E. Shepherd, *J. Inorg. Nucl. Chem.*, *34*, 3203 (1972).

72Mc T. Marshall and Q. Fernando, *Analy. Chem.*, *44*, 1346 (1972).

72Md J. N. Mathur, *Indian J. Chem.*, *10*, 299 (1972).

72Me R. E. Mesmer, C. F. Baes, and F. H. Sweeton, *Inorg. Chem.*, *11*, 537 (1972).

72Mf V. E. Mironov, L. A. Nikulina and N. P. Kolobov, *Russ. J. Phys. Chem. (English Trans.)*, *46*, 1370 (1972). (Russian Page No. 2395.)

72Mg V. E. Mironov, G. K. Ragulin, I. E. Umova, Y. B. Solovev, V. P. Mikhailova and F. T. Dong, *Russ. J. Phys. Chem. (English Trans.)*, *46*, 155 (1972). (Russian Page No. 257.)

72Mh V. E. Mironov, Y. B. Solovev, V. M. Fadeev and N. P. Kolobov, *Russ. J. Phys. Chem. (English Trans.)*, *46*, 926 (1972). (Russian Page No. 1612.)

72Mi R. Morassi and A. Dei, *Inorg. Chim. Acta*, *6*, 314 (1972).

72Na B. Nelander and I. Noren, *Acta Chem. Scand.*, *26*, 809 (1972).

72Nb R. M. Nikolic and I. J. Gal, *J. C. S. Dalton*, 162 (1972).

72Nc T. Nowlin and K. Cohn, *Inorg. Chem.*, *11*, 560 (1972).

72Nd M. S. Nozari and R. S. Drago, *Inorg. Chem.*, *11*, 280 (1972).

72Pa A. N. Pant, R. N. Soni, and S. L. Gupta, *J. Inorg. Nucl. Chem.*, *34*, 763 (1972).

72Pb A. N. Pant, R. N. Soni, and S. L. Gupta, *J. Inorg. Nucl. Chem.*, *34*, 2951 (1972).

72Pc W. Partenheimer, *Inorg. Chem.*, *11*, 743 (1972).

72Pd S. Parthasarathy and S. Ambujavalli, *Electrochimica Acta*, *17*, 1219 (1972).

72Pe M. E. Peisakhova, I. P. Goldshtein, E. N. Guryanova and K. A. Kocheshkov, *Dokiady Chem. Proc. Acad. Sci., U.S.S.R. (English Trans.)*, *203*, 372 (1972). (Russian Page No. 1316.)

72Pf A. N. Petrov, O. N. Temkin, and M. I. Bogdanov, *Russ. J. Phys. Chem. (English Trans.)*, *46*, 1043 (1972). (Russian Page No. 1822.)

72Pg K. S. Pitzer, *Trans. Farad. Soc.*, *68*, 101 (1972).

72Ph H. K. J. Powell and G. H. Nancollas, *J. Amer. Chem. Soc.*, *94*, 2664 (1972).

72Pi A. J. Poe and C. P. J. Vuik, *J. C. S. Dalton*, 2250 (1972).

72Pj C. S. G. Prasad, K. Balchandran and S. K. Banerji, *J. Indian Chem. Soc.*, *49*, 3 (1972).

72Ra P. C. Rawat and C. M. Gupta, *J. Inorg. Nucl. Chem.*, *34*, 951 (1972).

72Rb P. C. Rawat and C. M. Gupta, *J. Inorg. Nucl. Chem.*, *34*, 1621 (1972).

72Rc G. Rialdi, J. Levy and R. Biltonen, *Biochem.*, *11*, 2472 (1972).

1972 (continued)

72Rd I. P. Romm, E. N. Kharlamova and E. N. Guryanova, *J. General Chem. U.S.S.R. (English Trans.)*, *42*, 2242 (1972). (Russian Page No. 2246.)

72Re T. Ryhl, *Acta Chem. Scand.*, *26*, 2961 (1972).

72Sa R. Sarin and K. N. Munshi, *Aust. J. Chem.*, *25*, 929 (1972).

72Sb R. Sarin and K. N. Munshi, *J. Inorg. Nucl. Chem.*, *34*, 581 (1972).

72Sc R. S. Saxena and U. S. Chaturvedi, *J. Indian Chem. Soc.*, *49*, 321 (1972).

72Sd R. S. Saxena and U. S. Chaturvedi, *J. Inorg. Nucl. Chem.*, *34*, 913 (1972).

72Se R. S. Saxena and U. S. Chaturvedi, *J. Inorg. Nucl. Chem.*, *34*, 2964 (1972).

72Sf R. S. Saxena and P. Singh, *J. Indian Chem. Soc.*, *49*, 325 (1972).

72Sg H. A. Skinner and E. W. Tipping, *Rev. Chim. Minerale*, *9*, 51 (1972).

72Sh J. Smid, *Angew. Chem. Internat. Edt.*, *11*, 112 (1972).

72Si A. Streitwieser, C. J. Chang, and D. M. E. Reuben, *J. Amer. Chem. Soc.*, *94*, 5730 (1972).

72Sj H. C. Stynes and J. A. Ibers, *J. Amer. Chem. Soc.*, *94*, 1559 (1972).

72Ta A. D. Taneja and K. P. Srivastava, *Indian J. Chem.*, *10*, 1029 (1972).

72Tb E. W. Tipping and H. A. Skinner, *J. Chem. Soc., Faraday Trans.*, *68*, 1764 (1972).

72Tc G. Torsi and G. Mamantov, *Inorg. Chem.*, *11*, 1439 (1972).

72Va L. C. VanPoucke, G. F. Thiers, and Z. Eeckhaut, *Bull. Soc. Chim. Belges*, *81*, 357 (1972).

72Vb V. P. Vasilev and A. A. Ikonnikov, *Russ. J. Inorg. Chem. (English Trans.)*, *17*, 1700 (1972). (Russian Page No. 3232.)

72Vc L. S. Vasilev, M. M. Vartanyan, V. S. Bogdanov, V. G. Kiselev, and B. M. Mikhailov, *Russ. J. General Chem. (English Trans.)*, *42*, 1533 (1972). (Russian Page No. 1540.)

72Vd G. S. Vigee and C. L. Walkins, *J. Inorg. Nucl. Chem.*, *34*, 3936 (1972).

72Wa C. F. Wells and D. Whatley, *Trans. Farad. Soc.*, *68*, 434 (1972).

72Wb R. West and B. Bichlmeir, *J. Amer. Chem. Soc.*, *94*, 1649 (1972).

72Wc D. R. Williams, *J. C. S. Dalton Trans.*, 790 (1972).

72Wd D. R. Williams and P. A. Yeo, *J. C. S. Dalton*, 1988 (1972).

72Ya A. Yingst, R. M. Izatt and J. J. Christensen, *J. C. S. Dalton*, 1199 (1972).

72Yb H. Yokoi, M. Sai, and T. Isobe, *Bull. Chem. Soc. Japan*, *45*, 1100 (1972).

72Za P. S. Zacharias and A. Chakravorty, *Inorg. Chim. Acta*, *6*, 623 (1972).

1971

71Aa Y. K. Agrawal and J. P. Shukla, *J. Indian Chem.*, *48*, 771 (1971).

71Ab S. Ahrland and L. Kullberg, *Acta Chem. Scand.*, *25*, 3471 (1971).

1971 (continued)

71Ac S. Ahrland and L. Kullberg, *Acta Chem. Scand.*, 25, 3677 (1971).

71Ad S. Ahrland and L. Kullberg, *Acta Chem. Scand.*, 25, 3692 (1971).

71Ae S. O. Ajayi and D. R. Coddard, *J. Chem. Soc.*, (A), 2673 (1971).

71Af G. Anderegg and F. Wenk, *Helv. Chim. Acta*, 54, 216 (1971).

71Ag E. M. Arnett and T. C. Moriarity, *J. Amer. Chem. Soc.*, 93, 4908 (1971).

71Ah B. K. Avinashi and S. K. Banerji, *J. Indian Chem. Soc.*, 48, 174 (1971).

71Ai A. Aziz, S. J. Lyle and J. E. Newbery, *J. Inorg. Nucl. Chem.*, 33, 1757 (1971).

71Ba A. R. Bailey and J. W. Larson, *J. Phys. Chem.*, 75, 2368 (1971).

71Bb J. R. Baker, A. Hermann, and R. M. Wing, *J. Amer. Chem. Soc.*, 93, 6486 (1971).

71Bc D. S. Barnes, G. J. Ford, L. D. Pettit, and C. Sherrington, *Chem. Commun.*, 690 (1971).

71Bd D. S. Barnes, G. J. Ford, L. D. Pettit, and C. Sherrington, *J. Chem. Soc.*, (A), 2883 (1971).

71Be D. S. Barnes and L. D. Pettit, *J. Inorg. Nucl. Chem.*, 33, 2177 (1971).

71Bf J. L. Bear and M. E. Clark, *J. Inorg. Nucl. Chem.*, 33, 3805 (1971).

71Bg J. T. Bell, R. D. Baybarz and D. M. Helton, *J. Inor. Nucl. Chem.*, 33, 3077 (1971).

71Bh B. D. Berezin, L. V. Klopova, and A. N. Drobysheva, *Russ. J. Inorg. Chem. (English Trans.)*, 16, 1096 (1971). (Russian Page No. 2053.)

71Bi V. V. Blokhin, V. F. Anufrienko, Yu. A. Makashev, and V. E. Mironov, *Russ. J. Phys. Chem. (English Trans.)*, 45, 1062 (1971). (Russian Page No. 1860.)

71Bj V. S. Bogdanov, T. K. Baryshnikova, V. G. Kiselev, and B. M. Mikhailov, *Russ. J. Gen. Chem. (English Trans.)*, 41, 1537 (1971). (Russian Page No. 1533.)

71Bk D. W. Bolen, M. Flogel and R. Biltonen, *Biochemistry*, 10, 4136 (1971).

71Ca A. Cecal and I. A. Schneider, *Inorg. Chim. Acta*, 5, 623 (1971).

71Cb R. W. Chlebek and M. W. Lister, *Can. J. Chem.*, 49, 2943 (1971).

71Cc G. Ciullo, C. Furlani, L. Sestili and A. Sgamellotti, *Inorg. Chim. Acta*, 5, 489 (1971).

71Cd S. J. Cole, G. C. Curthoys and E. A. Magnusson, *J. Amer. Chem. Soc.*, 93, 2153 (1971).

71Ce R. O. Cook, A. Davies and L. A. K. Staveley, *J. Chem. Thermodynamics*, 3, 907 (1971).

71Cf B. J. Corden and P. H. Rieger, *Inorg. Chem.*, 10, 263 (1971).

71Da A. Dadgar and R. Delorenzo, *J. Inor. Nucl. Chem.*, 33, 4155 (1971).

71Db D. R. Dakternieks and D. P. Graddon, *Aust. J. Chem.*, 24, 2077 (1971).

71Dc D. R. Dakternieks and D. P. Graddon, *Aust. J. Chem.*, 24, 2509 (1971).

1971 (continued)

71Dd R. C. Das, A. C. Dash, D. Satyanarayan, and U. N. Dash, *Thermochimica Acta*, *2*, 435 (1971).

71De N. I. Dorokhova, V. A. Kogan and O. A. Osipov, *Russ. J. Phys. Chem. (English Trans.)*, *45*, 537 (1971). (Russian Page No. 962.)

71Ea D. R. Eaton and K. Zaw, *Can. J. Chem.*, *49*, 3315 (1971).

71Eb K. G. Everett and D. A. Skoog, *Anal. Chem.*, *43*, 1541 (1971).

71Fa V. A. Fedorov, T. N. Kalosh, G. E. Chernikova and V. E. Mironov, *Russ. J. Phys. Chem. (English Trans.)*, *45*, 775 (1971). (Russian Page No. 1364.)

71Fb V. A. Fedorov, T. N. Kalosh and V. E. Mironov, *Russ. J. Inorg. Chem. (English Trans.)*, *16*, 539 (1971). (Russian Page No. 1014.)

71Fc J. H. Forsberg, T. M. Kubik, T. Moeller and K. Gucwa, *Inorg. Chem.*, *10*, 2656 (1971).

71Fd J. H. Forsberg and C. A. Wathen, *Inorg. Chem.*, *10*, 1379 (1971).

71Fe P. U. Fruh, J. T. Clerc and W. Simon, *Helv. Chim. Acta*, *54*, 1445 (1971).

71Ga G. Geier and U. Karlen, *Helv. Chim. Acta*, *54*, 135 (1971).

71Gb A. Gergely, I. Nagypal and B. Kiraly, *Acta Chim. Acad. Sci. Hung.*, *68*, 285 (1971).

71Gc A. Gergely, I. Nagypal and I. Sovago, *Acta Chim. Acad. Sci. Hung.*, *67*, 241 (1971).

71Gd A. M. Goeminne and Z. Eeckhaut, *Bull. Soc. Chim. Belges*, *80*, 605 (1971).

71Ge I. P. Goldshtein, E. N. Guryanova, and T. I. Perepelkova, *Theor. and Exp. Chemistry (English Trans.)*, *7*, 343 (1971). (Russian Page No. 410.)

71Gf J. G. Gordon, M. J. O'Connor, and R. H. Holm, *Inorg. Chim. Acta*, *5*, 381 (1971).

71Gg D. P. Graddon and K. B. Heng, *Aust. J. Chem.*, *24*, 1781 (1971).

71Gh D. P. Graddon and C. Y. Hsu, *Aust. J. Chem.*, *24*, 2267 (1971).

71Gi H. Grasdalen and I. Svare, *Acta Chem. Scand.*, *25*, 1089 (1971).

71Gj T. I. Grigor, V. A. Fedorov and V. E. Mironov, *Russ. J. Inorg. Chem. (English Trans.)*, *16*, 337 (1971). (Russian Page No. 633.)

71Gk G. Guiheneuf, C. Laurence and B. Wojtkowiak, *Bull. Soc. Chim., (France)*, 1157, (1971).

71Ha P. R. Hammond and R. R. Lake, *J. Chem. Soc. (A)*, 3800 (1971).

71Hb P. R. Hammond and R. R. Lake, *J. Chem. Soc. (A)*, 3806 (1971).

71Hc P. R. Hammond and R. R. Lake, *J. Chem. Soc. (A)*, 3819 (1971).

71Hd P. R. Hammond and W. S. McEwan, *J. Chem. Soc. (A)*, 3812 (1971).

71He L. D. Hansen and D. J. Temer, *Inorg. Chem.*, *10*, 1439 (1971).

71Hf R. W. Hay and P. J. Morris, *J. Chem. Soc. (A)*, 1518 (1971).

71Hg R. W. Hay and P. J. Morris, *J. Chem. Soc. (A)*, 3562 (1971).

71Hh D. A. House and H. K. J. Powell, *Inorg. Chem.*, *10*, 1583 (1971).

1971 (continued)

71Hi L. G. Hubert-Pfalzgraf, J. Guion, and J. G. Riess, *Bull. Soc. Chim. France*, No. 11, 3855 (1971).

71Hj J. R. Hutchison, J. G. Gordon and R. H. Holm, *Inorg. Chem.*, *10*, 1004 (1971).

71Ia J. Israeli and J. R. Cayouette, *Can. J. Chem.*, *49*, 199 (1971).

71Ib J. Israeli and R. Volpe, *Bull. Soc. Chim. France*, No. 9, 3119 (1971).

71Ic J. Israeli and R. Volpe, *Inorg. Nucl. Chem. Letters*, *7*, 1183 (1971).

71Id J. Israeli and R. Volpe, *J. Inorg. Nucl. Chem.*, *33*, 4358 (1971).

71Ie N. Ivicic and V. Simeon, *Thermochimica Acta*, *2*, 345 (1971).

71If R. M. Izatt, C. H. Bartholomew, C. E. Morgan, D. J. Eatough, and J. J. Christensen, *Thermochimica Acta*, *2*, 313 (1971).

71Ig R. M. Izatt, H. D. Johnston, D. J. Eatough, J. W. Hansen and J. J. Christensen, *Thermochimica Acta*, *2*, 77 (1971).

71Ih R. M. Izatt, D. P. Nelson, J. H. Rytting, B. L. Haymore and J. J. Christensen, *J. Amer. Chem. Soc.*, *93*, 1619 (1971).

71Ja A. D. Jones and D. R. Williams, *J. Chem. Soc. (A)*, 3159 (1971).

71Ka M. H. Keyes, M. Falley and R. Lumry, *J. Amer. Chem. Soc.*, *93*, 2035 (1971).

71Kb V. A. Kogan, N. I. Dorokhova and O. A. Osipov, *Russ. J. Inorg. Chem. (English Trans.)*, *16*, 94 (1971). (Russian Page No. 179.)

71Kc N. A. Kostromina and N. N. Tananaeva, *Russ. J. Inorg. Chem. (English Trans.)*, *16*, 1256 (1971). (Russian Page No. 2356.)

71Kd H. Krakauer, *Biopolymers*, *10*, 2459 (1971).

71Ke C. Kuehn and W. Knoche, *Trans. Faraday Soc.*, *67*, 2101 (1971).

71La R. P. Lang, *J. Amer. Chem. Soc.*, *93*, 5047 (1971).

71Lb N. Langerman and J. M. Sturtevant, *Biochemistry*, *10*, 2809 (1971).

71Lc J. M. Lehn and J. P. Sauvage, *Chem. Commun.*, 440 (1971).

71Ld M. C. Lim and G. H. Nancollas, *Inorg. Chem.*, *10*, 1957 (1971).

71Le C. T. Lin and J. L. Bear, *J. Phys. Chem.*, *75*, 3705 (1971).

71Lf J. C. Lockhart and W. J. Mossop, *Chem. Commun.*, 61 (1971).

71Lg V. A. Logachev and V. I. Dulova, *Russ. J. Inorg. Chem. (English Trans.)*, *16*, 145 (1971). (Russian Page No. 275.)

71Lh V. A. Logachev and V. I. Dulova, *Russ. J. Phys. Chem. (English Trans.)*, *45*, 1311 (1971). (Russian Page No. 2325.)

71Li W. K. Lutz, P. U. Fruh and W. Simon, *Helv. Chim. Acta*, *54*, 2767 (1971).

71Ma F. Maggio and V. Romano, *J. Inorg. Nucl. Chem.*, *33*, 3993 (1971).

71Mb D. L. Mathur and G. V. Bakore, *J. Indian Chem. Soc.*, *48*, 7 (1971).

71Mc R. E. Mesmer and C. F. Baes, *Inorg. Chem.*, *10*, 2290 (1971).

71Md B. M. Mikhailov, M. N. Bochkareva, V. S. Bogdanov, O. G. Boldyreva, and V. A. Dorokhov, *J. Gen. Chem. U.S.S.R. (English Trans.)*, *41*, 1550 (1971). (Russian Page No. 1546.)

1971 (continued)

71Me A. I. Mogilyanskii, T. G. Sukhova, O. N. Temkin, and R. M. Flid, *Russ. J. Inorg. Chem. (English Trans.)*, *16*, 1074 (1971). (Russian Page No. 2017.)

71Mf H. J. Moschler, H. G. Weder, and R. Schwyzer, *Helv. Chim. Acta*, *54*, 1437 (1971).

71Oa G. Olofsson, *Acta Chem. Scand.*, *25*, 691 (1971).

71Pa A. Pagliardini, G. Torri, L. Elegant, and M. Azzaro, *Bull. Soc. Chim. France*, 54 (1971).

71Pb A. N. Pant, R. N. Soni, and S. L. Gupta, *J. Inorg. Nucl. Chem.*, *33*, 3202 (1971).

71Pc S. Parthasarathy and V. S. K. Nair, *Thermochimica Acta*, *2*, 69 (1971).

71Ra B. Rao and H. B. Mathur, *J. Inorg. Nucl. Chem.*, *33*, 809 (1971).

71Rb B. Rao and H. B. Mathur, *J. Inorg. Nucl. Chem.*, *33*, 2919 (1971).

71Rc P. V. S. Rao and P. Kamannarayana, *Z. Phys. Chemie, Leipzig*, *248*, 267 (1971).

71Rd N. A. Rumbaut, *Bull. Soc. Chim. Belges*, *80*, 63 (1971).

71Re L. Rusnak and R. B. Jordan, *Inorg. Chem.*, *10*, 2199 (1971).

71Sa R. S. Saxena and U. S. Chaturved, *J. Inorg. Nucl. Chem.*, *33*, 3597 (1971).

71Sb K. K. SenGupta and A. K. Chatterjee, *Z. Anorg. Allg. Chem.*, *384*, 280 (1971).

71Sc E. Shchori, J. Jagur-Grodzinski, Z. Luz and M. Shporer, *J. Amer. Chem. Soc.*, *93*, 7133 (1971).

71Sd L. O. Spreer and E. L. King, *Inorg. Chem.*, *10*, 916 (1971).

71Ta U. Takaki, T. E. Hogen Esch and J. Smid, *J. Amer. Chem. Soc.*, *93*, 6760 (1971).

71Tc A. D. Taneja and K. P. Srivastava, *J. Inorg. Nucl. Chem.*, *33*, 2694 (1971).

71Td C. P. Trivedi, P. N. Mathur and O. P. Sunar, *J. Indian Chem. Soc.*, *48*, 270 (1971).

71Te C. P. Trivedi and O. P. Sunar, *J. Indian Chem. Soc.*, *48*, 803 (1971).

71Ua J. Ulstrup, *Acta Chem. Scand.*, *25*, 3397 (1971).

71Va V. P. Vasilev, G. A. Zaitseva and V. G. Malakhova, *Russ. J. Inorg. Chem. (English Trans.)*, *16*, 1144 (1971). (Russian Page No. 2124.)

71Vb L. Vaska, *Inorg. Chim. Acta*, *5*, 295 (1971).

71Wa J. B. Walker, C. R. Twine and G. R. Choppin, *J. Inorg. Nucl. Chem.*, *33*, 1813 (1971).

71Wb J. L. Wardell, *J. Chem. Soc. (A)*, 2628 (1971).

71Wc D. Waysbort and G. Navon, *Chem. Commun.*, 1410 (1971).

71Ya T. Yamamoto, A. Yamamoto and S. Ikeda, *J. Amer. Chem. Soc.*, *93*, 3360 (1971).

71Yb H. Yokoi, M. Otagiri and T. Isobe, *Bull. Chem. Soc. Japan*, *44*, 1445 (1971).

71Yc H. Yokoi, M. Otagiri and T. Isobe, *Bull. Chem. Soc. Japan*, *44*, 2402 (1971).

1971 (continued)

71Za P. S. Zacharias and A. Chakravorty, *Inorg. Chem.*, *10*, 1961 (1971).

71Zb A. H. Zeltmann and L. O. Morgan, *Inorg. Chem.*, *10*, 2739 (1971).

71Zc T. V. Zhurba and V. I. Dulova, *Russ. J. Inorg. Chem. (English Trans.)*, *16*, 1168 (1971). (Russian Page No. 2190.)

71Zd T. V. Zhurba and V. I. Dulova, *Russ. J. Inorg. Chem. (English Trans.)*, *17*, 1203 (1972). (Russian Page No. 2306.)

71Ze T. V. Zhurba and V. I. Dulova, *Russ. J. Inorg. Chem. (English Trans.)*, *16*, 1309 (1971). (Russian Page No. 2454.)

71Zf R. A. Zingaro, K. J. Irgolic, D. H. O'Brien and L. J. Edmonson, *J. Amer. Chem. Soc.*, *93*, 5677 (1971).

1970

70Aa G. Amiconi, M. Brunori, E. Antonini, G. Tauzher and G. Costa, *Nature*, *228*, 549 (1970).

70Ab J. N. Armor and H. Taube, *J. Amer. Chem. Soc.*, *92*, 6170 (1970).

70Ac R. Arnek, *Arkiv Kemi*, *32*, 55 (1970).

70Ad B. K. Avinashi and S. K. Banerji, *Indian J. Chem.*, *8*, 1140 (1970).

70Ba K. Balachandran and S. K. Banerji, *Indian J. Chem.*, *8*, 1136 (1970).

70Bb K. Balachandran and S. K. Banerji, *J. Indian Chem. Soc.*, *47*, 343 (1970).

70Bc K. Balachandran and S. K. Banerji, *J. Indian Chem. Soc.*, *47*, 353 (1970).

70Bd D. S. Barnes and L. D. Pettit, *Chem. Commun.*, 1000 (1970).

70Be E. W. Baumann, *J. Inorg. Nucl. Chem.*, *32*, 3823 (1970).

70Bf V. V. Blokhin, G. K. Ragulin, V. F. Anufrienko, Yu. A. Makashev, and V. E. Mironov, *Russ. J. Phys. Chem. (English Trans.)*, *44*, 1512 (1970). (Russian Page No. 2654.)

70Bg R. A. Butler, A. S. Carson, P. G. Laye and W. V. Steele, *J. Organometal. Chem.*, *24*, C11 (1970).

70Ca D. L. Campbell and T. Moeller, *J. Inor. Nucl. Chem.*, *32*, 945 (1970).

70Cb R. L. Carlin and D. B. Losee, *Inorg. Chem.*, *9*, 2087 (1970).

70Cc R. L. Carlin and D. B. Losee, *Inorg. Chem.*, *9*, 2087 (1970).

70Cd A. S. Carson, P. G. Laye and W. V. Steele, *J. Chem. Thermodynamics*, *2*, 757 (1970).

70Ce L. L. Chan, K. H. Wong, and J. Smid, *J. Amer. Chem. Soc.*, *92*, 1955 (1970).

70Cf I. D. Chawla and A. C. Andrews, *J. Inorg. Nucl. Chem.*, *32*, 91 (1970).

70Cg G. R. Choppin and J. K. Schneider, *J. Inorg. Nucl. Chem.*, *32*, 3283 (1970).

70Ch S. J. Cole, G. C. Curthoys and E. A. Magnusson, *J. Amer. Chem. Soc.*, *92*, 2991 (1970).

70Da G. Degischer and G. H. Nancollas, *Inorg. Chem.*, *9*, 1259 (1970).

1970 (continued)

70Ea D. J. Eatough, *Analy. Chem.*, *42*, 635 (1970).

70Eb W. J. Eilbeck, F. Holmes and G. Phillips, *J. Chem. Soc. (A)*, 729 (1970).

70Ec A. H. Ewald and J. A. Scudder, *Aust. J. Chem.*, *23*, 1939 (1970).

70Ed G. J. Ewing and S. Maestas, *J. Phys. Chem.*, *74*, 2341 (1970).

70Fa E. L. Farquhar, L. Rusnock and S. J. Gill, *J. Amer. Chem. Soc.*, *92*, 416 (1970).

70Fb V. A. Fedorov, G. E. Chernikova and V. E. Mironov, *Zhurnal Neorganicheskoi Khimii*, *15*, 2100 (1970).

70Fc Z. Finta, Cs. Varhelyi and J. Zsako, *J. Inorg. Nucl. Chem.*, *32*, 3013 (1970).

70Ga A. Gergely, B. Kiraly, I. Nagypal, and J. Mojzes, *Magyar Kemiai Folyoirat*, *76*, 452 (1970).

70Gb A. Gergely, B. Kiraly, I. Nagypal, and J. Mojzes, *Acta Chim. Acad. Sci. Hung.* *67*, 133 (1971).

70Gc A. Gergely, I. Nagypal and I. Sovago, *Magyar Kemiai Folyoirat*, *76*, 550 (1970).

70Gd R. Ghosh and V. S. K. Nair, *J. Inorg. Nucl. Chem.*, *32*, 3025 (1970).

70Ge R. Ghosh and V. S. K. Nair, *J. Inorg. Nucl. Chem.*, *32*, 3033 (1970).

70Gf R. Ghosh and V. S. K. Nair, *J. Inorg. Nucl. Chem.*, *32*, 3041 (1970).

70Gg A. Y. Girgis and R. C. Fay, *J. Amer. Chem. Soc.*, *92*, 7061 (1970).

70Gh I. P. Gol'dshtein, E. N. Gur'yanova and E. S. Shcherbakova, *Russ. J. Gen. Chem. (English Trans.)*, *40*, 166 (1970). (Russian Page No. 183.)

70Gi L. W. Gosser and C. A. Tolman, *Inorg. Chem.*, *9*, 2350 (1970).

70Gj A. F. Grand and M. Tamres, *J. Phys. Chem.*, *74*, 208 (1970).

70Gk T. R. Griffiths and R. K. Scarrow, *J. Chem. Soc. (A)*, 827 (1970).

70Ia R. M. Izatt, G. D. Watt, C. H. Bartholomew, and J. J. Christensen, *Inorg. Chem.*, *9*, 2019 (1970).

70Ja J. M. Jallon and M. Cohn, *Biochim. Biophys. Acta*, *222*, 542 (1970).

70Jb G. R. Janig, G. Gerber, K. Ruckpaul, S. Rapoport and F. Jung, *Eur. J. Biochem*, *17*, 441 (1970).

70Jc A. D. Jones and D. R. Williams, *J. Chem. Soc. (A)*. 3138 (1970).

70Ka T. S. Kannan and A. Chakravorty, *Inorg. Chem.*, *9*, 1153 (1970).

70Kb G. C. Kugler and G. H. Carey, *Talanta*, *17*, 907 (1970).

70La J. W. Larson, *J. Phys. Chem.*, *74*, 3392 (1970).

70Lb J. E. Letter and J. E. Bauman, *J. Amer. Chem. Soc.*, *92*, 437 (1970).

70Ma J. W. Macklin and R. A. Plane, *Inorg. Chem.*, *9*, 821 (1970).

70Mb A. Martin and E. Uhlig, *Z. Anorg. Allgem. Chemie*, *375*, 166 (1970).

70Mc A. Martin and E. Uhlig, *Z. Anorg. Allgem. Chemie.*, *376*, 282 (1970).

70Oa A. L. Oleinikova, O. N. Temkin, M. I. Bogdanov and R. M. Flid, *Russ. J. Phys. Chem. (English Trans.)*, *44*, 1373 (1970). (Russian Page No. 2418.)

1970 (continued)

70Pa W. Partenheimer and R. S. Drago, *Inorg. Chem.*, *9*, 47 (1970).

70Pb A. N. Petrov, O. N. Temkin, and M. I. Bogdanov, *Russ. J. Phys. Chem.* (*English Trans.*), *44*, 1574 (1970). (Russian Page No. 2766.)

70Pc L. H. Pignolet, W. DeW. Horrocks, and R. H. Holm, *J. Amer. Chem. Soc.*, *92*, 1855 (1970).

70Pd R. H. Provost and C. A. Wulff, *J. Chem. Thermodynamics*, *2*, 793 (1970).

70Sa Z. A. Schelly, D. J. Harward, P. Hemmes, and E. M. Eyring, *J. Phys. Chem.*, *74*, 3040 (1970).

70Sb M. S. Subramanian and S. A. Pai, *J. Inorg. Nucl. Chem.*, *32*, 3677 (1970).

70Va V. P. Vasilev and P. N. Vorobev, *Russ. J. Phys. Chem.* (*English Trans.*), *44*, 657 (1970). (Russian Page No. 1181.)

70Wa S. P. Wasik and W. Tsang, *J. Phys. Chem.*, *74*, 2970 (1970).

70Wb D. R. Williams, *J. Chem. Soc.* (*A*), 1550 (1970).

1969

69Aa R. Arnek, *Acta Chem. Scand.*, *23*, 1986 (1969).

69Ba J. E. Bailey, J. G. Beetlestone and D. H. Irvine, *J. Chem. Soc.* (*A*), 241 (1969).

69Bb K. Balachandran and S. K. Banerji, *J. Indian Chem. Soc.*, *46*, 198 (1969).

69Bc K. Balachandran and S. K. Banerji, *Z. Phys. Chem., Leipzig*, *241*, 267 (1969).

69Bd J. L. Bear and M. E. Clark, *J. Inorg. Nucl. Chem.*, *31*, 1517 (1969).

69Be J. G. Beetlestone and D. H. Irvine, *J. Chem. Soc.* (*A*), 735 (1969).

69Bf J. P. Belaich and J. C. Sari, *Proc. Nat. Acad. Sci., U.S.*, *64*, 763 (1969).

69Bg G. Berthon and C. Luca, *Bull. Soc. Chim., France*, No. 2, 432 (1969).

69Bh R. M. Bhatnagar and A. C. Chatterji, *Z. Phys. Chem., Leipzig*, *241*, 130 (1969).

69Bi U. C. Bhattacharyya, S. C. Lahiri and S. Aditya, *J. Indian Chem. Soc.*, *46*, 247 (1969).

69Bj U. C. Bhattacharyya, S. C. Lahiri and S. Aditya, *J. Indian Chem. Soc.*, *46*, 715 (1969).

69Ca A. S. Carson, P. G. Laye, and W. V. Steele, *J. Chem. Thermodynamics*, *1*, 515 (1969).

69Cb I. D. Chawla and A. C. Andrews, *J. Inorg. Nucl. Chem.*, *31*, 3809 (1969).

69Cc G. R. Choppin and L. A. Martinez-Perez, *Inorg. Chem.*, *7*, 2657 (1969).

69Da A. G. Desai and R. M. Milburn, *J. Amer. Chem. Soc.*, *91*, 1958 (1969).

69Ea T. Ellingsen and J. Smid, *J. Phys. Chem.*, *73*, 2712 (1969).

69Fa D. P. Fay and N. Purdie, *J. Phys. Chem.*, *73*, 3462 (1969).

69Fb V. A. Fedorov, N. P. Samsonova and V. E. Mironov, *Zhurnal Neorganicheskoi Khimii*, *14*, 3264 (1969).

1969 (continued)

69Fc J. H. Forsberg and T. Moeller, *Inorg. Chem.*, *8*, 889 (1969).

69Ga A. F. Garito and B. B. Wayland, *J. Amer. Chem. Soc.*, *91*, 866 (1969).

69Gb P. Gerding, *Acta Chem. Scand.*, *23*, 1695 (1969).

69Gc D. R. Goddard, B. D. Lodam, S. O. Ajayi and M. J. Campbell, *J. Chem. Soc.* *(A)*, 506 (1969).

69Gd A. F. Grand and M. Tamres, *Inorg. Chem.*, *8*, 2495 (1969).

69Ge I. Grenthe and E. Hansson, *Acta Chem. Scand.*, *23*, 611 (1969).

69Gf S. L. Gupta and R. N. Soni, *J. Indian Chem. Soc.*, *46*, 561 (1969).

69Gg S. L. Gupta and R. N. Soni, *J. Indian Chem. Soc.*, *46*, 761 (1969).

69Gh S. L. Gupta and R. N. Soni, *J. Indian Chem. Soc.*, *46*, 763 (1969).

69Ha H. J. Harries and G. Wright, *J. Inorg. Nucl. Chem.*, *31*, 3149 (1969).

69Hb J. M. Harvilchuck, D. A. Aikens and R. C. Murray, *Inorg. Chem.*, *8*, 539 (1969).

69Hc R. W. Henkens, G. D. Watt and J. M. Sturtevant, *Biochemistry*, *8*, 1874 (1969).

69Hd D. A. House and H. K. J. Powell, *Chem. Commun.*, 382 (1969).

69Ia R. M. Izatt, D. Eatough, J. J. Christensen and C. H. Bartholomew, *J. Chem. Soc. (A)*, 45 (1969).

69Ib R. M. Izatt, D. Eatough, J. J. Christensen and C. H. Bartholomew, *J. Chem. Soc. (A)*, 47 (1969).

69Ic R. M. Izatt, J. H. Rytting, D. P. Nelson, B. L. Haymore and J. J. Christensen, *Science*, *164*, 443 (1969).

69Ja A. D. Jones and G. R. Choppin, *J. Inorg. Nucl. Chem.*, *31*, 3523 (1969).

69Jb J. P. Jones and D. W. Margerum, *Inorg. Chem.*, *8*, 1486 (1969).

69Ka I. V. Kolosov and Z. F. Andreeva, *Russ. J. Inorg. Chem. (English Trans.)*, *14*, 346 (1969). (Russian Page No. 664.)

69La B. L. Libutti, B. B. Wayland and A. F. Garito, *Inorg. Chem.*, *8*, 1510 (1969).

69Lb M. Lindheimer, J. Rouviere, A. Monnier, and J. Salvinien, *C. R. Acad. Sc. Paris*, *269*, 364 (1969).

69Ma W. J. McKinney and A. I. Popov, *J. Amer. Chem. Soc.*, *91*, 5215 (1969).

69Mb R. E. Mesmer and C. F. Baes, Jr., *Inorg. Chem.*, *8*, 618 (1969).

69Mc P. M. Milyukov and N. V. Polenova, *Khimiia i Khimicheskaia Tekhnologiia*, *12*, 409 (1969).

69Md P. Mohr and W. Scheler, *Eur. J. Biochem.*, *8*, 444 (1969).

69Me D. F. C. Morris and L. Vandersyde, *Radiochem. Radioanal. Letters*, *1*, 123 (1969).

69Na G. H. Nancollas and D. J. Poulton, *Inorg. Chem.*, *8*, 680 (1969).

69Pa K. H. Pearson, J. R. Baker and J. D. Goodrich, *Anal. Letters*, *2*, 577 (1969).

69Pb B. I. Peshchevitskii and V. I. Belevantsev, *Russ. J. Inorg. Chem. (English Trans.)*, *14*, 1256 (1969). (Russian Page No. 2393.)

1969 (continued)

69Pc T. J. Pinnavaia and S. O. Nweke, *Inorg. Chem.*, *8*, 639 (1969).

69Pd C. J. Popp, J. H. Nelson and R. O. Ragsdale, *J. Amer. Chem. Soc.*, *91*, 610 (1969).

69Ra E. V. Raju and H. B. Mathur, *J. Inorg. Nucl. Chem.*, *31*, 425 (1969).

69Rb M. V. Reddy and P. K. Bhattacharya, *J. Indian Chem. Soc.*, *46*, 1058 (1969).

69Rc P. Rigo, G. Guastalla and A. Turco, *Inorg. Chem.*, *8*, 375 (1969).

69Rd T. Ryhl, *Acta Chem. Scand.*, *23*, 2667 (1969).

69Sa G. Sahu and B. Prasad, *J. Indian Chem. Soc.*, *46*, 233 (1969).

69Sb R. S. Saxena and K. C. Gupta, *J. Indian Chem. Soc.*, *46*, 190 (1969).

69Sc R. S. Saxena and K. C. Gupta, *J. Indian Chem. Soc.*, *46*, 258 (1969).

69Sd R. S. Saxena and K. C. Gupta, *J. Indian Chem. Soc.*, *46*, 303 (1969).

69Se R. S. Saxena and K. C. Gupta, *Z. Phys. Chem. Leipzig*, *241*, 169 (1969).

69Sf A. Schejter and I. Aviram, *Biochemistry*, *8*, 149 (1969).

69Sg J. J. R. F. Silva and M. L. S. Simoes, *Rev. Port. Quim.*, *11*, 54 (1969).

69Sh V. Simeon, B. Svigir and N. Paulic, *J. Inorg. Nucl. Chem.*, *31*, 2085 (1969).

69Si T. G. Sukhova, O. N. Temkin and R. M. Flid, *Russ. J. Inorg. Chem. (English Trans.)*, *14*, 483 (1969). (Russian Page No. 928.)

69Sj V. T. Sushchenko, I. F. Mavrin, R. Ya. Kulba and V. E. Mironov, *Uch. Zap. Leningrad Gos. Pedagog. Inst. in A. I. Gertsena*, *385*, 74 (1969).

69Ua E. Uhlig and A. Martin, *J. Inorg. Nucl. Chem.*, *31*, 2781 (1969).

69Wa G. D. Watt and J. M. Sturtevant, *Biochemistry*, *8*, 4567 (1969).

1968

68Aa G. Aksnes and P. Albriktsen, *Acta Chem. Scand.*, *22*, 1866 (1968).

68Ab A. C. Anusiem, J. G. Beetlestone and D. H. Irvine, *J. Chem. Soc. (A)*, 1337 (1968).

68Ac A. C. Anusiem, J. G. Beetlestone and D. H. Irvine, *J. Chem. Soc. (A)*, 960 (1968).

68Ad R. Arnek, *Acta Chem. Scand.*, *22*, 1102 (1968).

68Ae R. Arnek and C. C. Patel, *Acta Chem. Scand.*, *22*, 1097 (1968).

68Af R. Arnek and K. Schlyter, *Acta Chem. Scand.*, *22*, 1327 (1968).

68Ag R. Arnek and K. Schlyter, *Acta Chem. Scand.*, *22*, 1331 (1968).

68Ba K. Balachandran and S. K. Banerji, *Curr. Sci. (India)*, *37*, 528 (1968).

68Bb D. G. Brown, R. S. Drago and T. F. Bolles, *J. Amer. Chem. Soc.*, *90*, 5706 (1968).

68Bc A. P. Brunetti, M. C. Lim and G. H. Nancollas, *J. Amer. Chem. Soc.*, *90*, 5120 (1968).

1968 (continued)

68Ca R. L. Carroll and R. R. Irani, *J. Inorg. Nucl. Chem.*, *30*, 2971 (1968).

68Cb A. S. Carson, P. G. Laye and P. N. Smith, *J. Chem. Soc. (A)*, 141 (1968).

68Cc A. S. Carson, P. G. Laye and P. N. Smith, *J. Chem. Soc. (A)*, 527 (1968).

68Cd A. S. Carson, P. G. Laye and P. N. Smith, *J. Chem. Soc. (A)*, 1384 (1968).

68Ce A. Chughtai, R. Marshall and G. H. Nancollas, *J. Phys. Chem.*, *72*, 208 (1968).

68Ea G. W. Everett, Jr. and R. H. Holm, *Inorg. Chem.*, *7*, 776 (1968).

68Fa V. A. Fedorov, A. M. Robov, N. P. Samsonova and M. Ya. Kutuzova, *Problemy Sovremennoi Khimii Koordinatsionnykh*, *2*, 217 (1968).

68Fb V. A. Fedorov, N. P. Samsonova and V. E. Mironov, *Russ. J. Inorg. Chem. (English Trans.)*, *13*, 198 (1968). (Russian Page No. 382.)

68Fc R. A. Firth, H. A. O. Hill, B. E. Mann, J. M. Pratt, R. G. Thorp and R. J. P. Williams, *J. Chem. Soc. (A)*, 2419 (1968).

68Fd T. Fueno, O. Kajimoto, T. Okuyama, and J. Furukawa, *Bull. Chem. Soc. Japan*, *41*, 785 (1968).

68Fe J. Fuger, E. Merciny and G. Duyckaerts, *Bull. Soc. Chim. Belg.*, *77*, 455 (1968).

68Ga R. C. Gardner and R. O. Ragsdale, *Inorg. Chim. Acta*, *2*, 139 (1968).

68Gb P. S. Gentile and A. Dadgar, *J. Chem. Eng. Data*, *13*, 367 (1968).

68Gc P. Gerding and I. Jönsson, *Acta Chem. Scand.*, *22*, 2247 (1968).

68Gd P. Gerding and B. Johansson, *Acta Chem. Scand.*, *22*, 2255 (1968).

68Ge I. P. Gol'dshtein, E. N. Kharlamova, and E. N. Guryanova, *J. Gen. Chem. U.S.S.R. (English Trans.)*, *38*, 1925 (1968). (Russian Page No. 1984.)

68Gg R. W. Green and W. G. Goodwin, *Aust. J. Chem.*, *21*, 1165 (1968).

68Gh G. Gutnikov and H. Freiser, *Anal. Chem.*, *40*, 39 (1968).

68Ha F. M. Hall and S. J. Slater, *Aust. J. Chem.*, *21*, 2663 (1968).

68Hb P. C. Harris and T. E. Moore, *Inorg. Chem.*, *7*, 656 (1968).

68Hc C. H. Henrickson, D. Duffy and D. P. Eyman, *Inorg. Chem.*, *7*, 1047 (1968).

68Hd C. H. Henrickson, K. M. Nykerk and D. P. Eyman, *Inorg. Chem.*, *7*, 1028 (1968).

68He N. Hirota, R. Carraway and W. Schook, *J. Amer. Chem. Soc.*, *90*, 3611 (1968).

68Ia R. M. Izatt, D. Eatough and J. J. Christensen, *J. Phys. Chem.*, *72*, 2720 (1968).

68Ib R. M. Izatt, D. Eatough, J. J. Christensen and R. L. Snow, *J. Phys. Chem.*, *72*, 1208 (1968).

68Ic R. M. Izatt, G. D. Watt, C. H. Bartholomew and J. J. Christensen, *Inorg. Chem.*, *7*, 2236 (1968).

68La S. C. Lahiri and S. Aditya, *J. Inorg. Nucl. Chem.*, *30*, 2487 (1968).

68Lb G. C. Lalor and G. W. Bushnell, *J. Chem. Soc. (A)*, 2520 (1968).

1968 (continued)

68Lc R. P. Lang, *J. Phys. Chem.*, *72*, 2129 (1968).

68Le O. G. Levanda, *Russ. J. Inorg. Chem. (English Trans.)*, *13*, 1707 (1968). (Russian Page No. 3311.)

68Ma J. M. Malin and J. H. Swinehart, *Inorg. Chem.*, *7*, 250 (1968).

68Mb D. F. Martin and W. J. Randall, *J. Inorg. Nucl. Chem.*, *30*, 2909 (1968).

68Mc R. M. Milburn and L. M. Venanzi, *Inorg. Chim. Acta*, *2*, 97 (1968).

68Md D. F. C. Morris and S. D. Hammond, *Electrochimica Acta*, *13*, 545 (1968).

68Na G. H. Nancollas and A. C. Park, *Inorg. Chem.*, *7*, 58 (1968).

68Nb D. Nicholls, C. Sutphen and M. Szwarc, *J. Phys. Chem.*, *72*, 1021 (1968).

68Oa G. Olofsson, *Acta Chem. Scand.*, *22*, 1352 (1968).

68Pa E. Peltonen and P. Kivalo, *Suomen Kemistilehti*, *B 41*, 187 (1968).

68Pb R. Phillips and P. George, *Biochim. Biophys. Acta*, *162*, 73 (1968).

68Pc N. G. Podder, *Current Sci.*, *37*, 48 (1968).

68Ra E. V. Raju and H. B. Mathur, *J. Inorg. Nucl. Chem.*, *30*, 2181 (1968).

68Rb C. M. Rodriguez, R. Cetina and J. Gomez Lara, *Boletin Institute Quimica Univ. Nacional Autonoma Mexico*, *20*, 56 (1968).

68Rc I. P. Romm and E. N. Guryanova, *J. General Chem., U.S.S.R. (English Trans.)*, *38*, 1872 (1968). (Russian Page No. 1927.)

68Rd I. P. Romm, T. G. Sevastyanova, E. N. Guryanova, I. D. Kolli and R. A. Rodionov, *J. General Chem., U.S.S.R. (English Trans.)*, *38*, 1881 (1968). (Russian Page No. 1938.)

68Sa R. S. Saxena, K. C. Gupta and M. L. Mittal, *J. Inorg. Nucl. Chem.*, *30*, 189 (1968).

68Sb R. S. Saxena, K. C. Gupta and M. L. Mittal, *Aust. J. Chem.*, *21*, 641 (1968).

68Sc R. S. Saxena, K. C. Gupta and M. L. Mittal, *Can. J. Chem.*, *46*, 311 (1968).

68Sd R. S. Saxena, K. C. Gupta and M. L. Mittal, *Monatshefte fur Chemie*, *99*, 1779 (1968).

68Se T. W. Swaddle and G. Guastalla, *Inorg. Chem.*, *7*, 1915 (1968).

68Ta S. P. Tanner and G. R. Choppin, *Inorg. Chem.*, *7*, 2046 (1968).

68Tb S. P. Tanner, J. B. Walker and G. R. Choppin, *J. Inorg. Nucl. Chem.*, *30*, 2067 (1968).

68Va A. Vacca and P. Paoletti, *J. Chem. Soc. (A)*, 2378 (1968).

68Wa D. R. Williams, *J. Chem. Soc. (A)*, 2965 (1968).

68Wb P. T. T. Wong and D. G. Brewer, *Can. J. Chem.*, *46*, 131 (1968).

68Za A. H. Zeltmann, N. A. Matwiyoff, and L. O. Morgan, *J. Phys. Chem.*, *72*, 121 (1968).

1967

67Aa K. P. Anderson, W. O. Greenhalgh and E. A. Butler, *Inorg. Chem.*, *6*, 1056 (1967).

67Ab S. Ahrland, *Helv. Chim. Acta*, *50*, 306 (1967).

67Ac M. Ardon and N. Sutin, *Inorg. Chem.*, *6*, 2268 (1967).

67Ad R. Arnek and W. Kakolowicz, *Acta Chem. Scand.*, *21*, 1449 (1967).

67Ae R. Arnek and W. Kakolowicz, *Acta Chem. Scand.*, *21*, 2180 (1967).

67Af G. Atkinson and S. K. Kor, *J. Phys. Chem.*, *71*, 673 (1967).

67Ba B. R. Baker, N. Sutin and T. J. Welch, *Inorg. Chem.*, *6*, 1948 (1967).

67Bb H. L. Bott and A. J. Poe, *J. Chem. Soc. (A)*, 205 (1967).

67Bc S. Boyd, J. R. Brannan, H. S. Dunsmore and G. H. Nancollas, *J. Chem. Eng. Data*, *12*, 601 (1967).

67Bd T. Boschi, P. Rigs, C. Pecile and A. Turco, *Gass. Chim. Ital.*, *97*, 1391 (1967).

67Ca D. K. Cabbiness and E. S. Amis, *Bull. Chem. Soc. Japan*, *40*, 435 (1967).

67Cb R. G. de Carvalho and G. R. Choppin, *J. Inorg. Nucl. Chem.*, *29*, 737 (1967).

67Cc J. M. Conner and V. C. Bulgrin, *J. Inorg. Nucl. Chem.*, *29*, 1953 (1967).

67Cd R. Cramer, *J. Amer. Chem. Soc.*, *89*, 4621 (1967).

67Ce G. Curthoys, *J. Inorg. Nucl. Chem.*, *29*, 1176 (1967).

67Da R. G. Denning, F. R. Hartley and L. M. Venanzi, *J. Chem. Soc. (A)*, 324 (1967).

67Db R. G. Denning, F. R. Hartley and L. M. Venanzi, *J. Chem. Soc. (A)*, 328 (1967).

67Dc R. G. Denning and L. M. Venanzi, *J. Chem. Soc. (A)*, 336 (1967).

67Dd L. Drougge, L. I. Elding and L. Gustafson, *Acta Chem. Scand.*, *21*, 1647 (1967).

67De N. V. Duffy and J. E. Earley, *J. Amer. Chem. Soc.*, *89*, 272 (1967).

67Ea W. A. Eaton, P. George and G. I. H. Hanania, *J. Phys. Chem.*, *71*, 2016 (1967).

67Eb R. P. Eswein, E. S. Howald, R. A. Howald and D. P. Keeton, *J. Inorg. Nucl. Chem.*, *29*, 437 (1967).

67Ec R. Ettorre, G. Dolcetti and A. Peloso, *Gazz. Chim. Ital.*, *97*, 1681 (1967).

67Ed R. Ettorre, A. Peloso and G. Dolcetti, *Gazz. Chim. Ital.*, *97*, 968 (1967).

67Fa V. A. Fedorov, A. M. Robov, and V. E. Mironov, *Russ. J. Inorg. Chem. (English Trans.)*, *12*, 1750 (1967). (Russian Page No. 3307.)

67Fb V. A. Fedorov, A. M. Robov and V. E. Mironov, *Russ. J. Inorg. Chem. (English Trans.)*, *12*, 1750 (1967). (Russian Page No. 3307.)

67Ga P. Gerding, *Acta Chem. Scand.*, *21*, 2015 (1967).

67Gb D. R. Goddard, S. I. Nwankwo and L. A. K. Staveley, *J. Chem. Soc. (A)*, 1376 (1967).

67Gc I. Grenthe and D. R. Williams, *Acta Chem. Scand.*, *21*, 347 (1967).

1967 (continued)

67Gd S. L. Gupta and R. N. Soni, *J. Indian Chem. Soc.*, *44*, 195 (1967).

67Ha C. F. Hale and E. L. King, *J. Phys. Chem.*, *71*, 1779 (1967).

67Hb F. R. Hartley and L. M. Venanzi, *J. Chem. Soc. (A)*, 330 (1967).

67Hc C. H. Henrickson and D. P. Eyman, *Inorg. Chem.*, *6*, 1461 (1967).

67Hd D. J. Hewkin and A. J. Poe, *J. Chem. Soc. (A)*, 1884 (1967).

67He N. Hirota, *J. Phys. Chem.*, *71*, 127 (1967).

67Hf F. Holmes and D. R. Williams, *J. Chem. Soc. (A)*, 1702 (1967).

67Hg T. M. Hseu and S. C. Wu, *J. Chin. Chem. Soc. (Taipei)*, *14*, 56 (1967).

67Hh D. H. Huchital and R. G. Wilkins, *Inorg. Chem.*, *6*, 1022 (1967).

67Hi J. U. Hwang and Y. J. Jun, *J. Korean Chem. Soc.*, *11*, 63 (1967).

67Ia S. S. Ivanov and L. P. Gavryuchenkova, *Polymer Sci. U.S.S.R. (English Trans.)*, *A9(1)*, 112 (1967). (Russian Page No. 103.)

67Ib R. M. Izatt, H. D. Johnston, G. D. Watt and J. J. Christensen, *Inorg. Chem.*, *6*, 132 (1967).

67Ic R. M. Izatt, C. D. Watt, D. Eatough and J. J. Christensen, *J. Chem. Soc. (A)*, 1304 (1967).

67Ka M. M. T. Kahn and A. E. Martell, *J. Amer. Chem. Soc.*, *89*, 5585 (1967).

67La S. C. Lahiri and S. Aditya, *Z. Phys. Chem. (Frankfurt am Main)*, *55*, 6 (1967).

67Lb C. H. Langford and W. R. Muir, *J. Phys. Chem.*, *71*, 2602 (1967).

67Ma J. A. Maguire, A. Bramley and J. J. Banewicz, *Inorg. Chem.*, *6*, 1752 (1967).

67Mb W. L. Marshall, *J. Phys. Chem.*, *71*, 3584 (1967).

67Mc G. Matsubayashi, Y. Kawasaki and T. Tanaka, *Nippon Kagaku Zasshi*, *88*, 1251 (1967).

67Md I. F. Mavrin, F. Ya. Kul'ba and V. E. Mironov, *Russ. J. Inorg. Chem. (English Trans.)*, *12*, 167 (1967). (Russian Page No. 324.)

67Me I. F. Mavrin, F. Ya. Kul'ba and V. E. Mironov, *Russ. J. Inorg. Chem. (English Trans.)*, *41*, 886 (1967). (Russian Page No. 1659.)

67Mf A. McAuley, G. H. Nancollas and K. Torrance, *Inorg. Chem.*, *6*, 136 (1967).

67Mg J. D. McCullough and A. Brunner, *Inorg. Chem.*, *6*, 1251 (1967).

67Mh R. E. Mesmer and C. F. Baes, *Inorg. Chem.*, *6*, 1951 (1967).

67Mi R. M. Milburn, *J. Amer. Chem. Soc.*, *89*, 54 (1967).

67Na G. H. Nancollas and K. Torrance, *Inorg. Chem.*, *6*, 1567 (1967).

67Oa G. Olofsson, *Acta Chem. Scand.*, *21*, 93 (1967).

67Ob G. Olofsson, *Acta Chem. Scand.*, *21*, 1114 (1967).

67Oc G. Olofsson, *Acta Chem. Scand.*, *21*, 1887 (1967).

67Od G. Olofsson, *Acta Chem. Scand.*, *21*, 1892 (1967).

67Oe G. Olofsson, *Acta Chem. Scand.*, *21*, 2143 (1967).

1967 (continued)

67Of G. Olofsson, *Acta Chem. Scand.*, *21*, 2415 (1967).

67Pa A. J. Poë and K. Shaw, *Chem. Commun.*, 52 (1967).

67Pb A. J. Poë, K. Shaw and M. J. Wendt, *Inorg. Chim. Acta*, *1*, 371 (1967).

67Pc H. K. J. Powell and N. F. Curtis, *J. Chem. Soc. (A)*, 1441 (1967).

67Ra V. H. Rau and K. H. Thiele, *Z. Anorg. Allg. Chem.*, *335*, 253 (1967).

67Rb P. S. Relan and P. K. Bhattacharya, *J. Indian Chem. Soc.*, *44*, 536 (1967).

67Sa D. E. Scaife and K. P. Wood, *Inorg. Chem.*, *6*, 358 (1967).

67Sb R. V. Slates and M. Szwarc, *J. Amer. Chem. Soc.*, *89*, 6043 (1967).

67Sc W. F. Stack and H. A. Skinner, *Trans. Faraday Soc.*, *63*, 1136 (1967).

67Sd M. A. Suwyn and R. E. Hamm, *Inorg. Chem.*, *6*, 2150 (1967).

67Va V. P. Vasil'ev and G. A. Lobanov, *Russ. J. Inorg. Chem. (English Trans.)*, *12*, 463 (1967). (Russian Page No. 878.)

67Vb V. P. Vasil'ev and G. A. Lobanov, *Russ. J. Phys. Chem. (English Trans.)*, *41*, 1053 (1967). (Russian Page No. 1969.)

67Wa J. B. Walker and G. R. Choppin, in *Advances in Chemistry Series*, *71*, 127 (1967). Amer. Chem. Soc., Washington, D.C., R. F. Gould, editor.

67Wb R. M. Wallace, *J. Phys. Chem.*, *71*, 1271 (1967).

67Wc C. Wu, R. J. Witonsky, P. George and R. J. Rutman, *J. Amer. Chem. Soc.*, *89*, 1987 (1967).

67Ya Yu. B. Yakovlev, F. Ya. Kulba and V. E. Mironov, *Russ. J. Phys. Chem. (English Trans.)*, *41*, 618 (1967). (Russian Page No. 1168.)

1966

66Aa K. P. Anderson, W. O. Greenhalgh and R. M. Izatt, *Inorg. Chem.*, *5*, 2106 (1966).

66Ab K. P. Anderson, D. A. Newell and R. M. Izatt, *Inorg. Chem.*, *5*, 62 (1966).

66Ac A. C. Anusiem, J. G. Beetlestone and D. H. Irvine, *J. Chem. Soc. (A)*, 357 (1966).

66Ad I. G. Arzamanova and E. N. Guryanova, *J. Gen. Chem. U.S.S.R. (English Trans.)* *36*, 1171 (1966). (Russian Page No. 1157.)

66Ba A. V. Belyaev and B. V. Ptitsyn, *Russ. J. Inorg. Chem. (English Trans.)*, *11*, 836 (1966). (Russian Page No. 1565.)

66Bb L. Z. Benet and J. E. Goyan, *J. Pharm. Sci.*, *55*, 1184 (1966).

66Bc K. R. Bhaskar, S. N. Bhat, A. S. N. Murthy and C. N. R. Rao, *Trans. Faraday Soc.*, *62*, 788 (1966).

66Be K. R. Bhaskar, R. K. Gosavi and C. N. R. Rao, *Trans. Faraday Soc.*, *62*, 29 (1966).

66Bg T. F. Bolles and R. S. Drago, *J. Amer. Chem. Soc.*, *88*, 3921 (1966).

66Bh T. F. Bolles and R. S. Drago, *J. Amer. Chem. Soc.*, *88*, 5730 (1966).

1966 (continued)

66Bi E. J. Bounsall and A. J. Poë, *J. Chem. Soc. (A)*, 286 (1966).

66Ca J. de O. Cabral, H. C. A. King, S. M. Nelson, T. M. Shepherd and E. Koros, *J. Chem. Soc. (A)*, 1348 (1966).

66Cb P. Chang, R. V. Slates and M. Szwarc, *J. Phys. Chem.*, *70*, 3180 (1966).

66Cc G. R. Choppin and H. G. Friedman, *Inorg. Chem.*, *5*, 1599 (1966).

66Da S. P. Datta and A. K. Grzybowski, *J. Chem. Soc. (A)*, 1059 (1966).

66Db G. Dolcetti, A. Peloso and L. Sindellari, *Gazz. Chim. Ital.*, *96*, 1648 (1966).

66Dc R. S. Drago, T. F. Bolles and R. J. Niedzielski, *J. Amer. Chem. Soc.*, *88*, 2717 (1966).

66Ga P. Gerding, *Acta Chem. Scand.*, *20*, 79 (1966).

66Gb P. Gerding, *Acta Chem. Scand.*, *20*, 2771 (1966).

66Gc J. Gohring and B. P. Susz, *Helv. Chim. Acta*, *49*, 486 (1966).

66Gd R. N. Goldberg, R. G. Riddell, M. R. Wingard, H. P. Hopkins, C. A. Wulff and L. G. Hepler, *J. Phys. Chem.*, *70*, 706 (1966).

66Ge D. M. L. Goodgame and M. A. Hitchman, *Inorg. Chem.*, *5*, 1303 (1966).

66Ha G. I. H. Hanania, D. H. Irving and M. V. Irving, *J. Chem. Soc. (A)*, 296 (1966).

66Hb J. O. Hill and R. J. Irving, *J. Chem. Soc. (A)*, 971 (1966).

66Hc T. E. Hogen-Esch and J. Smid, *J. Amer. Chem. Soc.*, *88*, 307 (1966).

66Ka M. M. T. Kahn and A. E. Martell, *J. Amer. Chem. Soc.*, *88*, 668 (1966).

66Kb M. B. Kennedy and M. W. Lister, *Can. J. Chem.*, *44*, 1709 (1966).

66Kc N. I. Kulevsky, *J. Chem. Eng. Data*, *11*, 492 (1966).

66La I. R. Lantzke and D. W. Watts, *Aust. J. Chem.*, *19*, 969 (1966).

66Ma R. Matejec, *Ber. Bunsenges. Phys. Chem.*, *70*, 703 (1966).

66Mb G. Matsubayashi, Y. Kawasaki, T. Tanaka and R. Okawara, *J. Inorg. Nucl. Chem.*, *28*, 2937 (1966).

66Mc A. McAuley, G. H. Nancollas and K. Torrance, *J. Inorg. Nucl. Chem.*, *28*, 917 (1966).

66Md W. A. Millen and D. W. Watts, *Aust. J. Chem.*, *19*, 43 (1966).

66Me V. E. Mironov and Yu. I. Rutkovskii, *Russ. J. Inorg. Chem. (English Trans.)*, *11*, 958 (1966). (Russian Page No. 1792.)

66Mf T. Moeller and S. Chu, *J. Inorg. Nucl. Chem.*, *28*, 153 (1966).

66Mg G. E. Mont and A. E. Martell, *J. Amer. Chem. Soc.*, *88*, 1387 (1966).

66Na G. H. Nancollas, *Interactions in Electroylte Solutions*, Elsevier, New York, N. Y., 1966.

66Nb B. Nelander, *Acta Chem. Scand.*, *20*, 2289 (1966).

66Pa P. Paoletti, F. Nuzzi and A. Vacca, *J. Chem. Soc. (A)*, 1385 (1966).

1966 (continued)

66Pb P. Paoletti, A. Vacca and D. Arenare, *Coordin. Chem. Rev.*, *1*, 280 (1966).

66Pc P. Paoletti, A. Vacca and D. Arenare, *J. Phys. Chem.*, *70*, 193 (1966).

66Pd J. A. Partridge, J. J. Christensen and R. M. Izatt, *J. Amer. Chem. Soc.*, *88*, 1649 (1966).

66Pe R. C. Phillips, P. George and R. J. Rutman, *J. Amer. Chem. Soc.*, *88*, 2631 (1966).

66Pf T. J. Pinnavaia and R. C. Fay, *Inorg. Chem.*, *5*, 233 (1966).

66Sa L. Sacconi and I. Bertini, *J. Amer. Chem. Soc.*, *88*, 5180 (1966).

66Sb V. I. Sklenskaya and A. A. Biryukov, *Russ. J. Inorg. Chem. (English Trans.)*, *11*, 28 (1966). (Russian Page No. 54.)

66Sc E. V. Sklenskaya and M. Kh. Karapet'yants, *Russ. J. Inorg. Chem. (English Trans.)*, *11*, 1102 (1966). (Russian Page No. 2061.)

66Sd C. B. Storm, A. H. Corwin, R. R. Arellano, M. Martz and R. Weintraub, *J. Amer. Chem. Soc.*, *88*, 2525 (1966).

66Se J. H. Swinehart and P. A. Rock, *Inorg. Chem.*, *5*, 573 (1966).

66Ta J. Y. Tong and R. L. Johnson, *Inorg. Chem.*, *5*, 1902 (1966).

66Va G. R. Van Hecke and W. DeW. Horrocks, *Inorg. Chem.*, *5*, 1968 (1966).

66Vb J. Vaissermann, *C. R. H. Acad. Sci., Ser. C*, *262*, 692 (1966).

66Vc J. Vaissermann and M. Quintin, *J. Chim. Phys. Physicochim. Biol.*, *63*, 731 (1966).

66Wa J. L. Walter and S. M. Rosalie, *J. Inorg. Nucl. Chem.*, *28*, 2969 (1966).

66Wb B. B. Wayland and R. H. Gold, *Inorg. Chem.*, *5*, 154 (1966).

1965

65Aa G. Anderegg, *Helv. Chim. Acta*, *48*, 1712 (1965).

65Ab G. Anderegg, *Helv. Chim. Acta*, *48*, 1718 (1965).

65Ac G. Anderegg, *Helv. Chim. Acta*, *48*, 1722 (1965).

65Ad A. C. Andrews and D. M. Zebolsky, *J. Chem. Soc. (London)*, 742 (1965).

65Ae R. Arnek, *Ark. Kemi*, *24*, 531 (1965).

65Af E. Augdahl and P. Klaeboe, *Acta Chem. Scand.*, *19*, 807 (1965).

65Ba C. F. Baes, N. J. Meyer and C. E. Roberts, *Inorg. Chem.*, *4*, 518 (1965).

65Bb T. F. Bolles and R. S. Drago, *J. Amer. Chem. Soc.*, *87*, 5015 (1965).

65Bc S. Boyd, A. Bryson, G. H. Nancollas and K. Torrance, *J. Chem. Soc. (London)*, 7353 (1965).

65Ca R. L. Carlin and F. A. Walker, *J. Amer. Chem. Soc.*, *87*, 2128 (1965).

65Cb C. Carvajal, K. J. Tolle, J. Smid, and M. Szwarc, *J. Amer. Chem. Soc.*, *87*, 5548 (1965).

65Cc L. Chi, T. Ts'ao, C. Lo and K. Huang, *K'o Hsueh T'ung Pao*, *16*, 352 (1965).

1965 (continued)

65Cd G. R. Choppin and A. J. Graffeo, *Inorg. Chem.*, *4*, 1254 (1965).

65Ce G. R. Choppin and J. Ketels, *J. Inorg. Nucl. Chem.*, *27*, 1335 (1965).

65Cf G. R. Choppin and W. F. Strazik, *Inorg. Chem.*, *4*, 1250 (1965).

65Cg J. J. Christensen, R. M. Izatt and D. Eatough, *Inorg. Chem.*, *4*, 1278 (1965).

65Ch E. Coates, B. Rigg, B. Saville, and D. Skelton, *J. Chem. Soc. (London)*, 5613 (1965).

65Ci R. J. Cvetanovic, F. J. Duncan, W. E. Falconer and R. S. Irwin, *J. Amer. Chem. Soc.*, *87*, 1827 (1965).

65Da R. C. Das, *Indian J. Chem.*, *3*, 179 (1965).

65Db R. L. Davies and K. W. Dunning, *J. Chem. Soc. (London)*, 4168 (1965).

65Fa T. Fueno, T. Okuyama, T. Deguchi and J. Furukawa, *J. Amer. Chem. Soc.*, *87*, 170 (1965).

65Fb J. Fuger and B. B. Cunningham, *J. Inorg. Nucl. Chem.*, *27*, 1079 (1965).

65Ga I. P. Gol'dshtein, E. N. Gur'yanova and I. R. Karpovich, *Russ. J. Phys. Chem. (English Trans.)*, *39*, 491 (1965). (Russian Page No. 932.)

65Gb I. P. Gol'dshtein, E. N. Gur'yanova, and K. A. Kocheshkov, *Proc. Doklady Akad. Nauk Chem. (English Trans.)*, *161*, 233 (1965). (Russian Page No. 111.)

65Ha S. Hamada, Y. Ishikawa and T. Shirai, *Nippon Kagaku Zasshi*, *86*, 1042 (1965).

65Hb H. P. Hopkins and C. A. Wulff, *J. Phys. Chem.*, *69*, 6 (1965).

65Hc H. P. Hopkins and C. A. Wulff, *J. Phys. Chem.*, *69*, 9 (1965).

65Hd T. M. Hseu, S. C. Wu and T. Chuang, *J. Inorg. Nucl. Chem.*, *27*, 1655 (1965).

65Ia R. M. Izatt, J. J. Christensen, J. W. Hansen and G. D. Watt, *Inorg. Chem.*, *4*, 718 (1965).

65Ib R. M. Izatt, J. W. Wrathall and K. P. Anderson, *J. Phys. Chem.*, *65*, 1914 (1965).

65Ka V. A. Korobova and G. A. Prik, *Russ. J. Inorg. Chem. (English Trans.)*, *10*, 456 (1965). (Russian Page No. 992.)

65Kb E. E. Kriss and K. B. Yatsimirskii, *Russ. J. Inorg. Chem. (English Trans.)*, *10*, 1326 (1965). (Russian Page No. 2436.)

65Kc T. Kubota, *J. Amer. Chem. Soc.*, *87*, 458 (1965).

65Kd F. Ya. Kul'ba, V. E. Mironov and I. F. Mavrin, *Russ. J. Phys. Chem. (English Trans.)*, *39*, 1387 (1965). (Russian Page No. 2595.)

65Ke F. Ya. Kul'ba, V. E. Mironov, I. F. Mavrin and Y. B. Yakovlev, *Russ. J. Inorg. Chem. (English Trans.)*, *10*, 1117 (1965). (Russian Page No. 2053.)

65La S. C. Lahiri, *J. Indian Chem. Soc.*, *42*, 715 (1965).

65Lb D. W. Larsen and A. L. Allred, *J. Amer. Chem. Soc.*, *87*, 1216 (1965).

65Lc D. W. Larsen and A. L. Allred, *J. Amer. Chem. Soc.*, *87*, 1219 (1965).

65Ld D. W. Larsen and A. L. Allred, *J. Phys. Chem.*, *69*, 2400 (1965).

1965 (continued)

65Le C. Luca, V. Magearu and G. Popa, *Analele Univ. "C. I. Parhon" Bucuresti, Ser. Stiint. Nat.*, *14*, 109 (1965).

65Ma J. B. Martinez, R. Ferrus, F. Puerta and L. L. Mateo, *An. Real Soc. Expan. Fis. Quim., Ser. B*, *61*, 1227 (1965).

65Mb V. E. Mironov and Y. I. Rutkovskii, *Russ. J. Inorg. Chem. (English Trans.)*, *10*, 1450 (1965). (Russian Page No. 2670.)

65Na V. S. K. Nair, *J. Chem. Soc. (London)*, 1450 (1965).

65Nb R. N. Nanda, R. C. Das and R. K. Nanda, *Indian J. Chem.*, *3*, 278 (1965).

65Nc S. M. Nelson and T. M. Shepherd, *J. Chem. Soc. (London)*, 3284 (1965).

65Oa G. Olofsson, *Acta Chem. Scand.*, *19*, 2155 (1965).

65Pa J. A. Partridge, R. M. Izatt and J. J. Christensen, *J. Chem. Soc. (London)*, 4231 (1965).

65Pb S. V. Pestrikov, I. I. Moiseev and T. N. Romanova, *Russ. J. Inorg. Chem. (English Trans.)*, *10*, 1199 (1965). (Russian Page No. 2203.)

65Pc E. Plahte, J. Grundnes and P. Klaeboe, *Acta Chem. Scand.*, *19*, 1897 (1965).

65Ra R. C. Roberts and M. Szwarc, *J. Amer. Chem. Soc.*, *87*, 5542 (1965).

65Sa G. Schwarzenbach and M. Schellenberg, *Helv. Chim. Acta*, *48*, 28 (1965).

65Sb V. S. Sharma and H. B. Mathur, *Indian J. Chem.*, *3*, 475 (1965).

65Sc D. C. Shuman and B. Sen, *Anal. Chim. Acta*, *38*, 487 (1965).

65Sd E. V. Sklenskaya and M. Kh. Karapet'yants, *Tr. Mosk. Khim. Tekhnol. Inst.*, *No. 48*, 11 (1965).

65Se T. W. Swaddle and E. L. King, *Inorg. Chem.*, *4*, 532 (1965).

65Ta Y. A. Treger, R. M. Flid, L. V. Antonova and S. S. Spektor, *Russ. J. Phys. Chem. (English Trans.)*, *39*, 1515 (1965). (Russian Page No. 2831.)

65Va J. Vaissermann and M. Quintin, *J. Chim. Phys.*, *63*, 731 (1965).

65Wa G. D. Watt, J. J. Christensen and R. M. Izatt, *Inorg. Chem.*, *4*, 220 (1965).

65Wb C. F. Wells and G. Davies, *Nature (London)*, *205*, 692 (1965).

65Wc D. Wobschall and D. A. Norton, *J. Amer. Chem. Soc.*, *87*, 3559 (1965).

65Wd D. L. Wright, J. H. Holloway and C. N. Reilley, *Anal. Chem.*, *37*, 884 (1965).

1964

64Aa G. Anderegg, *Experientia, Supplementum*, *9*, 75 (1964).

64Ab G. Anderegg, *Helv. Chim. Acta*, *47*, 1801 (1964).

64Ac A. C. Andrews and J. K. Romary, *Kansas Acad. Sci., Trans.*, *67*, 630 (1964).

64Ad A. C. Andrews and J. K. Romary, *J. Chem. Soc. (London)*, 405 (1964).

64Ba E. W. Baker, M. S. Brookhart and A. H. Corwin, *J. Amer. Chem. Soc.*, *86*, 4587 (1964).

64Bb R. J. Baltisberger and E. L. King, *J. Amer. Chem. Soc.*, *86*, 795 (1964).

1964 (continued)

64Bc J. E. Bauman and J. C. Wang, *Inorg. Chem.*, *3*, 368 (1964).

64Bd J. R. Brannan, H. S. Dunsmore and G. H. Nancollas, *J. Chem. Soc.* (*London*), 304 (1964).

64Be J. A. Broomhead, F. Basolo and R. G. Pearson, *Inorg. Chem.*, *3*, 826 (1964).

64Ca R. L. Carlin, J. S. Dubnoff and W. T. Huntress, *Proc. Chem. Soc.*, 228 (1964).

64Cb R. L. Carlin, J. S. Dubnoff and W. T. Huntress, *Proc. Chem. Soc.*, 228 (1964).

64Cc J. J. Christensen, R. M. Izatt, L. D. Hansen and J. D. Hale, *Inorg. Chem.*, *3*, 130 (1964).

64Da R. C. Das and S. Aditya, *J. Indian Chem. Soc.*, *41*, 765 (1964).

64Ea I. Eliezer, *J. Phys. Chem.*, *68*, 2722 (1964).

64Eb E. Eyal and A. Treinin, *J. Amer. Chem. Soc.*, *86*, 4287 (1964).

64Ga P. George, G. I. H. Hanania, D. H. Irvine and I. Abu-Issa, *J. Chem. Soc.* (*London*), 5689 (1964).

64Gb I. Grenthe, *Acta Chem. Scand.*, *18*, 283 (1964).

64Gc F. H. Guzzetta and W. B. Hadley, *Inorg. Chem.*, *3*, 259 (1964).

64Ha G. P. Haight, D. C. Richardson and N. H. Coburn, *Inorg. Chem.*, *3*, 1777 (1964).

64Hb G. I. H. Hanania and D. H. Irvine, *J. Chem. Soc.* (*London*), 5694 (1964).

64Hc J. A. Hull, R. H. Davies and L. A. K. Staveley, *J. Chem. Soc.* (*London*), 5422 (1964).

64Ia R. M. Izatt, J. J. Christensen and V. Kothari, *Inorg. Chem.*, *3*, 1565 (1964).

64Ja D. S. Jain and J. N. Gaur, *Indian J. Chem.*, *2*, 503 (1964).

64Ka H. C. A. King, E. Koros and S. M. Nelson, *J. Chem. Soc.* (*London*), 4832 (1964).

64La J. Laane and T. L. Brown, *Inorg. Chem.*, *3*, 148 (1964).

64Lb S. C. Lahiri and S. Aditya, *J. Indian Chem. Soc.*, *41*, 517 (1964).

64Lc S. C. Lahiri and S. Aditya, *Z. Phys. Chem.* (*Frankfurt am Main*), *41*, 173 (1964).

64Ld S. C. Lahiri and S. Aditya, *Z. Phys. Chem.* (*Frankfurt am Main*), *43*, 282 (1964).

64Le T. J. Lane, A. J. Kandathil and S. M. Rosalie, *Inorg. Chem.*, *3*, 487 (1964).

64Lf I. Leden and T. Ryhl, *Acta Chem. Scand.*, *18*, 1196 (1964).

64Lg D. L. Leussing and D. C. Shultz, *J. Amer. Chem. Soc.*, *86*, 4846 (1964).

64Ma R. L. Middaugh, R. S. Drago and R. J. Niedzielski, *J. Amer. Chem. Soc.*, *86*, 388 (1964).

64Mb V. E. Mironov, F. Ya. Kul'ba and V. A. Fedorov, *Russ. J. Inorg. Chem.* (*English Trans.*), *9*, 888 (1964). (Russian Page No. 1641.)

1964 (continued)

64Mc V. E. Mironov, F. Y. Kulba, A. V. Fokina, V. S. Golubeva, and V. A. Nazarov, *Russ. J. Inorg. Chem. (English Trans.)*, *9*, 1152 (1964). (Russian Page No. 2133.)

64Na R. J. Niedzielski, R. S. Drago and R. L. Middaugh, *J. Amer. Chem. Soc.*, *86*, 1694 (1964).

64Nb G. Nord and J. Ulstrup, *Acta Chem. Scand.*, *18*, 307 (1964).

64Oa G. Olofsson, *Acta Chem. Scand.*, *18*, 11 (1964).

64Ob G. Olofsson, *Acta Chem. Scand.*, *18*, 1022 (1964).

64Pa P. Paoletti and A. Vacca, *J. Chem. Soc. (London)*, 5051 (1964).

64Pb V. K. Potapov, Y. S. Shabarov and R. Ya. Levina, *Russ. J. General Chem. (English Trans.)*, *34*, 2535 (1964). (Russian Page No. 2512.)

64Pc P. L. E. De La Praudiere and L. A. K. Staveley, *J. Inorg. Nucl. Chem.*, *26*, 1713 (1964).

64Ra I. G. Ryss and N. F. Kulish, *Russ. J. Inorg. Chem. (English Trans.)*, *9*, 603 (1964). (Russian Page No. 1103.)

64Rb I. G. Ryss and N. F. Kulish, *Russ. J. Inorg. Chem. (English Trans.)*, *9*, 752 (1964). (Russian Page No. 1382.)

64Sa L. Sacconi, M. Ciampolini, N. Nardi, *J. Amer. Chem. Soc.*, *86*, 819 (1964).

64Sb L. Sacconi, M. Ciampolini and N. Nardi, *J. Amer. Chem. Soc.*, *86*, 819 (1964).

64Sc L. Sacconi, P. Paoletti and M. Ciampolini, *J. Chem. Soc. (London)*, 5046 (1964).

64Sd C. D. Schmulbach and D. M. Hart, *J. Amer. Chem. Soc.*, *86*, 2347 (1964).

64Se V. S. Sharma, H. B. Mathur and A. B. Biswas, *J. Inorg. Nucl. Chem.*, *26*, 382 (1964).

64Sf J. C. Sullivan, *Inorg. Chem.*, *3*, 315 (1964).

64Ta M. Tamres, S. Searles and J. M. Goodenow, *J. Amer. Chem. Soc.*, *86*, 3934 (1964).

64Tb B. G. Torre-Mori, D. Janjic and B. P. Susz, *Helv. Chim. Acta*, *47*, 1172 (1964).

64Va V. P. Vasil'ev, *Russ. J. Inorg. Chem. (English Trans.)*, *9*, 354 (1964). (Russian Page No. 641.)

64Wa M. J. M. Woods, P. K. Gallagher, Z. Z. Hugus and E. L. King, *Inorg. Chem.*, *3*, 1313 (1964).

64Wb J. C. Wang, J. E. Bauman and R. K. Murmann, *J. Phys. Chem.*, *68*, 2296 (1964).

1963

63Aa K. M. J. Al-Komser and B. Sen, *Inorg. Chem.*, *2*, 1219 (1963).

63Ab G. Anderegg, *Helv. Chim. Acta*, *46*, 1833 (1963).

63Ac G. Anderegg, *Helv. Chim. Acta*, *46*, 2813 (1963).

1963 (continued)

63Ad A. C. Andrews and J. K. Romary, *Inorg. Chem.*, *2*, 1060 (1963).

63Ae G. Atkinson and J. E. Bauman, *Inorg. Chem.*, *2*, 64 (1963).

63Ba F. Becker, J. Barthel, N. G. Schmahl and H. M. Luschow, *Z. Phys. Chem. (Frankfurt am Main)*, *37*, 52 (1963).

63Bb M. Bjorkman and L. G. Sillen, *Trans. Roy. Inst. Technol., Stockholm*, No. 199, 1963.

63Bc J. A. Bolzan and A. J. Arvia, *Electrochim. Acta*, *8*, 375 (1963).

63Bd J. A. Bolzan, E. A. Jauregui and A. J. Arvia, *Electrochim. Acta*, *8*, 841 (1963).

63Be J. A. Bolzan, J. J. Podesta and A. J. Arvia, *Anales Asociacion Quimica, Argentina*, *51*, 43 (1963).

63Ca R. L. Carlson and R. S. Drago, *J. Amer. Chem. Soc.*, *85*, 505 (1963).

63Cb G. R. Choppin and P. J. Unrein, *J. Inorg. Nucl. Chem.*, *25*, 387 (1963).

63Cc J. J. Christensen, R. M. Izatt, J. D. Hale, R. T. Pack and G. D. Watt, *Inorg. Chem.*, *2*, 337 (1963).

63Da R. S. Drago, B. B. Wayland and R.L. Carlson, *J. Amer. Chem. Soc.*, *85*, 3125 (1963).

63Db J. E. Dubois and H. Herzog, *Bull. Soc. Chim. Fr.*, 57 (1963).

63Ea A. G. Evans and B. J. Tabner, *J. Chem. Soc.*, 4613 (1963).

63Fa R. C. Fay and T. S. Piper, *J. Amer. Chem. Soc.*, *85*, 500 (1963).

63Ga P. George, R. C. Phillips, S. J. Rutman, and R. J. Rutman, *Biochem.*, *2*, 508 (1963).

63Gb D. C. Giedt and C. J. Nyman, *J. Phys. Chem.*, *67*, 2491 (1963).

63Gc I. Grenthe, *Acta Chem. Scand.*, *17*, 2487 (1963).

63Gd R. L. Gustafson and A. E. Martell, *J. Phys. Chem.*, *67*, 576 (1963).

63Ha R. P. Held and D. E. Goldberg, *Inorg. Chem.*, *2*, 585 (1963).

63Hb R. R. Holmes, W. P. Gallagher and R. P. Carter, *Inorg. Chem.*, *2*, 437 (1963).

63Ia H. Irving and J. J. R. F. da Silva, *J. Chem. Soc. (London)*, 448 (1963).

63Ka H. C. A. King, E. Körös and S. M. Nelson, *J. Chem. Soc. (London)*, 5449 (1963).

63Kb F. Ya. Kul'ba, V. E. Mironov, V. A. Fedorov and V. A. Baevskii, *Russ. J. Inorg. Chem. (English Trans.)*, *8*, 1012 (1963). (Russian Page No. 1945.)

63La T. J. Lane and H. B. Durham, *Inorg. Chem.*, *2*, 632 (1963).

63Ma W. R. May and M. M. Jones, *J. Inorg. Nucl. Chem. (London)*, *25*, 507 (1963).

63Mb A. McAuley and G. H. Nancollas, *J. Chem. Soc. (London)*, 989 (1963).

63Mc V. E. Mironov, *Russ. J. Inorg. Chem. (English Trans.)*, *8*, 388 (1963). (Russian Page No. 764.)

63Md V. E. Mironov, V. A. Fedorov and V. A. Nazarov, *Russ. J. Inorg. Chem. (English Trans.)*, *8*, 1102 (1963). (Russian Page No. 2109.)

1963 (continued)

63Me V. E. Mironov, F. Ya. Kul'ba, V. A. Fedorov and T. F. Nikitenko, *Russ. J. Inorg. Chem. (English Trans.)*, *8*, 964 (1963). (Russian Page No. 1852.)

63Mf V. E. Mironov and V. A. Nazarov, *Russ. J. Inorg. Chem. (English Trans.)*, *8*, 967 (1963). (Russian Page No. 1857.)

63Mg V. E. Mironov, F. Ya. Kul'ba, V. A. Fedorov and T. F. Nikitenko, *Russ. J. Inorg. Chem. (English Trans.)*, *8*, 1215 (1963). (Russian Page No. 2318.)

63Oa G. Olofsson, I. Lindquist and S. Sunner, *Acta Chem. Scand.*, *17*, 259 (1963).

63Pa P. Paoletti and M. Ciampolini, *Ric. Sci.*, *3*, 399 (1963).

63Pb P. Paoletti, M. Ciampolini and L. Sacconi, *J. Chem. Soc. (London)*, 3589 (1963).

63Pc W. B. Person, W. C. Golton and A. I. Popov, *J. Amer. Chem. Soc.*, *85*, 891 (1963).

63Pd G. A. Prik, *Russ. J. Inorg. Chem. (English Trans.)*, *8*, 1097 (1963). (Russian Page No. 2099.)

63Sa A. Ya. Sychev, A. P. Gerbeleu and P. K. Migal, *Russ. J. Inorg. Chem. (English Trans.)*, *8*, 1081 (1963). (Russian Page No. 2070.)

63Va J. VanderVeen and W. Stevens, *Rec. Trav. Chim.*, *82*, 287 (1963).

63Wa R. L. S. Willix, *Trans. Faraday Soc.*, *59*, 1315 (1963).

1962

62Aa A. C. Andrews, T. D. Lyons and T. D. O'Brien, *J. Chem. Soc. (London)*, 1776 (1962).

62Ab G. Atkinson and J. E. Bauman, *Inorg. Chem.*, *1*, 900 (1962).

62Ac J. M. Austin and A. D. Mair, *J. Phys. Chem.*, *66*, 519 (1962).

62Ba C. F. Baes and N. J. Meyer, *Inorg. Chem.*, *1*, 780 (1962).

62Bb J. A. Bolzan and A. J. Arvia, *Electrochimica Acta*, *7*, 589 (1962).

62Bc O. D. Bonner, H. Dolyniuk, C. F. Jordan and G. B. Hanson, *J. Inorg. Nucl. Chem.*, *24*, 689 (1962).

62Ca S. Cabani and M. Landucci, *J. Chem. Soc. (London)*, 278 (1962).

62Cb S. Cabani, G. Moretti and E. Scrocco, *J. Chem. Soc. (London)*, 88 (1962).

62Cc B. Carell and A. Olin, *Acta Chem. Scand.*, *16*, 2350 (1962).

62Cd B. Carell and A. Olin, *Acta Chem. Scand.*, *16*, 2357 (1962).

62Ce R. L. Carlson and R. S. Drago, *J. Amer. Chem. Soc.*, *84*, 2320 (1962).

62Cf E. Coates and B. Rigg, *Trans. Faraday Soc.*, *58*, 88 (1962).

62Da I. R. Desai and V. S. K. Nair, *J. Chem. Soc. (London)*, 2360 (1962).

62Db R. S. Drago and D. A. Wenz, *J. Amer. Chem. Soc.*, *84*, 526 (1962).

62Dc R. S. Drago, D. A. Wenz and R. L. Carlson, *J. Amer. Chem. Soc.*, *84*, 1106 (1962).

1962 (continued)

62Dd J. F. Duncan and D. L. Kepert, *J. Chem. Soc. (London)*, 205 (1962).

62Ea H. R. Ellison, J. O. Edwards and E. L. Healy, *J. Amer. Chem. Soc.*, *84*, 1820 (1962).

62Fa V. A. Fedorov, V. E. Mironov and F. Ya. Kul'ba, *Russ. J. Inorg. Chem. (English Trans.)*, *7*, 1311 (1962). (Russian Page No. 2528.)

62Fb N. Fogel, J. M. J. Tai and J. Yarborough, *J. Amer. Chem. Soc.*, *84*, 1145 (1962).

62Ga T. Gramstad and W. J. Fuglevik, *Acta Chem. Scand.*, *16*, 2368 (1962).

62Ha G. P. Haight, *J. Inorg. Nucl. Chem.*, *24*, 663 (1962).

62Hb G. I. H. Hanania and D. H. Irvine, *J. Chem. Soc. (London)*, 2745 (1962).

62Ia R. R. Irani and K. Moedritzer, *J. Phys. Chem.*, *66*, 1349 (1962).

62Ka H. C. A. King, E. Koros, and S. M. Nelson, *Nature (London)*, *196*, 573 (1962).

62Kb P. Kivalo and S. N. R. v. Schalien, *Suom. Kemistilehti B*, *35*, 7 (1962).

62Kc F. Ya. Kul'ba and Yu. A. Makashev, *Russ. J. Inorg. Chem. (English Trans.)*, *7*, 661 (1962). (Russian Page No. 1280.)

62La R. P. Lang, *J. Amer. Chem. Soc.*, *84*, 1185 (1962).

62Lb C. Lewis and D. A. Skoog, *J. Amer. Chem. Soc.*, *84*, 1101 (1962).

62Ma J. L. Mackey, J. E. Powell and F. H. Spedding, *J. Amer. Chem. Soc.*, *84*, 2047 (1962).

62Mb W. R. May and M. M. Jones, *J. Inorg. Nucl. Chem.*, *24*, 511 (1962).

62Mc J. D. McCullough and I. C. Zimmermann, *J. Phys. Chem.*, *66*, 1198 (1962).

62Md P. M. Milyukov, *Izv. Vyssh. Ucheb. Zaved., Khim. Khim. Tekhnol.*, *5*, 249 (1962).

62Me V. E. Mironov and V. A. Fedorov, *Russ. J. Inorg. Chem. (English Trans.)*, *7*, 1309 (1962). (Russian Page No. 2524.)

62Mf T. Moeller and R. Ferrus, *Inorg. Chem.*, *1*, 49 (1962).

62Mg T. Moeller and T. M. Hseu, *J. Inorg. Nucl. Chem.*, *24*, 1635 (1962).

62Mh T. Moeller and L. C. Thomson, *J. Inorg. Nucl. Chem.*, *24*, 499 (1962).

62Mi R. L. Montgomery, *U. S. Bur. Mines Rep. Invest.*, 6146 (1962).

62Na R. Näsänen, P. Meriläinen and M. Koskinen, *Suom. Kemistilehti B*, *35*, 59 (1962).

62Nb J. Nixon and R. A. Plane, *J. Amer. Chem. Soc.*, *84*, 4445 (1962).

62Pa A. D. Paul, *J. Phys. Chem.*, *66*, 1248 (1962).

62Pb D. D. Perrin, *J. Chem. Soc. (London)*, 4500 (1962).

62Sa K. Schlyter, *Trans. Roy. Inst. Technol., Stockholm*, No. 195 (1962).

62Sb K. Schlyter, *Trans. Roy. Inst. Technol., Stockholm*, No. 196 (1962).

62Sc A. Ya. Sychev and A. P. Gerbeleu, *Russ. J. Inorg. Chem. (English Trans.)*, *7*, 138 (1962). (Russian Page No. 269.)

1962 (continued)

62Ta M. Tamres and S. Searles, *J. Phys. Chem.*, *66*, 1099 (1962).

62Tb N. Tanaka, K. Ebata and T. Murayama, *Bull. Chem. Soc. Japan*, *35*, 124 (1962).

62Tc O. N. Temkin, R. M. Flid and A. I. Malakhov, *Kinet Katal. (English Trans.)*, *3*, 799 (1962). (Russian Page No. 915.)

62Va V. P. Vasil'ev, *Russ. J. Inorg. Chem. (English Trans.)*, *7*, 283 (1962). (Russian Page No. 555.)

62Wa M. J. M. Woods, P. K. Gallagher and E. L. King, *Inorg. Chem.*, *1*, 55 (1962).

62Ya K. B. Yatsimirskii and G. A. Prik, *Russ. J. Inorg. Chem. (English Trans.)*, *7*, 30 (1962). (Russian Page No. 62.)

1961

61Aa L. J. Andrews and R. M. Keefer, in *Advances in Inorganic Chemistry and Radiochemistry* (H. J. Emeleus and A. G. Sharpe, Eds.), Vol. 3, Academic Press, New York, N. Y., 1961.

61Ab N. M. Atherton and S. I. Weissman, *J. Amer. Chem. Soc.*, *83*, 1330 (1961).

61Ba D. Bunn, F. S. Dainton, and S. Duckworth, *Trans. Faraday Soc.*, *57*, 1131 (1961).

61Ca G. R. Choppin and J. A. Chopoorian, *J. Inorg. Nucl. Chem.*, *22*, 97 (1961).

61Cb M. Ciampolini, P. Paoletti and L. Sacconi, *J. Chem. Soc. (London)*, 2994 (1961).

61Cc R. E. Connick and A. D. Paul, *J. Phys. Chem.*, *65*, 1216 (1961).

61Cd R. E. Connick and A. D. Paul, *J. Phys. Chem.*, *65*, 1216 (1961).

61Da P. A. D. De Maine and P. Carapellucci, *J. Mol. Spectrosc.*, *7*, 83 (1961).

61Db R. S. Drago, R. L. Carlson, N. J. Rose and D. A. Wenz, *J. Amer. Chem. Soc.*, *83*, 3572 (1961).

61Dc F. R. Duke and H. M. Garfinkel, *J. Phys. Chem.*, *65*, 461 (1961).

61Ga M. Good, A. Major, J. Nag-Chaudhuri and S. P. McGlynn, *J. Amer. Chem. Soc.*, *83*, 4329 (1961).

61Ia R. R. Irani and C. F. Callis, *J. Phys. Chem.*, *65*, 296 (1961).

61Ib R. M. Izatt, J. W. Wrathall and K. P. Anderson, *J. Phys. Chem.*, *65*, 1914 (1961).

61Ja G. Junghähnel, *Z. Chem.*, *1*, 91 (1961).

61Ka L. I. Katzin, *J. Chem. Phys.*, *35*, 467 (1961).

61Kb G. A. Krestov and K. B. Yatsimirskii, *Russ. J. Inorg. Chem. (English Trans.)*, *6*, 1170 (1961). (Russian Page No. 2304.)

61La S. C. Levy and J. E. Powell, *U. S. At. Energy Comm.*, IS-421 (1961).

61Ma G. N. Malcolm, H. N. Parton and I. D. Watson, *J. Phys. Chem.*, *65*, 1900 (1961).

61Mb A. McAuley and G. H. Nancollas, *J. Chem. Soc. (London)*, 2215 (1961).

1961 (continued)

61Mc A. McAuley and G. H. Nancollas, *J. Chem. Soc. (London)*, 4458 (1961).

61Md J. D. McCullough and I. C. Zimmermann, *J. Phys. Chem.*, *65*, 888 (1961).

61Me P. M. Milyukov, *Izv. Vyssh. Ucheb. Zaved., Khim. Khim. Tekhnol.*, *4*, 212 (1961).

61Mf T. Moeller and R. Ferrus, *J. Inorg. Nucl. Chem.*, *20*, 261 (1961).

61Mg P. P. Mohapatra, R. C. Das and S. Aditya, *J. Indian Chem. Soc.*, *38*, 845 (1961).

61Na V. S. K. Nair and G. H. Nancollas, *J. Chem. Soc. (London)*, 4367 (1961).

61Nb W. C. Nicholas and W. C. Fernelius, *J. Phys. Chem.*, *65*, 1047 (1961).

61Pa J. M. Pagano, D. E. Goldberg and W. C. Fernelius, *J. Phys. Chem.*, *65*, 1062 (1961).

61Pb A. D. Paul, L. S. Gallo and J. B. Van Camp, *J. Phys. Chem.*, *65*, 441 (1961).

61Ra S. W. Rabideau and B. J. Masters, *J. Phys. Chem.*, *65*, 1256 (1961).

61Rb R. E. Reichard and W. C. Fernelius, *J. Phys. Chem.*, *65*, 380 (1961).

61Rc D. K. Roe, D. B. Masson and C. J. Nyman, *Anal. Chem.*, *33*, 1464 (1961).

61Rd E. M. Ryzhkov and A. M. Sukhotin, *Russ. J. Phys. Chem. (English Trans.)*, *35*, 648 (1961). (Russian Page No. 1327.)

61Sa L. Sacconi, P. Paoletti and M. Ciampolini, *J. Chem. Soc. (London)*, 5115 (1961).

61Sb K. Schiller and E. Thilo, *Z. Anorg. Allg. Chem.*, *310*, 261 (1961).

61Sc K. Schlyter, *Trans. Roy. Inst. Technol., Stockholm*, No. 182 (1961).

61Sd S. A. Shchukarev and O. A. Lobaneva, *Russ. J. Inorg. Chem. (English Trans.)*, *6*, 410 (1961). (Russian Page No. 804.)

61Se N. Sutin, J. K. Rowley and R. W. Dodson, *J. Phys. Chem.*, *65*, 1248 (1961).

61Ta H. Tsubomura and R. P. Lang, *J. Amer. Chem. Soc.*, *83*, 2085 (1961).

61Wa R. M. Wallace and E. K. Dukes, *J. Phys. Chem.*, *65*, 2094 (1961).

61Ya R. G. Yalman, *J. Amer. Chem. Soc.*, *83*, 4142 (1961). 4142 (1961).

1960

60Aa G. Aksnes and T. Gramstad, *Acta Chem. Scand.*, *14*, 1485 (1960).

60Bb J. A. Bolzan, *Rev. Fac. Cienc. Quim., Univ. Nac. La Plata*, *33*, 67 (1960).

60Bc M. Brandon, M. Tamres and S. Searles, *J. Amer. Chem. Soc.*, *82*, 2129 (1960).

60Ca M. Ciampolini, P. Paoletti and L. Sacconi, *J. Chem. Soc. (London)*, 4553 (1960).

60Cb M. Ciampolini, P. Paoletti and L. Sacconi, *Nature (London)*, *186*, 880 (1960).

60Cc M. Ciampolini, P. Paoletti and L. Sacconi, *Ric. Sci.*, *30*, 2148 (1960).

60Da G. Daniele, *Gazz. Chim., Ital.*, *90*, 1082 (1960).

1960 (continued)

60Db G. Daniele, *Gazz. Chim. Ital.*, *90*, 1585 (1960).

60Dc F. A. D. De Maine and J. Peone, *J. Mol. Spectrosc.*, *4*, 262 (1960).

60Ea J. H. Espenson and E. L. King, *J. Phys. Chem.*, *64*, 380 (1960).

60Ga P. K. Gallagher and E. L. King, *J. Amer. Chem. Soc.*, *82*, 3510 (1960).

60Gb P. George, D. H. Irvine and S. C. Glauser, *Ann. N. Y. Acad. Sci.*, *88*, 393 (1960).

60Ia R. R. Irani and C. F. Callis, *J. Phys. Chem.*, *64*, 1398 (1960).

60Ja J. Jordan, J. Meier, E. J. Billingham and J. Pendergrast, *Nature (London)*, *187*, 318 (1960).

60Ka T. Kuge and S. Ono, *Bull. Chem. Soc. Jap.*, *33*, 1273 (1960).

60Kb T. Kumai, *Nippon Kagaku Zasshi*, *81*, 1687 (1960).

60La D. L. Leussing and P. K. Gallagher, *J. Phys. Chem.*, *64*, 1631 (1960).

60Lb D. L. Leussing, R. E. Laramy and G. S. Alberts, *J. Amer. Chem. Soc.*, *82*, 4826 (1960).

60Lc M. H. Lietzke and R. W. Stoughton, *J. Phys. Chem.*, *64*, 816 (1960).

60Ld M. W. Lister and P. Rosenblum, *Can. J. Chem.*, *38*, 1827 (1960).

60Ma J. D. McCullough and I. C. Zimmermann, *J. Phys. Chem.*, *64*, 1084 (1960).

60Mb R. L. McCullough, L. H. Jones and R. A. Penneman, *J. Inorg. Nucl. Chem.*, *13*, 286 (1960).

60Na K. Naito, *Bull. Chem. Soc. Jap.*, *33*, 363 (1960).

60Pa P. Paoletti, M. Ciampolini and L. Sacconi, *Ric. Sci.*, *30*, 1791 (1960).

60Pb S. Pelletier, *J. Chim. Phys. Physicochim. Biol.*, *57*, 295 (1960).

60Pc S. Pelletier, *J. Chim. Phys. Physicochim. Biol.*, *57*, 301 (1960).

60Pd S. Pelletier, *J. Chim. Phys. Physicochim. Biol.*, *57*, 311 (1960).

60Pe S. Pelletier, *J. Chim. Phys. Physicochim. Biol.*, *57*, 318 (1960).

60Pf A. J. Poe and M. S. Vaidya, *J. Chem. Soc. (London)*, 3431 (1960).

60Ra S. W. Rabideau and R. J. Kline, *J. Phys. Chem.*, *64*, 680 (1960).

60Rb D. H. Richards and K. W. Sykes, *J. Chem. Soc. (London)*, 3626 (1960).

60Sa L. Sacconi and G. Lombardo, *J. Amer. Chem. Soc.*, *82*, 6266 (1960).

60Sb L. Sacconi, G. Lombardo and R. Ciofalo, *J. Amer. Chem. Soc.*, *82*, 4182 (1960).

60Sc L. Sacconi, P. Paoletti and M. Ciampolini, *Ric. Sci.*, *30*, 1610 (1960).

60Sd L. Sacconi, G. Lombardo and P. Paoletti, *J. Amer. Chem. Soc.*, *82*, 4185 (1960).

60Se S. N. R. v. Schalien, *Suom. Kemistilehti B*, *33*, 5 (1960).

60Sf C. D. Schmulbach and R. S. Drago, *J. Amer. Chem. Soc.*, *82*, 4484 (1960).

60Sg R. W. Stoughton and M. H. Lietzke, *J. Phys. Chem.*, *64*, 133 (1960).

1960 (continued)

60Ta M. Tamres and M. Brandon, *J. Amer. Chem. Soc.*, *82*, 2134 (1960).

60Tb V. F. Toropova and L. S. Kirillova, *Russ. J. Inorg. Chem.* (*English Trans.*), *5*, 276 (1960). (Russian Page No. 575.)

60Tc H. Tsubomura, *J. Amer. Chem. Soc.*, *82*, 40 (1960).

60Td H. Tsubomura and J. M. Kliegman, *J. Amer. Chem. Soc.*, *82*, 1314 (1960).

60Wa H. R. Weimer and W. C. Fernelius, *J. Phys. Chem.*, *64*, 1951 (1960).

60Ya H. Yada, J. Tanaka and S. Nagakura, *Bull. Chem. Soc. Japan*, *33*, 1660 (1960).

1959

59Aa J. M. Austin, R. A. Matheson and H. N. Parton, in *The Structure of Electrolytic Solutions* (W. J. Hamer, Ed.), J. Wiley & Sons, Inc., New York, N. Y., 1959. Chapter 24, p. 365-379.

59Ba R. H. Betts and O. F. Dahlinger, *Can. J. Chem.*, *37*, 91 (1959).

59Bb P. Brandt, *Acta Chem. Scand.*, *13*, 1639 (1959).

59Bc K. Burton, *Biochem. J.*, *71*, 388 (1959).

59Ca H. Coll, R. V. Nauman and P. W. West, *J. Amer. Chem. Soc.*, *81*, 1284 (1959).

59Cb R. E. Connick and C. P. Coppel, *J. Amer. Chem. Soc.*, *81*, 6389 (1959).

59Da S. P. Datta and A. K. Grzybowski, *J. Chem. Soc.* (*London*), 1091 (1959).

59Fa D. Fleischer and H. Freiser, *J. Phys. Chem.*, *63*, 260 (1959).

59Ga P. George, G. I. H. Hanania, and D. H. Irvine, *J. Chem. Soc.* (*London*), 2548 (1959).

59Gb D. E. Goldberg and W. C. Fernelius, *J. Phys. Chem.*, *63*, 1246 (1959).

59Gc D. E. Goldberg and W. C. Fernelius, *J. Phys. Chem.*, *63*, 1328 (1959).

59Gd A. M. Golub and Yu. V. Kosmatyi, *Russ. J. Inorg. Chem.* (*English Trans.*), *4*, 606 (1959). (Russian Page No. 1347.)

59Ge R. L. Gustafson and A. E. Martell, *J. Amer. Chem. Soc.*, *81*, 525 (1959).

59Ha T. R. Harkins and H. Freiser, *J. Phys. Chem.*, *63*, 309 (1959).

59Ka E. L. King and P. K. Gallagher, *J. Phys. Chem.*, *63*, 1073 (1959).

59Kb J. W. Kury, A. D. Paul, L. G. Hepler and R. E. Connick, *J. Amer. Chem. Soc.*, *81*, 4185 (1959).

59La T. J. Lane and J. M. Daly, *J. Amer. Chem. Soc.*, *81*, 2953 (1959).

59Lb J. R. Lotz, B. P. Block and W. C. Fernelius, *J. Phys. Chem.*, *63*, 541 (1959).

59Lc E. Lange, in *The Structure of Electrolyte Solutions* (W. J. Hamer, Ed.), J. Wiley & Sons, Inc. New York, N. Y., 1959.

59Ma G. H. McIntyre, B. P. Block and W. C. Fernelius, *J. Amer. Chem. Soc.*, *81*, 529 (1959).

1959 (continued)

59Na V. S. K. Nair and G. H. Nancollas, *J. Chem. Soc. (London)*, 3934 (1959).

59Pa A. J. Poe and M. S. Vaidya, *Nature, London*, *184*, 1139 (1959).

59Sa S. N. R. v. Schalien, *Suom. Kemistilehti B*, *32*, 148 (1959).

59Sb R. L. Seth and A. K. Day, *Z. Phys. Chem. (Leipzig)*, *210*, 108 (1959).

59Ta J. G. Traynham and J. R. Olechowski, *J. Amer. Chem. Soc.*, *81*, 571 (1959).

59Va V. P. Vasil'ev, *Termodin. Str. Rastvorov, Tr. Soveshch. M.*, 140 (1958), (Pub. 1959).

59Ya S. Yamashita, *Bull. Chem. Soc. Japan*, *32*, 1212 (1959).

59Yb K. B. Yatsimirskii and G. A. Karacheva, *Russ. J. Inorg. Chem. (English Trans.)*, *4*, 127 (1959). (Russian Page No. 294.)

59Za A. J. Zielen, *J. Amer. Chem. Soc.*, *81*, 5022 (1959).

1958

58Aa G. Anderegg, G. Schwarzenbach, M. Padmoyo and O. F. Borg, *Helv. Chim. Acta*, *41*, 988 (1958).

58Ba C. R. Bertsch, W. C. Fernelius and B. P. Block, *J. Phys. Chem.*, *62*, 444 (1958).

58Bb C. R. Bertsch, W. C. Fernelius and B. P. Block, *J. Phys. Chem.*, *62*, 503 (1958).

58Bc R. E. Buckles and W. D. Womer, *J. Amer. Chem. Soc.*, *80*, 5055 (1958).

58Ca S. Cabani and E. Scrocco, *Ann. Chim. (Rome)*, *48*, 99 (1958).

58Cb S. Cabani and E. Scrocco, *J. Inorg. Nucl. Chem.*, *8*, 332 (1958).

58Cc S. Cabani and E. Scrocco, *Ann. Chim. (Rome)*, *48*, 85 (1958).

58Cd R. E. Connick and A. D. Paul, *J. Amer. Chem. Soc.*, *80*, 2069 (1958).

58Ea T. S. Elleman, J. W. Reishus and D. S. Martin, *J. Amer. Chem. Soc.*, *80*, 536 (1958).

58Fa B. F. Freasier, A. G. Oberg and W. W. Wendlandt, *J. Phys. Chem.*, *62*, 700 (1958).

58Ga H. S. Gates and E. L. King, *J. Amer. Chem. Soc.*, *80*, 5011 (1958).

58Gb A. D. Gelman, N. N. Matorina and A. I. Moskvin, *Proc. Acad. Sci. U.S.S.R., Chem. Sect. (English Trans.)*, *117*, 973 (1958). (Russian Page No. 88.)

58Gc A. M. Golub and G. D. Ivanchenko, *Russ. J. Inorg. Chem. (English Trans.)*, *3*, No. 2, 130 (1958). (Russian Page No. 333.)

58Ja H. B. Jonassen and J. R. Oliver, *J. Amer. Chem. Soc.*, *80*, 2347 (1958).

58Jb H. B. Jonassen, A. Schaafsma and L. Westerman, *J. Phys. Chem.*, *62*, 1022 (1958).

58Ma P. E. Martin and A. G. White, *J. Chem. Soc. (London)*, 2490 (1958).

58Na S. Nagakura, *J. Amer. Chem. Soc.*, *80*, 520 (1958).

1958 (continued)

58Nb V. S. K. Nair and G. H. Nancollas, *J. Chem. Soc. (London)*, 3706 (1958).

58Pa M. H. Panckhurst and K. G. Woolmington, *Proc. Roy. Soc., Ser. A*, *244*, 124 (1958).

58Sa L. Sacconi, G. Lombardo and P. Paoletti, *J. Chem. Soc. (London)*, 848 (1958).

58Sb D. B. Scaife and H. J. V. Tyrrell, *J. Chem. Soc. (London)*, 386 (1958).

58Sc D. B. Scaife and H. J. V. Tyrrell, *J. Chem. Soc. (London)*, 392 (1958).

58Sd K. Schug and E. L. King, *J. Amer. Chem. Soc.*, *80*, 1089 (1958).

58Se L. A. K. Staveley and T. Randall, *Discuss. Faraday Soc.*, *26*, 157 (1958).

58Ya K. B. Yatsimirskii and G. A. Karacheva, *Izv. Vyssh. Ucheb. Zaved., Khim. Khim. Tekhnol.*, *1*, 13 (1958).

1957

57Ba W. D. Bale, E. W. Davies, D. B. Morgans and C. B. Monk, *Discuss. Faraday Soc.*, *24*, 94 (1957).

57Bb J. H. Baxendale and C. F. Wells, *Trans. Faraday Soc.*, *53*, 800 (1957).

57Ca H. Chateau, B. Hervier, and J. Pouradier, *J. Chimie. Physique*, *54*, 246 (1957).

57Cb H. Chateau, B. Hervier and J. Pouradier, *J. Phys. Chem.*, *61*, 250 (1957).

57Da P. A. D. De Maine, *J. Chem. Phys.*, *26*, 1036 (1957).

57Db P. A. D. De Maine, *J. Chem. Phys.*, *26*, 1042 (1957).

57Dc P. A. D. De Maine, *J. Chem. Phys.*, *26*, 1192 (1957).

57Dd P. A. D. De Maine, *J. Chem. Phys.*, *26*, 1199 (1957).

57De H. A. Droll, B. P. Block and W. C. Fernelius, *J. Phys. Chem.*, *61*, 1000 (1957).

57Ga A. D. Gelman, N. N. Matorina and A. I. Moskvin, *Akademiia Nauk S.S.S.R., Leningrad Comptes Rendus (Doklady)*, *117*, 88 (1957). (English 973.)

57Gb A. M. Golub and O. G. Bilyk, *Russ. J. Inorg. Chem. (English Trans.)*, *2*, 58 (1957). (Russian Page No. 2723).

57Gc A. M. Golub and V. M. Samoilenko, *Ukr. Khim. Zh.*, *23*, 17 (1957).

57Ha J. A. Hearne and A. G. White, *J. Chem. Soc. (London)*, 2168 (1957).

57Hb G. K. Helmkamp, F. L. Carter and H. J. Lucas, *J. Amer. Chem. Soc.*, *79*, 1306 (1957).

57Ja H. B. Johassen, J. A. Bertrand, F. R. Groves, Jr. and R. I. Stearns, *J. Amer. Chem. Soc.*, *79*, 4279 (1957).

57Jb H. B. Jonassen and L. Westerman, *J. Phys. Chem.*, *61*, 1006 (1957).

57Jc J. Jordan and T. G. Alleman, *Anal. Chem.*, *29*, 9 (1957).

57Ka T. Kubota, *Nippon Kagaku Zasshi*, *78*, 196 (1957).

1957 (continued)

57La M. S. Lewis and H. A. Saroff, *J. Amer. Chem. Soc.*, *79*, 2112 (1957).

57Ma R. M. Milburn, *J. Amer. Chem. Soc.*, *79*, 537 (1957).

57Mb C. B. Murphy and A. E. Martell, *J. Biol. Chem.*, *226*, 37 (1957).

57Na V. S. K. Nair and G. H. Nancollas, *J. Chem. Soc. (London)*, 318 (1957).

57Nb T. W. Nakagawa, L. J. Andrews and R. M. Keefer, *J. Amer. Chem. Soc.*, *61*, 1007 (1957).

57Nc L. B. Nanninga, *J. Phys. Chem.*, *61*, 1144 (1957).

57Nd Y. Nozaki, F. R. Gurd, R. F. Chen and J. T. Edsall, *J. Amer. Chem. Soc.*, *79*, 2123 (1957).

57Pa S. Pelletier, *C. R. H. Acad. Sci.*, *245*, 160 (1957).

57Pb A. L. Phipps and R. A. Plane, *J. Amer. Chem. Soc.*, *79*, 2458 (1957).

57Sa H. Schmied and R. W. Fink, *J. Chem. Phys.*, *27*, 1034 (1957).

57Sb R. L. Seth and A. K. Dey, *Proc. Nat. Acad. Sci., India, Sect. A*, *26*, 312 (1957).

57Sc S. A. Shchukarev, L. S. Lilich and V. A. Latysheva, *Uch. Zap. Leningrad. Gos. Univ., Ser. Khim. Nauk*, *15*, 17 (1957).

57Ta T. Takahashi and K. Sasaki, *Denki Kagaku*, *25*, 118 (1957).

57Ua E. Uusitalo, *Ann. Acad. Sci. Fenn., Ser. A*, *2*, No. 87 (1957).

57Va V. P. Vasil'ev, *Zh. Fiz. Khim.*, *31*, 692 (1957).

57Wa S. Wilska, *Suomen Kemistilehti*, *B*, *30*, 185 (1957).

57Ya K. B. Yatsimirskii and L. V. Gus'kova, *Russ. J. Inorg. Chem. (English Trans.)*, *2*, No. 9, 91 (1957). (Russian Page No. 2039.)

57Yb K. B. Yatsimirskii and P. M. Milyukov, *Russ. J. Inorg. Chem. (English Trans.)*, *2*, No. 5, 98 (1957). (Russian Page No. 1046.)

57Yc K. B. Yatsimirskii and P. M. Milyukov, *Zh. Fiz. Khim.*, *31*, 842 (1957).

1956

56Aa A. Agren, *Sv. Kem. Tidskr.*, *68*, 181 (1956).

56Ab Y. Amako, *Sci. Rep. Tohoku Univ., Ser. 1*, *40*, 147 (1956).

56Ba H. C. Brown and D. Gintis, *J. Amer. Chem. Soc.*, *78*, 5378 (1956).

56Bb H. C. Brown, D. Gintis and Harold Podall, *J. Amer. Chem. Soc.*, *78*, 5375 (1956).

56Bc H. C. Brown and R. R. Holmes, *J. Amer. Chem. Soc.*, *78*, 2173 (1956).

56Ca R. A. Care and L. A. K. Staveley, *J. Chem. Soc. (London)*, 4571 (1956).

56Cb J. Chatt and R. G. Wilkins, *J. Chem. Soc., (London)*, 525 (1956).

56Cc H. B. Clarke and S. P. Datta, *Biochem. J.*, *64*, 604 (1956).

56Cd R. E. Connick, L. G. Hepler, Z. Z. Hugus, J. W. Kury, W. M. Latimer and M. A. Tsao, *J. Amer. Chem. Soc.*, *78*, 1827 (1956).

1956 (continued)

56Da P. A. D. De Maine, *J. Chem. Phys.*, *24*, 1091 (1956).

56Fa S. N. Flengas and E. Rideal, *Proc. Roy. Soc., Ser. A.*, *233*, 443 (1956).

56Ga E. Gelles and G. H. Nancollas, *Trans. Faraday Soc.*, *52*, 680 (1956).

56Gb A. M. Golub, *J. Gen. Chem., U.S.S.R. (English Trans.)*, *26*, 2049 (1956). (Russian Page No. 1837.)

56Ha G. B. Hares, W. C. Fernelius and B. E. Douglas, *J. Amer. Chem. Soc.*, *78*, 1816 (1956).

56Hb V. L. Hughes and A. E. Martell, *J. Amer. Chem. Soc.*, *78*, 1319 (1956).

56Ka E. V. Kiseleva and S. M. Khodeeva, *Tr. Mosk. Khim. Tekhnol. Inst.*, *22*, 89 (1956).

56Kb E. Klein, *Z. Elektrochem.*, *60*, 1003 (1956).

56La G. S. Lawrence, *Trans. Faraday Soc.*, *52*, 236 (1956).

56Lb N. C. Li, J. M. White and R. L. Yoest, *J. Amer. Chem. Soc.*, *78*, 5218 (1956).

56Ma A. E. Martell, *Rec. Trav. Chim. Pays-Bas*, *75*, 781 (1956).

56Na G. H. Nancollas, *J. Chem. Soc. (London)*, 744 (1956).

56Oa N. Ogimachi, L. J. Andrews and R. M. Keefer, *J. Amer. Chem. Soc.*, *78*, 2210 (1956).

56Pa F. A. Posey and H. Taube, *J. Amer. Chem. Soc.*, *78*, 15 (1956).

56Ra J. J. Renier and D. S. Martin, *J. Amer. Chem. Soc.*, *78*, 1833 (1956).

56Sa S. A. Shchukarev, L. S. Lilich and V. A. Latysheva, *Russ. J. Inorg. Chem. (English Trans.)*, *1*, No. 2, 36 (1956). (Russian Page No. 225.)

56Sb R. M. Smith and R. A. Alberty, *J. Amer. Chem. Soc.*, *78*, 2376 (1956).

56Sc R. M. Smith and R. A. Alberty, *J. Phys. Chem.*, *60*, 180 (1956).

56Sd L. H. Sutcliffe and J. R. Weber, *Trans. Faraday Soc.*, *52*, 1225 (1956).

56Ta V. F. Toropova, *Zh. Neorg. Khim.*, *1*, 243 (1956).

56Tb V. F. Toropova, *Zh. Neorg. Khim.*, *1*, 930 (1956).

56Tc J. G. Traynham and M. F. Sehnert, *J. Amer. Chem. Soc.*, *78*, 4024 (1956).

56Td R. Trujillo and F. Brito, *Real Soc. Espan. Fis. Quim., Ser. B*, *52*, 4024 (1956).

56Ya K. B. Yatsimirskii and V. P. Vasil'ev, *Zh. Fiz. Khim.*, *30*, 901 (1956).

1955

55Ba R. H. Betts, *Can. J. Chem.*, *33*, 1775 (1955).

55Bb H. C. Brown and R. H. Horowitz, *J. Amer. Chem. Soc.*, *77*, 1730 (1955).

55Ca H. Chateau and J. Pouradier, *Comptes Rendus (France)*, *240*, 1882 (1955).

55Cb D. Cohen, J. C. Sullivan and J. C. Hindman, *J. Amer. Chem. Soc.*, *77*, 4964 (1955).

1955 (continued)

55Cc F. A. Cotton and F. E. Harris, *J. Phys. Chem.*, *59*, 1203 (1955).

55Da R. A. Day, R. N. Wilhoit and F. D. Hamilton, *J. Amer. Chem. Soc.*, *77*, 3180 (1955).

55Fa W. S. Fyfe, *J. Chem. Soc.* (*London*), 1347 (1955).

55Ga P. George and G. I. H. Hanania, *Nature* (*London*), *175*, 1034 (1955).

55Gb F. G. R. Gimblett and C. B. Monk, *Trans. Faraday Soc.*, *51*, 793 (1955).

55Gc L. F. Grantham, T. S. Elleman and D. S. Martin, *J. Amer. Chem. Soc.*, *77*, 2965 (1955).

55Ha L. J. Heidt and J. Berestecki, *J. Amer. Chem. Soc.*, *77*, 2049 (1955).

55Ia R. M. Izatt, W. C. Fernelius and B. P. Block, *J. Phys. Chem.*, *59*, 235 (1955).

55Ka L. I. Katzin and E. Gebert, *J. Amer. Chem. Soc.*, *77*, 5814 (1955).

55Kb R. M. Keefer and L. J. Andrews, *J. Amer. Chem. Soc.*, *77*, 2164 (1955).

55Kc K. A. Kraus and F. Nelson, *J. Amer. Chem. Soc.*, *77*, 3721 (1955).

55La M. W. Lister and D. E. Rivington, *Can. J. Chem.*, *33*, 1572 (1955).

55Lb E. M. Loebl, L. B. Luttinger and H. P. Gregor, *J. Phys. Chem.*, *59*, 559 (1955).

55Ma B. L. Mickel and A. C. Andrews, *J. Amer. Chem. Soc.*, *77*, 5291 (1955).

55Mb L. N. Mulay and P. W. Selwood, *J. Amer. Chem. Soc.*, *77*, 2693 (1955).

55Na G. H. Nancollas, *J. Chem. Soc.* (*London*), 1458 (1955).

55Oa N. Ogimachi, L. J. Andrews and R. M. Keefer, *J. Amer. Chem. Soc.*, *77*, 4202 (1955).

55Pa C. Postmus and E. L. King, *J. Phys. Chem.*, *59*, 1208 (1955).

55Pb I. Poulsen and J. Bjerrum, *Acta Chem. Scand.*, *9*, 1407 (1955).

55Ra S. W. Rabideau and H. D. Cowan, *J. Amer. Chem. Soc.*, *77*, 6145 (1955).

55Sa W. Scheler, A. Salewski and F. Jung, *Biochem. Z.*, *326*, 288 (1955).

1954

54Ba F. Basolo, Y. T. Chen and R. K. Murmann, *J. Amer. Chem. Soc.*, *76*, 956 (1954).

54Bb F. Basolo and R. K. Murmann, *J. Amer. Chem. Soc.*, *76*, 211 (1954).

54Ca F. F. Carini and A. E. Martell, *J. Amer. Chem. Soc.*, *76*, 2153 (1954).

54Cb R. G. Charles, *J. Amer. Chem. Soc.*, *76*, 5854 (1954).

54Cc R. G. Charles and H. Frieser, *Anal. Chim. Acta*, *11*, 101 (1954).

54Cd H. B. Clarke, D. C. Cusworth and S. P. Datta, *Biochem. J.*, *58*, 146 (1954).

54Ce R. E. Connick and M. S. Tsao, *J. Amer. Chem. Soc.*, *76*, 5311 (1954).

54Da T. Davies, S. S. Singer and L. A. K. Staveley, *J. Chem. Soc.* (*London*), 2304 (1954).

1954 (continued)

54Db R. A. Day and R. M. Powers, *J. Amer. Chem. Soc.*, *76*, 3895 (1954).

54Ea J. T. Edsall, G. Felsenfeld, D. S. Goodman and F. R. N. Gurd, *J. Amer. Chem. Soc.*, *76*, 3054 (1954).

54Ga F. G. R. Gimblett and C. B. Monk, *Trans. Faraday Soc.*, *50*, 965 (1954).

54Gb E. Gonick, W. C. Fernelius and B. E. Douglas, *J. Amer. Chem. Soc.*, *76*, 4671 (1954).

54Ha L. G. Hepler, J. W. Kury and Z. Z. Hugus, *J. Phys. Chem.*, *58*, 26 (1954).

54Ja W. D. Johnston and H. Freiser, *Anal. Chim. Acta*, *11*, 201 (1954).

54Ka J. A. A. Ketelaar, *J. Phys. Radium*, *15*, 197 (1954).

54Kb M. Kilpatrick and L. Pokras, *J. Electrochem. Soc.*, *101*, 39 (1954).

54La N. C. Li, J. M. White and E. Doody, *J. Amer. Chem. Soc.*, *76*, 6219 (1954).

54Ma L. N. Mulay and P. W. Selwood, *J. Amer. Chem. Soc.*, *76*, 6207 (1954).

54Pa F. M. Page, *Trans. Faraday Soc.*, *50*, 120 (1954).

54Pb K. G. Poulsen, J. Bjerrum and I. Poulsen, *Acta Chem. Scand.*, *8*, 921 (1954).

54Pc J. Pouradier, A. M. Venet and H. Chateau, *J. Chim. Phys. Physicochim. Biol.*, *51*, 375 (1954).

54Ra C. Reid and R. S. Mulliken, *J. Amer. Chem. Soc.*, *76*, 3869 (1954).

54Sa J. C. Sullivan and J. C. Hindman, *J. Amer. Chem. Soc.*, *76*, 5931 (1954).

54Wa R. J. P. Williams, *J. Phys. Chem.*, *58*, 121 (1954).

54Ya K. B. Yatsimirskii and A. A. Shutov, *Zh. Fiz. Khim.*, *28*, 30 (1954).

54Yb K. B. Yatsimirskii and E. K. Zolotarev, *Zh. Fiz. Khim.*, *28*, 1292 (1954).

1953

53Ba R. P. Bell and J. H. B. George, *Trans. Faraday Soc.*, *49*, 619 (1953).

53Bb R. H. Betts and F. S. Dainton, *J. Amer. Chem. Soc.*, *75*, 5721 (1953).

53Ea M. G. Evans and G. H. Nancollas, *Trans. Faraday Soc.*, *49*, 363 (1953).

53Fa H. Freiser, *Rec. Chem. Progr.*, *14*, 199 (1953).

53Ga D. M. Gruen and J. J. Katz, *J. Amer. Chem. Soc.*, *75*, 3772 (1953).

53Ha J. Hudis and A. 1. Wahl, *J. Amer. Chem. Soc.*, *75*, 4153 (1953).

53Ka E. L. King, *J. Chem. Educ.*, *30*, 71 (1953).

53Kb I. M. Klotz and W. C. Loh Ming, *J. Amer. Chem. Soc.*, *75*, 4159 (1953).

53La W. M. Latimer and W. L. Jolly, *J. Amer. Chem. Soc.*, *75*, 1548 (1953).

53Lb M. W. Lister, *Can J. Chem.*, *31*, 638 (1953).

53Na A. C. Nanda and S. Pani, *Curr. Sci.*, *22*, 300 (1953).

53Nb T. W. Newton and G. M. Arcand, *J. Amer. Chem. Soc.*, *75*, 2449 (1953).

1953 (continued)

63Sa C. G. Spike and R. W. Parry, *J. Amer. Chem. Soc.*, *75*, 2726 (1953).

53Sb C. G. Spike and R. W. Parry, *J. Amer. Chem. Soc.*, *75*, 3770 (1953).

53Ta C. Tanford and M. L. Wagner, *J. Amer. Chem. Soc.*, *75*, 434 (1953).

53Tb H. Taube and F. A. Posey, *J. Amer. Chem. Soc.*, *75*, 1463 (1953).

53Va C. E. Vanderzee and H. J. Dawson, *J. Amer. Chem. Soc.*, *75*, 5659 (1953).

53Ya R. G. Yalman and A. B. Lamb, *J. Amer. Chem. Soc.*, *75*, 1521 (1953).

1952

52Ba F. Basolo and R. K. Murmann, *J. Amer. Chem. Soc.*, *74*, 5243 (1952).

52Ca G. H. Cartledge, *J. Amer. Chem. Soc.*, *74*, 6015 (1952).

52Cb J. Chatt and R. G. Wilkins, *J. Chem. Soc.*, *London*, 4300 (1952).

52Da M. Davies and E. Gwynne, *J. Amer. Chem. Soc.*, *74*, 2748 (1952).

52Ea J. I. Evans and C. B. Monk, *Trans. Faraday Soc.*, *48*, 934 (1952).

52Fa W. S. Fyfe, *J. Chem. Soc.* (*London*), 2023 (1952).

52Ga P. George and C. L. Tsou, *Biochem. J.*, *50*, 440 (1952).

52Ja H. B. Jonassen, G. G. Hurst, R. B. LeBlanc and A. W. Meibohn, *J. Phys. Chem.*, *56*, 16 (1952).

52Jb H. W. Jones and C. B. Monk, *Trans. Faraday Soc.*, *48*, 929 (1952).

52Jc J. H. Jonte and D. S. Martin, *J. Amer. Chem. Soc.*, *74*, 2052 (1952).

52Ka J. A. A. Ketelaar, C. Van De Stolpe, A. Gousmit and W. Dzcubas, *Rec. Trav. Chim. Pays-Bas*, *71*, 1104 (1952).

52Ma J. R. Miller and G. D. Dorough, *J. Amer. Chem. Soc.*, *74*, 3977 (1952).

52Ua N. Uri, *Chem. Rev.*, *50*, 375 (1952).

52Va C. E. Vanderzee, *J. Amer. Chem. Soc.*, *74*, 4806 (1952).

52Vb C. E. Vanderzee and D. E. Rhodes, *J. Amer. Chem. Soc.*, *74*, 3552 (1952).

52Ya K. B. Yatsimirskii and A. A. Astasheva, *Zh. Fiz. Khim.*, *26*, 239 (1952).

1951

51Aa A. D. Awtrey and R. E. Connick, *J. Amer. Chem. Soc.*, *73*, 1842 (1951).

51Fa S. C. Furman and C. S. Garner, *J. Amer. Chem. Soc.*, *73*, 4528 (1951).

51Ha T. J. Hardwick and E. Robertson, *Can. J. Chem.*, *29*, 818 (1951).

51Ja E. L. Jahn and H. Staude, *Z. Naturforsch. A*, *6*, 385 (1951).

51Jb J. C. James, *J. Chem. Soc.* (*London*), 153 (1951).

51Jc H. B. Jonassen and A. W. Meibohn, *J. Phys. Colloid Chem.*, *55*, 726 (1951).

51Ka J. A. A. Ketelaar, C. Van De Stolpe and H. R. Gersmann, *Rec. Trav. Chim. Pays-Bas*, *70*, 499 (1951).

1950

50Ba J. H. Baxendale and P. George, *Trans. Faraday Soc.*, *46*, 55 (1950).

50Bb J. H. Baxendale and P. George, *Trans. Faraday Soc.*, *46*, 736 (1950).

50Ca T. M. Cromwell and R. L. Scott, *J. Amer. Chem. Soc.*, *72*, 3825 (1950).

50Fa S. C. Furman and C. S. Garner, *J. Amer. Chem. Soc.*, *72*, 1785 (1950).

50Ha K. Hartley and H. A. Skinner, *Trans. Faraday Soc.*, *46*, 621 (1950).

50Ja J. C. James and C. B. Monk, *Trans. Faraday Soc.*, *46*, 1041 (1950).

50Jb H. B. Jonassen, R. B. LeBlanc and R. M. Rogan, *J. Amer. Chem. Soc.*, *72*, 4968 (1950).

50Ma H. McConnell and N. Davidson, *J. Amer. Chem. Soc.*, *72*, 3164 (1950).

50Sa J. Schubert, E. R. Russell and L. S. Myers, *J. Biol. Chem.*, *185*, 387 (1950).

50Va M. G. Vladimirova and I. A. Kakovskii, *J. Appl. Chem. (Leningrad), (English Trans.)*, *23*, 615 (1950). (Russian Page No. 580.)

1949

49Da C. W. Davies and P. A. H. Wyatt, *Trans. Faraday Soc.*, *45*, 770 (1949).

49Db M. H. Dilke and D. D. Eley, *J. Chem. Soc. (London)*, 2601 (1949).

49Ea M. G. Evans, P. George and N. Uri, *Trans. Faraday Soc.*, *45*, 230 (1949).

49Ha J. Z. Hearon, D. Burk and A. L. Schade, *J. Nat. Cancer Inst.*, *9*, 337 (1949).

49Ja W. Jaenicke and K. Hauffe, *Z. Naturforsch. A*, *4*, 363 (1949).

49Ka E. L. King, *J. Amer. Chem. Soc.*, *71*, 319 (1949).

49Kb I. A. Korshunov and V. A. Osipova, *J. Gen. Chem., U.S.S.R. (English Trans.)*, *19*, 269 (1949). (Russian Page No. 1816.)

49Kc P. Krumholz, *J. Amer. Chem. Soc.*, *71*, 3654 (1949).

49Ma J. D. McCullough and M. K. Barsh, *J. Amer. Chem. Soc.*, *71*, 3029 (1949).

1948

48Ba J. Bjerrum and E. J. Nielsen, *Acta Chem. Scand.*, *2*, 297 (1948).

48Da C. W. Davies and J. C. James, *Proc. Roy. Soc. Ser. A*, *195*, 116 (1948).

48Ka I. M. Klotz and H. G. Curme, *J. Amer. Chem. Soc.*, *70*, 939 (1948).

1946

46Ca M. Calvin and R. H. Bailes, *J. Amer. Chem. Soc.*, *68*, 949 (1946).

1945

45Ma J. D. McCullough and B. A. Eckerson, *J. Amer. Chem. Soc.*, *67*, 707 (1945).

1944

44La F. Lindstrand, *Svensk Kemisk Tidskrift*, *56*, 251 (1944).

1942

42Na K. Nozaki and R. A. Ogg, *J. Amer. Chem. Soc.*, *64*, 697 (1942).

42Ra E. Rabinowitch and W. H. Stockmayer, *J. Amer. Chem. Soc.*, *64*, 335 (1942).

1941

41Ea W. Erber and A. Schühly, *J. Prakt. Chem.*, *158*, 176 (1941).

1938

38Oa B. B. Owen and S. R. Brinkley, *J. Amer. Chem. Soc.*, *60*, 2233 (1938).

38Wa S. Winstein and H. J. Lucas, *J. Amer. Chem. Soc.*, *60*, 836 (1938).

1937

37La F. H. Li and H. C. Chen, *Naking Journal*, *7*, 353 (1937).

37Sa W. V. Smith, O. L. I. Brown and K. S. Pitzer, *J. Amer. Chem. Soc.*, *59*, 1213 (1937).

1936

36Ca M. Chatelet, *J. Chim. Phys. Physicochim. Biol.*, *33*, 313 (1936).

36La W. M. Latimer, *Chem. Rev.*, *18*, 349 (1936).

36F. H. Lee and K. H. Lee, *J. Chin. Chem. Soc. (Peiping)*, *4*, 126 (1936).

1935

35Ra F. J. W. Roughton, *Biochem. J.*, *29*, 2604 (1935).

1934

34La W. M. Latimer, P. W. Shutz and J. F. G. Hicks, *J. Chem. Phys.*, *2*, 82 (1934).

1933

33La W. M. Latimer, J. F. G. Hicks and P. W. Schutz, *J. Chem. Phys.*, *1*, 424 (1933).

1931

31Ca E. Carriere and E. Raulet, *Comptes Rendus (Paris)*, *192*, 423 (1931).

1925

25Ma G. T. Morgan, S. R. Carter and W. F. Harrison, *J. Chem. Soc. (London)*, *127*, 1917 (1925).

1910

10Ba N. Bjerrum, *Z. Phys. Chem. (Leipzig)*, *73*, 724 (1910).

1909

09Ha J. H. Hildebrand and B. L. Glascock, *J. Amer. Chem. Soc.*, *31*, 26 (1909).